Introductory Algebra

6TH EDITION

Introductory Algebra

6
TH EDITION

Mervin L. Keedy

Purdue University

Marvin L. Bittinger

Indiana University—Purdue University at Indianapolis

ADDISON-WESLEY PUBLISHING COMPANY

Reading, Massachusetts • Menlo Park, California • New York
Don Mills, Ontario • Wokingham, England • Amsterdam • Bonn
Sydney • Singapore • Tokyo • Madrid • San Juan

Sponsoring Editor	Elizabeth Burr
Managing Editor	Karen Guardino
Production Supervisor	Jack Casteel
Design, Editorial, and Production Services	Quadrata, Inc.
Illustrator	ST Associates, Inc., and Scientific Illustrators
Art Consultant	Loretta Bailey
Manufacturing Supervisor	Roy Logan
Cover Design and Photograph	Marshall Henrichs

PHOTO CREDITS   **1,** NASA   **41,** Bohdan Hrynewych/Stock, Boston
109, Bill Bachman/Photo Researchers, Inc.   **147,** Federal Bureau of Investigation
173, Peter Menzel/Stock, Boston   **247,** Coco McCoy, Rainbow Pictures
309, Peter Menzel/Stock, Boston   **377,** Mark Antman/Stock, Boston
433, 457, Denny's Restaurants   **481,** David Frazier/Photo Researchers, Inc.
527, Bohdan Hrynewych/Stock, Boston

Library of Congress Cataloging-in-Publication Data

Keedy, Mervin Laverne.
    Introductory algebra/Mervin L. Keedy, Marvin L. Bittinger. —
6th ed.
     p.   cm.
    ISBN 0-201-19671-9
    1. Algebra. I. Bittinger, Marvin L. II. Title.
QA152.2.K43 1991
512.9—dc20   90-372
                                             CIP

Reprinted with corrections, May 1991

3 4 5 6 7 8 9 10 - DO - 95 94 93 92 91

Contents

R   Prealgebra Review   1

Pretest   2
R.1   Factoring and LCM's   3
R.2   Fractional Notation   9
R.3   Decimal Notation   19
R.4   Percent Notation   27
R.5   Exponential Notation and Order of Operations   31
Summary and Review   37
Test   39

1   Introduction to Real Numbers and Algebraic Expressions   41

Pretest   42
1.1   Introduction to Algebra   43
1.2   The Real Numbers   49
1.3   Addition of Real Numbers   59
1.4   Subtraction of Real Numbers   65
1.5   Multiplication of Real Numbers   73
1.6   Division of Real Numbers   79
1.7   Properties of Real Numbers   85

1.8   Simplifying Expressions; Order of Operations   97
        Summary and Review   105
        Test   107

2   Solving Equations and Inequalities   109

        Pretest   110
2.1   Solving Equations: The Addition Principle   111
2.2   Solving Equations: The Multiplication Principle   117
2.3   Using the Principles Together   123
2.4   Solving Problems   133
2.5   Solving Percent Problems   143
2.6   Formulas   149
2.7   Solving Inequalities   153
2.8   Solving Problems Using Inequalities   163
        Summary and Review   167
        Test   169

        Cumulative Review: Chapters 1–2   171

3   Polynomials: Operations   173

        Pretest   174
3.1   Integers as Exponents   175
3.2   Exponents and Scientific Notation   185
3.3   Introduction to Polynomials   193
3.4   Addition and Subtraction of Polynomials   203
3.5   Multiplication of Polynomials   211
3.6   Special Products   217
3.7   Operations with Polynomials in Several Variables   227
3.8   Division of Polynomials   235
        Summary and Review   241
        Test   243

        Cumulative Review: Chapters 1–3   245

4   Polynomials: Factoring   247

        Pretest   248
4.1   Introduction to Factoring   249
4.2   Factoring Trinomials of the Type $x^2 + bx + c$ 255
4.3   Factoring Trinomials of the Type $ax^2 + bx + c, a \neq 1$ 261
4.4   Factoring $ax^2 + bx + c, a \neq 1$, Using Grouping   267
4.5   Factoring Trinomial Squares and Differences of Squares   271
4.6   Factoring: A General Strategy   281
4.7   Solving Quadratic Equations by Factoring   289
4.8   Solving Problems   295
        Summary and Review   303
        Test   305

        Cumulative Review: Chapters 1–4   307

5 Rational Expressions and Equations 309

Pretest   310
5.1 Multiplying and Simplifying Rational Expressions   311
5.2 Division and Reciprocals   321
5.3 Least Common Multiples and Denominators   325
5.4 Adding Rational Expressions   329
5.5 Subtracting Rational Expressions   337
5.6 Solving Rational Equations   345
5.7 Solving Problems and Proportions   351
5.8 Formulas   361
5.9 Complex Rational Expressions   365
Summary and Review   371
Test   373

Cumulative Review: Chapters 1–5   375

6 Graphs of Equations and Inequalities 377

Pretest   378
6.1 Graphs and Equations   379
6.2 Graphing Linear Equations   387
6.3 More on Graphing Linear Equations   395
6.4 Slope and Equations of Lines   401
6.5 Parallel and Perpendicular Lines   409
6.6 Direct and Inverse Variation   413
6.7 Graphing Inequalities in Two Variables   421
Summary and Review   427
Test   429

Cumulative Review: Chapters 1–6   431

7 Systems of Equations 433

Pretest   434
7.1 Systems of Equations in Two Variables   435
7.2 The Substitution Method   441
7.3 The Elimination Method   447
7.4 More on Solving Problems   457
7.5 Motion Problems   469
Summary and Review   475
Test   477

Cumulative Review: Chapters 1–7   479

8 Radical Expressions and Equations 481

Pretest   482
8.1 Introduction to Square Roots and Radical Expressions   483
8.2 Multiplying and Simplifying with Radical Expressions   489

8.3   Quotients Involving Square Roots   497
8.4   Addition, Subtraction, and More Multiplication   505
8.5   Radical Equations   511
8.6   Right Triangles and Applications   517
      Summary and Review   521
      Test   523

Cumulative Review: Chapters 1–8   525

9   Quadratic Equations   527

      Pretest   528
9.1   Introduction to Quadratic Equations   529
9.2   Solving Quadratic Equations by Completing the Square   535
9.3   The Quadratic Formula   543
9.4   Formulas   549
9.5   Solving Problems   553
9.6   Graphs of Quadratic Equations   561
      Summary and Review   567
      Test   569

Cumulative Review: Chapters 1–9   571

FINAL EXAMINATION   575

TABLES

Table 1   Fractional and Decimal Equivalents   581
Table 2   Squares and Square Roots   582

ANSWERS   A-1

INDEX   I-1

Preface

Intended for students who have not studied algebra but have a firm background in basic mathematics, this text is appropriate for a one-term course in introductory algebra. It is the second in a series of texts that includes the following:

Keedy/Bittinger: *Basic Mathematics*, Sixth Edition,
Keedy/Bittinger: *Introductory Algebra*, Sixth Edition,
Keedy/Bittinger: *Intermediate Algebra*, Sixth Edition.

    Introductory Algebra, Sixth Edition, is a significant revision of the Fifth Edition, with respect to content, pedagogy, and an expanded supplements package. Its unique approach, which has been developed over many years, is designed to help today's students both learn *and* retain mathematical concepts. The Sixth Edition is accompanied by a comprehensive supplements package that has been integrated with the text to provide maximum support for both instructor and student.

    Following are some distinctive features of the approach and pedagogy that we feel will help meet some of the challenges all instructors face teaching developmental mathematics.

APPROACH

CAREFUL DEVELOPMENT OF CONCEPTS We have divided each section into discrete and manageable learning objectives. Within the presentation of each objective, there is a careful buildup of difficulty through a series of developmental and followup examples. These enable students to thoroughly understand the mathematical concepts involved at each step. Each objective is constructed in a similar way, which gives students a high level of comfort with both the text and their learning process.

FOCUS ON "WHY" Throughout the text, we present the appropriate mathematical rationale for a topic, rather than mathematical "shortcuts." For example, when manipulating rational expressions, we remove factors of 1 rather than cancel, although cancellation is mentioned with appropriate cautions. This helps prevent student errors in cancellation and other incorrectly remembered shortcuts in later courses.

PROBLEM SOLVING We include real-life applications and problem-solving techniques throughout the text to motivate students and encourage them to think about how mathematics can be used. We also introduce a five-step problem-solving process early in the text and use the basic steps of this process (Familiarize, Translate, Solve, Check, and State the Answer) whenever a problem is solved.

PEDAGOGY

INTERACTIVE WORKTEXT APPROACH   The pedagogy of this text is designed to provide students with a clear set of learning objectives, and involve them with the development of the material, providing immediate and continual reinforcement.

Section objectives are keyed to appropriate sections of the text, exercises, and answers, so that students can easily find appropriate review material if they are unable to do an exercise.

Numerous *margin exercises* throughout the text provide immediate reinforcement of concepts covered in each section.

STUDY AID REFERENCES   Many valuable study aids accompany this text. Each section is referenced to appropriate videotape, audiotape, and software diskette numbers to make it easy for students to find and use the correct support materials.

Important rules
and definitions
in color boxes

Students
encouraged to
do margin
exercises as they
work through
material

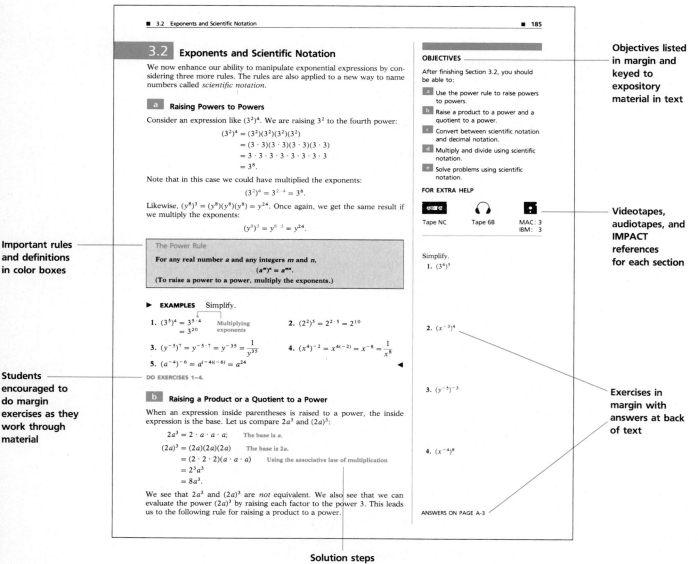

Objectives listed
in margin and
keyed to
expository
material in text

Videotapes,
audiotapes, and
IMPACT
references
for each section

Exercises in
margin with
answers at back
of text

Solution steps
in color

Exercises on tearout sheets for each section

Exercises keyed to objectives and material in text

Answer space provided for quick and easy grading

NAME   SECTION   DATE

Exercise Set 3.2 ■ 191

EXERCISE SET 3.2

a **b** Simplify.

1. $(2^3)^2$ 2. $(3^4)^3$ 3. $(5^2)^{-3}$ 4. $(9^3)^{-4}$

5. $(x^{-3})^{-4}$ 6. $(a^{-5})^{-6}$ 7. $(4x^3)^2$ 8. $4(x^3)^2$

9. $(x^4 y^5)^{-3}$ 10. $(t^5 x^3)^{-4}$ 11. $(x^{-6} y^{-2})^{-4}$ 12. $(x^{-2} y^{-7})^{-5}$

ANSWERS
1. _____
2. _____
3. _____
4. _____
5. _____
6. _____
7. _____
8. _____
9. _____

VERBALIZATION SKILLS AND "THINKING IT THROUGH"   Students' perception that mathematics is a foreign language is a significant barrier to their ability to think mathematically and is a major cause of math anxiety. In the Sixth Edition we have encouraged students to think through mathematical situations, synthesize concepts, and verbalize mathematics whenever possible.

"*Thinking it Through*" exercises at the end of each chapter encourage students to both think and write about key mathematical ideas that they have encountered in the chapter.

"*Synthesis Exercises*" at the end of most exercise sets require students to synthesize several learning objectives or to think through and provide insight into the present material.

In addition, many important definitions, such as the laws of exponents, are presented verbally as well as symbolically, to help students learn to read mathematical notation.

Skill Maintenance exercises at the end of most exercise sets review concepts from earlier chapters.

Synthesis exercises require students to synthesize objectives and provide insight into the material.

Thinking It Through exercises at the end of each chapter require students to think and write about key mathematical ideas.

48. The measure of the second angle of a triangle is 50° more than that of the first. The measure of the third angle is 10° less than twice the first. Find the measures of the angles.

49. Your quiz grades are 71, 75, 82, and 86. What is the lowest grade you can get on the next quiz and still have an average of at least 80?

50. The length of a rectangle is 43 cm. What widths will make the perimeter greater than 120 cm?

SKILL MAINTENANCE

51. Convert to decimal notation: $\frac{17}{12}$.

52. Divide: $12.42 \div 5.4$.

53. Add: $-12 + 10 + (-19) + (-24)$.

54. Remove parentheses and simplify: $5x - 8(6x - y)$.

SYNTHESIS

55. The total length of the Nile and Amazon Rivers is 13,108 km. If the Amazon were 234 km longer, it would be as long as the Nile. Find the length of each river.

56. Consumer experts advise us never to pay the sticker price for a car. A rule of thumb is to pay the sticker price minus 20% of the sticker price, plus $200. A car is purchased for $11,520 using the rule. What was the sticker price?

Solve.

57. $2|n| + 4 = 50$ 58. $|3n| = 60$ 59. $y = 2a - ab + 3$, for a

❖ THINKING IT THROUGH

Explain all possible errors in each of the following.

1. Solve: $4 - 3x = 5$
                   $3x = 9$
                   $x = 3$.

2. Solve: $2(x - 5) = 7$
                   $2x - 5 = 7$
                   $2x = 12$
                   $x = 6$.

3. Explain the difference in using the multiplication principle for solving equations and for solving inequalities.

SKILL MAINTENANCE   Because retention of skills is critical to students' future success, skill maintenance is a major emphasis of the Sixth Edition.

Each chapter begins with a *"Points to Remember"* box, which highlights key formulas and definitions from previous chapters.

In addition, we include *Skill Maintenance Exercises* at the end of most exercise sets. These review skills and concepts from earlier sections of the text.

At the end of each chapter, our *Summary and Review* summarizes important properties and formulas and includes extensive review exercises.

Each *Chapter Test* tests four review objectives from preceding chapters as well as the chapter objectives.

We also include a *Cumulative Review* at the end of each chapter; this reviews material from all preceding chapters.

At the back of the text are answers to all review exercises, together with section and objective references, so that students know exactly what material to restudy if they miss a review exercise.

TESTING AND SKILL ASSESSMENT   Accurate assessment of student comprehension is an important factor in a student's long-term success. In the Sixth Edition, we have provided many assessment opportunities.

A *Diagnostic Pretest* at the beginning of the text can place students in the appropriate chapter for their skill level, and identifies both familiar material and specific trouble areas later in the text.

Chapter Pretests diagnose student skills and place the students appropriately within each chapter, allowing them to concentrate on topics with which they have particular difficulty.

Chapter Tests at the end of each chapter allow students to review and test comprehension of chapter skills.

Answers to each question on all tests are included at the back of the text.

For additional testing options, we have developed a printed test bank with many alternative forms of each chapter test in both open-ended and multiple-choice formats. For a greater degree of flexibility in creating chapter tests, the text is also accompanied by extensive computerized testing programs for IBM, MAC, and Apple computers.

Key properties and skills from preceding material summarized at beginning of chapter

❖ POINTS TO REMEMBER: CHAPTER 2

Identity Properties of 0 and 1:	$a + 0 = a, \quad a \cdot 1 = a$
Simple-Interest Formula:	$I = Prt$
Sum of the Angles of a Triangle $= 180°$	
Perimeter of a Rectangle:	$P = 2l + 2w$
Consecutive Integers:	$x, \quad x + 1, \quad x + 2, \quad x + 3,$ etc.
Consecutive Even Integers:	$x, \quad x + 2, \quad x + 4, \quad x + 6,$ etc.
Consecutive Odd Integers:	$x, \quad x + 2, \quad x + 4, \quad x + 6,$ etc.

Chapter Pretest evaluates student's strengths and weaknesses in upcoming material

PRETEST: CHAPTER 2

Solve.

1. $-7x = 49$ **2.** $4y + 9 = 2y + 7$ **3.** $6a - 2 = 10$

4. $4 + x = 12$ **5.** $7 - 3(2x - 1) = 40$ **6.** $\dfrac{4}{9}x - 1 = \dfrac{7}{8}$

FLEXIBILITY OF TEACHING MODES

The flexible worktext format of *Introductory Algebra* allows the book to be used in many ways.

- **In a standard lecture.** To use the book in a lecture format, the instructor lectures in a conventional manner and encourages students to do the margin exercises while studying on their own. This greatly enhances the readability of the text.

- **For a modified lecture.** To bring student-centered activity into the class, the instructor stops lecturing and has the students do margin exercises.

- **For a no-lecture class.** The instructor makes assignments that students do on their own, including working the margin exercises. During the class period following the assignment, the instructor answers questions, and students have an extra day or two to polish their work before handing it in. In the meantime, they are working on the next assignment. This method provides individualization while keeping a class together. It also minimizes the number of instructor hours required and has been found to work well with large classes.

- **In a learning laboratory.** Because this text is highly readable and easy to understand, it can be used in a learning laboratory or any other self-study situation.

KEY CONTENT CHANGES

In response to both extensive user comments and reviewer feedback, there have been many organizational changes and revisions to the Sixth Edition. Detailed information about the changes made in this material is available in the form of a Conversion Guide. Please ask your local Addison-Wesley sales representative for more information. Following is a list of the major organizational changes for this revision:

- Where possible, short sections have been combined to streamline the presentation and reduce the overall number of sections.

- More exercises have been added throughout the text, increasing the overall number by approximately 10%.

- A five-step problem-solving process is now introduced early in the text, and these steps are used throughout the text whenever a problem is solved.

- The first three chapters have been *substantially* reorganized to separate the arithmetic review into a review chapter, streamline the presentation of operations and properties for the real numbers, introduce all equation solving and problem solving together in Chapter 2, and include material on solving inequalities with material on solving equations.

- Division of polynomials has been moved from the chapter on rational expressions to the chapter on polynomial operations.

- An introduction to polynomials in several variables is now included in the chapter on polynomial operations. Examples with polynomials in several variables are then integrated throughout the chapter on factoring.

- The chapter on rational expressions has been moved earlier in the text, so that it now immediately follows the chapter on factoring polynomials.

- Material on direct and inverse variation has been combined, and is also now included in the chapter on graphing.
- The chapter on inequalities has been split up and integrated into material on solving equations where appropriate. *Solving inequalities* is now covered in the chapter on solving equations, and *graphing inequalities* is now covered in the chapter on graphing equations. The set terminology from this chapter has been either integrated into the presentation of the real-number system and inequalities or omitted.

SUPPLEMENTS

This text is accompanied by a comprehensive supplements package. Below is a brief list of these supplements, followed by a detailed description of each one.

For the Instructor

Teacher's Edition
Instructor's Solutions Manual
Instructor's Resource Guide
Printed Test Bank
Lab Resource Manual
Answer Book
Computerized Testing

For the Student

Student's Solutions Manual
Videotapes
Audiotapes
The Math Hotline
Comprehensive Tutorial Software
Drill and Practice Software

SUPPLEMENTS FOR THE INSTRUCTOR

All supplements for the instructor are free upon adoption of this text.

Teacher's Edition

This is a specially bound version of the student text with exercise answers printed in a third color. It also includes additional information on the skill maintenance exercises, suggested syllabi for different length courses, and some information about the teaching aids that accompany the text.

Instructor's Solutions Manual

This manual by Judith A. Penna contains worked-out solutions to all even-numbered exercises and discussions of the "Thinking It Through" sections.

Instructor's Resource Guide

This guide contains the following:

- Additional "Thinking It Through" exercises.
- Extra practice problems for some of the most challenging topics in the text.
- Teaching essays on math anxiety and study skills.
- Indexes to the videotapes, the audiotapes, and the software that accompany the text.
- Number lines and grids for test preparation.
- Conversion guide that cross-references the Fifth Edition to the Sixth Edition.
- Black-line transparency masters including a selection of key definitions, procedures, problem-solving strategies, graphs, and figures to use in class.

Printed Test Bank

This is an extensive collection of alternative chapter test forms, including the following:

*   5 alternative test forms for each chapter with questions in the same topic order as the objectives presented in the chapter.

*   5 alternative forms for each chapter with the questions in a different order.

*   3 multiple-choice test forms for each chapter.

*   2 cumulative review tests for each chapter.

*   9 alternative forms of the final examination, 3 with questions organized by chapter, 3 with questions scrambled, and 3 with multiple-choice questions.

Lab Resource Manual

This manual contains a selection of essays on setting up learning labs, including information on running large testing centers and setting up mastery learning programs. It also includes a directory of learning lab coordinators who are available to answer questions.

Answer Book

The Answer Book will contain answers to all the exercises in the text for you to make available to your students.

Computerized Testing

OmniTest (IBM PC), AWTest (Apple II series)

This text is accompanied by algorithm-driven testing systems for both IBM and Apple. With both machine versions, it is easy to create up to 99 variations of a customized test with just a few keystrokes, choosing from over 300 open-ended and multiple-choice test items. Instructors can also print out tests in chapter-test format.

The IBM testing program, OmniTest, also allows users to enter their own test items and edit existing items in an easy-to-use WYSIWYG format.

LXR·TEST™ (MACINTOSH)

This is a versatile and flexible test-item bank of more than 1200 multiple-choice and open-ended test items with complete math graphics and full editing capabilities. Tests can be created by selecting specific test items or by requesting the computer to select items randomly from designated objectives. LXR·TEST can create multiple test versions by scrambling the order of multiple-choice distractors or the order of the questions themselves.

SUPPLEMENTS FOR THE STUDENT

Student's Solutions Manual

This manual by Judith A. Penna contains completely worked-out solutions with step-by-step annotations for all the odd-numbered exercises in the text. It is free to adopting instructors and may be purchased by your students from Addison-Wesley Publishing Company.

Videotapes

Using the chalkboard and manipulative aids, Donna DeSpain lectures in detail, works out exercises, and solves problems from most sections in the text on 20 70-minute videotapes. These tapes are ideal for students who have missed a lecture or who need extra help. Each section in the text is referenced to the appropriate tape number and section, underneath the icon ▣. A complete set of videotapes is free to qualifying adopters.

Audiotapes ◠

The audiotapes are designed to lead students through the material in each text section. Bill Saler explains solution steps to examples, cautions students about common errors, and instructs them to stop the tape and do exercises in the margin. He then reviews the margin-exercise solutions, pointing out potential errors. Each section in the text is referenced to the appropriate tape number and section, underneath the icon ◠. Audiotapes are free to qualifying adopters.

The Math Hotline

This telephone hotline is open 24 hours a day for students to receive detailed hints for exercises that have been developed by Larry Bittinger. Exercises covered include all the odd-numbered exercises in the exercise sets with the exception of the skill maintenance and synthesis exercises.

Tutorial Software

A variety of tutorial software packages is available to accompany this text. Please contact your Addison-Wesley representative for a software sampler that contains demonstration disks for these packages and a summary of our distribution policy.

Comprehensive Tutorials

IMPACT: An Interactive Mathematics Tutorial ▣ by Wayne Mackey and Doug Proffer, Collin County Community College (IBM PC or MACINTOSH).

This software was developed exclusively for Addison-Wesley and is keyed section by section to this text. Icons at the beginning of each section reference the appropriate disk number. The disk menus correspond to the text's section numbers.

IMPACT is designed to generate practice exercises based on the exercise sets in this book. If students are having trouble with a particular exercise, they can ask to see an example or a step-by-step solution to the problem they are working on. Each step of the step-by-step solutions is treated interactively to keep students involved in the solution of the problem and help them identify precisely where they are having trouble. *IMPACT* also keeps detailed records of students' scores.

Instructional Software for Algebra (Apple II series).

This software covers selected algebra topics. It also gives students brief explanations and examples, followed by practice exercises with interactive feedback for student error.

Drill and Practice Packages

The Math Lab by Chris Avery and Chris Barker, DeAnza College (Apple II series, IBM PC, or Macintosh).

Students choose the topic, level of difficulty, and number of exercises. If they get a wrong answer, *The Math Lab* will prompt them with the first step of the solution. This software also keeps detailed records of student scores.

Professor Weissman's Software by Martin Weissman, Essex County College
(IBM PC or compatible).

Professor Weissman's Software generates exercises based on the student's
selection of topic and level of difficulty. If they get a wrong answer, the software gives them a step-by-step solution. The level of difficulty increases if
students are successful.

In the back of this text is a coupon for *Professor Weissman's Software*
that allows students to buy the software directly from Martin Weissman at
a discount.

The Algebra Problem Solver by Michael Hoban and Kathirgama Nathan,
La Guardia Community College (IBM PC).

After selecting the topic and exercise type, students can enter their own
exercises or request an exercise from the computer. In each case, *The Algebra Problem Solver* will give the student detailed, annotated, step-by-step
solutions.

ACKNOWLEDGMENTS

Many of you who teach developmental mathematics have helped to shape
the Sixth Edition of this text by reviewing, answering surveys, participating
in focus groups, filling out questionnaires, and spending time with us on your
campuses. Our heartfelt thanks to all of you, and many apologies to anyone
we have missed on the following list.

TEXTBOOK REVIEWERS

Don Albers, *Menlo College;* Vickie Aldrich, *New Mexico State University—
Dona Ana Branch;* Joann Brossenbroek, *Columbus State Community
College;* Ellen E. Church, *Golden West College;* Bobbi Parrino Cook,
Indian River Community College; Stephen I. Gendler, *Clarion University;*
Gary A. Getchell, *Cape Cod Community College;* Wayne D. Gibson,
Rancho Santiago Community College; Helen M. Hancock, *Shoreline
Community College;* Mary Lou Hart, *Brevard Community College;* Phyllis
H. Jore, *Valencia Community College;* Sue Korsak, *New Mexico State
University;* Nenette Loftsgaarden, *University of Montana;* Larry Orelia,
Nassau Community College; Donald Perry, *Lee College*

FORMAL AND INFORMAL FOCUS GROUP PARTICIPANTS

Geoff Akst, *Borough of Manhattan Community College;* Betty Jo Baker,
Lansing Community College; Gene Beuthin, *Saginaw Valley State
University;* Rheta Beaver, *Valencia Community College;* Roy Boersema,
Front Range Community College; Dale Boye, *Schoolcraft College;* Jim
Brenner, *Black Hawk College;* Ben Cheatham, *Valencia Community
College;* Karen Clark, *Tacoma Community College;* Tom Clark, *Lane
Community College;* Sally Copeland, *Johnson County Community College;*
Ernie Danforth, *Corning Community College;* Sarah Evangelista, *Temple
University;* Bill Freed, *Concordia College;* Sally Glover-Richard, *Pierce
Community College;* Valerie Hayward, *Orange Coast College;* Eric Heinz,
Catonsville Community College; Bruce Hoelter, *Raritan Valley
Community College;* Lou Hoezle, *Bucks County Community College;*
Linda Horner, *Broward Community College;* Mary Indelicato,

Normandale Community College; Tom Jebson, *Pierce Community College;* Jeff Jones, *County College of Morris;* Judith Jones, *Valencia Community College;* Virginia Keen, *West Michigan University;* Roxanne King, *Prince Georges Community College;* Lee Marva Lacy, *Glendale Community College;* Ginny Licata, *Camden County College;* Randy Liefson, *Pierce Community College;* Charlie Luttrell, *Frederick Community College;* Marilyn MacDonald, *Red Deer College;* Sharon MacKendrick, *New Mexico State University;* Annette Magyar, *Southwestern Michigan College;* Bob Malena, *Community College of Allegheny County;* Marilyn Masterson, *Lansing Community College;* Don McNair, *Lane Community College;* John Pazdar, *Greater Hartford Community College;* Donald Perry, *Lee College;* Jeanne Romeo, *Delta College;* Jack Rotman, *Lansing Community College;* Winona Sathre, *Valencia Community College;* Billie Stacey, *Sinclair Community College;* John Steele, *Lane Community College;* Dave Steinfort, *Grand Rapids Junior College;* Betty Swift, *Cerritos College;* Bill Wittinfeld, *Tacoma Community College; Faculty of St. Petersburg Junior College*

QUESTIONNAIRE AND SURVEY RESPONDEES

Tony Abruzzo, *University of New Mexico;* Carol Alspach, *Reading Area Community College;* Boyd Benson, *Rio Hondo College;* Mary Jean Brod, *University of Montana;* Elaine Craft, *Chesterfield–Marlboro Technical College;* Mary Ann Fuerk, *Rosary College;* Masato Hayashi, *Irvine Valley College;* Nenette Loftsgaarden, *University of Montana;* Linda Long, *Ricks College;* Sharon MacKendrick, *New Mexico State University;* Ron Putthoff, *University of Southern Mississippi;* Rollin Quinn, *Flathad Valley Community College;* Richard Vaughn, *Jefferson College*

We also wish to thank the many people without whose committed efforts our work could not have been completed. In particular, we would like to thank Judy Beecher, Barbara Johnson, and Judy Penna for their work on proofreading the manuscript and overseeing the production process. We would also like to thank Pat Pasternak who did a marvelous job typing the text manuscript and answer section, and John Baumgart, Larry Bittinger, John Irons, and Jennifer Alsop, who did a thorough and conscientious job of checking the manuscript.

M.L.K.
M.L.B.

To The Student

This text has many features that can help you succeed in introductory algebra. To familiarize yourself with these, you might read the preface that starts on page ix and study the annotated pages that are included. Following are a few suggestions on how to use these features to enhance your learning process.

BEFORE YOU START THE TEXT

If you are in a classroom setting, your instructor might ask you to take the diagnostic pretest at the beginning of the text, checking your answers at the back of the text, to find out what material you already know and what material you need to spend time on. You can also use this pretest to skip material that you already know from an independent learning situation.

BEFORE YOU START A CHAPTER

The chapter opening page gives you an idea of the material that you are about to study and how it can be used. The chapter opening introduction also tells you what sections you will need to review in order to do the skill maintenance exercises on the chapter test. It's a good idea to restudy these sections to keep the material fresh in your mind for the midterm or final examination.

The first page of each chapter lists "Points to Remember" that will be needed to work certain examples and exercises in the chapter. You should try to review any skills listed here before beginning the chapter and learn any formulas or definitions.

This same page also includes a chapter pretest. You can work through this and check your answers at the back of the text to identify sections that you might skip or sections that give you particular difficulty and need extra concentration.

WORKING THROUGH A SECTION

First you should read the learning objectives for the section. The symbol next to an objective (a , b , c) appears next to the text, exercises, and answers that correspond to that objective, so you can always refer back to the appropriate material when you need to review a topic.

You will also notice that there are references to the audiotapes, video-tapes, and software that are available for extra help for the section underneath the objective listing. The software referenced is a program called *IMPACT: An Interactive Mathematics Tutorial*.

As you work through a section, you will see an instruction to "Do Exercises x–xx." This refers to the exercises in the margin of the page. You should always stop and do these to practice what you have just studied because they greatly enhance the readability of the text. Answers to the margin exercises are at the back of the text.

After you have completed a section, you should do the assigned exercises in the exercise set. The exercises are keyed to the section objectives, so that if you get an incorrect answer, you know that you should restudy the text section that corresponds to the symbol.

Answers to all the odd-numbered exercises are at the back of the text. A solutions manual with complete worked-out solutions to all the odd-numbered exercises is available from Addison-Wesley Publishing Company.

PREPARING FOR A CHAPTER TEST

To prepare for a chapter test, you can review your homework and restudy sections that were particularly difficult. You should also learn the "Important Properties and Formulas" that begin the chapter's summary and review and study the review sections that are listed at the beginning of the review exercises.

After studying, you might set aside a block of time to work through the summary and review as if it were a test. You can check your answers at the back of the text after you are done. The answers are coded to sections and objectives, so you can restudy any areas in which you are having trouble. You can also take the chapter test as practice, again checking your answers at the back of the text.

If you are still having difficulties with a topic, you might try either going to see your instructor or working with the videotapes, audiotapes, or tutorial software that are referenced at the beginning of the text sections. Be sure to start studying in time to get extra help before you must take the test.

PREPARING FOR A MIDTERM OR FINAL EXAMINATION

To keep material fresh in your mind for a midterm or final examination, you can work through the cumulative reviews at the end of each chapter. You can also use these as practice midterms or finals. In addition, there is a final examination at the end of the text. The answers to all the exercises in the cumulative reviews and the final examination are at the back of the text.

OTHER STUDY TIPS

There is a saying in the real-estate business: "The three most important things to consider when buying a house are *location, location, location*." When trying to learn mathematics, the three most important things are *time, time, time*. Try to carefully analyze your situation. Be sure to allow yourself *time* to do the lesson. Are you taking too many courses? Are you working so much that you do not have *time* to study? Are you taking *time* to maintain daily preparation? Other study tips are provided on pages marked "Side-lights" in the text.

Introductory Algebra

6 TH EDITION

DIAGNOSTIC PRETEST

Chapter R
Perform the indicated operations and simplify if possible.

1. $\dfrac{8}{9} \cdot \dfrac{3}{5}$

2. $\dfrac{1}{4} + \dfrac{2}{3}$

3. $4.94 \div 0.19$

4. $12.04 - 1.057$

Chapter 1
Compute and simplify.

5. $3.8 + (-4.62) - (-2)$

6. $-9(1.3)$

7. Your total assets are \$135.97. You borrow \$145.90 to fix your car. What are your total assets now?

8. Remove parentheses and simplify:
$$3[11(a - 2) - 2(3 - a)].$$

Chapter 2
Solve.

9. $2(x - 1) = 4(x + 2)$

10. $4 - 13x \le 10x - 5$

11. A 36-in. string is cut into two pieces. One piece is three times as long as the other. How long are the pieces?

12. A family spent \$270 one month on clothing. This was 18% of its income. What was their monthly income?

Chapter 3
Simplify.

13. $\dfrac{x^2 y^2}{x^{-2} y^3}$

14. $(-2x)^3 (2x^4)^2$

15. Subtract: $(x^2 + 3x - 1) - (2x^2 - 5)$.

16. Multiply: $(2x^2 + 3)(2x^2 - 3)$.

Chapter 4
Factor completely.

17. $2x^2 - 162$

18. $5x^2 - 14x - 3$

Solve.

19. $x^2 + 3x = 10$

20. The width of a rectangle is 9 m less than the length. The area is 136 m^2. Find the width and the length.

Chapter 5

21. Divide and simplify:

$$\frac{2x^3 + 6x^2}{x^2 + 10x + 25} \div \frac{4x^3 - 36x}{x^2 + x - 20}.$$

Solve.

23. $\dfrac{2}{x+4} = \dfrac{1}{x}$

22. Add and simplify:

$$\frac{1-x}{x^2 + x} + \frac{x}{x^2 + 3x + 2}.$$

24. One car goes 15 mph faster than another. While one car goes 165 mi, the other goes 120 mi. How fast is each car?

Chapter 6

25. Graph: $y = -2x + 1$.

27. Find an equation of the line containing the pair of points $(3, 2)$ and $(4, -1)$.

26. Find the slope and y-intercept of $2x + 3y = 8$.

28. Graph: $x - 2y \leq 6$.

Chapter 7

Solve.

29. $x + y = 5,$
$2x + 3y = 7$

31. Solution A is 20% alcohol and solution B is 50% alcohol. How much of each should be used to make 50 L of a solution that is 35% alcohol?

30. $2x + 4y = 5,$
$3x - 2y = 9$

32. Two cars leave a service station at the same time going in the same direction. One travels 56 mph and the other travels 62 mph. In how many hours will they be 60 mi apart?

Chapter 8

33. Multiply and simplify:

$$\sqrt{2x^2y} \cdot \sqrt{6xy^3}.$$

35. Rationalize the denominator:

$$\frac{3}{2 - \sqrt{3}}.$$

34. Divide and simplify:

$$\frac{\sqrt{5x^3}}{\sqrt{45xy^2}}.$$

36. Solve: $\sqrt{2x + 4} - 1 = 8$.

Chapter 9

Solve.

37. $3x^2 + 2x = 1$

39. The hypotenuse of a right triangle is 34 m. One leg is 14 m longer than the other. Find the lengths of the legs.

38. $2x^2 + 10 = x$

40. Graph: $y = x^2 - 4x + 1$.

INTRODUCTION This chapter is a review of skills basic to a study of algebra. By taking the book pretest on p. xxiii, or the chapter pretest on the next page, you may be able to skip this chapter. Consult with your instructor if you need advice. ❖

R

Prealgebra Review

AN APPLICATION

It takes Jupiter 12 years to revolve around the sun and it takes Saturn 30 years. How often will Jupiter and Saturn appear in the same position?

THE MATHEMATICS

Least common multiples can be used to solve problems related to orbital planets. The least common multiple of 12, or 2 · 2 · 3, and 30, or 2 · 3 · 5, is 2 · 2 · 3 · 5, or 60. Thus the planets will be in the same position once every 60 years.

$$\text{Definition of Percent:} \quad n\% = n \times 0.01 = n \times \frac{1}{100} = \frac{n}{100}$$

Identity Property of 0: $\quad a + 0 = a$

Identity Property of 1: $\quad a \cdot 1 = a$

Exponential Notation: $\quad a^n = \underbrace{a \cdot a \cdots a}_{n \text{ factors}}$

PRETEST: CHAPTER R

1. Find the prime factorization of 248.

2. Find the LCM: 12, 24, 42.

3. Write an expression equivalent to $\frac{2}{3}$ using $\frac{5}{5}$ as a name for 1.

4. Write an equivalent expression to $\frac{11}{12}$ with a denominator of 48.

Simplify.

5. $\dfrac{46}{128}$

6. $\dfrac{28}{42}$

Compute and simplify.

7. $\dfrac{3}{5} \div \dfrac{6}{11}$

8. $\dfrac{3}{7} - \dfrac{1}{3}$

9. $\dfrac{3}{10} + \dfrac{1}{5}$

10. $\dfrac{4}{7} \cdot \dfrac{5}{12}$

11. Convert to fractional notation (do not simplify): 32.17.

12. Convert to decimal notation: $\dfrac{789}{10,000}$.

13. Add: $8.25 + 91 + 34.7862$.

14. Subtract: $230 - 17.95$.

15. Multiply: 34.78×10.08.

16. Divide: $78.12 \div 6.3$.

17. Convert to decimal notation: $\dfrac{13}{9}$.

18. Round to the nearest hundredth: 345.8395.

19. Round to the nearest tenth: 345.8395.

20. Evaluate: 2^3.

21. Evaluate: $(1.1)^2$.

22. Calculate: $9 \cdot 3 + 24 \div 4 - 5^2 + 10$.

23. Convert to decimal notation: 11.6%.

24. Convert to fractional notation: 87%.

25. Convert to percent notation: $\dfrac{7}{8}$.

26. Write exponential notation: $5 \cdot 5 \cdot 5 \cdot 5$.

R.1 Factoring and LCM's

a Factors and Prime Factorizations

We begin our review of prealgebra with *factoring*, which is a necessary skill for addition and subtraction with fractional notation. Factoring is also an important skill in algebra. You will eventually learn to factor algebraic expressions.

The numbers we will be factoring are from the set of **natural numbers:**

$$1, \quad 2, \quad 3, \quad 4, \quad 5, \text{ and so on.}$$

Consider the product $12 = 3 \cdot 4$. We say that 3 and 4 are **factors** of 12 and that $3 \cdot 4$ is a **factorization** of 12. Since $12 = 12 \cdot 1$, we also know that 12 and 1 are factors of 12 and that $12 \cdot 1$ is a factorization of 12.

> To *factor* a number N means to express N as a product. A *factor* of a number N is a number that can be used to express N as a product. A *factorization* of a number N is an expression that names the number as a product of natural numbers.

▶ **EXAMPLE 1** Factor the number 8. Find the factors.

The number 8 can be named as a product in several ways:

$$2 \cdot 4, \quad 1 \cdot 8, \quad 2 \cdot 2 \cdot 2.$$

The factors of 8 are 1, 2, 4, and 8. ◀

Note that the word "factor" is used both as a noun and as a verb. You **factor** when you express a number as a product. The numbers you multiply together to get the product are **factors.**

DO EXERCISES 1 AND 2 (IN THE MARGIN AT THE RIGHT).

▶ **EXAMPLE 2** Write several factorizations of the number 12.

$$1 \cdot 12, \quad 2 \cdot 6, \quad 3 \cdot 4, \quad 2 \cdot 2 \cdot 3 \qquad ◀$$

DO EXERCISES 3 AND 4 (IN THE MARGIN).

> A natural number that has *exactly two different* factors, itself and 1, is called a *prime number.*

▶ **EXAMPLE 3** Which of these numbers are prime? 7, 4, 11, 16, 1

7 is prime. It has exactly two different factors, 7 and 1.

4 is not prime. It has three different factors, 1, 2, and 4.

11 is prime. It has exactly two different factors, 11 and 1.

16 is not prime. It has factors 1, 2, 4, 8, and 16.

1 is not prime. It has only itself as a factor. ◀

OBJECTIVES

After finishing Section R.1, you should be able to:

a Factor numbers and find prime factorizations of numbers.

b Find the LCM of two or more numbers using prime factorizations.

FOR EXTRA HELP

Tape 1A   Tape 1A   MAC: R
                                  IBM: R

Factor the number. List all the factors.

1. 9

2. 16

Write several factorizations for the number.

3. 18

4. 20

ANSWERS ON PAGE A-1

5. Which of these numbers are prime?

     8,   6,   13,   14,   1

Find the prime factorization.

6. 48

7. 50

8. 770

The following is a table of the prime numbers from 2 to 157. There are more extensive tables, but these prime numbers will be the most helpful to you in this text.

> **A Table of Primes**
>
> **2, 3, 5, 7, 11, 13, 17, 19, 23, 29, 31, 37, 41, 43, 47, 53, 59, 61, 67, 71, 73, 79, 83, 89, 97, 101, 103, 107, 109, 113, 127, 131, 137, 139, 149, 151, 157**

DO EXERCISE 5.

If a natural number, other than 1, is not prime, we call it **composite.** Every composite number can be factored into a product of prime numbers. Such a factorization is called a **prime factorization.**

▶ **EXAMPLE 4**   Find the prime factorization of 36.

We begin by factoring 36 any way we can. One way is like this:

$$36 = 4 \cdot 9.$$

The factors 4 and 9 are not prime, so we factor them:

$$
\begin{aligned}
36 &= \quad 4 \quad \cdot \quad 9 \\
& \quad\; \downarrow \qquad\;\; \downarrow \\
&= \; 2 \cdot 2 \; \cdot \; 3 \cdot 3
\end{aligned}
$$

The factors in the last factorization are all prime, so we now have the *prime factorization* of 36. Often a prime factorization can be obtained in several different ways. Another way to find the prime factorization of 36 is

$$36 = 2 \cdot 18 = 2 \cdot 3 \cdot 6 = 2 \cdot 3 \cdot 2 \cdot 3.$$

In effect, we begin factoring any way we can think of and keep factoring until all factors are prime. Using a **factor tree** might also be helpful.

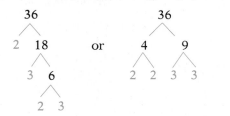

▶ **EXAMPLE 5**   Find the prime factorization of 60.

This time, we use our list of primes from the table. We go through the table until we find a prime that is a factor of 60. The first such prime is 2.

$$60 = 2 \cdot 30$$

We keep dividing by 2 until it is not possible to do so.

$$60 = 2 \cdot 2 \cdot 15.$$

Now we go to the next prime in the table that is a factor of 60. It is 3.

$$60 = 2 \cdot 2 \cdot 3 \cdot 5$$

Each factor in 2 · 2 · 3 · 5 is a prime. Thus this is the prime factorization. ◀

DO EXERCISES 6–8.

b   Least Common Multiples

9. Find several of the common multiples of 3 and 5 by making lists of multiples.

Least common multiples are used to add and subtract with fractional notation.

The **multiples** of a number all have that number as a factor. For example, the multiples of 2 are

$$2,\ 4,\ 6,\ 8,\ 10,\ 12,\ 14,\ 16,\ldots.$$

We could name each of them in such a way as to show 2 as a factor. For example, $14 = 2 \cdot 7$.

The multiples of 3 all have 3 as a factor. They are

$$3,\ 6,\ 9,\ 12,\ 15,\ 18,\ldots.$$

Two or more numbers always have many multiples in common. From lists of multiples, we can find common multiples.

▶ **EXAMPLE 6**   Find some of the multiples that 2 and 3 have in common.

We make lists of their multiples and circle the multiples that appear in both lists.

$$2, 4, ⑥, 8, 10, ⑫, 14, 16, ⑱, 20, 22, 24, 26, 28, 30, 32, 34, ㊱, \ldots;$$

$$3, ⑥, 9, ⑫, 15, ⑱, 21, ㉔, 27, ㉚, 33, ㊱, \ldots$$

The common multiples of 2 and 3 are 6,   12,   18,   24,   30,   36, ... ◀

DO EXERCISE 9.

In Example 6, we found many common multiples of 2 and 3. The *least*, or smallest, of those common multiples is 6. We abbreviate *least common multiple* as **LCM**.

There are several methods that work well for finding the LCM of several numbers. Some of these do not work well in algebra, especially when we consider expressions with variables such as $4ab$ and $12abc$. We now learn, or review, a method that will work in arithmetic *and in algebra as well*. The method uses factorizations. To see how it works, let's look at the prime factorizations of 9 and 15 in order to find the LCM:

$$9 = 3 \cdot 3, \qquad 15 = 3 \cdot 5.$$

Any multiple of 9 must have *two* 3's as factors. Any multiple of 15 must have *one* 3 and *one* 5 as factors. The smallest number satisfying all of these conditions is

┌──────── Two 3's; 9 is a factor
$3 \cdot 3 \cdot 5 = 45.$
└──────── One 3, one 5; 15 is a factor

The LCM must have all the factors of 9 and all the factors of 15, but the factors do not have to be repeated when they are common to both numbers.

To find the LCM of several numbers:

a) **Write the prime factorization of each number.**

b) **Form the LCM by writing the product of the different factors from step (a), using each the greatest number of times it occurs in any one factorization.**

ANSWERS ON PAGE A-1

Find the LCM by factoring.

10. 8 and 10

11. 18 and 40

12. Find the LCM of 24, 35, and 45.

Find the LCM.

13. 3,   18

14. 12,   24

Find the LCM.

15. 4,   9

16. 5,   6,   7

ANSWERS ON PAGE A-1

▶ **EXAMPLE 7**   Find the LCM of 40 and 100.

a) We find the prime factorizations:

$$40 = 2 \cdot 2 \cdot 2 \cdot 5,$$
$$100 = 2 \cdot 2 \cdot 5 \cdot 5.$$

b) We write 2 as a factor three times (the greatest number of times it occurs). We write 5 as a factor two times (the greatest number of times it occurs).

The LCM is $2 \cdot 2 \cdot 2 \cdot 5 \cdot 5$, or 200.   ◀

DO EXERCISES 10 AND 11.

▶ **EXAMPLE 8**   Find the LCM of 27, 90, and 84.

a) We factor:

$$27 = 3 \cdot 3 \cdot 3,$$
$$90 = 2 \cdot 3 \cdot 3 \cdot 5,$$
$$84 = 2 \cdot 2 \cdot 3 \cdot 7.$$

b) We write 2 as a factor two times, 3 three times, 5 one time, and 7 one time.

The LCM is $2 \cdot 2 \cdot 3 \cdot 3 \cdot 3 \cdot 5 \cdot 7$, or 3780.   ◀

DO EXERCISE 12.

▶ **EXAMPLE 9**   Find the LCM of 7 and 21.

Since 7 is prime, it has no prime factorization. We still need it as a factor, however:

$$7 = 7,$$
$$21 = 3 \cdot 7.$$

The LCM is $7 \cdot 3$, or 21.   ◀

> **If one number is a factor of another, then the LCM is the larger of the two numbers.**

DO EXERCISES 13 AND 14.

▶ **EXAMPLE 10**   Find the LCM of 8 and 9.

We have

$$8 = 2 \cdot 2 \cdot 2,$$
$$9 = 3 \cdot 3.$$

The LCM is $2 \cdot 2 \cdot 2 \cdot 3 \cdot 3$, or 72.   ◀

> **If two or more numbers have no common prime factor, then the LCM is the product of the numbers.**

DO EXERCISES 15 AND 16.

EXERCISE SET R.1

a Find some factorizations of the number. List all the factors.

1. 24   **2.** 18   **3.** 81   **4.** 72

Find the prime factorization of each number.

5. 14   **6.** 15   **7.** 33   **8.** 55

9. 9   **10.** 25   **11.** 49   **12.** 121

13. 18   **14.** 24   **15.** 40   **16.** 56

17. 90   **18.** 120   **19.** 210   **20.** 330

21. 91   **22.** 143   **23.** 119   **24.** 221

b Find the prime factorization of the numbers. Then find the LCM.

25. 3, 7   **26.** 18, 27   **27.** 20, 30   **28.** 24, 36

29. 3, 15   **30.** 20, 40   **31.** 30, 40   **32.** 50, 60

33. 13, 23   **34.** 12, 18   **35.** 18, 30   **36.** 45, 72

ANSWERS

1. _____
2. _____
3. _____
4. _____
5. _____
6. _____
7. _____
8. _____
9. _____
10. _____
11. _____
12. _____
13. _____
14. _____
15. _____
16. _____
17. _____
18. _____
19. _____
20. _____
21. _____
22. _____
23. _____
24. _____
25. _____
26. _____
27. _____
28. _____
29. _____
30. _____
31. _____
32. _____
33. _____
34. _____
35. _____
36. _____

ANSWERS

37. _____

38. _____

39. _____

40. _____

41. _____

42. _____

43. _____

44. _____

45. _____

46. _____

47. _____

48. _____

49. _____

50. _____

51. _____

52. _____

53. a) _____

b) _____

c) _____

d) _____

54. _____

55. _____

56. _____

57. _____

58. _____

37. 30, 36   **38.** 30, 50   **39.** 24, 30   **40.** 60, 70

41. 17, 29   **42.** 18, 24   **43.** 12, 28   **44.** 35, 45

45. 2, 3, 5   **46.** 3, 5, 7   **47.** 24, 36, 12   **48.** 8, 16, 22

49. 5, 12, 15   **50.** 12, 18, 40   **51.** 6, 12, 18   **52.** 18, 24, 30

SYNTHESIS

Synthesis Exercises are designed to allow you to combine objectives or skills studied in the chapter or preceding parts of the text.

53. Consider 8 and 12. Determine whether each of the following is the LCM of 8 and 12. Tell why or why not.

   a) $2 \cdot 2 \cdot 3 \cdot 3$ **b)** $2 \cdot 2 \cdot 3$ **c)** $2 \cdot 3 \cdot 3$ **d)** $2 \cdot 2 \cdot 2 \cdot 3$

▊ This symbol indicates exercises to be done using a calculator. Use your calculator to find the LCM of each pair of numbers.

54. 288, 324   **55.** 2700, 7800

Planet orbits and LCM's. The earth, Jupiter, Saturn, and Uranus all revolve around the sun. The earth takes 1 year, Jupiter takes 12 years, Saturn takes 30 years, and Uranus takes 84 years. On a certain night you look at all the planets, and you wonder how many years it will be before they have the same position again. To find out, you find the LCM of 12, 30, and 84. It will be that number of years.

56. How often will Jupiter and Saturn appear in the same position?

57. How often will Saturn and Uranus appear in the same position?

58. How often will Jupiter, Saturn, and Uranus appear in the same position?

R.2 Fractional Notation

We now review fractional notation and its use with multiplication, addition, subtraction, and division of numbers of arithmetic.

One feature of algebra that distinguishes it from arithmetic is the use of letters, or variables, to represent numbers. We will make some use of letters in this section.

a Equivalent Expressions and Fractional Notation

An example of **fractional notation** for a number is

$$\frac{2}{3} \begin{array}{l} \longleftarrow \text{Numerator} \\ \longleftarrow \text{Denominator} \end{array}$$

The top number is called the **numerator,** and the bottom number is called the **denominator.**

The **whole numbers** consist of the natural numbers and 0:

$$0, \quad 1, \quad 2, \quad 3, \quad 4, \quad 5, \ldots$$

The **numbers of ordinary arithmetic,** also called the **nonnegative rational numbers,** consist of the whole numbers and the fractions, such as $\frac{2}{3}$ and $\frac{9}{5}$. The numbers of arithmetic can also be described as follows:

> The *numbers of arithmetic* are the whole numbers and the fractions, such as $\frac{3}{4}$, $\frac{6}{8}$, or 8. All of these numbers can be named with fractional notation $\frac{a}{b}$, where a and b are whole numbers and $b \neq 0$.

Note that all whole numbers are also numbers of arithmetic. We can say this because we can name a whole number like 8 with fractional notation as $\frac{8}{1}$. Henceforth, another name for a number will be an **equivalent expression.**

Being able to find an equivalent expression is critical to a study of algebra. Some simple but powerful properties of numbers that allow us to find equivalent expressions are the identity properties of 0 and 1.

> The Identity Property of 0
>
> **For any number a,**
> $$a + 0 = a.$$
> (Adding 0 to any number gives that same number.)
>
> The Identity Property of 1
>
> **For any number a,**
> $$a \cdot 1 = a.$$
> (Multiplying any number by 1 gives that same number.)

OBJECTIVES

After finishing Section R.2, you should be able to:

a Find equivalent fractional expressions by multiplying by 1.

b Simplify fractional notation.

c Add, subtract, multiply, and divide using fractional notation.

FOR EXTRA HELP

Tape NC   Tape 1A   MAC: R
                                IBM: R

1. Write a fractional expression equivalent to $\frac{2}{3}$ with a denominator of 12.

2. Write a fractional expression equivalent to $\frac{3}{5}$ with a denominator of 35.

3. Multiply by 1 to find three different fractional expressions for $\frac{7}{8}$.

Here are some of the ways to name the number 1:

$$\frac{5}{5}, \quad \frac{3}{3}, \quad \text{and} \quad \frac{26}{26}.$$

The following property allows us to find equivalent fractional expressions, that is, find other names for numbers of arithmetic.

> **Equivalent Expressions for 1**
>
> **For any number a, $a \neq 0$,**
>
> $$\frac{a}{a} = 1.$$

We can use the identity property of 1 and the preceding result to find equivalent fractional expressions.

▶ **EXAMPLE 1**　Write a fractional expression equivalent to $\frac{2}{3}$ with a denominator of 15.

Note that $15 = 3 \cdot 5$. We want a denominator of 15, but it is missing a factor of 5. We multiply by 1, using $\frac{5}{5}$ as an expression for 1. Recall from arithmetic that to multiply with fractional notation, we multiply numerators and denominators:

$$\frac{2}{3} = \frac{2}{3} \cdot 1 \qquad \text{Using the identity property of 1}$$
$$= \frac{2}{3} \cdot \frac{5}{5} \qquad \text{Using } \frac{5}{5} \text{ for 1}$$
$$= \frac{10}{15}. \qquad \text{Multiplying numerators and denominators} \quad ◀$$

DO EXERCISES 1–3.

b　**Simplifying Expressions**

We know that $\frac{1}{2}$, $\frac{2}{4}$, $\frac{4}{8}$, and so on, all name the same number. Any number of arithmetic can be named in many ways. The **simplest fractional notation** is the notation that has the smallest numerator and denominator. We call the process of finding the simplest fractional notation **simplifying**. We reverse the process of Example 1 by first factoring the numerator and the denominator. Then we factor the fractional expression and remove a factor of 1 using the identity property of 1.

▶ **EXAMPLE 2**　Simplify: $\frac{10}{15}$.

$$\frac{10}{15} = \frac{2 \cdot 5}{3 \cdot 5} \qquad \text{Factoring the numerator and the denominator}$$
$$= \frac{2}{3} \cdot \frac{5}{5} \qquad \text{Factoring the fractional expression}$$
$$= \frac{2}{3} \cdot 1$$
$$= \frac{2}{3} \qquad \text{Using the identity property of 1 (removing a factor of 1)} \quad ◀$$

▶ **EXAMPLE 3** Simplify: $\dfrac{36}{24}$.

$$\frac{36}{24} = \frac{6 \cdot 6}{4 \cdot 6} \qquad \text{Factoring the numerator and the denominator}$$

$$= \frac{3 \cdot 2 \cdot 6}{2 \cdot 2 \cdot 6} \qquad \text{Factoring further}$$

$$= \frac{3}{2} \cdot \frac{2 \cdot 6}{2 \cdot 6} \qquad \text{Factoring the fractional expression}$$

$$= \frac{3}{2} \cdot 1$$

$$= \frac{3}{2} \qquad \text{Removing a factor of 1} \qquad ◀$$

It is always a good idea to check at the end to see if you have indeed factored out all the common factors of the numerator and denominator.

Canceling

Canceling is a shortcut that you may have used for removing a factor of 1 when working with fractional notation. With *great* concern, we mention it as a possibility of speeding up your work. You should use canceling only when removing common factors in numerators and denominators. Each common factor allows us to remove a factor of 1 in a product. Canceling *may not* be done in sums or when adding expressions together. Our concern is that "canceling" be done with care and understanding. Example 3 might have been done faster as follows:

$$\frac{36}{24} = \frac{3 \cdot \cancel{2 \cdot 6}}{2 \cdot \cancel{2 \cdot 6}} = \frac{3}{2}, \qquad \text{or} \qquad \frac{36}{24} = \frac{3 \cdot \cancel{12}}{2 \cdot \cancel{12}} = \frac{3}{2} \qquad \text{or} \qquad \frac{\overset{3}{\cancel{18}}\ \cancel{36}}{\underset{2}{\cancel{24}}\ \cancel{12}} = \frac{3}{2}$$

CAUTION! The difficulty with canceling is that it is often applied incorrectly in situations like these:

$$\frac{\cancel{2} + 3}{\cancel{2}} = 3, \qquad \frac{\cancel{4} + 1}{\cancel{4} + 2} = \frac{1}{2}, \qquad \frac{\cancel{15}}{\cancel{54}} = \frac{1}{4}.$$

$$\text{Wrong!} \qquad \text{Wrong!} \qquad \text{Wrong!}$$

$$\frac{2 + 3}{2} = \frac{5}{2} \qquad \frac{4 + 1}{4 + 2} = \frac{5}{6} \qquad \frac{15}{54} = \frac{5}{18}$$

In each of these situations, the expressions canceled out were *not* factors of 1. Factors are parts of products. For example, in $2 \cdot 3$, 2 and 3 are factors, but in $2 + 3$, 2 and 3 are *not* factors. **If you can't factor, you can't cancel! If in doubt, don't cancel!**

DO EXERCISES 4–6.

The number of factors in the numerator and the denominator may not always be the same. If not, we can always insert the number 1 as a factor. The identity property of 1 allows us to do that.

Simplify

4. $\dfrac{18}{27}$

5. $\dfrac{38}{18}$

6. $\dfrac{56}{49}$

ANSWERS ON PAGE A-1

Simplify.

7. $\dfrac{27}{54}$

8. $\dfrac{48}{12}$

Multiply and simplify.

9. $\dfrac{6}{5} \cdot \dfrac{25}{12}$

10. $\dfrac{3}{8} \cdot \dfrac{5}{3} \cdot \dfrac{7}{2}$

▶ **EXAMPLE 4**　Simplify: $\dfrac{18}{72}$.

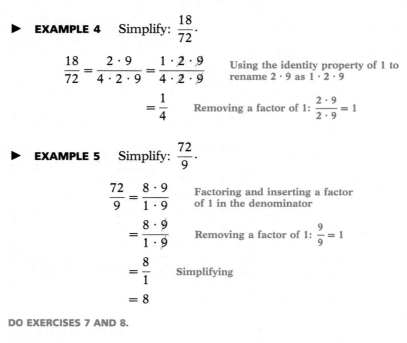

$$\frac{18}{72} = \frac{2 \cdot 9}{4 \cdot 2 \cdot 9} = \frac{1 \cdot 2 \cdot 9}{4 \cdot 2 \cdot 9}$$　Using the identity property of 1 to rename $2 \cdot 9$ as $1 \cdot 2 \cdot 9$

$$= \frac{1}{4}$$　Removing a factor of 1: $\dfrac{2 \cdot 9}{2 \cdot 9} = 1$　◀

▶ **EXAMPLE 5**　Simplify: $\dfrac{72}{9}$.

$$\frac{72}{9} = \frac{8 \cdot 9}{1 \cdot 9}$$　Factoring and inserting a factor of 1 in the denominator

$$= \frac{8 \cdot 9}{1 \cdot 9}$$　Removing a factor of 1: $\dfrac{9}{9} = 1$

$$= \frac{8}{1}$$　Simplifying

$$= 8$$　◀

DO EXERCISES 7 AND 8.

C　**Multiplication, Addition, Subtraction, and Division**

After we have performed an operation of multiplication, addition, subtraction, or division, the answer may or may not be simplified. We simplify, if at all possible. Let's continue as we have in the preceding examples.

Multiplication

▶ **EXAMPLE 6**　Multiply and simplify: $\dfrac{5}{6} \cdot \dfrac{9}{25}$.

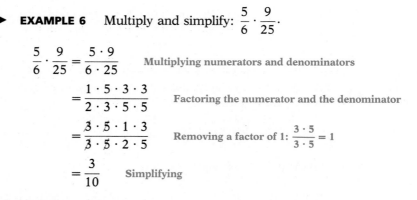

$$\frac{5}{6} \cdot \frac{9}{25} = \frac{5 \cdot 9}{6 \cdot 25}$$　Multiplying numerators and denominators

$$= \frac{1 \cdot 5 \cdot 3 \cdot 3}{2 \cdot 3 \cdot 5 \cdot 5}$$　Factoring the numerator and the denominator

$$= \frac{3 \cdot 5 \cdot 1 \cdot 3}{3 \cdot 5 \cdot 2 \cdot 5}$$　Removing a factor of 1: $\dfrac{3 \cdot 5}{3 \cdot 5} = 1$

$$= \frac{3}{10}$$　Simplifying　◀

DO EXERCISES 9 AND 10.

Addition

When denominators are the same, we can add by adding the numerators and keeping the same denominator.

▶ **EXAMPLE 7** Add and simplify: $\dfrac{4}{8} + \dfrac{5}{8}$.

The common denominator is 8. We add the numerators and keep the common denominator:

$$\frac{4}{8} + \frac{5}{8} = \frac{4+5}{8} = \frac{9}{8}. \qquad ◀$$

In arithmetic you usually write $1\frac{1}{8}$ rather than $\frac{9}{8}$. In algebra you will find that *improper* symbols such as $\frac{9}{8}$ are more useful and are quite *proper* for our purposes.

When denominators are different, we use the property of 1 and multiply to find a common denominator. The smallest such denominator is called the lowest or **least common denominator.** That number is the least common multiple of the original denominators. The least common denominator is often abbreviated **LCD.**

▶ **EXAMPLE 8** Add and simplify: $\dfrac{3}{8} + \dfrac{5}{12}$.

The LCD is 24. We multiply by 1 in each case to obtain the LCD:

$$\frac{3}{8} + \frac{5}{12} = \frac{3}{8} \cdot \frac{3}{3} + \frac{5}{12} \cdot \frac{2}{2} \quad \begin{array}{l}\textbf{Multiplying by 1.} \text{ Since } 3 \cdot 8 = 24, \text{ we multiply} \\ \text{the first number by } \frac{3}{3}. \text{ Since } 2 \cdot 12 = 24, \text{ we} \\ \text{multiply the second number by } \frac{2}{2}.\end{array}$$

$$= \frac{9}{24} + \frac{10}{24}$$

$$= \frac{19}{24}. \qquad ◀$$

DO EXERCISES 11–14.

Subtraction

▶ **EXAMPLE 9** Subtract and simplify: $\dfrac{9}{8} - \dfrac{4}{5}$.

$$\frac{9}{8} - \frac{4}{5} = \frac{9}{8} \cdot \frac{5}{5} - \frac{4}{5} \cdot \frac{8}{8} \qquad \text{The LCD is 40.}$$

$$= \frac{45}{40} - \frac{32}{40}$$

$$= \frac{13}{40} \qquad ◀$$

Add and simplify.

11. $\dfrac{4}{5} + \dfrac{3}{5}$

12. $\dfrac{5}{6} + \dfrac{7}{6}$

13. $\dfrac{5}{6} + \dfrac{7}{10}$

14. $\dfrac{1}{4} + \dfrac{1}{3}$

ANSWERS ON PAGE A-1

Subtract and simplify.

15. $\dfrac{4}{5} - \dfrac{4}{6}$

16. $\dfrac{5}{12} - \dfrac{2}{9}$

Find the reciprocal.

17. $\dfrac{4}{11}$ **18.** $\dfrac{15}{7}$

19. 5   **20.** $\dfrac{1}{3}$

21. Divide by multiplying by 1:

$$\dfrac{\frac{3}{5}}{\frac{4}{7}}.$$

▶ **EXAMPLE 10**   Subtract and simplify: $\dfrac{7}{10} - \dfrac{1}{5}$.

$$\dfrac{7}{10} - \dfrac{1}{5} = \dfrac{7}{10} - \dfrac{1}{5} \cdot \dfrac{2}{2} \qquad \text{The LCD is 10.}$$

$$= \dfrac{7}{10} - \dfrac{2}{10}$$

$$= \dfrac{5}{10} = \dfrac{1 \cdot \cancel{5}}{2 \cdot \cancel{5}} = \dfrac{1}{2} \qquad \text{Removing a factor of 1: } \dfrac{5}{5} = 1 \qquad ◀$$

DO EXERCISES 15 AND 16.

Reciprocals

Two numbers whose product is 1 are called **reciprocals,** or **multiplicative inverses,** of each other. All the numbers of arithmetic, except zero, have reciprocals.

▶ **EXAMPLES**

11. The reciprocal of $\frac{2}{3}$ is $\frac{3}{2}$ because $\frac{2}{3} \cdot \frac{3}{2} = \frac{6}{6} = 1$.

12. The reciprocal of 9 is $\frac{1}{9}$ because $9 \cdot \frac{1}{9} = \frac{9}{9} = 1$.

13. The reciprocal of $\frac{1}{4}$ is 4 because $\frac{1}{4} \cdot 4 = 1$.   ◀

DO EXERCISES 17–20.

Reciprocals and Division

The number 1 and reciprocals can be used to justify a fast way to divide numbers of arithmetic. We multiply by 1, choosing carefully the symbol for 1.

▶ **EXAMPLE 14**   Divide $\dfrac{2}{3}$ by $\dfrac{7}{5}$.

This is a symbol for 1.

$$\dfrac{\frac{2}{3}}{\frac{7}{5}} = \dfrac{\frac{2}{3}}{\frac{7}{5}} \cdot \dfrac{\frac{5}{7}}{\frac{5}{7}} \qquad \text{Multiplying by } \dfrac{\frac{5}{7}}{\frac{5}{7}}. \text{ We use } \tfrac{5}{7} \text{ because it is the reciprocal of } \tfrac{7}{5}.$$

$$= \dfrac{\frac{2}{3} \cdot \frac{5}{7}}{\frac{7}{5} \cdot \frac{5}{7}} \qquad \text{Multiplying numerators and denominators}$$

$$= \dfrac{\frac{10}{21}}{1} = \dfrac{10}{21} \qquad \text{Simplifying}$$

After multiplying we got 1 for a denominator. That was because we used $\frac{5}{7}$, the reciprocal of the divisor, for both the numerator and the denominator of the symbol for 1.   ◀

DO EXERCISE 21.

When multiplying by 1 to divide, we get a denominator of 1. What do we get in the numerator? In Example 14, we got $\frac{2}{3} \times \frac{5}{7}$. This is the product of $\frac{2}{3}$, the dividend, and $\frac{5}{7}$, the reciprocal of the divisor.

> **To divide, multiply by the reciprocal of the divisor:**
> $$\frac{a}{b} \div \frac{c}{d} = \frac{a}{b} \cdot \frac{d}{c}.$$

▶ **EXAMPLE 15** Divide $\frac{1}{2} \div \frac{3}{5}$ by multiplying by the reciprocal of the divisor.

$$\frac{1}{2} \div \frac{3}{5} = \frac{1}{2} \cdot \frac{5}{3} = \frac{5}{6} \qquad \frac{5}{3} \text{ is the reciprocal of } \frac{3}{5} \qquad ◀$$

After dividing, simplification is often possible and should be done.

▶ **EXAMPLE 16** Divide: $\frac{2}{3} \div \frac{4}{9}$.

$$\frac{2}{3} \div \frac{4}{9} = \frac{2}{3} \cdot \frac{9}{4} = \frac{2 \cdot 3 \cdot 3}{3 \cdot 2 \cdot 2} \qquad \text{Removing a factor of 1: } \frac{2 \cdot 3}{2 \cdot 3} = 1$$

$$= \frac{3}{2} \qquad ◀$$

DO EXERCISES 22–24.

▶ **EXAMPLE 17** Divide: $\frac{5}{6} \div 30$.

$$\frac{5}{6} \div 30 = \frac{5}{6} \div \frac{30}{1} = \frac{5}{6} \cdot \frac{1}{30} = \frac{5 \cdot 1}{6 \cdot 5 \cdot 6} = \frac{1}{36}$$

Removing a factor of 1: $\frac{5}{5} = 1$ ◀

▶ **EXAMPLE 18** Divide: $24 \div \frac{3}{8}$.

$$24 \div \frac{3}{8} = \frac{24}{1} \div \frac{3}{8} = \frac{24}{1} \cdot \frac{8}{3} = \frac{24 \cdot 8}{1 \cdot 3} = \frac{3 \cdot 8 \cdot 8}{1 \cdot 3} = \frac{8 \cdot 8}{1} = 64$$

Removing a factor of 1: $\frac{3}{3} = 1$ ◀

DO EXERCISES 25 AND 26.

Divide by multiplying by the reciprocal of the divisor.

22. $\frac{4}{3} \div \frac{7}{2}$

23. $\frac{5}{4} \div \frac{3}{2}$

24. $\frac{\frac{2}{9}}{\frac{5}{12}}$

Divide.

25. $\frac{7}{8} \div 56$

26. $36 \div \frac{4}{9}$

ANSWERS ON PAGE A-1

From time to time you will find some *"Sidelights"* like the one below. These are optional, but you may find them helpful and of interest. They will include such topics as study tips, career opportunities involving mathematics, applications, computer–calculator exercises, or other mathematical topics.

❖ SIDELIGHTS

Study Tips

Many students begin the study of a text by opening to the first section assigned by an instructor. There are many ways in which you can enhance your use of this book, and they have been outlined carefully in a page in the preface titled *To the student*. If you have not read that page, do so now before you start the exercise set on the next page.

There are some points on that page that bear repeating here.

- *Be sure to note the special symbols* a , b , c , *and so on, that correspond to the objectives you are to learn.* They appear many places throughout the text. The first time you will see them is in the headings for the section. The second time you will see them is in the exercise set, as follows. You will also see them in the answers to the Review Exercises, the Chapter Tests, and the Cumulative Reviews. These allow you to reference back when you need to review a topic.

- *Be sure to note also the symbols in the margin under the list of objectives at the beginning of the section.* These refer to the many distinctive study aids that accompany the book.

- *Be sure to stop and do the margin exercises as you study a section.* When our students come to us troubled about how they are doing in the course, the first question we ask them is "Are you doing the margin exercises when directed to do so?" This is one of the most effective ways to enhance your ability to learn mathematics from this text. Don't deprive yourself of its benefits!

- *When you study the book, don't mark points you think are important, but mark the points you do not understand!* The book is written with all kinds of processes that highlight important points. Use your efforts to mark where you are having trouble. Then when you go to class or a math lab or a tutoring session, you are prepared to ask questions that close in on your difficulties.

- *Try to keep one section ahead of your syllabus.* We have tried to write a book that is readable for students. If you study ahead of your lectures, you can concentrate on just the lectures, rather than trying to write everything down. You can then take notes only of special points or of questions related to what is happening in class.

EXERCISE SET R.2

a Write an equivalent expression for each of the following. Use the indicated name for 1.

1. $\dfrac{5}{6}$ $\left(\text{Use } \dfrac{8}{8} \text{ for } 1.\right)$

2. $\dfrac{9}{10}$ $\left(\text{Use } \dfrac{11}{11} \text{ for } 1.\right)$

3. $\dfrac{6}{7}$ $\left(\text{Use } \dfrac{100}{100} \text{ for } 1.\right)$

4. $\dfrac{7}{10}$ $\left(\text{Use } \dfrac{3}{3} \text{ for } 1.\right)$

5. $\dfrac{13}{20}$ $\left(\text{Use } \dfrac{6}{6} \text{ for } 1.\right)$

6. $\dfrac{11}{32}$ $\left(\text{Use } \dfrac{20}{20} \text{ for } 1.\right)$

Write an equivalent expression with the given denominator.

7. $\dfrac{7}{8}$ (Denominator: 24)

8. $\dfrac{5}{6}$ (Denominator: 48)

9. $\dfrac{x}{3}$ (Denominator: 15)

10. $\dfrac{4}{9}$ (Denominator: 54)

b Simplify.

11. $\dfrac{18}{45}$

12. $\dfrac{16}{56}$

13. $\dfrac{49}{14}$

14. $\dfrac{72}{27}$

15. $\dfrac{6}{42}$

16. $\dfrac{13}{104}$

17. $\dfrac{56}{7}$

18. $\dfrac{132}{11}$

19. $\dfrac{19}{76}$

20. $\dfrac{17}{51}$

21. $\dfrac{100}{20}$

22. $\dfrac{150}{25}$

23. $\dfrac{425}{525}$

24. $\dfrac{625}{325}$

25. $\dfrac{2600}{1400}$

26. $\dfrac{4800}{1600}$

27. $\dfrac{8 \cdot x}{6 \cdot x}$

28. $\dfrac{13 \cdot v}{39 \cdot v}$

1. _____
2. _____
3. _____
4. _____
5. _____
6. _____
7. _____
8. _____
9. _____
10. _____
11. _____
12. _____
13. _____
14. _____
15. _____
16. _____
17. _____
18. _____
19. _____
20. _____
21. _____
22. _____
23. _____
24. _____
25. _____
26. _____
27. _____
28. _____

C Compute and simplify.

29. $\dfrac{1}{4} \cdot \dfrac{1}{2}$　　　**30.** $\dfrac{11}{10} \cdot \dfrac{8}{5}$　　　**31.** $\dfrac{17}{2} \cdot \dfrac{3}{4}$　　　**32.** $\dfrac{11}{12} \cdot \dfrac{12}{11}$

33. $\dfrac{1}{2} + \dfrac{1}{2}$　　　**34.** $\dfrac{1}{2} + \dfrac{1}{4}$　　　**35.** $\dfrac{4}{9} + \dfrac{13}{18}$　　　**36.** $\dfrac{4}{5} + \dfrac{8}{15}$

37. $\dfrac{3}{10} + \dfrac{8}{15}$　　　**38.** $\dfrac{9}{8} + \dfrac{7}{12}$　　　**39.** $\dfrac{5}{4} - \dfrac{3}{4}$　　　**40.** $\dfrac{12}{5} - \dfrac{2}{5}$

41. $\dfrac{13}{18} - \dfrac{4}{9}$　　　**42.** $\dfrac{13}{15} - \dfrac{8}{45}$　　　**43.** $\dfrac{11}{12} - \dfrac{2}{5}$　　　**44.** $\dfrac{15}{16} - \dfrac{2}{3}$

45. $\dfrac{7}{6} \div \dfrac{3}{5}$　　　**46.** $\dfrac{7}{5} \div \dfrac{3}{4}$　　　**47.** $\dfrac{8}{9} \div \dfrac{4}{15}$　　　**48.** $\dfrac{3}{4} \div \dfrac{3}{7}$

49. $\dfrac{1}{4} \div \dfrac{1}{2}$　　　**50.** $\dfrac{1}{10} \div \dfrac{1}{5}$　　　**51.** $\dfrac{\frac{13}{12}}{\frac{39}{5}}$　　　**52.** $\dfrac{\frac{17}{6}}{\frac{3}{8}}$

53. $100 \div \dfrac{1}{5}$　　　**54.** $78 \div \dfrac{1}{6}$　　　**55.** $\dfrac{3}{4} \div 10$　　　**56.** $\dfrac{5}{6} \div 15$

SKILL MAINTENANCE

This heading indicates that the exercises that follow are *Skill Maintenance Exercises,* which review skills previously studied in the text. You can expect such exercises in almost every exercise set.

57. Find the prime factorization of 192.　　　**58.** Find the LCM of 28, 49, and 56.

SYNTHESIS

Simplify.

59. $\dfrac{128}{192}$　　　**60.** $\dfrac{p \cdot q}{r \cdot q}$　　　**61.** $\dfrac{33 \cdot b \cdot a}{11 \cdot b \cdot a}$

62. $\dfrac{4 \cdot 9 \cdot 16}{2 \cdot 8 \cdot 15}$　　　**63.** $\dfrac{36 \cdot (2 \cdot h)}{8 \cdot (9 \cdot h)}$

R.3 Decimal Notation

Let's say that the cost of a stereo system is

$$\$1768.95.$$

This is given in **decimal notation.** The following place-value chart shows the place value of each digit in 1768.95.

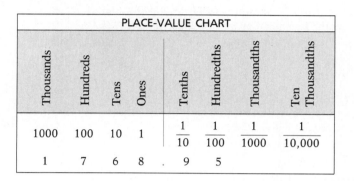

				PLACE-VALUE CHART			
Thousands	Hundreds	Tens	Ones	Tenths	Hundredths	Thousandths	Ten Thousandths
1000	100	10	1	$\frac{1}{10}$	$\frac{1}{100}$	$\frac{1}{1000}$	$\frac{1}{10,000}$
1	7	6	8	9	5		

OBJECTIVES

After finishing Section R.3, you should be able to:

a Convert between decimal notation and fractional notation.

b Add, subtract, multiply, and divide using decimal notation.

c Round numbers to a specified decimal place.

FOR EXTRA HELP

Tape 1B   Tape 1B   MAC: R
                                  IBM: R

a Converting Between Certain Decimal Notation and Fractional Notation

Decimals are defined in terms of fractions—for example,

$$0.1 = \frac{1}{10}, \qquad 0.6875 = \frac{6875}{10,000}, \qquad 53.47 = \frac{5347}{100}.$$

We see that the number of zeros in the denominator is the same as the number of decimal places in the number. From these examples, we obtain the following procedure for converting from this type of decimal notation to fractional notation.

Convert to fractional notation. Do not simplify.

1. 0.568

To convert from decimal notation to fractional notation:

a) **Count the number of decimal places.**   4.98
                                                  ↑
                                              2 places

b) **Move the decimal point that many places to the right.**   4.98
                                                Move 2 places

c) **Write the answer over a denominator with that number of zeros.** $\frac{498}{100}$
                                                  ↑
                                              2 zeros

2. 2.3

3. 89.04

► **EXAMPLE 1** Convert 0.876 to fractional notation. Do not simplify.

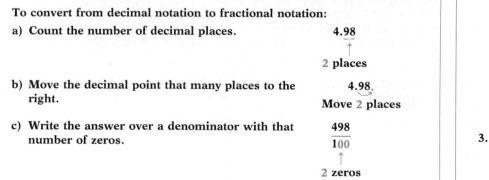

$$0.876 \qquad 0.876 \qquad 0.876 = \frac{876}{1000}$$

3 places   3 zeros

◄

Convert to decimal notation.

4. $\dfrac{4131}{1000}$

5. $\dfrac{4131}{10,000}$

6. $\dfrac{573}{100}$

Add.

7. $69 + 1.785 + 213.67$

8. $17.95 + 14.68 + 236$

▶ **EXAMPLE 2**   Convert 1.5018 to fractional notation. Do not simplify.

$$1.5018 \qquad 1.\underbrace{5018.}_{4 \text{ places}} \qquad 1.5018 = \dfrac{15,018}{\underset{\uparrow}{10,000}}$$

4 zeros ◀

DO EXERCISES 1–3 ON THE PRECEDING PAGE.

To convert from fractional notation to decimal notation when the denominator is a number like 10, 100, or 1000:

a) Count the number of zeros.

$$\dfrac{8679}{\underset{\uparrow}{1000}}$$

3 zeros

b) Move the decimal point that number of places to the left. Leave off the denominator.

$$8.\overset{\frown}{679}.$$

Move 3 places

▶ **EXAMPLE 3**   Convert to decimal notation: $\dfrac{123,067}{10,000}$.

$$\dfrac{123,067}{\underset{\uparrow}{10,000}} \qquad 12.\underbrace{3067.}_{4 \text{ places}} \qquad \dfrac{123,067}{10,000} = 12.3067$$

4 zeros ◀

DO EXERCISES 4–6.

---

b   **Addition, Subtraction, Multiplication, and Division**

Addition

To add with decimal notation:

Adding with decimal notation is similar to adding whole numbers. Add the thousandths, and then the hundredths, carrying if necessary. Then go on to the tenths, then the ones, and so on. To keep place values straight, line up the decimal points in a vertical column.

▶ **EXAMPLE 4**   Add: $74 + 26.46 + 0.998$.

We have

$$
\begin{array}{r}
\overset{1\;\;1\;\;1}{7\,4.} \\
2\,6.4\,6 \\
+\quad 0.9\,9\,8 \\
\hline
1\,0\,1.4\,5\,8
\end{array}
$$

You may put extra zeros to the right of any decimal point so there are the same number of decimal places, but this is not necessary. If you did, the preceding problem would look like this:

$$
\begin{array}{r}
7\,4.0\,0\,0 \\
2\,6.4\,6\,0 \\
+\quad 0.9\,9\,8 \\
\hline
1\,0\,1.4\,5\,8
\end{array}
$$

◀

DO EXERCISES 7 AND 8.

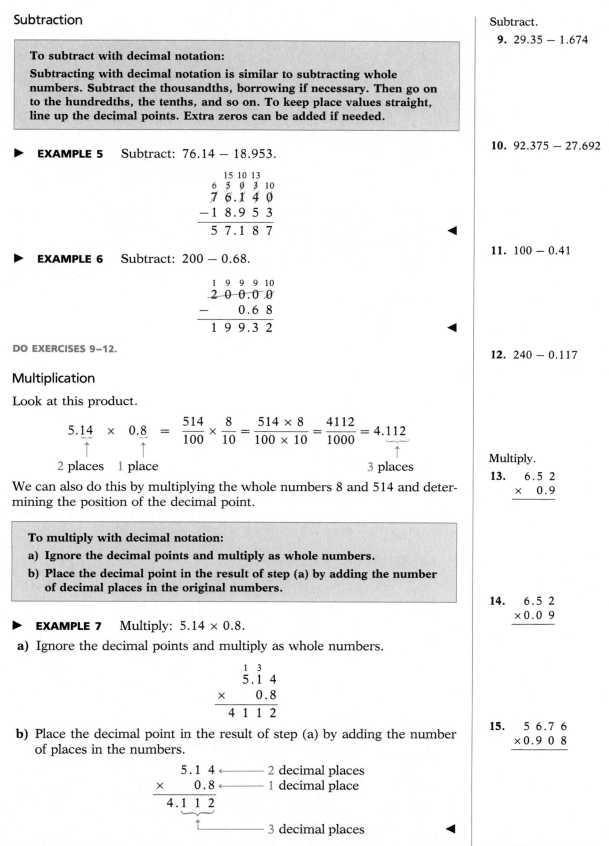

Subtraction

Subtract.
9. $29.35 - 1.674$

> **To subtract with decimal notation:**
>
> Subtracting with decimal notation is similar to subtracting whole numbers. Subtract the thousandths, borrowing if necessary. Then go on to the hundredths, the tenths, and so on. To keep place values straight, line up the decimal points. Extra zeros can be added if needed.

▶ **EXAMPLE 5**   Subtract: $76.14 - 18.953$.

$$
\begin{array}{r}
\overset{15\ 10\ 13}{\underset{6\ \ 5\ \ \cancel{0}\ \ \cancel{3}\ \ 10}{7\ 6.1\ 4\ 0}} \\
- 1\ 8.9\ 5\ 3 \\
\hline
5\ 7.1\ 8\ 7
\end{array}
$$
◀

10. $92.375 - 27.692$

▶ **EXAMPLE 6**   Subtract: $200 - 0.68$.

$$
\begin{array}{r}
\overset{1\ \ 9\ \ 9\ \ 9\ \ 10}{\cancel{2}\ \cancel{0}\ \cancel{0}.\cancel{0}\ \cancel{0}} \\
-\ \ \ \ \ \ 0.6\ 8 \\
\hline
1\ 9\ 9.3\ 2
\end{array}
$$
◀

DO EXERCISES 9–12.

11. $100 - 0.41$

Multiplication

Look at this product.

$$5.14 \times 0.8 = \frac{514}{100} \times \frac{8}{10} = \frac{514 \times 8}{100 \times 10} = \frac{4112}{1000} = 4.112$$

2 places 1 place   3 places

We can also do this by multiplying the whole numbers 8 and 514 and determining the position of the decimal point.

12. $240 - 0.117$

> **To multiply with decimal notation:**
>
> a) Ignore the decimal points and multiply as whole numbers.
> b) Place the decimal point in the result of step (a) by adding the number of decimal places in the original numbers.

Multiply.
13.
$$\begin{array}{r} 6.5\ 2 \\ \times\ \ \ 0.9 \\ \hline \end{array}$$

▶ **EXAMPLE 7**   Multiply: 5.14×0.8.

a) Ignore the decimal points and multiply as whole numbers.

$$\begin{array}{r} \overset{1\ \ 3}{5.1\ 4} \\ \times\ \ \ \ 0.8 \\ \hline 4\ 1\ 1\ 2 \end{array}$$

14.
$$\begin{array}{r} 6.5\ 2 \\ \times 0.0\ 9 \\ \hline \end{array}$$

b) Place the decimal point in the result of step (a) by adding the number of places in the numbers.

$$\begin{array}{r} 5.1\ 4 \longleftarrow 2 \text{ decimal places} \\ \times\ \ \ \ 0.8 \longleftarrow 1 \text{ decimal place} \\ \hline 4.1\ 1\ 2 \end{array}$$

3 decimal places
◀

15.
$$\begin{array}{r} 5\ 6.7\ 6 \\ \times 0.9\ 0\ 8 \\ \hline \end{array}$$

DO EXERCISES 13–15.

ANSWERS ON PAGE A-1

Divide.

16. 7)3 4 2.3

17. 1 6)2 5 3.1 2

Divide.

18. 2 5)3 2

19. 3 8)6 8 2.1

Division

Note that $37.6 \div 8 = 4.7$ because $8 \times 4.7 = 37.6$. If we write this as

```
         4.7
   8)3 7.6
     3 2
       5 6
       5 6
         0
```

we see how the following method can be used to divide by a whole number.

> **To divide by a whole number:**
> a) **Place the decimal point in the quotient directly above the decimal point in the dividend.**
> b) **Divide as whole numbers.**

▶ **EXAMPLE 8**   Divide: $216.75 \div 25$.

a)
```
            .
   2 5)2 1 6.7 5
```

b)
```
            8.6 7
   2 5)2 1 6.7 5
       2 0 0
       1 6 7
       1 5 0
         1 7 5
         1 7 5
             0
```
◀

DO EXERCISES 16 AND 17.

It is sometimes helpful to write extra zeros to the right of the decimal point. The answer is not changed. Remember that the decimal point for a whole number, though not normally written, is to the right of the number.

▶ **EXAMPLE 9**   Divide: $54 \div 8$.

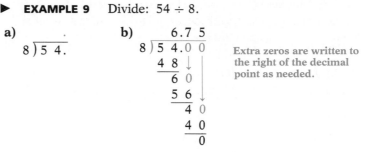

a)
```
         .
   8)5 4.
```

b)
```
         6.7 5
   8)5 4.0 0
     4 8
       6 0
       5 6
         4 0
         4 0
           0
```
Extra zeros are written to the right of the decimal point as needed.

◀

DO EXERCISES 18 AND 19.

> **To divide when the divisor is not a whole number:**
> a) **Move the decimal point in the divisor as many places to the right as it takes to make it a whole number. Move the decimal point in the dividend the same number of places to the right and place the decimal point in the quotient.**
> b) **Divide as whole numbers, adding zeros if necessary.**

► **EXAMPLE 10**   Divide: $83.79 \div 0.098$.

a)

$$0.098\overline{)83.790.}$$

b)

$$
\begin{array}{r}
8\ 5\ 5.\\
0.098_\wedge\overline{)83.790_\wedge}\\
7\ 8\ 4\\
\hline
5\ 3\ 9\\
4\ 9\ 0\\
\hline
4\ 9\ 0\\
4\ 9\ 0\\
\hline
0
\end{array}
$$

◄

DO EXERCISES 20 AND 21.

Divide.

20. $0.024\overline{)20.544}$

21. $4.6\overline{)3.91}$

Repeating Decimals

To convert from fractional notation to decimal notation when the denominator is not a number like 10, 100, or 1000, we divide the numerator by the denominator.

► **EXAMPLE 11**   Convert to decimal notation: $\frac{5}{16}$.

$$
\begin{array}{r}
0.3\ 1\ 2\ 5\\
16\overline{)5.0\ 0\ 0\ 0}\\
4\ 8\\
\hline
2\ 0\\
1\ 6\\
\hline
4\ 0\\
3\ 2\\
\hline
8\ 0\\
8\ 0\\
\hline
0
\end{array}
$$

If we get a remainder of 0, the decimal terminates. Decimal notation is 0.3125.   ◄

Convert to decimal notation.

22. $\frac{5}{8}$

23. $\frac{2}{3}$

24. $\frac{84}{11}$

► **EXAMPLE 12**   Convert to decimal notation: $\frac{7}{12}$.

$$
\begin{array}{r}
0.5\ 8\ 3\ 3\\
12\overline{)7.0\ 0\ 0\ 0}\\
6\ 0\\
\hline
1\ 0\ 0\\
9\ 6\\
\hline
4\ 0\\
3\ 6\\
\hline
4\ 0\\
3\ 6\\
\hline
4
\end{array}
$$

The number 4 repeats as a remainder, so the digits will repeat in the quotient. Therefore,

$$\frac{7}{12} = 0.583333\ldots.$$

ANSWERS ON PAGE A-1

Round to the nearest tenth.

25. 2.76

26. 13.85

27. 7.009

Round to the nearest hundredth.

28. 7.834

29. 34.675

30. 0.025

Round to the nearest thousandth.

31. 0.9434

32. 8.0038

33. 43.1119

34. 37.4005

Round 7459.3549 to the nearest:

35. thousandth.

36. hundredth.

37. tenth.

38. one.

39. ten.

ANSWERS ON PAGE A-1

Instead of dots, we often put a bar over the repeating part, in this case only the 3. Thus,

$$\frac{7}{12} = 0.58\overline{3}.$$ ◄

DO EXERCISES 22–24.

c   Rounding

When working with decimal notation, we often shorten notation by **rounding.** Although there are many ways to round, we will use the following.

> **To round to a certain place:**
> a) **Locate the digit in that place.**
> b) **Then consider the digit to its right.**
> c) **If the digit to the right is 5 or higher, round up; if the digit to the right is less than 5, round down.**

► **EXAMPLE 13**   Round 3872.2459 to the nearest tenth.

a) We locate the digit in the tenths place.

$$3 \; 8 \; 7 \; 2.2 \; 4 \; 5 \; 9$$
$$\uparrow$$

b) Then we consider the next digit to the right.

$$3 \; 8 \; 7 \; 2.2 \; 4 \; 5 \; 9$$
$$\uparrow$$

c) Since that digit is less than 5, we round down.

$$3 \; 8 \; 7 \; 2.2 \; \longleftarrow \text{ This is the answer.}$$ ◄

Note that 3872.3 is *not* a correct answer to Example 13. It is incorrect to round from the ten-thousandths place over as follows:

$$3872.246, \quad 3872.25, \quad 3872.3$$

► **EXAMPLE 14**   Round 3872.2459 to the nearest thousandth, hundredth, tenth, one, ten, hundred, and thousand.

thousandth:	3872.246
hundredth:	3872.25
tenth:	3872.2
one:	3872
ten:	3870
hundred:	3900
thousand:	4000

◄

DO EXERCISES 25–39.

We sometimes use the symbol ≈, meaning "is approximately equal to." Thus,

$$46.124 \approx 46.1.$$

NAME   SECTION   DATE

EXERCISE SET R.3

a Convert to fractional notation. Do not simplify.

1. 4.9   **2.** 1.3   **3.** 0.59   **4.** 0.81

5. 2.0007   **6.** 4.0008   **7.** 7889.8   **8.** 1122.3

Convert to decimal notation.

9. $\dfrac{1}{10}$ **10.** $\dfrac{1}{100}$ **11.** $\dfrac{1}{10,000}$ **12.** $\dfrac{1}{1000}$

13. $\dfrac{3079}{10}$ **14.** $\dfrac{1796}{10}$ **15.** $\dfrac{9999}{1000}$ **16.** $\dfrac{17}{1000}$

17. $\dfrac{39}{10,000}$ **18.** $\dfrac{4578}{10,000}$ **19.** $\dfrac{1}{100,000}$ **20.** $\dfrac{94}{100,000}$

b Add.

21. $\begin{array}{r} 4\,1\,5.7\,8 \\ +\ \ 2\,9.1\,6 \\ \hline \end{array}$ **22.** $\begin{array}{r} 7\,0\,8.9\,9 \\ +\ \ 7\,5.4\,8 \\ \hline \end{array}$ **23.** $\begin{array}{r} 2\,3\,4.0\,0\,0 \\ +1\,5\,6.6\,1\,7 \\ \hline \end{array}$ **24.** $\begin{array}{r} 1\,3\,4\,5.1\,2 \\ +\ \ 5\,6\,6.9\,8 \\ \hline \end{array}$

25. $85 + 67.95 + 2.774$ **26.** $119 + 43.74 + 18.876$

27. $17.95 + 16.99 + 28.85$ **28.** $14.59 + 16.79 + 19.95$

Subtract.

29. $\begin{array}{r} 7\,8.1\,1 \\ -4\,5.8\,7\,6 \\ \hline \end{array}$ **30.** $\begin{array}{r} 1\,4.0\,8 \\ -\ \ 9.1\,9\,9 \\ \hline \end{array}$ **31.** $\begin{array}{r} 3\,8.7\,0\,0 \\ -1\,1.8\,6\,5 \\ \hline \end{array}$ **32.** $\begin{array}{r} 3\,0\,0.0\,0\,0 \\ -\ \ 2\,4.6\,7\,7 \\ \hline \end{array}$

33. $57.86 - 9.95$ **34.** $2.6 - 1.08$ **35.** $3 - 1.0807$ **36.** $5 - 3.4051$

Multiply.

37. $\begin{array}{r} 6.2\,3 \\ \times\ \ \ 1.6 \\ \hline \end{array}$ **38.** $\begin{array}{r} 5.4\,4 \\ \times\ \ \ 3.2 \\ \hline \end{array}$ **39.** $\begin{array}{r} 0.5\,6 \\ \times 0.7\,8 \\ \hline \end{array}$ **40.** $\begin{array}{r} 0.0\,2\,4 \\ \times 6.8\,0\,7 \\ \hline \end{array}$

ANSWERS

1. _____
2. _____
3. _____
4. _____
5. _____
6. _____
7. _____
8. _____
9. _____
10. _____
11. _____
12. _____
13. _____
14. _____
15. _____
16. _____
17. _____
18. _____
19. _____
20. _____
21. _____
22. _____
23. _____
24. _____
25. _____
26. _____
27. _____
28. _____
29. _____
30. _____
31. _____
32. _____
33. _____
34. _____
35. _____
36. _____
37. _____
38. _____
39. _____
40. _____

41. $\begin{array}{r} 1\,7.9\,5 \\ \times\quad 1\,0 \\ \hline \end{array}$ **42.** $\begin{array}{r} 1\,7.9\,5 \\ \times\quad 1\,0\,0 \\ \hline \end{array}$ **43.** $\begin{array}{r} 1\,8.9\,4 \\ \times\quad 0.0\,1 \\ \hline \end{array}$ **44.** $\begin{array}{r} 1\,8.9\,4 \\ \times\quad 0.1 \\ \hline \end{array}$

45. $\begin{array}{r} 0.4\,5\,7 \\ \times\quad 3.0\,8 \\ \hline \end{array}$ **46.** $\begin{array}{r} 0.0\,0\,2\,4 \\ \times\quad 0.0\,1\,5 \\ \hline \end{array}$ **47.** $\begin{array}{r} 3.6\,4\,2 \\ \times\quad 0.9\,9 \\ \hline \end{array}$ **48.** $\begin{array}{r} 2\,8\,7.4 \\ \times\quad 1.0\,8 \\ \hline \end{array}$

Divide.

49. $7\,2\,)\overline{1\,6\,5.6}$ **50.** $5.2\,)\overline{4\,4.2}$ **51.** $8.5\,)\overline{4\,4.2}$

52. $7.8\,)\overline{7\,2.5\,4}$ **53.** $9.9\,)\overline{0.2\,2\,7\,7}$ **54.** $1\,0\,0\,)\overline{9\,5}$

55. $0.6\,4\,)\overline{1\,2}$ **56.** $1.6\,)\overline{7\,5}$ **57.** $1.0\,5\,)\overline{6\,9\,3}$

Convert to decimal notation.

58. $\dfrac{11}{32}$ **59.** $\dfrac{17}{32}$ **60.** $\dfrac{13}{11}$ **61.** $\dfrac{17}{12}$

62. $\dfrac{5}{9}$ **63.** $\dfrac{5}{6}$ **64.** $\dfrac{19}{9}$ **65.** $\dfrac{9}{11}$

c Round to the nearest hundredth, tenth, one, ten, and hundred.
66. 745.06534   **67.** 317.18565   **68.** 6780.50568   **69.** 840.15493

Round to the nearest cent and to the nearest dollar (nearest one).
70. $17.988   **71.** $20.492   **72.** $346.075   **73.** $4.718

Round to the nearest dollar.
74. $16.95   **75.** $17.50   **76.** $189.50   **77.** $567.24

Divide and round to the nearest ten-thousandth, thousandth, hundredth, tenth, and one.
78. $\dfrac{1000}{81}$ **79.** $\dfrac{23}{17}$

R.4 Percent Notation

a Converting from Percent Notation to Decimal Notation

The average family spends 26% of its income for food. What does this mean? It means that out of every $100 earned, $26 is spent for food. Thus 26% is a ratio of 26 to 100.

INCOME

Food
26%

The percent symbol % means "per hundred." We can regard the percent symbol as part of a name for a number. For example,

26% is defined to mean 26×0.01 or $26 \times \dfrac{1}{100}$ or $\dfrac{26}{100}$.

In general,

> $n\%$ means $n \times 0.01$, or $n \times \dfrac{1}{100}$, or $\dfrac{n}{100}$.

▶ **EXAMPLE 1** Convert 78.5% to decimal notation.

$$78.5\% = 78.5 \times 0.01 \qquad \text{Replacing \% by} \times 0.01$$
$$= 0.785 \qquad ◀$$

> To convert from percent notation to decimal notation, move the decimal point two places to the left and drop the percent symbol.

▶ **EXAMPLE 2** Convert 43.67% to decimal notation.

43.67%   0.43.67   43.67% = 0.4367

Move the decimal point two places to the left.   ◀

DO EXERCISES 1 AND 2.

b Converting from Percent Notation to Fractional Notation

▶ **EXAMPLE 3** Convert 88% to fractional notation.

$$88\% = 88 \times \frac{1}{100} \qquad \text{Replacing \% by} \times \frac{1}{100}$$

$$= \frac{88}{100} \qquad \text{You need not simplify.} \qquad ◀$$

OBJECTIVES

After finishing Section R.4, you should be able to:

a Convert from percent notation to decimal notation.

b Convert from percent notation to fractional notation.

c Convert from decimal notation to percent notation.

d Convert from fractional notation to percent notation.

FOR EXTRA HELP

Tape 1C   Tape 1B   MAC: R
                                  IBM: R

Convert to decimal notation.

1. 46.2%

2. 100%

Convert to fractional notation.

3. 67%

4. 45.6%

5. $\frac{1}{4}$%

Convert to percent notation.

6. 6.77

7. 0.9944

Convert to percent notation.

8. $\frac{1}{4}$

9. $\frac{3}{8}$

10. $\frac{2}{3}$

▶ **EXAMPLE 4**   Convert 34.7% to fractional notation.

$$34.7\% = 34.7 \times \frac{1}{100} \qquad \text{Replacing \% by } \times \frac{1}{100}$$

$$= \frac{34.7}{100}$$

$$= \frac{34.7}{100} \cdot \frac{10}{10} \qquad \text{Multiplying by 1 to get a whole number in the numerator}$$

$$= \frac{347}{1000} \qquad ◀$$

Table 1 at the back of the text contains many decimal and percent equivalents. It might be helpful to memorize some or all of them.

DO EXERCISES 3–5.

c   **Converting from Decimal Notation to Percent Notation**

By applying the definition of percent in reverse, we can convert from decimal notation to percent notation. We multiply by 1, expressing it as 100×0.01 and replacing $\times 0.01$ by %.

▶ **EXAMPLE 5**   Convert 0.93 to percent notation.

$$0.93 = 0.93 \times 1$$
$$= 0.93 \times (100 \times 0.01) \qquad \text{Expressing 1 as } 100 \times 0.01$$
$$= (0.93 \times 100) \times 0.01$$
$$= 93 \times 0.01$$
$$= 93\% \qquad \text{Replacing } \times 0.01 \text{ by \%} \qquad ◀$$

> To convert from decimal notation to percent notation, move the decimal point two places to the right and write a percent symbol.

▶ **EXAMPLE 6**   Convert 0.032 to percent notation.

$$0.032 \qquad 0.03\underset{\frown}{2} \qquad 0.032 = 3.2\%$$

Move the decimal point two places to the right.   ◀

DO EXERCISES 6 AND 7.

d   **Converting from Fractional Notation to Percent Notation**

We can also convert from fractional notation to percent notation by converting first to decimal notation.

▶ **EXAMPLE 7**   Convert $\frac{5}{8}$ to percent notation.

$$\frac{5}{8} = 0.625 = 62.5\% \qquad ◀$$

The result of Example 7 says that the ratio of 5 to 8 is the same as the ratio of 62.5 to 100.

DO EXERCISES 8–10.

EXERCISE SET R.4

a Convert to decimal notation.

1. 76%   **2.** 54%   **3.** 54.7%   **4.** 96.2%

5. 100%   **6.** 1%   **7.** 0.61%   **8.** 125%

9. 240%   **10.** 0.73%   **11.** 3.25%   **12.** 2.3%

b Convert to fractional notation.

13. 20%   **14.** 80%   **15.** 78.6%   **16.** 13.5%

17. $12\frac{1}{2}\%$ **18.** 120%   **19.** 0.042%   **20.** 0.68%

21. 250%   **22.** 3.2%   **23.** 3.47%   **24.** 12.557%

ANSWERS

1. _____
2. _____
3. _____
4. _____
5. _____
6. _____
7. _____
8. _____
9. _____
10. _____
11. _____
12. _____
13. _____
14. _____
15. _____
16. _____
17. _____
18. _____
19. _____
20. _____
21. _____
22. _____
23. _____
24. _____

Copyright © 1991 Addison-Wesley Publishing Co., Inc.

ANSWERS

25. _____

26. _____

27. _____

28. _____

29. _____

30. _____

31. _____

32. _____

33. _____

34. _____

35. _____

36. _____

37. _____

38. _____

39. _____

40. _____

41. _____

42. _____

43. _____

44. _____

45. _____

46. _____

47. _____

48. _____

49. _____

50. _____

51. _____

52. _____

53. _____

54. _____

55. _____

56. _____

57. _____

58. _____

59. _____

60. _____

61. _____

c Convert to percent notation.

25. 4.54   **26.** 1   **27.** 0.998   **28.** 0.73

29. 2   **30.** 0.0057   **31.** 0.072   **32.** 1.34

33. 9.2   **34.** 0.013   **35.** 0.0068   **36.** 0.675

d Convert to percent notation.

37. $\dfrac{1}{8}$ **38.** $\dfrac{1}{3}$ **39.** $\dfrac{17}{25}$ **40.** $\dfrac{11}{20}$

41. $\dfrac{17}{100}$ **42.** $\dfrac{119}{100}$ **43.** $\dfrac{7}{10}$ **44.** $\dfrac{8}{10}$

45. $\dfrac{3}{5}$ **46.** $\dfrac{17}{50}$ **47.** $\dfrac{2}{3}$ **48.** $\dfrac{3}{8}$

49. $\dfrac{7}{4}$ **50.** $\dfrac{7}{8}$ **51.** $\dfrac{3}{4}$ **52.** $\dfrac{99.4}{100}$

SYNTHESIS

Simplify. Express the answer in percent notation.

53. 12% + 14%   **54.** 84% − 16%   **55.** 1 − 10%

56. 81% − 10%   **57.** 12 × 100%   **58.** 42% − (1 − 58%)

59. 3(1 + 15%)   **60.** 7(1% + 13%)   **61.** $\dfrac{100\%}{40}$

R.5 Exponential Notation and Order of Operations

a Exponential Notation

Exponents provide a shorter way of writing products. A product in which the factors are the same is called a **power.** For

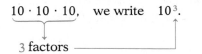

$$10 \cdot 10 \cdot 10, \quad \text{we write} \quad 10^3.$$

3 factors

This is read "ten to the third power." We call the number 3 an **exponent** and we say that 10 is the **base.** An exponent of 2 or greater tells how many times the base is used as a factor. For example,

$$a \cdot a \cdot a \cdot a = a^4.$$

Here the exponent is 4 and the base is a. An expression for a power is called **exponential notation.**

This is the exponent.

$$a^n$$

This is the base.

▶ **EXAMPLE 1** Write exponential notation for $10 \cdot 10 \cdot 10 \cdot 10 \cdot 10$.

$$10 \cdot 10 \cdot 10 \cdot 10 \cdot 10 = 10^5 \qquad ◀$$

DO EXERCISES 1–3.

b Evaluating Exponential Expressions

▶ **EXAMPLE 2** Evaluate: 5^2.

$$5^2 = 5 \cdot 5 = 25. \qquad ◀$$

▶ **EXAMPLE 3** Evaluate: 3^4.
We have

$$3^4 = 3 \cdot 3 \cdot 3 \cdot 3 = 9 \cdot 9 = 81.$$

We could also carry out the calculation as follows:

$$3^4 = 3 \cdot 3 \cdot 3 \cdot 3 = 9 \cdot 3 \cdot 3 = 27 \cdot 3 = 81. \qquad ◀$$

Exponential Notation

For any natural number n greater than or equal to 2,

n factors

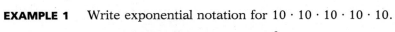

$$b^n = \overbrace{b \cdot b \cdot b \cdot b \cdots b}.$$

DO EXERCISES 4–6.

Write exponential notation.

1. $5 \cdot 5 \cdot 5$

2. $6 \cdot 6 \cdot 6 \cdot 6 \cdot 6$

3. 1.08×1.08

Evaluate.

4. 10^4

5. 8^3

6. $(1.1)^3$

Calculate.

7. $18 - 4 \times 3 + 7$

8. $(2 \times 5)^3$

9. 2×5^3

10. $8 + 2 \times 5^3 - 4 \cdot 20$

c Order of Operations

What does $5 \times 2 + 4$ mean? If we multiply 5 by 2 and add 4, we get 14. If we add 2 and 4 and multiply by 5, we get 30. Since our results are different, we see that the order in which we carry out operations is important. To tell which operation to do first, we can use grouping symbols such as parentheses ( ), or brackets [ ], or braces { }. For example,

$$(3 \times 5) + 6 = 15 + 6 = 21,$$

but

$$3 \times (5 + 6) = 3 \times 11 = 33.$$

Grouping symbols tell us what to do first. If there are no grouping symbols, we have agreements about the order in which operations should be done.

> Rules for Order of Operations
> 1. **Do all calculations within grouping symbols before operations outside.**
> 2. **Evaluate all exponential expressions.**
> 3. **Do all multiplications and divisions in order from left to right.**
> 4. **Do all additions and subtractions in order from left to right.**

▶ **EXAMPLE 4**   Calculate: $15 - 2 \times 5 + 3$.

$$15 - 2 \times 5 + 3 = 15 - 10 + 3 \qquad \text{Multiplying}$$
$$= 5 + 3 \qquad \text{Subtracting and adding from left to right}$$
$$= 8 \qquad \text{Adding} \qquad \blacktriangleleft$$

Always calculate within parentheses first. When there are exponents and no parentheses, simplify powers before multiplying or dividing.

▶ **EXAMPLE 5**   Calculate: $(3 \times 4)^2$.

$$(3 \times 4)^2 = (12)^2 \qquad \text{Working within parentheses first}$$
$$= 144 \qquad \blacktriangleleft$$

▶ **EXAMPLE 6**   Calculate: 3×4^2.
We have

$$3 \times 4^2 = 3 \times 16 \qquad \text{Simplifying the power}$$
$$= 48. \qquad \text{Multiplying} \qquad \blacktriangleleft$$

Note that $(3 \times 4)^2 \neq 3 \times 4^2$.

▶ **EXAMPLE 7**   Calculate: $7 + 3 \times 4^2 - 29$.

$$7 + 3 \times 4^2 - 29 = 7 + 3 \times 16 - 29 \qquad \text{There are no parentheses, so we find } 4^2 \text{ first.}$$
$$= 7 + 48 - 29 \qquad \text{Multiplying second}$$
$$= 55 - 29 \qquad \text{Adding}$$
$$= 26 \qquad \text{Subtracting} \qquad \blacktriangleleft$$

DO EXERCISES 7–10.

Sometimes combinations of grouping symbols are used, as in

$$5[4 + (8 - 2)].$$

The rules still apply. We begin with the innermost grouping symbols, in this case the parentheses, and work to the outside.

▶ **EXAMPLE 8** Calculate: $5[4 + (8 - 2)]$.

$$5[4 + (8 - 2)] = 5[4 + 6] \qquad \text{Subtracting within the parentheses first}$$
$$= 5[10] \qquad \text{Adding inside the brackets}$$
$$= 50 \qquad\qquad\qquad ◄$$

A fraction bar can play the role of a grouping symbol, although such a symbol is not as evident as the others.

▶ **EXAMPLE 9** Calculate: $\dfrac{12(9 - 7) + 4 \cdot 5}{3^4 + 2^3}$.

An equivalent expression with brackets as grouping symbols is

$$[12(9 - 7) + 4 \cdot 5] \div [3^4 + 2^3].$$

What this shows, in effect, is that we do the calculations in the numerator and then in the denominator, and divide the results:

$$\frac{12(9 - 7) + 4 \cdot 5}{3^4 + 2^3} = \frac{12(2) + 4 \cdot 5}{81 + 8}$$
$$= \frac{24 + 20}{89}$$
$$= \frac{44}{89}. \qquad ◄$$

DO EXERCISES 11–13.

Calculate.

11. $4[(8 - 3) + 7]$

12. $\dfrac{13(10 - 6) + 4.9}{5^2 - 3^2}$

13. $256 \div 16 \div 4$

ANSWERS ON PAGE A-2

❖ SIDELIGHTS

Factors and Sums

To *factor* a number is to express it as a product. Since $15 = 5 \cdot 3$, we say that 15 is *factored* and that 5 and 3 are *factors* of 15. In the table below, the top number has been factored in such a way that the sum of the factors is the bottom number. For example, in the first column, 56 has been factored as $7 \cdot 8$, and $7 + 8 = 15$, the bottom number. Such thinking will be important in a later chapter of the book.

Product	56	63	36	72	140	96		168	110			
Factor	7									9	24	3
Factor	8						8	8		10	18	
Sum	15	16	20	38	24	20	14		21			24

EXERCISE

Find the missing numbers in the table.

NAME   SECTION   DATE

EXERCISE SET R.5

a Write exponential notation.

1. $3 \times 3 \times 3 \times 3$ **2.** $2 \times 2 \times 2 \times 2 \times 2$ **3.** $10 \cdot 10 \cdot 10 \cdot 10 \cdot 10$

4. $10 \times 10 \times 10 \times 10$ **5.** $1 \cdot 1 \cdot 1 \cdot 1 \cdot 1$ **6.** $16 \cdot 16$

b Evaluate.

7. 5^2 **8.** 7^3 **9.** 9^5 **10.** 12^4

11. 10^2 **12.** 1^5 **13.** 1^4 **14.** $(1.8)^2$

15. $(2.3)^2$ **16.** $(0.1)^3$ **17.** $(0.2)^3$ **18.** $(14.8)^2$

19. $(20.4)^2$ **20.** $\left(\frac{4}{5}\right)^2$ **21.** $\left(\frac{3}{8}\right)^2$ **22.** 2^4

23. 5^3 **24.** $(1.4)^3$ **25.** $1000 \times (1.07)^3$ **26.** $2000 \times (1.08)^2$

c Calculate.

27. $7 + 2 \times 6$ **28.** $11 + 4 \times 4$ **29.** $8 \times 7 + 6 \times 5$

30. $10 \times 5 + 1 \times 1$ **31.** $19 - 5 \times 3 + 3$ **32.** $14 - 2 \times 6 + 7$

33. $9 \div 3 + 16 \div 8$ **34.** $32 - 8 \div 4 - 2$ **35.** $7 + 10 - 10 \div 2$

36. $(5 \cdot 4)^2$ **37.** $(6 \cdot 3)^2$ **38.** $3 \cdot 2^3$

ANSWERS

1. _____
2. _____
3. _____
4. _____
5. _____
6. _____
7. _____
8. _____
9. _____
10. _____
11. _____
12. _____
13. _____
14. _____
15. _____
16. _____
17. _____
18. _____
19. _____
20. _____
21. _____
22. _____
23. _____
24. _____
25. _____
26. _____
27. _____
28. _____
29. _____
30. _____
31. _____
32. _____
33. _____
34. _____
35. _____
36. _____
37. _____
38. _____

ANSWERS

39. _____

40. _____

41. _____

42. _____

43. _____

44. _____

45. _____

46. _____

47. _____

48. _____

49. _____

50. _____

51. _____

52. _____

53. _____

54. _____

55. _____

56. _____

57. _____

58. _____

59. _____

60. _____

61. _____

62. _____

63. _____

64. _____

65. _____

66. _____

67. _____

68. _____

69. _____

70. _____

71. _____

72. _____

73. _____

39. $4 \cdot 5^2$

40. $(8 + 2)^2$

41. $(5 + 3)^3$

42. $7 + 2^2$

43. $6 + 4^2$

44. $(5 - 2)^2$

45. $(3 - 2)^2$

46. $10 - 3^2$

47. $12 - 2^3$

48. $20 + 4^3 \div 8$

49. $2 \times 10^3 - 500$

50. $7 \times 3^4 + 18$

51. $6[9 + (3 + 4)]$

52. $8[(13 + 6) - 11]$

53. $8 + (7 + 9)$

54. $(8 + 7) + 9$

55. $15(4 + 2)$

56. $15 \cdot 4 + 15 \cdot 2$

57. $12 - (8 - 4)$

58. $(12 - 8) - 4$

59. $1000 \div 100 \div 10$

60. $256 \div 32 \div 4$

61. $\dfrac{80 - 6^2}{9^2 + 3^2}$

62. $\dfrac{5^2 + 4^3 - 3}{9^2 - 2^2 + 1^5}$

63. $\dfrac{3(6 + 7) - 5 \cdot 4}{6 \cdot 7 + 8(4 - 1)}$

64. $\dfrac{20(8 - 3) - 4(10 - 3)}{10(6 + 2) + 2(5 + 2)}$

SKILL MAINTENANCE

65. Find percent notation: $\dfrac{5}{16}$.

66. Simplify: $\dfrac{125}{325}$.

67. Find the prime factorization of 48.

68. Find the LCM of 12, 24, and 56.

SYNTHESIS

Write each of the following with a single exponent.

69. $\dfrac{10^5}{10^3}$

70. $\dfrac{10^7}{10^2}$

71. $\dfrac{5^4}{5^2}$

72. $\dfrac{2^8}{8^2}$

73. *Five five's.* We can use five 5's and any combination of grouping symbols to represent the numbers 0 through 10. For example,

$$0 = 5 \cdot 5 \cdot 5(5 - 5), \qquad 1 = \frac{5 + 5}{5} - \frac{5}{5}, \qquad 2 = \frac{5 \cdot 5 - 5}{5 + 5}.$$

Often more than one way to make a representation is possible. Use five 5's to represent the numbers 3 through 10.

SUMMARY AND REVIEW: CHAPTER R

IMPORTANT PROPERTIES AND FORMULAS

Identity Property of 0: $a + 0 = a$
Identity Property of 1: $a \cdot 1 = a$

Equivalent Expressions for 1: $\dfrac{a}{a} = 1$

Exponential Notation: $a^n = \underbrace{a \cdot a \cdot a \cdots a}_{n \text{ factors}}$

$$n\% = n \times 0.01 = n \times \frac{1}{100} = \frac{n}{100}$$

REVIEW EXERCISES

The review exercises that follow are for practice. Answers are at the back of the book. If you miss an exercise, restudy the section and objective indicated alongside the answer.

Find the prime factorization.

1. 92

2. 1400

Find the LCM.

3. 13, 32

4. 5, 18, 45

Write an equivalent expression using the indicated number for 1.

5. $\dfrac{2}{5}$ $\left(\text{Use } \dfrac{6}{6} \text{ for 1.}\right)$

6. $\dfrac{12}{23}$ $\left(\text{Use } \dfrac{8}{8} \text{ for 1.}\right)$

Write an equivalent expression with the given denominator.

7. $\dfrac{5}{8}$ (Denominator: 64)

8. $\dfrac{13}{12}$ (Denominator: 84)

Simplify.

9. $\dfrac{20}{48}$

10. $\dfrac{1020}{1820}$

Compute and simplify.

11. $\dfrac{4}{9} + \dfrac{5}{12}$

12. $\dfrac{3}{4} \div 3$

13. $\dfrac{2}{3} - \dfrac{1}{15}$

14. $\dfrac{9}{10} \cdot \dfrac{16}{5}$

15. Convert to fractional notation: 17.97.

16. Convert to decimal notation: $\dfrac{2337}{10{,}000}$.

Add.

17.
$$\begin{array}{r} 2\ 3\ 4\ 4.5\ 6 \\ +\ \ \ \ \ \ 9\ 8.3\ 4\ 5 \\ \hline \end{array}$$

18. $6.04 + 78 + 1.9898$

Subtract.

19. $20.4 - 11.058$

20.
$$\begin{array}{r} 7\ 8\ 9.0\ 3\ 2 \\ -6\ 5\ 5.7\ 6\ 8 \\ \hline \end{array}$$

21.
$$\begin{array}{r} 1\ 7.9\ 5 \\ \times\ \ \ \ \ 2\ 4 \\ \hline \end{array}$$

22.
$$\begin{array}{r} 5\ 6.9\ 5 \\ \times\ \ \ 1.9\ 4 \\ \hline \end{array}$$

Divide

23. $2.8\,)\overline{1\ 5\ 5.6\ 8}$

24. $5\ 2\,)\overline{2\ 3.4}$

25. Convert to decimal notation: $\dfrac{19}{12}$.

26. Round to the nearest tenth: 34.067.

27. Write exponential notation: $6 \cdot 6 \cdot 6$.

28. Evaluate: $(1.06)^2$.

Calculate and compare answers to Exercises 29–31.

29. $120 - 6^2 \div 4 + 8$

30. $(120 - 6^2) \div 4 + 8$

31. $(120 - 6^2) \div (4 + 8)$

32. Calculate: $\dfrac{4(18 - 8) + 7 \cdot 9}{9^2 - 8^2}$.

33. Convert to decimal notation: 4.7%.

34. Convert to fractional notation: 60%.

Convert to percent notation.

35. 0.886

36. $\dfrac{5}{8}$

37. $\dfrac{29}{25}$

38. For health reasons, there must be 6 parts per billion (ppb) of chlorine in a swimming pool. What percent of total volume is this?

TEST: CHAPTER R

1. Find the prime factorization of 300.

2. Find the LCM of 15, 24, and 60.

1. _____

3. Write an expression equivalent to $\frac{3}{7}$ using $\frac{7}{7}$ as a name for 1.

4. Write an equivalent expression with the given denominator:

$$\frac{11}{16}. \quad \text{(Denominator: 48)}$$

2. _____

3. _____

4. _____

Simplify.

5. $\frac{16}{24}$

6. $\frac{925}{1525}$

5. _____

6. _____

Compute and simplify.

7. $\frac{10}{27} \div \frac{8}{3}$

8. $\frac{9}{10} - \frac{5}{8}$

7. _____

8. _____

9. Convert to fractional notation (do not simplify): 6.78.

10. Convert to decimal notation: $\frac{1895}{1000}$.

9. _____

10. _____

11. Add: $7.14 + 89 + 2.8787$.

12. Subtract: $1800 - 3.42$.

11. _____

12. _____

13. _____

14. _____

15. _____

16. _____

17. _____

18. _____

19. _____

20. _____

21. _____

22. _____

23. _____

24. _____

13. Multiply:

$$\begin{array}{r} 1\ 2\ 3.6\ 8 \\ \times\ \ \ 3.5\ 2\ 1 \\ \hline \end{array}$$

14. Divide: $7.2\overline{)1\ 1.5\ 2.}$

15. Convert to decimal notation: $\dfrac{23}{11}$.

16. Round 234.7284 to the nearest tenth.

17. Round 234.7284 to the nearest thousandth.

18. Evaluate: 5^4.

19. Evaluate: $(1.2)^2$.

20. Calculate: $200 - 2^3 + 5 + 10$.

21. Convert to decimal notation: 0.7%.

22. Convert to fractional notation: 91%.

23. Convert to percent notation: $\dfrac{11}{25}$.

SYNTHESIS

24. Simplify: $\dfrac{13,860}{42,000}$.

INTRODUCTION In this chapter we consider the number system used most in algebra. It is called the real-number system. We will learn to add, subtract, multiply, and divide real numbers and to manipulate certain expressions. Such manipulation will be important when we solve equations and problems in Chapter 2.

The review sections to be tested in addition to the material in this chapter are R.1, R.2, R.4, and R.5. ❖

Introduction to Real Numbers and Algebraic Expressions

1

AN APPLICATION

In a football game, the quarterback attempted passes with the following results:

1st try: 13-yd gain;
2nd try: incomplete;
3rd try: 12-yd loss (tackled behind the line);
4th try: 21-yd gain;
5th try: 14-yd loss.

What is the total number of yards gained?

THE MATHEMATICS

We let $t =$ the total number of yards. Then t is given by

$$t = 13 + 0 + (-12) + 21 + (-14).$$

These are *negative* numbers.

Area of a Rectangle:	$A = l \cdot w$
Area of a Square:	$A = s^2$
Area of a Triangle:	$A = \frac{1}{2}b \cdot h$
Simple-Interest Formula:	$I = P \cdot r \cdot t$

PRETEST: CHAPTER 1

1. Evaluate $x/2y$ when $x = 5$ and $y = 8$.

2. Write an algebraic expression: Seventy-eight percent of some number.

3. Find the area of a rectangle when the length is 22.5 ft and the width is 16 ft.

4. Find $-x$ when $x = -12$.

Use either $<$ or $>$ for ▨ to write a true sentence.

5. 0 ▨ -5

6. 10 ▨ -5

7. -35 ▨ -45

8. $-\dfrac{2}{3}$ ▨ $\dfrac{4}{5}$

Find the absolute value.

9. $|-12|$

10. $|2.3|$

11. $|0|$

Find the opposite, or additive inverse.

12. 5.4

13. $-\dfrac{2}{3}$

Compute and simplify.

14. $-9 + (-8)$

15. $20.2 - (-18.4)$

16. $-\dfrac{5}{6} - \dfrac{3}{10}$

17. $-11.5 + 6.5$

18. $-9(-7)$

19. $\dfrac{5}{8}\left(-\dfrac{2}{3}\right)$

20. $-19.6 \div 0.2$

21. $-56 \div (-7)$

22. $12 - (-6) + 14 - 8$

23. $20 - 10 \div 5 + 2^3$

Multiply.

24. $9(z - 2)$

25. $-2(2a + b - 5c)$

Factor.

26. $4x - 12$

27. $6y - 9z - 18$

Simplify.

28. $3y - 7 - 2(2y + 3)$

29. $\{2[3(y + 1) - 4] - [5(y - 3) - 5]\}$

30. Write an inequality with the same meaning as $x > 12$.

1.1 Introduction to Algebra

Many kinds of problems require the use of equations in order to be solved effectively. The study of algebra involves the use of equations to solve problems. Equations are constructed from algebraic expressions. The purpose of this section is to introduce you to the types of expressions encountered in algebra.

a Algebraic Expressions

In arithmetic, you have worked with expressions such as

$$37 + 86, \qquad 7 \times 8, \qquad 19 - 7, \quad \text{and} \quad \frac{3}{8}.$$

In algebra, we use certain letters for numbers and work with *algebraic expressions* such as

$$x + 86, \qquad 7 \times t, \qquad 19 - y, \quad \text{and} \quad \frac{a}{b}.$$

Sometimes a letter can stand for various numbers. In that case we call the letter a **variable.** Sometimes a letter can stand for just one number. In that case we call the letter a **constant.** Let b = your date of birth. Then b is a constant. Let a = your age. Then a is a variable since a changes from year to year.

How do algebraic expressions arise? Most often they occur in problem-solving situations. For example, consider the following chart, which you might see in a magazine.

Starting Pay

Starting salaries of police in 1987 are up about 25% since 1982.

Police: 1982	$15,635
1987	$19,544

Suppose we wanted to know how much more the 1987 salary was than the 1982 salary.

In algebra, we translate the problem into an equation. It might be done as follows.

Salary in 1982	plus	How much?	is	Salary in 1987
$15,635	+	x	=	$19,544

Note that we have an algebraic expression on the left. To find the number x, we can subtract $15,635 on both sides of the equation:

$$\$15,635 + x - \$15,635 = \$19,544 - \$15,635$$
$$x = \$19,544 - \$15,635.$$

Then we carry out the subtraction and obtain the answer $3909.

In arithmetic, you probably would do this subtraction right away without considering an equation. In algebra, you will find most problems difficult without first solving an equation.

DO EXERCISE 1.

OBJECTIVES

After finishing Section 1.1, you should be able to:

a Evaluate algebraic expressions by substitution.

b Translate phrases to algebraic expressions.

FOR EXTRA HELP

Tape 2A	Tape 2A	MAC: 1 IBM: 1

1. Translate this problem to an equation. Use the graph below. How many more flights are there on the Dallas–Houston route than on the New York–Boston route?

Taking flight in traffic
Here's how many flights are made monthly on the busiest air routes:

Dallas–Houston	2,866
Los Angeles–San Francisco	2,822
New York–Chicago	2,710
New York–Washington	2,442
New York–Boston	2,128

ANSWER ON PAGE A-2

2. Evaluate $a + b$ for $a = 38$ and $b = 26$.

An **algebraic expression** consists of variables, numerals, and operation signs. When we replace a variable by a number, we say that we are **substituting** for the variable. This process is called **evaluating the expression.**

▶ **EXAMPLE 1**   Evaluate $x + y$ for $x = 37$ and $y = 29$.

We substitute 37 for x and 29 for y and carry out the addition:

$$x + y = 37 + 29 = 66.$$

The number 66 is called the **value** of the expression.   ◀

3. Evaluate $x - y$ for $x = 57$ and $y = 29$.

Algebraic expressions involving multiplication can be written in several ways. For example, "8 times a" can be written as $8 \times a$, $8 \cdot a$, $8(a)$, or simply $8a$. Two letters written together without a symbol, such as ab, also indicates a multiplication.

▶ **EXAMPLE 2**   Evaluate $3y$ for $y = 14$.

$$3y = 3(14) = 42$$ ◀

DO EXERCISES 2–4.

4. Evaluate $4t$ for $t = 15$.

▶ **EXAMPLE 3**   The area A of a rectangle of length l and width w is given by the formula $A = lw$. Find the area when l is 24.5 in. and w is 16 in.

We substitute 24.5 in. for l and 16 in. for w and carry out the multiplication:

$$
\begin{aligned}
A = lw &= (24.5 \text{ in.})(16 \text{ in.}) \\
&= (24.5)(16)(\text{in.})(\text{in.}) \\
&= 392 \text{ in}^2, \quad \text{or } 392 \text{ square inches.}
\end{aligned}
$$

◀

DO EXERCISE 5.

5. Find the area of a rectangle when l is 24 ft and w is 8 ft.

Algebraic expressions involving division can also be written in several ways. For example, "8 divided by t" can be written as $8 \div t$, or $\dfrac{8}{t}$, where the fraction bar is a division symbol.

▶ **EXAMPLE 4**   The time needed for a satellite to orbit the earth is determined by the height of the satellite above the earth's surface and the speed, or velocity, of the satellite. If a satellite is orbiting 300 mi above the earth's surface, it travels about 27,000 mi in one orbit. The time t, in hours, that it takes to orbit the earth one time is given by

$$t = \frac{27,000}{v},$$

6. Find the orbiting time of the satellite in Example 4 when the velocity is 8000 mph.

where v is the velocity of the satellite in miles per hour. Find the orbiting time of the satellite when the velocity v is 10,000 mph.

We substitute 10,000 for v and carry out the division:

$$t = \frac{27,000}{v} = \frac{27,000}{10,000} = 2.7 \text{ hr.}$$ ◀

DO EXERCISE 6.

▶ **EXAMPLE 5** Evaluate $\dfrac{a}{b}$ for $a = 63$ and $b = 9$.

We substitute 63 for a and 9 for b and carry out the division:

$$\frac{a}{b} = \frac{63}{9} = 7.$$ ◀

▶ **EXAMPLE 6** Evaluate $\dfrac{12m}{n}$ for $m = 8$ and $n = 16$.

$$\frac{12m}{n} = \frac{12 \cdot 8}{16} = \frac{96}{16} = 6$$ ◀

DO EXERCISES 7 AND 8.

b Translating to Algebraic Expressions

In algebra, we translate problems to equations. The different parts of an equation are translations of word phrases to algebraic expressions. It is easier to translate if we know that certain words translate to certain operation symbols.

KEY WORDS			
Addition (+)	**Subtraction (−)**	**Multiplication (·)**	**Division (÷)**
add	subtract	multiply	divide
sum	difference	product	quotient
plus	minus	times	divided by
more than	less than	twice	
increased by	decreased by	of	
	take from		

▶ **EXAMPLE 7** Translate to an algebraic expression:

Twice (or two times) some number.

Think of some number, say 8. What number is twice 8? It is 16. How did you get 16? You multiplied by 2. Do the same thing using a variable. We can use any variable we wish, such as x, y, m, or n. Let's use y to stand for some number. Multiply by 2. We get an expression

$$y \times 2, \quad 2 \times y, \quad 2 \cdot y, \quad \text{or} \quad 2y.$$ ◀

▶ **EXAMPLE 8** Translate to an algebraic expression:

Seven less than some number.

We let

x represent the number.

Now if the number were 23, then the translation would be $23 - 7$. If we knew the number to be 345, then the translation would be $345 - 7$. If the number is x, then the translation is

$$x - 7.$$ ◀

7. Evaluate a/b for $a = 200$ and $b = 8$.

8. Evaluate $10p/q$ when $p = 40$ and $q = 25$.

ANSWERS ON PAGE A-2

Translate to an algebraic expression.

9. Twelve less than some number

10. Twelve more than some number

11. Four less than some number

12. Half of some number

13. Six more than eight times some number

14. The difference of two numbers

15. Fifty-nine percent of some number

16. Two hundred less than the product of two numbers

17. The sum of two numbers

Note that $7 - x$ is *not* a correct translation of the expression in Example 8. The expression $7 - x$ is a translation of "seven minus some number" or "some number less than seven."

▶ **EXAMPLE 9**   Translate to an algebraic expression:

Eighteen more than a number.

We let

$$t = \text{the number.}$$

Now if the number were 26, then the translation would be $18 + 26$. If we knew the number to be 174, then the translation would be $18 + 174$. If the number is t, then the translation is

$$18 + t.$$ ◀

▶ **EXAMPLE 10**   Translate to an algebraic expression:

A number divided by 5.

We let

$$m = \text{the number.}$$

Now if the number were 76, then the translation would be $76 \div 5$, or $76/5$, or $\frac{76}{5}$. If the number were 213, then the translation would be $213 \div 5$, or $213/5$, or $\frac{213}{5}$. If the number is m, then the translation is

$$m \div 5, \qquad m/5, \quad \text{or} \quad \frac{m}{5}.$$ ◀

▶ **EXAMPLE 11**   Translate each of the following phrases to an algebraic expression.

Phrase	Algebraic expression
Five more than some number	$5 + n$, or $n + 5$
Half of a number	$\frac{1}{2}t$, or $\frac{t}{2}$
Five more than three times some number	$5 + 3p$, or $3p + 5$
The difference of two numbers	$x - y$
Six less than the product of two numbers	$mn - 6$
Seventy-six percent of some number	$76\%z$, or $0.76z$

◀

DO EXERCISES 9–17.

NAME   SECTION   DATE

EXERCISE SET 1.1

a Substitute to find values of the expressions.

1. Theresa is 6 yr younger than her husband Frank. Suppose the variable x stands for Frank's age. Then $x - 6$ stands for Theresa's age. How old is Theresa when Frank is 29? 34? 47?

2. Employee A took five times as long to do a job as employee B. Suppose t stands for the time it takes B to do a job. Then $5t$ stands for the time it takes A. How long did it take A if B took 30 sec? 90 sec? 2 min?

3. The area A of a parallelogram with base b and height h is given by $A = bh$. Find the area of the parallelogram when the height is 15.4 cm (centimeters) and the base is 6.5 cm.

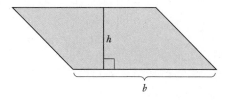

4. The area A of a triangle with base b and height h is given by $A = \frac{1}{2}bh$. Find the area when $b = 45$ m (meters) and $h = 86$ m.

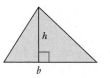

5. A driver who drives at a speed of r mph for t hr will travel a distance d mi given by $d = rt$ mi. How far will a driver travel at a speed of 55 mph for 4 hr?

6. *Simple interest.* The simple interest I on a principal of P dollars at interest rate r for time t, in years, is given by $I = Prt$. Find the simple interest on a principal of \$4800 at 9% for 2 yr. (*Hint:* 9% = 0.09.)

Evaluate.

7. $6x$ for $x = 7$

8. $7y$ for $y = 7$

9. $\dfrac{x}{y}$ for $x = 9$ and $y = 3$

10. $\dfrac{m}{n}$ for $m = 14$ and $n = 2$

11. $\dfrac{3p}{q}$ for $p = 2$ and $q = 6$

12. $\dfrac{5y}{z}$ for $y = 15$ and $z = 25$

13. $\dfrac{x + y}{5}$ for $x = 10$ and $y = 20$

14. $\dfrac{p + q}{2}$ for $p = 2$ and $q = 16$

15. $\dfrac{x - y}{8}$ for $x = 20$ and $y = 4$

16. $\dfrac{m - n}{5}$ for $m = 16$ and $n = 6$

1. _____
2. _____
3. _____
4. _____
5. _____
6. _____
7. _____
8. _____
9. _____
10. _____
11. _____
12. _____
13. _____
14. _____
15. _____
16. _____

ANSWERS

17. _____

18. _____

19. _____

20. _____

21. _____

22. _____

23. _____

24. _____

25. _____

26. _____

27. _____

28. _____

29. _____

30. _____

31. _____

32. _____

33. _____

34. _____

35. _____

36. _____

37. _____

38. _____

39. _____

40. _____

41. _____

42. _____

43. _____

44. _____

45. _____

46. _____

47. _____

48. _____

b   Translate to an algebraic expression.

17. 6 more than b

18. 8 more than t

19. 9 less than c

20. 4 less than d

21. 6 increased by q

22. 11 increased by z

23. b more than a

24. c more than d

25. x less than y

26. c less than h

27. x added to w

28. s added to t

29. m subtracted from n

30. p subtracted from q

31. The sum of r and s

32. The sum of d and f

33. Twice x

34. Three times p

35. 5 multiplied by t

36. The product of 3 and b

37. The product of 97% and some number

38. 43% of some number

39. A student had d dollars before going to the bookstore. The student bought a book for $29.95. How much did the student have after the purchase?

40. A driver drove at a speed of 65 mph for t hr. How far did the driver go?

SKILL MAINTENANCE

Find the prime factorization.

41. 54

42. 192

Find the LCM.

43. 6, 18

44. 6, 24, 32

SYNTHESIS

Translate to an algebraic expression.

45. Some number x plus three times y

46. Some number a plus 2 plus b

47. A number that is 3 less than twice x

48. Your age in 5 years, if you are a years old now

1.2 The Real Numbers

A **set** is a collection of objects. For our purposes we will most often be considering sets of numbers. One way to name a set uses what is called **roster notation.** For example, the set containing the numbers 0, 2, and 5 can be named {0, 2, 5}.

Sets that are part of other sets are called **subsets.** In this section, we become acquainted with the set of *real numbers* and its various subsets.

Two important subsets of the real numbers are listed below using roster notation.

> **Natural numbers = {1, 2, 3, . . .}. These are the numbers used for counting.**
>
> **Whole numbers = {0, 1, 2, 3, . . .}. This is the set of natural numbers with 0 included.**

We can represent these sets on a number line. The natural numbers are those to the right of 0.

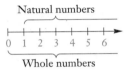

We create a new set, called the *integers*, by starting with the whole numbers, 0, 1, 2, 3, and so on. For each natural number 1, 2, 3, and so on, we obtain a new number to the left of 0 on the number line:

For the number 1, there will be an *opposite* number −1 (negative 1).

For the number 2, there will be an *opposite* number −2 (negative 2).

For the number 3, there will be an *opposite* number −3 (negative 3), and so on.

The **integers** consist of the whole numbers and these new numbers. We picture them on a number line as follows.

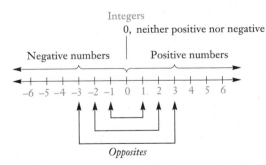

We call the newly obtained numbers **negative integers.** The natural numbers are called **positive integers.** Zero is neither positive nor negative. We call −1 and 1 opposites of each other. Similarly, −2 and 2 are opposites, −3 and 3 are opposites, −100 and 100 are opposites, and 0 is its own opposite. This gives us the set of integers, which extends infinitely on the number line to the left and right of 0.

> **The set of integers = {. . . , −5, −4, −3, −2, −1, 0, 1, 2, 3, 4, 5, . . .}.**

Tell which integers correspond to the given situation.

1. The halfback gained 8 yd on the first down. The quarterback was sacked for a 5-yd loss on second down.

2. The highest temperature ever recorded in the United States was 134° in Death Valley on July 10, 1913. The coldest temperature ever recorded in the United States was 76° below zero in Tanana, Alaska, in January of 1886.

3. At 10 sec before liftoff, ignition occurs. At 148 sec after liftoff, the first stage is detached from the rocket.

4. A student owes $137 to the bookstore. The student has $289 in a savings account.

a　Integers and the Real World

Integers can be associated with many real-world problems and situations. The following examples will help you get ready to translate problem situations to mathematical language.

▶ **EXAMPLE 1**　Tell which integer corresponds to this situation: The temperature is 3 degrees below zero.

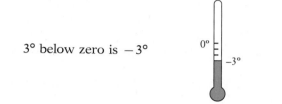

3° below zero is −3°

◀

▶ **EXAMPLE 2**　Tell which integer corresponds to this situation: Losing 21 points in a card game.

Losing 21 points in a card game gives you −21 points.

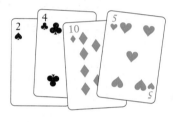

◀

▶ **EXAMPLE 3**　Tell which integer corresponds to this situation: Death Valley is 280 ft below sea level.

The integer −280 corresponds to the situation. The elevation is −280 ft.

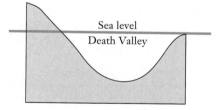

◀

▶ **EXAMPLE 4**　Tell which integers correspond to this situation: A salesperson made $78 on Monday, but lost $57 on Tuesday.

The integers 78 and −57 correspond to the situation. The integer 78 corresponds to the profit on Monday and −57 corresponds to the loss on Tuesday.　◀

DO EXERCISES 1–4.

b　The Rational Numbers

We created the set of integers by obtaining a negative number for each natural number. To create a larger number system, called the set of **rational numbers,** we consider quotients of integers with nonzero divisors. The following are rational numbers.

$$\frac{2}{3}, \quad -\frac{2}{3}, \quad \frac{7}{1}, \quad 4, \quad -3, \quad 0, \quad \frac{23}{-8}, \quad 2.4, \quad -0.17.$$

The number $-\frac{2}{3}$ (read "negative two-thirds") can also be named $\frac{-2}{3}$ or $\frac{2}{-3}$. The number 2.4 can be named $\frac{24}{10}$ or $\frac{12}{5}$, and -0.17 can be named $-\frac{17}{100}$.

Note that this new set of numbers, the rational numbers, contains the whole numbers, the integers, and the numbers of arithmetic (also called the nonnegative rational numbers). We can describe the set of rational numbers using set notation as follows.

The set of rational numbers $= \left\{ \dfrac{a}{b} \,\middle|\, a \text{ and } b \text{ are integers and } b \neq 0 \right\}.$

$\left(\text{This is read "the set of numbers } \dfrac{a}{b}, \text{ where } a \text{ and } b \text{ are integers and } b \neq 0." \right)$

We picture the rational numbers on a number line as follows. There is a point on the line for every rational number.

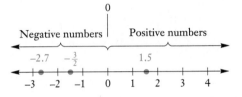

To **graph** a number means to find and mark its point on the line. Some numbers are graphed in the preceding figure.

▶ **EXAMPLE 5** Graph: $\frac{5}{2}$.

The number $\frac{5}{2}$ can be named $2\frac{1}{2}$, or 2.5. Its graph is halfway between 2 and 3.

◀

▶ **EXAMPLE 6** Graph: -3.2.

The graph of -3.2 is $\frac{2}{10}$ of the way from -3 to -4.

◀

▶ **EXAMPLE 7** Graph: $\frac{13}{8}$.

The number $\frac{13}{8}$ can be named $1\frac{5}{8}$, or 1.625. The graph is about $\frac{6}{10}$ of the way from 1 to 2.

◀

DO EXERCISES 5–7.

Graph on a number line.

5. $-\dfrac{7}{2}$

6. 1.4

7. $-\dfrac{11}{4}$

Convert to decimal notation.

8. $-\dfrac{3}{8}$

9. $-\dfrac{6}{11}$

10. $\dfrac{4}{3}$

 Notation for Rational Numbers

The rational numbers can be named using fractional or decimal notation.

▶ **EXAMPLE 8**   Convert to decimal notation: $-\frac{5}{8}$.

We first find decimal notation for $\frac{5}{8}$. Since $\frac{5}{8}$ means $5 \div 8$, we divide.

$$
\begin{array}{r}
0.6\;2\;5 \\
8\,)\overline{5.0\;0\;0} \\
4\;8 \\
\hline
2\;0 \\
1\;6 \\
\hline
4\;0 \\
4\;0 \\
\hline
0
\end{array}
$$

Thus, $\frac{5}{8} = 0.625$, so $-\frac{5}{8} = -0.625$.   ◀

Decimal notation for $-\frac{5}{8}$ is -0.625. We consider -0.625 to be a **terminating decimal.** Decimal notation for some numbers repeats.

▶ **EXAMPLE 9**   Convert to decimal notation: $\frac{7}{11}$.

We divide.

$$
\begin{array}{r}
0.6\;3\;6\;3\ldots \\
1\,1\,)\overline{7.0\;0\;0\;0} \\
6\;6 \\
\hline
4\;0 \\
3\;3 \\
\hline
7\;0 \\
6\;6 \\
\hline
4\;0 \\
3\;3 \\
\hline
7
\end{array}
$$

$$\frac{7}{11} = 0.\overline{63}$$

We can abbreviate repeating decimal notation by writing a bar over the repeating part, in this case, $0.\overline{63}$.   ◀

DO EXERCISES 8–10.

d   **The Real Numbers and Order**

Every rational number has a point on the number line. However, there are some points on the line for which there is no rational number. These points correspond to what are called **irrational numbers.**

What kinds of numbers correspond to points that are irrational numbers? One example is the number π, which is used in finding the area and circumference of a circle: $A = \pi r^2$ and $C = 2\pi r$.

Another example of an irrational number is the square root of 2, named $\sqrt{2}$.

It is the length of the diagonal of a square with sides of length 1. It is also the number that when multiplied by itself gives 2. There is no rational number that can be multiplied by itself to get 2. But the following are rational *approximations*:

1.4 is an approximation of $\sqrt{2}$ because $(1.4)^2 = 1.96$;

1.41 is a better approximation because $(1.41)^2 = 1.988$;

1.4142 is an even better approximation because $(1.4142)^2 = 1.99996164$.

We can find rational approximations for square roots using a calculator.

Decimal notation for rational numbers *either* terminates *or* repeats. Decimal notation for irrational numbers *neither* terminates *nor* repeats. Some other examples of irrational numbers are π, $\sqrt{3}$, $-\sqrt{8}$, $\sqrt{11}$, and $0.121221222122221\ldots$. Whenever we take the square root of a number that is not a perfect square we will get an irrational number.

The rational numbers and the irrational numbers together correspond to all the points on a number line and make up what is called the **real-number system.**

> **The set of real numbers = The set of all numbers corresponding to points on the number line.**

The real numbers consist of the rational numbers and the irrational numbers. The following figure shows the relationship between various kinds of numbers.

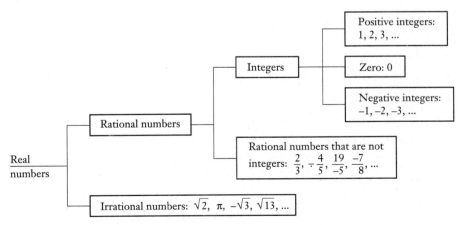

Use either < or > for ▨ to write a true sentence.

11. −3 ▨ 7

12. −8 ▨ −5

13. 7 ▨ −10

14. 3.1 ▨ −9.5

15. $-\frac{2}{3}$ ▨ −1

16. $-\frac{11}{8}$ ▨ $\frac{23}{15}$

17. $-\frac{2}{3}$ ▨ $-\frac{5}{9}$

18. −4.78 ▨ −5.01

▶ **EXAMPLE 10**   Graph the real number $\sqrt{3}$ on a number line.

We use a calculator or Table 2 at the back of the book and approximate $\sqrt{3} \approx 1.732$ ("≈" means "approximately equal to"). Then we locate this number on a number line.

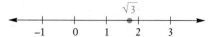

◀

Real numbers are named in order on the number line, with larger numbers named further to the right. For any two numbers on the line, the one to the left is less than the one to the right.

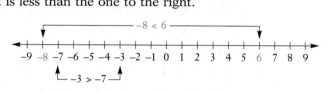

We use the symbol **<** to mean "**is less than.**" The sentence −8 < 6 means "−8 is less than 6." The symbol **>** means "**is greater than.**" The sentence −3 > −7 means "−3 is greater than −7."

▶ **EXAMPLES**   Use either < or > for ▨ to write a true sentence.

11. 2 ▨ 9   Since 2 is to the left of 9, 2 is less than 9, so 2 < 9.

12. −7 ▨ 3   Since −7 is to the left of 3, we have −7 < 3.

13. 6 ▨ −12   Since 6 is to the right of −12, then 6 > −12.

14. −18 ▨ −5   Since −18 is to the left of −5, we have −18 < −5.

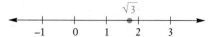

15. −2.7 ▨ $-\frac{3}{2}$ The answer is $-2.7 < -\frac{3}{2}$.

16. 1.5 ▨ −2.7   The answer is 1.5 > −2.7.

17. 1.38 ▨ 1.83   The answer is 1.38 < 1.83.

18. −3.45 ▨ 1.32   The answer is −3.45 < 1.32.

19. $\frac{5}{8}$ ▨ $\frac{7}{11}$ We convert to decimal notation: $\frac{5}{8} = 0.625$ and $\frac{7}{11} = 0.6363 \ldots$. Thus, $\frac{5}{8} < \frac{7}{11}$.

◀

DO EXERCISES 11–18.

Note that both $-8 < 6$ and $6 > -8$ are true. These are **inequalities.** Every true inequality yields another true inequality when we interchange the numbers or variables and reverse the direction of the inequality sign.

$a < b$ also has the meaning $b > a$.

▶ **EXAMPLES** Write another inequality with the same meaning.

20. $a < -5$ The inequality $-5 > a$ has the same meaning.

21. $-3 > -8$ The inequality $-8 < -3$ has the same meaning. ◀

A helpful mental device is to think of an inequality sign as an "arrow" with the arrow pointing to the smaller number.

DO EXERCISES 19 AND 20.

Note that all positive real numbers are greater than zero and all negative real numbers are less than zero.

If x is a positive real number, then $x > 0$.
If x is a negative real number, then $x < 0$.

Expressions like $a \leq b$ and $b \geq a$ are also inequalities. We read $a \leq b$ as "*a* **is less than or equal to** *b*." We read $a \geq b$ as "*a* **is greater than or equal to** *b*."

▶ **EXAMPLES** Write true or false for each statement.

22. $-3 \leq 5$ True since $-3 < 5$ is true

23. $-3 \leq -3$ True since $-3 = -3$ is true

24. $-5 \geq 4$ False since neither $-5 > 4$ nor $-5 = 4$ is true ◀

DO EXERCISES 21–23.

e   Absolute Value

From the number line, we see that numbers like 4 and -4 are the same distance from zero. Distance is always a nonnegative number. We call the distance from zero on a number line the **absolute value** of the number.

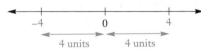

The *absolute value* of a number is its distance from zero on a number line. We use the symbol $|x|$ to represent the absolute value of a number x.

Write another inequality with the same meaning.

19. $-5 < 7$

20. $x > 4$

Write true or false.

21. $-4 \leq -6$

22. $7 \geq 7$

23. $-2 \leq 3$

ANSWERS ON PAGE A-2

Find the absolute value.

24. $|8|$

25. $|0|$

26. $|-9|$

27. $\left|-\dfrac{2}{3}\right|$

28. $|5.6|$

To find absolute value:

1. **If a number is negative, make it positive.**
2. **If a number is positive or zero, leave it alone.**

► **EXAMPLES**   Find the absolute value.

25. $|-7|$ The distance of -7 from 0 is 7, so $|-7|$ is 7.

26. $|12|$ The distance of 12 from 0 is 12, so $|12|$ is 12.

27. $|0|$ The distance of 0 from 0 is 0, so $|0|$ is 0.

28. $\left|\dfrac{3}{2}\right| = \dfrac{3}{2}$

29. $|-2.73| = 2.73$ ◄

DO EXERCISES 24–28.

EXERCISE SET 1.2

a Tell which real numbers correspond to the situation.

1. The temperature on Wednesday was 18° above zero. On Thursday it was 2° below zero.

2. The Dead Sea, between Jordan and Israel, is 1286 ft below sea level, whereas Mt. Everest is 29,028 ft above sea level.

3. A student deposited $750 in a savings account. Two weeks later, the student withdrew $125.

4. During a certain time period, the United States had a deficit of $3 million in foreign trade.

5. During a video game, a player intercepted a missile worth 20 points, lost a starship worth 150 points, and captured a base worth 300 points.

6. 3 seconds before liftoff of a rocket occurs. 128 seconds after the liftoff of a rocket occurs.

b Graph the number on the number line.

7. $\dfrac{10}{3}$

8. $-\dfrac{17}{5}$

9. -4.3

10. 3.87

Write an inequality with the same meaning.

11. $-6 > x$ **12.** $x < 8$ **13.** $-10 \leq y$ **14.** $12 \geq t$

c Convert to decimal notation.

15. $-\dfrac{3}{8}$ **16.** $-\dfrac{1}{8}$ **17.** $\dfrac{5}{3}$ **18.** $\dfrac{5}{6}$

19. $\dfrac{7}{6}$ **20.** $\dfrac{5}{12}$ **21.** $\dfrac{2}{3}$ **22.** $\dfrac{1}{4}$

23. $-\dfrac{1}{2}$ **24.** $\dfrac{5}{8}$ **25.** $\dfrac{1}{10}$ **26.** $-\dfrac{7}{20}$

1. _____
2. _____
3. _____
4. _____
5. _____
6. _____
7. See graph.
8. See graph.
9. See graph.
10. See graph.
11. _____
12. _____
13. _____
14. _____
15. _____
16. _____
17. _____
18. _____
19. _____
20. _____
21. _____
22. _____
23. _____
24. _____
25. _____
26. _____

d Use either < or > for ▦ to write a true sentence.

27. 5 ▦ 0   **28.** 9 ▦ 0   **29.** −9 ▦ 5   **30.** 8 ▦ −8

31. −6 ▦ 6   **32.** 0 ▦ −7   **33.** −8 ▦ −5   **34.** −4 ▦ −3

35. −5 ▦ −11   **36.** −3 ▦ −4   **37.** −6 ▦ −5   **38.** −10 ▦ −14

39. 2.14 ▦ 1.24   **40.** −3.3 ▦ −2.2   **41.** −14.5 ▦ 0.011

42. 17.2 ▦ −1.67   **43.** −12.88 ▦ −6.45   **44.** −14.34 ▦ −17.88

45. $\dfrac{5}{12}$ ▦ $\dfrac{11}{25}$ **46.** $-\dfrac{14}{17}$ ▦ $-\dfrac{27}{35}$

Write true or false.

47. $-3 \geq -11$ **48.** $5 \leq -5$ **49.** $0 \geq 8$ **50.** $-5 \leq 7$

e Find the absolute value.

51. $|-3|$ **52.** $|-7|$ **53.** $|10|$ **54.** $|11|$

55. $|0|$ **56.** $|-4|$ **57.** $|-24|$ **58.** $|325|$

59. $\left|-\dfrac{2}{3}\right|$ **60.** $\left|-\dfrac{10}{7}\right|$ **61.** $\left|\dfrac{0}{4}\right|$ **62.** $|14.8|$

SYNTHESIS

List in order from the least to the greatest.

63. $-\dfrac{2}{3}$, $\dfrac{1}{2}$, $-\dfrac{3}{4}$, $-\dfrac{5}{6}$, $\dfrac{3}{8}$, $\dfrac{1}{6}$

64. 7^1, -5, $|-6|$, 4, $|3|$, -100, 0, 1^7, $\dfrac{14}{4}$

1.3 Addition of Real Numbers

In this section, we consider addition of real numbers. First, to gain an understanding, we add using a number line. Then we consider rules for addition.

Addition of numbers can be illustrated on a number line. To do the addition $a + b$, we start at a, and then move according to b.

a) If b is positive, we move to the right.

b) If b is negative, we move to the left.

c) If b is 0, we stay at a.

OBJECTIVES

After finishing Section 1.3, you should be able to:

a Add real numbers without using a number line.

b Find the additive inverse, or opposite, of a real number.

FOR EXTRA HELP

Tape 2C   Tape 2B   MAC: 1
                                  IBM: 1

▶ **EXAMPLE 1** Add: $3 + (-5)$.

$$3 + (-5) = -2$$

Move 5 units to the left.   Start at 3.

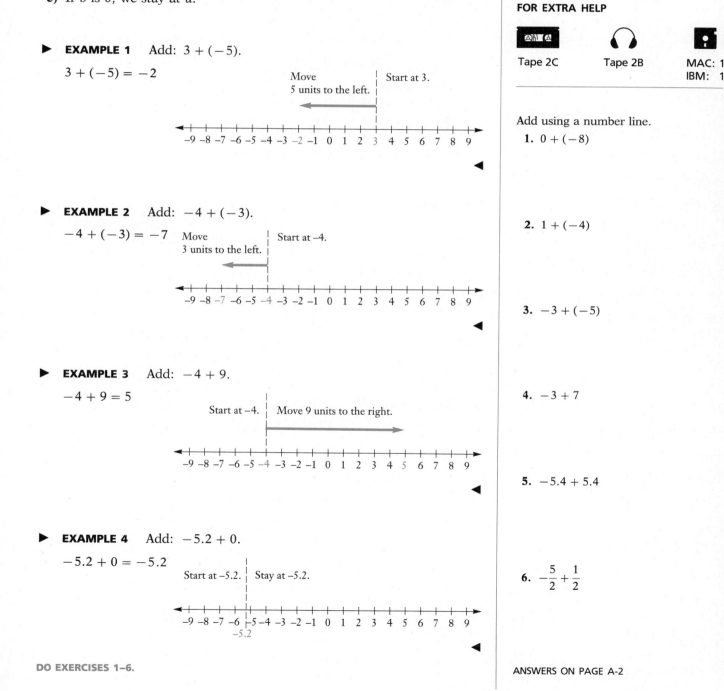

◀

▶ **EXAMPLE 2** Add: $-4 + (-3)$.

$$-4 + (-3) = -7$$

Move 3 units to the left.   Start at –4.

◀

▶ **EXAMPLE 3** Add: $-4 + 9$.

$$-4 + 9 = 5$$

Start at –4.   Move 9 units to the right.

◀

▶ **EXAMPLE 4** Add: $-5.2 + 0$.

$$-5.2 + 0 = -5.2$$

Start at –5.2.   Stay at –5.2.

◀

Add using a number line.

1. $0 + (-8)$

2. $1 + (-4)$

3. $-3 + (-5)$

4. $-3 + 7$

5. $-5.4 + 5.4$

6. $-\dfrac{5}{2} + \dfrac{1}{2}$

DO EXERCISES 1–6.

ANSWERS ON PAGE A-2

Add without using a number line.

7. $-5 + (-6)$ **8.** $-9 + (-3)$

9. $-4 + 6$ **10.** $-7 + 3$

11. $5 + (-7)$ **12.** $-20 + 20$

13. $-11 + (-11)$ **14.** $10 + (-7)$

15. $-0.17 + 0.7$ **16.** $-6.4 + 8.7$

17. $-4.5 + (-3.2)$ **18.** $-8.6 + 2.4$

19. $\dfrac{5}{9} + \left(-\dfrac{7}{9}\right)$ **20.** $-\dfrac{1}{5} + \left(-\dfrac{3}{4}\right)$

a   Adding Without a Number Line

You may have noticed some patterns in the preceding examples. These lead us to rules for adding without using a number line that are more efficient for adding larger or more complicated numbers.

> **Rules for Addition of Real Numbers**
>
> 1. *Positive numbers:* Add the same as numbers of arithmetic. The answer is positive.
> 2. *Negative numbers:* Add absolute values. The answer is negative.
> 3. *A positive and a negative number:* Subtract absolute values. Then:
>   a) If the positive number has the greater absolute value, the answer is positive.
>   b) If the negative number has the greater absolute value, the answer is negative.
>   c) If the numbers have the same absolute value, the answer is 0.
> 4. *One number is zero:* The sum is the other number.

Rule 4 is known as the **Identity Property of 0.** It says that for any real number a, $a + 0 = a$.

▶ **EXAMPLES**   Add without using a number line.

5. $-12 + (-7) = -19$ Two negatives. *Think:* Add the absolute values, getting 19. Make the answer *negative*, -19.

6. $-1.4 + 8.5 = 7.1$ The absolute values are 1.4 and 8.5. The difference is 7.1. The positive number has the larger absolute value, so the answer is *positive*, 7.1.

7. $-36 + 21 = -15$ The absolute values are 36 and 21. The difference is 15. The negative number has the larger absolute value, so the answer is *negative*, -15.

8. $1.5 + (-1.5) = 0$ The numbers have the same absolute value. The sum is 0.

9. $-\dfrac{7}{8} + 0 = -\dfrac{7}{8}$ One number is zero. The sum is $-\frac{7}{8}$.

10. $-9.2 + 3.1 = -6.1$

11. $-\dfrac{3}{2} + \dfrac{9}{2} = \dfrac{6}{2} = 3$

12. $-\dfrac{2}{3} + \dfrac{5}{8} = -\dfrac{16}{24} + \dfrac{15}{24} = -\dfrac{1}{24}$ ◀

DO EXERCISES 7–20.

Add.

21. $(-15) + (-37) + 25 + 42 + (-59) + (-14)$

22. $42 + (-81) + (-28) + 24 + 18 + (-31)$

23. $-2.5 + (-10) + 6 + (-7.5)$

24. -35
17
14
-27
31
-12
‾‾‾‾

Find the opposite.

25. -4 **26.** 8.7

27. -7.74 **28.** $-\dfrac{8}{9}$

29. 0 **30.** 12

Suppose we want to add several numbers, some positive and some negative, as follows. How can we proceed?

$$15 + (-2) + 7 + 14 + (-5) + (-12)$$

We can change grouping and order as we please when adding. For instance, we can group the positive numbers together and the negative numbers together and add them separately. Then we add the two results.

▶ **EXAMPLE 13**   Add: $15 + (-2) + 7 + 14 + (-5) + (-12)$.

a) $15 + 7 + 14 = 36$ Adding the positive numbers
b) $-2 + (-5) + (-12) = -19$ Adding the negative numbers
c) $36 + (-19) = 17$ Adding the results

We can also add the numbers in any other order we wish, say from left to right as follows:

$$
\begin{aligned}
15 + (-2) + 7 + 14 + (-5) + (-12) &= 13 + 7 + 14 + (-5) + (-12) \\
&= 20 + 14 + (-5) + (-12) \\
&= 34 + (-5) + (-12) \\
&= 29 + (-12) \\
&= 17 \quad ◀
\end{aligned}
$$

DO EXERCISES 21–24.

b   Opposites and Additive Inverses

Suppose we add two numbers that are opposites, such as 6 and -6. The result is 0. When opposites are added, the result is always 0. Such numbers are also called **opposites**, or **additive inverses**. Every real number has an opposite.

> Two numbers whose sum is 0 are called *opposites*, or *additive inverses* of each other.

▶ **EXAMPLES**   Find the opposite of each number.

14. 34 The opposite of 34 is -34 because $34 + (-34) = 0$.

15. -8 The opposite of -8 is 8 because $-8 + 8 = 0$.

16. 0 The opposite of 0 is 0 because $0 + 0 = 0$.

17. $-\dfrac{7}{8}$ The opposite of $-\dfrac{7}{8}$ is $\dfrac{7}{8}$ because $-\dfrac{7}{8} + \dfrac{7}{8} = 0$.   ◀

DO EXERCISES 25–30.

To name the opposite, we use the symbol $-$, as follows.

> The opposite, or additive inverse, of a number a can be named $-a$ (read "the opposite of a" or "the additive inverse of a").

Note that if we take a number, say 8, and find its opposite, -8, and then find the opposite of the result, we will have the original number, 8, again.

Find $-x$ and $-(-x)$ when x is:

31. 14   **32.** 1

33. -19 **34.** -1.6

35. $\dfrac{2}{3}$ **36.** $-\dfrac{9}{8}$

Change the sign. (Find the opposite.)

37. -4 **38.** -13.4

39. 0   **40.** $\dfrac{1}{4}$

> The opposite of the opposite of a number is the number itself. The additive inverse of the additive inverse of a number is the number itself. That is, for any number a,
>
> $$-(-a) = a.$$

▶ **EXAMPLE 18**   Find $-x$ and $-(-x)$ when $x = 16$.

a) If $x = 16$, then $-x = -16$.   The opposite of 16 is -16.

b) If $x = 16$, then $-(-x) = -(-16) = 16$.   The opposite of the opposite of 16 is 16.   ◄

▶ **EXAMPLE 19**   Find $-x$ and $-(-x)$ when $x = -3$.

a) If $x = -3$, then $-x = -(-3) = 3$.

b) If $x = -3$, then $-(-x) = -(-(-3)) = -3$.   ◄

Note in Example 19 that an extra set of parentheses is used to show that we are substituting the negative number -3 for x. Symbolism like $--x$ is not considered meaningful.

DO EXERCISES 31–36.

A symbol such as -8 is usually read "negative 8." It could be read "the additive inverse of 8," because the additive inverse of 8 is negative 8. It could also be read "the opposite of 8," because the opposite of 8 is -8. Thus a symbol like -8 can be read in more than one way. A symbol like $-x$, which has a variable, should be read "the opposite of x" or "the additive inverse of x" and *not* "negative x," because we do not know whether x represents a positive number, a negative number, or 0. Check this out by referring to the preceding examples.

We can use the symbolism $-a$ to restate the definition of opposite, or additive inverse.

> For any real number a, the *opposite*, or *additive inverse*, of a, $-a$, is such that
>
> $$a + (-a) = (-a) + a = 0.$$

Signs of Numbers

A negative number is sometimes said to have a "negative sign." A positive number is said to have a "positive sign." When we replace a number by its opposite, we can say that we have "changed its sign."

▶ **EXAMPLES**   Change the sign. (Find the opposite.)

20. -3 $-(-3) = 3$ The opposite of -3 is 3.

21. -10 $-(-10) = 10$

22. 0 $-(0) = 0$

23. 14 $-(14) = -14$ ◄

DO EXERCISES 37–40.

EXERCISE SET 1.3

a Add. Do not use a number line except as a check.

1. $-9 + 2$ **2.** $2 + (-5)$ **3.** $-10 + 6$ **4.** $8 + (-3)$

5. $-8 + 8$ **6.** $6 + (-6)$ **7.** $-3 + (-5)$ **8.** $-4 + (-6)$

9. $-7 + 0$ **10.** $-13 + 0$ **11.** $0 + (-27)$ **12.** $0 + (-35)$

13. $17 + (-17)$ **14.** $-15 + 15$ **15.** $-17 + (-25)$ **16.** $-24 + (-17)$

17. $18 + (-18)$ **18.** $-13 + 13$ **19.** $-18 + 18$ **20.** $11 + (-11)$

21. $8 + (-5)$ **22.** $-7 + 8$ **23.** $-4 + (-5)$ **24.** $10 + (-12)$

25. $13 + (-6)$ **26.** $-3 + 14$ **27.** $-25 + 25$ **28.** $40 + (-40)$

29. $63 + (-18)$ **30.** $85 + (-65)$ **31.** $-6.5 + 4.7$ **32.** $-3.6 + 1.9$

33. $-2.8 + (-5.3)$ **34.** $-7.9 + (-6.5)$ **35.** $-\dfrac{3}{5} + \dfrac{2}{5}$ **36.** $-\dfrac{4}{3} + \dfrac{2}{3}$

ANSWERS

1. _____
2. _____
3. _____
4. _____
5. _____
6. _____
7. _____
8. _____
9. _____
10. _____
11. _____
12. _____
13. _____
14. _____
15. _____
16. _____
17. _____
18. _____
19. _____
20. _____
21. _____
22. _____
23. _____
24. _____
25. _____
26. _____
27. _____
28. _____
29. _____
30. _____
31. _____
32. _____
33. _____
34. _____
35. _____
36. _____

ANSWERS

37. _____

38. _____

39. _____

40. _____

41. _____

42. _____

43. _____

44. _____

45. _____

46. _____

47. _____

48. _____

49. _____

50. _____

51. _____

52. _____

53. _____

54. _____

55. _____

56. _____

57. _____

58. _____

59. _____

60. _____

61. _____

62. _____

63. _____

64. _____

65. _____

66. _____

67. _____

68. _____

69. _____

70. _____

37. $-\dfrac{3}{7}+\left(-\dfrac{5}{7}\right)$ **38.** $-\dfrac{4}{9}+\left(-\dfrac{6}{9}\right)$ **39.** $-\dfrac{5}{8}+\dfrac{1}{4}$ **40.** $-\dfrac{5}{6}+\dfrac{2}{3}$

41. $-\dfrac{3}{7}+\left(-\dfrac{2}{5}\right)$ **42.** $-\dfrac{5}{8}+\left(-\dfrac{1}{3}\right)$ **43.** $-\dfrac{7}{16}+\dfrac{7}{8}$ **44.** $-\dfrac{3}{28}+\dfrac{5}{42}$

45. $75 + (-14) + (-17) + (-5)$ **46.** $28 + (-44) + 17 + 31 + (-94)$

47. $-44 + \left(-\dfrac{3}{8}\right) + 95 + \left(-\dfrac{5}{8}\right)$ **48.** $24 + 3.1 + (-44) + (-8.2) + 63$

49. $98 + (-54) + 113 + (-998) + 44 + (-612)$

50. $-455 + (-123) + 1026 + (-919) + 213$

b Find the opposite, or additive inverse.

51. 24 **52.** -64 **53.** -26.9 **54.** 48.2

Find $-x$ when x is:

55. 9 **56.** -26 **57.** $-\dfrac{14}{3}$ **58.** $\dfrac{1}{328}$

Find $-(-x)$ when x is:

59. -65 **60.** 29 **61.** $\dfrac{5}{3}$ **62.** -9.1

Change the sign. (Find the opposite.)

63. -14 **64.** -22.4 **65.** 10 **66.** $-\dfrac{7}{8}$

SYNTHESIS

67. For what numbers x is $-x$ negative?   **68.** For what numbers x is $-x$ positive?

Tell whether the sum is positive, negative, or zero.

69. If n is positive and m is negative, $-n + m$ is _____.

70. If $n = m$ and n and m are negative, $-n + (-m)$ is _____.

1.4 Subtraction of Real Numbers

a Subtraction

We now consider subtraction of real numbers. Subtraction is defined as follows.

> **The difference $a - b$ is the number that when added to b gives a.**

For example, $45 - 17 = 28$ because $28 + 17 = 45$. Let us consider an example whose answer is a negative number.

▶ **EXAMPLE 1**   Subtract: $5 - 8$.
Think: $5 - 8$ is the number that when added to 8 gives 5. What number can we add to 8 to get 5? The number must be negative. The number is -3:

$$5 - 8 = -3.$$

That is, $5 - 8 = -3$ because $5 = -3 + 8$.   ◀

DO EXERCISES 1–3.

The definition above does *not* provide the most efficient way to do subtraction. From that definition, however, we can develop a faster way to subtract. Look for a pattern in the following examples.

Subtractions	*Adding an Opposite*
$5 - 8 = -3$	$5 + (-8) = -3$
$-6 - 4 = -10$	$-6 + (-4) = -10$
$-7 - (-10) = 3$	$-7 + 10 = 3$
$-7 - (-2) = -5$	$-7 + 2 = -5$

DO EXERCISES 4–7.

Perhaps you have noticed that we can subtract by adding the opposite of the number being subtracted. This can always be done.

> **For any real numbers a and b,**
> $$a - b = a + (-b).$$
> **(To subtract, add the opposite, or additive inverse, of the number being subtracted.)**

This is the method normally used for quick subtraction of real numbers.

▶ **EXAMPLES**   Subtract. Check by addition.

2. $2 - 6 = 2 + (-6) = -4$ The opposite of 6 is -6. We change the subtraction to addition and add the opposite. *Check:* $-4 + 6 = 2$.

3. $4 - (-9) = 4 + 9 = 13$ The opposite of -9 is 9. We change the subtraction to addition and add the opposite. *Check:* $13 + (-9) = 4$.

OBJECTIVES

After finishing Section 1.4, you should be able to:

a Subtract real numbers and simplify combinations of additions and subtractions.

b Solve problems involving additions and subtractions of real numbers.

FOR EXTRA HELP

Tape 3A   Tape 3A   MAC: 1
                                  IBM: 1

Subtract.

1. $-6 - 4$ (*Think:* $-6 - 4$ is the number that when added to 4 gives -6. What number can be added to 4 to get -6?)

2. $-7 - (-10)$ (*Think:* $-7 - (-10)$ is the number that when added to -10 gives -7. What number can be added to -10 to get -7?)

3. $-7 - (-2)$ (*Think:* $-7 - (-2)$ is the number that when added to -2 gives -7. What number can be added to -2 to get -7?)

Complete each addition and compare with the subtraction.

4. $4 - 6 = -2$;

   $4 + (-6) = $ _____

5. $-3 - 8 = -11$;

   $-3 + (-8) = $ _____

6. $-5 - (-9) = 4$;

   $-5 + 9$ _____

7. $-5 - (-3) = -2$;

   $-5 + 3 = $ _____

ANSWERS ON PAGE A-2

Subtract. Check by addition.

8. $2 - 8$

9. $-6 - 10$

10. $12.4 - 5.3$

11. $-8 - (-11)$

12. $-8 - (-8)$

13. $\dfrac{2}{3} - \left(-\dfrac{5}{6}\right)$

Read each of the following. Then subtract by adding the opposite of the number being subtracted.

14. $3 - 11$

15. $12 - 5$

16. $-12 - (-9)$

17. $-12.4 - 10.9$

18. $-\dfrac{4}{5} - \left(-\dfrac{4}{5}\right)$

Simplify.

19. $-6 - (-2) - (-4) - 12 + 3$

20. $9 - (-6) + 7 - 11 - 14 - (-20)$

21. $-9.6 + 7.4 - (-3.9) - (-11)$

ANSWERS ON PAGE A-2

4. $-4.2 - (-3.6) = -4.2 + 3.6 = -0.6$ Adding the opposite.
                                            Check: $-0.6 + (-3.6) = -4.2.$

5. $-\dfrac{1}{2} - \left(-\dfrac{3}{4}\right) = -\dfrac{1}{2} + \dfrac{3}{4} = \dfrac{1}{4}$ Adding the opposite.
                                       Check: $\frac{1}{4} + (-\frac{3}{4}) = -\frac{1}{2}.$ ◄

DO EXERCISES 8–13.

► **EXAMPLES** Read each of the following. Then subtract by adding the opposite of the number being subtracted.

6. $3 - 5;$ Read "three minus 5"
    $3 - 5 = 3 + (-5) = -2$ Adding the opposite

7. $\dfrac{1}{8} - \dfrac{7}{8};$ Read "one-eighth minus seven-eighths"

    $\dfrac{1}{8} - \dfrac{7}{8} = \dfrac{1}{8} + \left(-\dfrac{7}{8}\right) = -\dfrac{6}{8},$ or $-\dfrac{3}{4}$

8. $-4.6 - (-9.8);$ Read "negative four point six minus negative
                             nine point eight"
    $-4.6 - (-9.8) = -4.6 + 9.8 = 5.2$

9. $-\dfrac{3}{4} - \dfrac{7}{5};$ Read "negative three-fourths minus seven-fifths"

    $-\dfrac{3}{4} - \dfrac{7}{5} = -\dfrac{15}{20} + \left(-\dfrac{28}{20}\right) = -\dfrac{43}{20}$ ◄

DO EXERCISES 14–18.

When several additions and subtractions occur together, we can make them all additions.

► **EXAMPLES** Simplify.

10. $8 - (-4) - 2 - (-4) + 2 = 8 + 4 + (-2) + 4 + 2$
                              $= 16$

11. $8.2 - (-6.1) + 2.3 - (-4) = 8.2 + 6.1 + 2.3 + 4$
                               $= 20.6$ ◄

DO EXERCISES 19–21.

b   Problem Solving

Let us see how we can use subtraction of real numbers to solve problems.

▶ **EXAMPLE 12**   The lowest point in Asia is the Dead Sea, which is 400 m below sea level. The lowest point in the United States is Death Valley, which is 86 m below sea level. How much higher is Death Valley than the Dead Sea?

It is helpful to draw a picture of the situation.

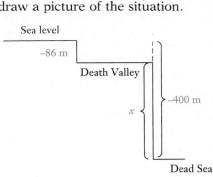

We see that −86 is the higher altitude at Death Valley and −400 is the lower altitude at the Dead Sea. To find how much higher Death Valley is, we subtract:

$$-86 - (-400) = -86 + 400 = 314.$$

Death Valley is 314 m higher than the Dead Sea.   ◀

DO EXERCISES 22 AND 23.

Solve.

22. A small business made a profit of $18 on Monday. There was a loss of $7 on Tuesday. On Wednesday there was a loss of $5, and on Thursday there was a profit of $11. Find the total profit or loss.

23. In Churchill, Manitoba, Canada, the average daily low temperature in January is −31°C. The average daily low temperature in Key West, Florida, is 19°C. How much higher is the average daily low temperature in Key West, Florida?

❖ SIDELIGHTS

Careers and Their Uses of Mathematics

Students typically ask the question "Why do we have to study mathematics?" This is a question with a complex set of answers. Certainly, one answer is that you will use this mathematics in the next course. While it is a correct answer, it sometimes frustrates students, because this answer can be given in the next mathematics course, and the next one, and so on. Sometimes an answer can be given by applications like those you have seen or will see in this book. Another answer is that you are living in a society in which mathematics becomes more and more critical with each passing day. Evidence of this was provided recently by a nationwide symposium sponsored by the National Research Council's Mathematical Sciences Education Board. Results showed that "Other than demographic factors, the *strongest* predictor of earnings nine years after high school is the number of mathematics courses taken." This is a significant testimony to the need for you to take as many mathematics courses as possible.

We try to provide other answers to "Why do we have to study mathematics?" in what follows. We have listed several occupations that are attractive and popular to students. Below each occupation are listed various kinds of mathematics that are useful in that occupation.

Accountant and businessperson	Travel agent
Computer skills	Whole-number skills
Calculator skills	Fraction/decimal skills
Equations	Estimation
Systems of equations	Percent notation
Formulas	Equations
Probability	Calculator skills
Statistics	Computer skills
Ratio and proportion	
Percent notation	
Estimation	

Librarian	Machinist
Whole-number skills	Whole-number skills
Fraction/decimal skills	Fraction/decimal skills
Estimation	Estimation
Percent notation	Percent notation
Ratio and proportion	Length, area, volume, and perimeter
Area and perimeter	Angle measures
Formulas	Geometry
Calculator skills	Pythagorean theorem
Computer skills	Square roots
	Equations
	Formulas
	Graphing
	Calculator skills
	Computer skills

Doctor	Lawyer
Equations	Equations
Percent notation	Percent notation
Graphing	Graphing
Statistics	Probability
Geometry	Statistics
Measurement	Ratio and proportion
Estimation	Area and volume
Exponents	Negative numbers
Logic	Formulas
	Calculator skills

Pilot	Firefighter
Equations	Percent notation
Percent notation	Graphing
Graphing	Estimation
Trigonometry	Formulas
Angles and geometry	Angles and geometry
Calculator skills	Probability
Computer skills	Statistics
Ratio and proportion	Area and geometry
Vectors	Square roots
	Exponents
	Pythagorean theorem

Nurse	Police officer
Whole-number skills	Whole-number skills
Fraction/decimal skills	Fraction/decimal skills
Estimation	Estimation
Percent notation	Percent notation
Ratio and proportion	Ratio and proportion
Equations	Geometry
English/Metric measurement	Negative numbers
Probability	Probability
Statistics	Statistics
Formulas	Calculator skills
Exponents and scientific notation	
Calculator skills	
Computer skills	

EXERCISE SET 1.4

a Subtract.

1. $3 - 7$ **2.** $4 - 9$ **3.** $0 - 7$ **4.** $0 - 10$

5. $-8 - (-2)$ **6.** $-6 - (-8)$ **7.** $-10 - (-10)$ **8.** $-6 - (-6)$

9. $12 - 16$ **10.** $14 - 19$ **11.** $20 - 27$ **12.** $30 - 4$

13. $-9 - (-3)$ **14.** $-7 - (-9)$ **15.** $-40 - (-40)$ **16.** $-9 - (-9)$

17. $7 - 7$ **18.** $9 - 9$ **19.** $7 - (-7)$ **20.** $4 - (-4)$

21. $8 - (-3)$ **22.** $-7 - 4$ **23.** $-6 - 8$ **24.** $6 - (-10)$

25. $-4 - (-9)$ **26.** $-14 - 2$ **27.** $2 - 9$ **28.** $2 - 8$

29. $-6 - (-5)$ **30.** $-4 - (-3)$ **31.** $8 - (-10)$ **32.** $5 - (-6)$

33. $0 - 5$ **34.** $0 - 6$ **35.** $-5 - (-2)$ **36.** $-3 - (-1)$

ANSWERS

1.
2.
3.
4.
5.
6.
7.
8.
9.
10.
11.
12.
13.
14.
15.
16.
17.
18.
19.
20.
21.
22.
23.
24.
25.
26.
27.
28.
29.
30.
31.
32.
33.
34.
35.
36.

37. _____

38. _____

39. _____

40. _____

41. _____

42. _____

43. _____

44. _____

45. _____

46. _____

47. _____

48. _____

49. _____

50. _____

51. _____

52. _____

53. _____

54. _____

55. _____

56. _____

57. _____

58. _____

59. _____

60. _____

61. _____

62. _____

63. _____

64. _____

65. _____

66. _____

67. _____

68. _____

37. $-7 - 14$

38. $-9 - 16$

39. $0 - (-5)$

40. $0 - (-1)$

41. $-8 - 0$

42. $-9 - 0$

43. $7 - (-5)$

44. $8 - (-3)$

45. $2 - 25$

46. $18 - 63$

47. $-42 - 26$

48. $-18 - 63$

49. $-71 - 2$

50. $-49 - 3$

51. $24 - (-92)$

52. $48 - (-73)$

53. $-50 - (-50)$

54. $-70 - (-70)$

55. $\dfrac{3}{8} - \dfrac{5}{8}$

56. $\dfrac{3}{9} - \dfrac{9}{9}$

57. $\dfrac{3}{4} - \dfrac{2}{3}$

58. $\dfrac{5}{8} - \dfrac{3}{4}$

59. $-\dfrac{3}{4} - \dfrac{2}{3}$

60. $-\dfrac{5}{8} - \dfrac{3}{4}$

61. $-\dfrac{5}{8} - \left(-\dfrac{3}{4}\right)$

62. $-\dfrac{3}{4} - \left(-\dfrac{2}{3}\right)$

63. $6.1 - (-13.8)$

64. $1.5 - (-3.5)$

65. $-3.2 - 5.8$

66. $-2.7 - 5.9$

67. $0.99 - 1$

68. $0.87 - 1$

69. $-79 - 114$ **70.** $-197 - 216$ **71.** $0 - (-500)$ **72.** $500 - (-1000)$

73. $-2.8 - 0$ **74.** $6.04 - 1.1$ **75.** $7 - 10.53$ **76.** $8 - (-9.3)$

77. $\dfrac{1}{6} - \dfrac{2}{3}$ **78.** $-\dfrac{3}{8} - \left(-\dfrac{1}{2}\right)$ **79.** $-\dfrac{4}{7} - \left(-\dfrac{10}{7}\right)$ **80.** $\dfrac{12}{5} - \dfrac{12}{5}$

81. $-\dfrac{7}{10} - \dfrac{10}{15}$ **82.** $-\dfrac{4}{18} - \left(-\dfrac{2}{9}\right)$ **83.** $\dfrac{1}{13} - \dfrac{1}{12}$ **84.** $-\dfrac{1}{7} - \left(-\dfrac{1}{6}\right)$

Simplify.

85. $18 - (-15) - 3 - (-5) + 2$ **86.** $22 - (-18) + 7 + (-42) - 27$

87. $-31 + (-28) - (-14) - 17$ **88.** $-43 - (-19) - (-21) + 25$

89. $-34 - 28 + (-33) - 44$ **90.** $39 + (-88) - 29 - (-83)$

91. $-93 - (-84) - 41 - (-56)$ **92.** $84 + (-99) + 44 - (-18) - 43$

93. $-5 - (-30) + 30 + 40 - (-12)$ **94.** $14 - (-50) + 20 - (-32)$

95. $132 - (-21) + 45 - (-21)$ **96.** $81 - (-20) - 14 - (-50) + 53$

ANSWERS

69. _____
70. _____
71. _____
72. _____
73. _____
74. _____
75. _____
76. _____
77. _____
78. _____
79. _____
80. _____
81. _____
82. _____
83. _____
84. _____
85. _____
86. _____
87. _____
88. _____
89. _____
90. _____
91. _____
92. _____
93. _____
94. _____
95. _____
96. _____

Copyright © 1991 Addison-Wesley Publishing Co., Inc.

ANSWERS

97. _____

98. _____

99. _____

100. _____

101. _____

102. _____

103. _____

104. _____

105. _____

106. _____

107. _____

108. _____

109. _____

110. _____

111. _____

112. _____

b   Solve.

97. Your total assets are $619.46. You borrow $950 for the purchase of a stereo system. What are your total assets now?

98. You owe a friend $420. The friend decides to cancel $156 of the debt. How much do you owe now?

99. You are in debt $215.50. How much money will you need to make your total assets y dollars?

100. On a winter night, the temperature dropped from $-5°C$ to $-12°C$. How many degrees did it drop?

101. The lowest point in Africa is Lake Assal, which is 156 m below sea level. The lowest point in South America is the Valdes Peninsula, which is 40 m below sea level. How much lower is Lake Assal than the Valdes Peninsula?

102. The deepest point in the Pacific Ocean is the Marianas Trench with a depth of 10,415 m. The deepest point in the Atlantic Ocean is the Puerto Rico Trench with a depth of 8648 m. How much higher is the Puerto Rico Trench than the Marianas Trench?

SKILL MAINTENANCE

103. Evaluate: 5^3.

104. Find the prime factorization of 864.

SYNTHESIS

Subtract.

105. ▤ $123,907 - 433,789$

106. ▤ $23,011 - (-60,432)$

Tell whether the statement is true or false for all integers m and n. If false, give a counterexample.

107. $n - 0 = 0 - n$

108. $0 - n = n$

109. If $m \neq n$, then $m - n \neq 0$.

110. If $m = -n$, then $m + n = 0$.

111. If $m + n = 0$, then m and n are additive inverses.

112. If $m - n = 0$, then $m = -n$.

1.5 Multiplication of Real Numbers

a Multiplication

Multiplication of real numbers is very much like multiplication of numbers of arithmetic. The only difference is that we must determine whether the answer is positive or negative.

Multiplication of a Positive Number and a Negative Number

To see how to multiply a positive number and a negative number, consider the pattern of the following.

$$
\begin{array}{ll}
\text{This number decreases} \rightarrow 4 \cdot 5 = & 20 \leftarrow \text{This number decreases} \\
\text{by 1 each time.} \qquad\quad 3 \cdot 5 = & 15 \quad \text{by 5 each time.} \\
\qquad\qquad\qquad\quad 2 \cdot 5 = & 10 \\
\qquad\qquad\qquad\quad 1 \cdot 5 = & 5 \\
\qquad\qquad\qquad\quad 0 \cdot 5 = & 0 \\
\qquad\qquad\quad -1 \cdot 5 = & -5 \\
\qquad\qquad\quad -2 \cdot 5 = & -10 \\
\qquad\qquad\quad -3 \cdot 5 = & -15
\end{array}
$$

DO EXERCISE 1.

According to this pattern, it looks as though the product of a negative number and a positive number is negative. That is the case, and we have the first part of the rule for multiplying numbers.

> **To multiply a positive number and a negative number, multiply their absolute values. The answer is negative.**

▶ **EXAMPLES** Multiply.

1. $8(-5) = -40$
2. $-\dfrac{1}{3} \cdot \dfrac{5}{7} = -\dfrac{5}{21}$
3. $(-7.2)5 = -36$ ◀

DO EXERCISES 2–7.

Multiplication of Two Negative Numbers

How do we multiply two negative numbers? Again, we look for a pattern.

$$
\begin{array}{ll}
\text{This number decreases} \rightarrow 4 \cdot (-5) = & -20 \leftarrow \text{This number increases} \\
\text{by 1 each time.} \qquad\quad 3 \cdot (-5) = & -15 \quad \text{by 5 each time.} \\
\qquad\qquad\qquad\quad 2 \cdot (-5) = & -10 \\
\qquad\qquad\qquad\quad 1 \cdot (-5) = & -5 \\
\qquad\qquad\qquad\quad 0 \cdot (-5) = & 0 \\
\qquad\qquad\quad -1 \cdot (-5) = & 5 \\
\qquad\qquad\quad -2 \cdot (-5) = & 10 \\
\qquad\qquad\quad -3 \cdot (-5) = & 15
\end{array}
$$

DO EXERCISE 8.

OBJECTIVE

After finishing Section 1.5, you should be able to:

a Multiply real numbers.

FOR EXTRA HELP

Tape 3B   Tape 3A   MAC: 1
                          IBM: 1

1. Complete, as in the example.
$$
\begin{array}{l}
4 \cdot 10 = 40 \\
3 \cdot 10 = 30 \\
2 \cdot 10 = \\
1 \cdot 10 = \\
0 \cdot 10 = \\
-1 \cdot 10 = \\
-2 \cdot 10 = \\
-3 \cdot 10 =
\end{array}
$$

Multiply.

2. $-3 \cdot 6$

3. $20 \cdot (-5)$

4. $4 \cdot (-20)$

5. $-\dfrac{2}{3} \cdot \dfrac{5}{6}$

6. $-4.23(7.1)$

7. $\dfrac{7}{8}\left(-\dfrac{4}{5}\right)$

8. Complete, as in the example.
$$
\begin{array}{l}
3 \cdot (-10) = -30 \\
2 \cdot (-10) = -20 \\
1 \cdot (-10) = \\
0 \cdot (-10) = \\
-1 \cdot (-10) = \\
-2 \cdot (-10) = \\
-3 \cdot (-10) =
\end{array}
$$

ANSWERS ON PAGE A-2

Multiply.

9. $-3 \cdot (-4)$

10. $-16 \cdot (-2)$

11. $-7 \cdot (-5)$

12. $-\dfrac{4}{7}\left(-\dfrac{5}{9}\right)$

13. $-\dfrac{3}{2}\left(-\dfrac{4}{9}\right)$

14. $-3.25(-4.14)$

Multiply.

15. $5(-6)$

16. $(-5)(-6)$

17. $(-3.2) \cdot 0$

18. $\left(-\dfrac{4}{5}\right)\left(\dfrac{10}{3}\right)$

ANSWERS ON PAGE A-2

According to the pattern, it appears that the product of two negative numbers is positive. That is actually so, and we have the second part of the rule for multiplying real numbers.

> **To multiply two negative numbers, multiply their absolute values. The answer is positive.**

DO EXERCISES 9–14.

The following is an alternative way to consider the rules we have for multiplication.

> **To multiply two real numbers:**
>
> 1. **Multiply the absolute values.**
> 2. **If the signs are the same, the answer is positive.**
> 3. **If the signs are different, the answer is negative.**

Multiplication by Zero

The only case that we have not considered is multiplying by zero. As with other numbers, the product of any real number and 0 is 0.

> **The Multiplicative Property of Zero**
>
> **For any real number a,**
> $$a \cdot 0 = 0.$$
> **(The product of 0 and any real number is 0.)**

▶ **EXAMPLES** Multiply.

4. $(-3)(-4) = 12$

5. $-1.6(2) = -3.2$

6. $-19 \cdot 0 = 0$

7. $\left(-\dfrac{5}{6}\right)\left(-\dfrac{1}{9}\right) = \dfrac{5}{54}$ ◀

DO EXERCISES 15–18.

Multiplying More Than Two Numbers

When multiplying more than two real numbers, we can choose order and grouping as we please.

▶ **EXAMPLES** Multiply.

8. $-8 \cdot 2(-3) = -16(-3)$ Multiplying the first two numbers
$= 48$ Multiplying the results

9. $-8 \cdot 2(-3) = 24 \cdot 2$ Multiplying the negatives. Every pair of negative numbers gives a positive product.
$= 48$

10. $-3(-2)(-5)(4) = 6(-5)(4)$ Multiplying the first two numbers
$= (-30)4$
$= -120$

11. $\left(-\dfrac{1}{2}\right)(8)\left(-\dfrac{2}{3}\right)(-6) = (-4)4$ Multiplying the first two numbers and the last two numbers

$\qquad\qquad\qquad\qquad = -16$

12. $-5 \cdot (-2) \cdot (-3) \cdot (-6) = 10 \cdot 18$

$\qquad\qquad\qquad\qquad\quad = 180$

13. $(-3)(-5)(-2)(-3)(-6) = (-30)(18) = -540$ ◀

We can see the following pattern in the results of Examples 12 and 13.

> **The product of an even number of negative numbers is positive.**
> **The product of an odd number of negative numbers is negative.**

DO EXERCISES 19–24.

▶ **EXAMPLE 14**   Evaluate $(-x)^2$ and $-x^2$ when $x = 5$.

$(-x)^2 = (-5)^2 = (-5)(-5) = 25;$ Substitute 5 for x. Then evaluate the power.

$-x^2 = -(5)^2 = -25$ Substitute 5 for x. Evaluate the power. Then find the opposite. ◀

Note that the expressions $(-x)^2$ and $-x^2$ are *not* equivalent. That is, they do not have the same value for every replacement of the variable by a real number. To find $(-x)^2$, we take the opposite and then square. To find $-x^2$, we find the square and then take the opposite.

▶ **EXAMPLE 15**   Evaluate $2x^2$ when $x = 3$ and $x = -3$.

$2x^2 = 2(3)^2 = 2(9) = 18;$

$2x^2 = 2(-3)^2 = 2(9) = 18$ ◀

DO EXERCISES 25–27.

Multiply.

19. $5 \cdot (-3) \cdot 2$

20. $-3 \times (-4.1) \times (-2.5)$

21. $-\dfrac{1}{2} \cdot \left(-\dfrac{4}{3}\right) \cdot \left(-\dfrac{5}{2}\right)$

22. $-2 \cdot (-5) \cdot (-4) \cdot (-3)$

23. $(-4)(-5)(-2)(-3)(-1)$

24. $(-1)(-1)(-2)(-3)(-1)(-1)$

25. Evaluate $(-x)^2$ and $-x^2$ when $x = 2$.

26. Evaluate $(-x)^2$ and $-x^2$ when $x = 3$.

27. Evaluate $3x^2$ when $x = 4$ and $x = -4$.

ANSWERS ON PAGE A-2

❖ SIDELIGHTS

Study Tips: Studying for Tests and Making the Most of Tutoring Sessions

As has been stated, we will often present some tips and guidelines to enhance your learning abilities. Sometimes these tips will be focused on mathematics, but sometimes they will be more general, as is the case here where we consider test preparation and tutoring.

TEST-TAKING TIPS

Many test-taking tips have been covered on the *To the student* page at the beginning of the book. If you have not read that material, do so now. We now provide some other test-taking tips.

- *Make up your own test questions as you study.* You have probably noted by now the section and objective codes that appear throughout the book. After you have done your homework over a particular objective, write one or two questions on your own that you think might be on a test. You will be amazed at the insight this will provide. You are actually carrying out a task similar to what a teacher does in preparing an exam.

- *Ask former students for old exams.* Working such exams can be very helpful and allows you to see what various professors think is important.

- *When taking a test, read each question carefully and try to do all the questions the first time through, but pace yourself.* Answer all the questions, and mark those to recheck if you have time at the end. Very often your first hunch will be correct.

- *Try to write your test in a neat and orderly manner.* Very often your instructor tries to give you partial credit when grading an exam. If your test paper is sloppy and disorderly, it is difficult to verify the partial credit. Doing your homework in a neat and orderly manner can ease such a task on an exam. Try using an erasable pen to make your writing darker and therefore more readable.

MAKING THE MOST OF TUTORING AND HELP SESSIONS

Often you will determine that a tutoring session may be helpful. The following comments may help you to make the most of such situations.

- *Work on the topics before you go to the help or tutoring session. Do not go to such sessions with the view of yourself as an empty cup and the tutor as a magician who will instantly pour in the learning.* The primary source of your ability to learn is within you. We have seen so many students over the years go to help or tutoring sessions with no advanced preparation. You are often wasting your time and perhaps your money if you are paying for such sessions. Go to class, study the textbook, and mark trouble spots. Then use the help and tutoring sessions to deal with these difficulties most efficiently.

- *Do not be afraid to ask questions in these help and tutoring sessions!* The more you relate to your tutor, the more the tutor can help you with your difficulties.

- *Try being a tutor yourself.* Explaining a topic to someone else is often the best way to learn it.

- *What about the student who says "I could do the work at home, but on the test I made silly mistakes"?* Yes, all of us, including instructors, make silly computational mistakes in class, on homework, and on tests. But your instructor, if he or she has taught for some time, is probably aware that 90% of students who make such comments in truth do not have the depth of knowledge of the subject matter, and such silly mistakes very often are a sign that the student has not mastered the material. There is no way we can make that analysis for you. It will have to be unraveled by some careful soul searching on your part or by a conference with your instructor.

EXERCISE SET 1.5

a Multiply.

1. $-8 \cdot 2$

2. $-2 \cdot 5$

3. $-7 \cdot 6$

4. $-9 \cdot 2$

5. $8 \cdot (-3)$

6. $9 \cdot (-5)$

7. $-9 \cdot 8$

8. $-10 \cdot 3$

9. $-8 \cdot (-2)$

10. $-2 \cdot (-5)$

11. $-7 \cdot (-6)$

12. $-9 \cdot (-2)$

13. $15 \cdot (-8)$

14. $-12 \cdot (-10)$

15. $-14 \cdot 17$

16. $-13 \cdot (-15)$

17. $-25 \cdot (-48)$

18. $39 \cdot (-43)$

19. $-3.5 \cdot (-28)$

20. $97 \cdot (-2.1)$

21. $9 \cdot (-8)$

22. $7 \cdot (-9)$

23. $4 \cdot (-3.1)$

24. $3 \cdot (-2.2)$

25. $-6 \cdot (-4)$

26. $-5 \cdot (-6)$

27. $-7 \cdot (-3.1)$

28. $-4 \cdot (-3.2)$

29. $\dfrac{2}{3} \cdot \left(-\dfrac{3}{5}\right)$

30. $\dfrac{5}{7} \cdot \left(-\dfrac{2}{3}\right)$

31. $-\dfrac{3}{8} \cdot \left(-\dfrac{2}{9}\right)$

32. $-\dfrac{5}{8} \cdot \left(-\dfrac{2}{5}\right)$

33. -6.3×2.7

34. -4.1×9.5

35. $-\dfrac{5}{9} \cdot \dfrac{3}{4}$

36. $-\dfrac{8}{3} \cdot \dfrac{9}{4}$

37. $7 \cdot (-4) \cdot (-3) \cdot 5$

38. $9 \cdot (-2) \cdot (-6) \cdot 7$

39. $-\dfrac{2}{3} \cdot \dfrac{1}{2} \cdot \left(-\dfrac{6}{7}\right)$

40. $-\dfrac{1}{8} \cdot \left(-\dfrac{1}{4}\right) \cdot \left(-\dfrac{3}{5}\right)$

41. $-3 \cdot (-4) \cdot (-5)$

ANSWERS

1. _____
2. _____
3. _____
4. _____
5. _____
6. _____
7. _____
8. _____
9. _____
10. _____
11. _____
12. _____
13. _____
14. _____
15. _____
16. _____
17. _____
18. _____
19. _____
20. _____
21. _____
22. _____
23. _____
24. _____
25. _____
26. _____
27. _____
28. _____
29. _____
30. _____
31. _____
32. _____
33. _____
34. _____
35. _____
36. _____
37. _____
38. _____
39. _____
40. _____
41. _____

ANSWERS

42. _____

43. _____

44. _____

45. _____

46. _____

47. _____

48. _____

49. _____

50. _____

51. _____

52. _____

53. _____

54. _____

55. _____

56. _____

57. _____

58. _____

59. _____

60. _____

61. _____

62. _____

63. _____

64. _____

65. _____

66. _____

67. _____

68. _____

69. _____

70. _____

71. _____

72. _____

73. _____

74. _____

75. a) _____

b) _____

c) _____

76. _____

42. $-2 \cdot (-5) \cdot (-7)$

43. $-2 \cdot (-5) \cdot (-3) \cdot (-5)$

44. $-3 \cdot (-5) \cdot (-2) \cdot (-1)$

45. $\frac{1}{5}\left(-\frac{2}{9}\right)$

46. $-\frac{3}{5}\left(-\frac{2}{7}\right)$

47. $-7 \cdot (-21) \cdot 13$

48. $-14 \cdot (34) \cdot 12$

49. $-4 \cdot (-1.8) \cdot 7$

50. $-8 \cdot (-1.3) \cdot (-5)$

51. $-\frac{1}{9}\left(-\frac{2}{3}\right)\left(\frac{5}{7}\right)$

52. $-\frac{7}{2}\left(-\frac{5}{7}\right)\left(-\frac{2}{5}\right)$

53. $4 \cdot (-4) \cdot (-5) \cdot (-12)$

54. $-2 \cdot (-3) \cdot (-4) \cdot (-5)$

55. $0.07 \cdot (-7) \cdot 6 \cdot (-6)$

56. $80 \cdot (-0.8) \cdot (-90) \cdot (-0.09)$

57. $\left(-\frac{5}{6}\right)\left(\frac{1}{8}\right)\left(-\frac{3}{7}\right)\left(-\frac{1}{7}\right)$

58. $\left(\frac{4}{5}\right)\left(-\frac{2}{3}\right)\left(-\frac{15}{7}\right)\left(\frac{1}{2}\right)$

59. $(-14) \cdot (-27) \cdot 0$

60. $7 \cdot (-6) \cdot 5 \cdot (-4) \cdot 3 \cdot (-2) \cdot 1 \cdot 0$

61. $(-8)(-9)(-10)$

62. $(-7)(-8)(-9)(-10)$

63. $(-6)(-7)(-8)(-9)(-10)$

64. $(-5)(-6)(-7)(-8)(-9)(-10)$

65. Evaluate $(-3x)^2$ and $-3x^2$ when $x = 7$.

66. Evaluate $(-2x)^2$ and $-2x^2$ when $x = 3$.

67. Evaluate $5x^2$ when $x = 2$ and $x = -2$.

68. Evaluate $2x^2$ when $x = 5$ and $x = -5$.

SYNTHESIS

Simplify. Keep in mind the rules for order of operations in Section R.5.

69. $-6[(-5) + (-7)]$

70. $-3[(-8) + (-6)]\left(-\frac{1}{7}\right)$

71. $-(3^5) \cdot [-(2^3)]$

72. $4(2^4) \cdot [-(3^3)] \cdot 6$

73. $|(-2)^3 + 4^2| - (2 - 7)^2$

74. $|-11(-3)^2 - 5^3 - 6^2 - (-4)^2|$

75. What must be true of m and n if $-mn$ is to be (a) positive? (b) zero? (c) negative?

76. Evaluate $-6(3x - 5y) + z$ when $x = -2$, $y = -4$, and $z = 5$.

1.6 Division of Real Numbers

We now consider division of real numbers. The definition of division results in rules for division very much like those for multiplication.

a Division of Integers

> The quotient $\dfrac{a}{b}$ (or $a \div b$) is the number, if there is one, that when multiplied by b gives a.

Let us use the definition to divide integers.

▶ **EXAMPLES** Divide, if possible. Check your answer.

1. $14 \div (-7) = -2$ We look for a number that when multiplied by -7 gives 14. That number is -2. *Check:* $(-2)(-7) = 14$.

2. $\dfrac{-32}{-4} = 8$ We look for a number that when multiplied by -4 gives -32. That number is 8. *Check:* $8(-4) = -32$.

3. $\dfrac{-10}{7} = -\dfrac{10}{7}$ We look for a number that when multiplied by 7 gives -10. That number is $-\frac{10}{7}$. *Check:* $-\frac{10}{7} \cdot 7 = -10$.

4. $\dfrac{-17}{0}$ is undefined.   We look for a number that when multiplied by 0 gives -17. There is no such number because the product of 0 and *any* number is 0.   ◀

The rules for division are the same as those for multiplication. We state them together.

> **To multiply or divide two real numbers:**
> 1. **Multiply or divide the absolute values.**
> 2. **If the signs are the same, the answer is positive.**
> 3. **If the signs are different, the answer is negative.**

DO EXERCISES 1–8.

Division by Zero

Example 4 shows why we cannot divide -17 by 0. We can use the same argument to show why we cannot divide any nonzero number b by 0. Consider $b \div 0$. We look for a number that when multiplied by 0 gives b. There is no such number because the product of 0 and any number is 0. Thus we cannot divide a nonzero number b by 0.

On the other hand, if we divide 0 by 0, we look for a number r such that $0 \cdot r = 0$. But, $0 \cdot r = 0$ for any number r. Thus it appears that $0 \div 0$ could be any number we choose. Getting any answer we want when we divide 0 by 0 would be very confusing. Thus we agree that division by zero is undefined.

> **Division by zero is undefined.** That is, $a \div 0$ is undefined for all real numbers a. But, $0 \div a = 0$, when a is nonzero. That is, 0 divided by a nonzero number is 0.

OBJECTIVES

After finishing Section 1.6, you should be able to:

a Divide integers.

b Find the reciprocal of a real number.

c Divide real numbers.

FOR EXTRA HELP

Tape 3C   Tape 3B   MAC: 1
                                  IBM: 1

Divide.

1. $6 \div (-3)$

2. $\dfrac{-15}{-3}$

3. $-24 \div 8$

4. $\dfrac{-32}{-4}$

5. $\dfrac{30}{-5}$

6. $\dfrac{30}{-7}$

7. $\dfrac{-5}{0}$

8. $\dfrac{0}{-3}$

ANSWERS ON PAGE A-2

Find the reciprocal.

9. $\dfrac{2}{3}$

10. $-\dfrac{5}{4}$

11. -3

12. $-\dfrac{1}{5}$

13. 5.78

14. $\dfrac{1}{2/3}$

---

b   **Reciprocals**

When two numbers like $\frac{1}{2}$ and 2 are multiplied, the result is 1. Such numbers are called **reciprocals** of each other. Every nonzero real number has a reciprocal, also called a **multiplicative inverse.**

> Two numbers whose product is 1 are called *reciprocals* of each other.

► **EXAMPLES**   Find the reciprocal of each number.

5. $\dfrac{7}{8}$ The reciprocal of $\dfrac{7}{8}$ is $\dfrac{8}{7}$ because $\dfrac{7}{8} \cdot \dfrac{8}{7} = 1$.

6. -5 The reciprocal of -5 is $-\dfrac{1}{5}$ because $-5\left(-\dfrac{1}{5}\right) = 1$.

7. 3.9 The reciprocal of 3.9 is $\dfrac{1}{3.9}$ because $3.9\left(\dfrac{1}{3.9}\right) = 1$.

8. $-\dfrac{1}{2}$ The reciprocal of $-\dfrac{1}{2}$ is -2 because $\left(-\dfrac{1}{2}\right)(-2) = 1$.

9. $-\dfrac{2}{3}$ The reciprocal of $-\dfrac{2}{3}$ is $-\dfrac{3}{2}$ because $\left(-\dfrac{2}{3}\right)\left(-\dfrac{3}{2}\right) = 1$.

10. $\dfrac{1}{3/4}$ The reciprocal of $\dfrac{1}{3/4}$ is $\dfrac{3}{4}$ because $\left(\dfrac{1}{3/4}\right)\left(\dfrac{3}{4}\right) = 1$.   ◄

> For $a \neq 0$, the reciprocal of a can be named $\dfrac{1}{a}$ and the reciprocal of $\dfrac{1}{a}$ is a.
>
> The reciprocal of a nonzero number $\dfrac{a}{b}$ can be named $\dfrac{b}{a}$.
>
> The number 0 has no reciprocal.

DO EXERCISES 9–14.

The reciprocal of a positive number is also a positive number, because their product must be the positive number 1. The reciprocal of a negative number is also a negative number, because their product must be the positive number 1.

> The reciprocal of a number has the same sign as the number itself.

It is important *not* to confuse *opposite* with *reciprocal*. Keep in mind that the opposite, or additive inverse, of a number is what we add to the number to get 0, whereas a reciprocal is what we multiply the number by to get 1. Compare the following.

Number	Opposite (Change the sign.)	Reciprocal (Invert but do not change the sign.)
$-\dfrac{3}{8}$	$\dfrac{3}{8}$	$-\dfrac{8}{3}$
19	-19	$\dfrac{1}{19}$
$\dfrac{18}{7}$	$-\dfrac{18}{7}$	$\dfrac{7}{18}$
-7.9	7.9	$-\dfrac{1}{7.9}$, or $-\dfrac{10}{79}$
0	0	Undefined

$$\left(-\frac{3}{8}\right)\left(-\frac{8}{3}\right)=1$$

$$-\frac{3}{8}+\frac{3}{8}=0$$

DO EXERCISE 15.

c **Division of Real Numbers**

We know that we can subtract by adding an opposite. Similarly, we can divide by multiplying by a reciprocal.

> **For any real numbers *a* and *b*, *b* ≠ 0,**
> $$\frac{a}{b}=a\cdot\frac{1}{b}.$$
> **(To divide, we can multiply by the reciprocal of the divisor.)**

▶ **EXAMPLES** Rewrite each division as a multiplication.

11. $-4\div 3$ $-4\div 3$ is the same as $-4\cdot\dfrac{1}{3}$

12. $\dfrac{6}{-7}$ $\dfrac{6}{-7}=6\left(-\dfrac{1}{7}\right)$

13. $\dfrac{x+2}{5}$ $\dfrac{x+2}{5}=(x+2)\dfrac{1}{5}$ Parentheses are necessary here.

14. $\dfrac{-17}{1/b}$ $\dfrac{-17}{1/b}=-17\cdot b$

15. $\dfrac{3}{5}\div\left(-\dfrac{9}{7}\right)$ $\dfrac{3}{5}\div\left(-\dfrac{9}{7}\right)=\dfrac{3}{5}\left(-\dfrac{7}{9}\right)$ ◀

DO EXERCISES 16–20.

When actually doing division calculations, we sometimes multiply by a reciprocal and we sometimes divide directly. With fractional notation, it is usually better to multiply by a reciprocal. With decimal notation, it is usually better to divide directly.

15. Complete the following table.

Number	Opposite	Reciprocal
$\dfrac{2}{3}$		
$-\dfrac{5}{4}$		
0		
1		
-4.5		

Rewrite the division as a multiplication.

16. $\dfrac{4}{7}\div\left(-\dfrac{3}{5}\right)$

17. $\dfrac{5}{-8}$

18. $\dfrac{a-b}{7}$

19. $\dfrac{-23}{1/a}$

20. $-5\div 7$

ANSWERS ON PAGE A-2

Divide by multiplying by the reciprocal of the divisor.

21. $\dfrac{4}{7} \div \left(-\dfrac{3}{5}\right)$

22. $-\dfrac{8}{5} \div \dfrac{2}{3}$

23. $-\dfrac{12}{7} \div \left(-\dfrac{3}{4}\right)$

24. $21.7 \div (-3.1)$

Find two equal expressions for the number with negative signs in different places.

25. $\dfrac{-5}{6}$

26. $-\dfrac{8}{7}$

27. $\dfrac{10}{-3}$

▶ **EXAMPLES**   Divide by multiplying by the reciprocal of the divisor.

16. $\dfrac{2}{3} \div \left(-\dfrac{5}{4}\right) = \dfrac{2}{3} \cdot \left(-\dfrac{4}{5}\right) = -\dfrac{8}{15}$

17. $-\dfrac{5}{6} \div \left(-\dfrac{3}{4}\right) = -\dfrac{5}{6} \cdot \left(-\dfrac{4}{3}\right) = \dfrac{20}{18} = \dfrac{10 \cdot 2}{9 \cdot 2} = \dfrac{10}{9} \cdot \dfrac{2}{2} = \dfrac{10}{9}$

> Be careful not to change the sign when taking a reciprocal!

18. $-\dfrac{3}{4} \div \dfrac{3}{10} = -\dfrac{3}{4} \cdot \left(\dfrac{10}{3}\right) = -\dfrac{30}{12} = -\dfrac{5}{2} \cdot \dfrac{6}{6} = -\dfrac{5}{2}$ ◀

With decimal notation, it is easier to carry out long division than to multiply by the reciprocal.

▶ **EXAMPLES**   Divide.

19. $-27.9 \div (-3) = \dfrac{-27.9}{-3} = 9.3$ Do the long division $3 \overline{)27.9}$. The answer is positive. $\left(\dfrac{9.3}{}\right)$

20. $-6.3 \div 2.1 = -3$ Do the long division $2.1 \overline{)6.30}$. The answer is negative.   ◀

DO EXERCISES 21–24.

Consider the following:

$$\dfrac{2}{3} = \dfrac{2}{3} \cdot 1 = \dfrac{2}{3} \cdot \dfrac{-1}{-1} = \dfrac{2(-1)}{3(-1)} = \dfrac{-2}{-3},$$

$$-\dfrac{2}{3} = -1 \cdot \dfrac{2}{3} = \dfrac{-1}{1} \cdot \dfrac{2}{3} = \dfrac{-1 \cdot 2}{1 \cdot 3} = \dfrac{-2}{3},$$

and

$$\dfrac{-2}{3} = \dfrac{-2}{3} \cdot 1 = \dfrac{-2}{3} \cdot \dfrac{-1}{-1} = \dfrac{-2(-1)}{3(-1)} = \dfrac{2}{-3}.$$

We can use the following properties to make sign changes in fractional notation.

> For any numbers a and b, $b \neq 0$:
>
> **1.** $\dfrac{-a}{b} = \dfrac{a}{-b} = -\dfrac{a}{b}$
>
> (The opposite of a number a divided by another number b is the same as the number a divided by the opposite of another number b, and both are the same as the opposite of a *divided by b*.)
>
> **2.** $\dfrac{-a}{-b} = \dfrac{a}{b}$
>
> (The opposite of a number a divided by the opposite of another number b is the same as the quotient of the two numbers a and b.)

DO EXERCISES 25–27.

EXERCISE SET 1.6

a Divide, if possible. Check each answer.

1. $36 \div (-6)$

2. $\dfrac{28}{-7}$

3. $\dfrac{26}{-2}$

4. $26 \div (-13)$

5. $\dfrac{-16}{8}$

6. $-22 \div (-2)$

7. $\dfrac{-48}{-12}$

8. $-63 \div (-9)$

9. $\dfrac{-72}{9}$

10. $\dfrac{-50}{25}$

11. $-100 \div (-50)$

12. $\dfrac{-200}{8}$

13. $-108 \div 9$

14. $\dfrac{-64}{-7}$

15. $\dfrac{200}{-25}$

16. $-300 \div (-13)$

17. $\dfrac{75}{0}$

18. $\dfrac{0}{-5}$

19. $\dfrac{88}{-9}$

20. $\dfrac{-145}{-5}$

b Find the reciprocal.

21. $\dfrac{15}{7}$

22. $\dfrac{3}{8}$

23. $-\dfrac{47}{13}$

24. $-\dfrac{31}{12}$

25. 13

26. -10

27. 4.3

28. -8.5

29. $-\dfrac{1}{7.1}$

30. $\dfrac{1}{-4.9}$

31. $\dfrac{p}{q}$

32. $\dfrac{s}{t}$

33. $\dfrac{1}{4y}$

34. $\dfrac{-1}{8a}$

35. $\dfrac{2a}{3b}$

36. $\dfrac{-4y}{3x}$

ANSWERS

1. 2. 3. 4. 5. 6. 7. 8. 9. 10. 11. 12. 13. 14. 15. 16. 17. 18. 19. 20. 21. 22. 23. 24. 25. 26. 27. 28. 29. 30. 31. 32. 33. 34. 35. 36.

c Rewrite the division as a multiplication.

37. $3 \div 19$

38. $4 \div (-9)$

39. $\dfrac{6}{-13}$

40. $-\dfrac{12}{41}$

41. $\dfrac{13.9}{-1.5}$

42. $-\dfrac{47.3}{21.4}$

43. $\dfrac{x}{\frac{1}{y}}$

44. $\dfrac{13}{x}$

45. $\dfrac{3x+4}{5}$

46. $\dfrac{4y-8}{-7}$

47. $\dfrac{5a-b}{5a+b}$

48. $\dfrac{2x+x^2}{x-5}$

Divide.

49. $\dfrac{3}{4} \div \left(-\dfrac{2}{3}\right)$

50. $\dfrac{7}{8} \div \left(-\dfrac{1}{2}\right)$

51. $-\dfrac{5}{4} \div \left(-\dfrac{3}{4}\right)$

52. $-\dfrac{5}{9} \div \left(-\dfrac{5}{6}\right)$

53. $-\dfrac{2}{7} \div \left(-\dfrac{4}{9}\right)$

54. $-\dfrac{3}{5} \div \left(-\dfrac{5}{8}\right)$

55. $-\dfrac{3}{8} \div \left(-\dfrac{8}{3}\right)$

56. $-\dfrac{5}{8} \div \left(-\dfrac{6}{5}\right)$

57. $-6.6 \div 3.3$

58. $-44.1 \div (-6.3)$

59. $\dfrac{-11}{-13}$

60. $\dfrac{-1.9}{20}$

61. $\dfrac{48.6}{-3}$

62. $\dfrac{-17.8}{3.2}$

63. $\dfrac{-9}{17-17}$

64. $\dfrac{-8}{-5+5}$

SKILL MAINTENANCE

65. Simplify: $\dfrac{264}{468}$.

66. Convert to decimal notation: 47.7%.

67. Simplify: $2^3 - 5 \cdot 3 + 8 \cdot 10 \div 2$.

68. Add and simplify: $\dfrac{2}{3} + \dfrac{5}{6}$.

SYNTHESIS

69. ▦ Find the reciprocal of -10.5.

70. Determine those real numbers that are their own reciprocals.

71. Determine those real numbers a for which the additive inverse of a is the same as the reciprocal of a.

72. ▦ What should happen if you enter a number on a calculator and press the reciprocal key twice? Why?

Tell whether the expression represents a positive number or a negative number when m and n are negative.

73. $\dfrac{-n}{m}$

74. $\dfrac{-n}{-m}$

75. $-\left(\dfrac{-n}{m}\right)$

76. $-\left(\dfrac{n}{-m}\right)$

77. $-\left(\dfrac{-n}{-m}\right)$

1.7 Properties of Real Numbers

a Equivalent Expressions

In solving equations and doing other kinds of work in algebra, we manipulate expressions in various ways. For example, instead of

$$x + x,$$

we might write

$$2x,$$

knowing that the two expressions represent the same number for any meaningful replacement of x. In that sense, the expressions $x + x$ and $2x$ are **equivalent.**

> Two expressions that have the same value for all meaningful replacements are called *equivalent.*

The expressions $x + 3x$ and $5x$ are *not* equivalent.

DO EXERCISES 1 AND 2.

We will consider several laws of real numbers in this section which will allow us to find equivalent expressions. The first two laws are *identity properties of 0 and 1.*

> **The Identity Property of 0**
>
> **For any real number a, $a + 0 = 0 + a = a$. (The number 0 is the *additive identity*.)**
>
> **The Identity Property of 1**
>
> **For any real number a, $a \cdot 1 = 1 \cdot a = a$. (The number 1 is the *multiplicative identity*.)**

We often refer to the use of the identity property of 1 as "multiplying by 1."

▶ **EXAMPLE 1** Use multiplying by 1 to find an expression equivalent to $\frac{2}{3}$ with a denominator of $3x$.

We multiply by 1, using x/x as a name for 1:

$$\frac{2}{3} = \frac{2}{3} \cdot 1 = \frac{2}{3} \cdot \frac{x}{x} = \frac{2x}{3x}.$$ ◀

Note that the expressions $2/3$ and $2x/3x$ are equivalent. They have the same value for any meaningful expression. Note that 0 is not a meaningful replacement in $2x/3x$, but for all nonzero real numbers the expressions $2/3$ and $2x/3x$ have the same value.

DO EXERCISE 3.

OBJECTIVES

After finishing Section 1.7, you should be able to:

a Find equivalent fractional expressions and simplify fractional expressions by multiplying by 1.

b Use the commutative and associative laws to find equivalent expressions.

c Use the distributive laws to multiply expressions like 8 and $x - y$.

d Use the distributive laws to factor expressions like $4x - 12$.

e Collect like terms.

FOR EXTRA HELP

Tape 4A	Tape 3B	MAC: 1
		IBM: 1

Complete each of the following tables by evaluating each expression for the given values.

1.

	$x + x$	$2x$
$x = 3$		
$x = -6$		
$x = 4.8$		

2.

	$x + 3x$	$5x$
$x = 2$		
$x = -6$		
$x = 4.8$		

3. Use multiplying by 1 to find an expression equivalent to $\frac{3}{4}$ with a denominator of $4y$.

ANSWERS ON PAGE A-2

Simplify.

4. $\dfrac{3y}{4y}$

5. $-\dfrac{16m}{12m}$

6. Evaluate $x + y$ and $y + x$ when $x = -2$ and $y = 3$.

7. Evaluate xy and yx when $x = -2$ and $y = 5$.

ANSWERS ON PAGE A-2

In algebra, we consider an expression like 2/3 to be "simplified" from $2x/3x$. To find such simplified expressions, we use the identity property of 1 to remove a factor of 1.

▶ **EXAMPLE 2**   Simplify: $-\dfrac{20x}{12x}$.

$$-\frac{20x}{12x} = -\frac{5 \cdot 4x}{3 \cdot 4x}$$ We look for the largest common factor of the numerator and the denominator and factor each.

$$= -\frac{5}{3} \cdot \frac{4x}{4x}$$ Factoring the fractional expression

$$= -\frac{5}{3} \cdot 1 \quad \frac{4x}{4x} = 1$$

$$= -\frac{5}{3}$$ Removing a factor of 1 using the identity property of 1   ◀

DO EXERCISES 4 AND 5.

b   **The Commutative and Associative Laws**

The Commutative Laws

Let us examine the expressions $x + y$ and $y + x$, as well as xy and yx.

▶ **EXAMPLE 3**   Evaluate $x + y$ and $y + x$ for $x = 4$ and $y = 3$.
We substitute 4 for x and 3 for y in both expressions:

$$x + y = 4 + 3 = 7; \qquad y + x = 3 + 4 = 7.$$ ◀

▶ **EXAMPLE 4**   Evaluate xy and yx for $x = 23$ and $y = 12$.
We substitute 23 for x and 12 for y in both expressions:

$$xy = 23 \cdot 12 = 276; \qquad yx = 12 \cdot 23 = 276.$$ ◀

DO EXERCISES 6 AND 7.

Note that the expressions

$$x + y \quad \text{and} \quad y + x$$

have the same values no matter what the variables stand for. Thus they are equivalent. Therefore, when we add two numbers, the order in which we add does not matter. Similarly, the expressions xy and yx are equivalent. They also have the same values, no matter what the variables stand for. Therefore, when we multiply two numbers, the order in which we multiply does not matter.

The following are examples of general patterns or laws.

The Commutative Laws

Addition.   **For any numbers *a* and *b*,**

$$a + b = b + a.$$

(We can change the order when adding without affecting the answer.)

Multiplication.   **For any numbers *a* and *b*,**

$$ab = ba.$$

(We can change the order when multiplying without affecting the answer.)

Using a commutative law, we know that $x + 2$ and $2 + x$ are equivalent. Similarly, $3x$ and $x(3)$ are equivalent. Thus, in an algebraic expression, we can replace one by the other and the result will be equivalent to the original expression.

▶ **EXAMPLE 5**   Use the commutative laws to write an expression equivalent to $y + 5$, xy, and $7 + ab$.

An expression equivalent to $y + 5$ is $5 + y$ by the commutative law of addition.

An expression equivalent to xy is yx by the commutative law of multiplication.

An expression equivalent to $7 + ab$ is $ab + 7$ by the commutative law of addition. Another expression equivalent to $7 + ab$ is $7 + ba$ by the commutative law of multiplication.   ◀

DO EXERCISES 8–10.

The Associative Laws

Now let us examine the expressions $a + (b + c)$ and $(a + b) + c$. Note that these expressions involve parentheses as *grouping* symbols, and they also involve three numbers. Calculations within parentheses are to be done first.

▶ **EXAMPLE 6**   Calculate and compare: $3 + (8 + 5)$ and $(3 + 8) + 5$.

$$3 + (8 + 5) = 3 + 13 \quad \text{Calculating within parentheses first; adding the 8 and 5}$$

$$= 16;$$

$$(3 + 8) + 5 = 11 + 5 \quad \text{Calculating within parentheses first; adding the 3 and 8}$$

$$= 16 \qquad ◀$$

Use a commutative law to write an equivalent expression.

8. $x + 9$

9. pq

10. $xy + t$

ANSWERS ON PAGE A-2

11. Calculate and compare:
$8 + (9 + 2)$ and $(8 + 9) + 2$.

12. Calculate and compare:
$10 \cdot (5 \cdot 3)$ and $(10 \cdot 5) \cdot 3$.

Use an associative law to write an equivalent expression.

13. $a + (b + 2)$

14. $3(vw)$

The two expressions in Example 6 name the same number. Moving the parentheses to group the additions differently did not affect the value of the expression.

▶ **EXAMPLE 7** Calculate and compare: $3 \cdot (4 \cdot 2)$ and $(3 \cdot 4) \cdot 2$.

$$3 \cdot (4 \cdot 2) = 3 \cdot 8 \qquad (3 \cdot 4) \cdot 2 = 12 \cdot 2$$
$$= 24; \qquad\qquad = 24 \quad ◀$$

DO EXERCISES 11 AND 12.

You may have noted that when only addition is involved, parentheses can be placed any way we please without affecting the answer. When only multiplication is involved, parentheses also can be placed any way we please without affecting the answer.

> **The Associative Laws**
>
> **Addition.** For any numbers a, b, and c,
> $$a + (b + c) = (a + b) + c.$$
> **(Numbers can be grouped in any manner for addition.)**
> **Multiplication.** For any numbers a, b, and c,
> $$a \cdot (b \cdot c) = (a \cdot b) \cdot c.$$
> **(Numbers can be grouped in any manner for multiplication.)**

The associative laws say parentheses may be placed any way we please when only additions or only multiplications are involved. So we often omit them. For example,

$$x + (y + 2) \quad \text{means} \quad x + y + 2, \quad \text{and} \quad (lw)h \quad \text{means} \quad lwh.$$

▶ **EXAMPLE 8** Use an associative law to write an expression equivalent to $(y + z) + 3$.

An equivalent expression is
$$y + (z + 3)$$
by the associative law of addition. ◀

▶ **EXAMPLE 9** Use an associative law to write an expression equivalent to $8(xy)$.

An equivalent expression is
$$(8x)y$$
by the associative law of multiplication. ◀

DO EXERCISES 13 AND 14.

Using the Commutative and Associative Laws Together

▶ **EXAMPLE 10** Use the commutative and associative laws to write at least three expressions equivalent to $(x + 5) + y$.

a) $(x + 5) + y = x + (5 + y)$

   Using the associative law first and then using the commutative law

   $= x + (y + 5)$

b) $(x + 5) + y = y + (x + 5)$

   Using the commutative law first and then the commutative law again

   $= y + (5 + x)$

c) $(x + 5) + y = 5 + (x + y)$

   Using the commutative law first and then the associative law ◀

▶ **EXAMPLE 11** Use the commutative and associative laws to write at least three expressions equivalent to $(3x)y$.

a) $(3x)y = 3(xy)$

   Using the associative law first and then using the commutative law

   $= 3(yx)$

b) $(3x)y = y(x3)$

   Using the commutative law twice

c) $(3x)y = x(y3)$

   Using the commutative law, and then the associative law, and then the commutative law again ◀

DO EXERCISES 15 AND 16.

c The Distributive Laws

The *distributive laws* are the basis of many procedures in both arithmetic and algebra. These are probably the most important laws that we use to manipulate algebraic expressions. The distributive law of multiplication over addition involves two operations: addition and multiplication.

   Let us begin by considering a multiplication problem from arithmetic:

$$\begin{array}{r} 4\ 5 \\ \times\quad 7 \\ \hline 3\ 5 \\ 2\ 8\ 0 \\ \hline 3\ 1\ 5 \end{array}$$
   3 5 ← This is $7 \cdot 5$.
   2 8 0 ← This is $7 \cdot 40$.
   3 1 5 ← This is the sum $7 \cdot 40 + 7 \cdot 5$.

To carry out the multiplication, we actually added two products. That is,

$$7 \cdot 45 = 7(40 + 5) = 7 \cdot 40 + 7 \cdot 5.$$

   Let us examine this further. If we wish to multiply a sum of several numbers by a factor, we can either add and then multiply, or multiply and then add.

Use the commutative and associative laws to write at least three equivalent expressions.

15. $4(tu)$

16. $r + (2 + s)$

17. a) $7 \cdot (3 + 6)$

b) $(7 \cdot 3) + (7 \cdot 6)$

18. a) $2 \cdot (10 + 30)$

b) $(2 \cdot 10) + (2 \cdot 30)$

19. a) $(2 + 5) \cdot 4$

b) $(2 \cdot 4) + (5 \cdot 4)$

Calculate.

20. a) $4(5 - 3)$

b) $4 \cdot 5 - 4 \cdot 3$

21. a) $-2 \cdot (5 - 3)$

b) $-2 \cdot 5 - (-2) \cdot 3$

22. a) $5 \cdot (2 - 7)$

b) $5 \cdot 2 - 5 \cdot 7$

What are the terms of the expression?

23. $5x - 4y + 3$

24. $-4y - 2x + 3z$

▶ **EXAMPLE 12**   Compute in two ways: $5 \cdot (4 + 8)$.

a) $5 \cdot \underbrace{(4 + 8)}$ Adding within parentheses first, and then multiplying

$$= 5 \cdot \quad 12$$
$$= 60$$

b) $\underbrace{(5 \cdot 4)} + \underbrace{(5 \cdot 8)}$ Distributing the multiplication to terms within parentheses first and then adding

$$= \quad 20 \quad + \quad 40$$
$$= \quad 60$$ ◀

DO EXERCISES 17–19.

The Distributive Law of Multiplication Over Addition

For any numbers a, b, and c,

$$a(b + c) = ab + ac.$$

In the statement of the distributive law, we know that in an expression such as $ab + ac$, the multiplications are to be done first according to our rules for order of operations. So, instead of writing $(4 \cdot 5) + (4 \cdot 7)$, we can write $4 \cdot 5 + 4 \cdot 7$. However, in $a(b + c)$, we cannot omit the parentheses. If we did, we would have $ab + c$, which means $(ab) + c$. For example, $3(4 + 2) = 18$, but $3 \cdot 4 + 2 = 14$.

There is another distributive law that relates multiplication and subtraction. This law says that to multiply by a difference, we can either subtract and then multiply or multiply and then subtract.

The Distributive Law of Multiplication Over Subtraction

For any numbers a, b, and c,

$$a(b - c) = ab - ac.$$

We often refer to "*the* distributive law" when we mean *either* of these laws.

DO EXERCISES 20–22.

What do we mean by the *terms* of an expression? **Terms** are separated by addition signs. If there are subtraction signs, we can find an equivalent expression that uses addition signs.

▶ **EXAMPLE 13**   What are the terms of $3x - 4y + 2z$?

We have

$$3x - 4y + 2z = 3x + (-4y) + 2z. \quad \text{Separating parts with + signs}$$

The terms are $3x$, $-4y$, and $2z$. ◀

DO EXERCISES 23 AND 24.

The distributive laws are a basis for a procedure in algebra called **multiplying.** In an expression like $8(a + 2b - 7)$, we multiply each term inside the parentheses by 8:

$$8(a + 2b - 7) = 8 \cdot a + 8 \cdot 2b - 8 \cdot 7 = 8a + 16b - 56.$$

► **EXAMPLES** Multiply.

14. $9(x - 5) = 9x - 9(5)$ Using the distributive law of multiplication over subtraction

$$= 9x - 45$$

15. $\frac{4}{3}(s - t + w) = \frac{4}{3}s - \frac{4}{3}t + \frac{4}{3}w$ Using both distributive laws

16. $-4(x - 2y + 3z) = -4 \cdot x - (-4)(2y) + (-4)(3z)$
$$= -4x - (-8y) + (-12z)$$
$$= -4x + 8y - 12z$$

We can also do this problem by first finding an equivalent expression with all plus signs and then multiplying:

$$-4(x - 2y + 3z) = -4[x + (-2y) + 3z]$$
$$= -4 \cdot x + (-4)(-2y) + (-4)(3z)$$
$$= -4x + 8y - 12z. \qquad ◄$$

DO EXERCISES 25–28.

d **Factoring**

Factoring is the reverse of multiplying. To factor, we can use the distributive laws in reverse: $ab + ac = a(b + c)$ and $ab - ac = a(b - c)$.

> To *factor* an expression is to find an equivalent expression that is a product.

Look at Example 14. To *factor* $9x - 45$, we find an equivalent expression that is a product, $9(x - 5)$. When all the terms of an expression have a factor in common, we can "factor it out" using the distributive laws. Note the following.

$9x$ has the factors 9, -9, 3, -3, 1, -1, x, $-x$, $3x$, $-3x$, $9x$, $-9x$;
-45 has the factors 1, -1, 3, -3, 5, -5, 9, -9, 15, -15, 45, -45

We usually remove the largest common factor. In this case, that factor is 9.

Remember that an expression is factored when we find an equivalent expression that is a product.

► **EXAMPLES** Factor.

17. $5x - 10 = 5 \cdot x - 5 \cdot 2$ Try to do this step mentally.
$$= 5(x - 2) \qquad \text{You can check by multiplying.}$$

18. $ax - ay + az = a(x - y + z)$

Multiply.

25. $3(x - 5)$

26. $5(x - y + 4)$

27. $-2(x - 3)$

28. $-5(x - 2y + 4z)$

ANSWERS ON PAGE A-2

Factor.

29. $6x - 12$

30. $3x - 6y + 9$

31. $bx + by - bz$

32. $16a - 36b + 42$

33. $\dfrac{3}{8}x - \dfrac{5}{8}y + \dfrac{7}{8}$

34. $-12x + 32y - 16z$

Collect like terms.

35. $6x - 3x$

36. $7x - x$

37. $x - 9x$

38. $x - 0.41x$

39. $5x + 4y - 2x - y$

40. $3x - 7x - 11 + 8y + 4 - 13y$

19. $9x + 27y - 9 = 9 \cdot x + 9 \cdot 3y - 9 \cdot 1$
$= 9(x + 3y - 1)$ ◀

CAUTION! Note that $3(3x + 9y - 3)$ is also equivalent to $9x + 27y - 9$, but it is *not* the desired form. However, we can complete the process by factoring out another factor of 3: $9x + 27y - 9 = 3(3x + 9y - 3) = 3 \cdot 3(x + 3y - 1) = 9(x + 3y - 1)$. Remember to factor out the largest common factor.

▶ **EXAMPLES**   Factor. Try to write just the answer if you can.

20. $5x - 5y = 5(x - y)$

21. $-3x + 6y - 9z = -3(x - 2y + 3z)$

We might also factor the expression in Example 21 as follows:
$$-3x + 6y - 9z = 3(-x + 2y - 3z).$$

We usually factor out a negative when the first term is negative. The way we factor can depend on the situation in which we are working.

22. $18z - 12x - 24 = 6(3z - 2x - 4)$

23. $\frac{1}{2}x + \frac{3}{2}y - \frac{1}{2} = \frac{1}{2}(x + 3y - 1)$

> *Remember:* An expression is factored when it is written as a product. ◀

DO EXERCISES 29–34.

e Collecting Like Terms

Terms such as $5x$ and $-4x$, whose variable factors are exactly the same, are called **like terms.** Similarly, $3y^2$ and $9y^2$ are like terms because the variables are raised to the same power. Terms such as $4y$ and $5y^2$ are not like terms, and $7x$ and $2y$ are not like terms.

The process of **collecting like terms** is also based on the distributive laws. We can apply the distributive law "on the right" because of the commutative law of multiplication.

▶ **EXAMPLES**   Collect like terms. Try to write just the answer if you can.

24. $4x + 2x = (4 + 2)x = 6x$ Factoring out the x using a distributive law

25. $2x + 3y - 5x - 2y = 2x - 5x + 3y - 2y$
$$= (2 - 5)x + (3 - 2)y = -3x + y$$

26. $3x - x = (3 - 1)x = 2x$

27. $x - 0.24x = 1 \cdot x - 0.24x = (1 - 0.24)x = 0.76x$

28. $x - 6x = 1 \cdot x - 6 \cdot x = (1 - 6)x = -5x$

29. $4x - 7y + 9x - 5 + 3y - 8 = 13x - 4y - 13$ ◀

DO EXERCISES 35–40.

EXERCISE SET 1.7

a Find an equivalent expression with the given denominator.

1. $\dfrac{2}{5}$; $5x$

2. $\dfrac{5}{6}$; $6t$

3. $\dfrac{2}{3}$; $15y$

4. $\dfrac{7}{8}$; $16x$

Simplify.

5. $-\dfrac{24a}{16a}$

6. $-\dfrac{42t}{18t}$

7. $-\dfrac{42ab}{36ab}$

8. $-\dfrac{64pq}{48pq}$

b Write an equivalent expression. Use a commutative law.

9. $y + 8$

10. $x + 3$

11. mn

12. ab

13. $9 + xy$

14. $11 + ab$

15. $ab + c$

16. $rs + t$

Write an equivalent expression. Use an associative law.

17. $a + (b + 2)$

18. $3(vw)$

19. $(8x)y$

20. $(y + z) + 7$

21. $(a + b) + 3$

22. $(5 + x) + y$

23. $3(ab)$

24. $(6x)y$

Use the commutative and associative laws to write three equivalent expressions.

25. $(a + b) + 2$

26. $(3 + x) + y$

27. $5 + (v + w)$

28. $6 + (x + y)$

29. $(xy)3$

30. $(ab)5$

31. $7(ab)$

32. $5(xy)$

1. _____
2. _____
3. _____
4. _____
5. _____
6. _____
7. _____
8. _____
9. _____
10. _____
11. _____
12. _____
13. _____
14. _____
15. _____
16. _____
17. _____
18. _____
19. _____
20. _____
21. _____
22. _____
23. _____
24. _____
25. _____
26. _____
27. _____
28. _____
29. _____
30. _____
31. _____
32. _____

c Multiply.

33. $2(b + 5)$ **34.** $4(x + 3)$ **35.** $7(1 + t)$ **36.** $4(1 + y)$

37. $6(5x + 2)$ **38.** $9(6m + 7)$ **39.** $7(x + 4 + 6y)$ **40.** $4(5x + 8 + 3p)$

41. $7(4 - 3)$ **42.** $15(8 - 6)$ **43.** $-3(3 - 7)$ **44.** $1.2(5 - 2.1)$

45. $4.1(6.3 - 9.4)$ **46.** $-\dfrac{8}{9}\left(\dfrac{2}{3} - \dfrac{5}{3}\right)$ **47.** $7(x - 2)$ **48.** $5(x - 8)$

49. $-7(y - 2)$ **50.** $-9(y - 7)$

51. $-9(-5x - 6y + 8)$ **52.** $-7(-2x - 5y + 9)$

53. $-4(x - 3y - 2z)$ **54.** $8(2x - 5y - 8z)$

55. $3.1(-1.2x + 3.2y - 1.1)$ **56.** $-2.1(-4.2x - 4.3y - 2.2)$

Give the terms of the expression.

57. $4x + 3z$ **58.** $8x - 1.4y$

59. $7x + 8y - 9z$ **60.** $8a + 10b - 18c$

d Factor. Check by multiplying.

61. $2x + 4$ **62.** $5y + 20$ **63.** $30 + 5y$ **64.** $7x + 28$

65. $14x + 21y$

66. $18a + 24b$

67. $5x + 10 + 15y$

68. $9a + 27b + 81$

69. $8x - 24$

70. $10x - 50$

71. $32 - 4y$

72. $24 - 6m$

73. $8x + 10y - 22$

74. $9a + 6b - 15$

75. $ax - a$

76. $by - 9b$

77. $ax - ay - az$

78. $cx + cy - cz$

79. $18x - 12y + 6$

80. $-14x + 21y + 7$

e Collect like terms.

81. $9a + 10a$

82. $12x + 2x$

83. $10a - a$

84. $-16x + x$

85. $2x + 9z + 6x$

86. $3a - 5b + 7a$

87. $7x + 6y^2 + 9y^2$

88. $12m^2 + 6q + 9m^2$

89. $41a + 90 - 60a - 2$

90. $42x - 6 - 4x + 2$

91. $23 + 5t + 7y - t - y - 27$

92. $45 - 90d - 87 - 9d + 3 + 7d$

ANSWERS

65. _____

66. _____

67. _____

68. _____

69. _____

70. _____

71. _____

72. _____

73. _____

74. _____

75. _____

76. _____

77. _____

78. _____

79. _____

80. _____

81. _____

82. _____

83. _____

84. _____

85. _____

86. _____

87. _____

88. _____

89. _____

90. _____

91. _____

92. _____

Copyright © 1991 Addison-Wesley Publishing Co., Inc.

ANSWERS

93. _____

94. _____

95. _____

96. _____

97. _____

98. _____

99. _____

100. _____

101. _____

102. _____

103. _____

104. _____

105. _____

106. _____

107. _____

108. _____

109. _____

110. _____

111. _____

112. _____

113. _____

114. _____

115. _____

116. _____

117. _____

118. _____

119. _____

120. _____

121. _____

122. _____

93. $\dfrac{1}{2}b + \dfrac{1}{2}b$

94. $\dfrac{2}{3}x + \dfrac{1}{3}x$

95. $2y + \dfrac{1}{4}y + y$

96. $\dfrac{1}{2}a + a + 5a$

97. $11x - 3x$

98. $9t - 17t$

99. $6n - n$

100. $10t - t$

101. $y - 17y$

102. $3m - 9m + 4$

103. $-8 + 11a - 5b + 6a - 7b + 7$

104. $8x - 5x + 6 + 3y - 2y - 4$

105. $9x + 2y - 5x$

106. $8y - 3z + 4y$

107. $11x + 2y - 4x - y$

108. $13a + 9b - 2a - 4b$

109. $2.7x + 2.3y - 1.9x - 1.8y$

110. $6.7a + 4.3b - 4.1a - 2.9b$

111. $\dfrac{1}{5}x + \dfrac{4}{5}y + \dfrac{2}{5}x - \dfrac{1}{5}y$

112. $\dfrac{7}{8}x + \dfrac{5}{8}y + \dfrac{1}{8}x - \dfrac{3}{8}y$

SKILL MAINTENANCE

113. Add and simplify: $\dfrac{11}{12} + \dfrac{15}{16}$.

114. Subtract and simplify: $\dfrac{7}{8} - \dfrac{2}{3}$.

115. Find the LCM for 16, 18, and 24.

116. Convert to percent notation: $\dfrac{3}{10}$.

SYNTHESIS

Tell whether the following expressions are equivalent. Also, tell why.

117. $3t + 5$ and $3 \cdot 5 + t$

118. $4x$ and $x + 4$

119. $5m + 6$ and $6 + 5m$

120. $(x + y) + z$ and $z + (x + y)$

Collect like terms if possible and factor the result.

121. $q + qr + qrs + qrst$

122. $21x + 44xy + 15y - 16x - 8y - 38xy + 2y + xy$

1.8 Simplifying Expressions; Order of Operations

We now expand our ability to manipulate expressions by first considering opposites of sums and differences. Then we simplify expressions involving parentheses.

a Inverses of Sums

What happens when we multiply a real number by -1? Consider the following products:

$$-1(7) = -7, \qquad -1(-5) = 5, \qquad -1(0) = 0.$$

From these examples, it appears that when we multiply a number by -1, we get the additive inverse, or opposite, of that number.

> **The Property of -1**
>
> **For any real number a,**
>
> $$-1 \cdot a = -a.$$
>
> **(Negative one times a is the opposite of a.)**
> **(Negative one times a is the additive inverse of a.)**

The property of -1 enables us to find certain expressions equivalent to opposites of sums.

▶ **EXAMPLES**   Find an equivalent expression without parentheses.

1. $-(3 + x) = -1(3 + x)$ Using the property of -1
    $= -1 \cdot 3 + (-1)x$ Using a distributive law, multiplying each term by -1
    $= -3 + (-x)$ Using the property of -1
    $= -3 - x$

2. $-(3x + 2y + 4) = -1(3x + 2y + 4)$ Using the property of -1
    $= -1(3x) + (-1)(2y) + (-1)4$ Using a distributive law
    $= -3x - 2y - 4$ Using the property of -1 ◀

DO EXERCISES 1 AND 2.

Suppose we want to remove parentheses in an expression like

$$-(x - 2y + 5).$$

We can first find an equivalent expression in which the inside expression is separated by plus signs. Then taking the opposite of each term we get

$$-(x - 2y + 5) = -[x + (-2y) + 5]$$
$$= -x + 2y - 5.$$

The most efficient method for this is to replace each term in the parentheses by its opposite ("change the sign of every term"). Doing so for $-(x - 2y + 5)$, we obtain $-x + 2y - 5$ as an equivalent expression.

Find an equivalent expression without parentheses.

1. $-(x + 2)$

2. $-(5x + 2y + 8)$

ANSWERS ON PAGE A-2

Find an equivalent expression without parentheses. Try to do this in one step.

3. $-(6 - t)$

4. $-(x - y)$

5. $-(-4a + 3t - 10)$

6. $-(18 - m - 2n + 4z)$

Remove parentheses and simplify.

7. $5x - (3x + 9)$

8. $5y - 2 - (2y - 4)$

Remove parentheses and simplify.

9. $6x - (4x + 7)$

10. $8y - 3 - (5y - 6)$

11. $(2a + 3b - c) - (4a - 5b + 2c)$

▶ **EXAMPLES**   Find an equivalent expression without parentheses.

3. $-(5 - y) = -5 + y$ Changing the sign of each term

4. $-(2a - 7b - 6) = -2a + 7b + 6$

5. $-(-3x + 4y + z - 7w - 23) = 3x - 4y - z + 7w + 23$ ◀

DO EXERCISES 3–6.

b   **Removing Parentheses and Simplifying**

When a sum is added, as in $5x + (2x + 3)$, we can simply remove, or drop, the parentheses and collect like terms because of the associative law of addition:

$$5x + (2x + 3) = 5x + 2x + 3 = 7x + 3.$$

On the other hand, when a sum is subtracted, as in $3x - (4x + 2)$, no "associative" law applies. However, we can subtract by adding an opposite. We then remove parentheses by changing the sign of each term inside the parentheses and collecting like terms.

▶ **EXAMPLE 6**   Remove parentheses and simplify.

$$\begin{aligned} 3x - (4x + 2) &= 3x + [-(4x + 2)] &&\text{Adding the opposite of } (4x + 2) \\ &= 3x + (-4x - 2) &&\text{Changing the sign of each term} \\ & &&\text{inside the parentheses} \\ &= 3x - 4x - 2 \\ &= -x - 2 &&\text{Collecting like terms} \end{aligned}$$ ◀

DO EXERCISES 7 AND 8.

In practice, the first three steps of Example 6 are usually combined by changing the sign of each term in parentheses and then collecting like terms.

▶ **EXAMPLES**   Remove parentheses and simplify.

7. $5y - (3y + 4) = 5y - 3y - 4$ Removing parentheses by changing the sign of every term inside the parentheses

                $= 2y - 4$ Collecting like terms

8. $3y - 2 - (2y - 4) = 3y - 2 - 2y + 4$

                      $= y + 2$

9. $(3a + 4b - 5) - (2a - 7b + 4c - 8) = 3a + 4b - 5 - 2a + 7b - 4c + 8$

                                     $= a + 11b - 4c + 3$ ◀

DO EXERCISES 9–11.

Next, consider subtracting an expression consisting of several terms preceded by a number other than 1 or -1.

▶ **EXAMPLE 10** Remove parentheses and simplify.

$$\begin{aligned}
x - 3(x + y) &= x + [-3(x + y)] && \text{Adding the opposite of } 3(x + y) \\
&= x + [-3x - 3y] && \text{Multiplying } x + y \text{ by } -3 \\
&= x - 3x - 3y \\
&= -2x - 3y && \text{Collecting like terms} \qquad ◀
\end{aligned}$$

In practice, the first three steps of Example 10 are usually combined by multiplying each term in parentheses by -3 and then collecting like terms.

▶ **EXAMPLES** Remove parentheses and simplify.

11. $3y - 2(4y - 5) = 3y - 8y + 10$ Multiplying each term in parentheses by -2
$$= -5y + 10$$

12. $(2a + 3b - 7) - 4(-5a - 6b + 12) = 2a + 3b - 7 + 20a + 24b - 48$
$$= 22a + 27b - 55 \qquad ◀$$

DO EXERCISES 12–14.

c Parentheses Within Parentheses

Some expressions contain more than one kind of grouping symbol such as brackets [ ] and braces { }.

> **When more than one kind of grouping symbol occurs, do the computations in the innermost ones first. Then work from the inside out.**

▶ **EXAMPLES** Simplify.

13. $[3 - (7 + 3)] = [3 - 10]$ Computing $7 + 3$
$$= -7$$

14. $\{8 - [9 - (12 + 5)]\} = \{8 - [9 - 17]\}$ Computing $12 + 5$
$$\begin{aligned}
&= \{8 - [-8]\} && \text{Computing } 9 - 17 \\
&= 8 + 8 \\
&= 16
\end{aligned}$$

15. $[(-4) \div (-\frac{1}{4})] \div \frac{1}{4} = [(-4) \cdot (-4)] \div \frac{1}{4}$ Working with the innermost parentheses first: computing $(-4) \div (-\frac{1}{4})$

$$\begin{aligned}
&= 16 \div \frac{1}{4} \\
&= 16 \cdot 4 \\
&= 64
\end{aligned}$$

16. $4(2 + 3) - \{7 - [4 - (8 + 5)]\}$
$$\begin{aligned}
&= 4 \cdot 5 - \{7 - [4 - 13]\} && \text{Working with the innermost parentheses first} \\
&= 20 - \{7 - [-9]\} && \text{Computing } 4 \cdot 5 \text{ and } 4 - 13 \\
&= 20 - 16 && \text{Computing } 7 - [-9] \\
&= 4 && ◀
\end{aligned}$$

DO EXERCISES 15–18.

Remove parentheses and simplify.

12. $y - 9(x + y)$

13. $5a - 3(7a - 6)$

14. $4a - b - 6(5a - 7b + 8c)$

Simplify.

15. $12 - (8 + 2)$

16. $\{9 - [10 - (13 + 6)]\}$

17. $[24 \div (-2)] \div (-2)$

18. $5(3 + 4) - \{8 - [5 - (9 + 6)]\}$

ANSWERS ON PAGE A-2

19. Simplify:

$[3(x + 2) + 2x] - [4(y + 2) - 3(y - 2)]$

▶ **EXAMPLE 17**   Simplify.

$[5(x + 2) - 3x] - [3(y + 2) - 7(y - 3)]$

$\quad = [5x + 10 - 3x] - [3y + 6 - 7y + 21]$ Working with the innermost parentheses first

$\quad = [2x + 10] - [-4y + 27]$ Collecting like terms within brackets

$\quad = 2x + 10 + 4y - 27$ Removing brackets

$\quad = 2x + 4y - 17$ Collecting like terms   ◀

DO EXERCISE 19.

d   Order of Operations

When several operations are to be done in a calculation or a problem, we apply the same rules that we did in Section R.4. We repeat them here for review. (If you did not study that section earlier, you should do so now.)

> **Rules for Order of Operations**
>
> 1. **Do all calculations within parentheses before operations outside.**
> 2. **Evaluate all exponential expressions.**
> 3. **Do all multiplications and divisions in order from left to right.**
> 4. **Do all additions and subtractions in order from left to right.**

These rules are consistent with the way in which most computers perform calculations.

Simplify.

20. $23 - 42 \cdot 30$

▶ **EXAMPLE 18**   Simplify: $-34 \cdot 56 - 17$.

There are no parentheses or powers so we start with the third step.

$-34 \cdot 56 - 17 = -1904 - 17$ Carrying out all multiplications and divisions in order from left to right

$\quad\quad\quad\quad = -1921$ Carrying out all additions and subtractions in order from left to right   ◀

▶ **EXAMPLE 19**   Simplify: $2^4 + 51 \cdot 4 - (37 + 23 \cdot 2)$.

$2^4 + 51 \cdot 4 - (37 + 23 \cdot 2)$

$\quad = 2^4 + 51 \cdot 4 - (37 + 46)$ Carrying out all operations inside parentheses first, multiplying 23 by 2, and following the rules for order of operations within the parentheses

21. $52 \cdot 5 + 5^3 - (4^2 - 48 \div 4)$

$\quad = 2^4 + 51 \cdot 4 - 83$ Completing the addition inside parentheses

$\quad = 16 + 51 \cdot 4 - 83$ Evaluating exponential expressions

$\quad = 16 + 204 - 83$ Doing all multiplications

$\quad = 220 - 83$ Doing all additions and subtractions in order from left to right

$\quad = 137$ ◀

DO EXERCISES 20 AND 21.

ANSWERS ON PAGE A-3

NAME   SECTION   DATE

EXERCISE SET 1.8

a Find an equivalent expression without parentheses.

1. $-(2x + 7)$ **2.** $-(3x + 5)$ **3.** $-(5x - 8)$

4. $-(6x - 7)$ **5.** $-(4a - 3b + 7c)$ **6.** $-(5x - 2y - 3z)$

7. $-(6x - 8y + 5)$ **8.** $-(8x + 3y + 9)$ **9.** $-(3x - 5y - 6)$

10. $-(6a - 4b - 7)$ **11.** $-(-8x - 6y - 43)$ **12.** $-(-2a + 9b - 5c)$

b Remove parentheses and simplify.

13. $9x - (4x + 3)$ **14.** $7y - (2y + 9)$

15. $2a - (5a - 9)$ **16.** $11n - (3n - 7)$

17. $2x + 7x - (4x + 6)$ **18.** $3a + 2a - (4a + 7)$

19. $2x - 4y - 3(7x - 2y)$ **20.** $3a - 7b - 1(4a - 3b)$

21. $15x - y - 5(3x - 2y + 5z)$ **22.** $4a - b - 4(5a - 7b + 8c)$

ANSWERS

1. _____

2. _____

3. _____

4. _____

5. _____

6. _____

7. _____

8. _____

9. _____

10. _____

11. _____

12. _____

13. _____

14. _____

15. _____

16. _____

17. _____

18. _____

19. _____

20. _____

21. _____

22. _____

Copyright © 1991 Addison-Wesley Publishing Co., Inc.

ANSWERS

23. _____

24. _____

25. _____

26. _____

27. _____

28. _____

29. _____

30. _____

31. _____

32. _____

33. _____

34. _____

35. _____

36. _____

37. _____

38. _____

39. _____

40. _____

41. _____

42. _____

43. _____

44. _____

23. $(3x + 2y) - 2(5x - 4y)$

24. $(-6a - b) - 3(4b + a)$

25. $(12a - 3b + 5c) - 5(-5a + 4b - 6c)$

26. $(-8x + 5y - 12) - 6(2x - 4y - 10)$

c Simplify.

27. $[9 - 2(5 - 4)]$

28. $[6 - 5(8 - 4)]$

29. $8[7 - 6(4 - 2)]$

30. $10[7 - 4(7 - 5)]$

31. $[4(9 - 6) + 11] - [14 - (6 + 4)]$

32. $[7(8 - 4) + 16] - [15 - (7 + 3)]$

33. $[10(x + 3) - 4] + [2(x - 1) + 6]$

34. $[9(x + 5) - 7] + [4(x - 12) + 9]$

35. $[7(x + 5) - 19] - [4(x - 6) + 10]$

36. $[6(x + 4) - 12] - [5(x - 8) + 11]$

37. $3\{[7(x - 2) + 4] - [2(2x - 5) + 6]\}$

38. $4\{[8(x - 3) + 9] - [4(3x - 7) + 2]\}$

39. $4\{[5(x - 3) + 2] - 3[2(x + 5) - 9]\}$

40. $3\{[6(x - 4) + 5] - 2[5(x + 8) - 10]\}$

d Simplify.

41. $8 - 2 \cdot 3 - 9$

42. $8 - (2 \cdot 3 - 9)$

43. $(8 - 2 \cdot 3) - 9$

44. $(8 - 2)(3 - 9)$

45. $[(-24) \div (-3)] \div (-\frac{1}{2})$

46. $[32 \div (-2)] \div (-2)$

47. $16 \cdot (-24) + 50$

48. $10 \cdot 20 - 15 \cdot 24$

49. $2^4 + 2^3 - 10$

50. $40 - 3^2 - 2^3$

51. $5^3 + 26 \cdot 71 - (16 + 25 \cdot 3)$

52. $4^3 + 10 \cdot 20 + 8^2 - 23$

53. $4 \cdot 5 - 2 \cdot 6 + 4$

54. $4 \cdot (6 + 8)/(4 + 3)$

55. $4^3/8$

56. $5^3 - 7^2$

57. $8(-7) + 6(-5)$

58. $10(-5) + 1(-1)$

59. $19 - 5(-3) + 3$

60. $14 - 2(-6) + 7$

61. $9 \div (-3) + 16 \div 8$

62. $-32 - 8 \div 4 - (-2)$

63. $6 - 4^2$

64. $(2 - 5)^2$

65. $(3 - 8)^2$

66. $3 - 3^2$

ANSWERS

45. _____

46. _____

47. _____

48. _____

49. _____

50. _____

51. _____

52. _____

53. _____

54. _____

55. _____

56. _____

57. _____

58. _____

59. _____

60. _____

61. _____

62. _____

63. _____

64. _____

65. _____

66. _____

ANSWERS

67. _____

68. _____

69. _____

70. _____

71. _____

72. _____

73. _____

74. _____

75. _____

76. _____

77. _____

78. _____

79. _____

80. _____

81. _____

82. _____

83. _____

84. _____

85. _____

86. _____

87. _____

88. _____

89. _____

90. _____

67. $12 - 20^3$

68. $20 + 4^3 \div (-8)$

69. $2 \times 10^3 - 5000$

70. $-7(3^4) + 18$

71. $6[9 - (3 - 4)]$

72. $8[(6 - 13) - 11]$

73. $-1000 \div (-100) \div 10$

74. $256 \div (-32) \div (-4)$

75. $8 - (7 - 9)$

76. $(8 - 7) - 9$

77. $\dfrac{10 - 6^2}{9^2 + 3^2}$

78. $\dfrac{5^2 - 4^3 - 3}{9^2 - 2^2 - 1^5}$

79. $\dfrac{3(6 - 7) - 5 \cdot 4}{6 \cdot 7 - 8(4 - 1)}$

80. $\dfrac{20(8 - 3) - 4(10 - 3)}{10(2 - 6) - 2(5 + 2)}$

81. $\dfrac{2^3 - 3^2 + 12 \cdot 5}{-32 \div (-16) \div (-4)}$

82. $\dfrac{|3 - 5|^2 - |7 - 13|}{|12 - 9| + |11 - 14|}$

SYNTHESIS

Find an equivalent expression by enclosing the last three terms in parentheses preceded by a minus sign.

83. $6y + 2x - 3a + c$　　　**84.** $x - y - a - b$　　　**85.** $6m + 3n - 5m + 4b$

Simplify.

86. $z - \{2z - [3z - (4z - 5z) - 6z] - 7z\} - 8z$

87. $\{x - [f - (f - x)] + [x - f]\} - 3x$

88. $x - \{x - 1 - [x - 2 - (x - 3 - \{x - 4 - [x - 5 - (x - 6)]\})]\}$

89. Determine whether it is true that, for any real numbers a and b, $ab = (-a)(-b)$. Explain why or why not.

90. Determine whether it is true that, for any real numbers a and b, $-(ab) = (-a)b = a(-b)$. Explain why or why not.

SUMMARY AND REVIEW: CHAPTER 1

IMPORTANT PROPERTIES AND FORMULAS

Properties of the Real-Number System

Commutative Laws: $a + b = b + a$, $ab = ba$
Associative Laws: $a + (b + c) = (a + b) + c$, $a(bc) = (ab)c$
Identity Properties: For every real number a, $a + 0 = a$ and $a \cdot 1 = a$.
Opposite Properties: For each real number a, there is an opposite $-a$, such that $a + (-a) = 0$.

For each nonzero real number a, there is a reciprocal $\dfrac{1}{a}$, such that $a\left(\dfrac{1}{a}\right) = 1$.

Distributive Laws: $a(b + c) = ab + ac$, $a(b - c) = ab - ac$

REVIEW EXERCISES

The review sections and objectives to be tested in addition to the material in this chapter are [R.1a, b], [R.2b, c], [R.4a, d], and [R.5b, c].

1. Evaluate $\dfrac{x - y}{3}$ when $x = 17$ and $y = 5$.

2. Translate to an algebraic expression: Nineteen percent of some number.

3. Tell which integers correspond to this situation: Mike has a debt of $45 and Joe has $72 in his savings account.

4. Find: $|-38|$.

Graph the number on a number line.

5. -2.5

6. $\dfrac{8}{9}$

Use either $<$ or $>$ for ▮ to write a true sentence.

7. -3 ▮ 10

8. -1 ▮ -6

9. 0.126 ▮ -12.6

10. $-\dfrac{2}{3}$ ▮ $-\dfrac{1}{10}$

Find the opposite.

11. 3.8

12. $-\dfrac{3}{4}$

Find the reciprocal.

13. $\dfrac{3}{8}$

14. -7

15. Find $-x$ when x is -34.

16. Find $-(-x)$ when x is 5.

Compute and simplify.

17. $4 + (-7)$

18. $6 + (-9) + (-8) + 7$

19. $-3.8 + 5.1 + (-12) + (-4.3) + 10$

20. $-3 - (-7)$

21. $-\dfrac{9}{10} - \dfrac{1}{2}$

22. $-3.8 - 4.1$

23. $-9 \cdot (-6)$

24. $-2.7(3.4)$

25. $\dfrac{2}{3} \cdot \left(-\dfrac{3}{7}\right)$

26. $3 \cdot (-7) \cdot (-2) \cdot (-5)$

27. $35 \div (-5)$

28. $-5.1 \div 1.7$

29. $-\dfrac{3}{5} \div \left(-\dfrac{4}{5}\right)$

30. $|-3.4 - 12.2| - 8(-7)$

31. $|-12(-3) - 2^3 - (-9)(-10)|$

Solve.

32. On the first, second, and third downs, a football team had these gains and losses: 5-yd gain, 12-yd loss, and 15-yd gain. Find the total gain (or loss).

33. Your total assets are $170. You borrow $300. What are your total assets now?

Multiply.

34. $5(3x - 7)$

35. $-2(4x - 5)$

36. $10(0.4x + 1.5)$

37. $-8(3 - 6x)$

Factor.

38. $2x - 14$

39. $6x - 6$

40. $5x + 10$

41. $12 - 3x$

Collect like terms.

42. $11a + 2b - 4a - 5b$

43. $7x - 3y - 9x + 8y$

44. $6x + 3y - x - 4y$

45. $-3a + 9b + 2a - b$

Remove parentheses and simplify.

46. $2a - (5a - 9)$

47. $3(b + 7) - 5b$

48. $3[11 - 3(4 - 1)]$

49. $2[6(y - 4) + 7]$

50. $[8(x + 4) - 10] - [3(x - 2) + 4]$

51. $5\{[6(x - 1) + 7] - [3(3x - 4) + 8]\}$

Write true or false.

52. $-9 \le 11$

53. $-11 \ge -3$

54. Write another inequality with the same meaning as $-3 < x$.

SKILL MAINTENANCE

55. Divide and simplify: $\dfrac{11}{12} \div \dfrac{7}{10}$.

56. Compute and simplify: $\dfrac{5^3 - 2^4}{5 \cdot 2 + 2^3}$.

57. Find the prime factorization of 648.

58. Convert to percent notation: $\dfrac{5}{8}$.

59. Convert to decimal notation: 5.67%.

60. Find the LCM of 15, 27, and 30.

SYNTHESIS

61. Simplify: $-\left| \dfrac{7}{8} - \left(-\dfrac{1}{2}\right) - \dfrac{3}{4} \right|$.

62. Simplify: $(|2.7 - 3| + 3^2 - |-3|) \div (-3)$.

THINKING IT THROUGH

1. List three examples of rational numbers that are not integers.
2. Explain at least three uses of the distributive laws considered in this chapter.

TEST: CHAPTER 1

1. Evaluate $\dfrac{3x}{y}$ when $x = 10$ and $y = 5$.

2. Write an algebraic expression: Nine less than some number.

3. Find the area of a triangle when the height h is 30 ft and the base b is 16 ft.

Use either $<$ or $>$ for ▨ to write a true sentence.

4. -4 ▨ 0

5. -3 ▨ -8

6. -0.78 ▨ -0.87

7. $-\dfrac{1}{8}$ ▨ $\dfrac{1}{2}$

Find the absolute value.

8. $|-7|$

9. $\left|\dfrac{9}{4}\right|$

10. $|-2.7|$

Find the opposite.

11. $\dfrac{2}{3}$

12. -1.4

13. Find $-x$ when x is -8.

Find the reciprocal.

14. -2

15. $\dfrac{4}{7}$

Compute and simplify.

16. $3.1 - (-4.7)$

17. $-8 + 4 + (-7) + 3$

18. $-\dfrac{1}{5} + \dfrac{3}{8}$

19. $2 - (-8)$

20. $3.2 - 5.7$

21. $\dfrac{1}{8} - \left(-\dfrac{3}{4}\right)$

1. _____

2. _____

3. _____

4. _____

5. _____

6. _____

7. _____

8. _____

9. _____

10. _____

11. _____

12. _____

13. _____

14. _____

15. _____

16. _____

17. _____

18. _____

19. _____

20. _____

21. _____

ANSWERS

22. _____

23. _____

24. _____

25. _____

26. _____

27. _____

28. _____

29. _____

30. _____

31. _____

32. _____

33. _____

34. _____

35. _____

36. _____

37. _____

38. _____

39. _____

40. _____

41. _____

42. _____

43. _____

44. _____

45. _____

22. $4 \cdot (-12)$

23. $-\dfrac{1}{2} \cdot \left(-\dfrac{3}{8}\right)$

24. $-45 \div 5$

25. $-\dfrac{3}{5} \div \left(-\dfrac{4}{5}\right)$

26. $4.864 \div (-0.5)$

27. $-2(16) - \left|2(-8) - 5^3\right|$

28. Wendy has \$143 in her savings account. She withdraws \$25. Then she makes a deposit of \$30. How much is now in her savings account?

Multiply.

29. $3(6 - x)$

30. $-5(y - 1)$

Factor.

31. $12 - 22x$

32. $7x + 21 + 14y$

Simplify.

33. $6 + 7 - 4 - (-3)$

34. $5x - (3x - 7)$

35. $4(2a - 3b) + a - 7$

36. $4\{3[5(y - 3) + 9] + 2(y + 8)\}$

37. $256 \div (-16) \div 4$

38. $2^3 - 10[4 - (-2 + 18)3]$

39. Write an inequality with the same meaning as $x \le -2$.

SKILL MAINTENANCE

40. Evaluate: $(1.2)^3$.

41. Convert to percent notation: $\dfrac{1}{8}$.

42. Find the prime factorization of 280.

43. Find the LCM of 16, 20, and 30.

SYNTHESIS

44. Simplify: $\left|-27 - 3(4)\right| - \left|-36\right| + \left|-12\right|$.

45. Simplify: $a - \{3a - [4a - (2a - 4a)]\}$.

INTRODUCTION In this chapter we use the manipulations discussed in Chapter 1 to solve equations and inequalities. We then solve problems using equations and inequalities.

The review sections to be tested in addition to the material in this chapter are R.3, 1.1, 1.3, and 1.8. ❖

Solving Equations and Inequalities

2

AN APPLICATION

The state of Colorado is in the shape of a rectangle whose perimeter is 1300 mi. The length is 110 mi more than the width. Find the dimensions.

THE MATHEMATICS

Let w = the width of the state of Colorado. The problem can be translated to the following *equation:*

$$2(w + 110) + 2w = 1300.$$

Identity Properties of 0 and 1: $a + 0 = a, \quad a \cdot 1 = a$
Simple-Interest Formula: $I = Prt$
Sum of the Angles of a Triangle = 180°
Perimeter of a Rectangle: $P = 2l + 2w$
Consecutive Integers: $x, \quad x + 1, \quad x + 2, \quad x + 3$, etc.
Consecutive Even Integers: $x, \quad x + 2, \quad x + 4, \quad x + 6$, etc.
Consecutive Odd Integers: $x, \quad x + 2, \quad x + 4, \quad x + 6$, etc.

PRETEST: CHAPTER 2

Solve.

1. $-7x = 49$

2. $4y + 9 = 2y + 7$

3. $6a - 2 = 10$

4. $4 + x = 12$

5. $7 - 3(2x - 1) = 40$

6. $\dfrac{4}{9}x - 1 = \dfrac{7}{8}$

7. $1 + 2(a + 3) = 3(2a - 1) + 6$

8. $-3x \leq 18$

9. $y + 5 > 1$

10. $5 - 2a < 7$

11. $3x + 4 \geq 2x + 7$

12. $8y < -18$

13. Solve for G: $P = 3KG$.

14. Solve for a: $A = \dfrac{3a - b}{b}$.

Solve.

15. The perimeter of a rectangular field is 146 m. The width is 5 m less than the length. Find the dimensions.

16. Money is invested in a savings account at 9% simple interest. After one year, there is $708.50 in the account. How much was originally invested?

17. The sum of three consecutive integers is 246. Find the integers.

18. When 18 is added to six times a number, the result is less than 120. For what numbers is this possible?

Graph on a number line.

19. $x > -3$

20. $x \leq 4$

<table>
</table>

2.1 Solving Equations: The Addition Principle

a Equations and Solutions

In order to solve problems, we must learn to solve equations.

> An *equation* is a number sentence that says that the expressions on either side of the equals sign, $=$, represent the same number.

Here are some examples:

$$3 + 2 = 5, \quad 14 - 10 = 1 + 3, \quad x + 6 = 13, \quad 3x - 2 = 7 - x.$$

Equations have expressions on each side of the equals sign. The sentence "$14 - 10 = 1 + 3$" asserts that the expressions $14 - 10$ and $1 + 3$ name the same number.

Some equations are true. Some are false. Some are neither true nor false.

▶ **EXAMPLES** Determine whether the equation is true, false, or neither.

1. $3 + 2 = 5$ The equation is *true*.

2. $7 - 2 = 4$ The equation is *false*.

3. $x + 6 = 13$ The equation is *neither* true nor false, because we do not know what number x represents. ◀

DO EXERCISES 1–3.

> Any replacement for the variable that makes an equation true is called a *solution* of the equation. To solve an equation means to find *all* of its solutions.

One way to determine whether a number is a solution of an equation is to evaluate the algebraic expression on each side of the equation by substitution. If the values are the same, then the number is a solution.

▶ **EXAMPLE 4** Determine whether 7 is a solution of $x + 6 = 13$.

We have

$$\begin{array}{c|c} x + 6 = 13 & \text{Writing the equation} \\ \hline 7 + 6 & 13 \quad \text{Substituting 7 for } x \\ 13 & \text{TRUE} \end{array}$$

Since the left-hand and the right-hand sides are the same, we have a solution. No other number makes the equation true, so the only solution is the number 7. ◀

OBJECTIVES

After finishing Section 2.1, you should be able to:

a Determine whether a given number is a solution of a given equation.

b Solve equations using the addition principle.

FOR EXTRA HELP

Tape 5A   Tape 4A   MAC: 2
IBM: 2

Determine whether the equation is true, false, or neither.

1. $5 - 8 = -4$

2. $12 + 6 = 18$

3. $x + 6 = 7 - x$

ANSWERS ON PAGE A-3

Determine whether the given number is a solution of the given equation.

4. 8; $x + 4 = 12$

5. 0; $x + 4 = 12$

6. -3; $7 + x = -4$

▶ **EXAMPLE 5**   Determine whether 19 is a solution of $7x = 141$.
We have

$$
\begin{array}{c|l}
7x = 141 & \text{Writing the equation} \\
\hline
7(19) & 141 \quad \text{Substituting 19 for } x \\
133 & \text{FALSE}
\end{array}
$$

Since the left-hand and the right-hand sides are not the same, we do not have a solution.   ◀

DO EXERCISES 4–6.

b Using the Addition Principle

Consider the equation

$$x = 7.$$

We can easily see that the solution of this equation is 7. If we replace x by 7, we get

$$7 = 7, \quad \text{which is true.}$$

Now consider the equation of Example 4:

$$x + 6 = 13.$$

In Example 4, we discovered that the solution of this equation is also 7, but the fact that 7 is the solution is not as obvious. We now begin to consider principles that allow us to start with an equation and end up with an equation like $x = 7$, in which the variable is alone on one side and for which the solution is easy to find. The equations $x + 6 = 13$ and $x = 7$ are **equivalent.**

> Equations with the same solutions are called *equivalent equations.*

One of the principles that we use in solving equations concerns adding. An equation $a = b$ says that a and b stand for the same number. Suppose this is true, and we add a number c to the number a. We get the same answer if we add c to b, because a and b are the same number.

> **The Addition Principle**
>
> If an equation $a = b$ is true, then
> $$a + c = b + c$$
> is true for any number c.

When we use the addition principle, we sometimes say that we "add the same number on both sides of an equation." This is also true for subtraction, since we can express every subtraction as an addition. That is, since

$$a - c = b - c \quad \text{means} \quad a + (-c) = b + (-c),$$

the addition principle tells us that we can "subtract the same number on both sides of an equation."

▶ **EXAMPLE 6** Solve: $x + 5 = -7$.

We have

$$x + 5 = -7$$
$$x + 5 - 5 = -7 - 5 \quad \text{Using the addition principle: adding } -5 \text{ on both sides or subtracting 5 on both sides}$$
$$x + 0 = -12 \quad \text{Simplifying}$$
$$x = -12. \quad \text{Identity property of 0}$$

We can see that the solution of $x = -12$ is the number -12. To check the answer, we substitute -12 in the original equation.

Check:

$$\begin{array}{c|c} x + 5 = -7 \\ \hline -12 + 5 & -7 \\ -7 & \text{TRUE} \end{array}$$

The solution of the original equation is -12. ◀

In Example 6, to get x alone, we used the addition principle and subtracted 5 on both sides. This eliminated the 5 on the left. We started with $x + 5 = -7$, and, using the addition principle, we found a simpler equation $x = -12$ for which it was easy to "*see*" the solution. The equations $x + 5 = -7$ and $x = -12$ are equivalent.

DO EXERCISE 7.

Now we solve an equation with a subtraction using the addition principle.

▶ **EXAMPLE 7** Solve: $-6.5 = y - 8.4$.

We have

$$-6.5 = y - 8.4$$
$$-6.5 + 8.4 = y - 8.4 + 8.4 \quad \text{Using the addition principle: adding 8.4 to eliminate } -8.4 \text{ on the right}$$
$$1.9 = y$$

Check:

$$\begin{array}{c|c} -6.5 = y - 8.4 \\ \hline -6.5 & 1.9 - 8.4 \\ & -6.5 \quad \text{TRUE} \end{array}$$

The solution is 1.9. ◀

Note that equations are reversible. That is, if $a = b$ is true, then $b = a$ is true. Thus when we solve $-6.5 = y - 8.4$, we can reverse it and solve $y - 8.4 = -6.5$ if we wish.

DO EXERCISES 8 AND 9.

7. Solve using the addition principle:
$$x + 7 = 2.$$

Solve.

8. $8.7 = n - 4.5$

9. $y + 17.4 = 10.9$

ANSWERS ON PAGE A-3

Solve.

10. $x + \dfrac{1}{2} = -\dfrac{3}{2}$

▶ **EXAMPLE 8**   Solve: $x - \dfrac{2}{3} = \dfrac{5}{2}$.

We have

$$x - \frac{2}{3} = \frac{5}{2}$$

$$x - \frac{2}{3} + \frac{2}{3} = \frac{5}{2} + \frac{2}{3} \qquad \text{Adding } \frac{2}{3}$$

$$x = \frac{5}{2} + \frac{2}{3}$$

$$x = \frac{5}{2} \cdot \frac{3}{3} + \frac{2}{3} \cdot \frac{2}{2} \qquad \begin{array}{l}\text{Multiplying by 1 to obtain equivalent}\\ \text{fractional expressions with the}\\ \text{least common denominator 6}\end{array}$$

$$x = \frac{15}{6} + \frac{4}{6}$$

$$x = \frac{19}{6}$$

Check:

$$x - \frac{2}{3} = \frac{5}{2}$$

$$\begin{array}{c|c} \dfrac{19}{6} - \dfrac{2}{3} & \dfrac{5}{2} \\ \hline \dfrac{19}{6} - \dfrac{4}{6} & \\ \dfrac{15}{6} & \\ \dfrac{5}{2} & \text{TRUE} \end{array}$$

11. $t - \dfrac{13}{4} = \dfrac{5}{8}$

The solution is $\dfrac{19}{6}$.   ◀

DO EXERCISES 10 AND 11.

EXERCISE SET 2.1

a Determine whether the given number is a solution of the given equation.

1. 15; $x + 17 = 32$

2. 75; $y + 28 = 93$

3. 21; $x - 7 = 12$

4. 27; $y - 8 = 19$

5. 7; $6x = 54$

6. 9; $8y = 72$

7. 30; $\dfrac{x}{6} = 5$

8. 49; $\dfrac{y}{8} = 6$

9. 19; $5x + 7 = 107$

10. 9; $9x + 5 = 86$

11. 11; $7(y - 1) = 63$

12. 18; $x + 3 = 3 + x$

b Solve using the addition principle. Don't forget to check!

13. $x + 2 = 6$

14. $x + 5 = 8$

15. $x + 15 = -5$

16. $y + 9 = 43$

17. $x + 6 = -8$

18. $t + 9 = -12$

19. $x + 16 = -2$

20. $y + 25 = -6$

21. $x - 9 = 6$

22. $x - 8 = 5$

23. $x - 7 = -21$

24. $x - 3 = -14$

25. $5 + t = 7$

26. $8 + y = 12$

27. $-7 + y = 13$

ANSWERS

1. _____

2. _____

3. _____

4. _____

5. _____

6. _____

7. _____

8. _____

9. _____

10. _____

11. _____

12. _____

13. _____

14. _____

15. _____

16. _____

17. _____

18. _____

19. _____

20. _____

21. _____

22. _____

23. _____

24. _____

25. _____

26. _____

27. _____

28. $-9 + z = 15$

29. $-3 + t = -9$

30. $-6 + y = -21$

31. $r + \dfrac{1}{3} = \dfrac{8}{3}$

32. $t + \dfrac{3}{8} = \dfrac{5}{8}$

33. $m + \dfrac{5}{6} = -\dfrac{11}{12}$

34. $x + \dfrac{2}{3} = -\dfrac{5}{6}$

35. $x - \dfrac{5}{6} = \dfrac{7}{8}$

36. $y - \dfrac{3}{4} = \dfrac{5}{6}$

37. $-\dfrac{1}{5} + z = -\dfrac{1}{4}$

38. $-\dfrac{1}{8} + y = -\dfrac{3}{4}$

39. $7.4 = x + 2.3$

40. $9.3 = 4.6 + x$

41. $7.6 = x - 4.8$

42. $9.5 = y - 8.3$

43. $-9.7 = -4.7 + y$

44. $-7.8 = 2.8 + x$

45. $5\dfrac{1}{6} + x = 7$

46. $5\dfrac{1}{4} = 4\dfrac{2}{3} + x$

47. $q + \dfrac{1}{3} = -\dfrac{1}{7}$

48. $47\dfrac{1}{8} = -76 + z$

SKILL MAINTENANCE

49. Add: $-3 + (-8)$.

50. Subtract: $-3 - (-8)$.

51. Multiply: $-\dfrac{2}{3} \cdot \dfrac{5}{8}$.

52. Divide: $-\dfrac{3}{7} \div \left(-\dfrac{9}{7}\right)$.

SYNTHESIS

Solve.

53. ▦ $-356.788 = -699.034 + t$

54. $-\dfrac{4}{5} + \dfrac{7}{10} = x - \dfrac{3}{4}$

55. $x + \dfrac{4}{5} = -\dfrac{2}{3} - \dfrac{4}{15}$

56. $8 - 25 = 8 + x - 21$

57. $16 + x - 22 = -16$

58. $x + x = x$

59. $x + 3 = 3 + x$

60. $x + 4 = 5 + x$

61. $-\dfrac{3}{2} + x = -\dfrac{5}{17} - \dfrac{3}{2}$

62. $|x| = 5$

63. $|x| + 6 = 19$

2.2 Solving Equations: The Multiplication Principle

a Using the Multiplication Principle

Suppose that $a = b$ is true, and we multiply a by some number c. We get the same answer if we multiply b by c, because a and b are the same number.

> **The Multiplication Principle**
>
> **If an equation $a = b$ is true, then**
> $$a \cdot c = b \cdot c$$
> **is true for any number c.**

When using the multiplication principle, we sometimes say that we "multiply on both sides by the same number."

► **EXAMPLE 1**   Solve: $\dfrac{3}{8} = -\dfrac{5}{4}x$.

We have

$$\frac{3}{8} = -\frac{5}{4}x$$

The reciprocal of $-\dfrac{5}{4}$ is $-\dfrac{4}{5}$. There is no sign change.

$$-\frac{4}{5} \cdot \frac{3}{8} = -\frac{4}{5} \cdot \left(-\frac{5}{4}x\right)$$ Multiplying by $-\dfrac{4}{5}$ to get $1 \cdot x$ and eliminate $-\dfrac{5}{4}$ on the right

$$-\frac{3}{10} = 1 \cdot x$$ Simplifying

$$-\frac{3}{10} = x$$ Identity property of 1: $1 \cdot x = x$

Check:
$$\frac{3}{8} = -\frac{5}{4}x$$

$$\begin{array}{c|c} \dfrac{3}{8} & -\dfrac{5}{4}\left(-\dfrac{3}{10}\right) \\[2mm] & \dfrac{3}{8} \end{array}$$ TRUE

The solution is $-\dfrac{3}{10}$.   ◄

DO EXERCISES 1 AND 2.

OBJECTIVE

After finishing Section 2.2, you should be able to:

a Solve equations using the multiplication principle.

FOR EXTRA HELP

Tape 5B   Tape 4B   MAC: 2
                                     IBM: 2

Solve.

1. $\dfrac{2}{3} = -\dfrac{5}{6}y$

2. $4x = -7$

3. Solve: $5x = 40$.

In Example 1, to get x alone, we multiplied by the *multiplicative inverse*, or *reciprocal*, of $-\frac{5}{4}$. When we multiplied, we got the *multiplicative identity* 1 times x, or $1 \cdot x$, which simplified to x. This enabled us to eliminate the $-\frac{5}{4}$ on the right.

The multiplication principle also tells us that we can "divide on both sides by a nonzero number." This is because division is the same as multiplying by a reciprocal. That is,

$$\frac{a}{c} = \frac{b}{c} \quad \text{means} \quad a \cdot \frac{1}{c} = b \cdot \frac{1}{c}, \quad \text{when } c \neq 0.$$

In an expression like $3x$, the number 3 is called the **coefficient.** In practice it is usually more convenient to "divide" both sides of the equation if the coefficient of the variable is in decimal notation or is an integer. When the coefficient is in fractional notation, it is more convenient to "multiply" by a reciprocal.

▶ **EXAMPLE 2**   Solve: $3x = 9$.

We have

$$3x = 9$$

$$\frac{3x}{3} = \frac{9}{3} \qquad \text{Using the multiplication principle: multiplying by } \tfrac{1}{3} \text{ on both sides or dividing on both sides by 3}$$

$$1 \cdot x = 3 \qquad \text{Simplifying}$$

$$x = 3. \qquad \text{Identity property of 1}$$

It is easy to see that the solution of $x = 3$ is 3.

Check:
$$\begin{array}{c|c} \multicolumn{2}{c}{3x = 9} \\ \hline 3 \cdot 3 & 9 \\ 9 & \text{TRUE} \end{array}$$

The solution of the original equation is 3.   ◀

DO EXERCISE 3.

4. Solve: $-6x = 108$.

▶ **EXAMPLE 3**   Solve: $-4x = 92$.

$$-4x = 92$$

$$\frac{-4x}{-4} = \frac{92}{-4} \qquad \text{Using the multiplication principle. Dividing on both sides by } -4 \text{ is the same as multiplying by } -\tfrac{1}{4}.$$

$$\left. \begin{array}{l} 1 \cdot x = -23 \\ x = -23 \end{array} \right\} \qquad \text{Simplifying}$$

Check:
$$\begin{array}{c|c} \multicolumn{2}{c}{-4x = 92} \\ \hline -4(-23) & 92 \\ 92 & \text{TRUE} \end{array}$$

The solution is -23.   ◀

DO EXERCISE 4.

ANSWERS ON PAGE A-3

► **EXAMPLE 4**   Solve: $-x = 9$.

$$-x = 9$$

$$-1 \cdot x = 9 \qquad \text{Using the property of } -1$$

$$-1 \cdot (-1 \cdot x) = -1 \cdot 9 \qquad \text{Multiplying on both sides by } -1\text{, the}$$
$$\text{reciprocal of itself, or dividing by } -1$$

$$1 \cdot x = -9$$

$$x = -9$$

Check:

$$\begin{array}{c|c} -x = 9 \\ \hline -(-9) & 9 \\ 9 & \text{TRUE} \end{array}$$

The solution is -9.   ◄

DO EXERCISE 5.

5. Solve: $-x = -10$.

Now we solve an equation with a division using the multiplication principle. Consider an equation like $-y/9 = 14$. In Chapter 1, we learned that a division can be expressed as multiplication by the reciprocal of the divisor. Thus,

$$\frac{-y}{9} = \frac{1}{9}(-y).$$

The reciprocal of $\frac{1}{9}$ is 9. Then, using the multiplication principle, we multiply on both sides by 9. This is shown in the following example.

6. Solve: $-14 = \dfrac{-y}{2}$.

► **EXAMPLE 5**   Solve: $\dfrac{-y}{9} = 14$.

$$\frac{-y}{9} = 14$$

$$\frac{1}{9}(-y) = 14$$

$$9 \cdot \frac{1}{9}(-y) = 9 \cdot 14 \qquad \text{Multiplying by 9 on both sides}$$

$$-y = 126$$

$$y = -126 \qquad \text{Multiplying by } -1 \text{ on both sides}$$

Check:

$$\frac{-y}{9} = 14$$

$$\begin{array}{c|c} \dfrac{-(-126)}{9} & 14 \\[2ex] \dfrac{126}{9} & \\[2ex] 14 & \text{TRUE} \end{array}$$

The solution is -126.   ◄

DO EXERCISE 6.

ANSWERS ON PAGE A-3

Solve.

7. $1.12x = 8736$

▶ **EXAMPLE 6**   Solve: $1.16y = 9744$.

$$1.16y = 9744$$

$$\frac{1.16y}{1.16} = \frac{9744}{1.16} \qquad \text{Dividing by 1.16}$$

$$y = \frac{9744}{1.16}$$

$$y = 8400$$

Check:

$$\begin{array}{c|c} 1.16y = 9744 \\ \hline 1.16(\,8400\,) & 9744 \\ 9744 & \text{TRUE} \end{array}$$

The solution is 8400.   ◀

DO EXERCISES 7 AND 8.

Note that equations are reversible. That is, if $a = b$ is true, then $b = a$ is true. Thus, when we solve $15 = 3x$, we can reverse it and solve $3x = 15$ if we wish.

8. $6.3 = -2.1y$

ANSWERS ON PAGE A-3

EXERCISE SET 2.2

a Solve using the multiplication principle. Don't forget to check!

1. $6x = 36$ **2.** $3x = 39$ **3.** $5x = 45$

4. $9x = 72$ **5.** $84 = 7x$ **6.** $56 = 8x$

7. $-x = 40$ **8.** $100 = -x$ **9.** $-x = -1$

10. $-68 = -r$ **11.** $7x = -49$ **12.** $9x = -36$

13. $-12x = 72$ **14.** $-15x = 105$ **15.** $-21x = -126$

16. $-13x = -104$ **17.** $\dfrac{t}{7} = -9$ **18.** $\dfrac{y}{-8} = 11$

19. $\dfrac{3}{4}x = 27$ **20.** $\dfrac{4}{5}x = 16$ **21.** $\dfrac{-t}{3} = 7$

22. $\dfrac{-x}{6} = 9$ **23.** $-\dfrac{m}{3} = \dfrac{1}{5}$ **24.** $\dfrac{1}{9} = -\dfrac{z}{7}$

1. _____
2. _____
3. _____
4. _____
5. _____
6. _____
7. _____
8. _____
9. _____
10. _____
11. _____
12. _____
13. _____
14. _____
15. _____
16. _____
17. _____
18. _____
19. _____
20. _____
21. _____
22. _____
23. _____
24. _____

ANSWERS

25. _____

26. _____

27. _____

28. _____

29. _____

30. _____

31. _____

32. _____

33. _____

34. _____

35. _____

36. _____

37. _____

38. _____

39. _____

40. _____

41. _____

42. _____

43. _____

44. _____

45. _____

46. _____

47. _____

48. _____

49. _____

50. _____

25. $-\dfrac{3}{5}r = \dfrac{9}{10}$

26. $\dfrac{2}{5}y = -\dfrac{4}{15}$

27. $-\dfrac{3}{2}r = -\dfrac{27}{4}$

28. $-\dfrac{5}{7}x = -\dfrac{10}{14}$

29. $6.3x = 44.1$

30. $2.7y = 54$

31. $-3.1y = 21.7$

32. $-3.3y = 6.6$

33. $38.7m = 309.6$

34. $29.4m = 235.2$

35. $-\dfrac{2}{3}y = -10.6$

36. $-\dfrac{9}{7}y = 12.06$

SKILL MAINTENANCE

Collect like terms.

37. $3x + 4x$

38. $6x + 5 - 7x$

Remove parentheses and simplify.

39. $3x - (4 + 2x)$

40. $2 - 5(x + 5)$

SYNTHESIS

Solve.

41. ▓ $-0.2344m = 2028.732$

42. $0 \cdot x = 0$

43. $0 \cdot x = 9$

44. $4|x| = 48$

45. $2|x| = -12$

Solve for x.

46. $ax = 5a$

47. $3x = \dfrac{b}{a}$

48. $cx = a^2 + 1$

49. $\dfrac{a}{b}x = 4$

50. A student makes a calculation and gets an answer of 22.5. On the last step, the student multiplies by 0.3 when a division by 0.3 should have been done. What should the correct answer be?

2.3 Using the Principles Together

a Applying Both Principles

Consider the equation $3x + 4 = 13$. It is more complicated than those we discussed in the preceding two sections. In order to solve such an equation, we first isolate the x-term, $3x$, using the addition principle. Then we apply the multiplication principle to get x by itself.

▶ **EXAMPLE 1** Solve: $3x + 4 = 13$.

$$3x + 4 = 13$$
$$3x + 4 - 4 = 13 - 4 \qquad \text{Using the addition principle: subtracting 4 on both sides}$$
$$3x = 9 \qquad \text{Simplifying}$$
$$\frac{3x}{3} = \frac{9}{3} \qquad \text{Using the multiplication principle: dividing on both sides by 3}$$
$$x = 3 \qquad \text{Simplifying}$$

Check:
$$\begin{array}{c|c} 3x + 4 = 13 \\ \hline 3 \cdot 3 + 4 & 13 \\ 9 + 4 & \\ 13 & \text{TRUE} \end{array}$$

We use our rules for order of operations to carry out the check. We find the product $3 \cdot 3$. Then we add 4.

The solution is 3. ◀

DO EXERCISE 1.

▶ **EXAMPLE 2** Solve: $-5x - 6 = 16$.

$$-5x - 6 = 16$$
$$-5x - 6 + 6 = 16 + 6 \qquad \text{Adding 6 on both sides}$$
$$-5x = 22$$
$$\frac{-5x}{-5} = \frac{22}{-5} \qquad \text{Dividing on both sides by } -5$$
$$x = -\frac{22}{5}, \quad \text{or} \quad -4\frac{2}{5} \qquad \text{Simplifying}$$

Check:
$$\begin{array}{c|c} -5x - 6 = 16 \\ \hline -5\left(-\dfrac{22}{5}\right) - 6 & 16 \\ 22 - 6 & \\ 16 & \text{TRUE} \end{array}$$

The solution is $-\dfrac{22}{5}$. ◀

DO EXERCISES 2 AND 3.

OBJECTIVES

After finishing Section 2.3, you should be able to:

a Solve equations using both the addition and the multiplication principles.

b Solve equations in which like terms may need to be collected.

c Solve equations by first removing parentheses and collecting like terms.

FOR EXTRA HELP

Tape 6A   Tape 4B   MAC: 2
                                  IBM: 2

1. Solve: $9x + 6 = 51$.

Solve.

2. $8x - 4 = 28$

3. $-\dfrac{1}{2}x + 3 = 1$

ANSWERS ON PAGE A-3

4. Solve: $-18 - x = -57$.

Solve.

5. $-4 - 8x = 8$

6. $41.68 = 4.7 - 8.6y$

Solve.

7. $4x + 3x = -21$

8. $x - 0.09x = 728$

▶ **EXAMPLE 3**   Solve: $45 - x = 13$.

$$45 - x = 13$$
$$-45 + 45 - x = -45 + 13 \qquad \text{Adding } -45 \text{ on both sides}$$
$$-x = -32$$
$$-1 \cdot x = -32 \qquad \text{Using the property of } -1: \ -x = -1 \cdot x$$
$$\frac{-1 \cdot x}{-1} = \frac{-32}{-1} \qquad \begin{array}{l}\text{Dividing on both sides by } -1 \text{ (You could have}\\\text{multiplied on both sides by } -1 \text{ instead. That}\\\text{would also change the sign on both sides.)}\end{array}$$
$$x = 32$$

The number 32 checks and is the solution.   ◀

DO EXERCISE 4.

As we improve our equation-solving skills, we begin to shorten some of our writing. Thus we may not always write a number being added, subtracted, multiplied, or divided on both sides. We simply write it on the opposite side.

▶ **EXAMPLE 4**   Solve: $16.3 - 7.2y = -8.18$.

$$16.3 - 7.2y = -8.18$$
$$-7.2y = -16.3 + (-8.18) \qquad \begin{array}{l}\text{Adding } -16.3 \text{ on both sides. We write the}\\\text{addition of } -16.3 \text{ on the right side.}\end{array}$$
$$-7.2y = -24.48$$
$$y = \frac{-24.48}{-7.2} \qquad \begin{array}{l}\text{Dividing by } -7.2 \text{ on both sides. We write}\\\text{the division by } -7.2 \text{ on the right side.}\end{array}$$
$$y = 3.4$$

Check:
$$\begin{array}{c|c}\hline 16.3 - 7.2y = -8.18 \\\hline 16.3 - 7.2(3.4) & -8.18 \\ 16.3 - 24.48 & \\ -8.18 & \qquad \text{TRUE} \end{array}$$

The solution is 3.4.   ◀

DO EXERCISES 5 AND 6.

b   **Collecting Like Terms**

If there are like terms on one side of the equation, we collect them before using the addition or the multiplication principle.

▶ **EXAMPLE 5**   Solve: $3x + 4x = -14$.

$$3x + 4x = -14$$
$$7x = -14 \qquad \text{Collecting like terms}$$
$$x = \frac{-14}{7} \qquad \text{Dividing by 7 on both sides}$$
$$x = -2$$

The number -2 checks, so the solution is -2.   ◀

DO EXERCISES 7 AND 8.

If there are like terms on opposite sides of the equation, we get them on the same side by using the addition principle. Then we collect them. In other words, we get all terms with a variable on one side and all numbers on the other.

▶ **EXAMPLE 6**   Solve: $2x - 2 = -3x + 3$.

$$2x - 2 = -3x + 3$$
$$2x - 2 + 2 = -3x + 3 + 2 \qquad \text{Adding 2}$$
$$2x = -3x + 5 \qquad \text{Collecting like terms}$$
$$2x + 3x = -3x + 3x + 5 \qquad \text{Adding } 3x$$
$$5x = 5 \qquad \text{Simplifying}$$
$$\frac{5x}{5} = \frac{5}{5} \qquad \text{Dividing by 5}$$
$$x = 1 \qquad \text{Simplifying}$$

Check: $\dfrac{2x - 2 = -3x + 3}{\begin{array}{c|c} 2 \cdot 1 - 2 & -3 \cdot 1 + 3 \\ 2 - 2 & -3 + 3 \\ 0 & 0 \qquad \text{TRUE} \end{array}}$

The solution is 1.   ◀

DO EXERCISE 9.

In Example 6, we used the addition principle to get all terms with a variable on one side and all numbers on the other side. Then we collected like terms and proceeded as before. If there are like terms on one side at the outset, they should be collected before proceeding.

▶ **EXAMPLE 7**   Solve: $6x + 5 - 7x = 10 - 4x + 3$.

$$6x + 5 - 7x = 10 - 4x + 3$$
$$-x + 5 = 13 - 4x \qquad \text{Collecting like terms}$$
$$4x - x + 5 = 13 - 4x + 4x \qquad \text{Adding } 4x$$
$$3x + 5 = 13 \qquad \text{Simplifying}$$
$$3x + 5 - 5 = 13 - 5 \qquad \text{Subtracting 5}$$
$$3x = 8 \qquad \text{Simplifying}$$
$$\frac{3x}{3} = \frac{8}{3} \qquad \text{Dividing by 3}$$
$$x = \frac{8}{3} \qquad \text{Simplifying}$$

The number $\frac{8}{3}$ checks, so it is the solution.   ◀

DO EXERCISES 10–12.

---

9. Solve: $7y + 5 = 2y + 10$.

Solve.

10. $5 - 2y = 3y - 5$

11. $7x - 17 + 2x = 2 - 8x + 15$

12. $3x - 15 = 5x + 2 - 4x$

ANSWERS ON PAGE A-3

13. Solve: $\dfrac{7}{8}x - \dfrac{1}{4} + \dfrac{1}{2}x = \dfrac{3}{4} + x.$

Clearing of Fractions and Decimals

We have stated that we generally use the addition principle first. There are, however, some situations in which it is to our advantage to use the multiplication principle first. Consider, for example,

$$\frac{1}{2}x = \frac{3}{4}.$$

If we multiply by 4 on both sides, we get $2x = 3$, which has no fractions. We have "cleared of fractions." Consider

$$2.3x = 5.$$

If we multiply by 10 on both sides, we get $23x = 50$, which has no decimal points. We have "cleared of decimals." The equations are then easier to solve. It is your choice whether to clear of the fractions or decimals, but doing so often eases computations.

In what follows, we use the multiplication principle first to "clear of" or "eliminate" fractions or decimals. For fractions, the number we multiply by is the **least common multiple of all the denominators.**

▶ **EXAMPLE 8**   Solve:

$$\frac{2}{3}x - \frac{1}{6} + \frac{1}{2}x = \frac{7}{6} + 2x.$$

The number 6 is the least common multiple of all the denominators. We multiply by 6 on both sides.

$$6\left(\tfrac{2}{3}x - \tfrac{1}{6} + \tfrac{1}{2}x\right) = 6\left(\tfrac{7}{6} + 2x\right) \quad \text{Multiplying by 6 on both sides}$$

$$6\cdot\tfrac{2}{3}x - 6\cdot\tfrac{1}{6} + 6\cdot\tfrac{1}{2}x = 6\cdot\tfrac{7}{6} + 6\cdot 2x \quad \text{Using the distributive laws (\emph{Caution!} Be sure to multiply all the terms by 6.)}$$

$$4x - 1 + 3x = 7 + 12x \quad \text{Simplifying. Note that the fractions are cleared.}$$

$$7x - 1 = 7 + 12x \quad \text{Collecting like terms}$$

$$7x - 12x = 7 + 1 \quad \text{Subtracting 12x and adding 1 to get all terms with variables on one side and all constant terms on the other side}$$

$$-5x = 8 \quad \text{Collecting like terms}$$

$$x = -\tfrac{8}{5} \quad \text{Multiplying by } -\tfrac{1}{5} \text{ or dividing by } -5$$

The number $-\frac{8}{5}$ checks and is the solution. ◀

DO EXERCISE 13.

Here is a procedure for solving the types of equation discussed in this section.

An Equation-Solving Procedure

1. **Multiply on both sides to clear the equation of fractions or decimals. (This is optional, but it can ease computations.)**
2. **Collect like terms on each side, if necessary.**
3. **Get all terms with variables on one side and all constant terms on the other side, using the addition principle.**
4. **Collect like terms again, if necessary.**
5. **Multiply or divide to solve for the variable, using the multiplication principle.**
6. **Check all possible solutions in the original equation.**

We illustrate this by repeating Example 4, but we clear the equation of decimals first.

▶ **EXAMPLE 9**   Solve: $16.3 - 7.2y = -8.18$.

The greatest number of decimal places in any one number is *two*. Multiplying by 100, which has *two* 0's, will clear of decimals.

$$100(16.3 - 7.2y) = 100(-8.18) \qquad \text{Multiplying by 100 on both sides}$$
$$100(16.3) - 100(7.2y) = 100(-8.18) \qquad \text{Using a distributive law}$$
$$1630 - 720y = -818 \qquad \text{Simplifying}$$
$$-720y = -818 - 1630 \qquad \text{Subtracting 1630 on both sides}$$
$$-720y = -2448 \qquad \text{Collecting like terms}$$
$$y = \frac{-2448}{-720} = 3.4 \qquad \text{Dividing by } -720 \text{ on both sides}$$

The number 3.4 checks and is the solution.   ◀

DO EXERCISE 14.

C   Equations Containing Parentheses

To solve certain kinds of equations that contain parentheses, we use the distributive laws to first remove the parentheses. Then we proceed as before.

▶ **EXAMPLE 10**   Solve: $4x = 2(12 - 2x)$.

$$4x = 2(12 - 2x)$$
$$4x = 24 - 4x \qquad \text{Using a distributive law to multiply and remove parentheses}$$
$$4x + 4x = 24 \qquad \text{Adding } 4x \text{ to get all } x\text{-terms on one side}$$
$$8x = 24 \qquad \text{Collecting like terms}$$
$$\frac{8x}{8} = \frac{24}{8} \qquad \text{Dividing by 8}$$
$$x = 3$$

14. Solve: $41.68 = 4.7 - 8.6y$.

ANSWER ON PAGE A-3

Solve.

15. $2(2y + 3) = 14$

16. $5(3x - 2) = 35$

Solve.

17. $3(7 + 2x) = 30 + 7(x - 1)$

18. $4(3 + 5x) - 4 = 3 + 2(x - 2)$

Check:

$$\begin{array}{c|c} \multicolumn{2}{c}{4x = 2(12 - 2x)} \\ \hline 4 \cdot 3 & 2(12 - 2 \cdot 3) \\ 12 & 2(12 - 6) \\ & 2 \cdot 6 \\ & 12 \qquad \text{TRUE} \end{array}$$

We use the rules for order of operations to carry out the calculations on each side of the equation.

The solution is 3. ◄

DO EXERCISES 15 AND 16.

► **EXAMPLE 11**   Solve: $2 - 5(x + 5) = 3(x - 2) - 1$.

$$2 - 5(x + 5) = 3(x - 2) - 1$$

$2 - 5x - 25 = 3x - 6 - 1$ Using the distributive laws to multiply and remove parentheses

$-5x - 23 = 3x - 7$ Simplifying

$-23 + 7 = 3x + 5x$ Adding $5x$ and 7 to get all x-terms on one side and all other terms on the other side

$-16 = 8x$ Simplifying

$-2 = x$ Dividing by 8

Check:

$$\begin{array}{c|c} \multicolumn{2}{c}{2 - 5(x + 5) = 3(x - 2) - 1} \\ \hline 2 - 5(-2 + 5) & 3(-2 - 2) - 1 \\ 2 - 5(3) & 3(-4) - 1 \\ 2 - 15 & -12 - 1 \\ -13 & -13 \qquad \text{TRUE} \end{array}$$

The solution is -2. ◄

Note that the solution of $-2 = x$ is -2, which is also the solution of $x = -2$.

DO EXERCISES 17 AND 18.

EXERCISE SET 2.3

a   Solve. Don't forget to check!

1. $5x + 6 = 31$ **2.** $3x + 6 = 30$ **3.** $8x + 4 = 68$ **4.** $7z + 9 = 72$

5. $4x - 6 = 34$ **6.** $6x - 3 = 15$ **7.** $3x - 9 = 33$ **8.** $5x - 7 = 48$

9. $7x + 2 = -54$ **10.** $5x + 4 = -41$ **11.** $-45 = 3 + 6y$

12. $-91 = 9t + 8$ **13.** $-4x + 7 = 35$ **14.** $-5x - 7 = 108$

15. $-7x - 24 = -129$ **16.** $-6z - 18 = -132$

b   Solve.

17. $5x + 7x = 72$ **18.** $4x + 5x = 45$ **19.** $8x + 7x = 60$

20. $3x + 9x = 96$ **21.** $4x + 3x = 42$ **22.** $6x + 19x = 100$

23. $-6y - 3y = 27$ **24.** $-4y - 8y = 48$ **25.** $-7y - 8y = -15$

ANSWERS

1. _____
2. _____
3. _____
4. _____
5. _____
6. _____
7. _____
8. _____
9. _____
10. _____
11. _____
12. _____
13. _____
14. _____
15. _____
16. _____
17. _____
18. _____
19. _____
20. _____
21. _____
22. _____
23. _____
24. _____
25. _____

Copyright © 1991 Addison-Wesley Publishing Co., Inc.

ANSWERS

26. _____

27. _____

28. _____

29. _____

30. _____

31. _____

32. _____

33. _____

34. _____

35. _____

36. _____

37. _____

38. _____

39. _____

40. _____

41. _____

42. _____

43. _____

44. _____

45. _____

46. _____

47. _____

48. _____

49. _____

50. _____

26. $-10y - 3y = -39$

27. $10.2y - 7.3y = -58$

28. $6.8y - 2.4y = -88$

29. $x + \dfrac{1}{3}x = 8$

30. $x + \dfrac{1}{4}x = 10$

31. $8y - 35 = 3y$

32. $4x - 6 = 6x$

33. $8x - 1 = 23 - 4x$

34. $5y - 2 = 28 - y$

35. $2x - 1 = 4 + x$

36. $5x - 2 = 6 + x$

37. $6x + 3 = 2x + 11$

38. $5y + 3 = 2y + 15$

39. $5 - 2x = 3x - 7x + 25$

40. $10 - 3x = 2x - 8x + 40$

41. $4 + 3x - 6 = 3x + 2 - x$

42. $5 + 4x - 7 = 4x - 2 - x$

43. $4y - 4 + y + 24 = 6y + 20 - 4y$

44. $5y - 7 + y = 7y + 21 - 5y$

Solve. Clear of fractions or decimals first.

45. $\dfrac{7}{2}x + \dfrac{1}{2}x = 3x + \dfrac{3}{2} + \dfrac{5}{2}x$

46. $\dfrac{7}{8}x - \dfrac{1}{4} + \dfrac{3}{4}x = \dfrac{1}{16} + x$

47. $\dfrac{2}{3} + \dfrac{1}{4}t = \dfrac{1}{3}$

48. $-\dfrac{3}{2} + x = -\dfrac{5}{6} - \dfrac{4}{3}$

49. $\dfrac{2}{3} + 3y = 5y - \dfrac{2}{15}$

50. $\dfrac{1}{2} + 4m = 3m - \dfrac{5}{2}$

51. $\dfrac{5}{3} + \dfrac{2}{3}x = \dfrac{25}{12} + \dfrac{5}{4}x + \dfrac{3}{4}$

52. $1 - \dfrac{2}{3}y = \dfrac{9}{5} - \dfrac{y}{5} + \dfrac{3}{5}$

53. $2.1x + 45.2 = 3.2 - 8.4x$

54. $0.96y - 0.79 = 0.21y + 0.46$

55. $1.03 - 0.62x = 0.71 - 0.22x$

56. $1.7t + 8 - 1.62t = 0.4t - 0.32 + 8$

57. $\dfrac{2}{7}x - \dfrac{1}{2}x = \dfrac{3}{4}x + 1$

58. $\dfrac{5}{16}y + \dfrac{3}{8}y = 2 + \dfrac{1}{4}y$

c Solve.

59. $3(2y - 3) = 27$

60. $4(2y - 3) = 28$

61. $40 = 5(3x + 2)$

62. $9 = 3(5x - 2)$

63. $2(3 + 4m) - 9 = 45$

64. $3(5 + 3m) - 8 = 88$

65. $5r - (2r + 8) = 16$

66. $6b - (3b + 8) = 16$

67. $6 - 2(3x - 1) = 2$

68. $10 - 3(2x - 1) = 1$

69. $5(d + 4) = 7(d - 2)$

70. $3(t - 2) = 9(t + 2)$

71. $8(2t + 1) = 4(7t + 7)$

72. $7(5x - 2) = 6(6x - 1)$

ANSWERS

51. _____

52. _____

53. _____

54. _____

55. _____

56. _____

57. _____

58. _____

59. _____

60. _____

61. _____

62. _____

63. _____

64. _____

65. _____

66. _____

67. _____

68. _____

69. _____

70. _____

71. _____

72. _____

Copyright © 1991 Addison-Wesley Publishing Co., Inc.

ANSWERS

73. _____

74. _____

75. _____

76. _____

77. _____

78. _____

79. _____

80. _____

81. _____

82. _____

83. _____

84. _____

85. _____

86. _____

87. _____

88. _____

89. _____

90. _____

91. _____

92. _____

93. _____

94. _____

95. _____

96. _____

97. _____

98. _____

99. _____

100. _____

101. _____

73. $3(r - 6) + 2 = 4(r + 2) - 21$

74. $5(t + 3) + 9 = 3(t - 2) + 6$

75. $19 - (2x + 3) = 2(x + 3) + x$

76. $13 - (2c + 2) = 2(c + 2) + 3c$

77. $\frac{1}{3}(6x + 24) - 20 = -\frac{1}{4}(12x - 72)$

78. $\frac{1}{4}(8y + 4) - 17 = -\frac{1}{2}(4y - 8)$

79. $2[4 - 2(3 - x)] - 1 = 4[2(4x - 3) + 7] - 25$

80. $5[3(7 - t) - 4(8 + 2t)] - 20 = -6[2(6 + 3t) - 4]$

81. $\frac{2}{3}(2x - 1) = 10$

82. $\frac{4}{5}(3x + 4) = 20$

83. $\frac{3}{4}\left(3x - \frac{1}{2}\right) - \frac{2}{3} = \frac{1}{3}$

84. $\frac{2}{3}\left(\frac{7}{8} - 4x\right) - \frac{5}{8} = \frac{3}{8}$

85. $0.7(3x + 6) = 1.1 - (x + 2)$

86. $0.9(2x + 8) = 20 - (x + 5)$

87. $a + (a - 3) = (a + 2) - (a + 1)$

88. $0.8 - 4(b - 1) = 0.2 + 3(4 - b)$

SKILL MAINTENANCE

89. Divide: $-22.1 \div 3.4$.

90. Factor: $7x - 21 - 14y$.

91. Use $<$ or $>$ for ▨ to write a true sentence: -15 ▨ -13.

92. Find $-(-x)$ when $x = -14$.

SYNTHESIS

Solve.

93. ▤ $0.008 + 9.62x - 42.8 = 0.944x + 0.0083 - x$.

94. $\frac{y - 2}{3} = \frac{2 - y}{5}$

95. $0 = y - (-14) - (-3y)$

96. $3x = 4x$

97. $\frac{5 + 2y}{3} = \frac{25}{12} + \frac{5y + 3}{4}$

98. ▤ $0.05y - 1.82 = 0.708y - 0.504$

99. $-2y + 5y = 6y$

100. $\frac{4 - 3x}{7} = \frac{2 + 5x}{49} - \frac{x}{14}$

101. Solve the equation $4x - 8 = 32$ by first using the addition principle. Then solve it by first using the multiplication principle.

2.4 Solving Problems

a Five Steps for Solving Problems

We have studied many new equation-solving tools in this chapter. We now apply them to problem solving. The following five-step strategy can be very helpful in solving problems.

> **Five Steps for Problem Solving in Algebra**
>
> 1. *Familiarize* yourself with the problem situation.
> 2. *Translate* the problem to an equation.
> 3. *Solve* the equation.
> 4. *Check* the answer in the original problem.
> 5. *State* the answer to the problem clearly.

Of the five steps, the most important is probably the first one: becoming familiar with the problem situation. Here are some hints for familiarization.

> **To familiarize yourself with the problem situation:**
> 1. If a problem is given in words, read it carefully.
> 2. Reread the problem, perhaps aloud. Try to verbalize the problem to yourself.
> 3. List the information given and the questions to be answered. Choose a variable (or variables) to represent the unknown and clearly state what the variable represents. Be descriptive! For example, let L = length, d = distance, and so on.
> 4. Find further information. Look up a formula on the inside frontcover of this book or in a reference book. Talk to a reference librarian or an expert in the field.
> 5. Make a table of the given information and the information you have collected. Look for patterns that may help in the translation to an equation.
> 6. Make a drawing and label it with known information. Also, indicate unknown information, using specific units if given.
> 7. Guess or estimate the answer.

▶ **EXAMPLE 1** A 72-in. board is cut into two pieces. One piece is twice as long as the other. How long are the pieces?

1. *Familiarize.* We first draw a picture. We let

$$x = \text{the length of the shorter piece.}$$

Then $\qquad 2x = \text{the length of the longer piece.}$

(We can also let y = the length of the longer piece. Then $\frac{1}{2}y$ = the length of the shorter piece. This, however, introduces fractions and will make the solution somewhat more difficult.)

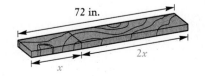

OBJECTIVE

After finishing Section 2.4, you should be able to:

a Solve problems by translating to equations.

FOR EXTRA HELP

Tape 6B   Tape 5A   MAC: 2
                                     IBM: 2

1. A 58-in. board is cut into two pieces. One piece is 2 in. longer than the other. How long are the pieces?

We can further familiarize ourselves with the problem by making some guesses. Suppose $x = 31$ in. Then $2x = 62$ in. and $x + 2x = 93$ in. This is not correct but does help us to become familiar with the problem.

2. *Translate.* From the figure, we can see that the lengths of the two pieces add up to 72 in. This gives us our translation.

$$\underbrace{\text{Length of one piece}}_{x} \text{ plus } \underbrace{\text{length of other}}_{2x} \text{ is } 72$$
$$x \quad + \quad 2x \quad = 72$$

3. *Solve.* We solve the equation:

$$x + 2x = 72$$
$$3x = 72 \qquad \text{Collecting like terms}$$
$$x = 24. \qquad \text{Dividing by 3}$$

4. *Check.* Do we have an answer to the *problem*? If one piece is 24 in. long, the other, to be twice as long, must be 48 in. long. The lengths of the pieces add up to 72 in. This checks.

5. *State.* One piece is 24 in. long, and the other is 48 in. long. ◀

DO EXERCISE 1.

▶ **EXAMPLE 2** Five plus three more than a number is nineteen. What is the number?

1. *Familiarize.* Let $x = $ the number. Then "three more than the number" translates to $x + 3$ and "5 more than $x + 3$" translates to $5 + (x + 3)$.

2. *Translate.* The familiarization leads us to the following translation:

$$\underbrace{\text{Five plus}}_{} \underbrace{\text{three more than a number}}_{} \underbrace{\text{is}}_{} \underbrace{\text{nineteen}}_{}$$
$$5 \quad + \quad (x + 3) \quad = \quad 19.$$

2. If 5 is subtracted from three times a certain number, the result is 10. What is the number?

3. *Solve.* We solve the equation:

$$5 + (x + 3) = 19$$
$$x + 8 = 19 \qquad \text{Collecting like terms}$$
$$x = 11. \qquad \text{Subtracting 8}$$

4. *Check.* Three more than 11 is 14. Adding 5 to 14, we get 19. This checks.

5. *State.* The number is 11. ◀

DO EXERCISE 2.

The following are examples of **consecutive integers**: 16, 17, 18, 19, 20; and $-31, -30, -29, -28$. Note that consecutive integers can be represented in the form $x, x + 1, x + 2$, and so on.

The following are examples of **consecutive even integers**: 16, 18, 20, 22, 24; and $-52, -50, -48, -46$. Note that consecutive even integers can be represented in the form $x, x + 2, x + 4$, and so on.

The following are examples of **consecutive odd integers**: 21, 23, 25, 27, 29; and $-71, -69, -67, -65$. Note that consecutive odd integers can be represented in the form $x, x + 2, x + 4$, and so on.

▶ **EXAMPLE 3** A book is opened. The sum of the page numbers on the facing pages is 233. Find the page numbers.

1. *Familiarize.* Look at your page numbers on the pages to which your book is now open. Note that they are consecutive positive integers. The numbers follow each other if we count by ones. Thus if we let $x =$ the smaller number, then $x + 1 =$ the larger number.

   To become more familiar with the problem, we can make a table. How do we get the entries to the table? First, we just guess a value for x. Then we find $x + 1$. Finally, we add the two numbers and see what happens. You might actually solve the problem this way, even though you need to practice using algebra.

x	$x + 1$	Sum of x and $x + 1$
14	15	29
24	25	49
102	103	205

2. *Translate.* We reword the problem and translate as follows.

   First integer + second integer = 233   *Rewording*

   x + $(x + 1)$ = 233   *Translating*

3. *Solve.* We solve the equation:

   $$x + (x + 1) = 233$$
   $$2x + 1 = 233 \quad \text{Collecting like terms}$$
   $$2x = 232 \quad \text{Subtracting 1}$$
   $$x = 116. \quad \text{Dividing by 2}$$

   If x is 116, then $x + 1$ is 117.

4. *Check.* Our possible answers are 116 and 117. These are consecutive integers. Their sum is 233, so the answers check in the *original problem*.

5. *State.* The page numbers are 116 and 117.   ◀

DO EXERCISE 3.

3. A book is opened. The sum of the page numbers on the facing pages is 457. Find the page numbers.

ANSWER ON PAGE A-3

4. Acme also rents compact cars at a daily rate of $34.95 plus 27 cents per mile. What mileage will allow the businessperson to stay within a budget of $100?

▶ **EXAMPLE 4**   Acme Rent-A-Car rents an intermediate-size car (such as a Chevrolet, Ford, or Plymouth) at a daily rate of $44.95 plus 29 cents a mile. A salesperson can spend $100 per day on car rental. How many miles can the person drive on the $100 budget?

1. *Familiarize.*   Suppose the businessperson drives 75 mi. Then the cost is

$$\underbrace{\text{Daily charge}}_{} \text{ plus } \underbrace{\text{mileage charge}}_{}$$

or

($44.95) plus (cost per mile) times (number of miles driven)
   $44.95   + $0.29   ·   75,

which is $44.95 + $21.75, or $66.70. This familiarizes us with the way in which a calculation is made. Note that we convert 29 cents to $0.29 so that we have the same units, dollars. Otherwise, we will not get a correct answer.

     Let m = the number of miles that can be driven on $100.

2. *Translate.*   We reword the problem and translate as follows.

Daily rate plus cost per mile times number of miles driven is cost
   $44.95   + $0.29   · m = $100

3. *Solve.*   We solve the equation:

$$44.95 + 0.29m = 100$$
$$100(44.95 + 0.29m) = 100(100) \quad \text{Multiplying by 100 on both sides to clear of the decimals}$$
$$100(44.95) + 100(0.29m) = 10,000 \quad \text{Using a distributive law}$$
$$4495 + 29m = 10,000$$
$$29m = 5505 \quad \text{Subtracting 4495}$$
$$m = \frac{5505}{29} \quad \text{Dividing by 29}$$
$$m \approx 189.8. \quad \text{Rounding to the nearest tenth. ``$\approx$'' means ``approximately equal to.''}$$

4. *Check.*   We check in the original problem. We multiply 189.8 by $0.29, getting $55.042. Then we add $55.042 to $44.95 and get $99.992, which is just about the $100 allotted.

5. *State.*   The person can drive about 189.8 mi on the car rental allotment of $100.   ◀

DO EXERCISE 4.

▶ **EXAMPLE 5** The state of Colorado is in the shape of a rectangle whose perimeter is 1300 mi. The length is 110 mi more than the width. Find the dimensions.

1. *Familiarize.* We first draw a picture. We let

$$w = \text{the width of the rectangle.}$$

Then $w + 110 = $ the length.

(We could also let $l = $ the length and $l - 110 = $ the width.)

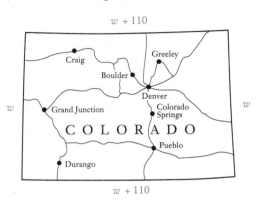

The perimeter *P* of a rectangle is the distance around it and is given by the formula $2l + 2w = P$, where $l = $ the length and $w = $ the width.

2. *Translate.* To translate the problem, we substitute $w + 110$ for l and 1300 for *P*, as follows:

$$2l + 2w = P$$
$$2(w + 110) + 2w = 1300.$$

3. *Solve.* We solve the equation:

$$2(w + 110) + 2w = 1300$$
$$2w + 220 + 2w = 1300$$
$$4w + 220 = 1300$$
$$4w = 1080$$
$$w = 270.$$

Possible dimensions are $w = 270$ mi and $w + 110 = 380$ mi.

4. *Check.* If the width is 270 mi and the length is 110 mi + 270 mi, or 380 mi, the perimeter is 2(380 mi) + 2(270 mi), or 1300 mi. This checks.

5. *State.* The width is 270 mi, and the length is 380 mi.   ◀

DO EXERCISE 5.

▶ **EXAMPLE 6** The second angle of a triangle is twice as large as the first. The measure of the third angle is 20° greater than that of the first angle. How large are the angles?

1. *Familiarize.* We draw a picture. We let.

measure of 1st angle = *x*.

Then   measure of 2nd angle = 2*x*,

and   measure of 3rd angle = *x* + 20.

5. A standard-sized rug has a perimeter of 42 ft. The length is 3 ft more than the width. Find the dimensions of the rug.

ANSWER ON PAGE A-3

6. The second angle of a triangle is three times as large as the first. The third angle measures 30° more than the first angle. Find the measures of the angles.

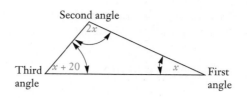

Second angle

Third angle $x + 20$ $2x$ x First angle

2. *Translate.* To translate we need to recall a geometric fact. (You might, as part of step 1, look it up in a geometry book or in the list of formulas at the back of this book.) The measures of the angles of any triangle add up to 180°.

$$\underbrace{\text{Measure of}}_{x} + \underbrace{\text{Measure of}}_{2x} + \underbrace{\text{Measure of}}_{(x+20)} = 180°$$

$$\text{first angle} \qquad \text{second angle} \qquad \text{third angle}$$

$$x \quad + \quad 2x \quad + \quad (x + 20) \quad = 180$$

3. *Solve.* We solve:

$$x + 2x + (x + 20) = 180$$
$$4x + 20 = 180$$
$$4x = 160$$
$$x = 40.$$

Possible measures for the angles are as follows:

First angle: $x = 40°$;

Second angle: $2x = 2(40) = 80°$;

Third angle: $x + 20 = 40 + 20 = 60°$.

4. *Check.* Consider 40°, 80°, and 60°. The second is twice the first, and the third is 20° greater than the first. The sum is 180°. These numbers check.

5. *State.* The measures of the angles are 40°, 80°, and 60°.   ◄

CAUTION!   Units are important in answers.

DO EXERCISE 6.

We close this section with some other tips to aid you in problem solving.

Problem-Solving Tips

1. To be good at problem solving, work lots of problems.

2. Look for patterns when solving problems. Each time you study an example in a text, you may observe a pattern for problems that you will see later in the exercise sets or some other practical situation.

3. When translating to an equation, or some other mathematical language, consider the dimensions of the variables and constants in the equation. The variables that represent length should all be in the same unit, those that represent money should all be in dollars or all in cents, and so on.

EXERCISE SET 2.4

a Solve.

1. What number added to 60 is 112?

2. Seven times what number is 2233?

3. When 42 is multiplied by a number, the result is 2352. Find the number.

4. When 345 is added to a number, the result is 987. Find the number.

5. A game board has 64 squares. If you win 35 squares and your opponent wins the rest, how many does your opponent win?

6. A consultant charges $80 an hour. How many hours did the consultant work to make $53,400?

7. In a recent year, the cost of four 12-oz boxes of Post^R Oat Flakes was $7.96. How much did one box cost?

8. The total amount spent on women's blouses in a recent year was $6.5 billion. This was $0.2 billion more than was spent on women's dresses. How much was spent on women's dresses?

9. When 18 is subtracted from six times a certain number, the result is 96. What is the number?

10. When 28 is subtracted from five times a certain number, the result is 232. What is the number?

11. If you double a number and then add 16, you get two fifths of the original number. What is the original number?

12. If you double a number and then add 85, you get three fourths of the original number. What is the original number?

ANSWERS

1. _____

2. _____

3. _____

4. _____

5. _____

6. _____

7. _____

8. _____

9. _____

10. _____

11. _____

12. _____

ANSWERS

13. _____

14. _____

15. _____

16. _____

17. _____

18. _____

19. _____

20. _____

21. _____

22. _____

23. _____

24. _____

13. If you add two fifths of a number to the number itself, you get 56. What is the number?

14. If you add one third of a number to the number itself, you get 48. What is the number?

15. A 180-m rope is cut into three pieces. The second piece is twice as long as the first. The third piece is three times as long as the second. How long is each piece of rope?

16. A 480-m wire is cut into three pieces. The second piece is three times as long as the first. The third piece is four times as long as the second. How long is each piece?

17. The sum of the page numbers on the facing pages of a book is 73. What are the page numbers?

18. The sum of the page numbers on the facing pages of a book is 81. What are the page numbers?

19. The sum of two consecutive even integers is 114. What are the integers?

20. The sum of two consecutive even integers is 106. What are the integers?

21. The sum of three consecutive integers is 108. What are the integers?

22. The sum of three consecutive integers is 126. What are the integers?

23. The sum of three consecutive odd integers is 189. What are the integers?

24. Three consecutive integers are such that the first plus one-half the second plus three less than twice the third is 964. Find the integers.

25. The top of the John Hancock Building in Chicago is a rectangle whose length is 60 ft more than the width. The perimeter is 520 ft. Find the width and the length of the rectangle. Find the area of the rectangle.

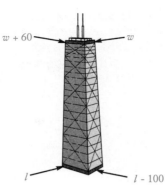

26. The ground floor of the John Hancock Building is a rectangle whose width is 100 ft less than the length. The perimeter is 860 ft. Find the width and the length of the rectangle. Find the area of the rectangle.

27. The perimeter of a standard-sized piece of typewriter paper is 99 cm. The width is 6.3 cm less than the length. Find the length and the width.

28. The perimeter of the state of Wyoming is 1280 mi. The width is 90 mi less than the length. Find the width and the length.

29. The second angle of a triangle is four times as large as the first. The third angle is 45° less than the sum of the other two angles. Find the measure of the first angle.

30. The second angle of a triangle is three times as large as the first. The third angle is 25° less than the sum of the other two angles. Find the measure of the first angle.

31. Badger Rent-A-Car rents an intermediate-size car at a daily rate of $34.95 plus 10 cents per mile. A businessperson is allotted $80 for car rental. How many miles can the businessperson travel on the $80 budget?

32. Badger also rents compact cars at $43.95 plus 10 cents per mile. A businessperson has a car rental allotment of $90. How many miles can the businessperson travel on the $90 budget?

ANSWERS

33. _____

34. _____

35. _____

36. _____

37. _____

38. _____

39. _____

40. _____

41. _____

42. _____

43. _____

44. _____

33. The second angle of a triangle is three times as large as the first. The measure of the third angle is 40° greater than that of the first angle. How large are the angles?

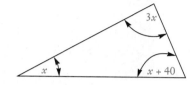

34. One angle of a triangle is 32 times as large as another. The measure of the third angle is 10° greater than that of the smallest angle. How large are the angles?

35. The equation

$$R = -0.028t + 20.8$$

can be used to predict the world record in the 200-m dash, where R stands for the record in seconds and t for the number of years since 1920. In what year will the record be 18.0 sec?

36. The equation

$$F = \frac{1}{4}N + 40$$

can be used to determine temperatures given how many times a cricket chirps per minute, where F represents temperature in degrees and N the number of chirps per minute. Determine the number of chirps per minute necessary in order for the temperature to be 80°.

SYNTHESIS

37. Abraham Lincoln's 1863 Gettysburg Address refers to the year 1776 as "Four *score* and seven years ago." Write an equation and find what number a score represents.

38. If the daily rental for a car is $18.90 plus a certain price per mile and a person must drive 190 mi and still stay within a $55.00 budget, what is the highest price per mile that the person can afford?

39. A student scored 78 on a test that had 4 seven-point fill-ins and 24 three-point multiple-choice questions. The student had one fill-in wrong. How many multiple-choice questions did the student answer correctly?

40. The width of a rectangle is three fourths of the length. The perimeter of the rectangle becomes 50 cm when the length and the width are each increased by 2 cm. Find the length and the width.

41. Apples are collected in a basket for six people. One third, one fourth, one eighth, and one fifth are given to four people, respectively. The fifth person gets ten apples with one apple remaining for the sixth person. Find the original number of apples in the basket.

42. A student has an average score of 82 on three tests. The student's average score on the first two tests is 85. What was the score on the third test?

43. A storekeeper goes to the bank to get $10 worth of change. The storekeeper requests twice as many quarters as half dollars, twice as many dimes as quarters, three times as many nickels as dimes, and no pennies or dollars. How many of each coin did the storekeeper get?

44. ▦ The area of this triangle is 2.9047 in². Find x.

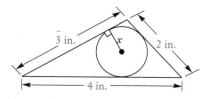

2.5 Solving Percent Problems

a Many problems involve percents. We can use our knowledge of equations and the problem-solving process to solve such problems.

▶ **EXAMPLE 1** What percent of 45 is 15?

1. *Familiarize.* This type of problem is stated so explicitly that we can proceed directly to the translation. We first let x = the percent.

2. *Translate.* We translate as follows:

$$\underbrace{\text{What percent}}_{x\%} \text{ of } \underbrace{45}_{\cdot\ 45} \underbrace{\text{is}}_{=} \underbrace{15?}_{15.}$$

3. *Solve.* We solve the equation:

$$x\% \cdot 45 = 15$$
$$x \times 0.01 \times 45 = 15$$
$$x(0.45) = 15$$
$$x = \frac{15}{0.45} \qquad \text{Dividing by 0.45}$$
$$x = \frac{15}{0.45} \times \frac{100}{100} = \frac{1500}{45}$$
$$x = 33\frac{1}{3}.$$

4. *Check.* We check by finding $33\frac{1}{3}\%$ of 45:

$$33\frac{1}{3}\% \cdot 45 = \frac{1}{3} \cdot 45 = 15. \qquad \text{See Table 1 at the back of the book.}$$

5. *State.* The answer is $33\frac{1}{3}\%$. ◀

DO EXERCISES 1 AND 2.

▶ **EXAMPLE 2** 3 is 16 percent of what?

1. *Familiarize.* This problem is stated so explicitly that we can proceed directly to the translation. We let y = the number that we are taking 16% of.

2. *Translate.* The translation is as follows:

$$\underbrace{3}_{3} \underbrace{\text{is}}_{=} \underbrace{16}_{16} \underbrace{\text{percent}}_{\%} \underbrace{\text{of}}_{\cdot} \underbrace{\text{what?}}_{y}$$

3. *Solve.* We solve the equation:

$$3 = 16\% \cdot y$$
$$3 = 0.16y$$
$$0.16y = 3$$
$$y = \frac{3}{0.16} \qquad \text{Dividing by 0.16}$$
$$y = 18.75.$$

Solve.

1. What percent of 50 is 16?

2. 15 is what percent of 60?

Solve.

3. 45 is 20 percent of what?

4. 120 percent of what is 60?

Solve.

5. What is 23% of 48?

6. The area of Arizona is 19% of the area of Alaska. The area of Alaska is 586,400 mi² (square miles). What is the area of Arizona?

ANSWERS ON PAGE A-3

4. *Check.*   We check by finding 16% of 18.75:

$$16\% \times 18.75 = 0.16 \times 18.75 = 3.$$

5. *State.*   The answer is 18.75.   ◄

DO EXERCISES 3 AND 4.

Perhaps you have noticed that to handle percents in problems such as those in Examples 1 and 2, you can convert to decimal notation before continuing.

▶ **EXAMPLE 3**   Blood is 90% water. The average adult has 5 quarts (qt) of blood. How much water is in the average adult's blood?

1. *Familiarize.*   We first write down the given information.

Blood: 90% water

Adult: Body contains 5 qt of blood.

We want to find the amount of water that is in the blood of an adult. We let $x =$ the amount of water in the blood of an adult. It seems reasonable that we take 90% of 5. This leads us to the rewording and translating of the problem.

2. *Translate.*

Rewording: 90% of 5 is what?

Translating: 90% · 5 = x

3. *Solve.*   We solve the equation:

$$90\% \cdot 5 = x$$
$$0.90 \times 5 = x \quad \text{Converting 90\% to decimal notation}$$
$$4.5 = x.$$

4. *Check.*   The check is actually the computation we use to solve the equation:

$$90\% \cdot 5 = 0.90 \times 5 = 4.5.$$

5. *State.*   The answer is that there are 4.5 qt of water in an adult who has 5 qt of blood.   ◄

DO EXERCISES 5 AND 6.

▶ **EXAMPLE 4**   An investment is made at 8% simple interest for 1 year. It grows to $783. How much was originally invested (the principal)?

1. *Familiarize.*   Suppose that $100 was invested. Recalling the formula for simple interest, $I = Prt$, we know that the interest for 1 year on $100 at 8% simple interest is given by $I = \$100 \cdot 8\% \cdot 1 = \$8$. Then, at the end of the year, the *amount* in the account is found by adding principal and interest:

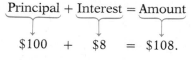

$$\text{Principal} + \text{Interest} = \text{Amount}$$
$$\$100 \quad + \quad \$8 \quad = \quad \$108.$$

In this problem we are working backward. We are trying to find the principal, which is the original investment. We let $x =$ the principal.

2. *Translate.* We reword the problem and then translate.

Rewording: $\underbrace{\text{Principal}} + \underbrace{\text{Interest}} = \underbrace{\text{Amount}}$

Translating: $ x + 8\%x = 783$ Interest is 8% of the principal.

3. *Solve.* We solve the equation:

$$x + 8\%x = 783$$
$$x + 0.08x = 783 \qquad \text{Converting}$$
$$1x + 0.08x = 783 \qquad \text{Identity property of 1}$$
$$1.08x = 783 \qquad \text{Collecting like terms}$$
$$x = \frac{783}{1.08} \qquad \text{Dividing by 1.08}$$
$$x = 725.$$

4. *Check.* We check by taking 8% of $725 and adding it to $725:

$$8\% \times \$725 = 0.08 \times 725 = \$58.$$

Then $725 + $58 = $783, so $725 checks.

5. *State.* The original investment was $725.   ◀

DO EXERCISE 7.

▶ **EXAMPLE 5** The price of an automobile was decreased to a sale price of $13,559. This was a 9% reduction. What was the former price?

BRAND–NEW 1992 CAVIAT
9% OFF
REDUCED FROM X?
TO ONLY $13,559

1. *Familiarize.* Suppose that the former price was $16,000. A 9% reduction can be found by taking 9% of $16,000, that is,

$$9\% \text{ of } \$16,000 = 0.09(\$16,000) = \$1440.$$

Then the sale price is found by subtracting the amount of the reduction:

$$\underbrace{\text{Former price}} - \underbrace{\text{Reduction}} = \underbrace{\text{Sale price}}$$

$$ \$16,000 - \$1440 = \$14,560.$$

Our guess of $16,000 was too high, but we are becoming familiar with the problem. We let $x =$ the former price of the automobile. It is reduced by 9%. So the sale price $= x - 9\%x$.

2. *Translate.* We reword and then translate:

$$\underbrace{\text{Former price}} - \underbrace{\text{Reduction}} = \underbrace{\text{Sale price}} \qquad \text{Rewording}$$

$$ x - 9\%x = \$13,559. \qquad \text{Translating}$$

7. An investment is made at 7% simple interest for 1 year. It grows to $8988. How much was originally invested (the principal)?

ANSWER ON PAGE A-3

8. The price of a suit was decreased to a sale price of $526.40. This was a 20% reduction. What was the former price?

3. *Solve.*   We solve the equation:

$$x - 9\%x = 13{,}559$$
$$x - 0.09x = 13{,}559 \quad \text{Converting to decimal notation}$$
$$1x - 0.09x = 13{,}559$$
$$(1 - 0.09)x = 13{,}559 \quad \text{Factoring out the } x$$
$$0.91x = 13{,}559 \quad \text{Collecting like terms}$$
$$x = \frac{13{,}559}{0.91} \quad \text{Dividing by 0.91}$$
$$x = 14{,}900.$$

4. *Check.*   To check, we find 9% of $14,900 and subtract:

$$9\% \times \$14{,}900 = 0.09 \times \$14{,}900 = \$1341$$
$$\$14{,}900 - \$1341 = \$13{,}559.$$

Since we get the sale price, $13,559, the $14,900 checks.

5. *State.*   The former price was $14,900.   ◀

This problem is easy with algebra. Without algebra it is not. A common error in a problem like this is to take 9% of the sale price and subtract or add. Note that 9% of the original price is not equal to 9% of the sale price!

DO EXERCISE 8.

NAME   SECTION   DATE

EXERCISE SET 2.5

a Solve.

1. What percent of 68 is 17?

2. What percent of 75 is 36?

3. What percent of 125 is 30?

4. What percent of 300 is 57?

5. 45 is 30% of what number?

6. 20.4 is 24% of what number?

7. 0.3 is 12% of what number?

8. 7 is 175% of what number?

9. What number is 65% of 840?

10. What number is 1% of one million?

11. What percent of 80 is 100?

12. What percent of 10 is 205?

13. What is 2% of 40?

14. What is 40% of 2?

15. 2 is what percent of 40?

16. 40 is 2% of what number?

17. The FBI annually receives 16,000 applicants to become agents. It accepts 600 of these applicants. What percent does it accept?

18. The U.S. Postal Service reports that we open and read 78% of the junk mail that we receive. A business sends out 9500 advertising brochures. How many of them can it expect to be opened and read?

19. It has been determined by sociologists that 17% of the population is left-handed. Each week 160 men enter a tournament conducted by the Professional Bowlers Association. How many of them would you expect to be left-handed? Round to the nearest one.

20. In a medical study, it was determined that if 800 people kiss someone else who has a cold, only 56 will actually catch the cold. What percent is this?

21. On a test of 88 items, a student got 76 correct. What percent were correct?

22. A baseball player had 13 hits in 25 times at bat. What percent were hits?

1. _____

2. _____

3. _____

4. _____

5. _____

6. _____

7. _____

8. _____

9. _____

10. _____

11. _____

12. _____

13. _____

14. _____

15. _____

16. _____

17. _____

18. _____

19. _____

20. _____

21. _____

22. _____

23. The cost of a Navy F-14 Tomcat jet is $24 million. Eventually each jet will need to be renovated at 2% of its original cost. How much is that cost?

24. The cost of a Navy A-6 Intruder attack bomber is $30 million. Eventually each bomber will need to be renovated at 2% of its original cost. How much is that cost?

25. A family spent $208 one month for food. This was 26% of its income. What was their monthly income?

26. The sales tax rate in New York City is 8%. How much would be charged on a purchase of $428.86? How much will the total cost of the purchase be?

27. Water volume increases 9% when water freezes. If 400 cubic centimeters of water is frozen, how much will its volume increase? What will be the volume of the ice?

28. An investment is made at 9% simple interest for 1 year. It grows to $8502. How much was originally invested?

29. Money is borrowed at 10% simple interest. After 1 year, $7194 pays off the loan. How much was originally borrowed?

30. Due to inflation the price of an item increased 12¢. This was an 8% increase. What was the old price? the new price?

31. After a 40% price reduction, a shirt is on sale at $9.60. What was the original price (that is, the price before reduction)?

32. After a 34% price reduction, a blouse is on sale at $9.24. What was the original price?

SYNTHESIS

33. In a basketball league, the Falcons won 15 of their first 20 games. How many more games will they have to play where they win only half the time in order to win 60% of the total number of games?

34. One number is 25% of another. The larger number is 12 more than the smaller. What are the numbers?

35. In one city, a sales tax of 9% was added to the price of gasoline as registered on the pump. Suppose a driver asked for $10 worth of gas. The attendant filled the tank until the pump read $9.10 and charged the driver $10. Something was wrong. Use algebra to correct the error.

36. The weather report is "a 60% chance of showers during the day, 30% tonight, and 5% tomorrow morning." What are the chances it won't rain during the day? tonight? tomorrow morning?

37. Twenty-seven people make a certain amount of money at a sale. What percentage does each receive if they share the profit equally?

38. If x is 160% of y, y is what percent of x?

39. Which of the following is higher, if either?

A. x is increased by 25%; then that amount is decreased by 25%.
B. x is decreased by 25%; then that amount is increased by 25%.

Explain.

2.6 Formulas

a Solving Formulas

A **formula** is a ''recipe'' for doing a certain type of calculation. Formulas are often given as equations. Here is an example of a formula that has to do with weather: $M = \frac{1}{5}n$. You see a flash of lightning. After a few seconds you hear the thunder associated with that flash. How far away was the lightning?

Your distance from the storm is M miles. You can find that distance by counting the number of seconds n that it takes the sound of the thunder to reach you and then multiplying by $\frac{1}{5}$.

▶ **EXAMPLE 1** Consider the formula $M = \frac{1}{5}n$. It takes 10 sec for the sound of thunder to reach you after you have seen a flash of lightning. How far away is the storm?

We substitute 10 for n and calculate M: $M = \frac{1}{5}n = \frac{1}{5}(10) = 2$. The storm is 2 mi away. ◀

DO EXERCISE 1.

Suppose that we know how far we are from the storm and want to calculate the number of seconds it would take the sound of the thunder to reach us. We could substitute a number for M, say 2, and solve for n:

$$2 = \frac{1}{5}n$$
$$10 = n. \quad \text{Multiplying by 5}$$

However, if we wanted to do this repeatedly, it might be easier to solve for n by getting it alone on one side. We ''solve'' the formula for n.

▶ **EXAMPLE 2** Solve for n: $M = \frac{1}{5}n$.

We have

$$M = \frac{1}{5}n \quad \text{We want this letter alone.}$$
$$5 \cdot M = 5 \cdot \frac{1}{5}n \quad \text{Multiplying on both sides by 5}$$
$$5M = n.$$

In the above situation for $M = 2$, $n = 5(2)$, or 10. ◀

DO EXERCISE 2.

To see how the addition and multiplication principles apply to formulas, compare the following.

A. Solve.

$$5x + 2 = 12$$
$$5x = 12 - 2$$
$$5x = 10$$
$$x = \frac{10}{5} = 2$$

B. Solve.

$$5x + 2 = 12$$
$$5x = 12 - 2$$
$$x = \frac{12 - 2}{5}$$

C. Solve for x.

$$ax + b = c$$
$$ax = c - b$$
$$x = \frac{c - b}{a}$$

In (A) we solved as we did before. In (B) we did not carry out the calculations. In (C) we could not carry out the calculations because we had unknown numbers.

OBJECTIVE

After finishing Section 2.6, you should be able to:

a Solve a formula for a specified letter.

FOR EXTRA HELP

Tape 7A   Tape 5B   MAC: 2
                                  IBM: 2

1. Suppose that it takes the sound of thunder 14 sec to reach you. How far away is the storm?

2. Solve for I: $E = IR$.

   (This is a formula from electricity relating voltage E, current I, and resistance R.)

ANSWERS ON PAGE A-3

3. Solve for D: $C = \pi D$.

(This is a formula for the circumference C of a circle of diameter D.)

4. Solve for c:

$$A = \frac{a + b + c + d}{4}.$$

5. Solve for I:

$$E = \frac{9R}{I}.$$

(This is a formula for computing the earned run average E of a pitcher who has given up R earned runs in I innings of pitching.)

▶ **EXAMPLE 3**   Solve for r: $C = 2\pi r$.

This is a formula for the circumference C of a circle of radius r.

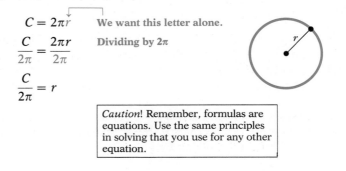

$C = 2\pi r$ **We want this letter alone.**

$$\frac{C}{2\pi} = \frac{2\pi r}{2\pi} \qquad \text{\textbf{Dividing by }} 2\pi$$

$$\frac{C}{2\pi} = r$$

> *Caution!* Remember, formulas are equations. Use the same principles in solving that you use for any other equation.

◀

DO EXERCISE 3.

With the formulas in this section, we can use a procedure like that described in Section 2.3.

> **To solve a formula for a given letter, identify the letter, and:**
> 1. **Multiply on both sides to clear of fractions or decimals, if that is needed.**
> 2. **Collect like terms on each side, if necessary.**
> 3. **Get all terms with the letter to be solved for on one side of the equation and all other terms on the other side.**
> 4. **Collect like terms again, if necessary.**
> 5. **Solve for the letter in question.**

▶ **EXAMPLE 4**   Solve for a: $A = \dfrac{a + b + c}{3}$.

This is a formula for the average A of three numbers a, b, and c.

$$A = \frac{a + b + c}{3} \qquad \text{\textbf{We want the letter } } a \text{ \textbf{alone.}}$$

$$3A = a + b + c \qquad \text{\textbf{Multiplying by 3 to clear of the fraction}}$$

$$3A - b - c = a$$

◀

DO EXERCISE 4.

▶ **EXAMPLE 5**   Solve for C: $Q = \dfrac{100M}{C}$.

This is a formula used in psychology for intelligence quotient Q, where M is mental age and C is chronological, or actual, age.

$$Q = \frac{100M}{C} \qquad \text{\textbf{We want the letter } } C \text{ \textbf{alone.}}$$

$$CQ = 100M \qquad \text{\textbf{Multiplying by } } C \text{ \textbf{ to clear of the fraction}}$$

$$C = \frac{100M}{Q} \qquad \text{\textbf{Dividing by } } Q$$

◀

DO EXERCISE 5.

NAME   SECTION   DATE

EXERCISE SET 2.6

a Solve for the given letter.

1. $A = bh$, for b
(Area of a parallelogram with base b
and height h)

2. $A = bh$, for h

3. $d = rt$, for r
(A distance formula, where d is
distance, r is speed, and t is time)

4. $d = rt$, for t

5. $I = Prt$, for P
(Simple-interest formula, where I is
interest, P is principal, r is interest
rate, and t is time)

6. $I = Prt$, for t

7. $F = ma$, for a
(A physics formula, where F is force,
m is mass, and a is acceleration)

8. $F = ma$, for m

9. $P = 2l + 2w$, for w
(Perimeter of a rectangle of length l
and width w)

10. $P = 2l + 2w$, for l

11. $A = \pi r^2$, for r^2
(Area of a circle with radius r)

12. $A = \pi r^2$, for π

13. $A = \dfrac{1}{2}bh$, for b
(Area of a triangle with base b and
height h)

14. $A = \dfrac{1}{2}bh$, for h

15. $E = mc^2$, for m
(A relativity formula)

16. $E = mc^2$, for c^2

17. $Q = \dfrac{c + d}{2}$, for d

18. $Q = \dfrac{p - q}{2}$, for p

19. $A = \dfrac{a + b + c}{3}$, for b

20. $A = \dfrac{a + b + c}{3}$, for c

ANSWERS

1. _____

2. _____

3. _____

4. _____

5. _____

6. _____

7. _____

8. _____

9. _____

10. _____

11. _____

12. _____

13. _____

14. _____

15. _____

16. _____

17. _____

18. _____

19. _____

20. _____

ANSWERS

21. _____

22. _____

23. _____

24. _____

25. _____

26. _____

27. _____

28. _____

29. _____

30. _____

31. _____

32. _____

33. _____

34. _____

35. _____

36. _____

37. _____

38. _____

39. _____

40. _____

41. _____

42. _____

21. $v = \dfrac{3k}{t}$, for t

22. $P = \dfrac{ab}{c}$, for c

23. $Ax + By = C$, for y

24. $Ax + By = C$, for x

25. $A = \dfrac{1}{2}ah + \dfrac{1}{2}bh$, for b; for h

26. $A = \dfrac{1}{2}ah - \dfrac{1}{2}bh$, for a; for h

27. $Q = 3a + 5ca$, for a

28. $P = 4m + 7mn$, for m

29. The formula
$$H = \frac{D^2N}{2.5}$$
is used to find the horsepower H of an N-cylinder engine. Solve for D^2.

30. Solve for N:
$$H = \frac{D^2N}{2.5}.$$

31. The area of a sector of a circle is given by
$$A = \frac{\pi r^2 S}{360},$$
where r is the radius and S is the angle measure of the sector. Solve for S.

32. Solve for r^2:
$$A = \frac{\pi r^2 S}{360}.$$

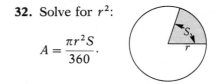

33. The formula
$$R = -0.0075t + 3.85$$
can be used to estimate the world record in the 1500-m run t years after 1930. Solve for t.

34. The formula
$$F = \frac{9}{5}C + 32$$
can be used to convert from Celsius, or Centigrade, temperature C to Fahrenheit temperature F. Solve for C.

SKILL MAINTENANCE

35. Convert to decimal notation: $\dfrac{23}{25}$.

36. Add: $-23 + (-67)$.

37. Subtract: $-45.8 - (-32.6)$.

38. Remove parentheses and simplify:
$$4a - 8b - 5(5a - 4b).$$

SYNTHESIS

39. In $A = lw$, l and w both double. What is the effect on A?

40. In $P = 2a + 2b$, P doubles. Do a and b necessarily both double?

41. In $A = \frac{1}{2}bh$, b increases by 4 units and h does not change. What happens to A?

42. Solve for F:
$$D = \frac{1}{E + F}.$$

2.7 Solving Inequalities

We now extend our equation-solving principles to the solving of inequalities.

a Solutions of Inequalities

In Section 1.2, we defined the symbols > (greater than), < (less than), ≥ (greater than or equal to), and ≤ (less than or equal to). For example, $3 \leq 4$ and $3 \leq 3$ are both true, but $-3 \leq -4$ and $0 \geq 2$ are both false.

An **inequality** is a number sentence with >, <, ≥, or ≤ as its verb— for example,

$$-4 > t, \quad x < 3, \quad 2x + 5 \geq 0, \quad \text{and} \quad -3y + 7 \leq -8.$$

Some replacements for a variable in an inequality make it true and some make it false. A replacement that makes an inequality true is called a **solution.** The set of all solutions is called the **solution set.** When we have found the set of all solutions of an inequality, we say that we have **solved** the inequality.

▶ **EXAMPLES** Determine whether the number is a solution of $x < 2$.

1. -2 Since $-2 < 2$ is true, -2 is a solution.

2. 2 Since $2 < 2$ is false, 2 is not a solution. ◀

▶ **EXAMPLES** Determine whether the number is a solution of $y \geq 6$.

3. 6 Since $6 \geq 6$ is true, 6 is a solution.

4. -4 Since $-4 \geq 6$ is false, -4 is not a solution. ◀

DO EXERCISES 1 AND 2.

b Graphs of Inequalities

Some solutions of $x < 2$ are 0.45, -8.9, $-\pi$, and so on. In fact, there are infinitely many real numbers that are solutions. Because we cannot list them all individually, it is helpful to make a drawing that represents all the solutions.

A **graph** of an inequality is a drawing that represents its solutions. An inequality in one variable can be graphed on a number line. An inequality in two variables can be graphed on a coordinate plane.

We first graph inequalities in one variable on a number line.

▶ **EXAMPLE 5** Graph: $x < 2$.

The solutions of $x < 2$ are those numbers less than 2. They are shown on the graph by shading all points to the left of 2. The open circle at 2 indicates that 2 is not part of the graph.

◀

OBJECTIVES

After finishing Section 2.7, you should be able to:

a Determine whether a given number is a solution of an inequality.

b Graph an inequality on a number line.

c Solve inequalities using the addition principle.

d Solve inequalities using the multiplication principle.

e Solve inequalities using the addition and multiplication principles together.

FOR EXTRA HELP

Tape 7B   Tape 5B   MAC: 2
IBM: 2

Determine whether each number is a solution of the inequality.

1. $x > 3$

  a) 2   **b)** 0

  c) -5 **d)** 15

  e) 3

2. $x \leq 6$

  a) 6   **b)** 0

  c) -4 **d)** 25

  e) -6

ANSWERS ON PAGE A-3

Graph.

3. $x < 4$

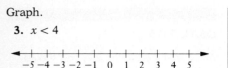

4. $y \geq -2$

5. $-2 \leq x < 4$

▶ **EXAMPLE 6**   Graph: $y \geq -3$.

The solutions of $y \geq -3$ are shown on the number line by shading the point for -3 and all points to the right of -3. The closed circle at -3 indicates that -3 *is* part of the graph.

◀

▶ **EXAMPLE 7**   Graph: $-2 < x \leq 3$.

The inequality $-2 < x \leq 3$ is read "-2 is less than x *and* x is less than or equal to 3," or "x is greater than -2 and less than or equal to 3." To be a solution of this inequality, a number must be a solution of both $-2 < x$ and $x \leq 3$. The number 1 is a solution, as are -0.5, 2, 2.5, and 3. The solution set is graphed as follows:

The open circle at -2 means that -2 is not part of the graph. The closed circle at 3 means that 3 is part of the graph. The other solutions are shaded.

◀

DO EXERCISES 3–5.

c   Solving Inequalities Using the Addition Principle

Consider the true inequality

$$3 < 7.$$

If we add 2 on both sides, we get another true inequality:

$$3 + 2 < 7 + 2, \quad \text{or} \quad 5 < 9.$$

Similarly, if we add -3 on both sides, we get another true inequality:

$$3 + (-3) < 7 + (-3), \quad \text{or} \quad 0 < 4.$$

> The Addition Principle for Inequalities
>
> **If the same number is added on both sides of a true inequality, we get another true inequality.**

Let's see how we use the addition principle to solve inequalities.

▶ **EXAMPLE 8**   Solve: $x + 2 > 8$. Then graph.

We use the addition principle, subtracting 2 on both sides:

$$x + 2 - 2 > 8 - 2$$
$$x > 6.$$

Using the addition principle, we get an inequality for which we can determine the solutions easily.

Any number greater than 6 makes the last sentence true and is a solution of that sentence. Any such number is also a solution of the original sentence. Thus the inequality is solved. The graph is as follows:

We cannot check all the solutions of an inequality by substitution, as we can check solutions of equations, because there are too many of them. A partial check can be done by substituting a number greater than 6, say 7, into the original inequality:

$$\frac{x + 2 > 8}{7 + 2 \mid 8}$$
$$9 \qquad \text{TRUE}$$

Since $9 > 8$ is true, 7 is a solution. Any number greater than 6 is a solution. ◀

When two inequalities have the same solutions, we say that they are **equivalent.** Whenever we use the addition principle with inequalities, the first and last sentences will be equivalent.

▶ **EXAMPLE 9**   Solve: $3x + 1 \le 2x - 3$. Then graph.

We have

$$3x + 1 \le 2x - 3$$
$$3x + 1 - 1 \le 2x - 3 - 1 \qquad \text{Subtracting 1}$$
$$3x \le 2x - 4 \qquad \text{Simplifying}$$
$$3x - 2x \le 2x - 4 - 2x \qquad \text{Subtracting } 2x$$
$$x \le -4. \qquad \text{Simplifying}$$

The graph is as follows:

Any number less than or equal to -4 is a solution. The following are some solutions:

$$-4, \qquad -5, \qquad -6, \qquad -4.1, \qquad -2045, \quad \text{and} \quad -18\pi.$$

Besides drawing a graph, we can also describe all the solutions of an inequality using **set notation.** We could just begin to list them in a set as follows:

$$\{-4, -5, -6, -4.1, -2045, -18\pi, \dots\}.$$

We can never list them all this way, however. Seeing this set without knowing the inequality makes it difficult for us to know what real numbers we are considering. There is another kind of notation used. It is

$$\{x \mid x \le -4\},$$

which is read:

"The set of all x such that x is less than or equal to -4."

This shorter notation for sets is called **set-builder notation.** From now on, you should use this notation when solving inequalities. ◀

DO EXERCISES 6–8.

Solve. Then graph.

6. $x + 3 > 5$

7. $x - 1 \le 2$

8. $5x + 1 < 4x - 2$

Solve.

9. $x + \dfrac{2}{3} \geq \dfrac{4}{5}$

▶ **EXAMPLE 10**   Solve: $x + \frac{1}{3} > \frac{5}{4}$.

We have

$$x + \tfrac{1}{3} > \tfrac{5}{4}$$
$$x + \tfrac{1}{3} - \tfrac{1}{3} > \tfrac{5}{4} - \tfrac{1}{3} \qquad \text{Subtracting } \tfrac{1}{3}$$
$$x > \tfrac{5}{4} \cdot \tfrac{3}{3} - \tfrac{1}{3} \cdot \tfrac{4}{4} \qquad \text{Multiplying by 1 to obtain a}$$
$$\phantom{x > \tfrac{5}{4} \cdot \tfrac{3}{3} - \tfrac{1}{3} \cdot \tfrac{4}{4}} \text{common denominator}$$
$$x > \tfrac{15}{12} - \tfrac{4}{12}$$
$$x > \tfrac{11}{12}.$$

Any number greater than $\frac{11}{12}$ is a solution. The solution set is

$$\{x \,|\, x > \tfrac{11}{12}\},$$

which is read: "The set of all x such that x is greater than $\frac{11}{12}$."   ◀

When solving inequalities, you may obtain an answer like $7 < x$. Recall from Chapter 1 that this has the same meaning as $x > 7$. Thus the solution set can be described as $\{x \,|\, 7 < x\}$ or as $\{x \,|\, x > 7\}$. The latter is used most often.

DO EXERCISES 9 AND 10.

10. $5y + 2 \leq -1 + 4y$

d Solving Inequalities Using the Multiplication Principle

There is a multiplication principle for inequalities similar to that for equations, but it must be modified when multiplying on both sides by a negative number. Consider the true inequality

$$3 < 7.$$

If we multiply on both sides by a positive number 2, we get another true inequality:

$$3 \cdot 2 < 7 \cdot 2,$$
or $$6 < 14. \qquad \text{True}$$

If we multiply on both sides by a negative number -3, we get the false inequality

$$3 \cdot (-3) < 7 \cdot (-3),$$
or $$-9 < -21. \qquad \text{False}$$

However, if we reverse the inequality symbol, we get a true inequality:

$$-9 > -21. \qquad \text{True}$$

Summarizing these results, we obtain the multiplication principle for inequalities.

> ### The Multiplication Principle for Inequalities
>
> **If we multiply (or divide) on both sides of a true inequality by a positive number, we get another true inequality. If we multiply (or divide) by a negative number and reverse the inequality symbol, we get another true inequality.**

▶ **EXAMPLE 11**   Solve: $4x < 28$. Then graph.

We have

$$\frac{4x}{4} < \frac{28}{4} \qquad \text{Dividing by 4}$$

The symbol stays the same.

$$x < 7. \qquad \text{Simplifying}$$

The solution set is $\{x \mid x < 7\}$. The graph is as follows:

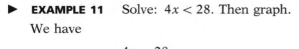

DO EXERCISES 11 AND 12.

▶ **EXAMPLE 12**   Solve: $-2y < 18$. Then graph.

We have

$$\frac{-2y}{-2} > \frac{18}{-2} \qquad \text{Dividing by } -2$$

The symbol must be reversed!

$$y > -9. \qquad \text{Simplifying}$$

The solution set is $\{y \mid y > -9\}$. The graph is as follows:

DO EXERCISES 13 AND 14.

e | Using the Principles Together

We use the addition and multiplication principles together in solving inequalities in much the same way as in solving equations. We generally use the addition principle first.

▶ **EXAMPLE 13**   Solve: $6 - 5y > 7$.

We have

$$-6 + 6 - 5y > -6 + 7 \qquad \text{Adding } -6$$

$$-5y > 1 \qquad \text{Simplifying}$$

$$\frac{-5y}{-5} < \frac{1}{-5} \qquad \text{Dividing by } -5$$

The symbol must be reversed.

$$y < -\frac{1}{5} \qquad \text{Simplifying}$$

The solution set is $\{y \mid y < -\frac{1}{5}\}$.

DO EXERCISE 15.

Solve. Then graph.

11. $8x < 64$

12. $5y \geq 160$

Solve.

13. $-4x \leq 24$

14. $-5y > 13$

15. Solve: $7 - 4x < 8$.

ANSWERS ON PAGE A-3

16. Solve: $24 - 7y \leq 11y - 14$.

17. Solve. Use a method like the one used in Example 15.

$$24 - 7y \leq 11y - 14$$

18. Solve:

$$3(7 + 2x) \leq 30 + 7(x - 1).$$

▶ **EXAMPLE 14**   Solve: $8y - 5 > 17 - 5y$.

We have

$$-17 + 8y - 5 > -17 + 17 - 5y \qquad \text{Adding } -17$$
$$8y - 22 > -5y \qquad \text{Simplifying}$$
$$-8y + 8y - 22 > -8y - 5y \qquad \text{Adding } -8y$$
$$-22 > -13y \qquad \text{Simplifying}$$
$$\frac{-22}{-13} < \frac{-13y}{-13} \qquad \text{Dividing by } -13$$

The symbol must be reversed.

$$\frac{22}{13} < y.$$

The solution set is $\{y | \frac{22}{13} < y\}$. Since $\frac{22}{13} < y$ has the same meaning as $y > \frac{22}{13}$, we can also describe the solution set as $\{y | y > \frac{22}{13}\}$. That is, $\frac{22}{13} < y$ and $y > \frac{22}{13}$ are equivalent. Answers are generally written, however, with the variable on the left. ◀

We can often solve inequalities in such a way as to avoid having to reverse the inequality symbol. We add so that after like terms have been collected, the coefficient of the variable term is positive. We show this by solving the inequality in Example 15 a different way.

▶ **EXAMPLE 15**   Solve: $8y - 5 > 17 - 5y$.

We note that if we add $5y$ on both sides, the coefficient of the y-term will be positive after like terms have been collected.

$$8y - 5 + 5y > 17 - 5y + 5y \qquad \text{Adding } 5y \text{ on both sides}$$
$$13y - 5 > 17 \qquad \text{Simplifying}$$
$$13y - 5 + 5 > 17 + 5 \qquad \text{Adding } 5 \text{ on both sides}$$
$$13y > 22 \qquad \text{Simplifying}$$
$$\frac{13y}{13} > \frac{22}{13} \qquad \text{Dividing by } 13 \text{ on both sides}$$
$$y > \frac{22}{13}$$

The solution set is $\{y | y > \frac{22}{13}\}$. ◀

DO EXERCISES 16 AND 17.

▶ **EXAMPLE 16**   Solve: $3(x - 2) - 1 < 2 - 5(x + 6)$.

$$3(x - 2) - 1 < 2 - 5(x + 6)$$
$$3x - 6 - 1 < 2 - 5x - 30 \qquad \text{Using the distributive laws to multiply and remove parentheses}$$
$$3x - 7 < -5x - 28 \qquad \text{Simplifying}$$
$$3x + 5x < -28 + 7 \qquad \text{Adding } 5x \text{ and } 7 \text{ to get all } x\text{-terms on one side and all other terms on the other side}$$
$$8x < -21 \qquad \text{Simplifying}$$
$$x < \frac{-21}{8}, \quad \text{or} \quad -\frac{21}{8} \qquad \text{Dividing by } 8$$

The solution set is $\{x | x < -\frac{21}{8}\}$. ◀

DO EXERCISE 18.

NAME   SECTION   DATE

EXERCISE SET 2.7

a Determine whether each number is a solution of the given inequality.

1. $x > -4$
   a) 4
   b) 0
   c) -4
   d) 6
   e) 5.6

2. $y < 5$
   a) 0
   b) 5
   c) -1
   d) -5
   e) $7\frac{1}{4}$

3. $x \geq 6$
   a) -6
   b) 0
   c) 6
   d) 8
   e) $-3\frac{1}{2}$

4. $x \leq 10$
   a) 4
   b) -10
   c) 0
   d) 11
   e) -4.7

b Graph on a number line.

5. $x > 4$

6. $y < 0$

7. $t < -3$

8. $y > 5$

9. $m \geq -1$

10. $p \leq 3$

11. $-3 < x \leq 4$

12. $-5 \leq x < 2$

13. $0 < x < 3$

14. $-5 \leq x \leq 0$

c Solve using the addition principle. Then graph.

15. $x + 7 > 2$

16. $x + 6 > 3$

17. $x + 8 \leq -10$

18. $x + 9 \leq -12$

Solve using the addition principle.

19. $y - 7 > -12$

20. $y - 10 > -16$

21. $2x + 3 > x + 5$

22. $2x + 4 > x + 7$

23. $3x + 9 \leq 2x + 6$

24. $3x + 10 \leq 2x + 8$

25. $5x - 6 < 4x - 2$

26. $6x - 8 < 5x - 9$

27. $-7 + c > 7$

1. _____

2. _____

3. _____

4. _____

5. See graph.

6. See graph.

7. See graph.

8. See graph.

9. See graph.

10. See graph.

11. See graph.

12. See graph.

13. See graph.

14. See graph.

15. _____

16. _____

17. _____

18. _____

19. _____

20. _____

21. _____

22. _____

23. _____

24. _____

25. _____

26. _____

27. _____

28. $-9 + c > 9$

29. $y + \dfrac{1}{4} \le \dfrac{1}{2}$

30. $y + \dfrac{1}{3} \le \dfrac{5}{6}$

31. $x - \dfrac{1}{3} > \dfrac{1}{4}$

32. $x - \dfrac{1}{8} > \dfrac{1}{2}$

d Solve using the multiplication principle. Then graph.

33. $5x < 35$

34. $8x \ge 32$

35. $-12x > -36$

36. $-16x > -64$

Solve using the multiplication principle.

37. $5y \ge -2$

38. $7x > -4$

39. $-2x \le 12$

40. $-3y \le 15$

41. $-4y \ge -16$

42. $-7x < -21$

43. $-3x < -17$

44. $-5y > -23$

45. $-2y > \dfrac{1}{7}$

46. $-4x \le \dfrac{1}{9}$

47. $-\dfrac{6}{5} \le -4x$

48. $-\dfrac{7}{8} > -56t$

e Solve using the addition and multiplication principles.

49. $4 + 3x < 28$

50. $5 + 4y < 37$

51. $3x - 5 \le 13$

52. $5y - 9 \le 21$

53. $13x - 7 < -46$

54. $8y - 4 < -52$

55. $30 > 3 - 9x$

56. $40 > 5 - 7y$

57. $4x + 2 - 3x \le 9$

58. $15x + 3 - 14x \le 7$

59. $-3 < 8x + 7 - 7x$

60. $-5 < 9x + 8 - 8x$

61. $6 - 4y > 4 - 3y$

62. $7 - 8y > 5 - 7y$

63. $5 - 9y \le 2 - 8y$

64. $6 - 13y \le 4 - 12y$

65. $19 - 7y - 3y < 39$

66. $18 - 6y - 9y < 63$

67. $2.1x + 45.2 > 3.2 - 8.4x$

68. $0.96y - 0.79 \le 0.21y + 0.46$

69. $\dfrac{x}{3} - 2 \le 1$

70. $\dfrac{2}{3} - \dfrac{x}{5} < \dfrac{4}{15}$

71. $\dfrac{y}{5} + 1 \le \dfrac{2}{5}$

72. $\dfrac{3x}{4} + \dfrac{7}{8} \ge -15$

73. $3(2y - 3) < 27$

74. $4(2y - 3) > 28$

75. $2(3 + 4m) - 9 \ge 45$

76. $3(5 + 3m) - 8 \le 88$

ANSWERS

55. _____

56. _____

57. _____

58. _____

59. _____

60. _____

61. _____

62. _____

63. _____

64. _____

65. _____

66. _____

67. _____

68. _____

69. _____

70. _____

71. _____

72. _____

73. _____

74. _____

75. _____

76. _____

77. $8(2t + 1) > 4(7t + 7)$

78. $7(5x - 2) < 6(6x - 1)$

79. $3(r - 6) + 2 < 4(r + 2) - 21$

80. $5(t + 3) + 9 > 3(t - 2) + 6$

81. $\frac{1}{4}(8y + 4) - 17 > -\frac{1}{2}(4y - 8)$

82. $\frac{1}{3}(6x + 24) - 20 < -\frac{1}{4}(12x - 72)$

83. $\frac{2}{3}(2x - 1) \geq 10$

84. $\frac{4}{5}(3x + 4) \leq 20$

85. $0.8(3x + 6) \geq 1.1 - (x + 2)$

86. $0.4(2x + 8) \geq 20 - (x + 5)$

87. $a + (a - 1) < (a + 2) - (a + 1)$

88. $0.8 - 4(b - 1) > 0.2 + 3(4 - b)$

SKILL MAINTENANCE

Add.

89. $-56 + (-18)$ **90.** $-2.3 + 7.1$ **91.** $-\frac{3}{4} + \frac{1}{8}$ **92.** $8.12 - 9.23$

SYNTHESIS

93. Suppose that $2x - 5 \geq 9$ is true for some value of x. Determine whether $2x - 5 \geq 8$ is true for that same value of x.

94. Determine whether each number is a solution of the inequality $|x| < 3$.
   a) 0   **b)** -2
   c) -3 **d)** 4
   e) 3   **f)** 1.7
   g) -2.8

95. Graph $|x| < 3$ on a number line.

96. Determine whether each number is a solution of the inequality $|x| \geq 4$.
   a) 0   **b)** -5
   c) 6   **d)** -3
   e) 3   **f)** -8
   g) 9.7

Solve.

97. $x + 3 \leq 3 + x$

98. $x + 4 < 3 + x$

99. Suppose we are considering *only* integer solutions to $x > 5$. Find an equivalent inequality involving $\geq$.

2.8 Solving Problems Using Inequalities

We can use inequalities to solve certain kinds of problems.

a Translating to Inequalities

Let us first practice translating sentences to inequalities.

▶ **EXAMPLES** Translate to an inequality.

1. A number is less than 5.

$$x < 5$$

2. A number is greater than or equal to $3\frac{1}{2}$.

$$y \geq 3\frac{1}{2}$$

3. My salary is at most $34,000.

$$S \leq \$34,000$$

4. The number of compact disc players is at least 2700.

$$C \geq 2700$$

5. 12 more than twice a number is less than 37.

$$12 + 2x < 37 \qquad ◀$$

DO EXERCISES 1–5.

b Solving Problems

▶ **EXAMPLE 6** A student is taking an introductory algebra course in which four tests are to be given. To get an A, the student must average at least 90 on the four tests. The student got scores of 91, 86, and 89 on the first three tests. Determine (in terms of an inequality) what scores on the last test will allow the student to get an A.

1. *Familiarize.* Let us try some guessing. Suppose the student gets a 92 on the last test. The average of the four scores is their sum divided by the number of tests, 4, and is given by

$$\frac{91 + 86 + 89 + 92}{4} = 89.5.$$

For this average to be *at least* 90, it must be greater than or equal to 90. In this case, we have $89.5 \geq 90$, which is not true. But there are scores that will give the A. To find them, we translate to an inequality and solve. Let $x =$ the student's score on the last test.

2. *Translate.* The average of the four scores must be *at least* 90. This means that it must be greater than or equal to 90. Thus we can translate the problem to the inequality

$$\frac{91 + 86 + 89 + x}{4} \geq 90.$$

Translate.

1. A number is less than or equal to 8.

2. A number is greater than -2.

3. The speed of that car is at most 180 mph.

4. The price of that car is at least $5800.

5. Twice a number minus 32 is greater than 5.

ANSWERS ON PAGE A-3

6. A student is taking a literature course in which four tests are to be given. To get a B, the student must average at least 80 on the four tests. The student got scores of 82, 76, and 78 on the first three tests. Determine (in terms of an inequality) what scores on the last test will allow the student to get at least a B.

3. *Solve.* We solve the inequality. We first multiply by 4 to clear of fractions.

$$4\left(\frac{91 + 86 + 89 + x}{4}\right) \geq 4 \cdot 90 \qquad \text{Multiplying by 4}$$

$$91 + 86 + 89 + x \geq 360$$

$$266 + x \geq 360 \qquad \text{Collecting like terms}$$

$$x \geq 94$$

The solution set is $\{x \,|\, x \geq 94\}$.

4. *Check.* We can obtain a partial check by substituting a number greater than or equal to 94. We leave it to the student to try 95 in a manner similar to what was done in the familiarization step.

5. *State.* Any score that is at least 94 will give the student an A in the course.   ◄

DO EXERCISE 6.

▶ **EXAMPLE 7** Butter stays solid at Fahrenheit temperatures below 88°. The formula

$$F = \tfrac{9}{5}C + 32$$

can be used to convert Celsius temperatures C to Fahrenheit temperatures F. Determine (in terms of an inequality) those Celsius temperatures for which butter stays solid.

1. *Familiarize.* Suppose we guess to see how we might consider a solution. We try a Celsius temperature of 40°. We substitute and find F:

$$F = \tfrac{9}{5}C + 32 = \tfrac{9}{5}(40) + 32 = 72 + 32 = 104°.$$

This is higher than 88°, so 40° is *not* a solution. To find the solutions, we need to solve an inequality.

7. Gold stays solid at Fahrenheit temperatures below 1945.4°. Determine (in terms of an inequality) those Celsius temperatures for which gold stays solid. Use the formula given in Example 7.

2. *Translate.* The Fahrenheit temperature F is to be less than 88. We have the inequality

$$F < 88.$$

To find the Celsius temperatures C that satisfy this condition, we substitute $\tfrac{9}{5}C + 32$ for F, which gives us the following inequality:

$$\tfrac{9}{5}C + 32 < 88.$$

3. *Solve.* We solve the inequality:

$$\tfrac{9}{5}C + 32 < 88$$

$$5(\tfrac{9}{5}C + 32) < 5(88) \qquad \text{Multiplying by 5 to clear of fractions}$$

$$5(\tfrac{9}{5}C) + 5(32) < 440 \qquad \text{Using a distributive law}$$

$$9C + 160 < 440 \qquad \text{Simplifying}$$

$$9C < 280 \qquad \text{Subtracting 160}$$

$$C < \frac{280}{9} \qquad \text{Dividing by 9}$$

$$C < 31.1. \qquad \text{Dividing and rounding to the nearest tenth}$$

The solution set of the inequality is $\{C \,|\, C < 31.1°\}$.

4. *Check.* The check is left to the student.

5. *State.* Butter stays solid at Celsius temperatures below 31.1°.   ◄

DO EXERCISE 7.

EXERCISE SET 2.8

ANSWERS

a Translate to an inequality.

1. A number is greater than 4.

2. A number is less than 7.

3. A number is less than or equal to −6.

4. A number is greater than or equal to 13.

5. The number of people is at least 1200.

6. The cost is at most $3457.95.

7. The amount of acid is not to exceed 500 liters.

8. The cost of gasoline is no less than 94 cents per gallon.

9. Two more than three times a number is less than 13.

10. Five less than one-half a number is greater than 17.

b Solve.

11. Your quiz grades are 73, 75, 89, and 91. Determine (in terms of an inequality) those scores that you can obtain on the last quiz in order to receive an average quiz grade of at least 85.

12. A human body is considered to be fevered when its temperature is higher than 98.6°F. Using the formula given in Example 7, determine (in terms of an inequality) those Celsius temperatures for which the body is fevered.

13. The formula
$$R = -0.075t + 3.85$$
can be used to predict the world record in the 1500-m run t years after 1930. Determine (in terms of an inequality) those years for which the world record will be less than 3.5 min.

14. The formula
$$R = -0.028t + 20.8$$
can be used to predict the world record in the 200-m dash t years after 1920. Determine (in terms of an inequality) those years for which the world record will be less than 19.0 sec.

15. Acme rents station wagons at a daily rate of $42.95 plus $0.46 per mile. A family wants to rent a wagon one day while on vacation, but must stay within a budget of $200. Determine (in terms of an inequality) those mileages that will allow the family to stay within budget. Round to the nearest tenth of a mile.

16. Atlas rents an intermediate-size car at a daily rate of $44.95 plus $0.39 per mile. A businessperson is not to exceed a daily car rental budget of $250. Determine (in terms of an inequality) those mileages that will allow the businessperson to stay within budget. Round to the nearest tenth of a mile.

1. _____

2. _____

3. _____

4. _____

5. _____

6. _____

7. _____

8. _____

9. _____

10. _____

11. _____

12. _____

13. _____

14. _____

15. _____

16. _____

17. _____

18. _____

19. _____

20. _____

21. _____

22. _____

23. _____

24. _____

25. _____

26. _____

27. _____

28. _____

29. _____

30. _____

17. Find all numbers such that the sum of the number and 15 is less than four times the number.

18. Find all numbers such that three times the number minus ten times the number is greater than or equal to eight times the number.

19. The width of a rectangle is fixed at 4 cm. Determine (in terms of an inequality) those lengths for which the area will be less than 86 cm².

20. The width of a rectangle is fixed at 16 yd. Determine (in terms of an inequality) those lengths for which the area will be greater than or equal to 264 yd².

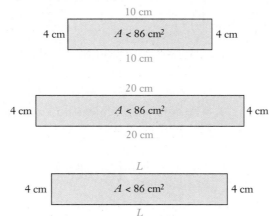

21. One side of a triangle is 2 cm shorter than the base. The other side is 3 cm longer than the base. What lengths of the base will allow the perimeter to be greater than 19 cm?

22. The perimeter of a rectangular swimming pool is not to exceed 70 ft. The length is to be twice the width. What widths will meet these conditions?

23. A salesperson made 18 customer calls last week and 22 calls this week. How many calls must be made next week in order to maintain an average of at least 20 for the three-week period?

24. George and Joan do volunteer work at a hospital. Joan worked 3 hr more than George, and together they worked more than 27 hr. What possible number of hours did each work?

25. A student is shopping for a new pair of jeans and two sweaters of the same kind. He is determined to spend no more than $120.00 for the outfit. He buys jeans for $21.95. What is the most the student can spend for each sweater?

26. The medium-size box of dog food weighs 1 lb more than the small size. The large size weighs 2 lb more than the small size. The total weight of the three boxes is at most 30 lb. What are the possible weights of the small box?

27. The width of a rectangle is 32 km. What lengths will make the area at least 2048 km²?

28. The height of a triangle is 20 cm. What lengths of the base will make the area at most 40 cm²?

SYNTHESIS

29. The area of a square can be no more than 64 cm². What lengths of a side will allow this?

30. The sum of two consecutive odd integers is less than 100. What is the largest possible pair of such integers?

SUMMARY AND REVIEW: CHAPTER 2

IMPORTANT PROPERTIES AND FORMULAS

The Addition Principles: If $a = b$ is true, then $a + c = b + c$ is true for any real number c. If the same number is added on both sides of an inequality, we get another true inequality.

The Multiplication Principles: If $a = b$ is true, then $ac = bc$ is true for any real number c. If we multiply on both sides of a true inequality by a positive number, we get another true inequality. If we multiply by a negative number and reverse the inequality symbol, we get another true inequality.

REVIEW EXERCISES

The review sections and objectives to be tested in addition to the material in this chapter are [R.3a, b], [1.1a, b], [1.3a], and [1.8b].

Solve.

1. $x + 5 = -17$

2. $-8x = -56$

3. $-\dfrac{x}{4} = 48$

4. $n - 7 = -6$

5. $15x = -35$

6. $x - 11 = 14$

7. $-\dfrac{2}{3} + x = -\dfrac{1}{6}$

8. $\dfrac{4}{5}y = -\dfrac{3}{16}$

9. $y - 0.9 = 9.09$

10. $5 - x = 13$

11. $5t + 9 = 3t - 1$

12. $7x - 6 = 25x$

13. $\dfrac{1}{4}x - \dfrac{5}{8} = \dfrac{3}{8}$

14. $14y = 23y - 17 - 10$

15. $0.22y - 0.6 = 0.12y + 3 - 0.8y$

16. $\dfrac{1}{4}x - \dfrac{1}{8}x = 3 - \dfrac{1}{16}x$

17. $4(x + 3) = 36$

18. $3(5x - 7) = -66$

19. $8(x - 2) = 5(x + 4)$

20. $-5x + 3(x + 8) = 16$

Determine whether the given number is a solution of the inequality $x \le 4$.

21. -3

22. 7

23. 4

Solve. Write set notation for the answers.

24. $y + \dfrac{2}{3} \ge \dfrac{1}{6}$

25. $9x \ge 63$

26. $2 + 6y > 14$

27. $7 - 3y \ge 27 + 2y$

28. $3x + 5 < 2x - 6$

29. $-4y < 28$

30. $3 - 4x < 27$

31. $4 - 8x < 13 + 3x$

32. $-3y \ge -21$

33. $-4x \le \dfrac{1}{3}$

Graph on a number line.

34. $4x - 6 < x + 3$

35. $-2 < x \le 5$

36. $y > 0$

Solve.

37. $C = \pi d$, for d

38. $V = \frac{1}{3}Bh$, for B

39. $A = \frac{a+b}{2}$, for a

40. A color television sold for $629 in May. This was $38 more than the cost in January. Find the cost in January.

41. Selma gets a $4 commission for each appliance that she sells. One week she got $108 in commissions. How many appliances did she sell?

42. An 8-m board is cut into two pieces. One piece is 2 m longer than the other. How long are the pieces?

43. If 14 is added to three times a certain number, the result is 41. Find the number.

44. The sum of two consecutive odd integers is 116. Find the integers.

45. The perimeter of a rectangle is 56 cm. The width is 6 cm less than the length. Find the width and the length.

46. After a 30% reduction, an item is on sale for $154. What was the marked price (the price before reducing)?

47. A businessperson's salary is $30,000. That is a 15% increase over the previous year's salary. What was the previous salary (to the nearest dollar)?

48. The measure of the second angle of a triangle is 50° more than that of the first. The measure of the third angle is 10° less than twice the first. Find the measures of the angles.

49. Your quiz grades are 71, 75, 82, and 86. What is the lowest grade you can get on the next quiz and still have an average of at least 80?

50. The length of a rectangle is 43 cm. What widths will make the perimeter greater than 120 cm?

SKILL MAINTENANCE

51. Convert to decimal notation: $\frac{17}{12}$.

52. Divide: $12.42 \div 5.4$.

53. Add: $-12 + 10 + (-19) + (-24)$.

54. Remove parentheses and simplify: $5x - 8(6x - y)$.

SYNTHESIS

55. The total length of the Nile and Amazon Rivers is 13,108 km. If the Amazon were 234 km longer, it would be as long as the Nile. Find the length of each river.

56. Consumer experts advise us never to pay the sticker price for a car. A rule of thumb is to pay the sticker price minus 20% of the sticker price, plus $200. A car is purchased for $11,520 using the rule. What was the sticker price?

Solve.

57. $2|n| + 4 = 50$

58. $|3n| = 60$

59. $y = 2a - ab + 3$, for a

❖ **THINKING IT THROUGH**

Explain all possible errors in each of the following.

1. Solve: $4 - 3x = 5$
$$3x = 9$$
$$x = 3.$$

2. Solve: $2(x - 5) = 7$
$$2x - 5 = 7$$
$$2x = 12$$
$$x = 6.$$

3. Explain the difference in using the multiplication principle for solving equations and for solving inequalities.

NAME   SECTION   DATE

TEST: CHAPTER 2

Solve.

1. $x + 7 = 15$

2. $t - 9 = 17$

3. $3x = -18$

4. $-\dfrac{4}{7}x = -28$

5. $3t + 7 = 2t - 5$

6. $\dfrac{1}{2}x - \dfrac{3}{5} = \dfrac{2}{5}$

7. $8 - y = 16$

8. $-\dfrac{2}{5} + x = -\dfrac{3}{4}$

9. $3(x + 2) = 27$

10. $-3x + 6(x + 4) = 9$

11. $0.4p + 0.2 = 4.2p - 7.8 - 0.6p$

Solve. Write set notation for the answers.

12. $x + 6 \leq 2$

13. $14x + 9 > 13x - 4$

14. $12x \leq 60$

15. $-2y \geq 26$

16. $-4y \leq -32$

17. $-5x \geq \dfrac{1}{4}$

18. $4 - 6x > 40$

19. $5 - 9x \geq 19 + 5x$

ANSWERS
1.
2.
3.
4.
5.
6.
7.
8.
9.
10.
11.
12.
13.
14.
15.
16.
17.
18.
19.

ANSWERS

20. _____

21. _____

22. . _____

23. _____

24. _____

25. _____

26. _____

27. _____

28. _____

29. _____

30. _____

31. _____

32. _____

33. _____

34. _____

35. _____

36. _____

37. _____

Graph on a number line.

20. $y \leq 9$

21. $6x - 3 < x + 2$

22. $-2 \leq x \leq 2$

Solve.

23. The perimeter of a rectangle is 36 cm. The length is 4 cm greater than the width. Find the width and the length.

24. If you triple a number and then subtract 14, you get two thirds of the original number. What is the original number?

25. The sum of three consecutive odd integers is 249. Find the integers.

26. Money is invested in a savings account at 12% simple interest. After 1 year, there is $840 in the account. How much was originally invested?

Solve the formulas for the given letter.

27. Solve $A = 2\pi rh$, for r.

28. Solve $w = \dfrac{P - 2l}{2}$, for l.

Solve.

29. Find all numbers such that six times the number is greater than the number plus 30.

30. The width of a rectangle is 96 yd. Find all possible lengths so that the perimeter of the rectangle will be at least 540 yd.

SKILL MAINTENANCE

31. Add: $\dfrac{2}{3} + \left(-\dfrac{8}{9}\right)$.

32. Evaluate $\dfrac{4x}{y}$ when $x = 2$ and $y = 3$.

33. Translate to an algebraic expression: Seventy-three percent of p.

34. Simplify: $2x - 3y - 5(4x - 8y)$.

SYNTHESIS

35. Solve $c = \dfrac{1}{a - d}$, for d.

36. Solve: $3|w| - 8 = 37$.

37. A movie theater had a certain number of tickets to give away. Five people got the tickets. The first got one third of the tickets, the second got one fourth of the tickets, and the third got one fifth of the tickets. The fourth person got eight tickets, and there were five tickets left for the fifth person. Find the total number of tickets given away.

CUMULATIVE REVIEW: CHAPTERS 1–2

Evaluate.

1. $\dfrac{y - x}{4}$ when $y = 12$ and $x = 6$ **2.** $\dfrac{3x}{y}$ when $x = 5$ and $y = 4$ **3.** $x - 3$ when $x = 3$

4. Translate to an algebraic expression: Four less than twice w.

Use $<$ or $>$ for ▦ to write a true sentence.

5. -4 ▦ -6 **6.** 0 ▦ -5 **7.** -8 ▦ 7

8. Find the opposite and the reciprocal of $\dfrac{2}{5}$.

Find the absolute value.

9. $|3|$ **10.** $\left| -\dfrac{3}{4} \right|$ **11.** $|0|$

Compute and simplify.

12. $-6.7 + 2.3$ **13.** $-\dfrac{1}{6} - \dfrac{7}{3}$ **14.** $-\dfrac{5}{8}\left(-\dfrac{4}{3} \right)$ **15.** $(-7)(5)(-6)(-0.5)$

16. $81 \div (-9)$ **17.** $-10.8 \div 3.6$ **18.** $-\dfrac{4}{5} \div -\dfrac{25}{8}$

Multiply.

19. $5(3x + 5y + 2z)$ **20.** $4(-3x - 2)$ **21.** $-6(2y - 4x)$

Factor.

22. $64 + 18x + 24y$ **23.** $16y - 56$ **24.** $5a - 15b + 25$

Collect like terms.

25. $9b + 18y + 6b + 4y$ **26.** $3y + 4 + 6z + 6y$

27. $-4d - 6a + 3a - 5d + 1$ **28.** $3.2x + 2.9y - 5.8x - 8.1y$

Simplify.

29. $7 - 2x - (-5x) - 8$ **30.** $-3x - (-x + y)$ **31.** $-3(x - 2) - 4x$

32. $10 - 2(5 - 4x)$

33. $[3(x + 6) - 10] - [5 - 2(x - 8)]$

Solve.

34. $x + 1.75 = 6.25$

35. $\dfrac{5}{2}y = \dfrac{2}{5}$

36. $-2.6 + x = 8.3$

37. $4\dfrac{1}{2} + y = 8\dfrac{1}{3}$

38. $-\dfrac{3}{4}x = 36$

39. $-2.2y = -26.4$

40. $5.8x = -35.96$

41. $-4x + 3 = 15$

42. $-3x + 5 = -8x - 7$

43. $4y - 4 + y = 6y + 20 - 4y$

44. $-3(x - 2) = -15$

45. $\dfrac{1}{3}x - \dfrac{5}{6} = \dfrac{1}{2} + 2x$

46. $-3.7x + 6.2 = -7.3x - 5.8$

47. $3x - 1 < 2x + 1$

48. $5 - y \le 2y - 7$

49. $3y + 7 > 5y + 13$

50. $A = \dfrac{1}{2}h(b + c)$, for h

51. $Q = \dfrac{p - q}{2}$, for q

Solve.

52. If 25 is subtracted from a certain number, the result is 129. Find the number.

53. Jane and Becky purchased identical dresses for a total of $107. Jane paid $17 more for her dress than Becky did. What did Becky pay?

54. Money is invested in a savings account at 12% simple interest. After 1 year, there is $1680 in the account. How much was originally invested?

55. A 143-m wire is cut into three pieces. The second is 3 m longer than the first. The third is four fifths as long as the first. How long is each piece?

56. Your test grades are 75, 82, 86, and 79. Determine, in terms of an inequality, what scores on the last test will allow you an average test score of at least 80.

57. After a 25% reduction, a tie is on sale for $5.85. What was the price before reduction?

SYNTHESIS

58. An engineer's salary at the end of a year is $26,780. This reflects a 4% salary increase and a later 3% cost-of-living adjustment during the year. What was the salary at the beginning of the year?

Solve.

59. $4|x| - 13 = 3$

60. $4(x + 2) = 4(x - 2) + 16$

61. $0(x + 3) + 4 = 0$

62. $\dfrac{2 + 5x}{4} = \dfrac{11}{28} + \dfrac{8x + 3}{7}$

63. $5(7 + x) = (x + 7)5$

64. $p = \dfrac{2}{m + Q}$, for Q

INTRODUCTION Algebraic expressions like $16t^2$, seen below, and
$3x^2 - 7x + 5$ are called *polynomials*. One of the most important parts of
introductory algebra is the study of polynomials. In this chapter, we learn to
add, subtract, multiply, and divide polynomials.

Of particular importance in this chapter is the study of fast ways to find
special products of polynomials, which will be helpful not only in this text
but also in more advanced mathematics.

The review sections to be tested in addition to the material in this
chapter are 1.4, 1.7, 2.3, and 2.4. ❖

Polynomials: Operations

3

AN APPLICATION

The distance s, in feet, traveled by a
body falling freely from rest in t seconds
is approximated by

$$16t^2.$$

An object is dropped and takes 3 sec to
hit the ground. From what height was it
dropped?

THE MATHEMATICS

To solve the problem, we substitute 3
for t and evaluate:

$$\underbrace{16t^2}_{} = 16(3)^2 = 144 \text{ ft.}$$

This is a *polynomial*.

Distributive Laws: $a(b + c) = ab + ac, \quad a(b - c) = ab - ac$

Definition of Exponents, $n \geq 2$: $a^n = \underbrace{a \cdot a \cdot a \cdots a}_{n \text{ factors}}$

PRETEST: CHAPTER 3

1. Multiply: $x^{-3}x^5$.

2. Divide: $\dfrac{x^{-2}}{x^5}$.

3. Simplify: $(-4x^2y^{-3})^2$.

4. Express using a positive exponent: p^{-3}.

5. Convert to scientific notation: 0.000347.

6. Convert to decimal notation: 3.4×10^6.

7. Identify the degree of each term and the degree of the polynomial:
$$2x^3 - 4x^2 + 3x - 5.$$

8. Collect like terms:
$$2a^3b - a^2b^2 + ab^3 + 9 - 5a^3b - a^2b^2 + 12b^3.$$

9. Add:
$$(5x^2 - 7x + 8) + (6x^2 + 11x - 19).$$

10. Subtract:
$$(5x^2 - 7x + 8) - (6x^2 + 11x - 19).$$

Multiply.

11. $5x^2(3x^2 - 4x + 1)$

12. $(x + 5)^2$

13. $(x - 5)(x + 5)$

14. $(x^3 + 6)(4x^3 - 5)$

15. $(2x - 3y)(2x - 3y)$

16. Divide: $(x^3 - x^2 + x + 2) \div (x - 2)$.

3.1 Integers as Exponents

We introduced integer exponents of 2 or higher in Section R.5. Here we consider 0 and 1, as well as negative integers, as exponents.

a Exponential Notation

An exponent of 2 or greater tells how many times the base is used as a factor. For example,

$$a \cdot a \cdot a \cdot a = a^4.$$

Here the **exponent** is 4 and the **base** is a. An expression for a power is called **exponential notation.**

$$a^n \leftarrow \text{This is an exponent.}$$
$$\uparrow$$
$$\text{This is the base.}$$

▶ **EXAMPLE 1** What is the meaning of 3^5? of n^4? of $(2n)^3$? of $50x^2$?

3^5 means $3 \cdot 3 \cdot 3 \cdot 3 \cdot 3$;

n^4 means $n \cdot n \cdot n \cdot n$;

$(2n)^3$ means $2n \cdot 2n \cdot 2n$;

$50x^2$ means $50 \cdot x \cdot x$ ◀

DO EXERCISES 1–4.

We read exponential notation as follows:

b^n is read the **nth power of b,** or simply **b to the nth,** or **b to the n.**

We often read x^2 as "**x-squared.**" The reason for this comes from the fact that the area of a square of side x is $x \cdot x$, or x^2. We often read x^3 as "**x-cubed.**" The reason for this comes from the fact that the volume of a cube with length, width, and height x is $x \cdot x \cdot x$, or x^3.

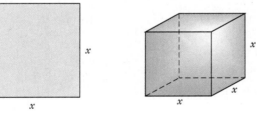

b One and Zero as Exponents

Look for a pattern in the following:

$$8 \cdot 8 \cdot 8 \cdot 8 = 8^4 \quad \text{We divide by 8 each time.}$$
$$8 \cdot 8 \cdot 8 = 8^3$$
$$8 \cdot 8 = 8^2$$
$$8 = 8^?$$
$$1 = 8^?.$$

The exponents decrease by 1 each time. To continue the pattern, we would say that

$$8 = 8^1$$

and

$$1 = 8^0.$$

OBJECTIVES

After finishing Section 3.1, you should be able to:

a Tell the meaning of exponential notation.

b Evaluate exponential expressions with exponents of 0 and 1.

c Evaluate algebraic expressions containing exponents.

d Use the product rule to multiply exponential expressions with like bases.

e Use the quotient rule to divide exponential expressions with like bases.

f Express an exponential expression involving negative exponents with positive exponents.

FOR EXTRA HELP

Tape 7C   Tape 6A   MAC: 3
                      IBM: 3

What is the meaning of each of the following?

1. 5^4

2. x^5

3. $(3t)^2$

4. $3t^2$

ANSWERS ON PAGE A-3

Evaluate.

5. 6^1

6. 7^0

7. $(8.4)^1$

8. 8654^0

We make the following definition.

> $b^1 = b$, for any number b;
> $b^0 = 1$, for any nonzero number b.

We consider 0^0 undefined. We will explain why later in this section.

▶ **EXAMPLE 2** Evaluate 5^1, 8^1, 3^0, and $(-7.3)^0$.

$5^1 = 5$;
$8^1 = 8$;
$3^0 = 1$;
$(-7.3)^0 = 1$ ◀

DO EXERCISES 5–8.

C Evaluating Algebraic Expressions

Algebraic expressions can involve exponential notation. For example, the following are algebraic expressions:

$$x^4, \qquad (3x)^3 - 2, \qquad a^2 + 2ab + b^2.$$

We evaluate algebraic expressions by replacing variables with numbers and following the rules for order of operations.

▶ **EXAMPLE 3** Evaluate x^4 when $x = 2$.

$$x^4 = 2^4 \quad \text{Substituting}$$
$$= 2 \cdot 2 \cdot 2 \cdot 2$$
$$= 16 \qquad\qquad ◀$$

▶ **EXAMPLE 4** The area of a circle is given by $A = \pi r^2$, where r is the radius. Find the area of a circle with a radius of 10 cm. Use 3.14 as an approximation for π.

$$A = \pi r^2 \approx 3.14 \times (10 \text{ cm})^2$$
$$= 3.14 \times 100 \text{ cm}^2$$
$$= 314 \text{ cm}^2$$

In Example 4, "cm²" means "square centimeters" and "$\approx$" means "approximately equal to."

▶ **EXAMPLE 5** Evaluate $m^1 + 5$ and $m^0 + 5$ when $m = 4$.

$$m^1 + 5 = 4^1 + 5 = 4 + 5 = 9;$$
$$m^0 + 5 = 4^0 + 5 = 1 + 5 = 6 \qquad ◀$$

▶ **EXAMPLE 6** Evaluate $(5x)^3$ when $x = -2$.

When we evaluate with a negative number, we often use extra parentheses to show the substitution.

$$(5x)^3 = [5 \cdot (-2)]^3 \quad \text{Substituting}$$
$$= [-10]^3 \quad \text{Multiplying within brackets first}$$
$$= -1000 \quad \text{Evaluating the power} \qquad ◀$$

▶ **EXAMPLE 7** Evaluate $5x^3$ when $x = -2$.

$$5x^3 = 5 \cdot (-2)^3 \qquad \text{Substituting}$$
$$= 5(-8) \qquad \text{Evaluating the power first}$$
$$= -40 \qquad \blacktriangleleft$$

Recall that two expressions are equivalent if they have the same value for all meaningful replacements. Note that Examples 6 and 7 show that $(5x)^3$ and $5x^3$ are *not* equivalent.

DO EXERCISES 9–13.

d Multiplying Powers with Like Bases

There are several rules for manipulating exponential notation to obtain equivalent expressions. We first consider multiplying powers with like bases:

$$a^3 \cdot a^2 = \underbrace{(a \cdot a \cdot a)}_{3 \text{ factors}}\underbrace{(a \cdot a)}_{2 \text{ factors}} = \underbrace{a \cdot a \cdot a \cdot a \cdot a}_{5 \text{ factors}} = a^5.$$

Since an integer exponent greater than 1 tells how many times we use a base as a factor, then $(a \cdot a \cdot a)(a \cdot a) = a \cdot a \cdot a \cdot a \cdot a = a^5$ by the associative law. Note that the exponent in a^5 is the sum of those in $a^3 \cdot a^2$. That is, $3 + 2 = 5$. Likewise,

$$b^4 \cdot b^3 = (b \cdot b \cdot b \cdot b)(b \cdot b \cdot b) = b^7, \quad \text{where} \quad 4 + 3 = 7.$$

Adding the exponents gives the correct result.

The Product Rule

For any number a and any positive integers m and n,

$$a^m \cdot a^n = a^{m+n}.$$

(When multiplying with exponential notation, if the bases are the same, keep the base and add the exponents.)

▶ **EXAMPLES** Multiply and simplify. By simplify, we mean write as one number to a nonnegative power.

8. $8^4 \cdot 8^3 = 8^{4+3} \qquad$ Adding exponents: $a^m \cdot a^n = a^{m+n}$
$$ = 8^7$$

9. $x^2 \cdot x^9 = x^{2+9}$
$$ = x^{11}$$

10. $m^5 m^{10} m^3 = m^{5+10+3}$
$$\phantom{m^5 m^{10} m^3} = m^{18}$$

11. $x \cdot x^8 = x^1 \cdot x^8 = x^{1+8}$
$$ = x^9$$

12. $(a^3 b^2)(a^3 b^5) = (a^3 a^3)(b^2 b^5)$
$$ = a^6 b^7 \qquad \blacktriangleleft$$

DO EXERCISES 14–18.

e Dividing Powers with Like Bases

The following suggests a rule for dividing powers with like bases, such as a^5/a^2:

$$\frac{a^5}{a^2} = \frac{a \cdot a \cdot a \cdot a \cdot a}{a \cdot a} = \frac{a \cdot a \cdot a \cdot a \cdot a}{1 \cdot a \cdot a} = \frac{a \cdot a \cdot a}{1} \cdot \frac{a \cdot a}{a \cdot a} = \frac{a \cdot a \cdot a}{1} \cdot 1$$
$$= a \cdot a \cdot a = a^3.$$

---

9. Evaluate t^3 when $t = 5$.

10. Find the area of a circle when $r = 32$ cm. Use 3.14 for π.

11. Evaluate $200 - a^4$ when $a = 3$.

12. Evaluate $t^1 - 4$ and $t^0 - 4$ when $t = 7$.

13. **a)** Evaluate $(4t)^2$ when $t = -3$.

b) Evaluate $4t^2$ when $t = -3$.

c) Determine whether $(4t)^2$ and $4t^2$ are equivalent.

Multiply and simplify.

14. $3^5 \cdot 3^5$

15. $x^4 \cdot x^6$

16. $p^4 p^{12} p^8$

17. $x \cdot x^4$

18. $(a^2 b^3)(a^7 b^5)$

ANSWERS ON PAGE A-3

Divide and simplify.

19. $\dfrac{4^5}{4^2}$

20. $\dfrac{y^6}{y^2}$

21. $\dfrac{p^{10}}{p}$

22. $\dfrac{a^7 b^6}{a^3 b^4}$

Note that the exponent in a^3 is the difference of those in $a^5 \div a^2$. If we subtract exponents, we get $5 - 2$, which is 3.

> **The Quotient Rule**
>
> **For any nonzero number a and any positive integers m and n,**
>
> $$\frac{a^m}{a^n} = a^{m-n}.$$
>
> **(When dividing with exponential notation, if the bases are the same, keep the base and subtract the exponent of the denominator from the exponent of the numerator.)**

▶ **EXAMPLES**   Divide and simplify. By simplify, we mean write as one number to a nonnegative power.

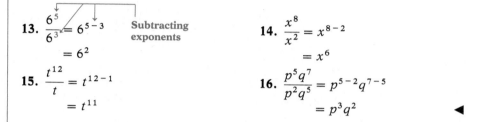

13. $\dfrac{6^5}{6^3} = 6^{5-3}$ Subtracting exponents

$\qquad = 6^2$

14. $\dfrac{x^8}{x^2} = x^{8-2}$

$\qquad = x^6$

15. $\dfrac{t^{12}}{t} = t^{12-1}$

$\qquad = t^{11}$

16. $\dfrac{p^5 q^7}{p^2 q^5} = p^{5-2} q^{7-5}$

$\qquad = p^3 q^2$ ◀

The quotient rule can also be used to explain the definition of 0 as an exponent. Consider the expression a^4/a^4, where a is nonzero:

$$\frac{a^4}{a^4} = \frac{a \cdot a \cdot a \cdot a}{a \cdot a \cdot a \cdot a} = 1.$$

This is true because the numerator and the denominator are the same. Now suppose we apply the rule for dividing powers with the same base:

$$\frac{a^4}{a^4} = a^{4-4} = a^0 = 1.$$

Since both expressions for a^4/a^4 are equivalent to 1, it follows that $a^0 = 1$, when $a \neq 0$.

We can explain why we do not define 0^0 using the quotient rule. We know that 0^0 is 0^{1-1}. But 0^{1-1} is also equal to $0/0$. We have already seen that division by 0 is undefined, so we also have 0^0 undefined.

DO EXERCISES 19–22.

f   Negative Integers as Exponents

We can use the rule for dividing powers with like bases to lead us to a definition of exponential notation when the exponent is a negative integer. Consider $5^3/5^7$ and first simplify it using procedures we have learned for working with fractions:

$$\frac{5^3}{5^7} = \frac{5 \cdot 5 \cdot 5}{5 \cdot 5 \cdot 5 \cdot 5 \cdot 5 \cdot 5 \cdot 5} = \frac{5 \cdot 5 \cdot 5 \cdot 1}{5 \cdot 5 \cdot 5 \cdot 5 \cdot 5 \cdot 5 \cdot 5}$$

$$= \frac{5 \cdot 5 \cdot 5}{5 \cdot 5 \cdot 5} \cdot \frac{1}{5 \cdot 5 \cdot 5 \cdot 5} = \frac{1}{5^4}.$$

Now we apply the rule for dividing powers with the same bases. Then

$$\frac{5^3}{5^7} = 5^{3-7} = 5^{-4}.$$

From these two expressions for $5^3/5^7$, it follows that

$$5^{-4} = \frac{1}{5^4}.$$

This leads to our definition of negative exponents:

For any real number a that is nonzero and any integer n,

$$a^{-n} = \frac{1}{a^n}.$$

(The numbers a^{-n} and a^n are reciprocals.)

▶ **EXAMPLES** Express using positive exponents. Then simplify.

17. $4^{-2} = \dfrac{1}{4^2} = \dfrac{1}{16}$

18. $(-3)^{-2} = \dfrac{1}{(-3)^2} = \dfrac{1}{(-3)(-3)} = \dfrac{1}{9}$

19. $m^{-3} = \dfrac{1}{m^3}$

20. $ab^{-1} = a\left(\dfrac{1}{b^1}\right) = a\left(\dfrac{1}{b}\right) = \dfrac{a}{b}$

21. $3c^{-5} = 3\left(\dfrac{1}{c^5}\right) = \dfrac{3}{c^5}$

22. $\dfrac{1}{x^{-3}} = x^3$ (x^{-3} and x^3 are reciprocals.) ◀

CAUTION! Note in Example 17 that

$$4^{-2} \neq 4(-2) \quad \text{and} \quad \frac{1}{4^2} \neq 4(-2).$$

Similarly, in Example 18,

$$(-3)^{-2} \neq (-3)(-2) \quad \text{and} \quad \frac{1}{(-3)^2} \neq (-3)(-2).$$

In particular, $a^{-n} \neq a(-n)$. The negative exponent also does not mean to multiply in the denominator. That is

$$4^{-2} = \frac{1}{16}, \quad not \quad \frac{1}{-16}.$$

DO EXERCISES 23–28.

The rules for multiplying and dividing powers with like bases still hold when exponents are 0 or negative. We will state them in a summary at the end of this section.

Express with positive exponents. Then simplify.

23. 4^{-3}

24. 5^{-2}

25. 2^{-4}

26. $(-2)^{-3}$

27. $4p^{-3}$

28. $\dfrac{1}{x^{-2}}$

ANSWERS ON PAGE A-3

Simplify.

29. $5^{-2} \cdot 5^4$

▶ **EXAMPLES** Simplify. By simplify, we mean write as one number to a nonnegative power.

23. $7^{-3} \cdot 7^6 = 7^{-3+6}$ Adding   **24.** $x^4 \cdot x^{-3} = x^{4+(-3)} = x^1 = x$
                    $= 7^3$ exponents

25. $\dfrac{5^4}{5^{-2}} = 5^{4-(-2)}$ Subtracting   **26.** $\dfrac{x}{x^7} = x^{1-7} = x^{-6} = \dfrac{1}{x^6}$
                       exponents

    $\phantom{\dfrac{5^4}{5^{-2}} } = 5^{4+2} = 5^6$

27. $\dfrac{b^{-4}}{b^{-5}} = b^{-4-(-5)} = b^1 = b$ ◀

30. $x^{-3} \cdot x^{-4}$

In Examples 23–27, it may help to think as follows: After writing the base, write the top exponent. Then write a subtraction sign. Then write the bottom exponent. Then do the subtraction by adding the opposite. For example,

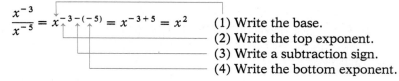

$$\frac{x^{-3}}{x^{-5}} = x^{-3-(-5)} = x^{-3+5} = x^2$$

(1) Write the base.
(2) Write the top exponent.
(3) Write a subtraction sign.
(4) Write the bottom exponent.

31. $\dfrac{7^{-2}}{7^3}$

DO EXERCISES 29–33.

The following is another way to explain the definition of negative exponents.

On this side, we divide by 5 at each step.

$$5 \cdot 5 \cdot 5 \cdot 5 = 5^4$$
$$5 \cdot 5 \cdot 5 = 5^3$$
$$5 \cdot 5 = 5^2$$
$$5 = 5^1$$
$$1 = 5^0$$
$$\frac{1}{5} = 5^?$$
$$\frac{1}{25} = 5^?$$

On this side, the exponents decrease by 1.

32. $\dfrac{b^{-2}}{b^{-3}}$

To continue the pattern, it should follow that

$$\frac{1}{5} = \frac{1}{5^1} = 5^{-1} \quad \text{and} \quad \frac{1}{25} = \frac{1}{5^2} = 5^{-2}.$$

The following is a summary of the definitions and rules for exponents that we have considered in this section.

33. $\dfrac{t}{t^{-5}}$

Definitions and Rules for Exponents

For any integers m and n,

1 as an exponent: $a^1 = a,$

0 as an exponent: $a^0 = 1, a \neq 0$

Negative integers as exponents: $a^{-n} = \dfrac{1}{a^n}$

Product Rule: $a^m \cdot a^n = a^{m+n}$

Quotient Rule: $\dfrac{a^m}{a^n} = a^{m-n}$

NAME   SECTION   DATE

EXERCISE SET 3.1

a   What is the meaning of each of the following?

1. 2^4 **2.** 5^3 **3.** $(1.4)^5$ **4.** m^6

5. $(7p)^2$ **6.** $(11c)^3$ **7.** $(19k)^4$ **8.** $(10pq)^2$

b   Evaluate.

9. t^0, $t \neq 0$ **10.** a^0, $a \neq 0$ **11.** a^1 **12.** q^1

13. 9.68^0 **14.** 9.68^1 **15.** $(ab)^1$ **16.** $(ab)^0$

c   Evaluate.

17. m^3 when $m = 3$ **18.** x^6 when $x = 2$

19. p^1 when $p = 19$ **20.** x^{19} when $x = 0$

21. x^4 when $x = 4$ **22.** y^{15} when $y = 1$

23. $y^2 - 7$ when $y = -10$ **24.** $z^5 + 5$ when $z = -2$

1. _____
2. _____
3. _____
4. _____
5. _____
6. _____
7. _____
8. _____
9. _____
10. _____
11. _____
12. _____
13. _____
14. _____
15. _____
16. _____
17. _____
18. _____
19. _____
20. _____
21. _____
22. _____
23. _____
24. _____

25. _____

26. _____

27. _____

28. _____

29. _____

30. _____

31. _____

32. _____

33. _____

34. _____

35. _____

36. _____

37. _____

38. _____

39. _____

40. _____

41. _____

42. _____

43. _____

44. _____

45. _____

46. _____

47. _____

48. _____

49. _____

50. _____

25. Find the area of a circle when $r = 34$ ft. Use 3.14 for π.

26. The area A of a square with sides of length s is given by $A = s^2$. Find the area of a square with sides of length 24 m (meters).

f   Express using positive exponents. Then simplify.

27. 3^{-2} **28.** 2^{-3} **29.** 10^{-4} **30.** 5^{-6}

31. 7^{-3} **32.** 5^{-2} **33.** a^{-3} **34.** x^{-2}

35. $\dfrac{1}{y^{-4}}$ **36.** $\dfrac{1}{t^{-7}}$ **37.** $\dfrac{1}{z^{-n}}$ **38.** $\dfrac{1}{h^{-n}}$

Express using negative exponents.

39. $\dfrac{1}{4^3}$ **40.** $\dfrac{1}{5^2}$ **41.** $\dfrac{1}{x^3}$ **42.** $\dfrac{1}{y^2}$

d , **f**   Multiply and simplify.

43. $2^4 \cdot 2^3$ **44.** $3^5 \cdot 3^2$ **45.** $8^5 \cdot 8^9$ **46.** $n^3 \cdot n^{20}$

47. $x^4 \cdot x^3$ **48.** $y^7 \cdot y^9$ **49.** $9^{17} \cdot 9^{21}$ **50.** $t^0 \cdot t^{16}$

51. $(3y)^4(3y)^8$ **52.** $(2t)^8(2t)^{17}$ **53.** $(7y)^1(7y)^{16}$ **54.** $(8x)^0(8x)^1$

55. $3^{-5} \cdot 3^8$ **56.** $5^{-8} \cdot 5^9$ **57.** $x^{-2} \cdot x$ **58.** $x \cdot x^{-1}$

59. $x^4 \cdot x^3$ **60.** $x^9 \cdot x^4$ **61.** $x^{-7} \cdot x^{-6}$ **62.** $y^{-5} \cdot y^{-8}$

63. $t^8 \cdot t^{-8}$ **64.** $m^{10} \cdot m^{-10}$

e , f Divide and simplify.

65. $\dfrac{7^5}{7^2}$ **66.** $\dfrac{4^7}{4^3}$ **67.** $\dfrac{8^{12}}{8^6}$ **68.** $\dfrac{9^{14}}{9^2}$

69. $\dfrac{y^9}{y^5}$ **70.** $\dfrac{x^{12}}{x^{11}}$ **71.** $\dfrac{16^2}{16^8}$ **72.** $\dfrac{5^4}{5^{10}}$

73. $\dfrac{m^6}{m^{12}}$ **74.** $\dfrac{p^4}{p^5}$ **75.** $\dfrac{(8x)^6}{(8x)^{10}}$ **76.** $\dfrac{(9t)^4}{(9t)^{11}}$

ANSWERS

51. _____

52. _____

53. _____

54. _____

55. _____

56. _____

57. _____

58. _____

59. _____

60. _____

61. _____

62. _____

63. _____

64. _____

65. _____

66. _____

67. _____

68. _____

69. _____

70. _____

71. _____

72. _____

73. _____

74. _____

75. _____

76. _____

77. $\dfrac{18^9}{18^9}$ **78.** $\dfrac{(6y)^7}{(6y)^7}$ **79.** $\dfrac{x}{x^{-1}}$ **80.** $\dfrac{x^6}{x}$

81. $\dfrac{x^7}{x^{-2}}$ **82.** $\dfrac{t^8}{t^{-3}}$ **83.** $\dfrac{z^{-6}}{z^{-2}}$ **84.** $\dfrac{y^{-7}}{y^{-3}}$

85. $\dfrac{x^{-5}}{x^{-8}}$ **86.** $\dfrac{y^{-4}}{y^{-9}}$ **87.** $\dfrac{m^{-9}}{m^{-9}}$ **88.** $\dfrac{x^{-8}}{x^{-8}}$

Simplify.

89. 8^2, 8^{-2}, $\left(\dfrac{1}{8}\right)^2$, $\left(\dfrac{1}{8}\right)^{-2}$, -8^2,   and $(-8)^2$

90. 5^2, 5^{-2}, $\left(\dfrac{1}{5}\right)^2$, $\left(\dfrac{1}{5}\right)^{-2}$, -5^2,   and $(-5)^2$

SKILL MAINTENANCE

91. Translate to an algebraic expression: Sixty-four percent of t.

92. Evaluate $3x/y$ when $x = 4$ and $y = 12$.

93. Divide: $1555.2 \div 24.3$.

94. Add: $1555.2 + 24.3$.

SYNTHESIS

95. Determine whether $(5y)^0$ and $5y^0$ are equivalent expressions.

Simplify.

96. $(y^{2x})(y^{3x})$ **97.** $a^{5k} \div a^{3k}$ **98.** $\dfrac{a^{6t}(a^{7t})}{a^{9t}}$

99. $\dfrac{\left(\dfrac{1}{2}\right)^4}{\left(\dfrac{1}{2}\right)^5}$ **100.** $\dfrac{(0.4)^5}{(0.4)^3(0.4)^2}$

101. Determine whether $(a + b)^2$ and $a^2 + b^2$ are equivalent. (*Hint:* Choose values for x and y and evaluate.)

Use $>$, $<$, or $=$ for ▨ to write a true sentence.

102. 3^5 ▨ 3^4 **103.** 4^2 ▨ 4^3 **104.** 4^3 ▨ 5^3 **105.** 4^3 ▨ 3^4

Find a value of the variable that shows that the two expressions are *not* equivalent.

106. $3x^2$; $(3x)^2$ **107.** $(a + 3)^2$; $a^2 + 3^2$

108. $\dfrac{x + 2}{2}$; x **109.** $\dfrac{y^6}{y^3}$; y^2

3.2 Exponents and Scientific Notation

We now enhance our ability to manipulate exponential expressions by considering three more rules. The rules are also applied to a new way to name numbers called *scientific notation*.

a Raising Powers to Powers

Consider an expression like $(3^2)^4$. We are raising 3^2 to the fourth power:

$$(3^2)^4 = (3^2)(3^2)(3^2)(3^2)$$
$$= (3 \cdot 3)(3 \cdot 3)(3 \cdot 3)(3 \cdot 3)$$
$$= 3 \cdot 3 \cdot 3 \cdot 3 \cdot 3 \cdot 3 \cdot 3 \cdot 3$$
$$= 3^8.$$

Note that in this case we could have multiplied the exponents:

$$(3^2)^4 = 3^{2 \cdot 4} = 3^8.$$

Likewise, $(y^8)^3 = (y^8)(y^8)(y^8) = y^{24}$. Once again, we get the same result if we multiply the exponents:

$$(y^8)^3 = y^{8 \cdot 3} = y^{24}.$$

> ### The Power Rule
>
> **For any real number a and any integers m and n,**
> $$(a^m)^n = a^{mn}.$$
> **(To raise a power to a power, multiply the exponents.)**

▶ **EXAMPLES** Simplify.

1. $(3^5)^4 = 3^{5 \cdot 4}$ Multiplying
      $= 3^{20}$ exponents

2. $(2^2)^5 = 2^{2 \cdot 5} = 2^{10}$

3. $(y^{-5})^7 = y^{-5 \cdot 7} = y^{-35} = \dfrac{1}{y^{35}}$

4. $(x^4)^{-2} = x^{4(-2)} = x^{-8} = \dfrac{1}{x^8}$

5. $(a^{-4})^{-6} = a^{(-4)(-6)} = a^{24}$ ◀

DO EXERCISES 1–4.

b Raising a Product or a Quotient to a Power

When an expression inside parentheses is raised to a power, the inside expression is the base. Let us compare $2a^3$ and $(2a)^3$:

$2a^3 = 2 \cdot a \cdot a \cdot a;$ The base is a.

$(2a)^3 = (2a)(2a)(2a)$ The base is $2a$.
$\quad = (2 \cdot 2 \cdot 2)(a \cdot a \cdot a)$ Using the associative law of multiplication
$\quad = 2^3 a^3$
$\quad = 8a^3.$

We see that $2a^3$ and $(2a)^3$ are *not* equivalent. We also see that we can evaluate the power $(2a)^3$ by raising each factor to the power 3. This leads us to the following rule for raising a product to a power.

OBJECTIVES

After finishing Section 3.2, you should be able to:

a Use the power rule to raise powers to powers.

b Raise a product to a power and a quotient to a power.

c Convert between scientific notation and decimal notation.

d Multiply and divide using scientific notation.

e Solve problems using scientific notation.

FOR EXTRA HELP

Tape NC   Tape 6B   MAC: 3
                                   IBM: 3

Simplify.
1. $(3^4)^5$

2. $(x^{-3})^4$

3. $(y^{-5})^{-3}$

4. $(x^{-4})^8$

ANSWERS ON PAGE A-3

Simplify.

5. $(2x^5y^{-3})^4$

6. $(5x^5y^{-6}z^{-3})^2$

7. $[(-x)^{37}]^2$

8. $(3y^{-2}x^{-5}z^8)^3$

Simplify.

9. $\left(\dfrac{x^6}{5}\right)^2$

10. $\left(\dfrac{2t^5}{w^4}\right)^3$

11. $\left(\dfrac{x^4}{3}\right)^{-2}$

> ### Raising a Product to a Power
>
> **For any real numbers a and b and any integer n,**
> $$(ab)^n = a^n b^n.$$
> **(To raise a product to the nth power, raise each factor to the nth power.)**

▶ **EXAMPLES**

6. $(4x^2)^3 = 4^3 \cdot (x^2)^3$ Raising each factor to the third power
$= 64x^6$

7. $(5x^3y^5z^2)^4 = 5^4(x^3)^4(y^5)^4(z^2)^4$ Raising each factor to the fourth power
$= 625x^{12}y^{20}z^8$

8. $(-5x^4y^3)^3 = (-5)^3(x^4)^3(y^3)^3$
$= -125x^{12}y^9$

9. $[(-x)^{25}]^2 = (-x)^{50}$
$= (-1 \cdot x)^{50}$ Using the property of -1 (Section 1.8)
$= (-1)^{50}x^{50}$
$= 1 \cdot x^{50}$ The product of an even number of negative factors is positive.
$= x^{50}$

10. $(5x^2y^{-2})^3 = 5^3(x^2)^3(y^{-2})^3 = 125x^6y^{-6}$ Be careful to raise *each* factor to the third power.
$= \dfrac{125x^6}{y^6}$

11. $(3x^3y^{-5}z^2)^4 = 3^4(x^3)^4(y^{-5})^4(z^2)^4$
$= 81x^{12}y^{-20}z^8$
$= \dfrac{81x^{12}z^8}{y^{20}}$

◀

DO EXERCISES 5–8.

There is a similar rule for raising a quotient to a power.

> ### Raising a Quotient to a Power
>
> **For any real numbers a and b, $b \neq 0$, and any integer n,**
> $$\left(\frac{a}{b}\right)^n = \frac{a^n}{b^n}.$$
> **(To raise a quotient to a power, raise the numerator to the power and divide by the denominator to the power.)**

▶ **EXAMPLES**   Simplify.

12. $\left(\dfrac{x^2}{4}\right)^3 = \dfrac{(x^2)^3}{4^3} = \dfrac{x^6}{64}$

13. $\left(\dfrac{3a^4}{b^3}\right)^2 = \dfrac{(3a^4)^2}{(b^3)^2} = \dfrac{3^2(a^4)^2}{b^{3\cdot2}} = \dfrac{9a^8}{b^6}$

14. $\left(\dfrac{y^3}{5}\right)^{-2} = \dfrac{(y^3)^{-2}}{5^{-2}} = \dfrac{y^{-6}}{5^{-2}} = \dfrac{\dfrac{1}{y^6}}{\dfrac{1}{5^2}} = \dfrac{1}{y^6} \div \dfrac{1}{5^2} = \dfrac{1}{y^6} \cdot \dfrac{5^2}{1} = \dfrac{25}{y^6}$

◀

DO EXERCISES 9–11.

c Scientific Notation

There are many kinds of symbols, or notation, for numbers. You are already familiar with fractional notation, decimal notation, and percent notation. Now we study another, **scientific notation,** which is especially useful when calculations involve very large or very small numbers. The following are examples of scientific notation:

The distance from the earth to the sun:
$$9.3 \times 10^7 \text{ mi} = 93,000,000 \text{ mi}$$

The mass of a hydrogen atom:
$$1.7 \times 10^{-24} \text{ gm} = 0.0000000000000000000000017 \text{ gm}$$

Scientific notation for a number is an expression of the type
$$N \times 10^n,$$
where 1 is less than or equal to N and N is less than 10 ($1 \leq N < 10$), and N is expressed in decimal notation. 10^n is also considered to be scientific notation when $N = 1$.

You should try to make conversions to scientific notation mentally as much as possible. Here is a handy mental device.

A positive exponent indicates a large number (greater than one) and a negative exponent indicates a small number (less than one).

▶ **EXAMPLES**

15. $78,000 = 7.8 \times 10^4$ 7.8,000.

                          4 places

Large number, so the exponent is positive.

16. $0.0000057 = 5.7 \times 10^{-6}$ 0.000005.7

                              6 places

Small number, so the exponent is negative. ◀

Each of the following is *not* scientific notation.

$$12.46 \times 10^7 \qquad\qquad 0.347 \times 10^{-5}$$

This number is greater than 10.   This number is less than 1.

DO EXERCISES 12 AND 13.

▶ **EXAMPLES**   Convert mentally to decimal notation.

17. $7.893 \times 10^5 = 789,300$ 7.89300.

                          5 places

Positive exponent, so the answer is a large number.

18. $4.7 \times 10^{-8} = 0.000000047$ 0.00000004.7

                              8 places

Negative exponent, so the answer is a small number. ◀

DO EXERCISES 14 AND 15.

Convert to scientific notation.

12. 0.000517

13. 523,000,000

Convert to decimal notation.

14. 6.893×10^{11}

15. 5.67×10^{-5}

ANSWERS ON PAGE A-3

Multiply and write scientific notation for the result.

16. $(1.12 \times 10^{-8})(5 \times 10^{-7})$

When using a calculator, we can express a number like 260,000,000 using scientific notation in a form like

$$2.6 \text{ E } 8, \quad \text{or} \quad 2.6 \quad 8,$$

or perhaps in other forms, depending on your calculator.

d   Multiplying and Dividing Using Scientific Notation

Multiplying

Consider the product

$$400 \cdot 2000 = 800,000.$$

In scientific notation, this would be

$$(4 \times 10^2) \cdot (2 \times 10^3) = (4 \cdot 2)(10^2 \cdot 10^3) = 8 \times 10^5.$$

By applying the commutative and associative laws, we can find this product by multiplying $4 \cdot 2$, to get 8, and $10^2 \cdot 10^3$, to get 10^5 (we do this by adding the exponents).

▶ **EXAMPLE 19**   Multiply: $(1.8 \times 10^6) \cdot (2.3 \times 10^{-4})$.

We apply the commutative and associative laws to get

$$
\begin{aligned}
(1.8 \times 10^6) \cdot (2.3 \times 10^{-4}) &= (1.8 \cdot 2.3) \times (10^6 \cdot 10^{-4}) \\
&= 4.14 \times 10^{6+(-4)} \quad \text{\small Adding exponents} \\
&= 4.14 \times 10^2. \quad\quad\quad\quad\quad ◀
\end{aligned}
$$

17. $(9.1 \times 10^{-17})(8.2 \times 10^3)$

▶ **EXAMPLE 20**   Multiply: $(3.1 \times 10^5) \cdot (4.5 \times 10^{-3})$.

We have

$$
\begin{aligned}
(3.1 \times 10^5) \cdot (4.5 \times 10^{-3}) &= (3.1 \times 4.5)(10^5 \cdot 10^{-3}) \\
&= 13.95 \times 10^2.
\end{aligned}
$$

The answer at this stage is

$$13.95 \times 10^2,$$

but this is *not* scientific notation, because 13.95 is not a number between 1 and 10. To find scientific notation, we convert 13.95 to scientific notation and simplify:

$$
\begin{aligned}
13.95 \times 10^2 &= (1.395 \times 10^1) \times 10^2 \quad \text{\small Substituting } 1.395 \times 10^1 \text{ for } 13.95 \\
&= 1.395 \times (10^1 \times 10^2) \quad \text{\small Associative law} \\
&= 1.395 \times 10^3. \quad\quad\quad \text{\small Adding exponents}
\end{aligned}
$$

The answer is

$$1.395 \times 10^3. \quad\quad\quad ◀$$

DO EXERCISES 16 AND 17.

Dividing

Consider the quotient

$$800,000 \div 400 = 2000.$$

In scientific notation, this is

$$(8 \times 10^5) \div (4 \times 10^2) = \frac{8 \times 10^5}{4 \times 10^2} = 2 \times 10^3.$$

We can find this product by dividing 8 by 4, to get 2, and 10^5 by 10^2, to get 10^3 (we do this by subtracting the exponents).

▶ **EXAMPLE 21**   Divide: $(3.41 \times 10^5) \div (1.1 \times 10^{-3})$.

$$(3.41 \times 10^5) \div (1.1 \times 10^{-3}) = \frac{3.41 \times 10^5}{1.1 \times 10^{-3}}$$
$$= \frac{3.41}{1.1} \times \frac{10^5}{10^{-3}}$$
$$= 3.1 \times 10^{5-(-3)}$$
$$= 3.1 \times 10^8 \qquad ◀$$

▶ **EXAMPLE 22**   Divide: $(6.4 \times 10^{-7}) \div (8.0 \times 10^6)$.
We have

$$(6.4 \times 10^{-7}) \div (8.0 \times 10^6) = \frac{6.4 \times 10^{-7}}{8.0 \times 10^6}$$
$$= \frac{6.4}{8.0} \times \frac{10^{-7}}{10^6}$$
$$= 0.8 \times 10^{-7-6}$$
$$= 0.8 \times 10^{-13}.$$

The answer at this stage is

$$0.8 \times 10^{-13},$$

but this is *not* scientific notation, because 0.8 is not a number between 1 and 10. To find scientific notation, we convert 0.8 to scientific notation and simplify:

$$0.8 \times 10^{-13} = (8.0 \times 10^{-1}) \times 10^{-13} \quad \text{Substituting } 8.0 \times 10^{-1} \text{ for } 0.8$$
$$= 8.0 \times (10^{-1} \times 10^{-13}) \quad \text{Associative law}$$
$$= 8.0 \times 10^{-14}. \quad \text{Adding exponents}$$

The answer is

$$8.0 \times 10^{-14}. \qquad ◀$$

DO EXERCISES 18 AND 19.

e   Solving Problems with Scientific Notation

▶ **EXAMPLE 23**   There are 3300 members in the Professional Bowlers Association. There are 244 million people in the United States. What part of the population are members of the Professional Bowlers Association? Write scientific notation for the answer.

Divide and write scientific notation for the result.

18. $\dfrac{4.2 \times 10^5}{2.1 \times 10^2}$

19. $\dfrac{1.1 \times 10^{-4}}{2.0 \times 10^{-7}}$

ANSWERS ON PAGE A-3

20. There are 300,000 words in the English language. The average person knows about 10,000 of them. What part of the total number of words does the average person know? Write scientific notation for the answer.

The part of the population that belongs to the Professional Bowlers Association is

$$\frac{3300}{244 \text{ million}}.$$

We know that 1 million = 1,000,000 = 10^6, so 244 million = 244×10^6, or 2.44×10^8. We also have 3300 = 3.3×10^3. We can now divide and write scientific notation for the answer:

$$\frac{3300}{244 \text{ million}} = \frac{3.3 \times 10^3}{2.44 \times 10^8}$$

$$\approx 1.3525 \times 10^{-5}. \qquad ◄$$

DO EXERCISE 20.

▶ **EXAMPLE 24**   Americans drink 3 million gallons of orange juice in one day. How much orange juice is consumed in this country in one year? Write scientific notation for the answer.

There are 365 days in a year, so the amount of orange juice consumed is

$$(365 \text{ days}) \cdot (3 \text{ million}) = (3.65 \times 10^2)(3 \times 10^6)$$

$$= 10.95 \times 10^8$$

$$= (1.095 \times 10^1) \times 10^8$$

$$= 1.095 \times 10^9.$$

There are 1.095×10^9 gallons of orange juice consumed in this country in one year. ◄

DO EXERCISE 21.

21. Americans eat 6.5 million gallons of popcorn each day. How much popcorn do they eat in one year? Write scientific notation for the answer.

The following is a summary of the definitions and rules for exponents that we have considered in this section and the preceding one.

Definitions and Rules for Exponents

$$a^1 = a, \qquad a^0 = 1, \qquad a \neq 0,$$

Negative exponents: $\qquad a^{-n} = \dfrac{1}{a^n}, \quad a \neq 0$

Product Rule: $\qquad a^m \cdot a^n = a^{m+n}$

Quotient Rule: $\qquad \dfrac{a^m}{a^n} = a^{m-n}$

Power Rule: $\qquad (a^m)^n = a^{mn}$

Raising a Product to a Power: $\quad (ab)^n = a^n b^n$

Raising a Quotient to a Power: $\quad \left(\dfrac{a}{b}\right)^n = \dfrac{a^n}{b^n}$

Scientific Notation: $\qquad N \times 10^n$, or 10^n, where N is a number such that $1 \leq N < 10$

NAME   SECTION   DATE

EXERCISE SET 3.2

a , **b** Simplify.

1. $(2^3)^2$

2. $(3^4)^3$

3. $(5^2)^{-3}$

4. $(9^3)^{-4}$

5. $(x^{-3})^{-4}$

6. $(a^{-5})^{-6}$

7. $(4x^3)^2$

8. $4(x^3)^2$

9. $(x^4y^5)^{-3}$

10. $(t^5x^3)^{-4}$

11. $(x^{-6}y^{-2})^{-4}$

12. $(x^{-2}y^{-7})^{-5}$

13. $(3x^3y^{-8}z^{-3})^2$

14. $(2a^2y^{-4}z^{-5})^3$

15. $\left(\dfrac{a^2}{b^3}\right)^4$

16. $\left(\dfrac{x^3}{y^4}\right)^5$

17. $\left(\dfrac{y^3}{2}\right)^2$

18. $\left(\dfrac{a^5}{3}\right)^3$

19. $\left(\dfrac{y^2}{2}\right)^{-3}$

20. $\left(\dfrac{a^4}{3}\right)^{-2}$

21. $\left(\dfrac{x^2y}{z}\right)^3$

22. $\left(\dfrac{m}{n^4p}\right)^3$

23. $\left(\dfrac{a^2b}{cd^3}\right)^{-2}$

24. $\left(\dfrac{2a^2}{3b^4}\right)^{-3}$

c Convert to scientific notation.

25. 78,000,000,000

26. 3,700,000,000,000

27. 907,000,000,000,000,000

28. 168,000,000,000,000

29. 0.00000374

30. 0.000000000275

31. 0.000000018

32. 0.00000000002

33. 100,000,000,000

34. 0.0000001

Convert to decimal notation.

35. 7.84×10^8

36. 1.35×10^7

37. 8.764×10^{-10}

38. 9.043×10^{-3}

39. 10^8

40. 10^4

41. 10^{-4}

42. 10^{-7}

1.
2.
3.
4.
5.
6.
7.
8.
9.
10.
11.
12.
13.
14.
15.
16.
17.
18.
19.
20.
21.
22.
23.
24.
25.
26.
27.
28.
29.
30.
31.
32.
33.
34.
35.
36.
37.
38.
39.
40.
41.
42.

d   Multiply or divide and write scientific notation for the result.

43. $(3 \times 10^4)(2 \times 10^5)$

44. $(1.9 \times 10^8)(3.4 \times 10^{-3})$

45. $(5.2 \times 10^5)(6.5 \times 10^{-2})$

46. $(7.1 \times 10^{-7})(8.6 \times 10^{-5})$

47. $(9.9 \times 10^{-6})(8.23 \times 10^{-8})$

48. $(1.123 \times 10^4) \times 10^{-9}$

49. $\dfrac{8.5 \times 10^8}{3.4 \times 10^{-5}}$

50. $\dfrac{5.6 \times 10^{-2}}{2.5 \times 10^5}$

51. $(3.0 \times 10^6) \div (6.0 \times 10^9)$

52. $(1.5 \times 10^{-3}) \div (1.6 \times 10^{-6})$

53. $\dfrac{7.5 \times 10^{-9}}{2.5 \times 10^{12}}$

54. $\dfrac{4.0 \times 10^{-3}}{8.0 \times 10^{20}}$

e   Solve. Write scientific notation for the answer.

55. About 250,000 people die per day in the world. How many die in one year?

56. The average discharge at the mouth of the Amazon River is 4,200,000 cubic feet per second. How much water is discharged from the Amazon River in one hour? in one year?

57. There are 300,000 words in the English language. The exceptional person knows about 20,000 of them. What part of the total number of words does the exceptional person know?

58. The mass of the earth is about 5.98×10^{24} kg. The mass of the planet Saturn is about 95 times the mass of the earth. Write scientific notation for the mass of Saturn.

SYNTHESIS

59. ▤ Carry out the indicated operations. Write scientific notation for the result.

$$\frac{(5.2 \times 10^6)(6.1 \times 10^{-11})}{1.28 \times 10^{-3}}$$

60. Find the reciprocal and express in scientific notation.

$$(6.25 \times 10^{-3})$$

61. Write $4^3 \cdot 8 \cdot 16$ as a power of 2.

62. Write $2^8 \cdot 16^3 \cdot 64$ as a power of 4.

Simplify.

63. $\dfrac{(5^{12})^2}{5^{25}}$

64. $\dfrac{a^{22}}{(a^2)^{11}}$

65. $\dfrac{(3^5)^4}{3^5 \cdot 3^4}$

66. $\dfrac{49^{18}}{7^{35}}$

67. $\left(\dfrac{1}{a}\right)^{-n}$

68. $\dfrac{(0.4)^5}{[(0.4)^3]^2}$

(*Hint:* Study Exercise 64.)

Determine whether each of the following is true for any pairs of integers m and n and any positive numbers x and y.

69. $x^m \cdot y^n = (xy)^{mn}$

70. $x^m \cdot y^m = (xy)^{2m}$

71. $(x - y)^m = x^m - y^m$

3.3 Introduction to Polynomials

We have already learned to evaluate and to manipulate certain kinds of algebraic expressions. We will now consider algebraic expressions called *polynomials*.

The following are examples of *monomials in one variable:*

$$3x^2, \quad 2x, \quad -5, \quad 37p^4, \quad 0.$$

Each expression is a constant or a constant times some variable to a non-negative integer power. More formally, a **monomial** is an expression of the type ax^n, where a is a real-number constant and n is a nonnegative integer.

Algebraic expressions like the following are **polynomials:**

$$\tfrac{3}{4}y^5, \quad -2, \quad 5y + 3, \quad 3x^2 + 2x - 5, \quad -7a^3 + \tfrac{1}{2}a, \quad 6x, \quad 37p^4, \quad x, \quad 0.$$

> A *polynomial* is a monomial or a combination of sums and/or differences of monomials.

The following algebraic expressions are *not* polynomials:

$$(1) \quad \frac{x+3}{x-4}, \qquad (2) \quad 5x^3 - 2x^2 + \frac{1}{x}, \qquad (3) \quad \frac{1}{x^3 - 2}.$$

Expressions (1) and (3) are not polynomials because they represent quotients, not sums. Expression (2) is not a polynomial because

$$\frac{1}{x} = x^{-1},$$

and this is not a monomial because the exponent is negative.

DO EXERCISE 1.

a Evaluating Polynomials and Applications

When we replace the variable in a polynomial by a number, the polynomial then represents a number called a **value** of the polynomial. Finding that number, or value, is called **evaluating the polynomial.** We evaluate a polynomial using our rules for order of operations (Section 1.8).

▶ **EXAMPLE 1** Evaluate the polynomial when $x = 2$.

a) $3x + 5 = 3 \cdot 2 + 5 = 6 + 5 = 11$

b) $2x^2 - 7x + 3 = 2 \cdot 2^2 - 7 \cdot 2 + 3 = 2 \cdot 4 - 14 + 3 = 8 - 14 + 3 = -3$ ◀

▶ **EXAMPLE 2** Evaluate the polynomial when $x = -5$.

a) $2 - x^3 = 2 - (-5)^3 = 2 - (-125) = 2 + 125 = 127$

b) $-x^2 - 3x + 1 = -(-5)^2 - 3(-5) + 1 = -25 + 15 + 1 = -9$ ◀

DO EXERCISES 2–5.

Polynomials occur in many real-world situations. The following examples are two such applications.

OBJECTIVES

After finishing Section 3.3, you should be able to:

a Evaluate a polynomial for a given value of the variable.

b Identify the terms of a polynomial.

c Identify the like terms of a polynomial.

d Identify the coefficients of a polynomial.

e Collect the like terms of a polynomial.

f Arrange a polynomial in descending order, or collect the like terms and then arrange in descending order.

g Identify the degree of each term of a polynomial and the degree of the polynomial.

h Identify the missing terms of a polynomial.

i Classify a polynomial as a monomial, binomial, trinomial, or none of these.

FOR EXTRA HELP

Tape 8A   Tape 6B   MAC: 3
                                  IBM: 3

---

1. Write three polynomials.

Evaluate the polynomial for $x = 3$.

2. $-4x - 7$

3. $-5x^3 + 7x + 10$

Evaluate the polynomial for $x = -4$.

4. $5x + 7$

5. $2x^2 + 5x - 4$

ANSWERS ON PAGE A-3

6. In the situation of Example 4, what is the total number of games to be played in a league of 12 teams?

7. The perimeter of a square of side x is given by the polynomial $4x$.

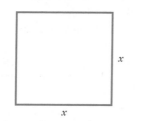

A baseball diamond is a square 90 ft on a side. Find the perimeter of a baseball diamond.

8. In the situation of Example 3, find the concentration after 3 hr.

Find an equivalent polynomial using only additions.

9. $-9x^3 - 4x^5$

10. $-2y^3 + 3y^7 - 7y$

▶ **EXAMPLE 3**   *Medical dosage.* The concentration, in parts per million, of a certain medication in the bloodstream after time t, in hours, is given by the polynomial

$$-0.05t^2 + 2t + 2.$$

Find the concentration after 2 hr.

To find the concentration after 2 hr, we evaluate the polynomial for $t = 2$:

$$\begin{aligned}
-0.05t^2 + 2t + 2 &= -0.05(2)^2 + 2(2) + 2 \\
&= -0.05(4) + 2(2) + 2 \\
&= -0.2 + 4 + 2 \\
&= -0.2 + 6 \\
&= 5.8.
\end{aligned}$$

Carrying out the calculation using rules for order of operations

The concentration after 2 hr is 5.8 parts per million.   ◀

▶ **EXAMPLE 4**   *Games in a sports league.* In a sports league of n teams in which each team plays every other team twice, the total number of games to be played is given by the polynomial

$$n^2 - n.$$

A women's slow-pitch softball league has 10 teams. What is the total number of games to be played?

We evaluate the polynomial for $n = 10$:

$$n^2 - n = 10^2 - 10 = 100 - 10 = 90.$$

The league plays 90 games.   ◀

DO EXERCISES 6–8.

b   Identifying Terms

As we saw in Section 1.4, subtractions can be rewritten as additions. For any polynomial that has some subtractions, we can find an equivalent polynomial using only additions.

▶ **EXAMPLES**   Find an equivalent polynomial using only additions.

5. $-5x^2 - x = -5x^2 + (-x)$

6. $4x^5 - 2x^6 - 4x + 7 = 4x^5 + (-2x^6) + (-4x) + 7$ ◀

DO EXERCISES 9 AND 10.

When a polynomial has only additions, the monomials being added are called **terms.** In Example 5, the terms are $-5x^2$ and $-x$. In Example 6, the terms are $4x^5$, $-2x^6$, $-4x$, and 7.

▶ **EXAMPLE 7**   Identify the terms of the polynomial

$$4x^7 + 3x + 12 + 8x^3 + 5x.$$

Terms: $4x^7$, $3x$, 12, $8x^3$, and $5x$.   ◀

If there are subtractions, you can *think* of them as additions without rewriting.

▶ **EXAMPLE 8** Identify the terms of the polynomial

$$3t^4 - 5t^6 - 4t + 2.$$

Terms: $3t^4$, $-5t^6$, $-4t$, and 2.   ◀

DO EXERCISES 11 AND 12.

c   Like Terms

When terms have the same variable and the variable is raised to the same power, we say that they are **like terms,** or **similar terms.**

▶ **EXAMPLES**   Identify the like terms in the polynomial.

9. $4x^3 + 5x - 4x^2 + 2x^3 + x^2$

Like terms: $4x^3$ and $2x^3$ Same variable and exponent
Like terms: $-4x^2$ and x^2 Same variable and exponent

$6 - 3a^2 + 8 - a - 5a$

10. Like terms: 6 and 8   Constant terms are like terms because
$6 = 6x^0$ and $8 = 8x^0$.

Like terms: $-a$ and $-5a$ ◀

DO EXERCISES 13 AND 14.

d   Coefficients

The coefficient of the term $5x^3$ is 5. In the following polynomial, the color numbers are the **coefficients:**

$$3x^5 - 2x^3 + 5x + 4.$$

▶ **EXAMPLE 11**   Identify the coefficient of each term in the polynomial

$$3x^4 - 4x^3 + 7x^2 + x - 8.$$

The coefficient of the first term is 3.
The coefficient of the second term is -4.
The coefficient of the third term is 7.
The coefficient of the fourth term is 1.
The coefficient of the fifth term is -8.   ◀

DO EXERCISE 15.

e   Collecting Like Terms

We can often simplify polynomials by **collecting like terms,** or **combining similar terms.** To do this, we use the distributive laws. We factor out the exponential expression and add or subtract the coefficients. We try to do this mentally as much as possible.

▶ **EXAMPLES**   Collect like terms.

12. $2x^3 - 6x^3 = (2 - 6)x^3 = -4x^3$ Using a distributive law

13. $5x^2 + 7 + 4x^4 + 2x^2 - 11 - 2x^4 = (5 + 2)x^2 + (4 - 2)x^4 + (7 - 11)$
$= 7x^2 + 2x^4 - 4$ ◀

Identify the terms of the polynomial.

11. $3x^2 + 6x + \dfrac{1}{2}$

12. $-4y^5 + 7y^2 - 3y - 2$

Identify the like terms in the polynomial.

13. $4x^3 - x^3 + 2$

14. $4t^4 - 9t^3 - 7t^4 + 10t^3$

15. Identify the coefficient of each term in the polynomial
$2x^4 - 7x^3 - 8.5x^2 + 10x - 4.$

ANSWERS ON PAGE A-3

Collect like terms.

16. $3x^2 + 5x^2$

17. $4x^3 - 2x^3 + 2 + 5$

18. $\frac{1}{2}x^5 - \frac{3}{4}x^5 + 4x^2 - 2x^2$

Collect like terms.

19. $24 - 4x^3 - 24$

20. $5x^3 - 8x^5 + 8x^5$

21. $-2x^4 + 16 + 2x^4 + 9 - 3x^5$

Collect like terms.

22. $7x - x$

23. $5x^3 - x^3 + 4$

24. $\frac{3}{4}x^3 + 4x^2 - x^3 + 7$

25. $8x^2 - x^2 + x^3 - 1 - 4x^2 + 10$

Arrange each polynomial in descending order.

26. $x + 3x^5 + 4x^3 + 5x^2 + 6x^7 - 2x^4$

27. $4x^2 - 3 + 7x^5 + 2x^3 - 5x^4$

28. $-14 + 7t^2 - 10t^5 + 14t^7$

　　Note that using the distributive laws in this manner allows us to collect like terms by adding or subtracting the coefficients. Often the middle step is omitted and we add or subtract mentally, just writing the answer. In collecting like terms, we may get 0.

▶ **EXAMPLES**　Collect like terms.

14. $5x^3 - 5x^3 = (5 - 5)x^3 = 0x^3 = 0$

15. $3x^4 + 2x^2 - 3x^4 + 8 = (3 - 3)x^4 + 2x^2 + 8$
$$= 0x^4 + 2x^2 + 8 = 2x^2 + 8 \quad ◀$$

DO EXERCISES 16–21.

　　Multiplying a term of a polynomial by 1 does not change the term, but it may make the polynomial easier to factor or add and subtract.

▶ **EXAMPLES**　Collect like terms.

16. $5x^2 + x^2 = 5x^2 + 1x^2$　　Replacing x^2 by $1x^2$
$$= (5 + 1)x^2 \qquad \text{Using a distributive law}$$
$$= 6x^2$$

17. $5x^4 - 6x^3 - x^4 = 5x^4 - 6x^3 - 1x^4 \qquad x^4 = 1x^4$
$$= (5 - 1)x^4 - 6x^3$$
$$= 4x^4 - 6x^3$$

18. $\frac{2}{3}x^4 - x^3 - \frac{1}{6}x^4 + \frac{2}{5}x^3 - \frac{3}{10}x^3 = (\frac{2}{3} - \frac{1}{6})x^4 + (-1 + \frac{2}{5} - \frac{3}{10})x^3$
$$= (\frac{4}{6} - \frac{1}{6})x^4 + (-\frac{10}{10} + \frac{4}{10} - \frac{3}{10})x^3$$
$$= \frac{3}{6}x^4 - \frac{9}{10}x^3$$
$$= \frac{1}{2}x^4 - \frac{9}{10}x^3 \quad ◀$$

DO EXERCISES 22–25.

f　Descending and Ascending Order

Note in the following polynomial that the exponents decrease. We say that the polynomial is arranged in **descending order:**

$$2x^4 - 8x^3 + 5x^2 - x + 3.$$

The term with the largest exponent is first. The term with the next largest exponent is second, and so on. The associative and commutative laws allow us to arrange the terms of a polynomial in descending order.

▶ **EXAMPLES**　Arrange the polynomial in descending order.

19. $6x^5 + 4x^7 + x^2 + 2x^3 = 4x^7 + 6x^5 + 2x^3 + x^2$

20. $\frac{2}{3} + 4x^5 - 8x^2 + 5x - 3x^3 = 4x^5 - 3x^3 - 8x^2 + 5x + \frac{2}{3} \quad ◀$

　　We usually arrange polynomials in descending order, but not always. The opposite order is called **ascending order.** Generally, if an exercise is written in a certain order, we give the answer in that same order.

DO EXERCISES 26–28.

▶ **EXAMPLE 21** Collect like terms and then arrange in descending order:

$$2x^2 - 4x^3 + 3 - x^2 - 2x^3.$$

We have

$$2x^2 - 4x^3 + 3 - x^2 - 2x^3 = x^2 - 6x^3 + 3 \qquad \text{Collecting like terms}$$
$$= -6x^3 + x^2 + 3 \qquad \text{Arranging in descending order} \quad ◀$$

DO EXERCISES 29 AND 30.

g Degrees

The **degree** of a term is the exponent of the variable. The degree of the term $5x^3$ is 3.

▶ **EXAMPLE 22** Identify the degree of each term of $8x^4 + 3x + 7$.

The degree of $8x^4$ is 4.

The degree of $3x$ is 1.   Recall that $x = x^1$.

The degree of 7 is 0.   Think of 7 as $7x^0$. Recall that $x^0 = 1$.   ◀

The **degree of a polynomial** is the largest of the degrees of the terms, unless it is the polynomial 0. The polynomial 0 is a special case. We agree that it has *no* degree either as a term or as a polynomial. This is because we can express 0 as $0 = 0x^5 = 0x^7$, and so on, using any exponent we wish.

▶ **EXAMPLE 23** Identify the degree of the polynomial $5x^3 - 6x^4 + 7$.

We have

$$5x^3 - 6x^4 + 7. \qquad \text{The largest exponent is 4.}$$

The degree of the polynomial is 4.   ◀

DO EXERCISE 31.

Let us summarize the terminology we have learned for the polynomial

$$3x^4 - 8x^3 + 5x^2 + 7x - 6$$

Term	Coefficient	Degree of the term	Degree of the polynomial
$3x^4$	3	4	4
$-8x^3$	-8	3	
$5x^2$	5	2	
$7x$	7	1	
-6	-6	0	

Collect like terms and then arrange in descending order.

29. $3x^2 - 2x + 3 - 5x^2 - 1 - x$

30. $-x + \dfrac{1}{2} + 14x^4 - 7x - 1 - 4x^4$

31. Identify the degree of each term and the degree of the polynomial:

$$-6x^4 + 8x^2 - 2x + 9.$$

ANSWERS ON PAGE A-4

Identify the missing terms in the polynomial.

32. $2x^3 + 4x^2 - 2$

33. $-3x^4$

34. $x^3 + 1$

35. $x^4 - x^2 + 3x + 0.25$

Classify the polynomial as a monomial, binomial, trinomial, or none of these.

36. $5x^4$

37. $4x^3 - 3x^2 + 4x + 2$

38. $3x^2 + x$

39. $3x^2 + 2x - 4$

h   Missing Terms

If a coefficient is 0, we usually do not write the term. We say that we have a **missing term.**

▶ **EXAMPLE 23**   Identify the missing terms in the polynomial

$$8x^5 - 2x^3 + 5x^2 + 7x + 8.$$

There is no term with x^4. We say that the x^4-term (or the *fourth-degree term*) is missing.   ◀

For certain skills or manipulations, we can write missing terms with zero coefficients or leave space. For example, we can write the polynomial $3x^2 + 9$ as

$$3x^2 + 0x + 9 \quad \text{or} \quad 3x^2 + \qquad 9.$$

DO EXERCISES 32–35.

i   Classifying Polynomials

Polynomials with just one term are called **monomials.** Polynomials with just two terms are called **binomials.** Those with just three terms are called **trinomials.** Those with more than three terms are usually not specified with a name.

▶ **EXAMPLE 24**

Monomials	Binomials	Trinomials	None of these
$4x^2$	$2x + 4$	$3x^3 + 4x + 7$	$4x^3 - 5x^2 + x - 8$
9	$3x^5 + 6x$	$6x^7 - 7x^2 + 4$	
$-23x^{19}$	$-9x^7 - 6$	$4x^2 - 6x - \frac{1}{2}$	

◀

DO EXERCISES 36–39.

NAME   SECTION   DATE

EXERCISE SET 3.3

a Evaluate the polynomial for $x = 4$.

1. $-5x + 2$ **2.** $-3x + 1$ **3.** $2x^2 - 5x + 7$

4. $3x^2 + x + 7$ **5.** $x^3 - 5x^2 + x$ **6.** $7 - x + 3x^2$

Evaluate the polynomial for $x = -1$.

7. $3x + 5$ **8.** $6 - 2x$ **9.** $x^2 - 2x + 1$

10. $5x - 6 + x^2$ **11.** $-3x^3 + 7x^2 - 3x - 2$ **12.** $-2x^3 - 5x^2 + 4x + 3$

Daily accidents. The daily number of accidents N (average number of accidents per day) involving drivers of age a is approximated by the polynomial

$$N = 0.4a^2 - 40a + 1039.$$

13. Evaluate the polynomial for $a = 18$ to find the daily number of accidents involving 18-year-old drivers.

14. Evaluate the polynomial for $a = 20$ to find the daily number of accidents involving 20-year-old drivers.

Falling distance. The distance s, in feet, traveled by a body falling freely from rest in t seconds is approximated by the polynomial

$$s = 16t^2.$$

$s = 16t^2$

15. A stone is dropped from a cliff and takes 8 sec to hit the ground. How high is the cliff?

16. A brick is dropped from a building and takes 3 sec to hit the ground. How high is the building?

Total revenue. An electronics firm is marketing a new kind of stereo. *Total revenue* is the total amount of money taken in. The firm determines that when it sells x stereos, it will take in

$$280x - 0.4x^2 \text{ dollars.}$$

17. What is the total revenue from the sale of 75 stereos?

18. What is the total revenue from the sale of 100 stereos?

Total cost. The electronics firm determines that the total cost of producing x stereos is given by

$$5000 + 0.6x^2 \text{ dollars.}$$

19. What is the total cost of producing 500 stereos?

20. What is the total cost of producing 650 stereos?

b Identify the terms of the polynomial.

21. $2 - 3x + x^2$ **22.** $2x^2 + 3x - 4$

1. _____

2. _____

3. _____

4. _____

5. _____

6. _____

7. _____

8. _____

9. _____

10. _____

11. _____

12. _____

13. _____

14. _____

15. _____

16. _____

17. _____

18. _____

19. _____

20. _____

21. _____

22. _____

c Identify the like terms in the polynomial.

23. $5x^3 + 6x^2 - 3x^2$ **24.** $3x^2 + 4x^3 - 2x^2$

25. $2x^4 + 5x - 7x - 3x^4$ **26.** $-3t + t^3 - 2t - 5t^3$

d Identify the coefficient of each term of the polynomial.

27. $-3x + 6$ **28.** $2x - 4$ **29.** $5x^2 + 3x + 3$

30. $3x^2 - 5x + 2$ **31.** $-7x^3 + 6x^2 + 3x + 7$ **32.** $5x^4 + x^2 - x + 2$

33. $-5x^4 + 6x^3 - 3x^2 + 8x - 2$ **34.** $7x^3 - 4x^2 - 4x + 5$

e Collect like terms.

35. $2x - 5x$ **36.** $2x^2 + 8x^2$ **37.** $x - 9x$

38. $x - 5x$ **39.** $5x^3 + 6x^3 + 4$ **40.** $6x^4 - 2x^4 + 5$

41. $5x^3 + 6x - 4x^3 - 7x$ **42.** $3a^4 - 2a + 2a + a^4$ **43.** $6b^5 + 3b^2 - 2b^5 - 3b^2$

44. $2x^2 - 6x + 3x + 4x^2$ **45.** $\frac{1}{4}x^5 - 5 + \frac{1}{2}x^5 - 2x - 37$

46. $\frac{1}{3}x^3 + 2x - \frac{1}{6}x^3 + 4 - 16$ **47.** $6x^2 + 2x^4 - 2x^2 - x^4 - 4x^2$

48. $8x^2 + 2x^3 - 3x^3 - 4x^2 - 4x^2$ **49.** $\frac{1}{4}x^3 - x^2 - \frac{1}{6}x^2 + \frac{3}{8}x^3 + \frac{5}{16}x^3$

50. $\frac{1}{5}x^4 + \frac{1}{5} - 2x^2 + \frac{1}{10} - \frac{3}{15}x^4 + 2x^2 - \frac{3}{10}$

f Arrange the polynomial in descending order.

51. $x^5 + x + 6x^3 + 1 + 2x^2$ **52.** $3 + 2x^2 - 5x^6 - 2x^3 + 3x$

53. $5y^3 + 15y^9 + y - y^2 + 7y^8$ **54.** $9p - 5 + 6p^3 - 5p^4 + p^5$

Collect like terms and then arrange in descending order.

55. $3x^4 - 5x^6 - 2x^4 + 6x^6$

56. $-1 + 5x^3 - 3 - 7x^3 + x^4 + 5$

57. $-2x + 4x^3 - 7x + 9x^3 + 8$

58. $-6x^2 + x - 5x + 7x^2 + 1$

59. $3x + 3x + 3x - x^2 - 4x^2$

60. $-2x - 2x - 2x + x^3 - 5x^3$

61. $-x + \dfrac{3}{4} + 15x^4 - x - \dfrac{1}{2} - 3x^4$

62. $2x - \dfrac{5}{6} + 4x^3 + x + \dfrac{1}{3} - 2x$

g Identify the degree of each term of the polynomial and the degree of the polynomial.

63. $2x - 4$

64. $6 - 3x$

65. $3x^2 - 5x + 2$

66. $5x^3 - 2x^2 + 3$

67. $-7x^3 + 6x^2 + 3x + 7$

68. $5x^4 + x^2 - x + 2$

69. $x^2 - 3x + x^6 - 9x^4$

70. $8x - 3x^2 + 9 - 8x^3$

71. For the polynomial $-7x^4 + 6x^3 - 3x^2 + 8x - 2$, complete the following table.

Term	Coefficient	Degree of the term	Degree of the polynomial
$6x^3$	6		
		2	
$8x$		1	
	-2		

ANSWERS

55.
56.
57.
58.
59.
60.
61.
62.
63.
64.
65.
66.
67.
68.
69.
70.
71.

72. For the polynomial $3x^2 + 8x^5 - 46x^3 + 6x - 2.4 - \frac{1}{2}x^4$, complete the following table.

Term	Coefficient	Degree of the term	Degree of the polynomial
		5	
$-\frac{1}{2}x^4$		4	
	-46		
$3x^2$		2	
	6		
-2.4			

h   Identify the missing terms in the polynomial.

73. $x^3 - 27$ **74.** $x^5 + x$ **75.** $x^4 - x$

76. $5x^4 - 7x + 2$ **77.** $2x^3 - 5x^2 + x - 3$ **78.** $-6x^3$

i   Classify the polynomial as a monomial, binomial, trinomial, or none of these.

79. $x^2 - 10x + 25$ **80.** $-6x^4$ **81.** $x^3 - 7x^2 + 2x - 4$

82. $x^2 - 9$ **83.** $4x^2 - 25$ **84.** $2x^4 - 7x^3 + x^2 + x - 6$

85. $40x$ **86.** $4x^2 + 12x + 9$

SKILL MAINTENANCE

87. Three tired campers stopped for the night. All they had to eat was a bag of apples. During the night, one awoke and ate one third of the apples. Later, a second camper awoke and ate one third of the apples that remained. Much later, the third camper awoke and ate one third of those apples yet remaining after the other two had eaten. When they got up the next morning, 8 apples were left. How many did they have to begin with?

88. A family spent $2011 to drive a car one year, during which the car was driven 7400 mi. The family spent $972 for insurance and $114 for a license registration fee. The only other cost was for gasoline. How much did gasoline cost per mile?

SYNTHESIS

Combine like terms.

89. $\frac{9}{2}x^8 + \frac{1}{9}x^2 + \frac{1}{2}x^9 + \frac{9}{2}x^1 + \frac{9}{2}x^9 + \frac{8}{9}x^2 + \frac{1}{2}x - \frac{1}{2}x^8$

90. $(3x^2)^3 + 4x^2 \cdot 4x^4 - x^4(2x)^2 + ((2x)^2)^3 - 100x^2(x^2)^2$

91. Construct a polynomial in x (meaning that x is the variable) of degree 5 with four terms and coefficients that are integers.

92. What is the degree of $(5m^5)^2$?

93. A polynomial in x has degree 3. The coefficient of x^2 is three less than the coefficient of x^3. The coefficient of x is three times the coefficient of x^2. The remaining coefficient is two more than the coefficient of x^3. The sum of the coefficients is -4. Find the polynomial.

3.4 Addition and Subtraction of Polynomials

a Addition

To add two polynomials, we can think of writing a plus sign between them and then collecting like terms. Depending on the situation, you may see polynomials written in descending order, ascending order, or neither. Generally, if an exercise is written in a particular order, we write the answer in that same order.

▶ **EXAMPLE 1** Add: $(-3x^3 + 2x - 4) + (4x^3 + 3x^2 + 2)$.

$(-3x^3 + 2x - 4) + (4x^3 + 3x^2 + 2)$
$= (-3 + 4)x^3 + 3x^2 + 2x + (-4 + 2)$ Collecting like terms (*No* signs are changed.)
$= x^3 + 3x^2 + 2x - 2$ ◀

▶ **EXAMPLE 2** Add:

$$\left(\frac{2}{3}x^4 + 3x^2 - 2x + \frac{1}{2}\right) + \left(-\frac{1}{3}x^4 + 5x^3 - 3x^2 + 3x - \frac{1}{2}\right).$$

$\left(\frac{2}{3}x^4 + 3x^2 - 2x + \frac{1}{2}\right) + \left(-\frac{1}{3}x^4 + 5x^3 - 3x^2 + 3x - \frac{1}{2}\right)$
$= \left(\frac{2}{3} - \frac{1}{3}\right)x^4 + 5x^3 + (3 - 3)x^2 + (-2 + 3)x + \left(\frac{1}{2} - \frac{1}{2}\right)$ Collecting like terms
$= \frac{1}{3}x^4 + 5x^3 + x$ ◀

We can add polynomials as we do because they represent numbers. After some practice, you will be able to add mentally.

DO EXERCISES 1–4.

▶ **EXAMPLE 3** Add: $(3x^2 - 2x + 2) + (5x^3 - 2x^2 + 3x - 4)$.

$(3x^2 - 2x + 2) + (5x^3 - 2x^2 + 3x - 4)$
$= 5x^3 + (3 - 2)x^2 + (-2 + 3)x + (2 - 4)$ You might do this step mentally.
$= 5x^3 + x^2 + x - 2$ Then you would write only this. ◀

DO EXERCISES 5 AND 6.

We can also add polynomials by writing like terms in columns.

▶ **EXAMPLE 4** Add: $9x^5 - 2x^3 + 6x^2 + 3$ and $5x^4 - 7x^2 + 6$ and $3x^6 - 5x^5 + x^2 + 5$.

We arrange the polynomials with like terms in columns.

$$
\begin{array}{l}
9x^5 \qquad\quad - 2x^3 + 6x^2 + 3 \\
\qquad\quad 5x^4 \qquad\qquad - 7x^2 + 6 \\
3x^6 - 5x^5 \qquad\qquad\quad + x^2 + 5 \\
\hline
3x^6 + 4x^5 + 5x^4 - 2x^3 \qquad\quad + 14
\end{array}
$$

We leave spaces for missing terms.

Adding

We write the answer as $3x^6 + 4x^5 + 5x^4 - 2x^3 + 14$ without the missing space. ◀

OBJECTIVES

After finishing Section 3.4, you should be able to:

a Add polynomials.

b Find the opposite of a polynomial.

c Subtract polynomials.

d Solve problems using addition and subtraction of polynomials.

FOR EXTRA HELP

Tape 8B   Tape 7A   MAC: 3   IBM: 3

Add.

1. $(3x^2 + 2x - 2) + (-2x^2 + 5x + 5)$

2. $(-4x^5 + x^3 + 4) + (7x^4 + 2x^2)$

3. $(31x^4 + x^2 + 2x - 1) + (-7x^4 + 5x^3 - 2x + 2)$

4. $(17x^3 - x^2 + 3x + 4) + \left(-15x^3 + x^2 - 3x - \frac{2}{3}\right)$

Add mentally. Try to write just the answer.

5. $(4x^2 - 5x + 3) + (-2x^2 + 2x - 4)$

6. $(3x^3 - 4x^2 - 5x + 3) + \left(5x^3 + 2x^2 - 3x - \frac{1}{2}\right)$

ANSWERS ON PAGE A-4

Add.

7.
$$-2x^3 + 5x^2 - 2x + 4$$
$$x^4 \qquad + 6x^2 + 7x - 10$$
$$-9x^4 + 6x^3 + x^2 \qquad - 2$$

8. $-3x^3 + 5x + 2$ and
$x^3 + x^2 + 5$ and
$x^3 - 2x - 4$

Find two equivalent expressions for the opposite of the polynomial.

9. $12x^4 - 3x^2 + 4x$

10. $-4x^4 + 3x^2 - 4x$

11. $-13x^6 + 2x^4 - 3x^2 + x - \dfrac{5}{13}$

12. $-7y^3 + 2y^2 - y + 3$

Simplify.

13. $-(4x^3 - 6x + 3)$

14. $-(5x^4 + 3x^2 + 7x - 5)$

15. $-\left(14x^{10} - \dfrac{1}{2}x^5 + 5x^3 - x^2 + 3x\right)$

It is sometimes easier to visualize the addition if we add in columns.

DO EXERCISES 7 AND 8.

b   **Opposites of Polynomials**

We now look at subtraction of polynomials. To do so, we first consider the opposite, or additive inverse, of a polynomial.

We know that two numbers are opposites of each other if their sum is zero. For example, 5 and -5 are opposites, since $5 + (-5) = 0$. The same definition holds for polynomials.

> Two polynomials are *opposites*, or *additive inverses*, of each other if their sum is zero.

To find a way to determine an opposite, look for a pattern in the following examples:

a) $2x + (-2x) = 0$;
b) $-6x^2 + 6x^2 = 0$;
c) $(5t^3 - 2) + (-5t^3 + 2) = 0$;
d) $(7x^3 - 6x^2 - x + 4) + (-7x^3 + 6x^2 + x - 4) = 0$.

Since $(5t^3 - 2) + (-5t^3 + 2) = 0$, we know that the opposite of $(5t^3 - 2)$ is $(-5t^3 + 2)$. To say the same thing with purely algebraic symbolism, consider

The opposite of $(5t^3 - 2)$ is $-5t^3 + 2$.

$$-(5t^3 - 2) = -5t^3 + 2.$$

> We can find an equivalent polynomial for the opposite, or additive inverse, of a polynomial by replacing each term by its opposite—that is, *changing the sign of every term.*

▶ **EXAMPLE 5** Find two equivalent expressions for the opposite of
$$4x^5 - 7x^3 - 8x + \frac{5}{6}.$$

a) $-\left(4x^5 - 7x^3 - 8x + \dfrac{5}{6}\right)$

b) $-4x^5 + 7x^3 + 8x - \dfrac{5}{6}$ Changing the sign of every term

Thus, $-(4x^5 - 7x^3 - 8x + \frac{5}{6})$ is equivalent to $-4x^3 + 7x^3 + 8x - \frac{5}{6}$, and each is the opposite of the original polynomial $4x^5 - 7x^3 - 8x + \frac{5}{6}$. ◀

DO EXERCISES 9–12.

▶ **EXAMPLE 6** Simplify: $-\left(-7x^4 - \dfrac{5}{9}x^3 + 8x^2 - x + 67\right)$.

$$-\left(-7x^4 - \frac{5}{9}x^3 + 8x^2 - x + 67\right) = 7x^4 + \frac{5}{9}x^3 - 8x^2 + x - 67 \quad ◀$$

DO EXERCISES 13–15.

C Subtraction of Polynomials

Recall that we can subtract a real number by adding its opposite, or additive inverse: $a - b = a + (-b)$. This allows us to find an equivalent expression for the difference of two polynomials.

▶ **EXAMPLE 7** Subtract:
$$(9x^5 + x^3 - 2x^2 + 4) - (2x^5 + x^4 - 4x^3 - 3x^2).$$

We have

$(9x^5 + x^3 - 2x^2 + 4) - (2x^5 + x^4 - 4x^3 - 3x^2)$
$= 9x^5 + x^3 - 2x^2 + 4 + [-(2x^5 + x^4 - 4x^3 - 3x^2)]$ Adding the opposite
$= 9x^5 + x^3 - 2x^2 + 4 - 2x^5 - x^4 + 4x^3 + 3x^2$ Finding the opposite by changing the sign of *each* term
$= 7x^5 - x^4 + 5x^3 + x^2 + 4$ Collecting like terms ◀

DO EXERCISES 16 AND 17.

As with similar work in Section 1.8, we combine steps by changing the sign of each term of the polynomial being subtracted and collecting like terms. Try to do this mentally as much as possible.

▶ **EXAMPLE 8** Subtract: $(9x^5 + x^3 - 2x) - (-2x^5 + 5x^3 + 6)$.

$(9x^5 + x^3 - 2x) - (-2x^5 + 5x^3 + 6)$
$= 11x^5 - 4x^3 - 2x - 6$ Change signs and collect like terms. Try to do so mentally. ◀

DO EXERCISES 18 AND 19.

We can use columns to subtract. We replace coefficients by their opposites, as shown in Example 7.

▶ **EXAMPLE 9** Write in columns and subtract:
$$(5x^2 - 3x + 6) - (9x^2 - 5x - 3).$$

a) $5x^2 - 3x + 6$
   $-(9x^2 - 5x - 3)$ Writing similar terms in columns

b) $5x^2 - 3x + 6$
   $-9x^2 + 5x + 3$ Changing signs

c) $5x^2 - 3x + 6$
   $-9x^2 + 5x + 3$
   $\overline{-4x^2 + 2x + 9}$ Adding ◀

If you can do so without error, you can arrange the polynomials in columns and write just the answer.

Subtract.
16. $(7x^3 + 2x + 4) - (5x^3 - 4)$

17. $(-3x^2 + 5x - 4) - (-4x^2 + 11x - 2)$

Subtract mentally. Try to write just the answer.
18. $(-6x^4 + 3x^2 + 6) - (2x^4 + 5x^3 - 5x^2 + 7)$

19. $\left(\frac{3}{2}x^3 - \frac{1}{2}x^2 + 0.3\right) - \left(\frac{1}{2}x^3 + \frac{1}{2}x^2 + \frac{4}{3}x + 1.2\right)$

ANSWERS ON PAGE A-4

Write in columns and subtract.

20. $(4x^3 + 2x^2 - 2x - 3) -$
$(2x^3 - 3x^2 + 2)$

21. $(2x^3 + x^2 - 6x + 2) -$
$(x^5 + 4x^3 - 2x^2 - 4x)$

22. Find a polynomial for the sum of
the areas of the rectangles.

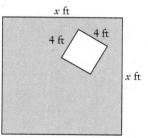

23. Find a polynomial for the shaded
area.

▶ **EXAMPLE 10**   Write in columns and subtract:

$$(x^3 + x^2 + 2x - 12) - (-2x^3 + x^2 - 3x).$$

We have

$$
\begin{array}{r}
x^3 + x^2 + 2x - 12 \\
-2x^3 + x^2 - 3x \\
\hline
3x^3 \qquad + 5x - 12.
\end{array}
$$ ◀

DO EXERCISES 20 AND 21.

d   **Solving Problems**

▶ **EXAMPLE 11**   Find a polynomial for the sum of the areas of these rectangles.

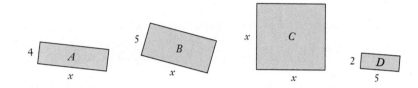

Recall that the area of a rectangle is the product of the length and the width. The sum of the areas is a sum of products. We find these products and then collect like terms.

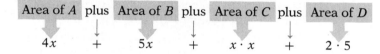

We collect like terms:

$$4x + 5x + x^2 + 10 = x^2 + 9x + 10.$$ ◀

DO EXERCISE 22.

▶ **EXAMPLE 12**   A 4-ft by 4-ft sandbox is placed on a square lawn x ft on a side. Find a polynomial for the remaining area.

We draw a picture of the situation as shown here.

We reword the problem and write the polynomial as follows.

$$\underbrace{\text{Area of lawn}} - \underbrace{\text{Area of sandbox}} = \text{Area left over}$$
$$x \cdot x \quad - \quad 4 \cdot 4 \quad = \text{Area left over}$$

Then

$$x^2 - 16 = \text{Area left over.}$$ ◀

DO EXERCISE 23.

EXERCISE SET 3.4

a Add.

1. $(3x + 2) + (-4x + 3)$

2. $(6x + 1) + (-7x + 2)$

3. $(-6x + 2) + (x^2 + x - 3)$

4. $(x^2 - 5x + 4) + (8x - 9)$

5. $(x^2 - 9) + (x^2 + 9)$

6. $(x^3 + x^2) + (2x^3 - 5x^2)$

7. $(3x^2 - 5x + 10) + (2x^2 + 8x - 40)$

8. $(6x^4 + 3x^3 - 1) + (4x^2 - 3x + 3)$

9. $(1.2x^3 + 4.5x^2 - 3.8x) +$
$(-3.4x^3 - 4.7x^2 + 23)$

10. $(0.5x^4 - 0.6x^2 + 0.7) +$
$(2.3x^4 + 1.8x - 3.9)$

11. $(1 + 4x + 6x^2 + 7x^3) +$
$(5 - 4x + 6x^2 - 7x^3)$

12. $(3x^4 - 6x - 5x^2 + 5) +$
$(6x^2 - 4x^3 - 1 + 7x)$

13. $(9x^8 - 7x^4 + 2x^2 + 5) +$
$(8x^7 + 4x^4 - 2x)$

14. $(4x^5 - 6x^3 - 9x + 1) +$
$(6x^3 + 9x^2 + 9x)$

15. $(\frac{1}{4}x^4 + \frac{2}{3}x^3 + \frac{5}{8}x^2 + 7) +$
$(-\frac{3}{4}x^4 + \frac{3}{8}x^2 - 7)$

16. $(\frac{1}{3}x^9 + \frac{1}{5}x^5 - \frac{1}{2}x^2 + 7) +$
$(-\frac{1}{5}x^9 + \frac{1}{4}x^4 - \frac{3}{5}x^5 + \frac{3}{4}x^2 + \frac{1}{2})$

17. $0.02x^5 - 0.2x^3 + x + 0.08$ and
$-0.01x^5 + x^4 - 0.8x - 0.02$

18. $(0.03x^6 + 0.05x^3 + 0.22x + 0.05) +$
$(\frac{7}{100}x^6 - \frac{3}{100}x^3 + 0.5)$

19. $-3x^4 + 6x^2 + 2x - 1$
$\underline{-3x^2 + 2x + 1}$

20. $-4x^3 + 8x^2 + 3x - 2$
$\underline{-4x^2 + 3x + 2}$

21. $\begin{aligned}
&0.15x^4 + 0.10x^3 - 0.9x^2 \\
&-0.01x^3 + 0.01x^2 + x \\
&1.25x^4 + 0.11x^2 + 0.01 \\
&0.27x^3 + 0.99 \\
&\underline{-0.35x^4 + 15x^2 - 0.03}
\end{aligned}$

22. $\begin{aligned}
&0.05x^4 + 0.12x^3 - 0.5x^2 \\
&-0.02x^3 + 0.02x^2 + 2x \\
&1.5x^4 + 0.01x^2 + 0.15 \\
&0.25x^3 + 0.85 \\
&\underline{-0.25x^4 + 10x^2 - 0.04}
\end{aligned}$

1. _____

2. _____

3. _____

4. _____

5. _____

6. _____

7. _____

8. _____

9. _____

10. _____

11. _____

12. _____

13. _____

14. _____

15. _____

16. _____

17. _____

18. _____

19. _____

20. _____

21. _____

22. _____

b   Find two equivalent expressions for the opposite of the polynomial.

23. $-5x$ **24.** $x^2 - 3x$ **25.** $-x^2 + 10x - 2$

26. $-4x^3 - x^2 - x$ **27.** $12x^4 - 3x^3 + 3$ **28.** $4x^3 - 6x^2 - 8x + 1$

Simplify.

29. $-(3x - 7)$ **30.** $-(-2x + 4)$

31. $-(4x^2 - 3x + 2)$ **32.** $-(-6a^3 + 2a^2 - 9a + 1)$

33. $-(-4x^4 + 6x^2 + \frac{3}{4}x - 8)$ **34.** $-(-5x^4 + 4x^3 - x^2 + 0.9)$

c   Subtract.

35. $(3x + 2) - (-4x + 3)$ **36.** $(6x + 1) - (-7x + 2)$

37. $(-6x + 2) - (x^2 + x - 3)$ **38.** $(x^2 - 5x + 4) - (8x - 9)$

39. $(x^2 - 9) - (x^2 + 9)$ **40.** $(x^3 + x^2) - (2x^3 - 5x^2)$

41. $(6x^4 + 3x^3 - 1) - (4x^2 - 3x + 3)$ **42.** $(-4x^2 + 2x) - (3x^3 - 5x^2 + 3)$

43. $(1.2x^3 + 4.5x^2 - 3.8x) -$
$(-3.4x^3 - 4.7x^2 + 23)$ **44.** $(0.5x^4 - 0.6x^2 + 0.7) -$
$(2.3x^4 + 1.8x - 3.9)$

45. $(\frac{5}{8}x^3 - \frac{1}{4}x - \frac{1}{3}) - (-\frac{1}{8}x^3 + \frac{1}{4}x - \frac{1}{3})$ **46.** $(\frac{1}{5}x^3 + 2x^2 - 0.1) -$
$(-\frac{2}{5}x^3 + 2x^2 + 0.01)$

47. $(0.08x^3 - 0.02x^2 + 0.01x) -$
$(0.02x^3 + 0.03x^2 - 1)$ **48.** $(0.8x^4 + 0.2x - 1) - (\frac{7}{10}x^4 + \frac{1}{5}x - 0.1)$

Subtract.

49. $x^2 + 5x + 6$
      $x^2 + 2x$

50. $x^3 \qquad + 1$
      $x^3 + x^2$

51. $\quad 5x^4 + 6x^3 - 9x^2$
      $-6x^4 - 6x^3 \qquad + 8x + 9$

52. $5x^4 \qquad + 6x^2 - 3x + 6$
      $\quad 6x^3 + 7x^2 - 8x - 9$

53. $x^5 \qquad\qquad - 1$
      $x^5 - x^4 + x^3 - x^2 + x - 1$

54. $x^5 + x^4 - x^3 + x^2 - x + 2$
      $x^5 - x^4 + x^3 - x^2 - x + 2$

[d] Solve.

55. Find a polynomial for the sum of the areas of these rectangles.

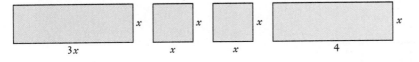

Find a polynomial for the perimeter of the figure.

56.

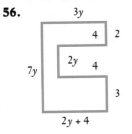

57.

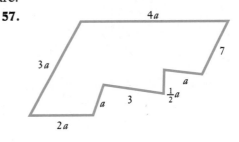

Find two algebraic expressions for the area of the figure.

58.

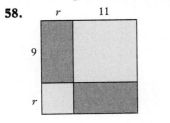

59.

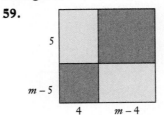

ANSWERS

49. _____

50. _____

51. _____

52. _____

53. _____

54. _____

55. _____

56. _____

57. _____

58. _____

59. _____

60. Find $(x + 3)^2$ using the four areas of the square shown here.

60. _____

Find a polynomial for the shaded area.

61. _____

61.

62. _____

62.

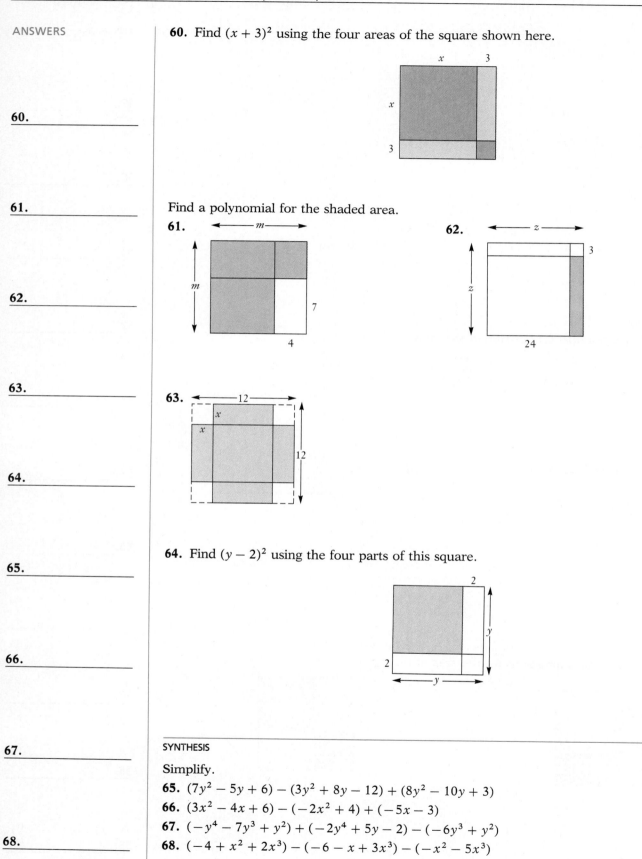

63. _____

63.

64. _____

64. Find $(y - 2)^2$ using the four parts of this square.

65. _____

66. _____

Copyright © 1991 Addison-Wesley Publishing Co., Inc.

67. _____

SYNTHESIS

Simplify.

65. $(7y^2 - 5y + 6) - (3y^2 + 8y - 12) + (8y^2 - 10y + 3)$

66. $(3x^2 - 4x + 6) - (-2x^2 + 4) + (-5x - 3)$

67. $(-y^4 - 7y^3 + y^2) + (-2y^4 + 5y - 2) - (-6y^3 + y^2)$

68. _____

68. $(-4 + x^2 + 2x^3) - (-6 - x + 3x^3) - (-x^2 - 5x^3)$

3.5 Multiplication of Polynomials

We now multiply polynomials using techniques based, for the most part, on the distributive laws, but also on the associative and commutative laws. As we proceed in this chapter, we will develop special ways to find certain products.

a Multiplying Monomials

Consider $(3x)(4x)$. We multiply as follows:

$$(3x)(4x) = 3 \cdot x \cdot 4 \cdot x \quad \text{By the associative law of multiplication}$$
$$= 3 \cdot 4 \cdot x \cdot x \quad \text{By the commutative law of multiplication}$$
$$= (3 \cdot 4) \cdot x \cdot x \quad \text{By the associative law}$$
$$= 12x^2. \quad \text{Using the product rule for exponents}$$

> **To find an equivalent expression for the product of two monomials, multiply the coefficients and then multiply the variables using the product rule for exponents.**

▶ **EXAMPLES** Multiply.

1. $(5x)(6x) = (5 \cdot 6)(x \cdot x) \quad \text{Multiplying the coefficients}$
$= 30x^2 \quad \text{Simplifying}$

2. $(3x)(-x) = (3x)(-1x)$
$= (3)(-1)(x \cdot x)$
$= -3x^2$

3. $(-7x^5)(4x^3) = (-7 \cdot 4)(x^5 \cdot x^3)$
$= -28x^{5+3} \quad \text{Adding the exponents}$
$= -28x^8 \quad \text{Simplifying}$ ◀

After some practice, you can do this mentally. Multiply the coefficients and then the variables by keeping the base and adding the exponents. Write only the answer.

DO EXERCISES 1–8.

b Multiplying a Monomial and Any Polynomial

To find an equivalent expression for the product of a monomial, such as $2x$, and a binomial, such as $5x + 3$, we use a distributive law.

▶ **EXAMPLE 4** Multiply: $2x$ and $5x + 3$.

$$(2x)(5x + 3) = (2x)(5x) + (2x)(3) \quad \text{Using a distributive law}$$
$$= 10x^2 + 6x \quad \text{Multiplying the monomials}$$ ◀

OBJECTIVES

After finishing Section 3.5, you should be able to:

a Multiply monomials.

b Multiply a monomial and any polynomial.

c Multiply two binomials.

d Multiply any two polynomials.

FOR EXTRA HELP

Tape 8C   Tape 7A   MAC: 3
                                   IBM: 3

Multiply.

1. $3x$ and -5

2. $-x$ and x

3. $-x$ and $-x$

4. $-x^2$ and x^3

5. $3x^5$ and $4x^2$

6. $4y^5$ and $-2y^6$

7. $-7y^4$ and $-y$

8. $7x^5$ and 0

ANSWERS ON PAGE A-4

Multiply.

9. $4x$ and $2x + 4$

10. $3t^2(-5t + 2)$

11. $5x^3(x^3 + 5x^2 - 6x + 8)$

Multiply.

12. $x + 8$ and $x + 5$

13. $(x + 5)(x - 4)$

Multiply.

14. $5x + 3$ and $x - 4$

15. $(2x - 3)(3x - 5)$

▶ **EXAMPLE 5**   Multiply: $5x(2x^2 - 3x + 4)$.

$$5x(2x^2 - 3x + 4) = (5x)(2x^2) - (5x)(3x) + (5x)(4)$$
$$= 10x^3 - 15x^2 + 20x \qquad ◀$$

> **To multiply a monomial and a polynomial, multiply each term of the polynomial by the monomial.**

▶ **EXAMPLE 6**   Multiply: $2x^2(x^3 - 7x^2 + 10x - 4)$.

$$2x^2(x^3 - 7x^2 + 10x - 4) = 2x^5 - 14x^4 + 20x^3 - 8x^2 \qquad ◀$$

DO EXERCISES 9–11.

C **Multiplying Two Binomials**

To find an equivalent expression for the product of two binomials, we use the distributive laws more than once. In Example 7, we use a distributive law three times.

▶ **EXAMPLE 7**   Multiply: $x + 5$ and $x + 4$.

$$(x + 5)(x + 4) = x(x + 4) + 5(x + 4) \qquad \text{Using a distributive law}$$
$$= x \cdot x + x \cdot 4 + 5 \cdot x + 5 \cdot 4 \qquad \text{Using a distributive law on each part}$$
$$= x^2 + 4x + 5x + 20 \qquad \text{Multiplying the monomials}$$
$$= x^2 + 9x + 20 \qquad \text{Collecting like terms} \qquad ◀$$

DO EXERCISES 12 AND 13.

▶ **EXAMPLE 8**   Multiply: $4x + 3$ and $x - 2$.

$$(4x + 3)(x - 2) = 4x(x - 2) + 3(x - 2) \qquad \text{Using a distributive law}$$
$$= 4x \cdot x - 4x \cdot 2 + 3 \cdot x - 3 \cdot 2 \qquad \text{Using a distributive law on each part}$$
$$= 4x^2 - 8x + 3x - 6 \qquad \text{Multiplying the monomials}$$
$$= 4x^2 - 5x - 6 \qquad \text{Collecting like terms} \qquad ◀$$

DO EXERCISES 14 AND 15.

Multiply.

d ❚ Multiplying Any Polynomials

Let us consider the product of a binomial and a trinomial. We again use a distributive law three times. You may see ways to skip some steps and do the work mentally.

16. $(x^2 + 3x - 4)(x^2 + 5)$

▶ **EXAMPLE 9**   Multiply: $(x^2 + 2x - 3)(x^2 + 4)$.

$$(\, x^2 + 2x - 3 \,)(x^2 + 4) = (\, x^2 + 2x - 3 \,)x^2 + (\, x^2 + 2x - 3 \,)4$$
$$= x^2 \cdot x^2 + 2x \cdot x^2 - 3 \cdot (x^2) + x^2(4) + 2x \cdot 4 - 3 \cdot 4$$
$$= x^4 + 2x^3 - 3x^2 + 4x^2 + 8x - 12$$
$$= x^4 + 2x^3 + x^2 + 8x - 12 \qquad ◀$$

DO EXERCISES 16 AND 17.

Perhaps you have discovered the following in the preceding examples.

> To multiply two polynomials P and Q, select one of the polynomials, say P. Then multiply each term of P by every term of Q and collect like terms.

We can use columns for long multiplications. We multiply each term at the top by every term at the bottom. We write like terms in columns, and then we add the results. Such multiplication is like multiplying with whole numbers:

$$
\begin{array}{r}
4\ 5\ 7 \\
\times \quad 6\ 3 \\
\hline
1\ 3\ 7\ 1 \\
2\ 7\ 4\ 2\ 0 \\
\hline
2\ 8\ 7\ 9\ 1
\end{array}
\qquad
\begin{array}{r}
4\ 5\ 7 \\
\times \qquad\quad 6\ 3 \\
\hline
1200 + 150 + 21 \\
2400 + 3000 + 420 \\
\hline
2400 + 4200 + 570 + 21
\end{array}
$$

$= 400 + 50 + 7$
$= 60 + 3$
$= 3(457) = 3(400 + 50 + 7)$
$= 60(457) = 60(400 + 50 + 7)$
$= 28{,}971$

17. $(3y^2 - 7)(2y^3 - 2y + 5)$

▶ **EXAMPLE 10**   Multiply: $(4x^2 - 2x + 3)(x + 2)$.

$$
\begin{array}{r}
4x^2 - 2x + 3 \\
x + 2 \\
\hline
8x^2 - 4x + 6 \\
4x^3 - 2x^2 + 3x \\
\hline
4x^3 + 6x^2 - \ x + 6
\end{array}
$$

Multiplying the top row by 2
Multiplying the top row by x
Collecting like terms

Line up like terms in columns. ◀

ANSWERS ON PAGE A-4

Multiply.

18. $3x^2 - 2x + 4$
            $x + 5$

19. $-5x^2 + 4x + 2$
       $-4x^2 - 8$

20. Multiply.

$$3x^2 - 2x - 5$$
$$2x^2 + x - 2$$

ANSWERS ON PAGE A-4

▶ **EXAMPLE 11**   Multiply: $(5x^3 - 3x + 4)(-2x^2 - 3)$.

When missing terms occur, it helps to leave spaces for them and align like terms as we multiply.

$$
\begin{array}{r}
5x^3 \quad\quad - 3x + 4 \\
-2x^2 \quad\quad\quad - 3 \\
\hline
-15x^3 \quad\quad + 9x - 12 \\
-10x^5 + 6x^3 - 8x^2 \quad\quad\quad \\
\hline
-10x^5 - 9x^3 - 8x^2 + 9x - 12 \\
\end{array}
$$

Multiplying by -3
Multiplying by $-2x^2$
Collecting like terms ◀

DO EXERCISES 18 AND 19.

▶ **EXAMPLE 12**   Multiply: $(2x^2 + 3x - 4)(2x^2 - x + 3)$.

$$
\begin{array}{r}
2x^2 + 3x - 4 \\
2x^2 - x + 3 \\
\hline
6x^2 + 9x - 12 \\
-2x^3 - 3x^2 + 4x \quad\quad \\
4x^4 + 6x^3 - 8x^2 \quad\quad\quad \\
\hline
4x^4 + 4x^3 - 5x^2 + 13x - 12 \\
\end{array}
$$

Multiplying by 3
Multiplying by $-x$
Multiplying by $2x^2$
Collecting like terms ◀

DO EXERCISE 20.

❖ SIDELIGHTS

Factors and Sums

To *factor* a number is to express it as a product. Since $12 = 4 \cdot 3$, we say that 12 is *factored* and that 4 and 3 are *factors* of 12. In the table below, the top number has been factored in such a way that the sum of the factors is the bottom number. For example, in the first column 40 has been factored as $5 \cdot 8$, and $5 + 8 = 13$, the bottom number. Such thinking is important in algebra when we factor trinomials of the type $x^2 + bx + c$.

Product	40	63	36	72	−140	−96	48	168	110			
Factor	5									−9	−24	−3
Factor	8									−10	18	
Sum	13	16	−20	−38	−4	4	−14	−29	−21			18

EXERCISES

Find the missing numbers in the table.

EXERCISE SET 3.5

a Multiply.

1. $(6x^2)(7)$

2. $(5x^2)(-2)$

3. $(-x^3)(-x)$

4. $(-x^4)(x^2)$

5. $(7x^5)(4x^3)$

6. $(10a^2)(3a^2)$

7. $(-0.1x^6)(0.2x^4)$

8. $(0.3x^3)(-0.4x^6)$

9. $(-\frac{1}{5}x^3)(-\frac{1}{3}x)$

10. $(-\frac{1}{4}x^4)(\frac{1}{5}x^8)$

11. $(-4x^2)(0)$

12. $(-4m^5)(-1)$

13. $(3x^2)(-4x^3)(2x^6)$

14. $(-2y^5)(10y^4)(-3y^3)$

b Multiply.

15. $3x(-x+5)$

16. $2x(4x-6)$

17. $-3x(x-1)$

18. $-5x(-x-1)$

19. $x^2(x^3+1)$

20. $-2x^3(x^2-1)$

21. $3x(2x^2-6x+1)$

22. $-4x(2x^3-6x^2-5x+1)$

23. $(-6x^2)(x^2+x)$

24. $(-4x^2)(x^2-x)$

25. $(3y^2)(6y^4+8y^3)$

26. $(4y^4)(y^3-6y^2)$

c Multiply.

27. $(x+6)(x+3)$

28. $(x+5)(x+2)$

29. $(x+5)(x-2)$

30. $(x+6)(x-2)$

31. $(x-4)(x-3)$

32. $(x-7)(x-3)$

33. $(x+3)(x-3)$

34. $(x+6)(x-6)$

35. $(5-x)(5-2x)$

36. $(3+x)(6+2x)$

37. $(2x+5)(2x+5)$

38. $(3x-4)(3x-4)$

39. $(x-\frac{5}{2})(x+\frac{2}{5})$

40. $(x+\frac{4}{3})(x+\frac{3}{2})$

ANSWERS

1. _____
2. _____
3. _____
4. _____
5. _____
6. _____
7. _____
8. _____
9. _____
10. _____
11. _____
12. _____
13. _____
14. _____
15. _____
16. _____
17. _____
18. _____
19. _____
20. _____
21. _____
22. _____
23. _____
24. _____
25. _____
26. _____
27. _____
28. _____
29. _____
30. _____
31. _____
32. _____
33. _____
34. _____
35. _____
36. _____
37. _____
38. _____
39. _____
40. _____

d   Multiply.

41. $(x^2 + x + 1)(x - 1)$

42. $(x^2 - x + 2)(x + 2)$

43. $(2x + 1)(2x^2 + 6x + 1)$

44. $(3x - 1)(4x^2 - 2x - 1)$

45. $(y^2 - 3)(3y^2 - 6y + 2)$

46. $(3y^2 - 3)(y^2 + 6y + 1)$

47. $(x^3 + x^2)(x^3 + x^2 - x)$

48. $(x^3 - x^2)(x^3 - x^2 + x)$

49. $(-5x^3 - 7x^2 + 1)(2x^2 - x)$

50. $(-4x^3 + 5x^2 - 2)(5x^2 + 1)$

51. $(1 + x + x^2)(-1 - x + x^2)$

52. $(1 - x + x^2)(1 - x + x^2)$

53. $(2t^2 - t - 4)(3t^2 + 2t - 1)$

54. $(3a^2 - 5a + 2)(2a^2 - 3a + 4)$

55. $(x - x^3 + x^5)(x^2 - 1 + x^4)$

56. $(x - x^3 + x^5)(3x^2 + 3x^6 + 3x^4)$

57. $(x^3 + x^2 + x + 1)(x - 1)$

58. $(x + 2)(x^3 - x^2 + x - 2)$

SKILL MAINTENANCE

59. Subtract: $-\frac{1}{4} - \frac{1}{2}$.

60. Factor: $16x - 24y + 36$.

SYNTHESIS

61. Find a polynomial for the shaded area.

62. A box with a square bottom is to be made from a 12-in.-square piece of cardboard. Squares with side x are cut out of the corners and the sides are folded up. Find polynomials for the volume and the outside surface area of the box.

63. The height of a triangle is 4 ft longer than its base. Find a polynomial for the area.

Compute and simplify.

64. $(x + 3)(x + 6) + (x + 3)(x + 6)$

65. $(x - 2)(x - 7) - (x - 2)(x - 7)$

3.6 Special Products

We now consider a special way of multiplying any two binomials. Such a special technique is called a **special product,** meaning that we encounter certain products so often that it is helpful to have faster methods of computing.

a Products of Two Binomials

To multiply two binomials, we can select one binomial and multiply each term of that binomial by every term of the other. Then we collect like terms. Consider the product $(x + 5)(x + 4)$:

$$(x + 5)(x + 4) = x \cdot x + 5 \cdot x + x \cdot 4 + 5 \cdot 4$$
$$= x^2 + 5x + 4x + 20$$
$$= x^2 + 9x + 20.$$

We can rewrite the first line of this product to show a special technique for finding the product of two binomials:

First   Outside   Inside   Last
terms   terms   terms   terms

$$(x + 5)(x + 4) = x \cdot x \; + \; 4 \cdot x \; + \; 5 \cdot x \; + \; 5 \cdot 4.$$

To remember this method of multiplying, use the initials **FOIL.**

The FOIL Method

To multiply two binomials, $A + B$ and $C + D$, multiply the First terms AC, the Outside terms AD, the Inside terms BC, and then the Last terms BD. Then collect like terms, if possible.

$$(A + B)(C + D) = AC + AD + BC + BD$$

1. Multiply **F**irst terms: AC.
2. Multiply **O**utside terms: AD.
3. Multiply **I**nside terms: BC.
4. Multiply **L**ast terms: BD.

FOIL

▶ **EXAMPLE 1** Multiply: $(x + 8)(x^2 + 5)$.

We have

$$\overset{\text{F}}{}\quad\overset{\text{O}}{}\quad\overset{\text{I}}{}\quad\overset{\text{L}}{}$$
$$(x + 8)(x^2 + 5) = x^3 + 5x + 8x^2 + 40$$
$$= x^3 + 8x^2 + 5x + 40.$$

Since each of the original binomials is in descending order, we write the product in descending order, as is customary, but this is not a "must." ◀

OBJECTIVES

After finishing Section 3.6, you should be able to:

a Multiply two binomials mentally using the FOIL method.

b Multiply the sum and difference of two terms mentally.

c Square a binomial mentally.

d Find special products when they are mixed together.

FOR EXTRA HELP

Tape 8D   Tape 7B   MAC: 3
                                    IBM: 3

Multiply mentally. Write just the answer.

1. $(x + 3)(x + 4)$

2. $(x + 3)(x - 5)$

3. $(2x + 1)(x + 4)$

4. $(2x^2 - 3)(x - 2)$

5. $(6x^2 + 5)(2x^3 + 1)$

6. $(y^3 + 7)(y^3 - 7)$

7. $(2x^4 + x^2)(-x^3 + x)$

Multiply.

8. $(t + 5)(t + 3)$

9. $\left(x + \dfrac{4}{5}\right)\left(x - \dfrac{4}{5}\right)$

10. $(x^3 - 0.5)(x^2 + 0.5)$

11. $(2 + 3x^2)(4 - 5x^2)$

12. $(6x^3 - 3x^2)(5x^2 + 2x)$

Often we can collect like terms after we multiply.

▶ **EXAMPLES**   Multiply.

2. $(x + 6)(x - 6) = x^2 - 6x + 6x - 36$ Using FOIL
$$= x^2 - 36 \qquad \text{Collecting like terms}$$

3. $(y + 3)(y - 2) = y^2 - 2y + 3y - 6$
$$= y^2 + y - 6$$

4. $(x^3 + 5)(x^3 - 5) = x^6 - 5x^3 + 5x^3 - 25$
$$= x^6 - 25$$

5. $(4t^3 + 5)(3t^2 - 2) = 12t^5 - 8t^3 + 15t^2 - 10$ ◀

DO EXERCISES 1–7.

▶ **EXAMPLES**   Multiply.

6. $(x + 7)(x + 4) = x^2 + 4x + 7x + 28$
$$= x^2 + 11x + 28$$

7. $(x - \frac{2}{3})(x + \frac{2}{3}) = x^2 + \frac{2}{3}x - \frac{2}{3}x - \frac{4}{9}$
$$= x^2 - \frac{4}{9}$$

8. $(x^2 - 0.3)(x^2 - 0.3) = x^4 - 0.3x^2 - 0.3x^2 + 0.09$
$$= x^4 - 0.6x^2 + 0.09$$

9. $(3 - 4x)(7 - 5x^3) = 21 - 15x^3 - 28x + 20x^4$
$$= 21 - 28x - 15x^3 + 20x^4$$

(*Note:* If the original polynomials are in ascending order, it is natural to write the product in ascending order, but this is not a ''must.'')

10. $(5x^4 + 2x^3)(3x^2 - 7x) = 15x^6 - 35x^5 + 6x^5 - 14x^4$
$$= 15x^6 - 29x^5 - 14x^4$$ ◀

DO EXERCISES 8–12.

We can show the FOIL method geometrically as follows:

The area of the large rectangle is $(A + B)(C + D)$.

   The area of rectangle ① is AC.

   The area of rectangle ② is AD.

   The area of rectangle ③ is BC.

   The area of rectangle ④ is BD.

The area of the large rectangle is the sum of the areas of the smaller rectangles. Thus,

$$(A + B)(C + D) = AC + AD + BC + BD.$$

b Multiplying Sums and Differences of Two Terms

Consider the product of the sum and difference of the same two terms, such as

$$(x + 2)(x - 2).$$

Since this is the product of two binomials, we can use FOIL. This product occurs so often, however, that it will speed up our work to be able to use an even faster method. To find a faster way to compute such a product, look for a pattern in the following:

a) $(x + 2)(x - 2) = x^2 - 2x + 2x - 4 = x^2 - 4$;

b) $(3x - 5)(3x + 5) = 9x^2 + 15x - 15x - 25 = 9x^2 - 25$.

DO EXERCISES 13 AND 14.

Perhaps you discovered in each case that when you multiply the two binomials, two terms are opposites, or additive inverses, which add to 0 and "drop out."

> **The product of the sum and difference of the same two terms is the square of the first term minus the square of the second term:**
>
> $$(A + B)(A - B) = A^2 - B^2.$$

It is helpful to memorize this rule in both words and symbols. (If you do forget it, you can, of course, use FOIL.)

▶ **EXAMPLES** Multiply. (Carry out the rule and say the words as you go.)

$$(A + B)(A - B) = A^2 - B^2$$

11. $(x + 4)(x - 4) = x^2 - 4^2$ "The square of the first term, x^2, minus the square of the second, 4^2."

$$= x^2 - 16 \quad \text{Simplifying}$$

12. $(5 + 2w)(5 - 2w) = 5^2 - (2w)^2$
$$= 25 - 4w^2$$

13. $(3x^2 - 7)(3x^2 + 7) = (3x^2)^2 - 7^2$
$$= 9x^4 - 49$$

14. $(-4x - 10)(-4x + 10) = (-4x)^2 - 10^2$
$$= 16x^2 - 100 \qquad ◀$$

DO EXERCISES 15–18.

c Squaring Binomials

Consider the square of a binomial, such as $(x + 3)^2$. This can be expressed as $(x + 3)(x + 3)$. Since this is the product of two binomials, we can again use FOIL. But again, this product occurs so often that it will speed up our work to be able to use an even faster method. Look for a pattern in the following:

a) $(x + 3)^2 = (x + 3)(x + 3)$
$$= x^2 + 3x + 3x + 9 = x^2 + 6x + 9;$$

b) $(5 + 3p)^2 = (5 + 3p)(5 + 3p)$
$$= 25 + 15p + 15p + 9p^2 = 25 + 30p + 9p^2;$$

Multiply.

13. $(x + 5)(x - 5)$

14. $(2x - 3)(2x + 3)$

Multiply.

15. $(x + 2)(x - 2)$

16. $(x - 7)(x + 7)$

17. $(6 - 4y)(6 + 4y)$

18. $(2x^3 - 1)(2x^3 + 1)$

ANSWERS ON PAGE A-4

Multiply.

19. $(x + 8)(x + 8)$

c) $(x - 3)^2 = (x - 3)(x - 3)$
$$= x^2 - 3x - 3x + 9$$
$$= x^2 - 6x + 9;$$

d) $(3x - 5)^2 = (3x - 5)(3x - 5)$
$$= 9x^2 - 15x - 15x + 25$$
$$= 9x^2 - 30x + 25.$$

DO EXERCISES 19 AND 20.

20. $(x - 5)(x - 5)$

When squaring a binomial, we multiply a binomial by itself. Perhaps you noticed that two terms are the same and when added give twice their product. The other two terms are squares.

Multiply.

21. $(x + 2)^2$

> **The square of a sum or difference of two terms is the square of the first term, plus or minus twice the product of the two terms, plus the square of the last term:**
> $$(A + B)^2 = A^2 + 2AB + B^2;$$
> $$(A - B)^2 = A^2 - 2AB + B^2.$$

It is helpful to memorize this rule in both words and symbols.

22. $(a - 4)^2$

▶ **EXAMPLES** Multiply. (Carry out the rule and say the words as you go.)

$$(A + B)^2 = A^2 + 2 \cdot A \cdot B + B^2$$

15. $(x + 3)^2 = x^2 + 2 \cdot x \cdot 3 + 3^2$ "x^2 plus 2 times x times 3 plus 3^2."
$$= x^2 + 6x + 9$$

23. $(2x + 5)^2$

16. $(t - 5)^2 = t^2 - 2 \cdot t \cdot 5 + 5^2$ "t^2 minus 2 times t times 5 plus 5^2."
$$= t^2 - 10t + 25$$

17. $(2x + 7)^2 = (2x)^2 + 2 \cdot 2x \cdot 7 + 7^2$
$$= 4x^2 + 28x + 49$$

18. $(5x - 3x^2)^2 = (5x)^2 - 2 \cdot 5x \cdot 3x^2 + (3x^2)^2$
$$= 25x^2 - 30x^3 + 9x^4$$ ◀

24. $(4x^2 - 3x)^2$

DO EXERCISES 21–26.

25. $(7 + y)(7 + y)$

> CAUTION!   Note carefully in these examples that the square of a sum is *not* the sum of the squares:
>
> $$\underbrace{}_{\text{The middle term } 2AB \text{ is missing.}}$$
>
> $$(A + B)^2 \neq \qquad A^2 + B^2.$$
>
> To see this, note that
>
> $$(20 + 5)^2 = 25^2 = 625,$$
>
> but
>
> $$20^2 + 5^2 = 400 + 25 = 425 \neq 625.$$
>
> However, $20^2 + 2(20)(5) + 5^2 = 625.$

26. $(3x^2 - 5)(3x^2 - 5)$

We can look at our rule for finding $(A + B)^2$ geometrically as follows. The area of the large square is

$$(A + B)(A + B) = (A + B)^2.$$

This is equal to the sum of the areas of the smaller rectangles:

$$A^2 + AB + AB + B^2 = A^2 + 2AB + B^2.$$

Thus,

$$(A + B)^2 = A^2 + 2AB + B^2.$$

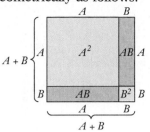

d Multiplications of Various Types

We have considered how to quickly multiply certain kinds of polynomials. Let us now try several kinds mixed together so that we can learn to sort them out. When you multiply, first see what kind of multiplication you have. Then use the best method. The formulas you should know and the questions you should ask yourself are as follows.

Multiplying Two Polynomials

1. **Is the product the square of a binomial? If so, use the following:**
$$(A + B)(A + B) = (A + B)^2 = A^2 + 2AB + B^2,$$
   or $\qquad (A - B)(A - B) = (A - B)^2 = A^2 - 2AB + B^2.$

   The square of a binomial is the square of the first term, plus or minus *twice* the product of the two terms, plus the square of the last term.

   [The answer has 3 terms.]

2. **Is it the product of the sum and difference of the *same* two terms? If so, use the following:**
$$(A + B)(A - B) = A^2 - B^2.$$
   The product of the sum and difference of the same two terms is the difference of squares.

   [The answer has 2 terms.]

3. **Is it the product of two binomials other than those above? If so, use FOIL.**

   [The answer will have 3 or 4 terms.]

4. **The product of a monomial and any polynomial is found by multiplying each term of the polynomial by the monomial.**

5. **Is it the product of two polynomials other than those above? If so, multiply each term of one by every term of the other. Use columns if you wish.**

   [The answer will have 2 or more terms, usually more than 2 terms.]

Note that FOIL will actually work instead of either of the first two rules, but those rules will make your work go faster.

Multiply.

27. $(x + 5)(x + 6)$

28. $(t - 4)(t + 4)$

29. $4x^2(-2x^3 + 5x^2 + 10)$

30. $(9x^2 + 1)^2$

31. $(2a - 5)(2a + 8)$

32. $\left(5x + \dfrac{1}{2}\right)^2$

33. $\left(2x - \dfrac{1}{2}\right)^2$

34. $(x^2 - x + 4)(x - 2)$

▶ **EXAMPLE 19**   Multiply: $(x + 3)(x - 3)$.

$$(x + 3)(x - 3) = x^2 - 9 \qquad$$ Using method 2 (the product of the sum and difference of two terms) ◀

▶ **EXAMPLE 20**   Multiply: $(t + 7)(t - 5)$.

$$(t + 7)(t - 5) = t^2 + 2t - 35 \qquad$$ Using method 3 (the product of two binomials, but neither the square of a binomial nor the product of the sum and difference of two terms) ◀

▶ **EXAMPLE 21**   Multiply: $(x + 7)(x + 7)$.

$$(x + 7)(x + 7) = x^2 + 14x + 49 \qquad$$ Using method 1 (the square of a binomial sum) ◀

▶ **EXAMPLE 22**   Multiply: $2x^3(9x^2 + x - 7)$.

$$2x^3(9x^2 + x - 7) = 18x^5 + 2x^4 - 14x^3 \qquad$$ Using method 4 (the product of a monomial and a trinomial; multiplying each term of the trinomial by the monomial) ◀

▶ **EXAMPLE 23**   Multiply: $(5x^3 - 7x)^2$.

$$(5x^3 - 7x)^2 = 25x^6 - 2(5x^3)(7x) + 49x^2 \qquad$$ Using method 1 (the square of a binomial difference)

$$= 25x^6 - 70x^4 + 49x^2 \qquad\qquad$$ ◀

▶ **EXAMPLE 24**   Multiply: $(3x + \frac{1}{4})^2$.

$$(3x + \tfrac{1}{4})^2 = 9x^2 + 2(3x)(\tfrac{1}{4}) + \tfrac{1}{16} \qquad$$ Using method 1 (the square of a binomial sum. To get the middle term, we multiply $3x$ by $\frac{1}{4}$ and double.)

$$= 9x^2 + \tfrac{3}{2}x + \tfrac{1}{16} \qquad\qquad$$ ◀

▶ **EXAMPLE 25**   Multiply: $(4x - \frac{3}{4})^2$.

$$(4x - \tfrac{3}{4})^2 = 16x^2 - 2(4x)(\tfrac{3}{4}) + \tfrac{9}{16} \qquad$$ Using method 1

$$= 16x^2 - 6x + \tfrac{9}{16} \qquad\qquad$$ ◀

▶ **EXAMPLE 26**   Multiply: $(p + 3)(p^2 + 2p - 1)$.

$$
\begin{array}{r}
p^2 + 2p - 1 \\
p + 3 \\
\hline
3p^2 + 6p - 3 \\
p^3 + 2p^2 - p \\
\hline
p^3 + 5p^2 + 5p - 3
\end{array}
$$

Using method 5

Multiplying by 3

Multiplying by p

◀

DO EXERCISES 27–34.

EXERCISE SET 3.6

a Multiply. Try to write only the answer. If you need more steps, by all means use them.

1. $(x + 1)(x^2 + 3)$

2. $(x^2 - 3)(x - 1)$

3. $(x^3 + 2)(x + 1)$

4. $(x^4 + 2)(x + 12)$

5. $(y + 2)(y - 3)$

6. $(a + 2)(a + 2)$

7. $(3x + 2)(3x + 3)$

8. $(4x + 1)(2x + 2)$

9. $(5x - 6)(x + 2)$

10. $(x - 8)(x + 8)$

11. $(3t - 1)(3t + 1)$

12. $(2m + 3)(2m + 3)$

13. $(4x - 2)(x - 1)$

14. $(2x - 1)(3x + 1)$

15. $(p - \frac{1}{4})(p + \frac{1}{4})$

16. $(q + \frac{3}{4})(q + \frac{3}{4})$

17. $(x - 0.1)(x + 0.1)$

18. $(x + 0.3)(x - 0.4)$

19. $(2x^2 + 6)(x + 1)$

20. $(2x^2 + 3)(2x - 1)$

21. $(-2x + 1)(x + 6)$

22. $(3x + 4)(2x - 4)$

23. $(a + 7)(a + 7)$

24. $(2y + 5)(2y + 5)$

25. $(1 + 2x)(1 - 3x)$

26. $(-3x - 2)(x + 1)$

27. $(x^2 + 3)(x^3 - 1)$

28. $(x^4 - 3)(2x + 1)$

29. $(3x^2 - 2)(x^4 - 2)$

30. $(x^{10} + 3)(x^{10} - 3)$

ANSWERS

1.
2.
3.
4.
5.
6.
7.
8.
9.
10.
11.
12.
13.
14.
15.
16.
17.
18.
19.
20.
21.
22.
23.
24.
25.
26.
27.
28.
29.
30.

31. $(3x^5 + 2)(2x^2 + 6)$ **32.** $(1 - 2x)(1 + 3x^2)$ **33.** $(8x^3 + 1)(x^3 + 8)$

34. $(4 - 2x)(5 - 2x^2)$ **35.** $(4x^2 + 3)(x - 3)$ **36.** $(7x - 2)(2x - 7)$

37. $(4y^4 + y^2)(y^2 + y)$ **38.** $(5y^6 + 3y^3)(2y^6 + 2y^3)$

b   Multiply mentally, if possible. If you need extra steps, by all means use them.

39. $(x + 4)(x - 4)$ **40.** $(x + 1)(x - 1)$ **41.** $(2x + 1)(2x - 1)$

42. $(x^2 + 1)(x^2 - 1)$ **43.** $(5m - 2)(5m + 2)$ **44.** $(3x^4 + 2)(3x^4 - 2)$

45. $(2x^2 + 3)(2x^2 - 3)$ **46.** $(6x^5 - 5)(6x^5 + 5)$ **47.** $(3x^4 - 4)(3x^4 + 4)$

48. $(t^2 - 0.2)(t^2 + 0.2)$ **49.** $(x^6 - x^2)(x^6 + x^2)$ **50.** $(2x^3 - 0.3)(2x^3 + 0.3)$

51. $(x^4 + 3x)(x^4 - 3x)$ **52.** $(\frac{3}{4} + 2x^3)(\frac{3}{4} - 2x^3)$ **53.** $(x^{12} - 3)(x^{12} + 3)$

54. $(12 - 3x^2)(12 + 3x^2)$ **55.** $(2y^8 + 3)(2y^8 - 3)$ **56.** $(m - \frac{2}{3})(m + \frac{2}{3})$

c   Multiply mentally, if possible.

57. $(x + 2)^2$ **58.** $(2x - 1)^2$ **59.** $(3x^2 + 1)^2$ **60.** $(3x + \frac{3}{4})^2$

61. $(a - \frac{1}{2})^2$ **62.** $(2a - \frac{1}{5})^2$ **63.** $(3 + x)^2$ **64.** $(x^3 - 1)^2$

65. $(x^2 + 1)^2$ **66.** $(8x - x^2)^2$ **67.** $(2 - 3x^4)^2$ **68.** $(6x^3 - 2)^2$

69. $(5 + 6t^2)^2$ **70.** $(3p^2 - p)^2$

d Multiply mentally, if possible.

71. $(3 - 2x^3)^2$ **72.** $(x - 4x^3)^2$ **73.** $4x(x^2 + 6x - 3)$

74. $8x(-x^5 + 6x^2 + 9)$ **75.** $(2x^2 - \frac{1}{2})(2x^2 - \frac{1}{2})$ **76.** $(-x^2 + 1)^2$

77. $(-1 + 3p)(1 + 3p)$ **78.** $(-3q + 2)(3q + 2)$ **79.** $3t^2(5t^3 - t^2 + t)$

80. $-6x^2(x^3 + 8x - 9)$ **81.** $(6x^4 + 4)^2$ **82.** $(8a + 5)^2$

83. $(3x + 2)(4x^2 + 5)$ **84.** $(2x^2 - 7)(3x^2 + 9)$ **85.** $(8 - 6x^4)^2$

86. $(\frac{1}{5}x^2 + 9)(\frac{3}{5}x^2 - 7)$ **87.** $(t - 1)(t^2 + t + 1)$ **88.** $(y + 5)(y^2 - 5y + 25)$

Compute each of the following and compare.

89. $3^2 + 4^2$;   **90.** $6^2 + 7^2$;   **91.** $9^2 - 5^2$;   **92.** $11^2 - 4^2$;
$\quad (3 + 4)^2$ $\quad (6 + 7)^2$ $\quad (9 - 5)^2$ $\quad (11 - 4)^2$

ANSWERS

65. _____

66. _____

67. _____

68. _____

69. _____

70. _____

71. _____

72. _____

73. _____

74. _____

75. _____

76. _____

77. _____

78. _____

79. _____

80. _____

81. _____

82. _____

83. _____

84. _____

85. _____

86. _____

87. _____

88. _____

89. _____

90. _____

91. _____

92. _____

SKILL MAINTENANCE

93. In an apartment, lamps, an air conditioner, and a television set are all operating at the same time. The lamps use 10 times as many watts as the television set, and the air conditioner uses 40 times as many watts as the television set. The total wattage used in the apartment is 2550 watts. How many watts are used by each appliance?

94. Solve: $3x - 8x = 4(7 - 8x)$.

SYNTHESIS

Multiply.

95. $4y(y + 5)(2y + 8)$

96. $8x(2x - 3)(5x + 9)$

97. $[(x + 1) - x^2][(x - 2) + 2x^2]$

98. $[(2x - 1)(2x + 1)](4x^2 + 1)$

Solve.

99. $(x + 2)(x - 5) = (x + 1)(x - 3)$

100. $(2x + 5)(x - 4) = (x + 5)(2x - 4)$

The height of a box is one more than its length l, and the length is one more than its width w. Find a polynomial for the volume V in terms of the following.

101. The width w　　　　**102.** The length l　　　　**103.** The height h

Find two expressions for the shaded area.

104.

105.

106.

107. ▦ Multiply: $(67.58x + 3.225)^2$.

Calculate as the difference of squares.

108. 18×22 [*Hint:* $(20 - 2)(20 + 2)$.]

109. 93×107

Multiply. (Do not collect like terms before multiplying.)

110. $[(3x - 2)(3x + 2)](9x^2 + 4)$

111. $[3a - (2a - 3)][3a + (2a - 3)]$

112. $(5t^2 - 3)^2(5t^2 + 3)^2$

113. A polynomial for the shaded area in this rectangle is $(A + B)(A - B)$.

 a) Find a polynomial for the area of the entire rectangle.

 b) Find a polynomial for the sum of the areas of the two small unshaded rectangles.

 c) Find a polynomial for the area in part (a) minus the area in part (b).

 d) Find a polynomial for the area of the shaded region and compare this with the polynomial found in part (c).

114. Find $(10x + 5)^2$. Use your result to show how to mentally square any two-digit number ending in 5.

3.7 Operations with Polynomials in Several Variables

The polynomials that we have been studying have only one variable. A **polynomial in several variables** is an expression like those you have already seen, but with more than one variable. Here are some examples:

$$3x + xy^2 + 5y + 4, \qquad 8xy^2z - 2x^3z - 13x^4y^2 + 15.$$

a Evaluating Polynomials

▶ **EXAMPLE 1** Evaluate the polynomial $4 + 3x + xy^2 + 8x^3y^3$ when $x = -2$ and $y = 5$.

We replace x by -2 and y by 5:

$$
\begin{aligned}
4 + 3x + xy^2 + 8x^3y^3 &= 4 + 3(-2) + (-2) \cdot 5^2 + 8(-2)^3 \cdot 5^3 \\
&= 4 - 6 - 50 - 8000 \\
&= -8052.
\end{aligned}
$$
◀

▶ **EXAMPLE 2** *Surface area of a right circular cylinder.* The surface area of a right circular cylinder is given by the polynomial

$$2\pi rh + 2\pi r^2,$$

where h is the height and r is the radius of the base. (This formula can be derived by cutting the cylinder apart, as shown on the right below, and adding the areas of the parts.) A 12-oz beverage can has a height of 4.7 in. and a radius of 1.2 in. To find the surface area, we can evaluate the polynomial when $h = 4.7$ and $r = 1.2$. Use 3.14 as an approximation for π.

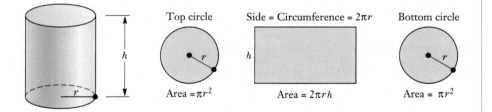

We evaluate the polynomial when $h = 4.7$, $r = 1.2$, and $\pi \approx 3.14$:

$$
\begin{aligned}
2\pi rh + 2\pi r^2 &\approx 2(3.14)(1.2)(4.7) + 2(3.14)(1.2)^2 \\
&= 2(3.14)(1.2)(4.7) + 2(3.14)(1.44) \\
&= 35.4192 + 9.0432 \\
&= 44.4624.
\end{aligned}
$$

The surface area is about 44.4624 in² (square inches.)   ◀

DO EXERCISES 1–3.

OBJECTIVES

After finishing Section 3.7, you should be able to:

a Evaluate a polynomial in several variables for given values of the variables.

b Identify the coefficients and the degrees of the terms of a polynomial and the degree of a polynomial.

c Collect like terms of a polynomial.

d Add polynomials.

e Subtract polynomials.

f Multiply polynomials.

FOR EXTRA HELP

Tape NC   Tape 7B   MAC: 3
                                  IBM: 3

1. Evaluate the polynomial
$$4 + 3x + xy^2 + 8x^3y^3$$
when $x = 2$ and $y = -5$.

2. Evaluate the polynomial
$$8xy^2 - 2x^3z - 13x^4y^2 + 5$$
when $x = -1$, $y = 3$, and $z = 4$.

3. For the situation of Example 2, find the surface area of a tank with $h = 20$ ft and $r = 3$ ft. Use 3.14 for π.

4. Identify the coefficient of each term:

$$-3xy^2 + 3x^2y - 2y^3 + xy + 2.$$

5. Identify the degree of each term and the degree of the polynomial

$$4xy^2 + 7x^2y^3z^2 - 5x + 2y + 4.$$

Collect like terms.

6. $4x^2y + 3xy - 2x^2y$

7. $-3pq - 5pqr^3 + 8pq + 5pqr^3 + 4$

b   **Coefficients and Degrees**

The **degree** of a term is the sum of the exponents of the variables. The **degree of a polynomial** is the degree of the term of highest degree.

▶ **EXAMPLE 3**   Identify the coefficient and the degree of each term and the degree of the polynomial

$$9x^2y^3 - 14xy^2z^3 + xy + 4y + 5x^2 + 7.$$

Term	Coefficient	Degree	Degree of the polynomial
$9x^2y^3$	9	5	
$-14xy^2z^3$	-14	6	6
xy	1	2	
$4y$	4	1	Think: $4y = 4y^1$
$5x^2$	5	2	
7	7	0	Think: $7 = 7x^0$, or $7x^0y^0z^0$ ◀

DO EXERCISES 4 AND 5.

c   **Collecting Like Terms**

Like terms (or **similar terms**) have exactly the same variables with exactly the same exponents. For example,

     $3x^2y^3$ and $-7x^2y^3$ are like terms;

     $9x^4z^7$ and $12x^4z^7$ are like terms.

But

     $13xy^5$ and $-2x^2y^5$ are *not* like terms, because the x-factors have different exponents;

and

     $3xyz^2$ and $4xy$ are *not* like terms, because there is no factor involving z in the second expression.

Collecting like terms is based on the distributive laws.

▶ **EXAMPLES**   Collect like terms.

   4. $5x^2y + 3xy^2 - 5x^2y - xy^2 = (5 - 5)x^2y + (3 - 1)xy^2 = 2xy^2$

   5. $7xy - 5xy^2 + 3xy^2 + 6x^3 + 9xy - 11x^3 + y - 1$
       $= -2xy^2 + 16xy - 5x^3 + y - 1$ ◀

DO EXERCISES 6 AND 7.

d Addition

We can find the sum of two polynomials in several variables by writing a plus sign between them and then collecting like terms.

▶ **EXAMPLE 6** Add: $(-5x^3 + 3y - 5y^2) + (8x^3 + 4x^2 + 7y^2)$.

$(-5x^3 + 3y - 5y^2) + (8x^3 + 4x^2 + 7y^2)$
$$= (-5 + 8)x^3 + 4x^2 + 3y + (-5 + 7)y^2$$
$$= 3x^3 + 4x^2 + 3y + 2y^2 \qquad ◀$$

▶ **EXAMPLE 7** Add:

$$(5xy^2 - 4x^2y + 5x^3 + 2) + (3xy^2 - 2x^2y + 3x^3y - 5).$$

We first look for like terms. They are $5xy^2$ and $3xy^2$, $-4x^2y$ and $-2x^2y$, and 2 and -5. We collect these. Since there are no more like terms, the answer is

$$8xy^2 - 6x^2y + 5x^3 + 3x^3y - 3. \qquad ◀$$

DO EXERCISES 8–10.

e Subtraction

We subtract a polynomial by adding its opposite, or additive inverse. The opposite of the polynomial

$$4x^2y - 6x^3y^2 + x^2y^2 - 5y$$

can be represented by

$$-(4x^2y - 6x^3y^2 + x^2y^2 - 5y).$$

We find an equivalent expression for the opposite of a polynomial by replacing each coefficient by its opposite, or by changing the sign of each term. Thus,

$$-(4x^2y - 6x^3y^2 + x^2y^2 - 5y) = -4x^2y + 6x^3y^2 - x^2y^2 + 5y.$$

▶ **EXAMPLE 8** Subtract:

$$(4x^2y + x^3y^2 + 3x^2y^3 + 6y) - (4x^2y - 6x^3y^2 + x^2y^2 - 5y).$$

We have

$(4x^2y + x^3y^2 + 3x^2y^3 + 6y) - (4x^2y - 6x^3y^2 + x^2y^2 - 5y)$
$$= 4x^2y + x^3y^2 + 3x^2y^3 + 6y - 4x^2y + 6x^3y^2 - x^2y^2 + 5y \qquad \text{Adding the opposite}$$
$$= 7x^3y^2 + 3x^2y^3 - x^2y^2 + 11y \qquad \text{Collecting like terms. (Try to write just the answer!)} \qquad ◀$$

DO EXERCISES 11 AND 12.

Add.

8. $4x^3 + 4x^2 - 8x - 3$ and $-8x^3 - 2x^2 + 4x + 5$

9. $(13x^3y + 3x^2y - 5y) + (x^3y + 4x^2y - 3xy + 3y)$

10. $(-5p^2q^4 + 2p^2q^2 + 3q) + (6pq^2 + 3p^2q + 5)$

Subtract.

11. $(-4s^4t + s^3t^2 + 2s^2t^3) - (4s^4t - 5s^3t^2 + s^2t^2)$

12. $(-5p^4q + 5p^3q^2 - 3p^2q^3 - 7q^4) - (4p^4q - 4p^3q^2 + p^2q^3 + 2q^4)$

ANSWERS ON PAGE A-4

Multiply.

13. $(x^2y^3 + 2x)(x^3y^2 + 3x)$

14. $(p^4q - 2p^3q^2 + 3q^3)(p + 2q)$

Multiply.

15. $(3xy + 2x)(x^2 + 2xy^2)$

16. $(x - 3y)(2x - 5y)$

17. $(4x + 5y)^2$

18. $(3x^2 - 2xy^2)^2$

19. $(2xy^2 + 3x)(2xy^2 - 3x)$

20. $(3xy^2 + 4y)(-3xy^2 + 4y)$

21. $(3y + 4 - 3x)(3y + 4 + 3x)$

22. $(2a + 5b + c)(2a - 5b - c)$

f **Multiplication**

To multiply polynomials in several variables, we can multiply each term of one by every term of the other. Where appropriate, we use the special products that we have learned.

▶ **EXAMPLE 9** Multiply: $(3x^2y - 2xy + 3y)(xy + 2y)$.

$$
\begin{array}{r}
3x^2y - 2xy + 3y \\
xy + 2y \\
\hline
6x^2y^2 - 4xy^2 + 6y^2 \\
3x^3y^2 - 2x^2y^2 + 3xy^2 \\
\hline
3x^3y^2 + 4x^2y^2 - xy^2 + 6y^2
\end{array}
$$

  Multiplying by $2y$
  Multiplying by xy
  Adding   ◀

DO EXERCISES 13 AND 14.

▶ **EXAMPLES** Multiply.

  F   O   I   L

10. $(x^2y + 2x)(xy^2 + y^2) = x^3y^3 + x^2y^3 + 2x^2y^2 + 2xy^2$

11. $(p + 5q)(2p - 3q) = 2p^2 - 3pq + 10pq - 15q^2$
  $= 2p^2 + 7pq - 15q^2$

$(A + B)^2 = A^2 + 2 \cdot A \cdot B + B^2$

12. $(3x + 2y)^2 = (3x)^2 + 2(3x)(2y) + (2y)^2$
  $= 9x^2 + 12xy + 4y^2$

$(A - B)^2 = A^2 - 2 \cdot A \cdot B + B^2$

13. $(2y^2 - 5x^2y)^2 = (2y^2)^2 - 2(2y^2)(5x^2y) + (5x^2y)^2$
  $= 4y^4 - 20x^2y^3 + 25x^4y^2$

$(A + B)(A - B) = A^2 - B^2$

14. $(3x^2y + 2y)(3x^2y - 2y) = (3x^2y)^2 - (2y)^2$
  $= 9x^4y^2 - 4y^2$

15. $(-2x^3y^2 + 5t)(2x^3y^2 + 5t) = (5t - 2x^3y^2)(5t + 2x^3y^2)$
  $= (5t)^2 - (2x^3y^2)^2$
  $= 25t^2 - 4x^6y^4$

$(A - B)(A + B) = A^2 - B^2$

16. $(2x + 3 - 2y)(2x + 3 + 2y) = (2x + 3)^2 - (2y)^2$
  $= 4x^2 + 12x + 9 - 4y^2$ ◀

DO EXERCISES 15-22.

EXERCISE SET 3.7

a Evaluate the polynomial when $x = 3$ and $y = -2$.

1. $x^2 - y^2 + xy$ **2.** $x^2 + y^2 - xy$

Evaluate the polynomial when $x = 2$, $y = -3$, and $z = -1$.

3. $xyz^2 + z$ **4.** $xy - xz + yz$

Interest compounded annually for two years. An amount of money P is invested at interest rate i. In 2 years, it will grow to an amount given by the polynomial

$$A = P(1 + i)^2.$$

5. Evaluate the polynomial when $P = 10,000$ and $i = 0.08$ to find the amount to which \$10,000 will grow at 8% interest for 2 years.

6. Evaluate the polynomial when $P = 10,000$ and $i = 0.07$ to find the amount to which \$10,000 will grow at 7% interest for 2 years.

Interest compounded annually for three years. An amount of money P is invested at interest rate i. In 3 years, it will grow to an amount given by the polynomial

$$A = P(1 + i)^3.$$

7. Evaluate the polynomial when $P = 10,000$ and $i = 0.08$ to find the amount to which \$10,000 will grow at 8% interest for 3 years.

8. Evaluate the polynomial when $P = 10,000$ and $i = 0.07$ to find the amount to which \$10,000 will grow at 7% interest for 3 years.

Surface area of a right circular cylinder. The area of a right circular cylinder is given by the polynomial

$$2\pi rh + 2\pi r^2,$$

where h is the height and r is the radius of the base.

9. A 26-oz coffee can has a height of 6.5 in. and a radius of 2.5 in. Evaluate the polynomial when $h = 6.5$ and $r = 2.5$ to find the area of the can. Use 3.14 for π.

10. A 16-oz beverage can has a height of 6.3 in. and a radius of 1.2 in. Evaluate the polynomial when $h = 6.3$ and $r = 1.2$ to find the area of the can. Use 3.14 for π.

b Identify the coefficient and the degree of each term of the polynomial. Then find the degree of the polynomial.

11. $x^3y - 2xy + 3x^2 - 5$ **12.** $5y^3 - y^2 + 15y + 1$

13. $17x^2y^3 - 3x^3yz - 7$ **14.** $6 - xy + 8x^2y^2 - y^5$

ANSWERS

1. _____

2. _____

3. _____

4. _____

5. _____

6. _____

7. _____

8. _____

9. _____

10. _____

11. _____

12. _____

13. _____

14. _____

c   Collect like terms.

15. $a + b - 2a - 3b$

16. $y^2 - 1 + y - 6 - y^2$

17. $3x^2y - 2xy^2 + x^2$

18. $m^3 + 2m^2n - 3m^2 + 3mn^2$

19. $2u^2v - 3uv^2 + 6u^2v - 2uv^2$

20. $3x^2 + 6xy + 3y^2 - 5x^2 - 10xy - 5y^2$

21. $6au + 3av + 14au + 7av$

22. $3x^2y - 2z^2y + 3xy^2 + 5z^2y$

d   Add.

23. $(2x^2 - xy + y^2) + (-x^2 - 3xy + 2y^2)$

24. $(2z - z^2 + 5) + (z^2 - 3z + 1)$

25. $(r - 2s + 3) + (2r + s) + (s + 4)$

26. $(b^3a^2 - 2b^2a^3 + 3ba + 4) + (b^2a^3 - 4b^3a^2 + 2ba - 1)$

27. $(2x^2 - 3xy + y^2) + (-4x^2 - 6xy - y^2) + (x^2 + xy - y^2)$

e   Subtract.

28. $(x^3 - y^3) - (-2x^3 + x^2y - xy^2 + 2y^3)$

29. $(xy - ab) - (xy - 3ab)$

30. $(3y^4x^2 + 2y^3x - 3y) - (2y^4x^2 + 2y^3x - 4y - 2x)$

31. $(-2a + 7b - c) - (-3b + 4c - 8d)$

32. Find the sum of $2a + b$ and $3a - b$. Then subtract $5a + 2b$.

f   Multiply.

33. $(3z - u)(2z + 3u)$

34. $(a - b)(a^2 + b^2 + 2ab)$

35. $(a^2b - 2)(a^2b - 5)$

36. $(xy + 7)(xy - 4)$

37. $(a + a^2 - 1)(a^2 + 1 - y)$

38. $(r + tx)(vx + s)$

39. $(a^3 + bc)(a^3 - bc)$

40. $(m^2 + n^2 - mn)(m^2 + mn + n^2)$

41. $(y^4x + y^2 + 1)(y^2 + 1)$

42. $(a - b)(a^2 + ab + b^2)$

43. $(3xy - 1)(4xy + 2)$

44. $(m^3n + 8)(m^3n - 6)$

45. $(3 - c^2d^2)(4 + c^2d^2)$

46. $(6x - 2y)(5x - 3y)$

47. $(m^2 - n^2)(m + n)$

48. $(pq + 0.2)(0.4pq - 0.1)$

49. $(xy + x^5y^5)(x^4y^4 - xy)$

50. $(x - y^3)(2y^3 + x)$

51. $(x + h)^2$

52. $(3a + 2b)^2$

53. $(r^3t^2 - 4)^2$

54. $(3a^2b - b^2)^2$

55. $(p^4 + m^2n^2)^2$

56. $(ab + cd)^2$

57. $(2a^3 - \frac{1}{2}b^3)^2$

58. $-5x(x + 3y)^2$

59. $3a(a - 2b)^2$

60. $(a^2 + b + 2)^2$

61. $(2a - b)(2a + b)$

62. $(x - y)(x + y)$

63. $(c^2 - d)(c^2 + d)$

64. $(p^3 - 5q)(p^3 + 5q)$

ANSWERS

39. _____

40. _____

41. _____

42. _____

43. _____

44. _____

45. _____

46. _____

47. _____

48. _____

49. _____

50. _____

51. _____

52. _____

53. _____

54. _____

55. _____

56. _____

57. _____

58. _____

59. _____

60. _____

61. _____

62. _____

63. _____

64. _____

65. _____

66. _____

67. _____

68. _____

69. _____

70. _____

71. _____

72. _____

73. _____

74. _____

75. _____

76. _____

77. _____

78. _____

79. _____

65. $(ab + cd^2)(ab - cd^2)$

66. $(xy + pq)(xy - pq)$

67. $(x + y - 3)(x + y + 3)$

68. $(p + q + 4)(p + q - 4)$

69. $[x + y + z][x - (y + z)]$

70. $[a + b + c][a - (b + c)]$

71. $(a + b + c)(a - b - c)$

72. $(3x + 2 - 5y)(3x + 2 + 5y)$

SYNTHESIS

Find a polynomial for the shaded area. (Leave results in terms of π where appropriate.)

73.

74.

75.

76.

77. _Lung capacity._ The polynomial

$$0.041h - 0.018A - 2.69$$

can be used to estimate the lung capacity, in liters, of a female with height h, in centimeters, and age A, in years. Find the lung capacity of a 29-year-old woman who is 138.7 cm tall.

78. _The magic number._ The Boston Red Sox are leading the New York Yankees for the Eastern Division championship of the American League. The magic number is 8. This means that any combination of Red Sox wins and Yankee losses that totals 8 will ensure the championship for the Red Sox. The magic number is given by the polynomial

$$G - P - L + 1,$$

where G is the number of games in the season, P is the number of games that the leading team has played, and L is the number of games ahead in the loss column.

Given the situation shown in the table and assuming a 162-game season, what is the magic number for the Philadelphia Phillies?

EASTERN DIVISION			
	W	**L**	**Pct.**	**GB**
Philadelphia	77	40	.658	—
Pittsburgh	65	53	.551	$12\frac{1}{2}$
New York	61	60	.504	18
Chicago	55	67	.451	$24\frac{1}{2}$
St. Louis	51	65	.440	$25\frac{1}{2}$
Montreal	41	73	.360	$34\frac{1}{2}$

79. Find a formula for $(A + B)^3$.

3.8 Division of Polynomials

In this section, we consider division of polynomials. You will see that such division is similar to what is done in arithmetic.

a Divisor a Monomial

We first consider division by a monomial. When we are dividing a monomial by a monomial, we can use our rules of exponents and subtract exponents when bases are the same. We studied this in Section 3.1. For example,

$$\frac{15x^{10}}{3x^4} = 5x^{10-4} = 5x^6; \qquad \frac{42a^2b^5}{-3ab^2} = \frac{42}{-3}a^{2-1}b^{5-2} = -14ab^3.$$

When we are dividing a monomial into a polynomial, we break up the division into an addition of quotients of monomials. To do this, we use the rule for addition using fractional notation. That is, since

$$\frac{A}{C} + \frac{B}{C} = \frac{A+B}{C},$$

we know that

$$\frac{A+B}{C} = \frac{A}{C} + \frac{B}{C}.$$

▶ **EXAMPLE 1** Divide and check: $x^3 + 10x^2 + 8x$ by $2x$.

We write the division as follows:

$$\frac{x^3 + 10x^2 + 8x}{2x}.$$

This is equivalent to

$$\frac{x^3}{2x} + \frac{10x^2}{2x} + \frac{8x}{2x}. \qquad \text{To see this, add and get the original expression.}$$

Next, we do the separate divisions:

$$\frac{x^3}{2x} + \frac{10x^2}{2x} + \frac{8x}{2x} = \frac{1}{2}x^{3-1} + \frac{10}{2}x^{2-1} + \frac{8}{2}x^{1-1} = \frac{1}{2}x^2 + 5x + 4.$$

We can check by multiplying the quotient by $2x$:

$$
\begin{array}{r}
\frac{1}{2}x^2 + 5x + 4 \\
2x \\
\hline
x^3 + 10x^2 + 8x
\end{array}
$$

We multiply.

The answer checks. ◀

DO EXERCISES 1 AND 2.

OBJECTIVES

After finishing Section 3.8, you should be able to:

a Divide a polynomial by a monomial and check the result.

b Divide a polynomial by a divisor that is not a monomial and, if there is a remainder, express the result in two ways.

FOR EXTRA HELP

Tape 9A   Tape 8A   MAC: 3
                              IBM: 3

Divide.

1. $\dfrac{2x^3 + 6x^2 + 4x}{2x}$

2. $(6x^2 + 3x - 2) \div 3$

Divide and check.

3. $(8x^2 - 3x + 1) \div 2$

▶ **EXAMPLE 2**   Divide and check: $(10a^5b^4 - 2a^3b^2 + 6a^2b) \div 2a^2b$.

$$\frac{10a^5b^4 - 2a^3b^2 + 6a^2b}{2a^2b} = \frac{10a^5b^4}{2a^2b} - \frac{2a^3b^2}{2a^2b} + \frac{6a^2b}{2a^2b}$$

$$= \frac{10}{2}a^{5-2}b^{4-1} - \frac{2}{2}a^{3-2}b^{2-1} + \frac{6}{2}$$

$$= 5a^3b^3 - ab + 3$$

Check:
$$\begin{array}{r} 5a^3b^3 - ab + 3 \\ 2a^2b \\ \hline 10a^5b^4 - 2a^3b^2 + 6a^2b \end{array}$$

We multiply.

The answer checks.   ◀

> **To divide a polynomial by a monomial, divide each term by the monomial.**

DO EXERCISES 3 AND 4.

b   **Divisor not a Monomial**

When the divisor is not a monomial, we use long division very much as we do in arithmetic. We write polynomials in descending order and write in missing terms.

4. $\dfrac{2x^4y^6 - 3x^3y^4 + 5x^2y^3}{x^2y^2}$

▶ **EXAMPLE 3**   Divide $x^2 + 5x + 6$ by $x + 2$.

We have

$$\begin{array}{r} x \\ x + 2 \overline{)\, x^2 + 5x + 6\,} \\ x^2 + 2x \\ \hline 3x \end{array}$$

Divide the first term by the first term: $x^2/x = x$.
Ignore the term 2.

Multiply x above by the divisor, $x + 2$.

Subtract: $(x^2 + 5x) - (x^2 + 2x) = x^2 + 5x - x^2 - 2x = 3x$.

We now "bring down" the next term of the dividend—in this case, 6.

$$\begin{array}{r} x \; + \; 3 \\ x + 2 \overline{)\, x^2 + 5x + 6\,} \\ x^2 + 2x \\ \hline 3x + 6 \\ 3x + 6 \\ \hline 0 \end{array}$$

Divide the first term by the first term: $3x/x = 3$.

The 6 has been "brought down."

Multiply 3 by the divisor, $x + 2$.

Subtract: $(3x + 6) - (3x + 6) = 3x + 6 - 3x - 6 = 0$.

The quotient is $x + 3$. The remainder is 0, usually expressed as R = 0. A remainder of 0 is generally not listed in an answer.

To check, we multiply the quotient by the divisor and add the remainder, if any, to see if we get the dividend:

Divisor   Quotient   Remainder   Dividend

$\overbrace{(x + 2)} \quad \overbrace{(x + 3)} + \quad \overbrace{0} \quad = x^2 + 5x + 6.$ The division checks.   ◀

▶ **EXAMPLE 4**   Divide and check: $(x^2 + 2x - 12) \div (x - 3)$.
We have

$$
\begin{array}{r}
x \\
x - 3 \overline{)\ x^2 + 2x - 12} \\
\underline{x^2 - 3x} \\
5x
\end{array}
$$

— Divide the first term by the first term: $x^2/x = x$.

— Multiply x above by the divisor, $x - 3$.

— Subtract: $(x^2 + 2x) - (x^2 - 3x) = x^2 + 2x - x^2 + 3x$
                  $= 5x$.

We now "bring down" the next term of the dividend—in this case, -12.

$$
\begin{array}{r}
x \ + \ 5 \\
x - 3 \overline{)\ x^2 + 2x - 12} \\
\underline{x^2 - 3x} \\
5x - 12 \\
\underline{5x - 15} \\
3
\end{array}
$$

— Divide the first term by the first term: $5x/x = 5$.

— Bring down the -12.

— Multiply 5 above by the divisor, $x - 3$.

— Subtract: $(5x - 12) - (5x - 15) = 5x - 12 - 5x + 15$
                   $= 3$.

The answer is $x + 5$ with R $= 3$, or

$$
\underbrace{x + 5}_{\text{Quotient}} + \frac{\overset{\displaystyle \longrightarrow \text{Remainder}}{3}}{\underset{\displaystyle \longrightarrow \text{Divisor}}{x - 3}}
$$

(This is the way answers will be given at the back of the book.)

Check: When the answer is given in the preceding form, we can check by multiplying the divisor by the quotient and adding, as follows:

$$(x - 3)(x + 5) + 3 = x^2 + 2x - 15 + 3$$
$$= x^2 + 2x - 12. \qquad ◀$$

When dividing, an answer may "come out even" (that is, have a remainder of 0, as in Example 1), or it may not (as in Example 2). If a remainder is not 0, we continue dividing until the degree of the remainder is less than the degree of the divisor. Check this in each of Examples 1 and 2.

DO EXERCISES 5 AND 6.

▶ **EXAMPLE 5**   Divide: $(x^3 + 1) \div (x + 1)$.

$$
\begin{array}{r}
x^2 - \ x \ + \ 1 \\
x + 1 \overline{)\ x^3 + 0x^2 + 0x + 1} \\
\underline{x^3 + \ x^2} \\
- \ x^2 + 0x \\
\underline{- \ x^2 - \ x} \\
x + 1 \\
\underline{x + 1} \\
0
\end{array}
$$

— Fill in the missing terms.

This subtraction is $x^3 - (x^3 + x^2)$.

This subtraction is $-x^2 - (-x^2 - x)$.

The answer is $x^2 - x + 1$.

Check: $(x + 1)(x^2 - x + 1) + 0 = x^2 - x + 1 + x^3 - x^2 + x + 0$
                                          $= x^3 + 1.$ ◀

Divide and check.

5. $(x^2 + x - 6) \div (x + 3)$

6. $x - 2 \overline{)\ x^2 + 2x - 8}$

ANSWERS ON PAGE A-4

Divide and check.

7. $x + 3 \overline{\smash{)}\, x^2 + 7x + 10}$

8. $(x^3 - 1) \div (x - 1)$

▶ **EXAMPLE 6**   Divide: $(x^4 - 3x^2 + 1) \div (x - 4)$.

$$
\begin{array}{r}
x^3 + 4x^2 + 13x \;+ 52 \\
x - 4 \overline{\smash{)}\, x^4 + 0x^3 - \;3x^2 + \;0x + \;\;1} \\
\underline{x^4 - 4x^3} \\
4x^3 - \;3x^2 \\
\underline{4x^3 - 16x^2} \\
13x^2 + \;0x \\
\underline{13x^2 - 52x} \\
52x + \;\;1 \\
\underline{52x - 208} \\
209
\end{array}
$$

⟵ Fill in the missing terms.

$x^4 - (x^4 - 4x^3)$

$(4x^3 - 3x^2) - (4x^3 - 16x^2)$

The answer is $x^3 + 4x^2 + 13x + 52$, with R $= 209$, or

$$
x^3 + 4x^2 + 13x + 52 + \frac{209}{x - 4}.
$$

Check: $(x - 4)(x^3 + 4x^2 + 13x + 52) + 209$

$$= -4x^3 - 16x^2 - 52x - 208 + x^4 + 4x^3 + 13x^2 + 52x + 209$$

$$= x^4 - 3x^2 + 1 \qquad\qquad ◀$$

DO EXERCISES 7 AND 8.

ANSWERS ON PAGE A-4

NAME   SECTION   DATE

EXERCISE SET 3.8

a   Divide and check.

1. $\dfrac{24x^4 - 4x^3 + x^2 - 16}{8}$

2. $\dfrac{12a^4 - 3a^2 + a - 6}{6}$

3. $\dfrac{u - 2u^2 - u^5}{u}$

4. $\dfrac{50x^5 - 7x^4 + x^2}{x}$

5. $(15t^3 + 24t^2 - 6t) \div 3t$

6. $(25t^3 + 15t^2 - 30t) \div 5t$

7. $(20x^6 - 20x^4 - 5x^2) \div (-5x^2)$

8. $(24x^6 + 32x^5 - 8x^2) \div (-8x^2)$

9. $(24x^5 - 40x^4 + 6x^3) \div (4x^3)$

10. $(18x^6 - 27x^5 - 3x^3) \div (9x^3)$

11. $\dfrac{18x^2 - 5x + 2}{2}$

12. $\dfrac{15x^2 + 30x - 4}{3}$

13. $\dfrac{12x^3 + 26x^2 + 8x}{2x}$

14. $\dfrac{2x^4 - 3x^3 + 5x^2}{x^2}$

15. $\dfrac{9r^2s^2 + 3r^2s - 6rs^2}{3rs}$

16. $\dfrac{4x^4y - 8x^6y^2 + 12x^8y^6}{4x^4y}$

b   Divide.

17. $(x^2 + 4x + 4) \div (x + 2)$

18. $(x^2 - 6x + 9) \div (x - 3)$

19. $(x^2 - 10x - 25) \div (x - 5)$

20. $(x^2 + 8x - 16) \div (x + 4)$

21. $(x^2 + 4x - 14) \div (x + 6)$

22. $(x^2 + 5x - 9) \div (x - 2)$

ANSWERS

1. _____
2. _____
3. _____
4. _____
5. _____
6. _____
7. _____
8. _____
9. _____
10. _____
11. _____
12. _____
13. _____
14. _____
15. _____
16. _____
17. _____
18. _____
19. _____
20. _____
21. _____
22. _____

23. $\dfrac{x^2 - 9}{x + 3}$

24. $\dfrac{x^2 - 25}{x + 5}$

25. $\dfrac{x^5 + 1}{x + 1}$

26. $\dfrac{x^5 - 1}{x - 1}$

27. $\dfrac{8x^3 - 22x^2 - 5x + 12}{4x + 3}$

28. $\dfrac{2x^3 - 9x^2 + 11x - 3}{2x - 3}$

29. $(x^6 - 13x^3 + 42) \div (x^3 - 7)$

30. $(x^6 + 5x^3 - 24) \div (x^3 - 3)$

31. $(x^4 - 16) \div (x - 2)$

32. $(x^4 - 81) \div (x - 3)$

33. $(t^3 - t^2 + t - 1) \div (t - 1)$

34. $(t^3 - t^2 + t - 1) \div (t + 1)$

SKILL MAINTENANCE

35. Subtract: $-2.3 - (-9.1)$.

36. Factor: $4x - 12 + 24y$.

37. The perimeter of a rectangle is 640 ft. The length is 15 ft more than the width. Find the area of the rectangle.

38. Solve: $-6(2 - x) + 10(5x - 7) = 10$.

SYNTHESIS

Divide.

39. $(x^4 + 9x^2 + 20) \div (x^2 + 4)$

40. $(y^4 + a^2) \div (y + a)$

41. $(5a^3 + 8a^2 - 23a - 1) \div (5a^2 - 7a - 2)$

42. $(15y^3 - 30y + 7 - 19y^2) \div (3y^2 - 2 - 5y)$

43. $(6x^5 - 13x^3 + 5x + 3 - 4x^2 + 3x^4) \div (3x^3 - 2x - 1)$

44. $(5x^7 - 3x^4 + 2x^2 - 10x + 2) \div (x^2 - x + 1)$

45. $(a^6 - b^6) \div (a - b)$

46. $(x^5 + y^5) \div (x + y)$

If the remainder is 0 when one polynomial is divided by another, the divisor is a *factor* of the dividend. Find the value(s) of c for which $x - 1$ is a factor of the polynomial.

47. $x^2 + 4x + c$

48. $2x^2 + 3cx - 8$

49. $c^2x^2 - 2cx + 1$

SUMMARY AND REVIEW: CHAPTER 3

IMPORTANT PROPERTIES AND FORMULAS

FOIL: $(A + B)(C + D) = AC + AD + BC + BD,$ $(A + B)(A + B) = (A + B)^2 = A^2 + 2AB + B^2$

$(A - B)(A - B) = (A - B)^2 = A^2 - 2AB + B^2,$ $(A + B)(A - B) = A^2 - B^2$

Definitions and Rules for Exponents
See p. 190.

REVIEW EXERCISES

The review sections and objectives to be tested in addition to the material in this chapter are [1.4a], [1.7d], [2.3b, c], and [2.4a].

Multiply.

1. $7^2 \cdot 7^{-4}$ **2.** $y^7 \cdot y^3 \cdot y$ **3.** $(3x)^5 \cdot (3x)^9$ **4.** $t^8 \cdot t^0$

Divide.

5. $\dfrac{4^5}{4^2}$ **6.** $\dfrac{a^5}{a^8}$ **7.** $\dfrac{(7x)^4}{(7x)^4}$

Simplify.

8. $(3t^4)^2$ **9.** $(2x^3)^2(-3x)^2$ **10.** $\left(\dfrac{2x}{y}\right)^{-3}$

11. Express using a negative exponent: $\dfrac{1}{t^5}$.

12. Express using a positive exponent: y^{-4}.

13. Convert to scientific notation: 0.0000328.

14. Convert to decimal notation: 8.3×10^6.

Multiply or divide and write scientific notation for the result.

15. $(3.8 \times 10^4)(5.5 \times 10^{-1})$

16. $\dfrac{1.28 \times 10^{-8}}{2.5 \times 10^{-4}}$

17. Each day Americans eat 170 million eggs. How many eggs are eaten in one year? Write scientific notation for the answer.

18. Evaluate the polynomial $x^2 - 3x + 6$ when $x = -1$.

19. Identify the terms of the polynomial $-4y^5 + 7y^2 - 3y - 2$.

20. Identify the missing terms in $x^3 + x$.

21. Identify the degree of each term and the degree of the polynomial $4x^3 + 6x^2 - 5x + \frac{5}{3}$.

Classify the polynomial as a monomial, binomial, trinomial, or none of these.

22. $4x^3 - 1$ **23.** $4 - 9t^3 - 7t^4 + 10t^2$ **24.** $7y^2$

Collect like terms and then arrange in descending order.

25. $3x^2 - 2x + 3 - 5x^2 - 1 - x$

26. $-x + \frac{1}{2} + 14x^4 - 7x^2 - 1 - 4x^4$

Add.

27. $(3x^4 - x^3 + x - 4) + (x^5 + 7x^3 - 3x^2 - 5) + (-5x^4 + 6x^2 - x)$

28. $(3x^5 - 4x^4 + x^3 - 3) + (3x^4 - 5x^3 + 3x^2) + (4x^5 + 4x^3) + (-5x^5 - 5x^2) + (-5x^4 + 2x^3 + 5)$

Subtract.

29. $(5x^2 - 4x + 1) - (3x^2 + 7)$

30. $(3x^5 - 4x^4 + 3x^2 + 3) - (2x^5 - 4x^4 + 3x^3 + 4x^2 - 5)$

31. The length of a rectangle is 4 m greater than its width. Find a polynomial for the perimeter and a polynomial for the area.

Multiply.

32. $(x + \frac{2}{3})(x + \frac{1}{2})$

33. $(7x + 1)^2$

34. $(4x^2 - 5x + 1)(3x - 2)$

35. $(3x^2 + 4)(3x^2 - 4)$

36. $5x^4(3x^3 - 8x^2 + 10x + 2)$

37. $(x + 4)(x - 7)$

38. $(3y^2 - 2y)^2$

39. $(2t^2 + 3)(t^2 - 7)$

40. Evaluate the polynomial $2 - 5xy + y^2 - 4xy^3 + x^6$ when $x = -1$ and $y = 2$.

41. Identify the coefficient and degree of each term of the polynomial $x^5y - 7xy + 9x^2 - 8$. Then find the degree of the polynomial.

Collect like terms.

42. $y + w - 2y + 8w - 5$

43. $m^6 - 2m^2n + m^2n^2 + n^2m - 6m^3 + m^2n^2 + 7n^2m$

44. Add:
$(5x^2 - 7xy + y^2) + (-6x^2 - 3xy - y^2) + (x^2 + xy - 2y^2)$.

45. Subtract:
$(6x^3y^2 - 4x^2y - 6x) - (-5x^3y^2 + 4x^2y + 6x^2 - 6)$.

Multiply.

46. $(p - q)(p^2 + pq + q^2)$

47. $(3a^4 - \frac{1}{3}b^3)^2$

Divide.

48. $(10x^3 - x^2 + 6x) \div 2x$

49. $(6x^3 - 5x^2 - 13x + 13) \div (2x + 3)$

SKILL MAINTENANCE

50. Factor: $25t - 50 + 100m$.

51. Solve: $7x + 6 - 8x = 11 - 5x + 4$.

52. Subtract: $-3.4 - 7.8$.

53. The perimeter of a rectangle is 540 m. The width is 19 m less than the length. Find the width and the length.

SYNTHESIS

54. Collect like terms:
$-3x^5 \cdot 3x^3 - x^6(2x)^2 + (3x^4)^2 + (2x^2)^4 - 40x^2(x^3)^2$.

55. Solve:
$(x - 7)(x + 10) = (x - 4)(x - 6)$.

❖ THINKING IT THROUGH

Explain the error(s) in each of the following.

1. $(a + 2)^2 = a^2 + 4$

2. $(p + 7)(p - 7) = p^2 + 14p - 49$

3. $(t - 3)^2 = t^2 - 9$

4. $2^{-3} = -6$

5. $\dfrac{a^2}{a^5} = a^3$

6. $m^{-2}m^5 = m^{-10}$

7. Explain why 0.23×10^5 is not scientific notation.

NAME   SECTION   DATE

TEST: CHAPTER 3

Multiply.

1. $6^{-2} \cdot 6^{-3}$ **2.** $x^6 \cdot x^2 \cdot x$ **3.** $(4a)^3 \cdot (4a)^8$

Divide.

4. $\dfrac{3^5}{3^2}$ **5.** $\dfrac{x^3}{x^8}$ **6.** $\dfrac{(2x)^5}{(2x)^5}$

Simplify.

7. $(x^3)^2$ **8.** $(-3y^2)^3$ **9.** $(2a^3b)^4$ **10.** $\left(\dfrac{ab}{c}\right)^3$

11. $(3x^2)^3(-2x^5)^3$ **12.** $3(x^2)^3(-2x^5)^3$ **13.** $2x^2(-3x^2)^4$ **14.** $(2x)^2(-3x^2)^4$

15. Express using a positive exponent: 5^{-3}.

16. Express using a negative exponent: $\dfrac{1}{y^8}$.

17. Convert to scientific notation: 3,900,000,000.

18. Convert to decimal notation: 5×10^{-8}.

Multiply or divide and write scientific notation for the answer.

19. $\dfrac{5.6 \times 10^6}{3.2 \times 10^{-11}}$ **20.** $(2.4 \times 10^5)(5.4 \times 10^{16})$

21. Each day Americans eat 170 million eggs. There are 243 million people in this country. How many eggs does each person eat in one year? Write scientific notation for the answer.

22. Evaluate the polynomial $x^5 + 5x - 1$ when $x = -2$.

23. Identify the coefficient of each term of the polynomial $\frac{1}{3}x^5 - x + 7$.

24. Identify the degree of each term and the degree of the polynomial $2x^3 - 4 + 5x + 3x^6$.

25. Classify the polynomial $7 - x$ as a monomial, binomial, trinomial, or none of these.

1. _____

2. _____

3. _____

4. _____

5. _____

6. _____

7. _____

8. _____

9. _____

10. _____

11. _____

12. _____

13. _____

14. _____

15. _____

16. _____

17. _____

18. _____

19. _____

20. _____

21. _____

22. _____

23. _____

24. _____

25. _____

26. _____	
27. _____	
28. _____	
29. _____	
30. _____	
31. _____	
32. _____	
33. _____	
34. _____	
35. _____	
36. _____	
37. _____	
38. _____	
39. _____	
40. _____	
41. _____	
42. _____	
43. _____	
44. _____	
45. _____	
46. _____	
47. _____	
48. _____	
49. _____	
50. _____	
51. _____	

Collect like terms.

26. $4a^2 - 6 + a^2$

27. $y^2 - 3y - y + \frac{3}{4}y^2$

28. Collect like terms and then arrange in descending order:
$$3 - x^2 + 2x^3 + 5x^2 - 6x - 2x + x^5.$$

Add.

29. $(3x^5 + 5x^3 - 5x^2 - 3) +$
$(x^5 + x^4 - 3x^3 - 3x^2 + 2x - 4)$

30. $\left(x^4 + \frac{2}{3}x + 5\right) + \left(4x^4 + 5x^2 + \frac{1}{3}x\right)$

Subtract.

31. $(2x^4 + x^3 - 8x^2 - 6x - 3) -$
$(6x^4 - 8x^2 + 2x)$

32. $(x^3 - 0.4x^2 - 12) -$
$(x^5 + 0.3x^3 + 0.4x^2 + 9)$

Multiply.

33. $-3x^2(4x^2 - 3x - 5)$

34. $\left(x - \frac{1}{3}\right)^2$

35. $(3x + 10)(3x - 10)$

36. $(3b + 5)(b - 3)$

37. $(x^6 - 4)(x^8 + 4)$

38. $(8 - y)(6 + 5y)$

39. $(2x + 1)(3x^2 - 5x - 3)$

40. $(5t + 2)^2$

41. Collect like terms: $x^3y - y^3 + xy^3 + 8 - 6x^3y - x^2y^2 + 11$.

42. Subtract: $(8a^2b^2 - ab + b^3) - (-6ab^2 - 7ab - ab^3 + 5b^3)$.

43. Multiply: $(3x^5 - 4y^5)(3x^5 + 4y^5)$.

Divide.

44. $(12x^4 + 9x^3 - 15x^2) \div 3x^2$

45. $(6x^3 - 8x^2 - 14x + 13) \div (3x + 2)$

SKILL MAINTENANCE

46. Solve: $7x - 4x - 2 = 37$.

47. Factor: $64t - 32m + 16$.

48. Subtract: $\frac{2}{5} - (-\frac{3}{4})$.

49. The first angle of a triangle is four times as large as the second. The measure of the third angle is 30° greater than that of the second. How large are the angles?

SYNTHESIS

50. The height of a box is one less than its length, and the length is two more than its width. Find the volume in terms of the length.

51. Solve: $(x - 5)(x + 5) = (x + 6)^2$.

CUMULATIVE REVIEW: CHAPTERS 1–3

1. Evaluate $\dfrac{x}{2y}$ when $x = 10$ and $y = 2$.

2. Evaluate $2x^3 + x^2 - 3$ when $x = -1$.

3. Evaluate $x^3y^2 + xy + 2xy^2$ when $x = -1$ and $y = 2$.

4. Find the absolute value: $|-4|$.

5. Find the reciprocal of 5.

Compute and simplify.

6. $-\dfrac{3}{5} + \dfrac{5}{12}$

7. $3.4 - (-0.8)$

8. $(-2)(-1.4)(2.6)$

9. $\dfrac{3}{8} \div \left(-\dfrac{9}{10}\right)$

10. $(1.1 \times 10^{10})(2 \times 10^{12})$

11. $(3.2 \times 10^{-10}) \div (8 \times 10^{-6})$

Simplify.

12. $\dfrac{-9x}{3x}$

13. $y - (3y + 7)$

14. $3(x - 1) - 2[x - (2x + 7)]$

15. $2 - [32 \div (4 + 2^2)]$

Add.

16. $(x^4 + 3x^3 - x + 7) + (2x^5 - 3x^4 + x - 5)$

17. $(x^2 + 2xy) + (y^2 - xy) + (2x^2 - 3y^2)$

Subtract.

18. $(x^3 + 3x^2 - 4) - (-2x^2 + x + 3)$

19. $\left(\dfrac{1}{3}x^2 - \dfrac{1}{4}x - \dfrac{1}{5}\right) - \left(\dfrac{2}{3}x^2 + \dfrac{1}{2}x - \dfrac{1}{5}\right)$

Multiply.

20. $3(4x - 5y + 7)$

21. $(-2x^3)(-3x^5)$

22. $2x^2(x^3 - 2x^2 + 4x - 5)$

23. $(y^2 - 2)(3y^2 + 5y + 6)$

24. $(2p^3 + p^2q + pq^2)(p - pq + q)$

25. $(2x + 3)(3x + 2)$

26. $(3x^2 + 1)^2$

27. $\left(t + \dfrac{1}{2}\right)\left(t - \dfrac{1}{2}\right)$

28. $(2y^2 + 5)(2y^2 - 5)$

29. $(2x^4 - 3)(2x^2 + 3)$

30. $(t - 2t^2)^2$

31. $(3p + q)(5p - 2q)$

Divide.

32. $(18x^3 + 6x^2 - 9x) \div 3x$

33. $(3x^3 + 7x^2 - 13x - 21) \div (x + 3)$

Solve.

34. $1.5 = 2.7 + x$

35. $\dfrac{2}{7}x = -6$

36. $5x - 9 = 36$

37. $\dfrac{2}{3} = \dfrac{-m}{10}$

38. $5.4 - 1.9x = 0.8x$

39. $x - \dfrac{7}{8} = \dfrac{3}{4}$

40. $2(2 - 3x) = 3(5x + 7)$

41. $\dfrac{1}{4}x - \dfrac{2}{3} = \dfrac{3}{4} + \dfrac{1}{3}x$

42. $y + 5 - 3y = 5y - 9$

43. $\dfrac{1}{4}x - 7 < 5 - \dfrac{1}{2}x$

44. $2(x + 2) \geq 5(2x + 3)$

45. $A = 2\pi r h + \pi r^2$ for h

46. A 6-ft by 3-ft raft is floating in a swimming pool of radius r. Find a polynomial for the area of the surface of the pool not covered by the raft.

Solve.

47. The sum of the page numbers on the facing pages of a book is 37. What are the page numbers?

48. The perimeter of a room is 88 ft. The width is 4 ft less than the length. Find the width and the length.

49. The second angle of a triangle is five times as large as the first. The third angle is twice the sum of the other two angles. Find the measure of the first angle.

50. If you triple a number and then add 99, you get $\frac{4}{5}$ of the original number. What is the original number?

51. A bookstore sells books at a price that is 80% higher than the price the store pays for the books. A book is priced for sale at $6.30. How much did the store pay for the book?

Simplify.

52. $y^2 \cdot y^{-6} \cdot y^8$

53. $\dfrac{x^6}{x^7}$

54. $(-3x^3 y^{-2})^3$

55. $\dfrac{x^3 x^{-4}}{x^{-5} x}$

56. Identify the coefficient of each term of the polynomial $\frac{2}{3}x^2 + 4x - 6$.

57. Identify the degree of each term and the degree of the polynomial $2x^4 + 3x^2 + 2x + 1$.

Classify the polynomial as a monomial, binomial, trinomial, or none of these.

58. $2x^2 + 1$

59. $2x^2 + x + 1$

SYNTHESIS

60. A picture frame is x inches square. The picture that it frames is 2 in. shorter than the frame in both length and width. Find a polynomial for the area of the frame.

Add.

61. $[(2x)^2 - (3x)^3 + 2x^2 x^3 + (x^2)^2] + [5x^2(2x^3) - ((2x)^2)^2]$

62. $(x - 3)^2 + (2x + 1)^2$

63. $[(3x^3 + 11x^2 + 11x + 15) \div (x + 3)] + [(2x^3 - 7x^2 + 2) \div (2x + 1)]$

Solve.

64. $(x + 3)(2x - 5) + (x - 1)^2 = (3x + 1)(x - 3)$

65. $(2x^2 + x - 6) \div (2x - 3) = (2x^2 - 9x - 5) \div (x - 5)$

66. $20 - 3|x| = 5$

67. $(x + 2)^2 = (x + 1)(x + 3)$

68. $(x - 3)(x + 4) = (x^3 - 4x^2 - 17x + 60) \div (x - 5)$

INTRODUCTION *Factoring* is the reverse of multiplying. To *factor* a polynomial, or other algebraic expression, is to find an equivalent expression that is a product. In this chapter we study factoring polynomials. To learn to factor quickly, we use the quick methods for multiplication learned in Chapter 3.

At the end of the chapter, we get the payoff for learning to factor. We have certain new equations containing second-degree polynomials that we can now solve. This then allows us to solve problems that we could not have solved before.

The review sections to be tested in addition to the material in this chapter are 1.6, 2.6, 2.7, and 3.6. ❖

Polynomials: Factoring

4

AN APPLICATION

A beverage can has height h and radius r. Its surface area is given by the polynomial

$$2\pi rh + 2\pi r^2.$$

THE MATHEMATICS

The polynomial can be *factored* as follows:

$$2\pi rh + 2\pi r^2 = 2\pi r(h + r).$$

Methods to find special products:	Chapter 3
Pythagorean Theorem:	If a and b are the lengths of the legs of a right triangle and c is the length of the hypotenuse, then $a^2 + b^2 = c^2$.
Equation-Solving Skills:	Sections 2.1–2.3

PRETEST: CHAPTER 4

1. Find three factorizations of $-20x^6$.

Factor.

2. $2x^2 + 4x + 2$

3. $x^2 + 6x + 8$

4. $8a^5 + 4a^3 - 20a$

5. $-6 + 5x^2 - 13x$

6. $81 - z^4$

7. $y^6 - 4y^3 + 4$

8. $3x^3 + 2x^2 + 12x + 8$

9. $p^2 - p - 30$

Solve.

10. $x^2 - 5x = 0$

11. $(x - 4)(5x - 3) = 0$

12. $3x^2 + 10x - 8 = 0$

Solve.

13. Six less than the square of a number is five times the number. Find all such numbers.

14. The height of a triangle is 3 cm longer than the base. The area of the triangle is 44 cm². Find the base and the height.

Factor.

15. $x^4y^2 - 64$

16. $2p^2 + 7pq - 4q^2$

4.1 Introduction to Factoring

To solve certain types of algebraic equations involving polynomials of second degree, we must learn to factor polynomials.

> To *factor* a polynomial is to find an equivalent expression that is a product.

When we factor, we do the reverse of multiplication.

a Factoring Monomials

To factor a monomial, we find two monomials whose product is equivalent to the original monomial. Compare.

Multiplying	*Factoring*
a) $(4x)(5x) = 20x^2$	$20x^2 = (4x)(5x)$
b) $(2x)(10x) = 20x^2$	$20x^2 = (2x)(10x)$
c) $(-4x)(-5x) = 20x^2$	$20x^2 = (-4x)(-5x)$
d) $(x)(20x) = 20x^2$	$20x^2 = (x)(20x)$

You can see that the monomial $20x^2$ has many factorizations. There are still other ways to factor $20x^2$.

DO EXERCISES 1 AND 2.

▶ **EXAMPLE 1** Find three factorizations of $15x^3$.

a) $15x^3 = (3 \cdot 5)(x \cdot x^2)$
$= (3x)(5x^2)$

b) $15x^3 = (3 \cdot 5)(x^2 \cdot x)$
$= (3x^2)(5x)$

c) $15x^3 = (-15)(-1)x^3$
$= (-15)(-x^3)$ ◀

DO EXERCISES 3–5.

OBJECTIVES

After finishing Section 4.1, you should be able to:

a Factor monomials.

b Factor polynomials when the terms have a common factor, factoring out the largest common factor.

c Factor certain expressions with four terms using factoring by grouping.

FOR EXTRA HELP

Tape 9B   Tape 8A   MAC: 4
                                  IBM: 4

1. a) Multiply: $(3x)(4x)$.

b) Factor: $12x^2$.

2. a) Multiply: $(2x)(8x^2)$.

b) Factor: $16x^3$.

Find three factorizations of the monomial.

3. $8x^4$

4. $21x^2$

5. $6x^5$

ANSWERS ON PAGE A-4

6. a) Multiply: $3(x + 2)$.

b) Factor: $3x + 6$.

7. a) Multiply: $2x(x^2 + 5x + 4)$.

b) Factor: $2x^3 + 10x^2 + 8x$.

Factoring When Terms Have a Common Factor

To factor polynomials quickly, we consider the special-product rules learned in Chapter 3, but we first factor out the largest common factor.

To multiply a monomial and a polynomial with more than one term, we multiply each term by the monomial using the distributive laws, $a(b + c) = ab + ac$ and $a(b - c) = ab - ac$. To factor, we do the reverse. We express a polynomial as a product using the distributive laws in reverse: $ab + ac = a(b + c)$ and $ab - ac = a(b - c)$. Compare.

Multiply

$3x(x^2 + 2x - 4)$

$\quad = 3x \cdot x^2 + 3x \cdot 2x - 3x \cdot 4$

$\quad = 3x^3 + 6x^2 - 12x$

Factor

$3x^3 + 6x^2 - 12x$

$\quad = 3x \cdot x^2 + 3x \cdot 2x - 3x \cdot 4$

$\quad = 3x(x^2 + 2x - 4)$

DO EXERCISES 6 AND 7.

CAUTION!   Consider the following:

$$3x^3 + 6x^2 - 12x = 3 \cdot x \cdot x \cdot x + 2 \cdot 3 \cdot x \cdot x - 2 \cdot 2 \cdot 3x.$$

The terms of the polynomial, $3x^3$, $6x^2$, and $-12x$, have been factored but the polynomial itself has not been factored. This is not a factorization. The *factorization* is

$$3x(x^2 + 2x - 4).$$

The expressions $3x$ and $x^2 + 2x - 4$ are *factors*.

To factor, we first try to find a factor common to all terms. There may not always be one other than 1. When there is, we generally use the factor with the largest possible coefficient and the largest possible exponent.

▶   **EXAMPLE 2**   Factor: $3x^2 + 6$.

We have

$$3x^2 + 6 = 3 \cdot x^2 + 3 \cdot 2 \qquad \text{Factoring each term}$$
$$= 3(x^2 + 2). \qquad \text{Factoring out the common factor, 3}$$

We can check by multiplying: $3(x^2 + 2) = 3 \cdot x^2 + 3 \cdot 2 = 3x^2 + 6$.   ◀

▶   **EXAMPLE 3**   Factor: $16x^3 + 20x^2$.

$$16x^3 + 20x^2 = (4x^2)(4x) + (4x^2)(5) \qquad \text{Factoring each term}$$
$$= 4x^2(4x + 5) \qquad \text{Factoring out } 4x^2 \qquad ◀$$

Suppose in Example 3 that you had not recognized the largest common factor and only removed part of it, as follows:

$$16x^3 + 20x^2 = (2x^2)(8x) + (2x^2)(10)$$
$$= 2x^2(8x + 10).$$

Note that $8x + 10$ still has a common factor of 2. You need not begin again. Just continue factoring out common factors, as follows, until finished:

$$= 2x^2[2(4x + 5)]$$
$$= 4x^2(4x + 5).$$

▶ **EXAMPLE 4** Factor: $15x^5 - 12x^4 + 27x^3 - 3x^2$.

$$15x^5 - 12x^4 + 27x^3 - 3x^2 = (3x^2)(5x^3) - (3x^2)(4x^2) + (3x^2)(9x) - (3x^2)(1)$$
$$= 3x^2(5x^3 - 4x^2 + 9x - 1) \qquad \text{Factoring out } 3x^2 \qquad ◀$$

| CAUTION! Don't forget the term -1. |

If you can spot the largest common factor without factoring each term, you can write just the answer.

▶ **EXAMPLES** Factor.

5. $8m^3 - 16m = 8m(m^2 - 2)$

6. $14p^2y^3 - 8py^2 + 2py = 2py(7py^2 - 4y + 1)$

7. $\dfrac{4}{5}x^2 + \dfrac{1}{5}x + \dfrac{2}{5} = \dfrac{1}{5}(4x^2 + x + 2)$ ◀

DO EXERCISES 8–12.

Below is one of the most important points to keep in mind as we study this chapter.

| Before doing any other kind of factoring, first try to factor out the largest common factor. |

Another tip is the following.

| You can always check the result of factoring by multiplying. |

Factor.

8. $x^2 + 3x$

9. $3y^6 - 5y^3 + 2y^2$

10. $9x^4 - 15x^3 + 3x^2$

11. $\dfrac{3}{4}t^3 + \dfrac{5}{4}t^2 + \dfrac{7}{4}t + \dfrac{1}{4}$

12. $35x^7 - 49x^6 + 14x^5 - 63x^3$

ANSWERS ON PAGE A-4

Factor.

13. $x^2(x + 7) + 3(x + 7)$

14. $x^2(a + b) + 2(a + b)$

Factor by grouping.

15. $x^3 + 7x^2 + 3x + 21$

16. $8t^3 + 2t^2 + 12t + 3$

17. $3m^5 - 15m^3 + 2m^2 - 10$

18. $4x^3 - 6x^2 - 6x + 9$

19. $y^4 - 2y^3 - 2y - 10$

c   Factoring by Grouping

Certain polynomials with four terms can be factored using a method called *factoring by grouping*.

▶ **EXAMPLE 8**   Factor: $x^2(x + 1) + 2(x + 1)$.

The binomial $x + 1$ is common to both terms:

$$x^2(x + 1) + 2(x + 1) = (x^2 + 2)(x + 1).$$

The factorization is $(x^2 + 2)(x + 1)$.   ◀

DO EXERCISES 13 AND 14.

Consider the four-term polynomial

$$x^3 + x^2 + 2x + 2.$$

There is no factor other than 1 that is common to all the terms. We can, however, factor $x^3 + x^2$ and $2x + 2$ separately:

$$x^3 + x^2 = x^2(x + 1); \qquad \text{Factoring } x^3 + x^2$$
$$2x + 2 = 2(x + 1). \qquad \text{Factoring } 2x + 2$$

We have grouped certain terms and factored each polynomial separately:

$$
\begin{aligned}
x^3 + x^2 + 2x + 2 &= (x^3 + x^2) + (2x + 2) \\
&= x^2(x + 1) + 2(x + 1) \\
&= (x^2 + 2)(x + 1),
\end{aligned}
$$

as in Example 8. This method is called **factoring by grouping.** We began with a polynomial with four terms. After grouping and removing common factors, we obtained a polynomial with two terms, each having a common factor $x + 1$. Not all polynomials with four terms can be factored by this method, but it does give us a method to try.

▶ **EXAMPLES**   Factor by grouping.

9. $6x^3 - 9x^2 + 4x - 6$
$$= (6x^3 - 9x^2) + (4x - 6)$$
$$= 3x^2(2x - 3) + 2(2x - 3) \qquad \text{Factoring each binomial}$$
$$= (3x^2 + 2)(2x - 3) \qquad \text{Factoring out the common factor}$$

10. $x^3 + x^2 + x + 1 = (x^3 + x^2) + (x + 1)$
$$= x^2(x + 1) + 1(x + 1) \qquad \text{Factoring each binomial}$$
$$= (x^2 + 1)(x + 1) \qquad \text{Factoring out the common factor}$$

11. $12x^5 + 20x^2 - 21x^3 - 35 = 12x^5 + 20x^2 - 21x^3 - 35$
$$= 4x^2(3x^3 + 5) - 7(3x^3 + 5)$$
$$= (4x^2 - 7)(3x^3 + 5)$$

12. $x^3 + x^2 + 2x - 2 = x^2(x + 1) + 2(x - 1)$

This polynomial is not factorable using factoring by grouping. It may be factorable, but not by methods that we will consider in this text.

      ◀

DO EXERCISES 15–19.

EXERCISE SET 4.1

a Find three factorizations for the monomial.

1. $6x^3$ **2.** $9x^4$ **3.** $-9x^5$ **4.** $-12x^6$ **5.** $24x^4$ **6.** $15x^5$

b Factor. Check by multiplying.

7. $x^2 - 4x$ **8.** $x^2 + 8x$ **9.** $2x^2 + 6x$

10. $3x^2 - 3x$ **11.** $x^3 + 6x^2$ **12.** $4x^4 + x^2$

13. $8x^4 - 24x^2$ **14.** $5x^5 + 10x^3$ **15.** $2x^2 + 2x - 8$

16. $6x^2 + 3x - 15$ **17.** $17x^5y^3 + 34x^3y^2 + 51xy$

18. $16x^6y^4 - 32x^5y^3 - 48xy^2$ **19.** $6x^4 - 10x^3 + 3x^2$

20. $5x^5 + 10x^2 - 8x$ **21.** $x^5y^5 + x^4y^3 + x^3y^3 - x^2y^2$

22. $x^9y^6 - x^7y^5 + x^4y^4 + x^3y^3$ **23.** $2x^7 - 2x^6 - 64x^5 + 4x^3$

24. $10x^3 + 25x^2 + 15x - 20$ **25.** $1.6x^4 - 2.4x^3 + 3.2x^2 + 6.4x$

1. _____
2. _____
3. _____
4. _____
5. _____
6. _____
7. _____
8. _____
9. _____
10. _____
11. _____
12. _____
13. _____
14. _____
15. _____
16. _____
17. _____
18. _____
19. _____
20. _____
21. _____
22. _____
23. _____
24. _____
25. _____

26. $2.5x^6 - 0.5x^4 + 5x^3 + 10x^2$

27. $\frac{5}{3}x^6 + \frac{4}{3}x^5 + \frac{1}{3}x^4 + \frac{1}{3}x^3$

28. $\frac{5}{7}x^7 + \frac{3}{7}x^5 - \frac{6}{7}x^3 - \frac{1}{7}x$

c Factor.

29. $x^2(x + 3) + 2(x + 3)$

30. $3z^2(2z + 1) + (2z + 1)$

Factor by grouping.

31. $x^3 + 3x^2 + 2x + 6$

32. $6z^3 + 3z^2 + 2z + 1$

33. $2x^3 + 6x^2 + x + 3$

34. $3x^3 + 2x^2 + 3x + 2$

35. $8x^3 - 12x^2 + 6x - 9$

36. $10x^3 - 25x^2 + 4x - 10$

37. $12x^3 - 16x^2 + 3x - 4$

38. $18x^3 - 21x^2 + 30x - 35$

39. $x^3 + 8x^2 - 3x - 24$

40. $2x^3 + 12x^2 - 5x - 30$

41. $2x^3 - 8x^2 - 9x + 36$

42. $20g^3 - 4g^2 - 25g + 5$

SKILL MAINTENANCE

Solve.

43. $-2x < 48$

44. $4x - 8x + 16 \geq 6(x - 2)$

45. Divide: $\dfrac{-108}{-4}$.

46. Solve $A = \dfrac{p + q}{2}$ for p.

Multiply.

47. $(y + 5)(y + 7)$

48. $(y + 7)^2$

49. $(y + 7)(y - 7)$

50. $(y - 7)^2$

SYNTHESIS

Factor.

51. $4x^5 + 6x^3 + 6x^2 + 9$

52. $x^6 + x^4 + x^2 + 1$

53. $x^{12} + x^7 + x^5 + 1$

54. $x^3 - x^2 - 2x + 5$

55. $p^3 + p^2 - 3p + 10$

56. Subtract $(x^2 + 1)^2$ from $x(x + 1)^2$ and factor the result.

4.2 Factoring Trinomials of the Type $x^2 + bx + c$

a   We now begin a study of the factoring of trinomials. We first try to factor trinomials like

$$x^2 + 5x + 6 \quad \text{and} \quad x^2 + 3x - 10$$

by *trial and error*. In this section, we restrict our attention to trinomials of the type $ax^2 + bx + c$, where $a = 1$. The coefficient a is often called the **leading coefficient.**

Constant Term Positive

Recall the FOIL method of multiplying two binomials:

$$\begin{array}{ccccc} \text{F} & \text{O} & \text{I} & \text{L} \\ (x + 2)(x + 5) = x^2 + 5x + 2x + 10 \end{array}$$

$$= x^2 + 7x + 10.$$

The product above is a trinomial. The term of highest degree, x^2, called the leading term, has a coefficient of 1. The constant term, 10, is positive. To factor $x^2 + 7x + 10$, we think of FOIL in reverse. We multiplied x times x to get the first term of the trinomial, so we know that the first term of each binomial factor is x. Next we look for numbers p and q such that

$$x^2 + 7x + 10 = (x + p)(x + q).$$

To get the middle term and the last term of the trinomial, we look for two numbers p and q whose product is 10 and whose sum is 7. Those numbers are 2 and 5. Thus the factorization is

$$(x + 2)(x + 5).$$

▶ **EXAMPLE 1**   Factor: $x^2 + 5x + 6$.

Think of FOIL in reverse. The first term of each factor is x:

$$(x + p)(x + q).$$

We then look for two numbers p and q whose product is 6 and whose sum is 5. Since both 5 and 6 are positive, we need consider only positive factors.

Pairs of factors	Sums of factors
1, 6	7
2, 3	5 ←

The numbers we need are 2 and 3.

The factorization is $(x + 2)(x + 3)$. We can check by multiplying to see whether we get the original trinomial.

Check: $(x + 2)(x + 3) = x^2 + 3x + 2x + 6 = x^2 + 5x + 6.$ ◀

DO EXERCISES 1 AND 2.

OBJECTIVE

After finishing Section 4.2, you should be able to:

a Factor trinomials of the type $x^2 + bx + c$ by examining the constant term c.

FOR EXTRA HELP

Tape 9C   Tape 8B   MAC: 4
                                  IBM: 4

Factor.

1. $x^2 + 7x + 12$

2. $x^2 + 13x + 36$

Factor.

3. $x^2 - 8x + 15$

4. $t^2 - 9t + 20$

Consider this multiplication:

$$\underset{\text{F}\quad\text{O}\quad\text{I}\quad\text{L}}{(x - 2)(x - 5) = x^2 - \underbrace{5x - 2x}_{} + 10}$$

$$= x^2 - 7x + 10.$$

> When the constant term of a trinomial is positive, we look for two numbers with the same sign. The sign is that of the middle term:
>
> $$(x^2 - 7x + 10) = (x - 2)(x - 5).$$

▶ **EXAMPLE 2**   Factor: $y^2 - 8y + 12$.

Since the constant term is positive and the coefficient of the middle term is negative, we look for a factorization of 12 in which both factors are negative. Their sum must be -8.

Pairs of factors	Sums of factors
$-1, -12$	-13
$-2, -6$	-8 ←
$-3, -4$	-7

The numbers we need are -2 and -6.

The factorization is $(y - 2)(y - 6)$.   ◀

DO EXERCISES 3 AND 4.

Constant Term Negative

Sometimes when we use FOIL, the product has a negative constant term. Consider these multiplications:

$$\textbf{a)}\;\underset{\text{F}\quad\text{O}\quad\text{I}\quad\text{L}}{(x - 5)(x + 2) = x^2 + \underbrace{2x - 5x}_{} - 10}$$

$$= x^2 - 3x - 10;$$

$$\textbf{b)}\;\underset{\text{F}\quad\text{O}\quad\text{I}\quad\text{L}}{(x + 5)(x - 2) = x^2 - \underbrace{2x + 5x}_{} - 10}$$

$$= x^2 + 3x - 10.$$

Reversing the signs of the factors changes the sign of the middle term.

> When the constant term is negative, we look for two factors whose product is negative. One of them must be positive and the other negative. Their sum must be the coefficient of the middle term.

▶ **EXAMPLE 3**   Factor: $x^3 - 8x^2 - 20x$.

Always look first for a common factor. This time there is one, x. We first factor it out:

$$x^3 - 8x^2 - 20x = x(x^2 - 8x - 20).$$

Now consider $x^2 - 8x - 20$. Since the constant term is negative, we look for a factorization of -20 in which one factor is positive and one factor is negative. The sum must be -8, so the negative factor must have the larger absolute value. Thus we consider only pairs of factors in which the negative factor has the larger absolute value.

Pairs of factors	Sums of factors
1, −20	−19
2, −10	−8 ←
4, −5	−1

The numbers we need are 2 and −10.

The numbers we need are 2 and -10. The factorization of $x^2 - 8x - 20$ is $(x + 2)(x - 10)$. But we must also remember to include the common factor. The factorization of the original polynomial is

$$x(x + 2)(x - 10). \quad ◀$$

▶ **EXAMPLE 4**   Factor: $t^2 - 24 + 5t$.

It helps to first write the trinomial in descending order: $t^2 + 5t - 24$. Since the constant term is negative, we look for a factorization of -24 in which one factor is positive and one factor is negative. Their sum must be 5, so the positive factor must have the larger absolute value. Thus we consider only pairs of factors in which the positive term has the larger absolute value.

Pairs of factors	Sums of factors
−1, 24	23
−2, 12	10
−3, 8	5 ←
−4, 6	2

The numbers we need are −3 and 8.

The factorization is $(t - 3)(t + 8)$. ◀

▶ **EXAMPLE 5**   Factor: $x^4 - x^2 - 110$.

Consider this trinomial as $(x^2)^2 - x^2 - 110$. We look for numbers p and q such that

$$x^4 - x^2 - 110 = (x^2 + p)(x^2 + q).$$

Since the constant term is negative, we look for a factorization of -110 in which one factor is positive and one factor is negative. Their sum must be -1. The middle-term coefficient, -1, is small compared to -110. This tells us that the desired factors are close to each other in absolute value. The numbers we want are 10 and -11. The factorization is

$$(x^2 + 10)(x^2 - 11). \quad ◀$$

Factor.

5. $x^3 + 4x^2 - 12x$

6. $y^2 - 12 - 4y$

7. $t^4 + 5t^2 - 14$

8. $p^2 - pq - 3pq^2$

9. $x^2 + 2x + 7$

10. Factor: $x^2 + 8x + 16$.

▶ **EXAMPLE 6**   Factor: $a^2 + 4ab - 21b^2$.

We consider the trinomial in the equivalent form

$$a^2 + 4ba - 21b^2.$$

We think of $4b$ as a "coefficient" of a. Then we look for factors of $-21b^2$ whose sum is $4b$. Those factors are $-3b$ and $7b$. The factorization is

$$(a - 3b)(a + 7b). \qquad ◀$$

There are polynomials that are not factorable.

▶ **EXAMPLE 7**   Factor: $x^2 - x + 5$.

Since 5 has very few factors, we can easily check all possibilities.

Pairs of factors	Sums of factors
5, 1	6
−5, −1	−6

There are no factors whose sum is -1. Thus the polynomial is *not* factorable into binomials. ◀

DO EXERCISES 5–9.

Can we factor a trinomial that is a perfect square using this method? The answer is "yes."

▶ **EXAMPLE 8**   Factor: $x^2 - 10x + 25$.

Since the constant term is positive and the coefficient of the middle term is negative, we look for a factorization of 25 in which both factors are negative. Their sum must be -10.

Pairs of factors	Sums of factors	
−25, −1	−26	
−5, −5	−10 ←	The numbers we need are −5 and −5.

The factorization is $(x - 5)(x - 5)$, or $(x - 5)^2$. ◀

DO EXERCISE 10.

The following is a summary of our procedure for factoring $x^2 + bx + c$.

To factor $x^2 + bx + c$:

1. **First arrange in descending order. Use a trial-and-error process that looks for factors of c whose sum is b.**
2. **If c is positive, the signs of the factors are the same as the sign of b.**
3. **If c is negative, one factor is positive and the other is negative. If the sum of two factors is the opposite of b, changing the sign of each factor will give the desired factors whose sum is b.**
4. **Check by multiplying.**

EXERCISE SET 4.2

a Factor. Remember that you can check by multiplying.

1. $x^2 + 8x + 15$ **2.** $x^2 + 5x + 6$ **3.** $x^2 + 7x + 12$

4. $x^2 + 9x + 8$ **5.** $x^2 - 6x + 9$ **6.** $y^2 + 11y + 28$

7. $x^2 + 9x + 14$ **8.** $a^2 + 11a + 30$ **9.** $b^2 + 5b + 4$

10. $x^2 - \dfrac{2}{5}x + \dfrac{1}{25}$ **11.** $x^2 + \dfrac{2}{3}x + \dfrac{1}{9}$ **12.** $z^2 - 8z + 7$

13. $d^2 - 7d + 10$ **14.** $x^2 - 8x + 15$ **15.** $y^2 - 11y + 10$

16. $x^2 - 2x - 15$ **17.** $x^2 + x - 42$ **18.** $x^2 + 2x - 15$

19. $x^2 - 7x - 18$ **20.** $y^2 - 3y - 28$ **21.** $x^3 - 6x^2 - 16x$

22. $x^3 - x^2 - 42x$ **23.** $y^2 - 4y - 45$ **24.** $x^2 - 7x - 60$

25. $-2x - 99 + x^2$ **26.** $x^2 - 72 + 6x$ **27.** $c^4 + c^2 - 56$

28. $b^4 + 5b^2 - 24$ **29.** $a^4 + 2a^2 - 35$ **30.** $2 - x^2 - x^4$

31. $x^2 + x + 1$ **32.** $x^2 + 2x + 3$ **33.** $7 - 2p + p^2$

34. $11 - 3w + w^2$ **35.** $x^2 + 20x + 100$ **36.** $x^2 + 20x + 99$

ANSWERS

1. _____
2. _____
3. _____
4. _____
5. _____
6. _____
7. _____
8. _____
9. _____
10. _____
11. _____
12. _____
13. _____
14. _____
15. _____
16. _____
17. _____
18. _____
19. _____
20. _____
21. _____
22. _____
23. _____
24. _____
25. _____
26. _____
27. _____
28. _____
29. _____
30. _____
31. _____
32. _____
33. _____
34. _____
35. _____
36. _____

37. $x^2 - 21x - 100$ **38.** $x^2 - 20x + 96$ **39.** $x^2 - 21x - 72$

40. $4x^2 + 40x + 100$ **41.** $x^2 - 25x + 144$ **42.** $y^2 - 21y + 108$

43. $a^2 + a - 132$ **44.** $a^2 + 9a - 90$ **45.** $120 - 23x + x^2$

46. $96 + 22d + d^2$ **47.** $108 - 3x - x^2$ **48.** $112 + 9y - y^2$

49. $y^2 - 0.2y - 0.08$ **50.** $t^2 - 0.3t - 0.10$ **51.** $p^2 + 3pq - 10q^2$

52. $a^2 - 2ab - 3b^2$ **53.** $m^2 + 5mn + 4n^2$ **54.** $x^2 - 11xy + 24y^2$

55. $s^2 - 2st - 15t^2$ **56.** $b^2 + 8bc - 20c^2$

SKILL MAINTENANCE

Multiply.

57. $8x(2x^2 - 6x + 1)$ **58.** $(7w + 6)(4w - 11)$ **59.** $(7w + 6)^2$

60. Simplify: $(3x^4)^3$.

SYNTHESIS

61. Find all integers m for which $y^2 + my + 50$ can be factored.

62. Find all integers b for which $a^2 + ba - 50$ can be factored.

Factor completely.

63. $x^2 - \frac{1}{2}x - \frac{3}{16}$ **64.** $x^2 - \frac{1}{4}x - \frac{1}{8}$ **65.** $x^2 + \frac{30}{7}x - \frac{25}{7}$

66. $\frac{1}{3}x^3 + \frac{1}{3}x^2 - 2x$ **67.** $b^{2n} + 7b^n + 10$ **68.** $a^{2m} - 11a^m + 28$

Find a polynomial in factored form for the shaded area. (Leave answers in terms of π.)

69.

70.

4.3 Factoring Trinomials of the Type $ax^2 + bx + c$, $a \neq 1$

In Section 4.2, we learned a trial-and-error method to factor trinomials of the type $x^2 + bx + c$. In this section, we factor trinomials in which the leading, or x^2, coefficient is not 1. The method we learn is the *standard* trial-and-error method. (In Section 4.4, we will consider an alternative method for the same kind of factoring. It involves *factoring by grouping*.)

OBJECTIVE

After finishing Section 4.3, you should be able to:

a Factor trinomials of the type $ax^2 + bx + c$, $a \neq 1$.

FOR EXTRA HELP

Tape 9D   Tape 8B   MAC: 4
                                  IBM: 4

a We want to factor trinomials of the type $ax^2 + bx + c$. Consider the following multiplication:

$$
\begin{array}{ccccccccc}
& & \text{F} & & \text{O} & & \text{I} & & \text{L} \\
(2x + 5)(3x + 4) & = & 6x^2 & + & 8x & + & 15x & + & 20 \\
& = & 6x^2 & + & & 23x & & + & 20
\end{array}
$$

F	O + I	L
$2 \cdot 3$	$2 \cdot 4 + 5 \cdot 3$	$5 \cdot 4$

To factor $6x^2 + 23x + 20$, we reverse the above multiplication. We look for two binomials $rx + p$ and $sx + q$ whose product is this trinomial. The product of the First terms must be $6x^2$. The product of the Outside terms plus the product of the Inside terms must be $23x$. The product of the Last terms must be 20. We know from the preceding discussion that the answer is

$$(2x + 5)(3x + 4).$$

Generally, however, finding such an answer is a trial-and-error process. It turns out that $(-2x - 5)(-3x - 4)$ is also a correct answer, but we usually choose an answer in which the first coefficients are positive.

We will use the following trial-and-error method.

To factor $ax^2 + bx + c$, $a \neq 1$, using the FOIL method:

1. **Factor out a common factor, if any.**

2. **Factor the term ax^2. This gives these possibilities for r and s:**
$$(rx + p)(sx + q).$$
$$rx \cdot sx = ax^2$$

3. **Factor the last term c. This gives these possibilities for p and q:**
$$(rx + p)(sx + q).$$
$$p \cdot q = c$$

4. **Look for combinations of factors from steps (2) and (3) for which the sum of their products is the middle term bx:**
$$rx \cdot q$$
$$(rx + p)(sx + q). \qquad rx \cdot q + p \cdot sx \overset{?}{=} bx$$
$$p \cdot sx$$

Factor.

1. $2x^2 - x - 15$

2. $12x^2 - 17x - 5$

▶ **EXAMPLE 1**　Factor: $3x^2 - 10x - 8$.

1) First, factor out a common factor, if any. There is none (other than 1 or −1).

2) Factor the first term, $3x^2$. The only possibility is $3x \cdot x$. The desired factorization is then of the form

$$(3x + \underline{})(x + \underline{}),$$

where we must determine the numbers in the blanks.

3) Factor the last term, -8, which is negative. The possibilities are

$$
\begin{aligned}
-8 &= (-8)(1); \\
&= 8(-1); \\
&= (-2)(4); \\
&= 2(-4).
\end{aligned}
$$

4) From steps (2) and (3), we see that there are $1 \cdot (2 \cdot 4)$, or 8 possibilities for factorizations. We look for combinations of factors from steps (2) and (3) such that the sum of their products is the middle term, $-10x$:

$$(3x - 8)(x + 1) = 3x^2 - 5x - 8;$$
Wrong middle term

$$(3x + 8)(x - 1) = 3x^2 + 5x - 8;$$
Wrong middle term

$$(3x - 2)(x + 4) = 3x^2 + 10x - 8;$$
Wrong middle term

$$(3x + 2)(x - 4) = 3x^2 - 10x - 8;$$
Correct middle term!

There are four other possibilities that we could try, but we need not since we have found a factorization. The factorization is $(3x + 2)(x - 4)$.

◀

DO EXERCISES 1 AND 2.

▶ **EXAMPLE 2**　Factor: $24x^2 - 76x + 40$.

1) Factor out a common factor, if any. This time there is one, 4. We factor it out:

$$4(6x^2 - 19x + 10).$$

Now we factor the trinomial $6x^2 - 19x + 10$.

2) Factor the first term, $6x^2$. These are $3x, 2x$, or $6x, x$. Then we have these as possibilities for factorizations:

$$(3x + \underline{})(2x + \underline{}) \quad \text{or} \quad (6x + \underline{})(x + \underline{}).$$

3) Factor the last term, 10, which is positive. The possibilities are

$$10, 1 \quad \text{and} \quad -10, -1 \quad \text{and} \quad 5, 2 \quad \text{and} \quad -5, -2.$$

4) From steps (2) and (3), we see that there are $2 \cdot (2 \cdot 4)$, or 16 possibilities for factorizations. Look for combinations of factors from steps (2) and (3) such that the sum of their products is the middle term, $-19x$. The sign of the middle term is negative, but the sign of the last term, 10, is positive. Thus the signs of both factors of the last term, 10, must be nega-

tive. From our list of factors in step (3), we can only use -10, -1 and -5, -2 as possibilities. This reduces the possibilities for factorizations to 8. We start by using these factors with $(3x + \underline{})(2x + \underline{})$. Should we not find the correct factorization, we will consider $(6x + \underline{})(x + \underline{})$.

$$\overset{\displaystyle -3x}{(3x - 10)(2x - 1)} = 6x^2 - 23x + 10;$$
$$\underset{\displaystyle -20x}{}$$
Wrong middle term

$$\overset{\displaystyle -30x}{(3x - 1)(2x - 10)} = 6x^2 - 32x + 10;$$
$$\underset{\displaystyle -2x}{}$$
Wrong middle term

$$\overset{\displaystyle -6x}{(3x - 5)(2x - 2)} = 6x^2 - 16x + 10;$$
$$\underset{\displaystyle -10x}{}$$
Wrong middle term

$$\overset{\displaystyle -15x}{(3x - 2)(2x - 5)} = 6x^2 - 19x + 10;$$
$$\underset{\displaystyle -4x}{}$$
Correct middle term!

We have a correct answer. We need not consider $(6x + \underline{})(x + \underline{})$.

Look again at the possibility $(3x - 5)(2x - 2)$. Without multiplying, we can reject such a possibility. Look at the following:

$$(3x - 5)(2x - 2) = 2(3x - 5)(x - 1).$$

The expression $2x - 2$ has a common factor, 2. But we removed the largest common factor before we began. If this expression were a factorization, then 2 would have to be a common factor in addition to the original 4. Thus, as we saw when we multiplied, $(3x - 5)(2x - 2)$ cannot be part of the factorization of the original trinomial.

> Given that we factored out the largest common factor at the outset, we can eliminate factorizations that have a common factor.

The factorization of $6x^2 - 19x + 10$ is $(3x - 2)(2x - 5)$. But do not forget the common factor! We must include it in order to get a factorization of the original trinomial:

$$24x^2 - 76x + 40 = 4(3x - 2)(2x - 5). \quad \blacktriangleleft$$

DO EXERCISES 3 AND 4.

▶ **EXAMPLE 3** Factor: $10x^2 + 37x + 7$.

1) First, factor out a common factor, if any. There is none (other than 1 or -1).

2) Factor the term $10x^2$: $10x$, x or $5x$, $2x$. We have these as possibilities for factorizations:

$$(10x + \underline{})(x + \underline{}) \quad \text{and} \quad (5x + \underline{})(2x + \underline{}).$$

Factor.

3. $3x^2 - 19x + 20$

4. $20x^2 - 46x + 24$

ANSWERS ON PAGE A-4

5. Factor: $6x^2 + 7x + 2$.

3) Factor the last term, 7. The possibilities are 1, 7 and -1, -7.

4) From steps (2) and (3), we see that there are 8 possibilities for factorizations. Look for factors from steps (2) and (3) such that the sum of their products is the middle term. In this case, all signs are positive, so we need consider only plus signs. The possibilities are

$$(10x + 1)(x + 7) = 10x^2 + 71x + 7,$$
$$(10x + 7)(x + 1) = 10x^2 + 17x + 7,$$
$$(5x + 7)(2x + 1) = 10x^2 + 19x + 7,$$
$$(5x + 1)(2x + 7) = 10x^2 + 37x + 7.$$

The factorization is $(5x + 1)(2x + 7)$. ◄

Tips for factoring $ax^2 + bx + c$, $a \neq 1$:

1. **If the largest common factor has been factored out of the original trinomial, then no binomial factor can have a common factor (other than 1 or -1).**

2. **If all the signs of all the terms are positive, then the signs of all the terms of the binomial factors are positive.**

3. **Be systematic about your trials. Keep track of those you have tried and those you have not.**

DO EXERCISE 5.

Factor.

6. $6a^2 - 5ab + b^2$

Keep in mind that this method of factoring trinomials of the type $ax^2 + bx + c$ involves trial and error. As you practice, you will find that you can make better and better guesses. Don't forget: When factoring any polynomial, always look first for a common factor. Failure to do so is such a common error that this caution bears repeating.

▶ **EXAMPLE 4**   Factor: $6p^2 - 13pq - 28q^2$.

1) Factor out a common factor, if any. There is none (other than 1 or -1).

2) Factor the first term, $6p^2$. Possibilities are $2p$, $3p$ and $6p$, p. We have these as possibilities for factorizations:

$$(2p + \underline{\quad})(3p + \underline{\quad}) \quad \text{and} \quad (6p + \underline{\quad})(p + \underline{\quad}).$$

7. $6x^2 + 15xy + 9y^2$

3) Factor the last term, $-28q^2$, which has a negative coefficient. The possibilities are $-14q$, $2q$ and $14q$, $-2q$; $-28q$, q and $28q$, $-q$; and $-7q$, $4q$ and $7q$, $-4q$.

4) The coefficient of the middle term is negative, so we look for combinations of factors from steps (2) and (3) such that the sum of their products has a negative coefficient. We try some possibilities:

$$(2p - 14q)(3p + 2q) = 6p^2 - 38pq - 28q^2,$$
$$(2p - 28q)(3p + q) = 6p^2 - 82pq - 28q^2,$$
$$(2p - 7q)(3p + 4q) = 6p^2 - 13pq - 28q^2.$$

The factorization of $6p^2 - 13pq - 28q^2$ is $(2p - 7q)(3p + 4q)$. ◄

DO EXERCISES 6 AND 7.

ANSWERS ON PAGE A-4

NAME   SECTION   DATE

EXERCISE SET 4.3

a Factor.

1. $2x^2 - 7x - 4$

2. $3x^2 - x - 4$

3. $5x^2 + x - 18$

4. $3x^2 - 4x - 15$

5. $6x^2 + 23x + 7$

6. $6x^2 + 13x + 6$

7. $3x^2 + 4x + 1$

8. $7x^2 + 15x + 2$

9. $4x^2 + 4x - 15$

10. $9x^2 + 6x - 8$

11. $2x^2 - x - 1$

12. $15x^2 - 19x - 10$

13. $9x^2 + 18x - 16$

14. $2x^2 + 5x + 2$

15. $3x^2 - 5x - 2$

16. $18x^2 - 3x - 10$

17. $12x^2 + 31x + 20$

18. $15x^2 + 19x - 10$

19. $14x^2 + 19x - 3$

20. $35x^2 + 34x + 8$

21. $9x^2 + 18x + 8$

22. $6 - 13x + 6x^2$

23. $49 - 42x + 9x^2$

24. $25x^2 + 40x + 16$

25. $24x^2 + 47x - 2$

26. $16a^2 + 78a + 27$

27. $35x^2 - 57x - 44$

28. $9a^2 + 12a - 5$

29. $20 + 6x - 2x^2$

30. $15 + x - 2x^2$

31. $12x^2 + 28x - 24$

32. $6x^2 + 33x + 15$

33. $30x^2 - 24x - 54$

34. $20x^2 - 25x + 5$

35. $4x + 6x^2 - 10$

36. $-9 + 18x^2 - 21x$

1. _____
2. _____
3. _____
4. _____
5. _____
6. _____
7. _____
8. _____
9. _____
10. _____
11. _____
12. _____
13. _____
14. _____
15. _____
16. _____
17. _____
18. _____
19. _____
20. _____
21. _____
22. _____
23. _____
24. _____
25. _____
26. _____
27. _____
28. _____
29. _____
30. _____
31. _____
32. _____
33. _____
34. _____
35. _____
36. _____

ANSWERS

37.
38.
39.
40.
41.
42.
43.
44.
45.
46.
47.
48.
49.
50.
51.
52.
53.
54.
55.
56.
57.
58.
59.
60.
61.
62.
63.
64.
65.
66.
67.
68.
69.
70.
71.
72.
73.
74.
75.
76.
77.
78.

37. $3x^2 - 4x + 1$ **38.** $6x^2 - 13x + 6$ **39.** $12x^2 - 28x - 24$

40. $6x^2 - 33x + 15$ **41.** $-1 + 2x^2 - x$ **42.** $-19x + 15x^2 + 6$

43. $9x^2 - 18x - 16$ **44.** $14x^2 + 35x + 14$ **45.** $15x^2 - 25x - 10$

46. $18x^2 + 3x - 10$ **47.** $12x^3 + 31x^2 + 20x$ **48.** $15x^3 + 19x^2 - 10x$

49. $14x^4 + 19x^3 - 3x^2$ **50.** $70x^4 + 68x^3 + 16x^2$ **51.** $168x^3 - 45x^2 + 3x$

52. $144x^5 + 168x^4 + 48x^3$ **53.** $15x^4 - 19x^2 + 6$ **54.** $9x^4 + 18x^2 + 8$

55. $25t^2 + 80t + 64$ **56.** $9x^2 - 42x + 49$ **57.** $6x^3 + 4x^2 - 10x$

58. $18x^3 - 21x^2 - 9x$ **59.** $25x^2 + 79x + 64$ **60.** $9y^2 - 42y + 47$

61. $x^2 + 3x - 7$ **62.** $x^2 + 13x - 12$ **63.** $12m^2 + mn - 20n^2$

64. $12a^2 + 17ab + 6b^2$ **65.** $6a^2 - ab - 15b^2$ **66.** $3p^2 - 16pq - 12q^2$

67. $9a^2 + 18ab + 8b^2$ **68.** $10s^2 + 4st - 6t^2$ **69.** $35p^2 + 34pq + 8q^2$

70. $30a^2 + 87ab + 30b^2$ **71.** $18x^2 - 6xy - 24y^2$ **72.** $15a^2 - 5ab - 20b^2$

SKILL MAINTENANCE

73. Solve $A = pq - 7$ for q.   **74.** Solve: $2x - 4(x + 3x) \geq 6x - 8 - 9x$.

SYNTHESIS

Factor.

75. $20x^{2n} + 16x^n + 3$ **76.** $-15x^{2m} + 26x^m - 8$

77. $3x^{6a} - 2x^{3a} - 1$ **78.** $x^{2n+1} - 2x^{n+1} + x$

4.4 Factoring $ax^2 + bx + c$, $a \neq 1$, Using Grouping

a Another method of factoring trinomials of the type $ax^2 + bx + c$, $a \neq 1$, is known as the **grouping method.** It involves factoring by grouping. We know how to factor the trinomial $x^2 + 5x + 6$. We look for factors of the constant term, 6, whose sum is the coefficient of the middle term, 5:

$$x^2 + 5x + 6.$$

(1) Factor: $6 = 2 \cdot 3$
(2) Sum: $2 + 3 = 5$

What happens when the leading coefficient is not 1? Consider the trinomial $3x^2 - 10x - 8$. The method we use is similar to what we used for the preceding trinomial, but we need two more steps. The method is outlined as follows.

To factor $ax^2 + bx + c$, $a \neq 1$, using the grouping method:

1. **Factor out a common factor, if any.**
2. **Multiply the leading coefficient a and the constant c.**
3. **Try to factor the product ac so that the sum of the factors is b. That is, find integers p and q such that $pq = ac$ and $p + q = b$.**
4. **Split the middle term. That is, write it as a sum using the factors found in step (3).**
5. **Then factor by grouping.**

▶ **EXAMPLE 1**   Factor: $3x^2 - 10x - 8$.

1) First, factor out a common factor, if any. There is none (other than 1 or -1).

2) Multiply the leading coefficient, 3, and the constant, -8:

$$3(-8) = -24.$$

3) Then look for a factorization of -24 in which the sum of the factors is the coefficient of the middle term, -10.

Pairs of factors	Sums of factors
$-1, 24$	23
$-2, 12$	10
$-3, \;\; 8$	5
$-4, \;\; 6$	2
$-6, \;\; 4$	-2
$-8, \;\; 3$	-5
$-12, \;\; 2$	$-10 \longleftarrow \quad -12 + 2 = -10$
$-24, \;\; 1$	-23

4) Next, split the middle term as a sum or difference using the factors found in step (3):

$$-10x = -12x + 2x.$$

OBJECTIVE

After finishing Section 4.4, you should be able to:

a Factor trinomials of the type $ax^2 + bx + c$, $a \neq 1$, by splitting the middle term and using grouping.

FOR EXTRA HELP

Tape 9E   Tape 9A   MAC: 4
                                  IBM: 4

Factor.

1. $6x^2 + 7x + 2$

2. $12x^2 - 17x - 5$

Factor.

3. $6x^2 + 15x + 9$

4. $20x^2 - 46x + 24$

5) Factor by grouping as follows:

$$3x^2 - 10x - 8 = 3x^2 - 12x + 2x - 8$$
$$= 3x(x - 4) + 2(x - 4) \quad \text{Factoring by grouping; see Section 4.1}$$
$$= (3x + 2)(x - 4).$$

It does not matter which way we split the middle term, so long as we split it correctly. We still get the same factorization, although the factors may be in a different order. Note the following:

$$3x^2 - 10x - 8 = 3x^2 + 2x - 12x - 8$$
$$= x(3x + 2) - 4(3x + 2)$$
$$= (x - 4)(3x + 2).$$

Check by multiplying: $(x - 4)(3x + 2) = 3x^2 - 10x - 8.$ ◄

DO EXERCISES 1 AND 2.

▶ **EXAMPLE 2**   Factor: $8x^2 + 8x - 6.$

1) First, factor out a common factor, if any. The number 2 is common to all three terms, so we factor it out:

$$2(4x^2 + 4x - 3).$$

2) Next, factor the trinomial $4x^2 + 4x - 3$. Multiply the leading coefficient and the constant, 4 and -3:

$$4(-3) = -12.$$

3) Try to factor -12 so that the sum of the factors is 4.

Pairs of factors	Sums of factors
$-3, \quad 4$	1
$3, \; -4$	-1
$-12, \quad 1$	-11
$12, \; -1$	11
$-6, \quad 2$	-4
$6, \; -2$	$4 \longleftarrow$ ┤ $6 + (-2) = 4$

4) Split the middle term, $4x$, as follows:

$$4x = 6x - 2x.$$

5) Factor by grouping:

$$4x^2 + 4x - 3 = 4x^2 + 6x - 2x - 3 \quad \text{Substituting } 6x - 2x \text{ for } 4x$$
$$= 2x(2x + 3) - 1(2x + 3) \quad \text{Factoring by grouping}$$
$$= (2x - 1)(2x + 3).$$

The factorization of $4x^2 + 4x - 3$ is $(2x - 1)(2x + 3)$. But don't forget the common factor! We must include it to get a factorization of the original trinomial:

$$8x^2 + 8x - 6 = 2(2x - 1)(2x + 3).$$ ◄

DO EXERCISES 3 AND 4.

NAME   SECTION   DATE

EXERCISE SET 4.4

a Factor. Note that the middle term has already been split.

1. $y^2 + 4y + y + 4$

2. $x^2 + 5x + 2x + 10$

3. $x^2 - 4x - x + 4$

4. $a^2 + 5a - 2a - 10$

5. $6x^2 + 4x + 9x + 6$

6. $3x^2 - 2x + 3x - 2$

7. $3x^2 - 4x - 12x + 16$

8. $24 - 18y - 20y + 15y^2$

9. $35x^2 - 40x + 21x - 24$

10. $8x^2 - 6x - 28x + 21$

11. $4x^2 + 6x - 6x - 9$

12. $2x^4 - 6x^2 - 5x^2 + 15$

13. $2x^4 + 6x^2 + 5x^2 + 15$

14. $4x^4 - 6x^2 - 6x^2 + 9$

Factor by grouping.

15. $2x^2 - 7x - 4$

16. $3x^2 - x - 4$

17. $5x^2 + x - 18$

18. $3x^2 - 4x - 15$

19. $6x^2 + 23x + 7$

20. $6x^2 + 13x + 6$

21. $3x^2 + 4x + 1$

22. $7x^2 + 15x + 2$

23. $4x^2 + 4x - 15$

1. _____
2. _____
3. _____
4. _____
5. _____
6. _____
7. _____
8. _____
9. _____
10. _____
11. _____
12. _____
13. _____
14. _____
15. _____
16. _____
17. _____
18. _____
19. _____
20. _____
21. _____
22. _____
23. _____

ANSWERS

24. _____

25. _____

26. _____

27. _____

28. _____

29. _____

30. _____

31. _____

32. _____

33. _____

34. _____

35. _____

36. _____

37. _____

38. _____

39. _____

40. _____

41. _____

42. _____

43. _____

44. _____

45. _____

46. _____

47. _____

48. _____

49. _____

50. _____

24. $9x^2 + 6x - 8$

25. $2x^2 - x - 1$

26. $15x^2 - 19x - 10$

27. $9x^2 + 18x - 16$

28. $2x^2 + 5x + 2$

29. $3x^2 - 5x - 2$

30. $18x^2 - 3x - 10$

31. $12x^2 + 31x + 20$

32. $15x^2 + 19x - 10$

33. $14x^2 + 19x - 3$

34. $35x^2 + 34x + 8$

35. $9x^2 + 18x + 8$

36. $6 - 13x + 6x^2$

37. $49 - 42x + 9x^2$

38. $25x^2 + 40x + 16$

39. $24x^2 + 47x - 2$

40. $16a^2 + 78a + 27$

41. $35x^5 - 57x^4 - 44x^3$

42. $18a^3 + 24a^2 - 10a$

43. $60x + 18x^2 - 6x^3$

44. $60x + 4x^2 - 8x^3$

SKILL MAINTENANCE

Solve.

45. $3x - 6x + 2(x - 4) > 2(9 - 4x)$

46. $-6(x - 4) + 8(4 - x) \le 3(x - 7)$

SYNTHESIS

Factor.

47. $9x^{10} - 12x^5 + 4$

48. $24x^{2n} + 22x^n + 3$

49. $16x^{10} + 8x^5 + 1$

50. $(a + 4)^2 - 2(a + 4) + 1$

4.5 Factoring Trinomial Squares and Differences of Squares

In this section, we first learn to factor trinomials that are squares of binomials. Then we factor binomials that are differences of squares.

a Recognizing Trinomial Squares

Some trinomials are squares of binomials. For example, the trinomial $x^2 + 10x + 25$ is the square of the binomial $x + 5$. To see this, we can calculate $(x + 5)^2$. It is $x^2 + 2 \cdot x \cdot 5 + 5^2$, or $x^2 + 10x + 25$. A trinomial that is the square of a binomial is called a **trinomial square.**

In Chapter 3, we considered squaring binomials as a special-product rule.

$$(A + B)^2 = A^2 + 2AB + B^2;$$
$$(A - B)^2 = A^2 - 2AB + B^2.$$

We can use these equations in reverse to factor trinomial squares.

$$A^2 + 2AB + B^2 = (A + B)^2;$$
$$A^2 - 2AB + B^2 = (A - B)^2$$

How can we recognize when an expression to be factored is a trinomial square? Look at $A^2 + 2AB + B^2$ and $A^2 - 2AB + B^2$. In order for an expression to be a trinomial square:

a) Two terms, A^2 and B^2, must be squares, such as

$$4, \quad x^2, \quad 25x^4, \quad 16t^2.$$

b) There must be no minus sign before A^2 or B^2.

c) If we multiply A and B (the square roots of A^2 and B^2) and double the result, we get the remaining term $2 \cdot A \cdot B$, or its opposite, $-2 \cdot A \cdot B$.

▶ **EXAMPLE 1**   Determine whether $x^2 + 6x + 9$ is a trinomial square.

a) We know that x^2 and 9 are squares.

b) There is no minus sign before x^2 or 9.

c) If we multiply the square roots, x and 3, and double the product, we get the remaining term: $2 \cdot x \cdot 3 = 6x$.

Thus, $x^2 + 6x + 9$ is the square of a binomial. In fact, $x^2 + 6x + 9 = (x + 3)^2$. ◀

▶ **EXAMPLE 2**   Determine whether $x^2 + 6x + 11$ is a trinomial square.
The answer is no, because only one term is a square. ◀

OBJECTIVES

After finishing Section 4.5, you should be able to:

a Recognize trinomial squares.

b Factor trinomial squares.

c Recognize differences of squares.

d Factor differences of squares, being careful to factor completely.

FOR EXTRA HELP

Tape 9F   Tape 9A   MAC: 4
                                  IBM: 4

Determine whether each is a trinomial square. Write "yes" or "no."

1. $x^2 + 8x + 16$

2. $25 - x^2 + 10x$

3. $t^2 - 12t + 4$

4. $25 + 20y + 4y^2$

5. $5x^2 + 16 - 14x$

6. $16x^2 + 40x + 25$

7. $p^2 + 6p - 9$

8. $25a^2 + 9 - 30a$

Factor.

9. $x^2 + 2x + 1$

10. $1 - 2x + x^2$

11. $4 + t^2 + 4t$

12. $25x^2 - 70x + 49$

13. $49 - 56y + 16y^2$

14. $48m^2 + 75 + 120m$

▶ **EXAMPLE 3**   Determine whether $16x^2 + 49 - 56x$ is a trinomial square.

It helps to first write the trinomial in descending order:

$$16x^2 - 56x + 49.$$

a) We know that $16x^2$ and 49 are squares.

b) There is no minus sign before $16x^2$ or 49.

c) If we multiply the square roots, $4x$ and 7, and double the product, we get the opposite of the remaining term: $2 \cdot 4x \cdot 7 = 56x$; $56x$ is the opposite of $-56x$.

Thus, $16x^2 + 49 - 56x$ is a trinomial square. In fact, $16x^2 - 56x + 49 = (4x - 7)^2$. ◀

DO EXERCISES 1–8.

b   Factoring Trinomial Squares

We can use the trial-and-error or grouping methods from Sections 4.2–4.4 to factor such trinomial squares, but there is a faster method using the following equations:

$$A^2 + 2AB + B^2 = (A + B)^2;$$
$$A^2 - 2AB + B^2 = (A - B)^2.$$

We use square roots of the squared terms and the sign of the remaining term.

▶ **EXAMPLE 4**   Factor: $x^2 + 6x + 9$.

$$x^2 + 6x + 9 = x^2 + 2 \cdot x \cdot 3 + 3^2 = (x + 3)^2$$

The sign of the middle term is positive.

$$A^2 + 2A \; B + B^2 = (A + B)^2 \qquad ◀$$

▶ **EXAMPLE 5**   Factor: $x^2 + 49 - 14x$.

$$x^2 + 49 - 14x = x^2 - 14x + 49 \qquad \text{Changing order}$$
$$= x^2 - 2 \cdot x \cdot 7 + 7^2 \qquad \text{The sign of the middle term is negative.}$$
$$= (x - 7)^2 \qquad ◀$$

▶ **EXAMPLE 6**   Factor: $16x^2 - 40x + 25$.

$$16x^2 - 40x + 25 = (4x)^2 - 2 \cdot 4x \cdot 5 + 5^2 = (4x - 5)^2$$

$$A^2 - 2A \; B + B^2 = (A - B)^2 \qquad ◀$$

DO EXERCISES 9–14.

▶ **EXAMPLE 7**   Factor: $t^4 + 20t^2 + 100$.

$$t^4 + 20t^2 + 100 = (t^2)^2 + 2(t^2)(10) + 10^2$$
$$= (t^2 + 10)^2 \qquad ◀$$

▶ **EXAMPLE 8**   Factor: $75m^3 + 210m^2 + 147m$.

Always look first for a common factor. This time there is one, $3m$:

$$75m^3 + 210m^2 + 147m = 3m[25m^2 + 70m + 49]$$
$$= 3m[(5m)^2 + 2(5m)(7) + 7^2]$$
$$= 3m(5m + 7)^2. \qquad ◀$$

▶ **EXAMPLE 9**   Factor: $4p^2 - 12pq + 9q^2$.

$$4p^2 - 12pq + 9q^2 = (2p)^2 - 2(2p)(3q) + (3q)^2 = (2p - 3q)^2 \qquad ◀$$

DO EXERCISES 15–17.

c   **Recognizing Differences of Squares**

The following polynomials are *differences of squares:*

$$x^2 - 9, \qquad 4t^2 - 49, \qquad a^2 - 25b^2.$$

To factor a difference of squares such as $x^2 - 9$, think about the formula we used in Chapter 3:

$$(A + B)(A - B) = A^2 - B^2.$$

Equations are reversible, so we also know that

$$A^2 - B^2 = (A + B)(A - B).$$

Thus,

$$x^2 - 9 = (x + 3)(x - 3).$$

To use this formula, we must be able to recognize when it applies. **A difference of squares** is an expression like the following:

$$A^2 - B^2.$$

How can we recognize such expressions? Look at $A^2 - B^2$. In order for a binomial to be a difference of squares:

a) There must be two expressions, both squares, such as

$$4x^2, \quad 9, \quad 25t^4, \quad 1, \quad x^6, \quad 49y^8.$$

When the coefficient is a perfect square and the power(s) of the variable(s) is (are) even, then the expression is a perfect square.

b) The terms must have different signs.

▶ **EXAMPLE 10**   Is $9x^2 - 64$ a difference of squares?

a) The first expression is a square: $9x^2 = (3x)^2$.
The second expression is a square: $64 = 8^2$.

b) The terms have different signs.

Thus we have a difference of squares, $(3x)^2 - 8^2$. ◀

Factor.

15. $p^4 + 18p^2 + 81$

16. $4z^5 - 20z^4 + 25z^3$

17. $9a^2 + 30ab + 25b^2$

ANSWERS ON PAGE A-5

Determine whether each is a difference of squares. Write ''yes'' or ''no.''

18. $x^2 - 25$

19. $t^2 - 24$

20. $y^2 + 36$

21. $4x^2 - 15$

22. $16x^4 - 49$

23. $9w^6 - 1$

24. $-49 + 25t^2$

▶ **EXAMPLE 11**   Is $25 - t^3$ a difference of squares?

a) The expression t^3 is not a square.

The expression is not a difference of squares.   ◀

▶ **EXAMPLE 12**   Is $-4x^2 + 16$ a difference of squares?

a) The expressions $4x^2$ and 16 are squares: $4x^2 = (2x)^2$ and $16 = 4^2$.

b) The terms have different signs.

Thus we have a difference of squares. We can also see this by rewriting in the equivalent form: $16 - 4x^2$.   ◀

DO EXERCISES 18–24.

d   Factoring Differences of Squares

To factor a difference of squares, we use the following equation:

$$A^2 - B^2 = (A + B)(A - B).$$

We consider 3 to be a square root of 9 because $3^2 = 9$. Similarly, A is a square root of A^2. To factor a difference of squares $A^2 - B^2$, we find A and B, which are square roots of the expressions A^2 and B^2. We then use A and B to form two factors. One is the sum $A + B$, and the other is the difference $A - B$.

▶ **EXAMPLE 13**   Factor: $x^2 - 4$.

$$x^2 - 4 = x^2 - 2^2 = (x + 2)(x - 2)$$
$$A^2 - B^2 = (A + B)(A - B)$$
    ◀

▶ **EXAMPLE 14**   Factor: $9 - 16t^4$.

$$9 - 16t^4 = 3^2 - (4t^2)^2 = (3 + 4t^2)(3 - 4t^2)$$
$$A^2 - B^2 = (A + B)(A - B)$$
    ◀

▶ **EXAMPLE 15**   Factor: $m^2 - 4p^2$.

$$m^2 - 4p^2 = m^2 - (2p)^2 = (m + 2p)(m - 2p)$$
    ◀

▶ **EXAMPLE 16**   Factor: $18x^2 - 50x^6$.

Always look first for a factor common to all terms. This time there is one, $2x^2$.

$$18x^2 - 50x^6 = 2x^2(9 - 25x^4)$$
$$= 2x^2[3^2 - (5x^2)^2]$$
$$= 2x^2(3 - 5x^2)(3 + 5x^2) \quad ◀$$

▶ **EXAMPLE 17**   Factor: $49x^4 - 9x^6$.

$$49x^4 - 9x^6 = x^4(49 - 9x^2) = x^4(7 + 3x)(7 - 3x) \quad ◀$$

DO EXERCISES 25–29.

CAUTION!   Note carefully in these examples that a difference of squares is *not* the square of the difference; that is,

$$A^2 - B^2 \neq (A - B)^2 = A^2 - 2AB + B^2.$$

For example,

$$(45 - 5)^2 = 40^2 = 1600,$$

but

$$45^2 - 5^2 = 2025 - 25 = 2000.$$

Factoring Completely

If a factor with more than one term can still be factored, you should do so. When no factor can be factored further, you have **factored completely.** Always factor completely whenever told to factor.

▶ **EXAMPLE 18**   Factor: $p^4 - 16$.

$$p^4 - 16 = (p^2)^2 - 4^2$$
$$= (p^2 + 4)(p^2 - 4) \qquad \text{Factoring a difference of squares}$$
$$= (p^2 + 4)(p + 2)(p - 2) \qquad \text{Factoring further. The factor } x^2 - 4 \text{ is a difference of squares.} \quad ◀$$

The polynomial $p^2 + 4$ cannot be factored further into polynomials with real coefficients.

▶ **EXAMPLE 19**   Factor: $y^4 - 16x^{12}$.

$$y^4 - 16x^{12} = (y^2 + 4x^6)(y^2 - 4x^6) \qquad \text{Factoring a difference of squares}$$
$$= (y^2 + 4x^6)(y + 2x^3)(y - 2x^3) \qquad \text{Factoring further. The factor } y^2 - 4x^6 \text{ is a difference of squares.} \quad ◀$$

Factor.

25. $x^2 - 9$

26. $64 - 4t^2$

27. $a^2 - 25b^2$

28. $64x^4 - 25x^6$

29. $5 - 20t^6$
[*Hint:* $1 = 1^2$, $t^6 = (t^3)^2$.]

ANSWERS ON PAGE A-5

Factor completely.

30. $81x^4 - 1$

Factoring Hints

1. **Always look first for a common factor. If there is one, factor it out!**
2. **Always factor completely.**
3. **Check by multiplying.**

CAUTION!   If the greatest common factor has been removed, then you cannot factor a sum of squares further. In particular,

$$(A + B)^2 \neq A^2 + B^2.$$

Consider $25x^2 + 100$. This is a case in which we have a sum of squares, but there is a common factor, 25. Factoring, we get $25(x^2 + 4)$. Now $x^2 + 4$ cannot be factored further.

DO EXERCISES 30 AND 31.

31. $49p^4 - 25q^6$

EXERCISE SET 4.5

ANSWERS

a Determine whether each of the following is a trinomial square.

1. $x^2 - 14x + 49$ **2.** $x^2 - 16x + 64$ **3.** $x^2 + 16x - 64$

4. $x^2 - 14x - 49$ **5.** $x^2 - 3x + 9$ **6.** $x^2 + 2x + 4$

7. $9x^2 - 36x + 24$ **8.** $36x^2 - 24x + 16$

b Factor completely. Remember to look first for a common factor and to check by multiplying.

9. $x^2 - 14x + 49$ **10.** $x^2 - 16x + 64$ **11.** $x^2 + 16x + 64$

12. $x^2 + 14x + 49$ **13.** $x^2 - 2x + 1$ **14.** $x^2 + 2x + 1$

15. $4 + 4x + x^2$ **16.** $4 + x^2 - 4x$ **17.** $y^4 + 6y^2 + 9$

18. $64 - 16p^2 + p^4$ **19.** $49 - 56y + 16y^2$ **20.** $120m + 75 + 48m^2$

21. $2x^2 - 4x + 2$ **22.** $2x^2 - 40x + 200$ **23.** $x^3 - 18x^2 + 81x$

24. $x^3 + 24x^2 + 144x$ **25.** $20x^2 + 100x + 125$ **26.** $12x^2 + 36x + 27$

1. _____

2. _____

3. _____

4. _____

5. _____

6. _____

7. _____

8. _____

9. _____

10. _____

11. _____

12. _____

13. _____

14. _____

15. _____

16. _____

17. _____

18. _____

19. _____

20. _____

21. _____

22. _____

23. _____

24. _____

25. _____

26. _____

ANSWERS

27. _____

28. _____

29. _____

30. _____

31. _____

32. _____

33. _____

34. _____

35. _____

36. _____

37. _____

38. _____

39. _____

40. _____

41. _____

42. _____

43. _____

44. _____

45. _____

46. _____

47. _____

48. _____

27. $49 - 42x + 9x^2$

28. $64 - 112x + 49x^2$

29. $5y^4 + 10y^2 + 5$

30. $a^4 + 14a^2 + 49$

31. $1 + 4x^4 + 4x^2$

32. $1 - 2a^5 + a^{10}$

33. $4p^2 + 12pq + 9q^2$

34. $25m^2 + 20mn + 4n^2$

35. $a^2 - 14ab + 49b^2$

36. $x^2 - 6xy + 9y^2$

37. $64m^2 + 16mn + n^2$

38. $81p^2 - 18pq + q^2$

39. $16s^2 - 40st + 25t^2$

40. $36a^2 + 96ab + 64b^2$

c Determine whether each of the following is a difference of squares.

41. $x^2 - 4$

42. $x^2 - 36$

43. $x^2 + 36$

44. $x^2 + 4$

45. $x^2 - 35$

46. $x^2 - 50y^2$

47. $16x^2 - 25y^2$

48. $-1 + 36x^2$

d Factor completely. Remember to look first for a common factor.

49. $y^2 - 4$ **50.** $x^2 - 36$ **51.** $p^2 - 9$ **52.** $q^2 - 1$

53. $-49 + t^2$ **54.** $-64 + m^2$ **55.** $a^2 - b^2$ **56.** $p^2 - q^2$

57. $25t^2 - m^2$ **58.** $w^2 - 49z^2$ **59.** $100 - k^2$ **60.** $81 - w^2$

61. $16a^2 - 9$ **62.** $25x^2 - 4$ **63.** $4x^2 - 25y^2$ **64.** $9a^2 - 16b^2$

65. $8x^2 - 98$ **66.** $24x^2 - 54$ **67.** $36x - 49x^3$ **68.** $16x - 81x^3$

69. $49a^4 - 81$ **70.** $25a^4 - 9$ **71.** $x^4 - 1$ **72.** $x^4 - 16$

73. $4x^4 - 64$ **74.** $5x^4 - 80$ **75.** $1 - y^8$ **76.** $x^8 - 1$

ANSWERS

49.
50.
51.
52.
53.
54.
55.
56.
57.
58.
59.
60.
61.
62.
63.
64.
65.
66.
67.
68.
69.
70.
71.
72.
73.
74.
75.
76.

77. $x^{12} - 16$ **78.** $x^8 - 81$ **79.** $y^2 - \dfrac{1}{16}$ **80.** $x^2 - \dfrac{1}{25}$

81. $25 - \dfrac{1}{49}x^2$ **82.** $4 - \dfrac{1}{9}y^2$ **83.** $16m^4 - t^4$ **84.** $1 - a^4 b^4$

SKILL MAINTENANCE

Divide.

85. $(-110) \div 10$ **86.** $-1000 \div (-2.5)$ **87.** $\left(-\dfrac{2}{3}\right) \div \dfrac{4}{5}$ **88.** $8.1 \div (-9)$

SYNTHESIS

Factor completely, if possible.

89. $49x^2 - 216$ **90.** $27x^3 - 13x$ **91.** $x^2 + 22x + 121$

92. $x^2 - 5x + 25$ **93.** $18x^3 + 12x^2 + 2x$ **94.** $162x^2 - 82$

95. $x^8 - 2^8$ **96.** $4x^4 - 4x^2$ **97.** $3x^5 - 12x^3$

98. $3x^2 - \dfrac{1}{3}$ **99.** $18x^3 - \dfrac{8}{25}x$ **100.** $x^2 - 2.25$

101. $0.49p - p^3$ **102.** $3.24x^2 - 0.81$ **103.** $0.64x^2 - 1.21$

104. $1.28x^2 - 2$ **105.** $(x + 3)^2 - 9$ **106.** $(y - 5)^2 - 36q^2$

107. $x^2 - \left(\dfrac{1}{x}\right)^2$ **108.** $a^{2n} - 49b^{2n}$ **109.** $81 - b^{4k}$

110. $9x^{18} + 48x^9 + 64$ **111.** $9b^{2n} + 12b^n + 4$ **112.** $(x + 7)^2 - 4x - 24$

113. $(y + 3)^2 + 2(y + 3) + 1$ **114.** $49(x + 1)^2 - 42(x + 1) + 9$

Find c so that the polynomial will be the square of a binomial.

115. $cy^2 + 6y + 1$ **116.** $cy^2 - 24y + 9$

4.6 Factoring: A General Strategy

a We now combine all of our factoring techniques and consider a general strategy for factoring polynomials. Here we will encounter polynomials of all the types we have considered, in random order, so you will have to determine which method to use.

To factor a polynomial:

a) **Always look first for a common factor.** If there is one, factor out the largest common factor.

b) **Then look at the number of terms.**

   Two terms: Determine whether you have a difference of squares. Do not try to factor a sum of squares: $A^2 + B^2$.

   Three terms: Determine whether the trinomial is a square. If so, you know how to factor. If not, try trial and error, using the standard method or grouping.

   Four terms: Try factoring by grouping.

c) **Always *factor completely.*** If a factor with more than one term can still be factored, you should factor it. When no factor can be factored further, you have finished.

▶ **EXAMPLE 1**   Factor: $5t^4 - 80$.

a) We look for a common factor:

$$5t^4 - 80 = 5(t^4 - 16).$$

b) The factor $t^4 - 16$ has only two terms. It is a difference of squares: $(t^2)^2 - 4^2$. We factor it, being careful to include the common factor:

$$5(t^2 + 4)(t^2 - 4).$$

We see that one of the factors is again a difference of squares. We factor it:

$$5(t^2 + 4)(t - 2)(t + 2).$$
       ↑
This is a sum of squares. It cannot be factored!

c) We have factored completely because no factor with more than one term can be factored further.   ◄

▶ **EXAMPLE 2**   Factor: $2x^3 + 10x^2 + x + 5$.

a) We look for a common factor. There isn't one.

b) There are four terms. We try factoring by grouping:

$$2x^3 + 10x^2 + x + 5$$
$$= (2x^3 + 10x^2) + (x + 5) \qquad \text{Separating into two binomials}$$
$$= 2x^2(x + 5) + 1(x + 5) \qquad \text{Factoring each binomial}$$
$$= (2x^2 + 1)(x + 5). \qquad \text{Factoring out the common factor, } x + 5$$

c) No factor with more than one term can be factored further, so we have factored completely.   ◄

OBJECTIVE

After finishing Section 4.6, you should be able to:

a Factor polynomials completely using any of the methods considered in this chapter.

FOR EXTRA HELP

Tape 10A   Tape 9B   MAC: 4
                                    IBM: 4

Factor.

1. $3m^4 - 3$

▶ **EXAMPLE 3**   Factor: $x^5 - 2x^4 - 35x^3$.

a) We look first for a common factor. This time there is one, x^3:

$$x^5 - 2x^4 - 35x^3 = x^3(x^2 - 2x - 35).$$

b) The factor $x^2 - 2x - 35$ has three terms, but it is not a trinomial square. We factor it using trial and error:

$$x^5 - 2x^4 - 35x^3 = x^3(x^2 - 2x - 35) = x^3(x - 7)(x + 5).$$

> Don't forget to include the common factor in the final answer!

We have studied two methods for such factoring: the standard trial-and-error method and grouping, which is also trial and error. Use the one that you prefer or follow the direction of your instructor.

c) No factor with more than one term can be factored further, so we have factored completely.   ◀

2. $x^6 + 8x^3 + 16$

▶ **EXAMPLE 4**   Factor: $x^4 - 10x^2 + 25$.

a) We look first for a common factor. There isn't one.

b) There are three terms. We see that this polynomial is a trinomial square. We factor it:

$$x^4 - 10x^2 + 25 = (x^2)^2 - 2 \cdot x^2 \cdot 5 + 5^2 = (x^2 - 5)^2.$$

c) No factor with more than one term can be factored further, so we have factored completely.   ◀

DO EXERCISES 1–5.

3. $2x^4 + 8x^3 + 6x^2$

▶ **EXAMPLE 5**   Factor: $6x^2y^4 - 21x^3y^5 + 3x^2y^6$.

a) We look first for a common factor:

$$6x^2y^4 - 21x^3y^5 + 3x^2y^6 = 3x^2y^4(2 - 7xy + y^2).$$

4. $3x^3 + 12x^2 - 2x - 8$

b) There are three terms in $2 - 7xy + y^2$. We determine whether the trinomial is a square. Since only y^2 is a square, we do not have a trinomial square. Can the trinomial be factored by trial and error? A key to the answer is that x is only in the term $-7xy$. The polynomial might be in a form like $(1 - y)(2 + y)$, but there would be no x in the middle term. Thus, $2 - 7xy + y^2$ cannot be factored.

c) Have we factored completely? Yes, because no factor with more than one term can be factored further.   ◀

5. $8x^3 - 200x$

ANSWERS ON PAGE A-5

▶ **EXAMPLE 6**   Factor: $(p + q)(x + 2) + (p + q)(x + y)$.

a) We look for a common factor:

$$(p + q)(x + 2) + (p + q)(x + y) = (p + q)[(x + 2) + (x + y)]$$
$$= (p + q)(2x + y + 2).$$

b) There are three terms in $2x + y + 2$, but this trinomial cannot be factored further.

c) No factor with more than one term can be factored further, so we have factored completely.   ◀

▶ **EXAMPLE 7**   Factor: $px + py + qx + qy$.

a) We look first for a common factor. There isn't one.

b) There are four terms. We try factoring by grouping:

$$px + py + qx + qy = p(x + y) + q(x + y)$$
$$= (p + q)(x + y).$$

c) Have we factored completely? Since no factor with more than one term can be factored further, we have factored completely.   ◀

▶ **EXAMPLE 8**   Factor: $25x^2 + 20xy + 4y^2$.

a) We look first for a common factor. There isn't one.

b) There are three terms. We determine whether the trinomial is a square. The first term and the last term are squares:

$$25x^2 = (5x)^2 \quad \text{and} \quad 4y^2 = (2y)^2.$$

Since twice the product of $5x$ and $2y$ is the other term,

$$2 \cdot 5x \cdot 2y = 20xy,$$

the trinomial is a perfect square.

  We factor by writing the square roots of the square terms and the sign of the middle term:

$$25x^2 + 20xy + 4y^2 = (5x + 2y)^2.$$

We can check by squaring $5x + 2y$.

c) No factor with more than one term can be factored further, so we have factored completely.   ◀

Factor.

6. $x^4y^2 + 2x^3y + 3x^2y$

7. $10p^6q^2 + 4p^5q^3 + 2p^4q^4$

8. $(a - b)(x + 5) + (a - b)(x + y^2)$

9. $ax^2 + ay + bx^2 + by$

10. $x^4 + 2x^2y^2 + y^4$

11. $x^2y^2 + 5xy + 4$

12. $p^4 - 81q^4$

▶ **EXAMPLE 9** Factor: $p^2q^2 + 7pq + 12$.

a) We look first for a common factor. There isn't one.

b) There are three terms. We determine whether the trinomial is a square. The first term is a square, but neither of the other terms is a square, so we do not have a trinomial square. We use the trial-and-error or grouping method, thinking of the product pq as a single variable. We consider this possibility for factorization:

$$(pq + \underline{\quad})(pq + \underline{\quad}).$$

We factor the last term, 12. All the signs are positive, so we consider only positive factors. Possibilities are 1, 12 and 2, 6 and 3, 4. The pair 3, 4 gives a sum of 7 for the coefficient of the middle term. Thus,

$$p^2q^2 + 7pq + 12 = (pq + 3)(pq + 4).$$

c) No factor with more than one term can be factored further, so we have factored completely. ◀

▶ **EXAMPLE 10** Factor: $8x^4 - 20x^2y - 12y^2$.

a) We look first for a common factor:

$$8x^4 - 20x^2y - 12y^2 = 4(2x^4 - 5x^2y - 3y^2).$$

b) There are three terms in $2x^4 - 5x^2y - 3y^2$. We determine whether the trinomial is a square. Since none of the terms is a square, we do not have a trinomial square. We use trial and error to factor $2x^4$. Possibilities are $2x^2$, x^2 and $2x$, x^3 and others. We also factor the last term, $-3y^2$. Possibilities are $3y$, $-y$ and $-3y$, y and others. We look for factors such that the sum of their products is the middle term. We try some possibilities:

$$(2x - y)(x^3 + 3y) = 2x^4 + 6xy - x^3y - 3y^2,$$
$$(2x^2 - y)(x^2 + 3y) = 2x^4 + 5x^2y - 3y^2,$$
$$(2x^2 + y)(x^2 - 3y) = 2x^4 - 5x^2y - 3y^2.$$

c) No factor with more than one term can be factored further, so we have factored completely. The factorization, including the common factor, is

$$4(2x^2 + y)(x^2 - 3y).$$ ◀

▶ **EXAMPLE 11** Factor: $a^4 - 16b^4$.

a) We look first for a common factor. There isn't one.

b) There are two terms. Since $a^4 = (a^2)^2$ and $16b^4 = (4b^2)^2$, we see that we do have a difference of squares. Thus,

$$a^4 - 16b^4 = (a^2 + 4b^2)(a^2 - 4b^2).$$

c) The last factor can be factored further. It is also a difference of squares. Thus,

$$a^4 - 16b^4 = (a^2 + 4b^2)(a + 2b)(a - 2b).$$ ◀

DO EXERCISES 6–12.

EXERCISE SET 4.6

ANSWERS

a Factor completely.

1. $2x^2 - 128$

2. $3t^2 - 27$

3. $a^2 + 25 - 10a$

4. $y^2 + 49 + 14y$

5. $2x^2 - 11x + 12$

6. $8y^2 - 18y - 5$

7. $x^3 + 24x^2 + 144x$

8. $x^3 - 18x^2 + 81x$

9. $x^3 + 3x^2 - 4x - 12$

10. $x^3 - 5x^2 - 25x + 125$

11. $24x^2 - 54$

12. $8x^2 - 98$

13. $20x^3 - 4x^2 - 72x$

14. $9x^3 + 12x^2 - 45x$

15. $x^2 + 4$

16. $t^2 + 25$

17. $x^4 + 7x^2 - 3x^3 - 21x$

18. $m^4 + 8m^3 + 8m^2 + 64m$

19. $x^5 - 14x^4 + 49x^3$

20. $2x^6 + 8x^5 + 8x^4$

21. $20 - 6x - 2x^2$

22. $45 - 3x - 6x^2$

1. _____

2. _____

3. _____

4. _____

5. _____

6. _____

7. _____

8. _____

9. _____

10. _____

11. _____

12. _____

13. _____

14. _____

15. _____

16. _____

17. _____

18. _____

19. _____

20. _____

21. _____

22. _____

ANSWERS

23. _____

24. _____

25. _____

26. _____

27. _____

28. _____

29. _____

30. _____

31. _____

32. _____

33. _____

34. _____

35. _____

36. _____

37. _____

38. _____

39. _____

40. _____

41. _____

42. _____

43. _____

44. _____

23. $x^2 + 3x + 1$

24. $x^2 + 5x + 2$

25. $4x^4 - 64$

26. $5x^5 - 80x$

27. $1 - y^8$

28. $t^8 - 1$

29. $x^5 - 4x^4 + 3x^3$

30. $x^6 - 2x^5 + 7x^4$

31. $36a^2 - 15a + \dfrac{25}{16}$

32. $\dfrac{1}{81}x^6 - \dfrac{8}{27}x^3 + \dfrac{16}{9}$

33. $12n^2 + 24n^3$

34. $ax^2 + ay^2$

35. $9x^2y^2 - 36xy$

36. $x^2y - xy^2$

37. $2\pi rh + 2\pi r^2$

38. $10p^4q^4 + 35p^3q^3 + 10p^2q^2$

39. $(a + b)(x - 3) + (a + b)(x + 4)$

40. $5c(a^3 + b) - (a^3 + b)$

41. $(x - 1)(x + 1) - y(x + 1)$

42. $x^2 + x + xy + y$

43. $n^2 + 2n + np + 2p$

44. $a^2 - 3a + ay - 3y$

45. $2x^2 - 4x + xz - 2z$

46. $6y^2 - 3y + 2py - p$

47. $x^2 + y^2 - 2xy$

48. $4b^2 + a^2 - 4ab$

49. $9c^2 + 6cd + d^2$

50. $16x^2 + 24xy + 9y^2$

51. $49m^4 - 112m^2n + 64n^2$

52. $4x^2y^2 + 12xyz + 9z^2$

53. $y^4 + 10y^2z^2 + 25z^4$

54. $0.01x^4 - 0.1x^2y^2 + 0.25y^4$

55. $\dfrac{1}{4}a^2 + \dfrac{1}{3}ab + \dfrac{1}{9}b^2$

56. $4p^2q + pq^2 + 4p^3$

57. $a^2 - ab - 2b^2$

58. $3b^2 - 17ab - 6a^2$

59. $2mn - 360n^2 + m^2$

60. $15 + x^2y^2 + 8xy$

61. $m^2n^2 - 4mn - 32$

62. $p^2q^2 + 7pq + 6$

63. $a^5b^2 + 3a^4b - 10a^3$

64. $m^2n^6 + 4mn^5 - 32n^4$

65. $a^5 + 4a^4b - 5a^3b^2$

66. $2s^6t^2 + 10s^3t^3 + 12t^4$

ANSWERS

45. _____

46. _____

47. _____

48. _____

49. _____

50. _____

51. _____

52. _____

53. _____

54. _____

55. _____

56. _____

57. _____

58. _____

59. _____

60. _____

61. _____

62. _____

63. _____

64. _____

65. _____

66. _____

67. $x^6 + x^3y - 2y^2$

68. $a^4 + a^2bc - 2b^2c^2$

69. $x^2 - y^2$

70. $p^2q^2 - r^2$

71. $7p^4 - 7q^4$

72. $a^4b^4 - 16$

73. $81a^4 - b^4$

74. $1 - 16x^{12}y^{12}$

75. $w^3 - 7w^2 - 4w + 28$

76. $y^3 + 8y^2 - y - 8$

SKILL MAINTENANCE

77. Divide: $\dfrac{7}{5} \div \left(-\dfrac{11}{10}\right)$.

78. Multiply: $(5x - t)^2$.

79. Solve $A = aX + bX - 7$ for X.

80. Solve: $4(x - 9) - 2(x + 7) < 14$.

SYNTHESIS

Factor completely.

81. $a^4 - 2a^2 + 1$

82. $x^4 + 9$

83. $12.25x^2 - 7x + 1$

84. $\dfrac{1}{5}x^2 - x + \dfrac{4}{5}$

85. $5x^2 + 13x + 7.2$

86. $x^3 - (x - 3x^2) - 3$

87. $18 + y^3 - 9y - 2y^2$

88. $-(x^4 - 7x^2 - 18)$

89. $a^3 + 4a^2 + a + 4$

90. $x^3 + x^2 - (4x + 4)$

91. $x^4 - 7x^2 - 18$

92. $3x^4 - 15x^2 + 12$

93. $x^3 - x^2 - 4x + 4$

94. $y^2(y + 1) - 4y(y + 1) - 21(y + 1)$

95. $y^2(y - 1) - 2y(y - 1) + (y - 1)$

96. $6(x - 1)^2 + 7y(x - 1) - 3y^2$

97. $(y + 4)^2 + 2x(y + 4) + x^2$

98. $2(a + 3)^2 - (a + 3)(b - 2) - (b - 2)^2$

99. Factor $x^{2k} - 2^{2k}$ when $k = 4$.

100. Factor: $a^4 - 81$.

4.7 Solving Quadratic Equations by Factoring

In this section, we introduce a new equation-solving method and use it along with factoring to solve certain equations like $x^2 + x - 156 = 0$.

> *A quadratic equation* is an equation equivalent to an equation of the type
> $$ax^2 + bx + c = 0, \quad \text{where } a > 0.$$
> The trinomial on the left is of second degree.

OBJECTIVES

After finishing Section 4.7, you should be able to:

a Solve equations (already factored) using the principle of zero factors.

b Solve quadratic equations by factoring and then using the principle of zero products.

FOR EXTRA HELP

Tape 10B   Tape 9B   MAC: 4
                               IBM: 4

a The Principle of Zero Products

The product of two numbers is 0 if one or both of the numbers is 0. Furthermore, *if any product is* 0, *then a factor must be* 0. For example, if $7x = 0$, then we know that $x = 0$. If $x(2x - 9) = 0$, we know that $x = 0$ or $2x - 9 = 0$. If $(x + 3)(x - 2) = 0$, we know that $x + 3 = 0$ or $x - 2 = 0$. In a product such as $ab = 24$, we cannot conclude that a is 24 or that b is 24.

▶ **EXAMPLE 1** Solve: $(x + 3)(x - 2) = 0$.

We have a product of 0. This equation will be true when either factor is 0. Hence it is true when

$$x + 3 = 0 \quad \text{or} \quad x - 2 = 0.$$

Here we have two simple equations that we know how to solve:

$$x = -3 \quad \text{or} \quad x = 2.$$

Each of the numbers -3 and 2 is a solution of the original equation, as we can see in the following checks.

Check: For -3:

$$
\begin{array}{c}
(x + 3)(x - 2) = 0 \\
\hline
(-3 + 3)(-3 - 2) \; | \; 0 \\
0(-5) \\
0 \quad | \quad \text{TRUE}
\end{array}
$$

For 2:

$$
\begin{array}{c}
(x + 3)(x - 2) = 0 \\
\hline
(2 + 3)(2 - 2) \; | \; 0 \\
5(0) \\
0 \quad | \quad \text{TRUE}
\end{array}
$$ ◀

We now have a principle to help in solving quadratic equations.

> **The Principle of Zero Products**
>
> An equation $ab = 0$ is true if and only if $a = 0$ or $b = 0$, or both. **(A product is 0 if and only if one or both of the factors is 0.)**

▶ **EXAMPLE 2** Solve: $(5x + 1)(x - 7) = 0$.

$$
\begin{array}{lll}
(5x + 1)(x - 7) = 0 \\
5x + 1 = 0 \quad \text{or} \quad x - 7 = 0 & \quad \text{Using the principle of zero products} \\
5x = -1 \quad \text{or} \quad x = 7 & \quad \text{Solving the two equations separately} \\
x = -\frac{1}{5} \quad \text{or} \quad x = 7
\end{array}
$$

Solve using the principle of zero products.

1. $(x - 3)(x + 4) = 0$

2. $(x - 7)(x - 3) = 0$

3. $(4t + 1)(3t - 2) = 0$

4. Solve: $y(3y - 17) = 0$.

Check: For $-\frac{1}{5}$:

$$\frac{(5x + 1)(x - 7) = 0}{(5(-\frac{1}{5}) + 1)(-\frac{1}{5} - 7) \mid 0}$$
$$(-1 + 1)(-7\tfrac{1}{5})$$
$$0(-7\tfrac{1}{5})$$
$$0 \mid \text{TRUE}$$

For 7:

$$\frac{(5x + 1)(x - 7) = 0}{(5(7) + 1)(7 - 7) \mid 0}$$
$$(35 + 1) \cdot 0$$
$$36 \cdot 0$$
$$0 \mid \text{TRUE}$$

The solutions are $-\frac{1}{5}$ and 7. ◀

When you solve an equation using the principle of zero products, you may wish to check by substitution as in Examples 1 and 2. Such a check will detect errors in solving.

DO EXERCISES 1–3.

When some factors have only one term, you can still use the principle of zero products.

▶ **EXAMPLE 3** Solve: $x(2x - 9) = 0$.

$$x(2x - 9) = 0$$
$$x = 0 \quad \text{or} \quad 2x - 9 = 0 \quad \text{Using the principle of zero products}$$
$$x = 0 \quad \text{or} \quad 2x = 9$$
$$x = 0 \quad \text{or} \quad x = \frac{9}{2}$$

The solutions are 0 and $\frac{9}{2}$. The check is left to the student. ◀

DO EXERCISE 4.

b Using Factoring to Solve Equations

Using factoring and the principle of zero products, we can solve some new kinds of equations. Thus we have extended our equation-solving abilities.

▶ **EXAMPLE 4** Solve: $x^2 + 5x + 6 = 0$.

Compare this equation to those that we know how to solve from Chapter 2. There are no like terms to collect, and we have a squared term. We first factor the polynomial. Then we use the principle of zero products:

$$x^2 + 5x + 6 = 0$$
$$(x + 2)(x + 3) = 0 \quad \text{Factoring}$$
$$x + 2 = 0 \quad \text{or} \quad x + 3 = 0 \quad \text{Using the principle of zero products}$$
$$x = -2 \quad \text{or} \quad x = -3.$$

Check: For -2:

$$x^2 + 5x + 6 = 0$$
$$\overline{(-2)^2 + 5(-2) + 6 \mid 0}$$
$$4 - 10 + 6$$
$$-6 + 6$$
$$0 \mid \text{TRUE}$$

For -3:

$$x^2 + 5x + 6 = 0$$
$$\overline{(-3)^2 + 5(-3) + 6 \mid 0}$$
$$9 - 15 + 6$$
$$-6 + 6$$
$$0 \mid \text{TRUE}$$

5. Solve: $x^2 - x - 6 = 0$.

The solutions are -2 and -3. The check is left to the student.   ◄

> CAUTION!   Keep in mind that you *must* have 0 on one side before you can use the principle of zero products. Get all nonzero terms on one side and 0 on the other.

DO EXERCISE 5.

Solve.

6. $x^2 - 3x = 28$

► **EXAMPLE 5**   Solve: $x^2 - 8x = -16$.

We first add 16 to get a 0 on one side:

$$x^2 - 8x = -16$$
$$x^2 - 8x + 16 = 0 \qquad \text{Adding 16}$$
$$(x - 4)(x - 4) = 0 \qquad \text{Factoring}$$
$$x - 4 = 0 \quad \text{or} \quad x - 4 = 0 \qquad \text{Using the principle of zero products}$$
$$x = 4 \quad \text{or} \qquad x = 4.$$

7. $x^2 = 6x - 9$

There is only one solution, 4. The check is left to the student.   ◄

DO EXERCISES 6 AND 7.

► **EXAMPLE 6**   Solve: $x^2 + 5x = 0$.

$$x^2 + 5x = 0$$
$$x(x + 5) = 0 \qquad \text{Factoring out a common factor}$$
$$x = 0 \quad \text{or} \quad x + 5 = 0 \qquad \text{Using the principle of zero products}$$
$$x = 0 \quad \text{or} \qquad x = -5$$

Solve.

8. $x^2 - 4x = 0$

The solutions are 0 and -5. The check is left to the student.   ◄

► **EXAMPLE 7**   Solve: $4x^2 = 25$.

$$4x^2 = 25$$
$$4x^2 - 25 = 0 \qquad \text{Subtracting 25 on both sides to get 0 on one side}$$
$$(2x - 5)(2x + 5) = 0 \qquad \text{Factoring a difference of squares}$$
$$2x - 5 = 0 \quad \text{or} \quad 2x + 5 = 0$$
$$2x = 5 \quad \text{or} \qquad 2x = -5$$
$$x = \frac{5}{2} \quad \text{or} \qquad x = -\frac{5}{2}$$

9. $9x^2 = 16$

The solutions are $\frac{5}{2}$ and $-\frac{5}{2}$.   ◄

DO EXERCISES 8 AND 9.

ANSWERS ON PAGE A-5

10. Solve: $(x + 1)(x - 1) = 8$.

▶ **EXAMPLE 8**   Solve: $(x + 2)(x - 2) = 5$.

    Be careful with an equation like this one! It might be tempting to set each factor equal to 5. We must have a 0 on one side. We first carry out the product on the left. Then we subtract 5 on both sides to get 0 on one side. Then we proceed with the principle of zero products.

$$(x + 2)(x - 2) = 5$$
$$x^2 - 4 = 5 \qquad \text{Multiplying on the left}$$
$$x^2 - 4 - 5 = 5 - 5 \qquad \text{Subtracting 5}$$
$$x^2 - 9 = 0$$
$$(x + 3)(x - 3) = 0 \qquad \text{Factoring}$$
$$x + 3 = 0 \quad \text{or} \quad x - 3 = 0 \qquad \text{Using the principle of zero products}$$
$$x = -3 \quad \text{or} \qquad x = 3$$

Check: For -3:

$$\frac{(x + 2)(x - 2) = 5}{\begin{array}{c|c} (-3 + 2)(-3 - 2) & 5 \\ (-1)(-5) & \\ 5 & \text{TRUE} \end{array}}$$

For 3:

$$\frac{(x + 2)(x - 2) = 5}{\begin{array}{c|c} (3 + 2)(3 - 2) & 5 \\ (5)(1) & \\ 5 & \text{TRUE} \end{array}}$$

The solutions are -3 and 3.   ◀

DO EXERCISE 10.

EXERCISE SET 4.7

a Solve using the principle of zero products.

1. $(x + 8)(x + 6) = 0$

2. $(x + 3)(x + 2) = 0$

3. $(x - 3)(x + 5) = 0$

4. $(x + 9)(x - 3) = 0$

5. $(x + 12)(x - 11) = 0$

6. $(x - 13)(x + 53) = 0$

7. $x(x + 5) = 0$

8. $y(y + 7) = 0$

9. $0 = y(y + 10)$

10. $0 = x(x - 21)$

11. $(2x + 5)(x + 4) = 0$

12. $(2x + 9)(x + 8) = 0$

13. $(5x + 1)(4x - 12) = 0$

14. $(4x + 9)(14x - 7) = 0$

15. $(7x - 28)(28x - 7) = 0$

16. $(12x - 11)(8x - 5) = 0$

17. $2x(3x - 2) = 0$

18. $75x(8x - 9) = 0$

19. $\frac{1}{2}x(\frac{2}{3}x - 12) = 0$

20. $\frac{5}{7}x(\frac{3}{4}x - 6) = 0$

21. $(\frac{1}{5} + 2x)(\frac{1}{9} - 3x) = 0$

22. $(\frac{7}{4}x - \frac{1}{12})(\frac{2}{3}x - \frac{12}{11}) = 0$

23. $(0.3x - 0.1)(0.05x - 1) = 0$

24. $(0.1x - 0.3)(0.4x - 20) = 0$

25. $9x(3x - 2)(2x - 1) = 0$

26. $(x - 5)(x + 55)(5x - 1) = 0$

b Solve by factoring and using the principle of zero products.

27. $x^2 + 6x + 5 = 0$

28. $x^2 + 7x + 6 = 0$

29. $x^2 + 7x - 18 = 0$

30. $x^2 + 4x - 21 = 0$

31. $x^2 - 8x + 15 = 0$

32. $x^2 - 9x + 14 = 0$

33. $x^2 - 8x = 0$

34. $x^2 - 3x = 0$

35. $x^2 + 19x = 0$

36. $x^2 + 12x = 0$

37. $x^2 = 16$

38. $100 = x^2$

ANSWERS

1. _____
2. _____
3. _____
4. _____
5. _____
6. _____
7. _____
8. _____
9. _____
10. _____
11. _____
12. _____
13. _____
14. _____
15. _____
16. _____
17. _____
18. _____
19. _____
20. _____
21. _____
22. _____
23. _____
24. _____
25. _____
26. _____
27. _____
28. _____
29. _____
30. _____
31. _____
32. _____
33. _____
34. _____
35. _____
36. _____
37. _____
38. _____

39. $9x^2 - 4 = 0$

40. $4x^2 - 9 = 0$

41. $0 = 6x + x^2 + 9$

42. $0 = 25 + x^2 + 10x$

43. $x^2 + 16 = 8x$

44. $1 + x^2 = 2x$

45. $5x^2 = 6x$

46. $7x^2 = 8x$

47. $6x^2 - 4x = 10$

48. $3x^2 - 7x = 20$

49. $12y^2 - 5y = 2$

50. $2y^2 + 12y = -10$

51. $x(x - 5) = 14$

52. $t(3t + 1) = 2$

53. $64m^2 = 81$

54. $100t^2 = 49$

55. $3x^2 + 8x = 9 + 2x$

56. $x^2 - 5x = 18 + 2x$

57. $10x^2 - 23x + 12 = 0$

58. $12x^2 - 17x - 5 = 0$

SKILL MAINTENANCE

Translate to an algebraic expression.

59. The square of the sum of a and b

60. The sum of the squares of a and b

61. Divide: $144 \div (-9)$.

62. Divide: $-24.3 \div 5.4$.

SYNTHESIS

Solve.

63. $b(b + 9) = 4(5 + 2b)$

64. $y(y + 8) = 16(y - 1)$

65. $(t - 3)^2 = 36$

66. $(t - 5)^2 = 2(5 - t)$

67. $x^2 - \frac{1}{64} = 0$

68. $x^2 - \frac{25}{36} = 0$

69. $\frac{5}{16}x^2 = 5$

70. $\frac{27}{25}x^2 = \frac{1}{3}$

71. Find an equation that has the given numbers as solutions. For example, 3 and -2 are solutions to $x^2 - x - 6 = 0$.

    a) $-3, 4$ **b)** $-3, -4$ **c)** $\frac{1}{2}, \frac{1}{2}$ **d)** $5, -5$ **e)** $0, 0.1, \frac{1}{4}$

72. Check the numbers found below. What's wrong with the methods used? Try to find the solutions.

    a) $(x - 3)(x + 4) = 0$
        $x = -3$ or $x = 4$

    b) $(x - 3)(x + 4) = 8$
        $x - 3 = 2$ or $x + 4 = 4$
            $x = 5$ or $x = 0$

4.8 Solving Problems

a We can now use the new method for solving quadratic equations and our five steps for solving problems.

▶ **EXAMPLE 1** One more than a number times one less than the number is 8. Find all such numbers.

1. *Familiarize.* Let's make a guess. Try 5. One more than 5 is 6. One less than the number is 4. The product of one more than the number and one less than the number is 6(4), or 24, which is too large. We could continue to guess, but let's use our algebraic skills to find the numbers. Let x = the number (there could be more than one).

2. *Translate.* From the familiarization, we can translate as follows:

$$\underbrace{\text{One more than a number}}_{(x + 1)} \quad \underbrace{\text{times}}_{\cdot} \quad \underbrace{\text{one less than that number}}_{(x - 1)} \quad \underbrace{\text{is}}_{=} \quad \underbrace{8.}_{8}$$

3. *Solve.* We solve the equation as follows:

$$(x + 1)(x - 1) = 8$$
$$x^2 - 1 = 8 \qquad \text{Multiplying}$$
$$x^2 - 1 - 8 = 0 \qquad \text{Subtracting 8 to get 0 on one side}$$
$$x^2 - 9 = 0$$
$$(x - 3)(x + 3) = 0 \qquad \text{Factoring}$$
$$x - 3 = 0 \quad \text{or} \quad x + 3 = 0 \qquad \text{Using the principle of zero products}$$
$$x = 3 \quad \text{or} \qquad x = -3.$$

4. *Check.* One more than 3 (this is 4) times one less than 3 (this is 2) is 8. Thus, 3 checks. One more than -3 (this is -2) times one less than -3 (this is -4) is 8. Thus, -3 also checks.

5. *State.* There are two such numbers, 3 and -3. ◀

DO EXERCISES 1 AND 2.

▶ **EXAMPLE 2** The square of a number minus twice the number is 48. Find all such numbers.

1. *Familiarize.* Let us again make a guess to help understand the problem and ease the translation. Try 6. The square of 6 is 36, and twice the number is 12. Then $36 - 12 = 24$, so 6 is not a number we want. We find the numbers using our algebraic skills. Let x = the number, or numbers.

2. *Translate.* We translate as follows:

$$\underbrace{\text{The square of a number}}_{x^2} \quad \underbrace{\text{minus}}_{-} \quad \underbrace{\text{twice the number}}_{2x} \quad \underbrace{\text{is}}_{=} \quad \underbrace{48.}_{48}$$

OBJECTIVE

After finishing Section 4.8, you should be able to:

a Solve problems involving quadratic equations that can be solved by factoring.

FOR EXTRA HELP

Tape 10C   Tape 10A   MAC: 4
                                     IBM: 4

1. One more than a number times one less than a number is 24. Find all such numbers.

2. Seven less than a number times eight less than the number is 0. Find all such numbers.

ANSWERS ON PAGE A-5

3. The square of a number minus the number is 20. Find all such numbers.

3. *Solve.*　We solve the equation as follows:

$$x^2 - 2x = 48$$
$$x^2 - 2x - 48 = 0 \qquad \text{Subtracting 48 to get 0 on one side}$$
$$(x - 8)(x + 6) = 0 \qquad \text{Factoring}$$
$$x - 8 = 0 \quad \text{or} \quad x + 6 = 0 \qquad \text{Using the principle of zero products}$$
$$x = 8 \quad \text{or} \qquad x = -6.$$

4. *Check.*　The square of 8 is 64, and twice the number 8 is 16. Then $64 - 16$ is 48, so 8 checks. The square of -6 is $(-6)^2$, or 36, and twice -6 is -12. Then $36 - (-12)$ is 48, so -6 checks.

5. *State.*　There are two such numbers, 8 and -6. ◄

DO EXERCISE 3.

▶ **EXAMPLE 3**　The height of a triangular sail is 7 ft more than the base. The area of the triangle is 30 ft². Find the height and the base.

1. *Familiarize.*　We first make a drawing. If you don't remember the formula for the area of a triangle, look it up, either in this book or a geometry book. The area is $\frac{1}{2}$(base)(height).
　　We let b = the base of the triangle. Then $b + 7$ = the height.

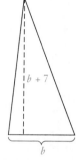

4. The width of a rectangle is 2 cm less than the length. The area is 15 cm². Find the length and the width.

2. *Translate.*　It helps to reword this problem before translating:

$$\frac{1}{2} \text{ times the base times the base plus 7 is 30.} \qquad \text{Rewording}$$

$$\frac{1}{2} \quad \cdot \quad b \quad \cdot \quad (b + 7) \quad = \quad 30 \qquad \text{Translating}$$

3. *Solve.*　We solve the equation as follows:

$$\frac{1}{2} \cdot b \cdot (b + 7) = 30$$
$$\frac{1}{2}(b^2 + 7b) = 30 \qquad \text{Multiplying}$$
$$b^2 + 7b = 60 \qquad \text{Multiplying by 2}$$
$$b^2 + 7b - 60 = 0 \qquad \text{Subtracting 60 to get 0 on one side}$$
$$(b + 12)(b - 5) = 0 \qquad \text{Factoring}$$
$$b + 12 = 0 \quad \text{or} \quad b - 5 = 0 \qquad \text{Using the principle of zero products}$$
$$b = -12 \quad \text{or} \qquad b = 5.$$

4. *Check.*　The base of a triangle cannot have a negative length, so -12 cannot be a solution. Suppose the base is 5 ft. Then the height is 7 ft more than the base, so the height is 12 ft and the area is $\frac{1}{2}(5)(12)$, or 30 ft². These numbers check in the original problem.

5. *State.*　The height is 12 ft and the base is 5 ft. ◄

DO EXERCISE 4.

▶ **EXAMPLE 4**   In a sports league of n teams in which each team plays every other team twice, the total number N of games to be played is given by

$$n^2 - n = N. \quad \textbf{(1)}$$

If a basketball league plays a total of 240 games, how many teams are in the league?

1., 2. *Familiarize* and *Translate.*   We are given that n = the number of teams in a league and N = the number of games. To familiarize yourself with this problem, reread Example 5 in Section 3.3 where we first considered it. To find the number of teams n in a league when 240 games are played, we substitute 240 for N in equation (1):

$$n^2 - n = 240. \quad \text{Substituting 240 for } N$$

3. *Solve.*   We solve the equation as follows:

$$n^2 - n = 240$$
$$n^2 - n - 240 = 0 \quad \text{Subtracting 240 to get 0 on one side}$$
$$(n - 16)(n + 15) = 0 \quad \text{Factoring}$$
$$n - 16 = 0 \quad \text{or} \quad n + 15 = 0 \quad \text{Using the principle of zero products}$$
$$n = 16 \quad \text{or} \quad n = -15.$$

4. *Check.*   The solutions of the equation are 16 and -15. Since the number of teams cannot be negative, -15 cannot be a solution. But 16 checks, since $16^2 - 16 = 256 - 16 = 240$.

5. *State.*   There are 16 teams in the league.   ◀

DO EXERCISE 5.

▶ **EXAMPLE 5**   The product of the numbers on two consecutive pages in a book is 156. Find the page numbers.

1. *Familiarize.*   The page numbers are consecutive integers. Recall that consecutive integers are next to each other, such as 49 and 50, or -6 and -5. Let x = the smaller integer; then $x + 1$ = the larger integer.

2. *Translate.*   It helps to reword the problem before translating:

First integer   times   second integer   is   156.   *Rewording*

$$x \quad \cdot \quad (x + 1) \quad = \quad 156 \quad \text{Translating}$$

3. *Solve.*   We solve the equation as follows:

$$x(x + 1) = 156$$
$$x^2 + x = 156 \quad \text{Multiplying}$$
$$x^2 + x - 156 = 0 \quad \text{Subtracting 156 to get 0 on one side}$$
$$(x - 12)(x + 13) = 0 \quad \text{Factoring}$$
$$x - 12 = 0 \quad \text{or} \quad x + 13 = 0 \quad \text{Using the principle of zero products}$$
$$x = 12 \quad \text{or} \quad x = -13.$$

4. *Check.*   The solutions of the equation are 12 and -13. When x is 12, then $x + 1$ is 13, and $12 \cdot 13 = 156$. The numbers 12 and 13 are consecutive integers that are solutions to the problem. When x is -13, then $x + 1$ is -12, and $(-13)(-12) = 156$. The numbers -13 and -12 are

5. Use $N = n^2 - n$ for the following.

a) A woman's volleyball league has 19 teams. What is the total number of games to be played?

b) A slow-pitch softball league plays a total of 72 games. How many teams are in the league?

ANSWERS ON PAGE A-5

6. The product of the page numbers on two facing pages of a book is 506. Find the page numbers.

also consecutive integers but they are not solutions of the problem since negative numbers are not used as page numbers.

5. *State.* The page numbers are 12 and 13. ◀

DO EXERCISE 6.

The following problem involves the Pythagorean theorem, which relates the lengths of the sides of a right triangle. A **right triangle** has a 90° angle. The side opposite the 90° angle is called the **hypotenuse.** The other sides are called **legs.**

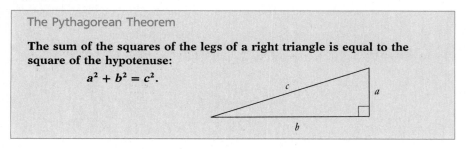

The Pythagorean Theorem

The sum of the squares of the legs of a right triangle is equal to the square of the hypotenuse:

$$a^2 + b^2 = c^2.$$

▶ **EXAMPLE 6** The length of one leg of a right triangle is 7 ft longer than the other. The length of the hypotenuse is 13 ft. Find the lengths of the legs.

1. *Familiarize.* We first make a drawing. We let

x = the length of one leg.

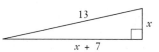

Since the other leg is 7 ft longer, we know that

$x + 7$ = the length of the other leg.

The hypotenuse has length 13 ft.

2. *Translate.* Applying the Pythagorean theorem, we obtain the following translation:

$$a^2 + b^2 = c^2$$
$$x^2 + (x + 7)^2 = 13^2.$$

7. The length of one leg of a right triangle is 1 m longer than the other. The length of the hypotenuse is 5 m. Find the lengths of the legs.

3. *Solve.* We solve the equation as follows:

$$x^2 + (x^2 + 14x + 49) = 169 \qquad \text{Squaring the binomial and 13}$$
$$2x^2 + 14x + 49 = 169 \qquad \text{Collecting like terms}$$
$$2x^2 + 14x - 120 = 0 \qquad \text{Subtracting 169 to get 0 on one side}$$
$$2(x^2 + 7x - 60) = 0 \qquad \text{Factoring out a common factor}$$
$$x^2 + 7x - 60 = 0 \qquad \text{Dividing by 2 on both sides, or multiplying by } \tfrac{1}{2}$$
$$(x + 12)(x - 5) = 0 \qquad \text{Factoring}$$
$$x + 12 = 0 \qquad \text{or} \quad x - 5 = 0$$
$$x = -12 \quad \text{or} \qquad x = 5.$$

4. *Check.* The integer -12 cannot be a length of a side because it is negative. When $x = 5$, $x + 7 = 12$, and $5^2 + 12^2 = 13^2$. So 5 and 12 check.

5. *State.* The lengths of the legs are 5 ft and 12 ft. ◀

DO EXERCISE 7.

EXERCISE SET 4.8

a Solve.

1. If you subtract a number from four times its square, the result is 3. Find all such numbers.

2. If 7 is added to the square of a number, the result is 32. Find all such numbers.

1. _____

3. Eight more than the square of a number is six times the number. Find all such numbers.

4. Fifteen more than the square of a number is eight times the number. Find all such numbers.

2. _____

3. _____

5. The product of the page numbers on two facing pages of a book is 210. Find the page numbers.

6. The product of the page numbers on two facing pages of a book is 156. Find the page numbers.

4. _____

5. _____

7. The product of two consecutive even integers is 168. Find the integers.

8. The product of two consecutive even integers is 224. Find the integers.

6. _____

7. _____

9. The product of two consecutive odd integers is 255. Find the integers.

10. The product of two consecutive odd integers is 143. Find the integers.

8. _____

9. _____

10. _____

11. The length of a rectangular garden is 4 m greater than the width. The area of the garden is 96 m². Find the length and the width.

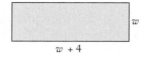

w

$w + 4$

12. The length of a rectangular calculator is 5 cm greater than the width. The area of the calculator is 84 cm². Find the length and the width.

11. _____

12. _____

13. The area of a square bookcase is five more than the perimeter. Find the length of a side.

14. The perimeter of a square porch is three more than the area. Find the length of a side.

13. _____

15. The base of a triangle is 10 cm greater than the height. The area is 28 cm². Find the height and the base.

h

$h + 10$

16. The height of a triangle is 8 m less than the base. The area is 10 m². Find the height and the base.

14. _____

15. _____

16. _____

17. If the sides of a square are lengthened by 3 km, the area becomes 81 km². Find the length of a side of the original square.

18. If the sides of a square are lengthened by 7 km, the area becomes 121 km². Find the length of a side of the original square.

17. _____

18. _____

19. The sum of the squares of two consecutive odd positive integers is 74. Find the integers.

20. The sum of the squares of two consecutive odd positive integers is 130. Find the integers.

19. _____

Use $n^2 - n = N$ for Exercises 21–24.

21. A women's volleyball league has 23 teams. What is the total number of games to be played?

22. A chess league has 14 teams. What is the total number of games to be played?

20. _____

21. _____

23. A woman's slow-pitch softball league plays a total of 132 games. How many teams are in the league?

24. The basketball league plays a total of 90 games. How many teams are in the league?

22. _____

23. _____

24. _____

The number of possible handshakes within a group of n people is given by $N = \frac{1}{2}(n^2 - n)$.

25. There are 40 people at a meeting. How many handshakes are possible?

26. There are 100 people at a party. How many handshakes are possible?

25. _____

27. Everyone shook hands at a party. There were 190 handshakes in all. How many were at the party?

28. Everyone shook hands at a meeting. There were 300 handshakes in all. How many were at the meeting?

26. _____

27. _____

28. _____

29. The length of one leg of a right triangle is 8 ft. The length of the hypotenuse is 2 ft longer than the other leg. Find the length of the hypotenuse and the other leg.

30. The length of one leg of a right triangle is 24 ft. The length of the other leg is 16 ft shorter than the hypotenuse. Find the length of the hypotenuse and the other leg.

SYNTHESIS

31. A cement walk of constant width is built around a 20-ft by 40-ft rectangular pool. The total area of the pool and walk is 1500 ft². Find the width of the walk.

32. A model rocket is launched with an initial velocity of 180 ft/sec. Its height h, in feet, after t seconds is given by the formula $h = 180t - 16t^2$.
    a) After how many seconds will the rocket first reach a height of 464 ft?
    b) After how many seconds will it be at that height again?

33. The one's digit of a number less than 100 is four greater than the ten's digit. The sum of the number and the product of the digits is 58. Find the number.

34. The total surface area of a closed box is 350 m². The box is 9 m high and has a square base and lid. Find the length of the side of the base.

35. A rectangular piece of cardboard is twice as long as it is wide. A 4-cm square is cut out of each corner, and the sides are turned up to make a box with an open top. The volume of the box is 616 cm³. Find the original dimensions of the cardboard.

36. An open rectangular gutter is made by turning up the sides of a piece of metal 20 in. wide. The area of the cross-section of the gutter is 50 in². Find the depth of the gutter.

37. The length of each side of a square is increased by 5 cm to form a new square. The area of the new square is $2\frac{1}{4}$ times the area of the original square. Find the area of each square.

SUMMARY AND REVIEW: CHAPTER 4

IMPORTANT PROPERTIES AND FORMULAS

Factoring Formulas: $\quad A^2 - B^2 = (A + B)(A - B),$
$$A^2 + 2AB + B^2 = (A + B)^2,$$
$$A^2 - 2AB + B^2 = (A - B)^2$$

REVIEW EXERCISES

The review sections and objectives to be tested in addition to the material in this chapter are [1.6a, c], [2.6a], [2.7e], and [3.6d].

Find three factorizations of the monomial.

1. $-10x^2$

2. $36x^5$

Factor completely.

3. $5 - 20x^6$

4. $x^2 - 3x$

5. $9x^2 - 4$

6. $x^2 + 4x - 12$

7. $x^2 + 14x + 49$

8. $6x^3 + 12x^2 + 3x$

9. $x^3 + x^2 + 3x + 3$

10. $6x^2 - 5x + 1$

11. $x^4 - 81$

12. $9x^3 + 12x^2 - 45x$

13. $2x^2 - 50$

14. $x^4 + 4x^3 - 2x - 8$

15. $16x^4 - 1$

16. $8x^6 - 32x^5 + 4x^4$

17. $75 + 12x^2 + 60x$

18. $x^2 + 9$

19. $x^3 - x^2 - 30x$

20. $4x^2 - 25$

21. $9x^2 + 25 - 30x$

22. $6x^2 - 28x - 48$

23. $x^2 - 6x + 9$

24. $2x^2 - 7x - 4$

25. $18x^2 - 12x + 2$

26. $3x^2 - 27$

27. $15 - 8x + x^2$

28. $25x^2 - 20x + 4$

Solve.

29. $(x - 1)(x + 3) = 0$

30. $x^2 + 2x - 35 = 0$

31. $x^2 + x - 12 = 0$

32. $3x^2 + 2 = 5x$

33. $2x^2 + 5x = 12$

34. $16 = x(x - 6)$

35. The square of a number is six more than the number. Find all such numbers.

36. The product of two consecutive even integers is 288. Find the integers.

37. The product of two consecutive odd integers is 323. Find the integers.

38. Twice the square of a number is ten more than the number. Find all such numbers.

Factor.

39. $49b^{10} + 4a^8 - 28a^4b^5$

40. $x^2y^2 + xy - 12$

41. $12a^2 + 84ab + 147b^2$

42. $m^2 + 5m + mt + 5t$

43. $32x^4 - 128y^4z^4$

SKILL MAINTENANCE

44. Divide: $-\dfrac{12}{25} \div \left(-\dfrac{21}{10}\right)$.

45. Solve: $20 - (3x + 2) \geq 2(x + 5) + x$.

46. Solve $A = a + 2b$ for b.

47. Multiply: $(2a + 3)^2$.

48. Multiply: $(2a - 3)(2a + 3)$.

49. Multiply: $(2a - 3)(5a + 7)$.

SYNTHESIS

Solve.

50. The pages of a book measure 15 cm by 20 cm. Margins of equal width surround the printing on each page and constitute one half of the area of the page. Find the width of the margins.

51. The cube of a number is the same as twice the square of the number. Find all such numbers.

52. The length of a rectangle is two times its width. When the length is increased by 20 and the width decreased by 1, the area is 160. Find the original length and width.

Solve.

53. $x^2 + 25 = 0$

54. $(x - 2)(x + 3)(2x - 5) = 0$

55. ▤ $(0.00005x + 0.1)(0.0097x + 0.5) = 0$

56. For each equation on the left, find an equivalent equation on the right.

a) $3x^2 - 4x + 8 = 0$

b) $(x - 6)(x + 3) = 0$

c) $x^2 + 2x + 9 = 0$

d) $(2x - 5)(x + 4) = 0$

e) $5x^2 - 5 = 0$

f) $x^2 + 10x - 2 = 0$

g) $4x^2 + 8x + 36 = 0$

h) $(2x + 8)(2x - 5) = 0$

i) $9x^2 - 12x + 24 = 0$

j) $(x + 1)(5x - 5) = 0$

k) $x^2 - 3x - 18 = 0$

l) $2x^2 + 20x - 4 = 0$

❖ THINKING IT THROUGH

1. Compare the type of equations we are able to solve after studying this chapter with those previously studied.

2. What is one procedure you can always use to check the result of factoring a polynomial?

3. Suppose we know that $(x - 5)(x + 2)$ would be a correct factorization of a trinomial $x^2 + bx + c$ except the sign of the middle term is incorrect. What would a correct factorization be?

Explain the error, if any, in each of the following when the direction is "Factor completely."

4. $x^2 + 9 = (x + 3)^2$

5. $x^2 - 6x + 8 = (x + 4)(x - 2)$

6. $p^2 - 9 = (p - 3)^2$

7. $a^2 + 6a - 9 = (a - 3)^2$

8. $16a^4 - 81 = (4a^2 - 9)(4a^2 + 9)$

9. $16m^2 - 80m + 100 = (4m - 10)^2$

TEST: CHAPTER 4

1. Find three factorizations of $4x^3$.

Factor completely.

2. $x^2 - 7x + 10$

3. $x^2 + 25 - 10x$

4. $6y^2 - 8y^3 + 4y^4$

5. $x^3 + x^2 + 2x + 2$

6. $x^2 - 5x$

7. $x^3 + 2x^2 - 3x$

8. $28x - 48 + 10x^2$

9. $4x^2 - 9$

10. $x^2 - x - 12$

11. $6m^3 + 9m^2 + 3m$

12. $3w^2 - 75$

13. $60x + 45x^2 + 20$

14. $3x^4 - 48$

15. $49x^2 - 84x + 36$

16. $5x^2 - 26x + 5$

ANSWERS

1. _____

2. _____

3. _____

4. _____

5. _____

6. _____

7. _____

8. _____

9. _____

10. _____

11. _____

12. _____

13. _____

14. _____

15. _____

16. _____

17. $x^4 + 2x^3 - 3x - 6$

18. $80 - 5x^4$

19. $4x^2 - 4x - 15$

20. $6t^3 + 9t^2 - 15t$

Solve.

21. $x^2 - x - 20 = 0$

22. $2x^2 + 7x = 15$

23. $x(x - 3) = 28$

24. The square of a number is 24 more than five times the number. Find all such numbers.

25. The length of a rectangle is 6 m more than the width. The area of the rectangle is 40 m². Find the length and the width.

26. Factor: $3m^2 - 9mn - 30n^2$.

SKILL MAINTENANCE

27. Divide: $\dfrac{5}{8} \div \left(-\dfrac{11}{16} \right)$.

28. Solve: $10(x - 3) < 4(x + 2)$.

29. Solve $I = PRT$ for T.

30. Multiply: $(5x^2 - 7)^2$.

SYNTHESIS

31. The length of a rectangle is five times its width. When the length is decreased by 3 and the width is increased by 2, the area of the new rectangle is 60. Find the original length and width.

32. Factor: $(a + 3)^2 - 2(a + 3) - 35$.

33. If $x^2 - 4 = (14)(18)$, then one possibility for x is which of the following?

   a) 12   b) 14
   c) 16   d) 18

34. If $x + y = 4$ and $x - y = 6$, then $x^2 - y^2 = ?$

   a) 2   b) 10
   c) 34   d) 24

CUMULATIVE REVIEW: CHAPTERS 1–4

Use either $<$ or $>$ for ■ to write a true sentence.

1. $\dfrac{2}{3}$ ■ $\dfrac{5}{7}$

2. $-\dfrac{4}{7}$ ■ $-\dfrac{8}{11}$

Compute and simplify.

3. $2.06 + (-4.79) - (-3.08)$

4. $5.652 \div (-3.6)$

5. $\left(\dfrac{2}{9}\right)\left(-\dfrac{3}{8}\right)\left(\dfrac{6}{7}\right)$

6. $\dfrac{21}{5} \div \left(-\dfrac{7}{2}\right)$

Simplify.

7. $[3x + 2(x - 1)] - [2x - (x + 3)]$

8. $1 - [14 + 28 \div 7 - (6 + 9 \div 3)]$

9. $(2x^2 y^{-1})^3$

10. $\dfrac{3x^5}{4x^3} \cdot \dfrac{-2x^{-3}}{9x^2}$

11. Add:

$$(2x^2 - 3x^3 + x - 4) + (x^4 - x - 5x^2).$$

12. Subtract:

$$(2x^2 y^2 + xy - 2xy^2) - (2xy - 2xy^2 + x^2 y).$$

13. Divide: $(x^3 + 2x^2 - x + 1) \div (x - 1)$.

Multiply.

14. $(2t - 3)^2$

15. $(x^2 - 3)(x^2 + 3)$

16. $(2x + 4)(3x - 4)$

17. $2x(x^3 + 3x^2 + 4x)$

18. $(2y - 1)(2y^2 + 3y + 4)$

19. $\left(x + \dfrac{2}{3}\right)\left(x - \dfrac{2}{3}\right)$

Factor.

20. $x^2 + 2x - 8$

21. $4x^2 - 25$

22. $3x^3 - 4x^2 + 3x - 4$

23. $x^2 - 26x + 169$

24. $75x^2 - 108y^2$

25. $6x^2 - 13x - 63$

26. $x^4 - 2x^2 - 3$

27. $4y^3 - 6y^2 - 4y + 6$

28. $6p^2 + pq - q^2$

29. $10x^3 + 52x^2 + 10x$

30. $49x^3 - 42x^2 + 9x$

31. $3x^2 + 5x - 4$

32. $75x^3 + 27x$

33. $3x^8 - 48y^8$

34. $14x^2 + 28 + 42x$

35. $x^5 - x^3 + x^2 - 1$

Solve.

36. $3x - 5 = 2x + 10$

37. $3y + 4 > 5y - 8$

38. $(x - 15)\left(x + \dfrac{1}{4}\right) = 0$

39. $-98x(x + 37) = 0$

40. $x^3 + x^2 = 25x + 25$

41. $2x^2 = 72$

42. $9x^2 + 1 = 6x$

43. $x^2 + 17x + 70 = 0$

44. $14y^2 = 21y$

45. $1.6 - 3.5x = 0.9$

46. $(x + 3)(x - 4) = 8$

47. $1.5x - 3.6 \leq 1.3x + 0.4$

48. $2x - [3x - (2x + 3)] = 3x + [4 - (2x + 1)]$

49. $y = mx + b$, for m

Solve.

50. The sum of two consecutive even integers is 102. Find the integers.

51. The product of two consecutive even integers is 360. Find the integers.

52. The length of a rectangular window is 3 ft longer than the height. The area of the window is 18 ft^2. Find the length and the height.

53. The length of a rectangular lot is 200 m longer than the width. The perimeter of the lot is 1000 m. Find the dimensions of the lot.

54. Money is borrowed at 12% simple interest. After 1 year, $7280 pays off the loan. How much was originally borrowed?

55. The length of one leg of a right triangle is 15 m. The length of the other leg is 9 m shorter than the length of the hypotenuse. Find the length of the hypotenuse.

56. A 100-m wire is cut into three pieces. The second piece is twice as long as the first piece. The third piece is one-third as long as the first piece. How long is each piece?

57. After a 25% price reduction, a pair of shoes is on sale for $21.75. What was the price before reduction?

58. The height of a triangle is 2 cm more than the base. The area of the triangle is 144 cm^2. Find the height and the base.

SYNTHESIS

Solve.

59. $(x + 3)(x - 5) \leq (x + 2)(x - 1)$

60. $\dfrac{x - 3}{2} - \dfrac{2x + 5}{26} = \dfrac{4x + 11}{13}$

61. $(x + 1)^2 = 25$

Factor.

62. $x^2(x - 3) - x(x - 3) - 2(x - 3)$

63. $4a^2 - 4a + 1 - 9b^2 - 24b - 16$

64. Find c so that the polynomial will be the square of a binomial: $cx^2 - 40x + 16$.

65. The length of the radius of a circle is increased by 2 cm to form a new circle. The area of the new circle is four times the area of the original circle. Find the length of the radius of the original circle.

INTRODUCTION In this chapter, we learn to manipulate rational expressions. We learn how to simplify rational expressions as well as to add, subtract, multiply, and divide them. Then we use these skills to solve equations, formulas, and problems.

The review sections to be tested in addition to the material in this chapter are 3.2, 3.4, 4.6, and 4.8. ❖

Rational Expressions and Equations

5

AN APPLICATION

One car travels 20 km/h faster than another. While one of them goes 240 km, the other goes 160 km. How can we use rational equations to find their speeds?

THE MATHEMATICS

Let r = the speed of the slow car. The problem translates to the following equation:

These are rational expressions.

$$\frac{160}{r} = \frac{240}{r + 20}.$$

This is a rational equation.

We now have a new type of equation that we must solve in order to solve the problem. This equation has a variable in a denominator and thus differs from the kinds of equations we have solved up to this point.

Motion Formulas: $\quad d = rt, \quad t = \dfrac{d}{r}, \quad r = \dfrac{d}{t}$

Equation-Solving Skills: Sections 2.1–2.3
Formula-Solving Skills: Section 2.6

PRETEST: CHAPTER 5

1. Find the LCM of $x^2 + 5x + 6$ and $x^2 + 6x + 9$.

Perform the indicated operations and simplify.

2. $\dfrac{b-1}{2-b} + \dfrac{b^2-3}{b^2-4}$

3. $\dfrac{4y-4}{y^2-y-2} - \dfrac{3y-5}{y^2-y-2}$

4. $\dfrac{4}{a+2} + \dfrac{3}{a}$

5. $\dfrac{x}{x+1} - \dfrac{x}{x-1} + \dfrac{2x^2}{x^2-1}$

6. $\dfrac{4x+8}{x+1} \cdot \dfrac{x^2-2x-3}{2x^2-8}$

7. Simplify: $\dfrac{\dfrac{1}{x} + \dfrac{1}{y}}{\dfrac{1}{x} - \dfrac{1}{y}}$.

Solve.

8. $\dfrac{1}{x+4} = \dfrac{5}{x}$

9. $\dfrac{3}{x-2} + \dfrac{x}{2} = \dfrac{6}{2x-4}$

10. Solve $R = \dfrac{1}{3}M(a-b)$ for M.

11. It takes 6 hr for a typist to address 200 envelopes. At this rate, how long would it take to address 350 envelopes?

12. One typist can type a report in 6 hr. Another typist can type the same report in 5 hr. How long would it take them to type the same report working together?

13. One car travels 20 mph faster than another. While one car travels 300 mi, the other travels 400 mi. Find their speeds.

5.1 Multiplying and Simplifying Rational Expressions

a Rational Expressions and Replacements

Rational numbers are quotients of integers. Some examples are

$$\frac{2}{3}, \quad \frac{4}{-5}, \quad \frac{-8}{17}, \quad \frac{563}{1}.$$

The following are called **rational expressions** or **fractional expressions.**
They are quotients of polynomials:

$$\frac{3}{4}, \quad \frac{5}{x+2}, \quad \frac{t^2 + 3t - 10}{7t^2 - 4}.$$

A rational expression is also a division. For example,

$$\frac{3}{4} \quad \text{means} \quad 3 \div 4 \quad \text{and} \quad \frac{x-8}{x+2} \quad \text{means} \quad (x-8) \div (x+2).$$

Because division is indicated by a rational expression, replacements of a
variable by a number that makes the denominator 0 cannot be allowed.
Such replacements are *not meaningful.* For example, in

$$\frac{x-8}{x+2},$$

-2 is not a meaningful replacement because it allows the denominator to
be 0, and division by 0 is not defined. The number 3 is a meaningful replacement because, as we see below, the denominator is nonzero and the
expression can be evaluated:

$$\frac{x-8}{x+2} = \frac{3-8}{3+2} = \frac{-5}{5} = -1.$$

The meaningful replacements are all real numbers except -2; that is, all x
such that $x \neq -2$.

▶ **EXAMPLE 1** Find the meaningful replacements in

$$\frac{x+7}{x^2 - 3x - 10}.$$

The meaningful replacements are all those real numbers for which the
denominator is not 0. To find them, we first find those that do make the
denominator 0. We set the denominator equal to 0 and solve:

$$x^2 - 3x - 10 = 0$$
$$(x-5)(x+2) = 0 \qquad \text{Factoring}$$
$$x - 5 = 0 \quad \text{or} \quad x + 2 = 0 \qquad \text{Using the principle of zero products}$$
$$x = 5 \quad \text{or} \qquad x = -2.$$

The meaningful replacements are all real numbers except 5 and -2; that
is, all x such that $x \neq 5$ and $x \neq -2$. ◄

DO EXERCISES 1 AND 2.

OBJECTIVES

After finishing Section 5.1, you should
be able to:

a Find the meaningful replacements
in a rational expression.

b Multiply a rational expression by 1,
using an expression such as *A/A.*

c Simplify rational expressions by
factoring the numerator and the
denominator and removing factors
of 1.

d Multiply rational expressions and
simplify.

FOR EXTRA HELP

Tape 10D	Tape 10A	MAC: 5
		IBM: 5

Find the meaningful replacements.

1. $\dfrac{16}{x-3}$

2. $\dfrac{2x-7}{x^2 + 5x - 24}$

ANSWERS ON PAGE A-5

Multiply.

3. $\dfrac{2x+1}{3x-2} \cdot \dfrac{x}{x}$

4. $\dfrac{x+1}{x-2} \cdot \dfrac{x+2}{x+2}$

5. $\dfrac{x-8}{x-y} \cdot \dfrac{-1}{-1}$

b Multiplying by 1

For rational expressions, multiplication is done as in arithmetic.

> **To multiply rational expressions, we multiply numerators and multiply denominators.**

For example,

$$\frac{x-2}{3} \cdot \frac{x+2}{x+7} = \frac{(x-2)(x+2)}{3(x+7)}.$$ Multiplying the numerators and the denominators

Note that we leave the numerator, $(x-2)(x+2)$, and the denominator, $3(x+7)$, in factored form because it is easier to simplify if we do not multiply. In order to learn to simplify, we first need to consider multiplying the rational expression by 1.

Any rational expression with the same numerator and denominator is a symbol for 1:

$$\frac{x+8}{x+8} = 1, \qquad \frac{3x^2-4}{3x^2-4} = 1, \qquad \frac{-1}{-1} = 1.$$

> **Expressions that have the same value for all meaningful replacements are called *equivalent expressions.***

We can multiply by 1 to obtain an equivalent expression.

▶ **EXAMPLES**   Multiply.

2. $\dfrac{3x+2}{x+1} \cdot \dfrac{2x}{2x} = \dfrac{(3x+2)2x}{(x+1)2x}$

3. $\dfrac{x+2}{x-7} \cdot \dfrac{x+3}{x+3} = \dfrac{(x+2)(x+3)}{(x-7)(x+3)}$

4. $\dfrac{2+x}{2-x} \cdot \dfrac{-1}{-1} = \dfrac{(2+x)(-1)}{(2-x)(-1)}$ ◀

DO EXERCISES 3–5.

c Simplifying Rational Expressions

We now consider simplifying rational expressions. What we do is similar to simplifying fractional expressions in arithmetic. But, instead of simplifying an expression like

$$\frac{16}{64},$$

we may simplify an expression like

$$\frac{x^2-16}{x+4}.$$

Just as factoring is important in simplifying in arithmetic, so too is it important in simplifying rational expressions. The factoring we use most is the factoring of polynomials, which we studied in Chapter 4.

To simplify, we can do the reverse of multiplying. We factor the numerator and the denominator and "remove" a factor of 1.

► **EXAMPLE 5** Simplify by removing a factor of 1: $\dfrac{8x^2}{24x}$.

$$\frac{8x^2}{24x} = \frac{8 \cdot x \cdot x}{3 \cdot 8 \cdot x} \qquad \text{Factoring the numerator and the denominator}$$

$$= \frac{8x}{8x} \cdot \frac{x}{3} \qquad \text{Factoring the rational expression}$$

$$= 1 \cdot \frac{x}{3} \qquad \frac{8x}{8x} = 1$$

$$= \frac{x}{3} \qquad \text{We removed a factor of 1.}$$

DO EXERCISES 6 AND 7.

► **EXAMPLES** Simplify by removing a factor of 1.

6. $\dfrac{5a + 15}{10} = \dfrac{5(a + 3)}{5 \cdot 2}$ Factoring the numerator and the denominator

$$= \frac{5}{5} \cdot \frac{a + 3}{2} \qquad \text{Factoring the rational expression}$$

$$= 1 \cdot \frac{a + 3}{2}$$

$$= \frac{a + 3}{2} \qquad \text{Removing a factor of 1: } \frac{5}{5} = 1$$

7. $\dfrac{6a + 12}{7a + 14} = \dfrac{6(a + 2)}{7(a + 2)}$ Factoring the numerator and the denominator

$$= \frac{6}{7} \cdot \frac{a + 2}{a + 2} \qquad \text{Factoring the rational expression}$$

$$= \frac{6}{7} \cdot 1$$

$$= \frac{6}{7} \qquad \text{Removing a factor of 1: } \frac{a + 2}{a + 2} = 1$$

8. $\dfrac{6x^2 + 4x}{2x^2 + 2x} = \dfrac{2x(3x + 2)}{2x(x + 1)}$ Factoring the numerator and the denominator

$$= \frac{2x}{2x} \cdot \frac{3x + 2}{x + 1} \qquad \text{Factoring the rational expression}$$

$$= 1 \cdot \frac{3x + 2}{x + 1}$$

$$= \frac{3x + 2}{x + 1} \qquad \text{Removing a factor of 1. Note in this step that you } \textit{cannot} \text{ remove the } x\text{'s because they are not factors of the entire numerator and the entire denominator.}$$

9. $\dfrac{x^2 + 3x + 2}{x^2 - 1} = \dfrac{(x + 2)(x + 1)}{(x + 1)(x - 1)}$

$$= \frac{x + 1}{x + 1} \cdot \frac{x + 2}{x - 1}$$

$$= 1 \cdot \frac{x + 2}{x - 1} = \frac{x + 2}{x - 1}$$

Simplify by removing a factor of 1.

6. $\dfrac{5y}{y}$

7. $\dfrac{9x^2}{36x}$

ANSWERS ON PAGE A-5

Simplify by removing a factor of 1.

8. $\dfrac{2x^2 + x}{3x^2 + 2x}$

9. $\dfrac{x^2 - 1}{2x^2 - x - 1}$

10. $\dfrac{7x + 14}{7}$

11. $\dfrac{12y + 24}{48}$

Simplify.

12. $\dfrac{x - 8}{8 - x}$

13. $\dfrac{a - b}{b - a}$

CANCELING. You may have encountered "canceling" when working with rational expressions. With great concern, we mention it as a possibility to speed up your work. Our concern is that canceling be done with care and understanding. Example 9 might have been done faster as follows:

$$\frac{x^2 + 3x + 2}{x^2 - 1} = \frac{(x + 2)(x + 1)}{(x + 1)(x - 1)} \quad \text{Factoring the numerator and the denominator}$$

$$= \frac{(x + 2)\cancel{(x + 1)}}{\cancel{(x + 1)}(x - 1)} \quad \begin{array}{l}\text{When a factor of 1 is noted, it is}\\[4pt]\text{"canceled" as shown: } \dfrac{x + 1}{x + 1} = 1.\end{array}$$

$$= \frac{x + 2}{x - 1} \quad \text{Simplifying}$$

> CAUTION! The difficulty with canceling is that it is applied incorrectly in situations such as the following:
>
> $$\frac{\cancel{2} + 3}{\cancel{2}} = 3; \qquad \frac{\cancel{4} + 1}{\cancel{4} + 2} = \frac{1}{2}; \qquad \frac{1\cancel{5}}{\cancel{5}4} = \frac{1}{4}.$$
>
>   Wrong!   Wrong!   Wrong!
>
> In each of these situations, the expressions canceled were *not* factors of 1. Factors are parts of products. For example, in $2 \cdot 3$, 2 and 3 are factors, but in $2 + 3$, 2 and 3 are *not* factors. If you can't factor, you can't cancel. If in doubt, don't cancel!

DO EXERCISES 8–11.

Consider

$$\frac{x - 4}{4 - x}.$$

At first glance the numerator and the denominator do not appear to have any common factors other than 1. But $x - 4$ and $4 - x$ are opposites, or additive inverses, of each other. Thus we can rewrite one as the inverse of the other.

▶ **EXAMPLE 10**   Simplify: $\dfrac{x - 4}{4 - x}$.

$$\frac{x - 4}{4 - x} = \frac{-1(-x + 4)}{4 - x}$$

$$= \frac{-1(4 - x)}{4 - x}$$

$$= -1 \cdot \frac{4 - x}{4 - x}$$

$$= -1 \cdot 1$$

$$= -1 \qquad\qquad ◀$$

DO EXERCISES 12 AND 13.

d ■ Multiplying and Simplifying

We try to simplify after we multiply. That is why we do not multiply out the numerator and the denominator too soon. We would need to factor them again anyway in order to simplify.

Multiply and simplify.

14. $\dfrac{a^2 - 4a + 4}{a^2 - 9} \cdot \dfrac{a + 3}{a - 2}$

▶ **EXAMPLE 11** Multiply and simplify: $\dfrac{5a^3}{4} \cdot \dfrac{2}{5a}$.

$$\frac{5a^3}{4} \cdot \frac{2}{5a} = \frac{5a^3(2)}{4(5a)} \qquad \text{Multiplying the numerators and the denominators}$$

$$= \frac{2 \cdot 5 \cdot a \cdot a \cdot a}{2 \cdot 2 \cdot 5 \cdot a} \qquad \text{Factoring the numerator and the denominator}$$

$$= \frac{2 \cdot 5 \cdot a \cdot a \cdot a}{2 \cdot 2 \cdot 5 \cdot a} \qquad \text{Removing a factor of 1: } \frac{2 \cdot 5 \cdot a}{2 \cdot 5 \cdot a} = 1$$

$$= \frac{a^2}{2} \qquad \text{Simplifying} \qquad \blacktriangleleft$$

▶ **EXAMPLE 12** Multiply and simplify: $\dfrac{x^2 + 6x + 9}{x^2 - 4} \cdot \dfrac{x - 2}{x + 3}$.

$$\frac{x^2 + 6x + 9}{x^2 - 4} \cdot \frac{x - 2}{x + 3} = \frac{(x^2 + 6x + 9)(x - 2)}{(x^2 - 4)(x + 3)} \qquad \text{Multiplying the numerators and the denominators}$$

$$= \frac{(x + 3)(x + 3)(x - 2)}{(x + 2)(x - 2)(x + 3)} \qquad \text{Factoring the numerator and the denominator}$$

$$= \frac{(x + 3)(x + 3)(x - 2)}{(x + 2)(x - 2)(x + 3)} \qquad \begin{array}{l} \text{Removing a factor of 1:} \\ \dfrac{(x + 3)(x - 2)}{(x + 3)(x - 2)} = 1 \end{array}$$

$$= \frac{x + 3}{x + 2} \qquad \text{Simplifying} \qquad \blacktriangleleft$$

15. $\dfrac{x^2 - 25}{6} \cdot \dfrac{3}{x + 5}$

▶ **EXAMPLE 13** Multiply and simplify: $\dfrac{x^2 + x - 2}{15} \cdot \dfrac{5}{2x^2 - 3x + 1}$.

$$\frac{x^2 + x - 2}{15} \cdot \frac{5}{2x^2 - 3x + 1} = \frac{(x^2 + x - 2)5}{15(2x^2 - 3x + 1)} \qquad \text{Multiplying the numerators and the denominators}$$

$$= \frac{(x + 2)(x - 1)5}{5(3)(x - 1)(2x - 1)} \qquad \text{Factoring the numerator and the denominator}$$

$$= \frac{(x + 2)(x - 1)5}{5(3)(x - 1)(2x - 1)} \qquad \begin{array}{l} \text{Removing a factor of 1:} \\ \dfrac{(x - 1)5}{(x - 1)5} = 1 \end{array}$$

$$= \underbrace{\frac{x + 2}{3(2x - 1)}}_{\uparrow} \qquad \text{Simplifying}$$

| You need not carry out this multiplication. |

$\blacktriangleleft$

DO EXERCISES 14 AND 15.

ANSWERS ON PAGE A-5

❖ SIDELIGHTS

Careers Involving Mathematics

You are about to finish this course in introductory algebra. If you have done well, you might be considering a career in mathematics or one that involves mathematics. If either is the case, the following information may be valuable to you.

CAREERS INVOLVING MATHEMATICS The following is the result of a survey conducted by *The Jobs Related Almanac*, published by the American References Inc., of Chicago. It used the criteria of salary, stress, work environment, outlook, security, and physical demands to rate the desirability of 250 jobs. The top 10 of the 250 jobs listed were:

1. Actuary
2. Computer programmer
3. Computer systems analyst   } The top 5 involve
4. Mathematician   mathematics.
5. Statistician
6. Hospital administrator
7. Industrial engineer
8. Physicist
9. Astrologer
10. Paralegal.

Two things are interesting to note. First, the top five rated professions involve a heavy use of mathematics. The top, actuary, involves the application of mathematics to insurance. The second point of interest is that choices like doctor, lawyer, and astronaut are *not* in the top ten.

Perhaps you might be interested in a career in teaching mathematics. This profession will be expanding increasingly in the next ten years. The field of mathematics will need well-qualified mathematics teachers in all areas from elementary to junior high to secondary to two-year college to college instruction. Some questions you might ask yourself in making a decision about a career in mathematics teaching are the following.

1. Do you find yourself carefully observing the strengths and weaknesses of your teachers?
2. Are you deeply interested in mathematics?
3. Are you interested in the ways of learning? If a student is struggling with a topic, would it be challenging

to you to discover two or three other ways to present the material so that the student might understand?
4. Are you able to put yourself in the place of the students in order to help them be successful in learning mathematics?

If you are interested in a career involving mathematics, the next courses you would take are *intermediate algebra, precalculus algebra and trigonometry* and *calculus*. You might want to seek out a counselor in the mathematics department at your college for further assistance.

WHAT KIND OF SALARIES ARE THERE IN VARIOUS FIELDS? The College Placement Council published the following comparisons of the average salaries of graduating students with bachelors degrees who were taking the following jobs:

Subject area	Annual salary
All engineering	$27,800
Computer science	$26,400
Mathematics	$25,900
Sciences other than math and computer science	$22,200
Humanities and social science	$21,800
Accounting	$21,700
All business	$21,300

Many people choose to go on to earn a masters degree. Here are salaries in the same fields for students just graduating with a masters degree:

Subject area	Annual salary
All engineering	$34,000
Computer science	$33,800
Mathematics	$27,900
Sciences other than math and computer science	$27,400
Humanities and social science	$22,300
Accounting	$26,000
Business administration	$30,700

NAME   SECTION   DATE

EXERCISE SET 5.1

a Find the meaningful replacements in the expression.

1. $\dfrac{-5}{2x}$

2. $\dfrac{14}{-5y}$

3. $\dfrac{a+7}{a-8}$

4. $\dfrac{a-8}{a+7}$

5. $\dfrac{3}{2y+5}$

6. $\dfrac{x^2-9}{4x-12}$

7. $\dfrac{x^2+11}{x^2-3x-28}$

8. $\dfrac{p^2-9}{p^2-7p+10}$

9. $\dfrac{m^3-2m}{m^2-25}$

10. $\dfrac{7-3x+x^2}{49-x^2}$

b Multiply. Do not simplify. Note that in each case you are multiplying by 1.

11. $\dfrac{3a}{3a}\cdot\dfrac{5a^2}{2c}$

12. $\dfrac{5x^2}{5x^2}\cdot\dfrac{6y^3}{3z^4}$

13. $\dfrac{2x}{2x}\cdot\dfrac{x-1}{x+4}$

14. $\dfrac{3y-1}{2y+1}\cdot\dfrac{y}{y}$

15. $\dfrac{-1}{-1}\cdot\dfrac{3-x}{4-x}$

16. $\dfrac{-1}{-1}\cdot\dfrac{x-5}{5-x}$

17. $\dfrac{y+6}{y+6}\cdot\dfrac{y-7}{y+2}$

18. $\dfrac{x-3}{x-3}\cdot\dfrac{x^2-4}{x^3+1}$

ANSWERS

1. _____

2. _____

3. _____

4. _____

5. _____

6. _____

7. _____

8. _____

9. _____

10. _____

11. _____

12. _____

13. _____

14. _____

15. _____

16. _____

17. _____

18. _____

19. _____

20. _____

21. _____

22. _____

23. _____

24. _____

25. _____

26. _____

27. _____

28. _____

29. _____

30. _____

31. _____

32. _____

33. _____

34. _____

35. _____

36. _____

37. _____

38. _____

C Simplify.

19. $\dfrac{8x^3}{32x}$

20. $\dfrac{6x^2}{18x}$

21. $\dfrac{48p^7q^5}{18p^5q^4}$

22. $\dfrac{-76x^8y^3}{-24x^4y^3}$

23. $\dfrac{4x - 12}{4x}$

24. $\dfrac{8y + 20}{8}$

25. $\dfrac{3m^2 + 3m}{6m^2 + 9m}$

26. $\dfrac{4y^2 - 2y}{5y^2 - 5y}$

27. $\dfrac{a^2 - 9}{a^2 + 5a + 6}$

28. $\dfrac{t^2 - 25}{t^2 + t - 20}$

29. $\dfrac{a^2 - 10a + 21}{a^2 - 11a + 28}$

30. $\dfrac{y^2 - 3y - 18}{y^2 - 2y - 15}$

31. $\dfrac{x^2 - 25}{x^2 - 10x + 25}$

32. $\dfrac{x^2 + 8x + 16}{x^2 - 16}$

33. $\dfrac{a^2 - 1}{a - 1}$

34. $\dfrac{t^2 - 1}{t + 1}$

35. $\dfrac{x^2 + 1}{x + 1}$

36. $\dfrac{y^2 + 4}{y + 2}$

37. $\dfrac{6x^2 - 54}{4x^2 - 36}$

38. $\dfrac{8x^2 - 32}{4x^2 - 16}$

39. $\dfrac{6t + 12}{t^2 - t - 6}$

40. $\dfrac{5y + 5}{y^2 + 7y + 6}$

41. $\dfrac{2t^2 + 6t + 4}{4t^2 - 12t - 16}$

42. $\dfrac{3a^2 - 9a - 12}{6a^2 + 30a + 24}$

43. $\dfrac{t^2 - 4}{(t + 2)^2}$

44. $\dfrac{(a - 3)^2}{a^2 - 9}$

45. $\dfrac{6 - x}{x - 6}$

46. $\dfrac{x - 8}{8 - x}$

47. $\dfrac{a - b}{b - a}$

48. $\dfrac{q - p}{-p + q}$

49. $\dfrac{6t - 12}{2 - t}$

50. $\dfrac{5a - 15}{3 - a}$

51. $\dfrac{a^2 - 1}{1 - a}$

52. $\dfrac{a^2 - b^2}{b^2 - a^2}$

d   Multiply and simplify.

53. $\dfrac{4x^3}{3x} \cdot \dfrac{14}{x}$

54. $\dfrac{32}{b^4} \cdot \dfrac{3b^2}{8}$

55. $\dfrac{3c}{d^2} \cdot \dfrac{4d}{6c^3}$

56. $\dfrac{3x^2y}{2} \cdot \dfrac{4}{xy^3}$

57. $\dfrac{x^2 - 3x - 10}{x^2 - 4x + 4} \cdot \dfrac{x - 2}{x - 5}$

58. $\dfrac{t^2}{t^2 - 4} \cdot \dfrac{t^2 - 5t + 6}{t^2 - 3t}$

ANSWERS

39. _____

40. _____

41. _____

42. _____

43. _____

44. _____

45. _____

46. _____

47. _____

48. _____

49. _____

50. _____

51. _____

52. _____

53. _____

54. _____

55. _____

56. _____

57. _____

58. _____

59. $\dfrac{a^2 - 9}{a^2} \cdot \dfrac{a^2 - 3a}{a^2 + a - 12}$

60. $\dfrac{x^2 + 10x - 11}{x^2 - 1} \cdot \dfrac{x + 1}{x + 11}$

61. $\dfrac{4a^2}{3a^2 - 12a + 12} \cdot \dfrac{3a - 6}{2a}$

62. $\dfrac{5v + 5}{v - 2} \cdot \dfrac{v^2 - 4v + 4}{v^2 - 1}$

63. $\dfrac{x^4 - 16}{x^4 - 1} \cdot \dfrac{x^2 + 1}{x^2 + 4}$

64. $\dfrac{t^4 - 1}{t^4 - 81} \cdot \dfrac{t^2 + 9}{t^2 + 1}$

65. $\dfrac{(t - 2)^3}{(t - 1)^3} \cdot \dfrac{t^2 - 2t + 1}{t^2 - 4t + 4}$

66. $\dfrac{(y + 4)^3}{(y + 2)^3} \cdot \dfrac{y^2 + 4y + 4}{y^2 + 8y + 16}$

67. $\dfrac{5a^2 - 180}{10a^2 - 10} \cdot \dfrac{20a + 20}{2a - 12}$

68. $\dfrac{2t^2 - 98}{4t^2 - 4} \cdot \dfrac{8t + 8}{16t - 112}$

SKILL MAINTENANCE

69. The product of two consecutive even integers is 360. Find the integers.

Factor.

70. $16 - t^4$ **71.** $2y^3 - 10y^2 + y - 5$ **72.** $x^5 - 2x^4 - 35x^3$

SYNTHESIS

Simplify.

73. $\dfrac{x^4 - 16y^4}{(x^2 + 4y^2)(x - 2y)}$

74. $\dfrac{(a - b)^2}{b^2 - a^2}$

75. $\dfrac{t^4 - 1}{t^4 - 81} \cdot \dfrac{t^2 - 9}{t^2 + 1} \cdot \dfrac{(t - 9)^2}{(t + 1)^2}$

76. $\dfrac{(t + 2)^3}{(t + 1)^3} \cdot \dfrac{t^2 + 2t + 1}{t^2 + 4t + 4} \cdot \dfrac{t + 1}{t + 2}$

77. $\dfrac{x^2 - y^2}{(x - y)^2} \cdot \dfrac{x^2 - 2xy + y^2}{x^2 - 4xy - 5y^2}$

78. $\dfrac{x - 1}{x^2 + 1} \cdot \dfrac{x^4 - 1}{(x - 1)^2} \cdot \dfrac{x^2 - 1}{x^4 - 2x^2 + 1}$

5.2 Division and Reciprocals

There is a similarity throughout this chapter between what we do with rational expressions and what we do with rational numbers. In fact, after replacements of variables by rational numbers, a rational expression represents a rational number.

a Finding Reciprocals

Two expressions are reciprocals of each other if their product is 1. The reciprocal of a rational expression is found by interchanging the numerator and the denominator.

► **EXAMPLES**

1. The reciprocal of $\frac{2}{5}$ is $\frac{5}{2}$. $\left(\text{This is because } \frac{2}{5} \cdot \frac{5}{2} = \frac{10}{10} = 1.\right)$

2. The reciprocal of $\frac{2x^2 - 3}{x + 4}$ is $\frac{x + 4}{2x^2 - 3}$.

3. The reciprocal of $x + 2$ is $\frac{1}{x + 2}$. $\left(\text{Think of } x + 2 \text{ as } \frac{x + 2}{1}.\right)$ ◄

DO EXERCISES 1–4.

b Division

> **To divide rational expressions, multiply by the reciprocal of the divisor. Then factor and simplify the result.**

► **EXAMPLES** Divide.

4. $\dfrac{3}{4} \div \dfrac{2}{5} = \dfrac{3}{4} \cdot \dfrac{5}{2}$ Multiplying by the reciprocal of the divisor

$= \dfrac{3 \cdot 5}{4 \cdot 2}$

$= \dfrac{15}{8}$

5. $\dfrac{x + 1}{x + 2} \div \dfrac{x - 1}{x + 3} = \dfrac{x + 1}{x + 2} \cdot \dfrac{x + 3}{x - 1}$ Multiplying by the reciprocal of the divisor

$= \dfrac{(x + 1)(x + 3)}{(x + 2)(x - 1)}$ ⎫ You need not carry out the multiplications in the numerator and the denominator. ◄

DO EXERCISES 5 AND 6.

OBJECTIVES

After finishing Section 5.2, you should be able to:

a Find the reciprocal of a rational expression.

b Divide rational expressions and simplify.

FOR EXTRA HELP

Tape 10E   Tape 10B   MAC: 5
                        IBM: 5

Find the reciprocal.

1. $\dfrac{7}{2}$

2. $\dfrac{x^2 + 5}{2x^3 - 1}$

3. $x - 5$

4. $\dfrac{1}{x^2 - 3}$

Divide.

5. $\dfrac{3}{5} \div \dfrac{7}{2}$

6. $\dfrac{x - 3}{x + 5} \div \dfrac{x + 5}{x - 2}$

ANSWERS ON PAGE A-5

Divide and simplify.

7. $\dfrac{x-3}{x+5} \div \dfrac{x+2}{x+5}$

8. $\dfrac{x^2-5x+6}{x+5} \div \dfrac{x+2}{x+5}$

9. $\dfrac{y^2-1}{y+1} \div \dfrac{y^2-2y+1}{y+1}$

▶ **EXAMPLE 6**   Divide and simplify: $\dfrac{x+1}{x^2-1} \div \dfrac{x+1}{x^2-2x+1}$.

$$\dfrac{x+1}{x^2-1} \div \dfrac{x+1}{x^2-2x+1}$$

$$= \dfrac{x+1}{x^2-1} \cdot \dfrac{x^2-2x+1}{x+1} \qquad \text{Multiplying by the reciprocal}$$

$$= \dfrac{(x+1)(x^2-2x+1)}{(x^2-1)(x+1)}$$

$$= \dfrac{(x+1)(x-1)(x-1)}{(x-1)(x+1)(x+1)} \qquad \begin{array}{l}\text{Factoring the numerator}\\\text{and the denominator}\end{array}$$

$$= \dfrac{(x+1)(x-1)(x-1)}{(x-1)(x+1)(x+1)} \qquad \text{Removing a factor of 1: } \dfrac{(x+1)(x-1)}{(x+1)(x-1)}=1$$

$$= \dfrac{x-1}{x+1} \qquad\qquad\qquad\qquad\qquad\qquad ◀$$

▶ **EXAMPLE 7**   Divide and simplify: $\dfrac{x^2-2x-3}{x^2-4} \div \dfrac{x+1}{x+5}$.

$$\dfrac{x^2-2x-3}{x^2-4} \div \dfrac{x+1}{x+5}$$

$$= \dfrac{x^2-2x-3}{x^2-4} \cdot \dfrac{x+5}{x+1} \qquad \text{Multiplying by the reciprocal}$$

$$= \dfrac{(x^2-2x-3)(x+5)}{(x^2-4)(x+1)}$$

$$= \dfrac{(x-3)(x+1)(x+5)}{(x-2)(x+2)(x+1)} \qquad \begin{array}{l}\text{Factoring the numerator}\\\text{and the denominator}\end{array}$$

$$= \dfrac{(x-3)(x+1)(x+5)}{(x-2)(x+2)(x+1)} \qquad \text{Removing a factor of 1: } \dfrac{x+1}{x+1}=1$$

$$= \dfrac{(x-3)(x+5)}{(x-2)(x+2)} \Big\}$$

> You need not carry out the multiplications in the numerator and the denominator.

$$\qquad\qquad\qquad\qquad\qquad\qquad\qquad\qquad ◀$$

DO EXERCISES 7–9.

ANSWERS ON PAGE A-5

EXERCISE SET 5.2

a Find the reciprocal.

1. $\dfrac{4}{x}$

2. $\dfrac{a+3}{a-1}$

3. $x^2 - y^2$

4. $\dfrac{1}{a+b}$

5. $\dfrac{x^2+2x-5}{x^2-4x+7}$

6. $\dfrac{x^2-3xy+y^2}{x^2+7xy-y^2}$

b Divide and simplify.

7. $\dfrac{2}{5} \div \dfrac{4}{3}$

8. $\dfrac{5}{6} \div \dfrac{2}{3}$

9. $\dfrac{2}{x} \div \dfrac{8}{x}$

10. $\dfrac{x}{2} \div \dfrac{3}{x}$

11. $\dfrac{x^2}{y} \div \dfrac{x^3}{y^3}$

12. $\dfrac{a}{b^2} \div \dfrac{a^2}{b^3}$

13. $\dfrac{a+2}{a-3} \div \dfrac{a-1}{a+3}$

14. $\dfrac{y+2}{4} \div \dfrac{y}{2}$

15. $\dfrac{x^2-1}{x} \div \dfrac{x+1}{x-1}$

16. $\dfrac{4y-8}{y+2} \div \dfrac{y-2}{y^2-4}$

17. $\dfrac{x+1}{6} \div \dfrac{x+1}{3}$

18. $\dfrac{a}{a-b} \div \dfrac{b}{a-b}$

19. $\dfrac{5x-5}{16} \div \dfrac{x-1}{6}$

20. $\dfrac{-4+2x}{8} \div \dfrac{x-2}{2}$

21. $\dfrac{-6+3x}{5} \div \dfrac{4x-8}{25}$

ANSWERS

1. ___
2. ___
3. ___
4. ___
5. ___
6. ___
7. ___
8. ___
9. ___
10. ___
11. ___
12. ___
13. ___
14. ___
15. ___
16. ___
17. ___
18. ___
19. ___
20. ___
21. ___

22. _____

23. _____

24. _____

25. _____

26. _____

27. _____

28. _____

29. _____

30. _____

31. _____

32. _____

33. _____

34. _____

35. _____

36. _____

37. _____

38. _____

39. _____

40. _____

22. $\dfrac{-12 + 4x}{4} \div \dfrac{-6 + 2x}{6}$

23. $\dfrac{a + 2}{a - 1} \div \dfrac{3a + 6}{a - 5}$

24. $\dfrac{t - 3}{t + 2} \div \dfrac{4t - 12}{t + 1}$

25. $\dfrac{x^2 - 4}{x} \div \dfrac{x - 2}{x + 2}$

26. $\dfrac{x + y}{x - y} \div \dfrac{x^2 + y}{x^2 - y^2}$

27. $\dfrac{x^2 - 9}{4x + 12} \div \dfrac{x - 3}{6}$

28. $\dfrac{x - b}{2x} \div \dfrac{x^2 - b^2}{5x^2}$

29. $\dfrac{c^2 + 3c}{c^2 + 2c - 3} \div \dfrac{c}{c + 1}$

30. $\dfrac{x - 5}{2x} \div \dfrac{x^2 - 25}{4x^2}$

31. $\dfrac{2y^2 - 7y + 3}{2y^2 + 3y - 2} \div \dfrac{6y^2 - 5y + 1}{3y^2 + 5y - 2}$

32. $\dfrac{x^2 - x - 20}{x^2 + 7x + 12} \div \dfrac{x^2 - 10x + 25}{x^2 + 6x + 9}$

33. $\dfrac{x^2 - 1}{4x + 4} \div \dfrac{2x^2 - 4x + 2}{8x + 8}$

34. $\dfrac{5x^2 + 5x - 30}{10x + 30} \div \dfrac{2x^2 - 8}{6x^2 + 36x + 54}$

SKILL MAINTENANCE

35. Sixteen more than the square of a number is eight times the number. Find the number.

36. Subtract:

$$(8x^3 - 3x^2 + 7) - (8x^2 + 3x - 5).$$

SYNTHESIS

Simplify.

37. $\dfrac{3a^2 - 5ab - 12b^2}{3ab + 4b^2} \div (3b^2 - ab)$

38. $\dfrac{3x^2 - 2xy - y^2}{x^2 - y^2} \div 3x^2 + 4xy + y^2$

39. $\dfrac{3x + 3y + 3}{9x} \div \left(\dfrac{x^2 + 2xy + y^2 - 1}{x^4 + x^2} \right)$

40. $\left(\dfrac{y^2 + 5y + 6}{y^2} \cdot \dfrac{3y^3 + 6y^2}{y^2 - y - 12} \right) \div \dfrac{y^2 - y}{y^2 - 2y - 8}$

5.3 Least Common Multiples and Denominators

a Least Common Multiples

To add when denominators are different, we first find a common denominator. For example, to add $\frac{5}{12}$ and $\frac{7}{30}$, we first look for the **least common multiple, LCM,** of both 12 and 30. That number becomes the **least common denominator, LCD.** To find the LCM of 12 and 30, we factor:

$$12 = 2 \cdot 2 \cdot 3;$$
$$30 = 2 \cdot 3 \cdot 5.$$

The LCM is the number that has 2 as a factor twice, 3 as a factor once, and 5 as a factor once:

$$\text{LCM} = 2 \cdot 2 \cdot 3 \cdot 5, \text{ or } 60.$$

> **To find the LCM, use each factor the greatest number of times that it appears in any one factorization.**

▶ **EXAMPLE 1** Find the LCM of 24 and 36.

$$\left.\begin{array}{l} 24 = 2 \cdot 2 \cdot 2 \cdot 3 \\ 36 = 2 \cdot 2 \cdot 3 \cdot 3 \end{array}\right\} \quad \text{LCM} = 2 \cdot 2 \cdot 2 \cdot 3 \cdot 3, \text{ or } 72 \qquad ◀$$

DO EXERCISES 1–4.

b Adding Using the LCD

Let us finish adding $\frac{5}{12}$ and $\frac{7}{30}$:

$$\frac{5}{12} + \frac{7}{30} = \frac{5}{2 \cdot 2 \cdot 3} + \frac{7}{2 \cdot 3 \cdot 5}.$$

The least common denominator, LCD, is $2 \cdot 2 \cdot 3 \cdot 5$. To get the LCD in the first denominator, we need a 5. To get the LCD in the second denominator, we need another 2. We get these numbers by multiplying by 1:

$$\frac{5}{12} + \frac{7}{30} = \frac{5}{2 \cdot 2 \cdot 3} \cdot \frac{5}{5} + \frac{7}{2 \cdot 3 \cdot 5} \cdot \frac{2}{2} \quad \text{Multiplying by 1}$$

$$= \frac{25}{2 \cdot 2 \cdot 3 \cdot 5} + \frac{14}{2 \cdot 3 \cdot 5 \cdot 2} \quad \begin{array}{l}\text{The denominators are}\\\text{now the LCD.}\end{array}$$

$$= \frac{39}{2 \cdot 2 \cdot 3 \cdot 5} \quad \begin{array}{l}\text{Adding the numerators}\\\text{and keeping the LCD}\end{array}$$

$$= \frac{3 \cdot 13}{2 \cdot 2 \cdot 3 \cdot 5}$$

$$= \frac{13}{20} \quad \text{Simplifying}$$

OBJECTIVES

After finishing Section 5.3, you should be able to:

a Find the LCM of several numbers by factoring.

b Add fractions, first finding the LCD.

c Find the LCM of algebraic expressions by factoring.

FOR EXTRA HELP

Tape 11A   Tape 10B   MAC: 5
                          IBM: 5

Find the LCM by factoring.

1. 16, 18

2. 6, 12

3. 2, 5

4. 24, 30, 20

Add, first finding the LCM of the denominators. Simplify if possible.

5. $\dfrac{3}{16} + \dfrac{1}{18}$

6. $\dfrac{1}{6} + \dfrac{1}{12}$

7. $\dfrac{1}{2} + \dfrac{3}{5}$

8. $\dfrac{1}{24} + \dfrac{1}{30} + \dfrac{3}{20}$

Find the LCM.

9. $12xy^2, \ 15x^3y$

10. $y^2 + 5y + 4, \ y^2 + 2y + 1$

11. $t^2 + 16, \ t - 2, \ 7$

12. $x^2 + 2x + 1, \ 3x^2 - 3x, \ x^2 - 1$

▶ **EXAMPLE 2**   Add: $\dfrac{5}{12} + \dfrac{11}{18}$.

$$\left.\begin{array}{l} 12 = 2 \cdot 2 \cdot 3 \\ 18 = 2 \cdot 3 \cdot 3 \end{array}\right\} \quad \text{LCD} = 2 \cdot 2 \cdot 3 \cdot 3, \text{ or } 36.$$

$$\frac{5}{12} + \frac{11}{18} = \frac{5}{2 \cdot 2 \cdot 3} \cdot \frac{3}{3} + \frac{11}{2 \cdot 3 \cdot 3} \cdot \frac{2}{2} = \frac{37}{2 \cdot 2 \cdot 3 \cdot 3} = \frac{37}{36} \quad ◀$$

DO EXERCISES 5–8.

⌐C⌐   LCMs of Algebraic Expressions

To find the LCM of two or more algebraic expressions, we factor them. Then we use each factor the greatest number of times it occurs in any one expression.

▶ **EXAMPLE 3**   Find the LCM of $12x$, $16y$, and $8xyz$.

$$\left.\begin{array}{l} 12x = 2 \cdot 2 \cdot 3 \cdot x \\ 16y = 2 \cdot 2 \cdot 2 \cdot 2 \cdot y \\ 8xyz = 2 \cdot 2 \cdot 2 \cdot x \cdot y \cdot z \end{array}\right\} \quad \begin{array}{l} \text{LCM} = 2 \cdot 2 \cdot 2 \cdot 2 \cdot 3 \cdot x \cdot y \cdot z \\ \qquad\quad = 48xyz \end{array} \quad ◀$$

▶ **EXAMPLE 4**   Find the LCM of $x^2 + 5x - 6$ and $x^2 - 1$.

$$\left.\begin{array}{r} x^2 + 5x - 6 = (x + 6)(x - 1) \\ x^2 - 1 = (x + 1)(x - 1) \end{array}\right\} \quad \text{LCM} = (x + 6)(x - 1)(x + 1) \quad ◀$$

▶ **EXAMPLE 5**   Find the LCM of $x^2 + 4$, $x + 1$, and 5.

These expressions are not factorable, so the LCM is their product:

$$5(x^2 + 4)(x + 1). \quad ◀$$

▶ **EXAMPLE 6**   Find the LCM of $x^2 - 25$ and $2x - 10$.

$$\left.\begin{array}{r} x^2 - 25 = (x + 5)(x - 5) \\ 2x - 10 = 2(x - 5) \end{array}\right\} \quad \text{LCM} = 2(x + 5)(x - 5) \quad ◀$$

▶ **EXAMPLE 7**   Find the LCM of $x^2 - 4y^2$, $x^2 - 4xy + 4y^2$, and $x - 2y$.

$$\left.\begin{array}{r} x^2 - 4y^2 = (x - 2y)(x + 2y) \\ x^2 - 4xy + 4y^2 = (x - 2y)(x - 2y) \\ x - 2y = x - 2y \end{array}\right\} \quad \begin{array}{l} \text{LCM} = (x + 2y)(x - 2y)(x - 2y) \\ \qquad\quad = (x + 2y)(x - 2y)^2 \end{array} \quad ◀$$

DO EXERCISES 9–12.

ANSWERS ON PAGE A-5

EXERCISE SET 5.3

a Find the LCM.

1. 12, 27

2. 10, 15

3. 8, 9

4. 12, 15

5. 6, 9, 21

6. 8, 36, 40

7. 24, 36, 40

8. 3, 4, 5

9. 28, 42, 60

10. 10, 100, 500

b Add, first finding the LCD. Simplify if possible.

11. $\dfrac{7}{24} + \dfrac{11}{18}$

12. $\dfrac{7}{60} + \dfrac{6}{75}$

13. $\dfrac{1}{6} + \dfrac{3}{40}$

14. $\dfrac{5}{24} + \dfrac{3}{20}$

15. $\dfrac{2}{15} + \dfrac{5}{9} + \dfrac{3}{20}$

16. $\dfrac{1}{20} + \dfrac{1}{30} + \dfrac{2}{45}$

c Find the LCM.

17. $6x^2,\ 12x^3$

18. $2a^2b,\ 8ab^2$

19. $2x^2,\ 6xy,\ 18y^2$

20. $c^2d,\ cd^2,\ c^3d$

21. $2(y-3),\ 6(y-3)$

22. $4(x-1),\ 8(x-1)$

23. $t,\ t+2,\ t-2$

24. $x,\ x+3,\ x-3$

25. $x^2-4,\ x^2+5x+6$

26. $x^2+3x+2,\ x^2-4$

ANSWERS

1. _____

2. _____

3. _____

4. _____

5. _____

6. _____

7. _____

8. _____

9. _____

10. _____

11. _____

12. _____

13. _____

14. _____

15. _____

16. _____

17. _____

18. _____

19. _____

20. _____

21. _____

22. _____

23. _____

24. _____

25. _____

26. _____

27. $t^3 + 4t^2 + 4t,\ t^2 - 4t$

28. $y^3 - y^2,\ y^4 - y^2$

29. $a + 1,\ (a - 1)^2,\ a^2 - 1$

30. $x^2 - y^2,\ 2x + 2y,\ x^2 + 2xy + y^2$

31. $m^2 - 5m + 6,\ m^2 - 4m + 4$

32. $2x^2 + 5x + 2,\ 2x^2 - x - 1$

33. $2 + 3x,\ 4 - 9x^2,\ 2 - 3x$

34. $3 - 2x,\ 9 - 4x^2,\ 3 + 2x$

35. $10v^2 + 30v,\ 5v^2 + 35v + 60$

36. $12a^2 + 24a,\ 4a^2 + 20a + 24$

37. $9x^3 - 9x^2 - 18x,\ 6x^5 - 24x^4 + 24x^3$

38. $x^5 - 4x^3,\ x^3 + 4x^2 + 4x$

39. $x^5 + 4x^4 + 4x^3,\ 3x^2 - 12,\ 2x + 4$

40. $x^5 + 2x^4 + x^3,\ 2x^3 - 2x,\ 5x - 5$

SKILL MAINTENANCE

Factor.

41. $x^2 - 6x + 9$

42. $6x^2 + 4x$

43. $x^2 - 9$

44. $x^2 + 4x - 21$

SYNTHESIS

Find the LCM.

45. 72, 90, 96

46. $8x^2 - 8,\ 6x^2 - 12x + 6,\ 10x - 10$

47. Two joggers leave the starting point of a circular course at the same time. One jogger completes one round in 6 min and the second jogger in 8 min. Assuming they continue to run at the same pace, after how many minutes will they meet again at the starting place?

48. If the LCM of two expressions is the same as one of the expressions, what is their relationship?

5.4 Adding Rational Expressions

a We add rational expressions as we do rational numbers.

> **To add when the denominators are the same, add the numerators and keep the same denominator.**

▶ **EXAMPLES**   Add.

1. $\dfrac{x}{x+1} + \dfrac{2}{x+1} = \dfrac{x+2}{x+1}$

2. $\dfrac{2x^2+3x-7}{2x+1} + \dfrac{x^2+x-8}{2x+1} = \dfrac{(2x^2+3x-7)+(x^2+x-8)}{2x+1}$

$$= \dfrac{3x^2+4x-15}{2x+1}$$

3. $\dfrac{x-5}{x^2-9} + \dfrac{2}{x^2-9} = \dfrac{(x-5)+2}{x^2-9} = \dfrac{x-3}{x^2-9}$

$$= \dfrac{x-3}{(x-3)(x+3)} \qquad \text{Factoring}$$

$$= \dfrac{\cancel{x-3}}{\cancel{(x-3)}(x+3)} \qquad \text{Removing a factor of 1: } \dfrac{x-3}{x-3}=1$$

$$= \dfrac{1}{x+3} \qquad \text{Simplifying} \qquad ◀$$

As in Example 3, simplifying should be done if possible after adding.

DO EXERCISES 1–3.

When denominators are not the same, we multiply by 1 to obtain equivalent expressions with the same denominator. When one denominator is the opposite of the other, we can first multiply either expression by $-1/-1$.

▶ **EXAMPLES**

4. $\dfrac{x}{2} + \dfrac{3}{-2} = \dfrac{x}{2} + \dfrac{-1}{-1} \cdot \dfrac{3}{-2}$ Multiplying by $\dfrac{-1}{-1}$

$$= \dfrac{x}{2} + \dfrac{-3}{2} \qquad \text{The denominators are now the same.}$$

$$= \dfrac{x+(-3)}{2} = \dfrac{x-3}{2}$$

5. $\dfrac{3x+4}{x-2} + \dfrac{x-7}{2-x} = \dfrac{3x+4}{x-2} + \dfrac{-1}{-1} \cdot \dfrac{x-7}{2-x}$

> We could have chosen to multiply this expression by $-1/-1$. We multiply only one expression, *not* both.

$$= \dfrac{3x+4}{x-2} + \dfrac{-x+7}{x-2} \qquad \textit{Note: } -1(2-x)=-2+x$$
$$\phantom{= \dfrac{3x+4}{x-2} + \dfrac{-x+7}{x-2} \qquad \textit{Note: }} = x-2$$

$$= \dfrac{(3x+4)+(-x+7)}{x-2} = \dfrac{2x+11}{x-2} \qquad ◀$$

DO EXERCISES 4 AND 5.

ANSWERS ON PAGE A-5

OBJECTIVE

After finishing Section 5.4, you should be able to:

a Add rational expressions.

FOR EXTRA HELP

Tape 11B   Tape 11A   MAC: 5
                                    IBM: 5

Add.

1. $\dfrac{5}{9} + \dfrac{2}{9}$

2. $\dfrac{3}{x-2} + \dfrac{x}{x-2}$

3. $\dfrac{4x+5}{x-1} + \dfrac{2x-1}{x-1}$

Add.

4. $\dfrac{x}{4} + \dfrac{5}{-4}$

5. $\dfrac{2x+1}{x-3} + \dfrac{x+2}{3-x}$

Add.

6. $\dfrac{3x}{16} + \dfrac{5x^2}{24}$

7. $\dfrac{3}{16x} + \dfrac{5}{24x^2}$

When denominators are different, we find the least common denominator, LCD. The procedure we will use is as follows.

> **To add rational expressions with different denominators:**
> 1. **Find the LCM of the denominators. This is the least common denominator (LCD).**
> 2. **For each rational expression, find an equivalent expression with the LCD. To do so, multiply by 1 using an expression for 1 made up of factors of the LCD missing from the original denominator.**
> 3. **Add the numerators. Write the sum over the LCD.**
> 4. **Simplify, if possible.**

▶ **EXAMPLE 6**   Add: $\dfrac{5x^2}{8} + \dfrac{7x}{12}$.

First, we find the LCD:

$$\left.\begin{array}{l} 8 = 2 \cdot 2 \cdot 2 \\ 12 = 2 \cdot 2 \cdot 3 \end{array}\right\} \quad \text{LCD} = 2 \cdot 2 \cdot 2 \cdot 3, \text{ or } 24.$$

The factor of the LCD missing from 8 is 3. The factor of the LCD missing from 12 is 2. We multiply by 1 to get the LCD in each expression, and then add and simplify, if possible.

$$\begin{aligned}
\frac{5x^2}{8} + \frac{7x}{12} &= \frac{5x^2}{2 \cdot 2 \cdot 2} + \frac{7x}{2 \cdot 2 \cdot 3} \\
&= \frac{5x^2}{2 \cdot 2 \cdot 2} \cdot \frac{3}{3} + \frac{7x}{2 \cdot 2 \cdot 3} \cdot \frac{2}{2} \quad \begin{array}{l}\text{\small\bfseries Multiplying by 1 to get}\\\text{\small\bfseries the same denominators}\end{array} \\
&= \frac{15x^2}{24} + \frac{14x}{24} = \frac{15x^2 + 14x}{24}.
\end{aligned}$$ ◀

▶ **EXAMPLE 7**   Add: $\dfrac{3}{8x} + \dfrac{5}{12x^2}$.

First, we find the LCD:

$$\left.\begin{array}{l} 8x = 2 \cdot 2 \cdot 2 \cdot x \\ 12x^2 = 2 \cdot 2 \cdot 3 \cdot x \cdot x \end{array}\right\} \quad \text{LCD} = 2 \cdot 2 \cdot 2 \cdot 3 \cdot x \cdot x, \text{ or } 24x^2.$$

The factors of the LCD missing from $8x$ are 3 and x. The factor of the LCD missing from $12x^2$ is 2. We multiply by 1 to get the LCD in each expression, and then add and simplify, if possible:

$$\begin{aligned}
\frac{3}{8x} + \frac{5}{12x^2} &= \frac{3}{8x} \cdot \frac{3 \cdot x}{3 \cdot x} + \frac{5}{12x^2} \cdot \frac{2}{2} \\
&= \frac{9x}{24x^2} + \frac{10}{24x^2} = \frac{9x + 10}{24x^2}.
\end{aligned}$$ ◀

DO EXERCISES 6 AND 7.

▶ **EXAMPLE 8** Add: $\dfrac{2a}{a^2 - 1} + \dfrac{1}{a^2 + a}$.

First, we find the LCD:

$$\left.\begin{array}{l} a^2 - 1 = (a - 1)(a + 1) \\ a^2 + a = a(a + 1) \end{array}\right\} \quad \text{LCD} = a(a - 1)(a + 1).$$

We multiply by 1 to get the LCD in each expression, and then add and simplify:

$$\dfrac{2a}{(a - 1)(a + 1)} \cdot \dfrac{a}{a} + \dfrac{1}{a(a + 1)} \cdot \dfrac{a - 1}{a - 1}$$

$$= \dfrac{2a^2}{a(a - 1)(a + 1)} + \dfrac{a - 1}{a(a - 1)(a + 1)}$$

$$= \dfrac{2a^2 + a - 1}{a(a - 1)(a + 1)}$$

$$= \dfrac{(a + 1)(2a - 1)}{a(a - 1)(a + 1)} \qquad \text{Factoring the numerator in order to simplify}$$

$$= \dfrac{(a + 1)(2a - 1)}{a(a - 1)(a + 1)} \qquad \text{Removing a factor of 1: } \dfrac{a + 1}{a + 1} = 1$$

$$= \dfrac{2a - 1}{a(a - 1)}. \qquad \blacktriangleleft$$

DO EXERCISE 8.

▶ **EXAMPLE 9** Add: $\dfrac{x + 4}{x - 2} + \dfrac{x - 7}{x + 5}$.

First, we find the LCD. It is just the product of the denominators:

$$\text{LCD} = (x - 2)(x + 5).$$

We multiply by 1 to get the LCD in each expression, and then add and simplify:

$$\dfrac{x + 4}{x - 2} \cdot \dfrac{x + 5}{x + 5} + \dfrac{x - 7}{x + 5} \cdot \dfrac{x - 2}{x - 2} = \dfrac{(x + 4)(x + 5)}{(x - 2)(x + 5)} + \dfrac{(x - 7)(x - 2)}{(x - 2)(x + 5)}$$

$$= \dfrac{x^2 + 9x + 20}{(x - 2)(x + 5)} + \dfrac{x^2 - 9x + 14}{(x - 2)(x + 5)}$$

$$= \dfrac{x^2 + 9x + 20 + x^2 - 9x + 14}{(x - 2)(x + 5)}$$

$$= \dfrac{2x^2 + 34}{(x - 2)(x + 5)}. \qquad \blacktriangleleft$$

DO EXERCISE 9.

8. Add:

$$\dfrac{3}{x^3 - x} + \dfrac{4}{x^2 + 2x + 1}.$$

9. Add:

$$\dfrac{x - 2}{x + 3} + \dfrac{x + 7}{x + 8}.$$

ANSWERS ON PAGE A-5

10. Add:

$$\frac{5}{x^2 + 17x + 16} + \frac{3}{x^2 + 9x + 8}.$$

▶ **EXAMPLE 10**   Add: $\dfrac{x}{x^2 + 11x + 30} + \dfrac{-5}{x^2 + 9x + 20}$.

$$\frac{x}{x^2 + 11x + 30} + \frac{-5}{x^2 + 9x + 20}$$

$$= \frac{x}{(x+5)(x+6)} + \frac{-5}{(x+5)(x+4)}$$ Factoring the denominators in order to find the LCM. The LCD is $(x+4)(x+5)(x+6)$.

$$= \frac{x}{(x+5)(x+6)} \cdot \frac{x+4}{x+4} + \frac{-5}{(x+5)(x+4)} \cdot \frac{x+6}{x+6}$$ Multiplying by 1

$$= \frac{x(x+4) + (-5)(x+6)}{(x+4)(x+5)(x+6)} = \frac{x^2 + 4x - 5x - 30}{(x+4)(x+5)(x+6)}$$

$$= \frac{x^2 - x - 30}{(x+4)(x+5)(x+6)}$$

$$\left. \begin{array}{c} = \dfrac{(x-6)(x+5)}{(x+4)(x+5)(x+6)} \\[2em] = \dfrac{(x-6)}{(x+4)(x+6)} \end{array} \right\} \longrightarrow$$ Always simplify at the end if possible: $\dfrac{x+5}{x+5} = 1$. ◀

DO EXERCISE 10.

Suppose that after we factor to find the LCD, we find factors that are opposites. There are several ways to handle this, but the easiest is to first go back and multiply by $-1/-1$ appropriately to change factors so they are not opposites.

11. Add:

$$\frac{x+3}{x^2 - 16} + \frac{5}{12 - 3x}.$$

▶ **EXAMPLE 11**   Add: $\dfrac{x}{x^2 - 25} + \dfrac{3}{10 - 2x}$.

First, we factor as though we are going to find the LCD:

$$x^2 - 25 = (x - 5)(x + 5);$$
$$10 - 2x = 2(5 - x).$$

We note that there is an $x - 5$ as one factor and a $5 - x$ as another factor. If the denominator of the second expression were $2x - 10$, this situation would not arise. To avoid this, we first multiply by $-1/-1$ and continue as before:

$$\frac{x}{x^2 - 25} + \frac{3}{10 - 2x} = \frac{x}{(x-5)(x+5)} + \frac{-1}{-1} \cdot \frac{3}{10 - 2x}$$

$$= \frac{x}{(x-5)(x+5)} + \frac{-3}{2x - 10}$$

$$= \frac{x}{(x-5)(x+5)} + \frac{-3}{2(x-5)}$$ LCD $= 2(x-5)(x+5)$

$$= \frac{x}{(x-5)(x+5)} \cdot \frac{2}{2} + \frac{-3}{2(x-5)} \cdot \frac{x+5}{x+5}$$

$$= \frac{2x - 3(x+5)}{2(x-5)(x+5)} = \frac{2x - 3x - 15}{2(x-5)(x+5)}$$ Collecting like terms

$$= \frac{-x - 15}{2(x-5)(x+5)}$$ ◀

DO EXERCISE 11.

NAME   SECTION   DATE

EXERCISE SET 5.4

a Add. Simplify, if possible.

1. $\dfrac{5}{12} + \dfrac{7}{12}$

2. $\dfrac{3}{14} + \dfrac{5}{14}$

3. $\dfrac{1}{3+x} + \dfrac{5}{3+x}$

4. $\dfrac{4x+1}{6x+5} + \dfrac{3x-7}{5+6x}$

5. $\dfrac{x^2+7x}{x^2-5x} + \dfrac{x^2-4x}{x^2-5x}$

6. $\dfrac{a}{x+y} + \dfrac{b}{y+x}$

7. $\dfrac{7}{8} + \dfrac{5}{-8}$

8. $\dfrac{11}{6} + \dfrac{5}{-6}$

9. $\dfrac{3}{t} + \dfrac{4}{-t}$

10. $\dfrac{5}{-a} + \dfrac{8}{a}$

11. $\dfrac{2x+7}{x-6} + \dfrac{3x}{6-x}$

12. $\dfrac{3x-2}{4x-3} + \dfrac{2x-5}{3-4x}$

13. $\dfrac{y^2}{y-3} + \dfrac{9}{3-y}$

14. $\dfrac{t^2}{t-2} + \dfrac{4}{2-t}$

15. $\dfrac{b-7}{b^2-16} + \dfrac{7-b}{16-b^2}$

16. $\dfrac{a-3}{a^2-25} + \dfrac{a-3}{25-a^2}$

17. $\dfrac{z}{(y+z)(y-z)} + \dfrac{y}{(z+y)(z-y)}$

[*Hint:* Multiply by $-1/-1$. Note that $(z+y)(z-y)(-1) = (z+y)(y-z)$.]

18. $\dfrac{a^2}{a-b} + \dfrac{b^2}{b-a}$

19. $\dfrac{x+3}{x-5} + \dfrac{2x-1}{5-x} + \dfrac{2(3x-1)}{x-5}$

20. $\dfrac{3(x-2)}{2x-3} + \dfrac{5(2x+1)}{2x-3} + \dfrac{3(x+1)}{3-2x}$

21. $\dfrac{2(4x+1)}{5x-7} + \dfrac{3(x-2)}{7-5x} + \dfrac{-10x-1}{5x-7}$

22. $\dfrac{5(x-2)}{3x-4} + \dfrac{2(x-3)}{4-3x} + \dfrac{3(5x+1)}{4-3x}$

23. $\dfrac{x+1}{(x+3)(x-3)} + \dfrac{4(x-3)}{(x-3)(x+3)} + \dfrac{(x-1)(x-3)}{(3-x)(x+3)}$

24. $\dfrac{2(x+5)}{(2x-3)(x-1)} + \dfrac{3x+4}{(2x-3)(1-x)} + \dfrac{x-5}{(3-2x)(x-1)}$

25. $\dfrac{2}{x} + \dfrac{5}{x^2}$

26. $\dfrac{4}{x} + \dfrac{8}{x^2}$

27. $\dfrac{5}{6r} + \dfrac{7}{8r}$

28. $\dfrac{2}{9t} + \dfrac{11}{6t}$

29. $\dfrac{4}{xy^2} + \dfrac{6}{x^2 y}$

30, $\dfrac{2}{c^2 d} + \dfrac{7}{cd^3}$

31. $\dfrac{2}{9t^3} + \dfrac{1}{6t^2}$

32. $\dfrac{-2}{3xy^2} + \dfrac{6}{x^2 y^3}$

33. $\dfrac{x+y}{xy^2} + \dfrac{3x+y}{x^2 y}$

34. $\dfrac{2c-d}{c^2 d} + \dfrac{c+d}{cd^2}$

35. $\dfrac{3}{x-2} + \dfrac{3}{x+2}$

36. $\dfrac{2}{x-1} + \dfrac{2}{x+1}$

37. $\dfrac{3}{x+1} + \dfrac{2}{3x}$

38. $\dfrac{2}{x+5} + \dfrac{3}{4x}$

39. $\dfrac{2x}{x^2-16} + \dfrac{x}{x-4}$

40. $\dfrac{4x}{x^2-25} + \dfrac{x}{x+5}$

41. $\dfrac{5}{z+4} + \dfrac{3}{3z+12}$

42. $\dfrac{t}{t-3} + \dfrac{5}{4t-12}$

43. $\dfrac{3}{x-1} + \dfrac{2}{(x-1)^2}$

44. $\dfrac{2}{x+3} + \dfrac{4}{(x+3)^2}$

45. $\dfrac{4a}{5a-10} + \dfrac{3a}{10a-20}$

46. $\dfrac{3a}{4a-20} + \dfrac{9a}{6a-30}$

47. $\dfrac{x+4}{x} + \dfrac{x}{x+4}$

48. $\dfrac{x}{x-5} + \dfrac{x-5}{x}$

49. $\dfrac{x}{x^2+2x+1} + \dfrac{1}{x^2+5x+4}$

50. $\dfrac{7}{a^2+a-2} + \dfrac{5}{a^2-4a+3}$

51. $\dfrac{x+3}{x-5} + \dfrac{x-5}{x+3}$

52. $\dfrac{3x}{2y-3} + \dfrac{2x}{3y-2}$

53. $\dfrac{a}{a^2-1} + \dfrac{2a}{a^2-a}$

54. $\dfrac{3x+2}{3x+6} + \dfrac{x-2}{x^2-4}$

ANSWERS

37. _____

38. _____

39. _____

40. _____

41. _____

42. _____

43. _____

44. _____

45. _____

46. _____

47. _____

48. _____

49. _____

50. _____

51. _____

52. _____

53. _____

54. _____

55. $\dfrac{6}{x-y} + \dfrac{4x}{y^2 - x^2}$

56. $\dfrac{a-2}{3-a} + \dfrac{4 - a^2}{a^2 - 9}$

57. $\dfrac{y+2}{y-7} + \dfrac{3-y}{49 - y^2}$

58. $\dfrac{4-p}{25 - p^2} + \dfrac{p+1}{p-5}$

59. $\dfrac{10}{x^2 + x - 6} + \dfrac{3x}{x^2 - 4x + 4}$

60. $\dfrac{2}{z^2 - z - 6} + \dfrac{3}{z^2 - 9}$

SKILL MAINTENANCE

Subtract.

61. $(x^2 + x) - (x + 1)$

62. $(4y^3 - 5y^2 + 7y - 24) - (-9y^3 + 9y^2 - 5y + 49)$

Simplify.

63. $(2x^4 y^3)^{-3}$

64. $\left(\dfrac{x^3}{5y}\right)^2$

SYNTHESIS

Find the perimeter and the area of the figure.

65.

66.

Add. Simplify, if possible.

67. $\dfrac{5}{z+2} + \dfrac{4z}{z^2 - 4} + 2$

68. $\dfrac{-2}{y^2 - 9} + \dfrac{4y}{(y-3)^2} + \dfrac{6}{3-y}$

69. $\dfrac{3z^2}{z^4 - 4} + \dfrac{5z^2 - 3}{2z^4 + z^2 - 6}$

70. Find an expression equivalent to

$$\dfrac{a - 3b}{a - b}$$

that is a sum of two fractional expressions. Answers can vary.

5.5 Subtracting Rational Expressions

a We subtract rational expressions as we do rational numbers.

> **To subtract when the denominators are the same, subtract the numerators and keep the same denominator.**

▶ **EXAMPLE 1** Subtract: $\dfrac{3x}{x+2} - \dfrac{x-2}{x+2}$.

$$\frac{3x}{x+2} - \frac{x-2}{x+2} = \frac{3x - (x-2)}{x+2}$$

> The parentheses are important to make sure that you subtract the entire numerator.

$$= \frac{3x - x + 2}{x+2} = \frac{2x+2}{x+2}$$ ◀

DO EXERCISES 1 AND 2.

When one denominator is the additive inverse of the other, we can first multiply one expression by $-1/-1$ to obtain a common denominator.

▶ **EXAMPLE 2** Subtract: $\dfrac{x}{5} - \dfrac{3x-4}{-5}$.

$$\frac{x}{5} - \frac{3x-4}{-5} = \frac{x}{5} - \frac{-1}{-1} \cdot \frac{3x-4}{-5}$$

Multiplying by $\dfrac{-1}{-1}$ ←

$$= \frac{x}{5} - \frac{(-1)(3x-4)}{(-1)(-5)}$$

> This is equal to 1 (not -1).

$$= \frac{x}{5} - \frac{4-3x}{5}$$

Remember the parentheses!

$$= \frac{x - (4-3x)}{5}$$

$$= \frac{x - 4 + 3x}{5} = \frac{4x-4}{5}$$ ◀

▶ **EXAMPLE 3** Subtract: $\dfrac{5y}{y-5} - \dfrac{2y-3}{5-y}$.

$$\frac{5y}{y-5} - \frac{2y-3}{5-y} = \frac{5y}{y-5} - \frac{-1}{-1} \cdot \frac{2y-3}{5-y}$$

$$= \frac{5y}{y-5} - \frac{(-1)(2y-3)}{(-1)(5-y)} = \frac{5y}{y-5} - \frac{3-2y}{y-5}$$

Remember the parentheses!

$$= \frac{5y - (3-2y)}{y-5}$$

$$= \frac{5y - 3 + 2y}{y-5}$$

$$= \frac{7y-3}{y-5}$$ ◀

OBJECTIVES

After finishing Section 5.5, you should be able to:

a Subtract rational expressions.

b Simplify combined additions and subtractions of rational expressions.

FOR EXTRA HELP

Tape 11C Tape 11A MAC: 5
                    IBM: 5

Subtract.

1. $\dfrac{7}{11} - \dfrac{3}{11}$

2. $\dfrac{2x^2 + 3x - 7}{2x+1} - \dfrac{x^2 + x - 8}{2x+1}$

ANSWERS ON PAGE A-5

Subtract.

3. $\dfrac{x}{3} - \dfrac{2x-1}{-3}$

4. $\dfrac{3x}{x-2} - \dfrac{x-3}{2-x}$

5. Subtract:

$$\dfrac{x-2}{3x} - \dfrac{2x-1}{5x}.$$

DO EXERCISES 3 AND 4.

To subtract rational expressions with different denominators, we use a procedure similar to what we used for addition, except that we subtract numerators and write the difference over the LCD.

To subtract rational expressions with different denominators:

1. **Find the LCM of the denominators. This is the least common denominator (LCD).**
2. **For each rational expression, find an equivalent expression with the LCD. To do so, multiply by 1 using a symbol for 1 made up of factors of the LCD missing from the original denominator.**
3. **Subtract the numerators. Write the difference over the LCD.**
4. **Simplify, if possible.**

▶ **EXAMPLE 4**   Subtract: $\dfrac{x+2}{x-4} - \dfrac{x+1}{x+4}$.

The LCM $= (x-4)(x+4)$.

$$\dfrac{x+2}{x-4} \cdot \dfrac{x+4}{x+4} - \dfrac{x+1}{x+4} \cdot \dfrac{x-4}{x-4}$$

$$= \dfrac{(x+2)(x+4)}{(x-4)(x+4)} - \dfrac{(x+1)(x-4)}{(x-4)(x+4)}$$

$$= \dfrac{x^2+6x+8}{(x-4)(x+4)} - \dfrac{x^2-3x-4}{(x-4)(x+4)}$$

$$= \dfrac{x^2+6x+8-(x^2-3x-4)}{(x-4)(x+4)}$$

Subtracting this numerator. Don't forget the parentheses.

$$= \dfrac{x^2+6x+8-x^2+3x+4}{(x-4)(x+4)}$$

$$= \dfrac{9x+12}{(x-4)(x+4)}$$ ◀

DO EXERCISE 5.

▶ **EXAMPLE 5**   Subtract: $\dfrac{x}{x^2+5x+6} - \dfrac{2}{x^2+3x+2}$.

$$\dfrac{x}{x^2+5x+6} - \dfrac{2}{x^2+3x+2}$$

$$= \dfrac{x}{(x+2)(x+3)} - \dfrac{2}{(x+2)(x+1)}$$

$$= \dfrac{x}{(x+2)(x+3)} \cdot \dfrac{x+1}{x+1} - \dfrac{2}{(x+2)(x+1)} \cdot \dfrac{x+3}{x+3}$$

LCD $= (x+1)(x+2)(x+3)$

$$= \dfrac{x^2+x}{(x+1)(x+2)(x+3)} - \dfrac{2x+6}{(x+1)(x+2)(x+3)}$$

Then

Subtracting this numerator.
Don't forget the parentheses.

$$= \frac{x^2 + x - (2x + 6)}{(x + 1)(x + 2)(x + 3)}$$

$$= \frac{x^2 + x - 2x - 6}{(x + 1)(x + 2)(x + 3)}$$

$$= \frac{x^2 - x - 6}{(x + 1)(x + 2)(x + 3)}$$

$$= \frac{(x + 2)(x - 3)}{(x + 1)(x + 2)(x + 3)}$$

$$= \frac{(x + 2)(x - 3)}{(x + 1)(x + 2)(x + 3)} \qquad \text{Simplifying by removing a factor of 1: } \frac{x + 2}{x + 2} = 1$$

$$= \frac{x - 3}{(x + 1)(x + 3)} \qquad \blacktriangleleft$$

DO EXERCISE 6.

Suppose that after we factor to find the LCD, we find factors that are opposites. Then we multiply by $-1/-1$ appropriately to change factors so they are not opposites.

▶ **EXAMPLE 6**   Subtract: $\dfrac{p}{64 - p^2} - \dfrac{5}{p - 8}$.

Factoring $64 - p^2$, we get $(8 - p)(8 + p)$. Note that the factor $8 - p$ in the first denominator and $p - 8$ in the second denominator are opposites. We multiply the first expression by $-1/-1$ to avoid this situation. Then we proceed as before.

$$\frac{p}{64 - p^2} - \frac{5}{p - 8} = \frac{-1}{-1} \cdot \frac{p}{64 - p^2} - \frac{5}{p - 8}$$

$$= \frac{-p}{p^2 - 64} - \frac{5}{p - 8}$$

$$= \frac{-p}{(p - 8)(p + 8)} - \frac{5}{p - 8} \qquad \text{LCD} = (p - 8)(p + 8)$$

$$= \frac{-p}{(p - 8)(p + 8)} - \frac{5}{p - 8} \cdot \frac{p + 8}{p + 8}$$

$$= \frac{-p}{(p - 8)(p + 8)} - \frac{5p + 40}{(p - 8)(p + 8)}$$

Subtracting this numerator.
Don't forget the parentheses.

$$= \frac{-p - (5p + 40)}{(p - 8)(p + 8)}$$

$$= \frac{-p - 5p - 40}{(p - 8)(p + 8)}$$

$$= \frac{-6p - 40}{(p - 8)(p + 8)}$$

DO EXERCISE 7.

6. Subtract:

$$\frac{x}{x^2 + 15x + 56} - \frac{6}{x^2 + 13x + 42}.$$

7. Subtract:

$$\frac{y}{16 - y^2} - \frac{7}{y - 4}.$$

ANSWERS ON PAGE A-5

8. Perform the indicated operations and simplify:

$$\frac{x+2}{x^2-9} - \frac{x-7}{9-x^2} + \frac{-8-x}{x^2-9}.$$

9. Perform the indicated operations and simplify:

$$\frac{1}{x} - \frac{5}{3x} + \frac{2x}{x+1}.$$

b **Combined Additions and Subtractions**

▶ **EXAMPLE 7**   Perform the indicated operations and simplify:

$$\frac{x+9}{x^2-4} + \frac{5-x}{4-x^2} - \frac{2+x}{x^2-4}.$$

We have

$$\frac{x+9}{x^2-4} + \frac{5-x}{4-x^2} - \frac{2+x}{x^2-4} = \frac{x+9}{x^2-4} + \frac{-1}{-1}\cdot\frac{5-x}{4-x^2} - \frac{2+x}{x^2-4}$$

$$= \frac{x+9}{x^2-4} + \frac{(-1)(5-x)}{(-1)(4-x^2)} - \frac{2+x}{x^2-4}$$

$$= \frac{(x+9)+(-5+x)-(2+x)}{x^2-4}$$

$$= \frac{x+9-5+x-2-x}{x^2-4}$$

$$= \frac{x+2}{x^2-4}$$

$$= \frac{(x+2)\cdot 1}{(x+2)(x-2)}$$

$$= \frac{1}{x-2}. \qquad \frac{x+2}{x+2}=1$$ ◀

DO EXERCISE 8.

▶ **EXAMPLE 8**   Perform the indicated operations and simplify:

$$\frac{1}{x} - \frac{1}{x^2} + \frac{2}{x+1}.$$

The LCD $= x \cdot x(x+1)$, or $x^2(x+1)$.

$$\frac{1}{x}\cdot\frac{x(x+1)}{x(x+1)} - \frac{1}{x^2}\cdot\frac{(x+1)}{(x+1)} + \frac{2}{x+1}\cdot\frac{x^2}{x^2}$$

$$= \frac{x(x+1)}{x^2(x+1)} - \frac{x+1}{x^2(x+1)} + \frac{2x^2}{x^2(x+1)}$$

$$= \frac{x(x+1)-(x+1)+2x^2}{x^2(x+1)}$$ Subtract this numerator. Don't forget the parentheses.

$$= \frac{x^2+x-x-1+2x^2}{x^2(x+1)}$$

$$= \frac{3x^2-1}{x^2(x+1)}$$ ◀

DO EXERCISE 9.

EXERCISE SET 5.5

ANSWERS

a Subtract. Simplify, if possible.

1. $\dfrac{7}{8} - \dfrac{3}{8}$

2. $\dfrac{5}{y} - \dfrac{7}{y}$

3. $\dfrac{x}{x-1} - \dfrac{1}{x-1}$

4. $\dfrac{x^2}{x+4} - \dfrac{16}{x+4}$

5. $\dfrac{x+1}{x^2-2x+1} - \dfrac{5-3x}{x^2-2x+1}$

6. $\dfrac{2x-3}{x^2+3x-4} - \dfrac{x-7}{x^2+3x-4}$

7. $\dfrac{11}{6} - \dfrac{5}{-6}$

8. $\dfrac{7}{8} - \dfrac{5}{-8}$

9. $\dfrac{5}{a} - \dfrac{8}{-a}$

10. $\dfrac{3}{t} - \dfrac{4}{-t}$

11. $\dfrac{x}{4} - \dfrac{3x-5}{-4}$

12. $\dfrac{2}{x-1} - \dfrac{2}{1-x}$

13. $\dfrac{3-x}{x-7} - \dfrac{2x-5}{7-x}$

14. $\dfrac{t^2}{t-2} - \dfrac{4}{2-t}$

1. _____

2. _____

3. _____

4. _____

5. _____

6. _____

7. _____

8. _____

9. _____

10. _____

11. _____

12. _____

13. _____

14. _____

Copyright © 1991 Addison-Wesley Publishing Co., Inc.

ANSWERS

15. _____

16. _____

17. _____

18. _____

19. _____

20. _____

21. _____

22. _____

23. _____

24. _____

25. _____

26. _____

27. _____

28. _____

15. $\dfrac{x-8}{x^2-16} - \dfrac{x-8}{16-x^2}$

16. $\dfrac{x-2}{x^2-25} - \dfrac{6-x}{25-x^2}$

17. $\dfrac{4-x}{x-9} - \dfrac{3x-8}{9-x}$

18. $\dfrac{3-x}{x-7} - \dfrac{2x-5}{7-x}$

19. $\dfrac{2(x-1)}{2x-3} - \dfrac{3(x+2)}{2x-3} - \dfrac{x-1}{3-2x}$

20. $\dfrac{3(x-2)}{2x-3} - \dfrac{5(2x+1)}{2x-3} - \dfrac{3(x-1)}{3-2x}$

Subtract. Simplify, if possible.

21. $\dfrac{x-2}{6} - \dfrac{x+1}{3}$

22. $\dfrac{a+2}{2} - \dfrac{a-4}{4}$

23. $\dfrac{4z-9}{3z} - \dfrac{3z-8}{4z}$

24. $\dfrac{x-1}{4x} - \dfrac{2x+3}{x}$

25. $\dfrac{4x+2t}{3xt^2} - \dfrac{5x-3t}{x^2t}$

26. $\dfrac{5x+3y}{2x^2y} - \dfrac{3x+4y}{xy^2}$

27. $\dfrac{5}{x+5} - \dfrac{3}{x-5}$

28. $\dfrac{2z}{z-1} - \dfrac{3z}{z+1}$

29. $\dfrac{3}{2t^2 - 2t} - \dfrac{5}{2t - 2}$

30. $\dfrac{8}{x^2 - 4} - \dfrac{3}{x + 2}$

31. $\dfrac{2s}{t^2 - s^2} - \dfrac{s}{t - s}$

32. $\dfrac{3}{12 + x - x^2} - \dfrac{2}{x^2 - 9}$

33. $\dfrac{y - 5}{y} - \dfrac{3y - 1}{4y}$

34. $\dfrac{3x - 2}{4x} - \dfrac{3x + 1}{6x}$

35. $\dfrac{a}{x + a} - \dfrac{a}{x - a}$

36. $\dfrac{t}{y - t} - \dfrac{y}{y + t}$

37. $\dfrac{8x}{16 - x^2} - \dfrac{5}{x - 4}$

38. $\dfrac{5x}{x^2 - 9} - \dfrac{4}{3 - x}$

39. $\dfrac{t^2}{2t^2 - 2t} - \dfrac{1}{2t - 2}$

40. $\dfrac{4}{5b^2 - 5b} - \dfrac{3}{5b - 5}$

41. $\dfrac{x}{x^2 + 5x + 6} - \dfrac{2}{x^2 + 3x + 2}$

42. $\dfrac{x}{x^2 + 11x + 30} - \dfrac{5}{x^2 + 9x + 20}$

29. _____

30. _____

31. _____

32. _____

33. _____

34. _____

35. _____

36. _____

37. _____

38. _____

39. _____

40. _____

41. _____

42. _____

Copyright © 1991 Addison-Wesley Publishing Co., Inc.

ANSWERS

43. _____

44. _____

45. _____

46. _____

47. _____

48. _____

49. _____

50. _____

51. _____

52. _____

53. _____

54. _____

55. _____

56. _____

57. _____

58. _____

59. _____

60. _____

b Perform the indicated operations and simplify.

43. $\dfrac{3(2x+5)}{x-1} - \dfrac{3(2x-3)}{1-x} + \dfrac{6x-1}{x-1}$

44. $\dfrac{2x-y}{x-y} + \dfrac{x-2y}{y-x} - \dfrac{3x-3y}{x-y}$

45. $\dfrac{x-y}{x^2-y^2} + \dfrac{x+y}{x^2-y^2} - \dfrac{2x}{x^2-y^2}$

46. $\dfrac{x+y}{2(x-y)} - \dfrac{2x-2y}{2(x-y)} + \dfrac{x-3y}{2(y-x)}$

47. $\dfrac{10}{2y-1} - \dfrac{6}{1-2y} + \dfrac{y}{2y-1} + \dfrac{y-4}{1-2y}$

48. $\dfrac{(x+1)(2x-1)}{(2x-3)(x-3)} - \dfrac{(x-3)(x+1)}{(3-x)(3-2x)} + \dfrac{(2x+1)(x+3)}{(3-2x)(x-3)}$

49. $\dfrac{4y}{y^2-1} - \dfrac{2}{y} - \dfrac{2}{y+1}$

50. $\dfrac{x+6}{4-x^2} - \dfrac{x+3}{x+2} + \dfrac{x-3}{2-x}$

51. $\dfrac{2z}{1-2z} + \dfrac{3z}{2z+1} - \dfrac{3}{4z^2-1}$

52. $\dfrac{1}{x+y} + \dfrac{1}{x-y} - \dfrac{2x}{x^2-y^2}$

53. $\dfrac{1}{x+y} - \dfrac{1}{x-y} + \dfrac{2x}{x^2-y^2}$

54. $\dfrac{1}{a-b} - \dfrac{1}{a+b} + \dfrac{2b}{a^2-b^2}$

SKILL MAINTENANCE

Simplify.

55. $\dfrac{x^8}{x^3}$

56. $3x^4 \cdot 10x^8$

57. $(a^2b^{-5})^{-4}$

58. $\dfrac{54x^{10}}{3x^7}$

SYNTHESIS

Subtract. Simplify, if possible.

59. $\dfrac{5}{3-2x} + \dfrac{3}{2x-3} - \dfrac{x-3}{2x^2-x-3}$

60. $\dfrac{2r}{r^2-s^2} + \dfrac{1}{r+s} - \dfrac{1}{r-s}$

5.6 Solving Rational Equations

a In Sections 5.1–5.5, we studied operations with rational expressions. These were expressions not having an equals sign. We cannot clear expressions of fractions other than occasionally when simplifying by removing a factor of 1. In this section, we are studying rational equations. Equations do have an equals sign, and we can clear of fractions as we did in Section 2.3.

A **rational**, or **fractional**, **equation** is an equation containing one or more rational expressions. Here are some examples:

$$\frac{2}{3} + \frac{5}{6} = \frac{x}{9}, \qquad x + \frac{6}{x} = -5, \qquad \frac{x^2}{x-1} = \frac{1}{x-1}.$$

> **To solve a rational equation, the first step is to clear the equation of fractions. To do this, multiply both sides of the equation by the LCM of all the denominators. Then carry out the equation-solving process as we learned it in Chapter 2.**

When clearing an equation of fractions, we use the terminology LCM instead of LCD because we are *not* adding or subtracting rational expressions.

▶ **EXAMPLE 1** Solve: $\frac{2}{3} + \frac{5}{6} = \frac{x}{9}$.

The LCM of all denominators is $2 \cdot 3 \cdot 3$, or 18. We multiply on both sides by 18:

$$\frac{2}{3} + \frac{5}{6} = \frac{x}{9}$$

$$18\left(\frac{2}{3} + \frac{5}{6}\right) = 18 \cdot \frac{x}{9} \qquad \text{Multiplying on both sides by the LCM}$$

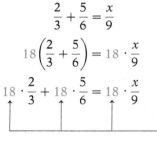
$$18 \cdot \frac{2}{3} + 18 \cdot \frac{5}{6} = 18 \cdot \frac{x}{9} \qquad \text{Multiplying to remove parentheses}$$

> When clearing an equation of fractions, be sure to multiply *each* rational expression in the equation by the LCM.

$$12 + 15 = 2x \qquad \text{Simplifying. Note that we have now cleared of fractions.}$$

$$27 = 2x$$

$$\frac{27}{2} = x. \qquad \blacktriangleleft$$

DO EXERCISE 1.

OBJECTIVE

After finishing Section 5.6, you should be able to:

a Solve rational equations.

FOR EXTRA HELP

Tape 11D   Tape 11B   MAC: 5
                                IBM: 5

1. Solve: $\frac{3}{4} + \frac{5}{8} = \frac{x}{12}$.

2. Solve: $\dfrac{1}{x} = \dfrac{1}{6 - x}$.

▶ **EXAMPLE 2**   Solve: $\dfrac{1}{x} = \dfrac{1}{4 - x}$.

The LCM is $x(4 - x)$. We multiply on both sides by $x(4 - x)$:

$$\frac{1}{x} = \frac{1}{4 - x}$$

$$x(4 - x) \cdot \frac{1}{x} = x(4 - x) \cdot \frac{1}{4 - x} \qquad \text{Multiplying on both sides by the LCM}$$

$$4 - x = x \qquad \text{Simplifying}$$

$$4 = 2x$$

$$x = 2.$$

Check:

$$\frac{1}{x} = \frac{1}{4 - x}$$

$$\begin{array}{c|c} \dfrac{1}{2} & \dfrac{1}{4 - 2} \\ \hline & \dfrac{1}{2} \quad \text{TRUE} \end{array}$$

This checks, so 2 is the solution. ◀

DO EXERCISE 2.

3. Solve: $\dfrac{x}{4} - \dfrac{x}{6} = \dfrac{1}{8}$.

▶ **EXAMPLE 3**   Solve: $\dfrac{x}{6} - \dfrac{x}{8} = \dfrac{1}{12}$.

The LCM is 24. We multiply on both sides by 24:

$$\frac{x}{6} - \frac{x}{8} = \frac{1}{12}$$

$$24\left(\frac{x}{6} - \frac{x}{8}\right) = 24 \cdot \frac{1}{12} \qquad \text{Multiplying on both sides by the LCM}$$

$$24 \cdot \frac{x}{6} - 24 \cdot \frac{x}{8} = 24 \cdot \frac{1}{12} \qquad \text{Multiplying to remove parentheses}$$

> Be sure to multiply *each* term by the LCM.

$$4x - 3x = 2 \qquad \text{Simplifying}$$

$$x = 2.$$

Check:

$$\frac{x}{6} - \frac{x}{8} = \frac{1}{12}$$

$$\begin{array}{c|c} \dfrac{2}{6} - \dfrac{2}{8} & \dfrac{1}{12} \\[2mm] \dfrac{1}{3} - \dfrac{1}{4} & \\[2mm] \dfrac{4}{12} - \dfrac{3}{12} & \\[2mm] \dfrac{1}{12} & \text{TRUE} \end{array}$$

This checks, so the solution is 2. ◀

DO EXERCISE 3.

▶ **EXAMPLE 4**   Solve: $\dfrac{2}{3x} + \dfrac{1}{x} = 10$.

The LCM is $3x$. We multiply on both sides by $3x$:

$$\frac{2}{3x} + \frac{1}{x} = 10$$

$$3x\left(\frac{2}{3x} + \frac{1}{x}\right) = 3x \cdot 10 \qquad \text{Multiplying on both sides by the LCM}$$

$$3x \cdot \frac{2}{3x} + 3x \cdot \frac{1}{x} = 3x \cdot 10 \qquad \text{Multiplying to remove parentheses}$$

$$2 + 3 = 30x \qquad \text{Simplifying}$$

$$5 = 30x$$

$$\frac{5}{30} = x$$

$$\frac{1}{6} = x.$$

We leave the check to the student. The solution is $\frac{1}{6}$.   ◀

DO EXERCISE 4.

▶ **EXAMPLE 5**   Solve: $x + \dfrac{6}{x} = -5$.

The LCM is x. We multiply on both sides by x:

$$x + \frac{6}{x} = -5$$

$$x\left(x + \frac{6}{x}\right) = -5x \qquad \text{Multiplying on both sides by } x$$

$$x \cdot x + x \cdot \frac{6}{x} = -5x \qquad \begin{array}{l}\text{Note that each rational expression} \\ \text{on the left is now multiplied by } x.\end{array}$$

$$x^2 + 6 = -5x \qquad \text{Simplifying}$$

$$x^2 + 5x + 6 = 0 \qquad \text{Subtracting } 5x \text{ to get a 0 on one side}$$

$$(x + 3)(x + 2) = 0 \qquad \text{Factoring}$$

$$x + 3 = 0 \quad \text{or} \quad x + 2 = 0 \qquad \text{Using the principle of zero products}$$

$$x = -3 \quad \text{or} \qquad x = -2.$$

Check: For -3:   For -2:

$$\begin{array}{c|c} x + \dfrac{6}{x} = -5 & x + \dfrac{6}{x} = -5 \\ \hline -3 + \dfrac{6}{-3} \ \bigg|\ -5 & -2 + \dfrac{6}{-2} \ \bigg|\ -5 \\ -3 - 2 & -2 - 3 \\ -5 \quad \text{TRUE} & -5 \quad \text{TRUE} \end{array}$$

Both of these check, so there are two solutions, -3 and -2.   ◀

DO EXERCISE 5.

4. Solve: $\dfrac{1}{2x} + \dfrac{1}{x} = -12$.

5. Solve: $x + \dfrac{1}{x} = 2$.

ANSWERS ON PAGE A-5

6. Solve: $\dfrac{x^2}{x+2} = \dfrac{4}{x+2}$.

When we multiply on both sides of an equation by the LCM, we might not get equivalent equations. Thus we must *always* check possible solutions in the original equation.

1. If you have carried out all algebraic procedures correctly, you need only check to see if a number is a meaningful replacement in all parts of the original equation.

2. To be sure that no computational errors have been made and that you indeed have a solution, a complete check is necessary, as we did in Chapter 2.

The next example will illustrate the importance of the preceding comments.

▶ **EXAMPLE 6**   Solve: $\dfrac{x^2}{x-1} = \dfrac{1}{x-1}$.

The LCM is $x - 1$. We multiply on both sides by $x - 1$:

$$\frac{x^2}{x-1} = \frac{1}{x-1}$$

$$(x-1)\cdot\frac{x^2}{x-1} = (x-1)\cdot\frac{1}{x-1} \qquad \text{Multiplying on both sides by } x-1$$

$$x^2 = 1 \qquad \text{Simplifying}$$

$$x^2 - 1 = 0 \qquad \text{Subtracting 1 to get a 0 on one side}$$

$$(x-1)(x+1) = 0 \qquad \text{Factoring}$$

$$x - 1 = 0 \quad \text{or} \quad x + 1 = 0 \qquad \text{Using the principle of zero products}$$

$$x = 1 \quad \text{or} \qquad x = -1.$$

The numbers 1 and -1 are possible solutions. We look at the original equation and see that 1 is not a meaningful replacement because it makes a denominator zero. The number -1 checks and is a solution.   ◀

DO EXERCISE 6.

7. Solve: $\dfrac{4}{x-2} + \dfrac{1}{x+2} = \dfrac{26}{x^2-4}$.

▶ **EXAMPLE 7**   Solve: $\dfrac{3}{x-5} + \dfrac{1}{x+5} = \dfrac{2}{x^2-25}$.

The LCM is $(x - 5)(x + 5)$. We multiply on both sides by $(x - 5)(x + 5)$:

$$(x-5)(x+5)\left(\frac{3}{x-5} + \frac{1}{x+5}\right) = (x-5)(x+5)\left(\frac{2}{x^2-25}\right)$$

$$\text{Multiplying on both sides by the LCM}$$

$$(x-5)(x+5)\cdot\frac{3}{x-5} + (x-5)(x+5)\cdot\frac{1}{x+5} = (x-5)(x+5)\cdot\frac{2}{x^2-25}$$

$$3(x+5) + (x-5) = 2 \qquad \text{Simplifying}$$

$$3x + 15 + x - 5 = 2 \qquad \text{Removing parentheses}$$

$$4x + 10 = 2$$

$$4x = -8$$

$$x = -2.$$

CAUTION! We have introduced a new use of the LCM in this section. You previously used the LCM in adding or subtracting rational expressions. *Now* we have equations with equals signs. We clear of fractions by multiplying on both sides of the equation by the LCM. This eliminates the denominators. Do *not* make the mistake of trying to "clear of fractions" when you do not have an equation.

The check is left to the student. The number -2 checks and is the solution.

DO EXERCISE 7.   ◀

NAME   SECTION   DATE

EXERCISE SET 5.6

a Solve.

1. $\dfrac{3}{8} + \dfrac{4}{5} = \dfrac{x}{20}$

2. $\dfrac{3}{5} + \dfrac{2}{3} = \dfrac{x}{9}$

3. $\dfrac{2}{3} - \dfrac{5}{6} = \dfrac{1}{x}$

4. $\dfrac{1}{8} - \dfrac{3}{5} = \dfrac{1}{x}$

5. $\dfrac{1}{6} + \dfrac{1}{8} = \dfrac{1}{t}$

6. $\dfrac{1}{8} + \dfrac{1}{10} = \dfrac{1}{t}$

7. $x + \dfrac{4}{x} = -5$

8. $x + \dfrac{3}{x} = -4$

9. $\dfrac{x}{4} - \dfrac{4}{x} = 0$

10. $\dfrac{x}{5} - \dfrac{5}{x} = 0$

11. $\dfrac{5}{x} = \dfrac{6}{x} - \dfrac{1}{3}$

12. $\dfrac{4}{x} = \dfrac{5}{x} - \dfrac{1}{2}$

13. $\dfrac{5}{3x} + \dfrac{3}{x} = 1$

14. $\dfrac{3}{4x} + \dfrac{5}{x} = 1$

15. $\dfrac{x-7}{x+2} = \dfrac{1}{4}$

16. $\dfrac{a-2}{a+3} = \dfrac{3}{8}$

17. $\dfrac{2}{x+1} = \dfrac{1}{x-2}$

18. $\dfrac{5}{x-1} = \dfrac{3}{x+2}$

19. $\dfrac{x}{6} - \dfrac{x}{10} = \dfrac{1}{6}$

20. $\dfrac{x}{8} - \dfrac{x}{12} = \dfrac{1}{8}$

21. $\dfrac{x+1}{3} - \dfrac{x-1}{2} = 1$

22. $\dfrac{x+2}{5} - \dfrac{x-2}{4} = 1$

23. $\dfrac{a-3}{3a+2} = \dfrac{1}{5}$

24. $\dfrac{x-1}{2x+5} = \dfrac{1}{4}$

ANSWERS

25. _____

26. _____

27. _____

28. _____

29. _____

30. _____

31. _____

32. _____

33. _____

34. _____

35. _____

36. _____

37. _____

38. _____

39. _____

40. _____

41. _____

42. _____

43. _____

44. _____

45. _____

46. _____

47. _____

48. _____

25. $\dfrac{x-1}{x-5} = \dfrac{4}{x-5}$

26. $\dfrac{x-7}{x-9} = \dfrac{2}{x-9}$

27. $\dfrac{2}{x+3} = \dfrac{5}{x}$

28. $\dfrac{3}{x+4} = \dfrac{4}{x}$

29. $\dfrac{x-2}{x-3} = \dfrac{x-1}{x+1}$

30. $\dfrac{2b-3}{3b+2} = \dfrac{2b+1}{3b-2}$

31. $\dfrac{1}{x+3} + \dfrac{1}{x-3} = \dfrac{1}{x^2-9}$

32. $\dfrac{4}{x-3} + \dfrac{2x}{x^2-9} = \dfrac{1}{x+3}$

33. $\dfrac{x}{x+4} - \dfrac{4}{x-4} = \dfrac{x^2+16}{x^2-16}$

34. $\dfrac{5}{y-3} - \dfrac{30}{y^2-9} = 1$

35. $\dfrac{-3}{y-7} = \dfrac{-10-y}{7-y}$

36. $\dfrac{4-m}{8-m} = \dfrac{4}{m-8}$

SKILL MAINTENANCE

Simplify.

37. $(a^2 b^5)^{-3}$

38. $(x^{-2} y^{-3})^{-4}$

39. $\left(\dfrac{2x}{t^2}\right)^4$

40. $\left(\dfrac{y^3}{w^2}\right)^{-2}$

SYNTHESIS

Solve.

41. $\dfrac{4}{y-2} - \dfrac{2y-3}{y^2-4} = \dfrac{5}{y+2}$

42. $\dfrac{x}{x^2+3x-4} + \dfrac{x+1}{x^2+6x+8} = \dfrac{2x}{x^2+x-2}$

43. $\dfrac{y}{y+0.2} - 1.2 = \dfrac{y-0.2}{y+0.2}$

44. $\dfrac{x^2}{x^2-4} = \dfrac{x}{x+2} - \dfrac{2x}{2-x}$

45. $4a - 3 = \dfrac{a+13}{a+1}$

46. $\dfrac{3x-9}{x-3} = \dfrac{5x-4}{2}$

47. $\dfrac{y^2-4}{y+3} = 2 - \dfrac{y-2}{y+3}$

48. $\dfrac{3a-5}{a^2+4a+3} + \dfrac{2a+2}{a+3} = \dfrac{a-3}{a+1}$

5.7 Solving Problems and Proportions

a Solving Problems

▶ **EXAMPLE 1** If 2 is subtracted from a number and then the reciprocal is found, the result is twice the reciprocal of the number itself. What is the number?

1. *Familiarize.* Let us try to guess such a number. Try 10: $10 - 2$ is 8, and the reciprocal of 8 is $\frac{1}{8}$. Two times the reciprocal of 10 is $2(\frac{1}{10})$, or $\frac{1}{5}$. Since $\frac{1}{8} \neq \frac{1}{5}$, the number 10 does not check, but the process helps us understand the translation. Let $x =$ the number.

2. *Translate.* From the familiarization step, we get the following translation. Subtracting 2 from the number gives us $x - 2$. Twice the reciprocal of the original number is $2(1/x)$.

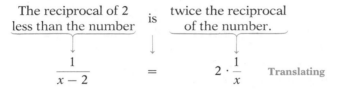

$$\frac{1}{x - 2} = 2 \cdot \frac{1}{x} \quad \textbf{Translating}$$

3. *Solve.* We solve the equation. The LCM is $x(x - 2)$.

$$x(x - 2) \cdot \frac{1}{x - 2} = x(x - 2) \cdot \frac{2}{x} \quad \textbf{Multiplying by the LCM}$$

$$x = 2(x - 2) \quad \textbf{Simplifying}$$

$$x = 2x - 4$$

$$-x = -4$$

$$x = 4.$$

4. *Check.* We go back to the original problem. The number to be checked is 4. Two from 4 is 2. The reciprocal of 2 is $\frac{1}{2}$. The reciprocal of the number itself is $\frac{1}{4}$. Since $\frac{1}{2}$ is twice $\frac{1}{4}$, the conditions are satisfied.

5. *State.* The number is 4. ◀

DO EXERCISE 1.

▶ **EXAMPLE 2** One car travels 20 km/h faster than another. While one car goes 240 km, the other goes 160 km. Find their speeds.

1. *Familiarize.* First we make a drawing. We really do not know the directions in which the cars are traveling, but it does not matter. Let $r =$ the speed of the slow car. Then $r + 20 =$ the speed of the fast car.

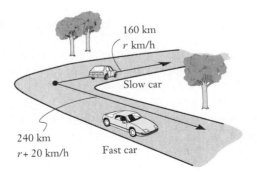

160 km
r km/h
Slow car

240 km
$r + 20$ km/h
Fast car

OBJECTIVES

After finishing Section 5.7, you should be able to:

a Solve applied problems using rational equations.

b Solve proportion problems.

FOR EXTRA HELP

Tape 12A   Tape 11B   MAC: 5
                                    IBM: 5

1. The reciprocal of two more than a number is three times the reciprocal of the number. Find the number.

ANSWER ON PAGE A-5

2. One car goes 10 km/h faster than another. While one car goes 120 km, the other goes 150 km. How fast does each car travel?

The cars travel for the same length of time, so we can just use t for time. We have the notions of distance, speed, and time in this problem. Are they related? Recall that we may need to find a formula that relates the parts of a problem. Indeed, you may need to look up such a formula. Actually we have considered this formula before; it is $d = rt$, *Distance = Speed · Time*. We can organize the information in a chart, as follows.

$$d \;=\; r \;\cdot\; t$$

	Distance	Speed	Time	
Slow car	160	r	t	$\longrightarrow 160 = rt$
Fast car	240	$r + 20$	t	$\longrightarrow 240 = (r + 20)t$

2. *Translate.* We can apply the formula $d = rt$ along the rows of the table to obtain two equations:

$$160 = rt, \qquad \textbf{(1)}$$
$$240 = (r + 20)t. \qquad \textbf{(2)}$$

The cars travel for the same length of time. Thus if we solve each equation for t and set the results equal, we get an equation in terms of r.

Solving $160 = rt$ for t: $t = \dfrac{160}{r}$

Solving $240 = (r + 20)t$: $t = \dfrac{240}{r + 20}$

Since the times are the same, we get the following equation:

$$\frac{160}{r} = \frac{240}{r + 20}.$$

3. *Solve.* To solve the equation, we first multiply on both sides by the LCM, which is $r(r + 20)$:

$$r(r + 20) \cdot \frac{160}{r} = r(r + 20) \cdot \frac{240}{r + 20} \qquad \text{\textbf{Multiplying on both sides by the LCM, } } r(r + 20)$$

$$160(r + 20) = 240r \qquad \text{\textbf{Simplifying}}$$

$$160r + 3200 = 240r \qquad \text{\textbf{Removing parentheses}}$$

$$3200 = 80r \qquad \text{\textbf{Subtracting } } 160r$$

$$\frac{3200}{80} = r \qquad \text{\textbf{Dividing by 80}}$$

$$40 = r.$$

We now have a possible solution. The speed of the slow car is 40 km/h, and the speed of the fast car is $r = 40 + 20$, or 60 km/h.

4. *Check.* We first reread the problem to see what we are to find. We check the speeds of 40 km/h for the slow car and 60 km/h for the fast car. The fast car does travel 20 km/h faster than the slow car. The fast car will travel farther than the slow car. If the fast car goes 240 km at 60 km/h, the time it has traveled is $\frac{240}{60}$, or 4 hr. If the slow car goes 160 km at 40 km/h, the time it travels is $\frac{160}{40}$, or 4 hr. Since the times are the same, the speeds check.

5. *State.* The slow car has a speed of 40 km/h, and the fast car has a speed of 60 km/h.   ◀

▶ **EXAMPLE 3**   The head of a secretarial pool examines work records and finds that it takes Helen 4 hr to type a certain report. It takes Willie 6 hr to type the same report. How long would it take them, working together, to type the report?

1. *Familiarize.*   We familiarize ourselves with the problem by considering two *incorrect* ways of translating the problem to mathematical language.

   a) A common incorrect way to translate the problem is just to add the two times:

   $$4 \text{ hr} + 6 \text{ hr} = 10 \text{ hr}.$$

   Now think about this. Helen can do the job alone in 4 hr. If Helen and Willie work together, whatever time it takes them should be *less* than 4 hr. Thus we reject 10 hr as a solution, but we do have a partial check on any answer we get. The answer should be less than 4 hr.

   b) Another incorrect way to translate the problem is as follows. Suppose the two people split up the typing job in such a way that Helen does half the typing and Willie does the other half. Then

   $$\text{Helen types } \frac{1}{2} \text{ the report in } \frac{1}{2}(4 \text{ hr}), \text{ or 2 hr,}$$

   and $$\text{Willie types } \frac{1}{2} \text{ the report in } \frac{1}{2}(6 \text{ hr}), \text{ or 3 hr.}$$

   But time is wasted since Helen would get done 1 hr earlier than Willie. In effect, they have not worked together to get the job done as fast as possible. If Helen helps Willie after completing her half, the entire job could be done in a time somewhere between 2 hr and 3 hr.

We proceed to a translation by considering how much of the job is finished in 1 hr, 2 hr, 3 hr, and so on. It takes Helen 4 hr to do the typing job alone. Then, in 1 hr, she can do $\frac{1}{4}$ of the job. It takes Willie 6 hr to do the job alone. Then, in 1 hr, he can do $\frac{1}{6}$ of the job. Working together, they can do

$$\frac{1}{4} + \frac{1}{6}, \quad \text{or } \frac{5}{12} \text{ of the job in 1 hr.}$$

In 2 hr, Helen can do $2(\frac{1}{4})$ of the job and Willie can do $2(\frac{1}{6})$ of the job. Working together in two hours, they can do

$$2\left(\frac{1}{4}\right) + 2\left(\frac{1}{6}\right), \quad \text{or } \frac{5}{6} \text{ of the job in 2 hr.}$$

Continuing this reasoning, we can form a table like the following one.

Time	Helen	Willie	Together
			Fraction of the job completed
1 hr	$\frac{1}{4}$	$\frac{1}{6}$	$\frac{1}{4} + \frac{1}{6}$, or $\frac{5}{12}$
2 hr	$2\left(\frac{1}{4}\right)$	$2\left(\frac{1}{6}\right)$	$2\left(\frac{1}{4}\right) + 2\left(\frac{1}{6}\right)$, or $\frac{5}{6}$
3 hr	$3\left(\frac{1}{4}\right)$	$3\left(\frac{1}{6}\right)$	$3\left(\frac{1}{4}\right) + 3\left(\frac{1}{6}\right)$, or $1\frac{1}{4}$
t hr	$t\left(\frac{1}{4}\right)$	$t\left(\frac{1}{6}\right)$	$t\left(\frac{1}{4}\right) + t\left(\frac{1}{6}\right)$

3. By checking work records, a contractor finds that it takes Red Bryck 6 hr to construct a wall of a certain size. It takes Lotta Mudd 8 hr to construct the same wall. How long would it take if they worked together?

4. Find the ratio of 145 km to 2.5 liters (L).

5. Recently, a baseball player got 7 hits in 25 times at bat. What was the rate, or batting average, in hits per times at bat?

6. Impulses in nerve fibers travel 310 km in 2.5 hr. What is the rate, or speed, in kilometers per hour?

7. A lake of area 550 yd² contains 1320 fish. What is the population density of the lake in fish per square yard?

From the table, we see that if they worked 3 hr, the fraction of the job that they get done is $1\frac{1}{4}$, which is more of the job than needs to be done. We also see that the answer is somewhere between 2 hr and 3 hr. What we want is a number t such that the fraction of the job that gets completed is 1; that is, the job is just completed—not more than $1\frac{1}{4}$ and not less than $\frac{5}{6}$.

2. *Translate.* From the table, we see that the time we want is some number t for which

$$t\left(\frac{1}{4}\right) + t\left(\frac{1}{6}\right) = 1, \quad \text{or} \quad \frac{t}{4} + \frac{t}{6} = 1,$$

where 1 represents the idea that the entire job is completed in time t.

3. *Solve.* We solve the equation:

$$\frac{t}{4} + \frac{t}{6} = 1$$

$$12\left(\frac{t}{4} + \frac{t}{6}\right) = 12 \cdot 1 \qquad \text{The LCM is } 2 \cdot 2 \cdot 3, \text{ or } 12.$$

$$12 \cdot \frac{t}{4} + 12 \cdot \frac{t}{6} = 12$$

$$3t + 2t = 12$$

$$5t = 12$$

$$t = \frac{12}{5}, \quad \text{or } 2\frac{2}{5} \text{ hr.}$$

4. *Check.* The check can be done by repeating the computations:

$$\frac{12}{5}\left(\frac{1}{4}\right) + \frac{12}{5}\left(\frac{1}{6}\right) = \frac{3}{5} + \frac{2}{5} = \frac{5}{5} = 1.$$

We also have another check in what we learned from our familiarization. The answer, $2\frac{2}{5}$ hr, is between 2 hr and 3 hr (see the table), and it is less than 4 hr, the time it takes Helen working alone.

5. *State.* It takes $2\frac{2}{5}$ hr for them to do the job working together.   ◀

> **The Work Principle**
>
> Suppose $a =$ the time it takes person A to do a job, $b =$ the time it takes person B to do the same job, and $t =$ the time it takes them to do the same job working together. Then
>
> $$\frac{t}{a} + \frac{t}{b} = 1.$$

DO EXERCISE 3.

b   Solving Proportion Problems

We now consider problems concerning proportions. A **proportion** involves ratios. A **ratio** of two quantities is their quotient. For example, 37% is the ratio of 37 to 100, $\frac{37}{100}$. The ratio of two different kinds of measure is called a **rate**. If you travel 400 mi in 7 hr, your rate, or **speed**, is

$$\frac{400 \text{ mi}}{7 \text{ hr}} \approx 57.1 \frac{\text{mi}}{\text{hr}}, \quad \text{or } 57.1 \text{ mph.}$$

DO EXERCISES 4–7.

> An equality of ratios, $A/B = C/D$, is called a *proportion*. The numbers named in a true proportion are said to be *proportional.*

Proportions can be used to solve applied problems by expressing a single ratio in two ways. For example, suppose it takes 9 gal of gas to drive 120 mi, and we wish to find how much will be required to go 550 mi. If we assume that the car uses gas at the same rate throughout the trip, the ratios are the same, and we can write a proportion.

$$\text{Gas} \longrightarrow \frac{9}{120} = \frac{x}{550} \longleftarrow \text{Gas}$$
$$\text{Miles} \longrightarrow \qquad\qquad \longleftarrow \text{Miles}$$

To solve this proportion, we multiply by 550 to get x alone on one side:

$$550 \cdot \frac{9}{120} = 550 \cdot \frac{x}{550}$$

$$\frac{550 \cdot 9}{120} = x$$

$$41.25 = x.$$

Thus, 41.25 gal will be required to go 550 mi. (Note that we could have multiplied by the LCM of 120 and 550, which is 66,000, but in this case, that would have been more complicated.) We can also use **cross products** to solve proportions:

$$\frac{9}{120} = \frac{x}{550}$$

$9 \cdot 550 = 120 \cdot x$ $\qquad$ $9 \cdot 550$ and $120 \cdot x$ are called *cross products.*

$$\frac{9 \cdot 550}{120} = x$$

$$41.25 = x.$$

This method can be verified using the multiplication principle, multiplying on both sides by 550 and then by 120, but we will not do so here.

▶ **EXAMPLE 4** A student is to read 32 essays. It takes the student 40 min to read 5 essays. At this rate, how long will it take to read all 32 essays?

1. *Familiarize.* The student reads 5 essays in 40 min, and we wish to find how long it will take to read all 32 essays. We can set up ratios. We let $t =$ the total time to read 32 essays.

2. *Translate.* If we assume that the student continues to read at the same rate, the ratios are the same, and we have an equation:

$$\text{Number of essays} \longrightarrow \frac{5}{40} = \frac{32}{t} \longleftarrow \text{Number of essays}$$
$$\text{Amount of time} \longrightarrow \qquad\qquad \longleftarrow \text{Amount of time}$$

3. *Solve.* We solve the equation:

$$40t \cdot \frac{5}{40} = 40t \cdot \frac{32}{t} \qquad \textbf{Multiplying by the LCM, } 40t$$

$$5t = 40 \cdot 32$$

$$t = \frac{40 \cdot 32}{5} = 256 \text{ min}, \quad \text{or } 4\frac{4}{15} \text{ hr.}$$

8. It takes 60 oz of grass seed to seed 3000 ft² of lawn. At this rate, how much would be needed for 5000 ft² of lawn?

9. A sample of 184 light bulbs contained 6 defective bulbs. How many would you expect to find in 1288 bulbs?

10. To determine the number of deer in a forest, a conservationist catches 612 deer, tags them, and lets them loose. Later, 244 deer are caught. Seventy-two of them are tagged. Estimate how many deer are in the forest.

4. *Check.*   We leave the check to the student.

5. *State.*   It will take the student 256 min, or $4\frac{4}{15}$ hr, to read 32 essays.   ◄

DO EXERCISES 8 AND 9.

▶   **EXAMPLE 5**   *Estimating wildlife populations.* To determine the number of fish in a lake, a conservationist catches 225 fish, tags them, and throws them back into the lake. Later, 108 fish are caught. Fifteen of them are found to be tagged. Estimate how many fish are in the lake.

1. *Familiarize.*   The ratio of fish tagged to the total number of fish in the lake, F, is $\frac{225}{F}$. Of the 108 fish caught later, 15 fish were tagged. The ratio of fish tagged to fish caught is $\frac{15}{108}$.

2. *Translate.*   Assuming the two ratios are the same, we can translate to a proportion.

$$\text{Fish tagged originally} \longrightarrow \frac{225}{F} = \frac{15}{108} \longleftarrow \text{Tagged fish caught later} \atop \longleftarrow \text{Fish caught later}$$

$$\text{Fish in lake} \longrightarrow$$

3. *Solve.*   We solve the proportion. We multiply by the LCM, which is $108F$:

$$108F \cdot \frac{225}{F} = 108F \cdot \frac{15}{108} \qquad \text{\textbf{Multiplying by 108}}F$$

$$108 \cdot 225 = F \cdot 15$$

$$\frac{108 \cdot 225}{15} = F \qquad \text{\textbf{Dividing by 15}}$$

$$1620 = F.$$

4. *Check.*   We leave the check to the student.

5. *State.*   We estimate that there are about 1620 fish in the lake.   ◄

DO EXERCISE 10.

NAME   SECTION   DATE

EXERCISE SET 5.7

a Solve.

1. The reciprocal of 4 plus the reciprocal of 5 is the reciprocal of what number?

2. The reciprocal of 3 plus the reciprocal of 8 is the reciprocal of what number?

3. One number is 5 more than another. The quotient of the larger divided by the smaller is $\frac{4}{3}$. Find the numbers.

4. One number is 4 more than another. The quotient of the larger divided by the smaller is $\frac{5}{2}$. Find the numbers.

5. One car travels 40 km/h faster than another. While one travels 150 km, the other goes 350 km. Find their speeds.

Complete this table and the equations as part of the familiarization.

$$d = r \cdot t$$

	Distance	Speed	Time
Slow car	150	r	
Fast car	350		t

→ 150 = $r($ $)$

→ 350 = $($ $)t$

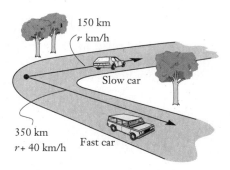

150 km
r km/h

Slow car

350 km
$r + 40$ km/h   Fast car

6. One car travels 30 km/h faster than another. While one goes 250 km, the other goes 400 km. Find their speeds.

7. The speed of a freight train is 14 km/h slower than the speed of a passenger train. The freight train travels 330 km in the same time that it takes the passenger train to travel 400 km. Find the speed of each train.

Complete this table and the equations as part of the familiarization.

$$d = r \cdot t$$

	Distance	Speed	Time
Freight	330		t
Passenger	400	r	

→ 330 = $($ $)t$

→ 400 = $r($ $)$

8. The speed of a freight train is 15 km/h slower than the speed of a passenger train. The freight train travels 390 km in the same time that it takes the passenger train to travel 480 km. Find the speed of each train.

1. _____

2. _____

3. _____

4. _____

5. _____

6. _____

7. _____

8. _____

9. _____

10. _____

11. _____

12. _____

13. _____

14. _____

15. _____

16. _____

17. _____

18. _____

9. A person traveled 120 mi in one direction. The return trip was accomplished at double the speed and took 3 hr less time. Find the speed going.

10. After making a trip of 126 mi, a person found that the trip would have taken 1 hr less time by increasing the speed by 8 mph. What was the actual speed?

11. It takes David 4 hr to paint a certain area of a house. It takes Sierra 5 hr to do the same job. How long would it take them, working together, to do the painting job?

12. By checking work records, a carpenter finds that Juanita can build a certain type of garage in 12 hr. Antoine can do the same job in 16 hr. How long would it take if they worked together?

13. By checking work records, a plumber finds that Rory can do a certain job in 12 hr. Mira can do the same job in 9 hr. How long would it take if they worked together?

14. A tank can be filled in 18 hr by pipe A alone and in 24 hr by pipe B alone. How long would it take to fill the tank if both pipes were working?

b Find the ratio of the following. Simplify, if possible.

15. 54 days, 6 days

16. 800 mi, 50 gal

17. A black racer snake travels 4.6 km in 2 hr. What is the speed in km/h?

18. Light travels 558,000 mi in 3 sec. What is the speed in mi/sec?

Solve.

19. The coffee beans from 14 trees are required to produce 7.7 kg of coffee (this is the average that each person in the United States drinks each year). How many trees are required to produce 320 kg of coffee?

20. Last season a minor-league baseball player got 240 hits in 600 times at bat. This season, his ratio of hits to number of times at bat is the same. He batted 500 times. How many hits has he had?

19. _____

20. _____

21. A student traveled 234 km in 14 days. At this same ratio, how far would the student travel in 42 days?

22. In a potato bread recipe, the ratio of milk to flour is $\frac{3}{13}$. If 5 cups of milk are used, how many cups of flour are used?

21. _____

23. 10 cm^3 of a normal specimen of human blood contains 1.2 g of hemoglobin. How many grams would 16 cm^3 of the same blood contain?

24. The winner of an election for class president won by a vote of 3 to 2, with 324 votes. How many votes did the loser get?

22. _____

23. _____

25. To determine the number of trout in a lake, a conservationist catches 112 trout, tags them, and throws them back into the lake. Later, 82 trout are caught; 32 of them are tagged. How many trout are in the lake?

26. To determine the number of deer in a game preserve, a conservationist catches 318 deer, tags them, and lets them loose. Later, 168 deer are caught; 56 of them are tagged. How many deer are in the preserve?

24. _____

25. _____

26. _____

27. The ratio of the weight of an object on the moon to the weight of an object on earth is 0.16 to 1.

   a) How much would a 12-ton rocket weigh on the moon?

   b) How much would a 180-lb astronaut weigh on the moon?

28. The ratio of the weight of an object on Mars to the weight of an object on earth is 0.4 to 1.

   a) How much would a 12-ton rocket weigh on Mars?

   b) How much would a 120-lb astronaut weigh on Mars?

29. Simplest fractional notation for a rational number is $\frac{9}{17}$. Find an equal ratio where the sum of the numerator and the denominator is 104.

30. A baseball team has 12 more games to play. They have won 25 out of the 36 games they have played. How many more games must they win in order to finish with a 0.750 record?

SYNTHESIS

31. The denominator of a fraction is 1 more than the numerator. If 2 is subtracted from both the numerator and the denominator, the resulting fraction is $\frac{1}{2}$. Find the original fraction.

32. Ann and Betty work together and complete a job in 4 hr. It would take Betty 6 hr longer, working alone, to do the job than it would Ann. How long would it take each of them to do the job working alone?

33. The speed of a boat in still water is 10 mph. It travels 24 mi upstream and 24 mi downstream in a total time of 5 hr. What is the speed of the current?

34. Express 100 as the sum of two numbers for which the ratio of one number, increased by 5, to the other number, decreased by 5, is 4.

35. In a proportion

$$\frac{A}{B} = \frac{C}{D},$$

the numbers A and D are often called extremes, whereas the numbers B and C are called the means. Write four true proportions.

36. Compare

$$\frac{A + B}{B} = \frac{C + D}{D}$$

with the proportion

$$\frac{A}{B} = \frac{C}{D}.$$

37. Rosina, Ng, and Oscar can complete a certain job in 3 days. Rosina can do the job in 8 days and Ng can do it in 10 days. How many days will it take Oscar to complete the job?

38. How soon after 5 o'clock will the hands on a clock first be together?

39. To reach an appointment 50 mi away, Dr. Wright allowed 1 hr. After driving 30 mi, she realized that her speed would have to be increased 15 mph for the remainder of the trip. What was her speed for the first 30 mi?

40. Together, Michelle, Sal, and Kristen can do a job in 1 hr and 20 min. To do the job alone, Michelle needs twice the time that Sal needs and 2 hr more than Kristen. How long would it take each to complete the job working alone?

5.8 Formulas

a The use of formulas is important in many applications of mathematics. For rational formulas, we use the following procedure to solve a formula for a letter.

To solve a formula for a given letter, identify the letter, and:

1. **Multiply on both sides to clear of fractions or decimals, if that is needed.**
2. **Multiply if necessary to remove parentheses.**
3. **Get all terms with the letter to be solved for on one side of the equation and all other terms on the other side, using the addition principle.**
4. **Collect like terms again, if necessary.**
5. **Solve for the letter in question, using the multiplication principle.**

▶ **EXAMPLE 1** *Gravitational force.* The gravitational force f between planets of mass M and m, at a distance d from each other, is given by

$$f = \frac{kMm}{d^2},$$

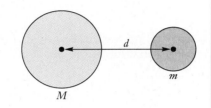

where k represents a fixed number constant. Solve for m.

We have

$$fd^2 = kMm \qquad \text{Multiplying by the LCM, } d^2$$

$$\frac{fd^2}{kM} = m. \qquad \text{Dividing by } kM \qquad ◀$$

DO EXERCISE 1.

▶ **EXAMPLE 2** *The area of a trapezoid.* The area A of a trapezoid is half the product of the height h and the sum of the lengths b_1 and b_2 of the parallel sides:

$$A = \frac{1}{2}(b_1 + b_2)h.$$

Solve for b_2.

We consider b_1 and b_2 to be different variables (or constants). The letter b_1 represents the length of the first parallel side and b_2 represents the length of the second parallel side. The small numbers 1 and 2 are called **subscripts.** Subscripts are used to identify different variables with related meanings.

$$2A = (b_1 + b_2)h \qquad \text{Multiplying by 2 to clear of fractions}$$

$$2A = b_1h + b_2h$$

$$2A - b_1h = b_2h \qquad \text{Subtracting } b_1h$$

$$\frac{2A - b_1h}{h} = b_2 \qquad \text{Dividing by } h \qquad ◀$$

DO EXERCISE 2.

OBJECTIVE

After finishing Section 5.8, you should be able to:

a Solve a formula for a letter.

FOR EXTRA HELP

Tape 12B   Tape 12A   MAC: 5
                                                   IBM: 5

1. Solve for M: $F = \dfrac{kMm}{d^2}$.

2. Solve for b_1: $A = \dfrac{1}{2}h(b_1 + b_2)$.

3. Solve for f: $\dfrac{1}{p} + \dfrac{1}{q} = \dfrac{1}{f}$.

This is an optics formula.

4. Solve for a: $Q = \dfrac{a - b}{2b}$.

▶ **EXAMPLE 3**   *A work formula.* The following work formula was considered in Section 5.7. Solve it for t.

$$\frac{t}{a} + \frac{t}{b} = 1$$

We multiply by the LCM, which is ab:

$$ab \cdot \left(\frac{t}{a} + \frac{t}{b} \right) = ab \cdot 1 \qquad \text{Multiplying by } ab$$

$$ab \cdot \frac{t}{a} + ab \cdot \frac{t}{b} = ab$$

$$\frac{abt}{a} + \frac{abt}{b} = ab$$

$$bt + at = ab \qquad \text{Simplifying}$$

$$(b + a)t = ab \qquad \text{Factoring out } t$$

$$t = \frac{ab}{b + a}. \qquad \text{Dividing by } b + a \qquad ◀$$

The answer to Example 3 can be used to find solutions to problems such as Example 3 in Section 5.7:

$$t = \frac{4 \cdot 6}{6 + 4} = \frac{24}{10} = 2\frac{2}{5}.$$

DO EXERCISE 3.

In Examples 1 and 2, the letter for which we solved was on the right side of the equation. In Example 3, the letter was on the left. The location of the letter is a matter of choice, since all equations are reversible.

Recall the following tip, which we also considered in Chapter 2.

> **Tip for Formula Solving**
>
> **The variable to be solved for should be alone on one side of the equation, with *no* occurrence of that variable on the other side.**

▶ **EXAMPLE 4**   Solve for b: $S = \dfrac{a + b}{3b}$.

We multiply by the LCM, which is $3b$:

$$3b \cdot S = 3b \cdot \frac{a + b}{3b} \qquad \text{Multiplying by } 3b$$

$$3bS = a + b \qquad \text{Simplifying}$$

> If we had divided by $3S$, we would have b alone on the left, but we would still have a term with b on the right.

$$3bS - b = a \qquad \text{Subtracting } b \text{ to get all terms involving } b \text{ on one side}$$

$$b(3S - 1) = a$$

$$b = \frac{a}{3S - 1}. \qquad ◀$$

DO EXERCISE 4.

NAME   SECTION   DATE

EXERCISE SET 5.8

a Solve.

1. $S = 2\pi rh$ for r

2. $A = P(1 + rt)$ for t
   (An interest formula)

3. $A = \dfrac{1}{2}bh$ for b
   (The area of a triangle)

4. $s = \dfrac{1}{2}gt^2$ for g

5. $S = 180(n - 2)$ for n

6. $S = \dfrac{n}{2}(a + l)$ for a

7. $V = \dfrac{1}{3}k(B + b + 4M)$ for b

8. $A = P + Prt$ for P
   (*Hint:* Factor the right-hand side.)

9. $S(r - 1) = rl - a$ for r

10. $T = mg - mf$ for m
    (*Hint:* Factor the right-hand side.)

11. $A = \dfrac{1}{2}h(b_1 + b_2)$ for h

12. $S = 2\pi r(r + h)$ for h
    (The area of a right circular cylinder)

13. $\dfrac{A - B}{AB} = Q$ for B

14. $L = \dfrac{Mt + g}{t}$ for t

15. $\dfrac{1}{p} + \dfrac{1}{q} = \dfrac{1}{f}$ for p

16. $\dfrac{1}{a} + \dfrac{1}{b} = \dfrac{1}{t}$ for b

1. _____

2. _____

3. _____

4. _____

5. _____

6. _____

7. _____

8. _____

9. _____

10. _____

11. _____

12. _____

13. _____

14. _____

15. _____

16. _____

17. $\dfrac{A}{P} = 1 + r$ for A

18. $\dfrac{2A}{h} = a + b$ for h

19. $\dfrac{1}{R} = \dfrac{1}{r_1} + \dfrac{1}{r_2}$ for R
(An electricity formula)

20. $\dfrac{1}{R} = \dfrac{1}{r_1} + \dfrac{1}{r_2}$ for r_1

21. $\dfrac{A}{B} = \dfrac{C}{D}$ for D

22. $\dfrac{A}{B} = \dfrac{C}{D}$ for C

23. $h_1 = q\left(1 + \dfrac{h_2}{p}\right)$ for h_2

24. $S = \dfrac{a - ar^n}{1 - r}$ for a

25. $C = \dfrac{Ka - b}{a}$ for a

26. $Q = \dfrac{Pt + h}{t}$ for t

SKILL MAINTENANCE

27. Subtract: $(5x^4 - 6x^3 + 23x^2 - 79x + 24) - (-18x^4 - 56x^3 + 84x - 17)$.

Factor.

28. $x^2 - 4$

29. $30y^4 + 9y^2 - 12$

30. $49m^2 - 112mn + 64n^2$

SYNTHESIS

Solve.

31. $u = -F\left(E - \dfrac{P}{T}\right)$ for T

32. $l = a + (n - 1)d$ for d

33. The formula

$$C = \frac{5}{9}(F - 32)$$

is used to convert Fahrenheit temperatures to Celsius temperatures. At what temperature are the Fahrenheit and Celsius readings the same?

34. In

$$N = \frac{a}{c},$$

what is the effect on N when c increases? when c decreases? Assume that a, c, and N are positive.

5.9 Complex Rational Expressions

a A **complex rational expression**, or **complex fractional expression**, is a rational expression that has one or more rational expressions within its numerator or denominator. Here are some examples:

$$\frac{1 + \frac{2}{x}}{3}, \quad \frac{\frac{x+y}{2}}{\frac{2x}{x+1}}, \quad \frac{\frac{1}{3} + \frac{1}{5}}{\frac{2}{x} - \frac{x}{y}}.$$ These are rational expressions within the complex rational expression.

There are two methods to simplify complex rational expressions. We will consider them both. Use the one that works best for you or the one that your instructor directs you to use.

Multiplying by the LCM of All the Denominators: Method 1

> **Method 1**
>
> **To simplify a complex rational expression:**
> 1. **First, find the LCM of all the denominators of all the rational expressions occurring *within* both the numerator and the denominator of the complex rational expression. Let a = the LCM.**
> 2. **Then multiply by 1 using a/a.**
> 3. **If possible, simplify by removing a factor of 1.**

▶ **EXAMPLE 1** Simplify: $\dfrac{\frac{1}{2} + \frac{3}{4}}{\frac{5}{6} - \frac{3}{8}}$.

We have

$$\frac{\frac{1}{2} + \frac{3}{4}}{\frac{5}{6} - \frac{3}{8}} \quad \begin{cases} \text{The denominators } \textit{within} \text{ the complex rational expression} \\ \text{are 2, 4, 6, and 8. The LCM of these denominators is 24.} \\ \text{We multiply by 1 using } \frac{24}{24}. \end{cases}$$

$$= \frac{\frac{1}{2} + \frac{3}{4}}{\frac{5}{6} - \frac{3}{8}} \cdot \frac{24}{24} \quad \text{Multiplying by 1}$$

$$= \frac{\left(\frac{1}{2} + \frac{3}{4}\right)24}{\left(\frac{5}{6} - \frac{3}{8}\right)24} \quad \begin{array}{l} \longleftarrow \text{Multiplying the numerator by 24} \\ \\ \longleftarrow \text{Multiplying the denominator by 24} \end{array}$$

OBJECTIVE

After finishing Section 5.9, you should be able to:

a Simplify complex rational expressions.

FOR EXTRA HELP

Tape 12C

Tape 12A

MAC: 5
IBM: 5

1. Simplify. Use Method 1.

$$\dfrac{\dfrac{1}{3}+\dfrac{4}{5}}{\dfrac{7}{8}-\dfrac{5}{6}}$$

Using the distributive laws, we carry out the multiplications.

$$=\dfrac{\dfrac{1}{2}(24)+\dfrac{3}{4}(24)}{\dfrac{5}{6}(24)-\dfrac{3}{8}(24)}$$

$$=\dfrac{12+18}{20-9}\qquad\text{Simplifying}$$

$$=\dfrac{30}{11}.\qquad\blacktriangleleft$$

Multiplying in this manner has the effect of clearing of fractions in both the top and bottom of the complex rational expression.

DO EXERCISE 1.

2. Simplify. Use Method 1.

$$\dfrac{\dfrac{x}{2}+\dfrac{2x}{3}}{\dfrac{1}{x}-\dfrac{x}{2}}$$

▶ **EXAMPLE 2**　Simplify: $\dfrac{\dfrac{3}{x}+\dfrac{1}{2x}}{\dfrac{1}{3x}-\dfrac{3}{4x}}$.

The denominators within the complex expression are x, $2x$, $3x$, and $4x$. The LCM of these denominators is $12x$. We multiply by 1 using $12x/12x$.

$$\dfrac{\dfrac{3}{x}+\dfrac{1}{2x}}{\dfrac{1}{3x}-\dfrac{3}{4x}}\cdot\dfrac{12x}{12x}=\dfrac{\left(\dfrac{3}{x}+\dfrac{1}{2x}\right)12x}{\left(\dfrac{1}{3x}-\dfrac{3}{4x}\right)12x}=\dfrac{\dfrac{3}{x}(12x)+\dfrac{1}{2x}(12x)}{\dfrac{1}{3x}(12x)-\dfrac{3}{4x}(12x)}$$

$$=\dfrac{36+6}{4-9}=-\dfrac{42}{5}\qquad\blacktriangleleft$$

DO EXERCISE 2.

3. Simplify. Use Method 1.

$$\dfrac{1+\dfrac{1}{x}}{1-\dfrac{1}{x^2}}$$

▶ **EXAMPLE 3**　Simplify: $\dfrac{1-\dfrac{1}{x}}{1-\dfrac{1}{x^2}}$.

The denominators within the complex expression are x and x^2. The LCM of these denominators is x^2. We multiply by 1 using x^2/x^2. Then, after obtaining a single rational expression, we simplify:

$$\dfrac{1-\dfrac{1}{x}}{1-\dfrac{1}{x^2}}\cdot\dfrac{x^2}{x^2}=\dfrac{\left(1-\dfrac{1}{x}\right)x^2}{\left(1-\dfrac{1}{x^2}\right)x^2}=\dfrac{1(x^2)-\dfrac{1}{x}(x^2)}{1(x^2)-\dfrac{1}{x^2}(x^2)}=\dfrac{x^2-x}{x^2-1}$$

$$=\dfrac{x(x-1)}{(x+1)(x-1)}=\dfrac{x}{x+1}.\qquad\blacktriangleleft$$

DO EXERCISE 3.

ANSWERS ON PAGE A-5

Adding in the Numerator and the Denominator: Method 2

> **Method 2**
>
> **To simplify a complex rational expression:**
>
> 1. **Add or subtract, as necessary, to get a single rational expression in the numerator.**
> 2. **Add or subtract, as necessary, to get a single rational expression in the denominator.**
> 3. **Divide the numerator by the denominator.**
> 4. **If possible, simplify by removing a factor of 1.**

We will redo Examples 1–3 using this method.

▶ **EXAMPLE 4**   Simplify: $\dfrac{\dfrac{1}{2} + \dfrac{3}{4}}{\dfrac{5}{6} - \dfrac{3}{8}}$.

We have

$$\frac{\dfrac{1}{2} + \dfrac{3}{4}}{\dfrac{5}{6} - \dfrac{3}{8}} = \frac{\dfrac{1}{2} \cdot \dfrac{2}{2} + \dfrac{3}{4}}{\dfrac{5}{6} \cdot \dfrac{4}{4} - \dfrac{3}{8} \cdot \dfrac{3}{3}}$$

$\left.\begin{array}{c}\\\end{array}\right\}$ ← **Multiplying the $\frac{1}{2}$ by 1 to get a common denominator**

$\left.\begin{array}{c}\\\end{array}\right\}$ ← **Multiplying the $\frac{5}{6}$ and the $\frac{3}{8}$ by 1 to get a common denominator**

$$= \frac{\dfrac{2}{4} + \dfrac{3}{4}}{\dfrac{20}{24} - \dfrac{9}{24}}$$

$$= \frac{\dfrac{5}{4}}{\dfrac{11}{24}} \qquad \textbf{Adding in the numerator; subtracting in the denominator}$$

$$= \frac{5}{4} \cdot \frac{24}{11} \qquad \textbf{Multiplying by the reciprocal of the divisor}$$

$$= \frac{5 \cdot 3 \cdot 2 \cdot 2 \cdot 2}{2 \cdot 2 \cdot 11} \qquad \textbf{Factoring}$$

$$= \frac{5 \cdot 3 \cdot 2 \cdot 2 \cdot 2}{2 \cdot 2 \cdot 11} \qquad \textbf{Removing a factor of 1: } \frac{2 \cdot 2}{2 \cdot 2} = 1$$

$$= \frac{30}{11}. \qquad\qquad\qquad\qquad ◀$$

DO EXERCISE 4.

4. Simplify. Use Method 2.

$$\frac{\dfrac{1}{3} + \dfrac{4}{5}}{\dfrac{7}{8} - \dfrac{5}{6}}$$

5. Simplify. Use Method 2.

$$\frac{\dfrac{x}{2} + \dfrac{2x}{3}}{\dfrac{1}{x} - \dfrac{x}{2}}$$

▶ **EXAMPLE 5**   Simplify: $\dfrac{\dfrac{3}{x} + \dfrac{1}{2x}}{\dfrac{1}{3x} - \dfrac{3}{4x}}$.

We have

$$\frac{\dfrac{3}{x} + \dfrac{1}{2x}}{\dfrac{1}{3x} - \dfrac{3}{4x}} = \frac{\dfrac{3}{x} \cdot \dfrac{2}{2} + \dfrac{1}{2x}}{\dfrac{1}{3x} \cdot \dfrac{4}{4} - \dfrac{3}{4x} \cdot \dfrac{3}{3}}$$

   } ← Finding the LCM, $2x$, and multiplying by 1 in the numerator

   } ← Finding the LCM, $12x$, and multiplying by 1 in the denominator

$$= \frac{\dfrac{6}{2x} + \dfrac{1}{2x}}{\dfrac{4}{12x} - \dfrac{9}{12x}} = \frac{\dfrac{7}{2x}}{\dfrac{-5}{12x}}$$

← Adding in the numerator and subtracting in the denominator

$$= \frac{7}{2x} \cdot \frac{12x}{-5}$$ Multiplying by the reciprocal of the divisor

$$= \frac{7}{2x} \cdot \frac{6(2x)}{-5}$$ Factoring

$$= \frac{7}{2x} \cdot \frac{6(2x)}{-5}$$ Removing a factor of 1: $\dfrac{2x}{2x} = 1$

$$= \frac{42}{-5} = -\frac{42}{5}.$$

DO EXERCISE 5.

6. Simplify. Use Method 2.

$$\frac{1 + \dfrac{1}{x}}{1 - \dfrac{1}{x^2}}$$

▶ **EXAMPLE 6**   Simplify: $\dfrac{1 - \dfrac{1}{x}}{1 - \dfrac{1}{x^2}}$.

We have

$$\frac{1 - \dfrac{1}{x}}{1 - \dfrac{1}{x^2}} = \frac{\dfrac{x}{x} - \dfrac{1}{x}}{\dfrac{x^2}{x^2} - \dfrac{1}{x^2}}$$

   } ← Finding the LCM, x, and multiplying by 1 in the numerator

   } ← Finding the LCM, x^2, and multiplying by 1 in the denominator

$$= \frac{\dfrac{x - 1}{x}}{\dfrac{x^2 - 1}{x^2}}$$

← Subtracting in the numerator and subtracting in the denominator

$$= \frac{x - 1}{x} \cdot \frac{x^2}{x^2 - 1}$$ Multiplying by the reciprocal of the divisor

$$= \frac{(x - 1)x \cdot x}{x(x - 1)(x + 1)}$$ Factoring

$$= \frac{(x - 1)x \cdot x}{x(x - 1)(x + 1)}$$ Removing a factor of 1: $\dfrac{x(x - 1)}{x(x - 1)} = 1$

$$= \frac{x}{x + 1}.$$

DO EXERCISE 6.

NAME   SECTION   DATE

EXERCISE SET 5.9

a Simplify.

1. $\dfrac{1 + \dfrac{9}{16}}{1 - \dfrac{3}{4}}$

2. $\dfrac{9 - \dfrac{1}{4}}{3 + \dfrac{1}{2}}$

3. $\dfrac{1 - \dfrac{3}{5}}{1 + \dfrac{1}{5}}$

4. $\dfrac{\dfrac{5}{27} - 5}{\dfrac{1}{3} + 1}$

5. $\dfrac{\dfrac{1}{2} + \dfrac{3}{4}}{\dfrac{5}{8} - \dfrac{5}{6}}$

6. $\dfrac{\dfrac{2}{3} - \dfrac{5}{6}}{\dfrac{3}{4} + \dfrac{7}{8}}$

7. $\dfrac{\dfrac{1}{x} + 3}{\dfrac{1}{x} - 5}$

8. $\dfrac{\dfrac{3}{s} + s}{\dfrac{s}{3} + s}$

9. $\dfrac{\dfrac{2}{y} + \dfrac{1}{2y}}{y + \dfrac{y}{2}}$

10. $\dfrac{4 - \dfrac{1}{x^2}}{2 - \dfrac{1}{x}}$

11. $\dfrac{8 + \dfrac{8}{d}}{1 + \dfrac{1}{d}}$

12. $\dfrac{2 - \dfrac{3}{b}}{2 - \dfrac{b}{3}}$

13. $\dfrac{\dfrac{x}{8} - \dfrac{8}{x}}{\dfrac{1}{8} + \dfrac{1}{x}}$

14. $\dfrac{\dfrac{2}{m} + \dfrac{m}{2}}{\dfrac{m}{3} - \dfrac{3}{m}}$

15. $\dfrac{1 + \dfrac{1}{y}}{1 - \dfrac{1}{y^2}}$

1. _____

2. _____

3. _____

4. _____

5. _____

6. _____

7. _____

8. _____

9. _____

10. _____

11. _____

12. _____

13. _____

14. _____

15. _____

16. $\dfrac{\dfrac{1}{q^2} - 1}{\dfrac{1}{q} + 1}$

17. $\dfrac{\dfrac{1}{5} - \dfrac{1}{a}}{\dfrac{5 - a}{5}}$

18. $\dfrac{2 - \dfrac{1}{x}}{\dfrac{2}{x}}$

19. $\dfrac{\dfrac{x}{x - y}}{\dfrac{x^2}{x^2 - y^2}}$

20. $\dfrac{\dfrac{x}{y} - \dfrac{y}{x}}{\dfrac{1}{y} + \dfrac{1}{x}}$

21. $\dfrac{x - 3 + \dfrac{2}{x}}{x - 4 + \dfrac{3}{x}}$

22. $\dfrac{1 + \dfrac{a}{b - a}}{\dfrac{a}{a + b} - 1}$

SKILL MAINTENANCE

23. Add: $(2x^3 - 4x^2 + x - 7) + (4x^4 + x^3 + 4x^2 + x)$.

24. The length of a rectangle is 3 yd greater than the width. The area of the rectangle is 10 yd^2. Find the perimeter.

SYNTHESIS

25. Find the reciprocal of $\dfrac{2}{x - 1} - \dfrac{1}{3x - 2}$.

Simplify.

26. $\dfrac{\dfrac{a}{b} + \dfrac{c}{d}}{\dfrac{b}{a} + \dfrac{d}{c}}$

27. $\dfrac{\dfrac{a}{b} - \dfrac{c}{d}}{\dfrac{b}{a} - \dfrac{d}{c}}$

28. $\left[\dfrac{\dfrac{x + 1}{x - 1} + 1}{\dfrac{x + 1}{x - 1} - 1} \right]^5$

29. $1 + \dfrac{1}{1 + \dfrac{1}{1 + \dfrac{1}{1 + \dfrac{1}{x}}}}$

30. $\dfrac{\dfrac{z}{1 - \dfrac{z}{2 + 2z}} - 2z}{\dfrac{2z}{5z - 2} - 3}$

SUMMARY AND REVIEW EXERCISES: CHAPTER 5

The review sections and objectives to be tested in addition to the material in this chapter are [3.2a, b], [3.4a, c], [4.6a], and [4.8a].

Determine the replacements that are not meaningful.

1. $\dfrac{3}{x}$

2. $\dfrac{4}{x-6}$

3. $\dfrac{x+5}{x^2-36}$

4. $\dfrac{x^2-3x+2}{x^2+x-30}$

5. $\dfrac{-4}{(x+2)^2}$

6. $\dfrac{x-5}{x^3-8x^2+15x}$

Simplify.

7. $\dfrac{4x^2-8x}{4x^2+4x}$

8. $\dfrac{14x^2-x-3}{2x^2-7x+3}$

9. $\dfrac{(y-5)^2}{y^2-25}$

Multiply or divide and simplify.

10. $\dfrac{a^2-36}{10a}\cdot\dfrac{2a}{a+6}$

11. $\dfrac{6t-6}{2t^2+t-1}\cdot\dfrac{t^2-1}{t^2-2t+1}$

12. $\dfrac{10-5t}{3}\div\dfrac{t-2}{12t}$

13. $\dfrac{4x^4}{x^2-1}\div\dfrac{2x^3}{x^2-2x+1}$

Find the LCM.

14. $3x^2,\quad 10xy,\quad 15y^2$

15. $a-2,\quad 4a-8$

16. $y^2-y-2,\quad y^2-4$

Add or subtract and simplify.

17. $\dfrac{x+8}{x+7}+\dfrac{10-4x}{x+7}$

18. $\dfrac{3}{3x-9}+\dfrac{x-2}{3-x}$

19. $\dfrac{6x-3}{x^2-x-12}-\dfrac{2x-15}{x^2-x-12}$

20. $\dfrac{3x-1}{2x}-\dfrac{x-3}{x}$

21. $\dfrac{x+3}{x-2}-\dfrac{x}{2-x}$

22. $\dfrac{2a}{a+1}+\dfrac{4a}{a^2-1}$

23. $\dfrac{d^2}{d-c}+\dfrac{c^2}{c-d}$

24. $\dfrac{1}{x^2-25}-\dfrac{x-5}{x^2-4x-5}$

25. $\dfrac{3x}{x+2}-\dfrac{x}{x-2}+\dfrac{8}{x^2-4}$

Simplify.

26. $\dfrac{\dfrac{1}{z}+1}{\dfrac{1}{z^2}-1}$

27. $\dfrac{\dfrac{c}{d}-\dfrac{d}{c}}{\dfrac{1}{c}+\dfrac{1}{d}}$

Solve.

28. $\dfrac{3}{y}-\dfrac{1}{4}=\dfrac{1}{y}$

29. $\dfrac{15}{x}-\dfrac{15}{x+2}=2$

Solve.

30. In checking records, a contractor finds that crew A can pave a certain length of highway in 9 hr. Crew B can do the same job in 12 hr. How long would it take if they worked together?

31. A manufacturer is testing two high-speed trains. One train travels 40 km/h faster than the other. While one train travels 70 km, the other travels 60 km. Find their speeds.

32. The reciprocal of one more than a number is twice the reciprocal of the number itself. What is the number?

33. A sample of 250 batteries contained 8 defective batteries. How many defective batteries would you expect among 5000 batteries?

34. One plane travels 80 km/h faster than another. While one of them travels 1750 km, the other travels 950 km. Find the speed of each plane.

Solve for the letter indicated.

35. $\dfrac{1}{r} + \dfrac{1}{s} = \dfrac{1}{t}$, for s

36. $F = \dfrac{9C + 160}{5}$, for C

SKILL MAINTENANCE

37. Factor: $5x^3 + 20x^2 - 3x - 12$.

38. Simplify: $(5x^3y^2)^{-3}$.

39. Subtract:
$$(5x^3 - 4x^2 + 3x - 4) - (7x^3 - 7x^2 - 9x + 14).$$

40. The width of a rectangle is 2 cm less than the length. The area is 15 cm². Find the dimensions and the perimeter of the rectangle.

SYNTHESIS

Simplify.

41. $\dfrac{2a^2 + 5a - 3}{a^2} \cdot \dfrac{5a^3 + 30a^2}{2a^2 + 7a - 4} \div \dfrac{a^2 + 6a}{a^2 + 7a + 12}$

42. $\dfrac{12a}{(a - b)(b - c)} - \dfrac{2a}{(b - a)(c - b)}$

43. a) Add: $\dfrac{6}{x - 3} + \dfrac{4}{x + 3}$.

  b) Solve: $\dfrac{6}{x - 3} + \dfrac{4}{x + 3} = \dfrac{7}{x^2 - 9}$.

  c) Explain the difference in the use of the LCM in (a) and (b).

❖ THINKING IT THROUGH

Carry out the direction for each of the following. Explain the use of the LCM in each problem.

1. Add: $\dfrac{4}{x - 2} + \dfrac{1}{x + 2}$.

2. Subtract: $\dfrac{4}{x - 2} - \dfrac{1}{x + 2}$.

3. Solve: $\dfrac{4}{x - 2} + \dfrac{1}{x + 2} = \dfrac{26}{x^2 - 4}$.

4. Simplify: $\dfrac{1 - \dfrac{2}{x}}{1 + \dfrac{x}{4}}$.

5. Explain the different uses of the LCM in each of Exercises 1–4.

NAME   SECTION   DATE

TEST: CHAPTER 5

Determine the replacements that are not meaningful.

1. $\dfrac{8}{2x}$

2. $\dfrac{5}{x+8}$

3. $\dfrac{x-7}{x^2-49}$

4. $\dfrac{x^2+x-30}{x^2-3x+2}$

5. $\dfrac{11}{(x-1)^2}$

6. $\dfrac{x+2}{x^3+8x^2+15x}$

7. Simplify:

$$\frac{6x^2+17x+7}{2x^2+7x+3}.$$

8. Multiply and simplify:

$$\frac{a^2-25}{6a}\cdot\frac{3a}{a-5}.$$

9. Divide and simplify:

$$\frac{25x^2-1}{9x^2-6x}\div\frac{5x^2+9x-2}{3x^2+x-2}.$$

10. Find the LCM:

$$y^2-9,\ y^2+10y+21,\ y^2+4y-21.$$

Add or subtract. Simplify, if possible.

11. $\dfrac{16+x}{x^3}+\dfrac{7-4x}{x^3}$

12. $\dfrac{5-t}{t^2+1}-\dfrac{t-3}{t^2+1}$

13. $\dfrac{x-4}{x-3}+\dfrac{x-1}{3-x}$

14. $\dfrac{x-4}{x-3}-\dfrac{x-1}{3-x}$

15. $\dfrac{5}{t-1}+\dfrac{3}{t}$

16. $\dfrac{1}{x^2-16}-\dfrac{x+4}{x^2-3x-4}$

17. $\dfrac{1}{x-1}+\dfrac{4}{x^2-1}-\dfrac{2}{x^2-2x+1}$

ANSWERS

1. _____

2. _____

3. _____

4. _____

5. _____

6. _____

7. _____

8. _____

9. _____

10. _____

11. _____

12. _____

13. _____

14. _____

15. _____

16. _____

17. _____

18. Simplify: $\dfrac{9 - \dfrac{1}{y^2}}{3 - \dfrac{1}{y}}$.

18. _____

19. _____

Solve.

19. $\dfrac{7}{y} - \dfrac{1}{3} = \dfrac{1}{4}$　　　　　20. $\dfrac{15}{x} - \dfrac{15}{x-2} = -2$

20. _____

21. _____

21. The reciprocal of three less than a number is four times the reciprocal of the number itself. What is the number?

22. A sample of 125 spark plugs contained 4 defective spark plugs. How many defective spark plugs would you expect among 500 spark plugs?

22. _____

23. _____

24. _____

23. One car travels 20 km/h faster than another. While one goes 225 km, the other goes 325 km. Find their speeds.

24. Solve $L = \dfrac{Mt - g}{t}$ for t.

25. _____

26. _____

SKILL MAINTENANCE

25. Factor: $16a^2 - 49$.

26. Simplify: $\left(\dfrac{3x^2}{y^3}\right)^{-4}$.

27. _____

27. Subtract:
$(5x^2 - 19x + 34) - (-8x^2 + 10x - 42)$.

28. A product of two consecutive integers is 462. Find the integers.

28. _____

SYNTHESIS

29. _____

29. Team A and team B work together and complete a job in $2\frac{6}{7}$ hr. It would take team B 6 hr longer, working alone, to do the job than it would team A. How long would it take each of them to do the job working alone?

30. Simplify: $1 + \dfrac{1}{1 + \dfrac{1}{1 + \dfrac{1}{a}}}$.

30. _____

CUMULATIVE REVIEW: CHAPTERS 1–5

Evaluate.

1. $\dfrac{2x + 5}{y - 10}$ when $x = 2$ and $y = 5$

2. $4 - x^3$ when $x = -2$

Simplify.

3. $x - [x - 2(x + 3)]$

4. $(2x^{-2})^{-2}(3x)^3$

5. $\dfrac{24x^8}{18x^{-2}}$

6. $\dfrac{2t^2 + 8t - 42}{2t^2 + 13t - 7}$

7. $\dfrac{\dfrac{2}{x} + 1}{\dfrac{x}{x + 2}}$

8. $\dfrac{a^2 - 16}{a^2 - 8a + 16}$

Add. Simplify, if possible.

9. $\dfrac{9}{14} + \left(-\dfrac{5}{21}\right)$

10. $\dfrac{2x + y}{x^2 y} + \dfrac{x + 2y}{xy^2}$

11. $\dfrac{z}{z^2 - 1} + \dfrac{2}{z + 1}$

12. $(2x^4 + 5x^3 + 4) + (3x^3 - 2x + 5)$

Subtract. Simplify, if possible.

13. $1.53 - (-0.8)$

14. $(x^2 - xy - y^2) - (x^2 - y^2)$

15. $\dfrac{3}{x^2 - 9} - \dfrac{x}{9 - x^2}$

16. $\dfrac{2x}{x^2 - x - 20} - \dfrac{4}{x^2 - 10x + 25}$

Multiply. Simplify, if possible.

17. $(1.3)(-0.5)(2)$

18. $3x^2(2x^2 + 4x - 5)$

19. $(3t + \tfrac{1}{2})(3t - \tfrac{1}{2})$

20. $(2p - q)^2$

21. $(3x + 5)(x - 4)$

22. $(2x^2 + 1)(2x^2 - 1)$

23. $\dfrac{6t + 6}{t^3 - 2t^2} \cdot \dfrac{t^3 - 3t^2 + 2t}{3t + 3}$

24. $\dfrac{a^2 - 1}{a^2} \cdot \dfrac{2a}{1 - a}$

Divide. Simplify, if possible.

25. $(3x^3 - 7x^2 + 9x - 5) \div (x - 1)$

26. $-\dfrac{21}{25} \div \dfrac{28}{15}$

27. $\dfrac{x^2 - x - 2}{4x^3 + 8x^2} \div \dfrac{x^2 - 2x - 3}{2x^2 + 4x}$

28. $\dfrac{3 - 3x}{x^2} \div \dfrac{x - 1}{4x}$

Factor completely.

29. $4x^3 + 12x^2 - 9x - 27$

30. $x^2 + 7x - 8$

31. $3x^2 - 14x - 5$

32. $16y^2 + 40xy + 25x^2$

33. $3x^3 + 24x^2 + 45x$

34. $2x^2 - 2$

35. $x^2 - 28x + 196$

36. $4y^3 + 10y^2 + 12y + 30$

Solve.

37. $2(x - 3) = 5(x + 3)$

38. $2x(3x + 4) = 0$

39. $x^2 = 8x$

40. $x^2 + 16 = 8x$

41. $x - 5 \leq 2x + 4$

42. $3x^2 = 27$

43. $\dfrac{1}{3}x - \dfrac{2}{5} = \dfrac{4}{5}x + \dfrac{1}{3}$

44. $\dfrac{x}{3} = \dfrac{3}{x}$

45. $\dfrac{x + 5}{2x + 1} = \dfrac{x - 7}{2x - 1}$

46. $\dfrac{1}{3}x\left(2x - \dfrac{1}{5}\right) = 0$

47. $\dfrac{3 - x}{x - 1} = \dfrac{2}{x - 1}$

48. $\dfrac{3}{2x + 5} = \dfrac{2}{5 - x}$

49. $\dfrac{1}{x} + \dfrac{1}{y} = \dfrac{1}{z}$, for z

50. $\dfrac{3N}{T} = D$, for N

Solve.

51. The sum of three consecutive integers is 99. What are the integers?

52. The speed of one bicyclist is 2 km/h faster than the speed of another bicyclist. The first bicyclist travels 60 km in the same time it takes the second to travel 50 km. Find the speed of each bicyclist.

53. A swimming pool can be filled in 5 hr by hose A alone and in 6 hr by hose B alone. How long would it take to fill the tank if both hoses were working?

54. The sum of the page numbers on the facing pages of a book is 67. What are the page numbers?

55. The product of the page numbers on two facing pages of a book is 272. Find the page numbers.

56. In a cake recipe, the ratio of flour to sugar is $\frac{6}{5}$. If 2 cups of flour are used, how many cups of sugar are used?

57. The area of a circle is 35π more than the circumference. Find the length of the radius.

58. The sum of the squares of two consecutive odd positive integers is 202. Find the integers.

SYNTHESIS

Solve.

59. $(2x - 1)^2 = (x + 3)^2$

60. $\dfrac{x + 2}{3x + 2} = \dfrac{1}{x}$

61. $\dfrac{2 + \dfrac{2}{x}}{x + 2 + \dfrac{1}{x}} = \dfrac{x + 2}{3}$

62. $\dfrac{x^6 x^4}{x^9 x^{-1}} = \dfrac{5^{14}}{25^6}$

63. Find the reciprocal of $\dfrac{1 - x}{x + 3} + \dfrac{x + 1}{2 - x}$.

64. Find the reciprocal of 2.0×10^{-8} and express in scientific notation.

We now study the graphs of equations in two variables. These variables will be raised only to the first power, and there will be no products or quotients involving variables. We also learn to graph such equations, called *linear*. The graphs of linear equations are straight lines. Later, we assign numbers called *slopes* to the way the lines slant. We will also consider variation and the graphing of inequalities in two variables.

The review sections to be tested in addition to the material in this chapter are 1.8, 4.7, 5.6, and 5.7. ❖

Graphs of Equations and Inequalities

6

AN APPLICATION

A road drops 422.4 ft vertically for every 5280 ft horizontally. What is the grade of the road?

THE MATHEMATICS

The slope of a line is a number related to the slant of a line. This sign shows the grade of a road. It is a number that represents how steeply the road slants. We find slope by computing the ratio of the vertical change to the horizontal change:

$$\text{Grade} = \text{Slope} = \frac{422.4}{5280} = 0.08 = 8\%.$$

Equation-Solving Skills: Sections 2.1–2.3
Inequality-Solving Skills: Section 2.7
Formula-Solving Skills: Section 2.6

PRETEST: CHAPTER 6

Graph on a plane.

1. $y = -x$

2. $x = -4$

3. $4x - 5y = 20$

4. $y = \frac{2}{3}x - 1$

5. In which quadrant is the point $(-4, -1)$ located?

6. Determine whether the ordered pair $(-2, 4)$ is a solution of the equation $2x - y = 0$.

Find the slope, if it exists, of the line.

7. $-4x + y = 6$

8. $y = 3$

9. Find the slope and the y-intercept of the line $x - 3y = 7$.

10. Find the slope, if it exists, of the line containing the points $(3, 0)$ and $(3, 6)$.

11. Find an equation of the line containing the points $(3, -1)$ and $(1, -3)$.

12. Find an equation of the line containing the point $(-1, 3)$ and having slope 4.

13. Find an equation of variation where y varies directly as x and $y = 10$ when $x = 4$.

14. Find an equation of variation where y varies inversely as x and $y = 10$ when $x = 4$.

Graph on a plane.

15. $y < x + 2$

16. $2y - 3x \geq 6$

Determine whether the graphs of the equations are parallel, perpendicular, or neither.

17. $y - 3x = 9,$
$y - 3x = 7$

18. $-x + 2y = 7,$
$2x + y = 4$

19. $y = \frac{2}{3}x - 5,$

$y = -\frac{3}{2}x + 4$

20. Determine whether the ordered pair $(-3, 4)$ is a solution of $2x + 5y < 17$.

6.1 Graphs and Equations

We often see graphs of various kinds in newspapers and magazines and on television. In the past decade, graphs have been more common because they are easy to prepare using a computer. We first solve some problems with commonly used graphs.

a   Problem Solving with Graphs

Bar Graphs

Graphs are used to illustrate information. A *bar graph* is convenient for showing comparisons. The bars may be vertical or horizontal. Typically, certain units, such as percent in the case of Example 1, are shown horizontally. With each horizontal number, there is associated a vertical number, or unit. In the case of Example 1, the vertical units are not numbers, but various "reasons" for dropping out of high school.

Let us solve some problems with this graph.

▶ **EXAMPLE 1** These reasons for dropping out of high school were given in a recent National Assessment of Educational Progress survey.

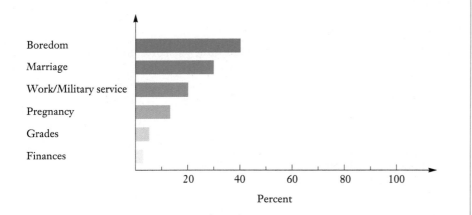

a) Approximately what percent in the survey dropped out because of pregnancy?

b) What reason was given least often for dropping out?

c) What reason was given by about 30% for dropping out?

a) We go to the right of the bar representing pregnancy and then go down to the percent scale. We find that approximately 12% dropped out because of pregnancy. Note that this is truly an estimate. We would need more detailed scaling to make a closer determination.

b) The shortest bar represents finances.

c) We go across to the 30% mark on the percent scale and then up until we reach a bar ending at approximately 30%. We then go across to the left and read the reason. The reason given by about 30% was marriage. ◀

DO EXERCISE 1.

OBJECTIVES

After finishing Section 6.1, you should be able to:

a Solve problems related to bar, line, and circle graphs.

b Plot points associated with ordered pairs of numbers.

c Determine the quadrant in which a point lies.

d Find the coordinates of a point on a graph.

FOR EXTRA HELP

Tape 12D   Tape 12B   MAC: 6
                                      IBM: 6

1. Consider the graph in Example 1.

   a) Approximately how many in the survey dropped out because of boredom?

   b) What reason was given by about 5% for dropping out?

2. Consider the graph in Example 2.

a) For which week was the DJIA closing about 2000?

Line Graphs

We often use *line graphs* to show change over time. Certain horizontal units are associated with vertical units. Each association determines a point. When the points are connected by line segments, we get a line graph.

▶ **EXAMPLE 2**   This line graph shows the closing Dow Jones Industrial Average (DJIA) for each of six weeks.

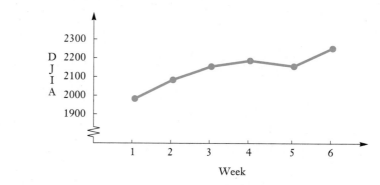

a) For which week was the DJIA closing the lowest?

b) Between which two weeks did the DJIA closing decrease?

c) For which week was the DJIA closing about 2200?

a) During the six weeks, the lowest closing was 2000 at the end of the first week.

b) Reading the graph from left to right, we see that the line went down only between the fourth and fifth weeks.

c) We locate 2200 on the DJIA scale and then move to the right until we reach the line. At that point, we move down to the "Week" scale and read the information we are seeking. For the fourth week, the DJIA closing was about 2200.   ◀

b) For which week was the DJIA closing about 2150?

DO EXERCISE 2.

Circle Graphs

Circle graphs are often used to show the percent of a quantity for each particular item in a group.

▶ **EXAMPLE 3**   This circle graph shows expenses as a percent of income by a family of four, according to the Bureau of Labor Statistics.

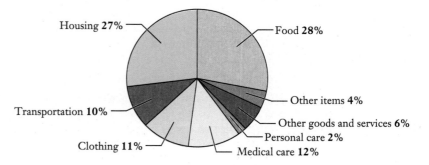

A family with $2000 monthly income would typically spend how much on housing per month?

1. *Familiarize.* The graph tells us that housing is 27% of income. We let y = the amount spent on housing.

2. *Translate.* We restate and translate the problem as follows:

Restate: What is 27% of income?

Translate: $y = 27\% \cdot \$2000$

3. *Solve.* We solve by carrying out the computation:

$$y = 0.27 \cdot \$2000 = \$540.$$

4. *Check.* We leave the check to the student.

5. *State.* The family would spend $540 on housing.

DO EXERCISE 3.

b Points and Ordered Pairs

We have graphed numbers on a line. To enable us to graph an equation that contains two variables, we now learn to graph number pairs on a plane.

On a number line each point is the graph of a number. On a plane, each point is the graph of a number pair. We use two perpendicular number lines called **axes.** They cross at a point called the **origin.** It has coordinates $(0, 0)$ but is usually labeled with the number 0. The arrows show the positive directions. Consider the ordered pair $(3, 4)$. The numbers in an ordered pair are called **coordinates.** In $(3, 4)$, the **first coordinate** is 3 and the **second coordinate** is 4. To plot $(3, 4)$, we start at the origin and move horizontally to the 3. Then we move up vertically 4 units and make a "dot." We can also plot this point by starting at the origin, going up 4 units vertically, and then moving to the right 3 units horizontally.

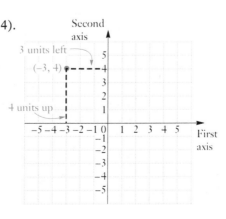

The point $(4, 3)$ is also plotted. Note that $(3, 4)$ and $(4, 3)$ give different points. The order of the numbers in the pair is indeed important. They are called **ordered pairs** because it makes a difference which number comes first.

▶ **EXAMPLE 4** Plot the point $(-3, 4)$.

The first number, -3, is negative. Starting at the origin, we go -3 units in the horizontal direction (3 units to the left). The second number, 4, is positive. We go 4 units in the vertical direction (up). We could also go up 4 units and then 3 units to the left.

3. Consider the graph in Example 3.

a) A family with a $2000 monthly income would typically spend how much on transportation per month?

b) What percent of the income is spent on food?

Plot these points on the graph below.

4. $(4, 5)$ 5. $(5, 4)$

6. $(-2, 5)$ 7. $(-3, -4)$

8. $(5, -3)$ 9. $(-2, -1)$

10. $(0, -3)$ 11. $(2, 0)$

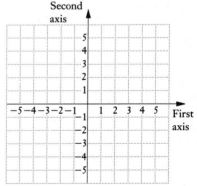

12. What can you say about the coordinates of a point in the third quadrant?

DO EXERCISES 4–11 ON THE PRECEDING PAGE.

c **Quadrants**

This figure shows some points and their coordinates. In region I (the *first quadrant*), both coordinates of any point are positive. In region II (the *second quadrant*), the first coordinate is negative and the second positive. In region III (the *third quadrant*), both coordinates are negative. In region IV (the *fourth quadrant*), the first coordinate is positive and the second is negative.

13. What can you say about the coordinates of a point in the fourth quadrant?

The point $(-4, 5)$ is in the second quadrant. The point $(5, -5)$ is in the fourth quadrant.

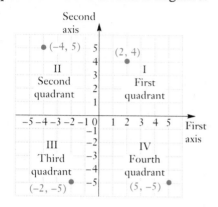

In which quadrant is the point located?

14. (5, 3)

DO EXERCISES 12–17.

15. $(-6, -4)$

d **Finding Coordinates**

To find the coordinates of a point, we see how far to the right or left of zero it is located and how far up or down.

16. $(10, -14)$

17. $(-13, 9)$

▶ **EXAMPLE 5**   Find the coordinates of points *A, B, C, D, E, F,* and *G.*

18. Find the coordinates of points *A, B, C, D, E, F,* and *G* on the graph below.

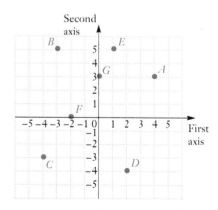

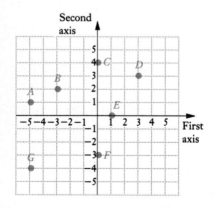

Point *A* is 4 units to the right (horizontal direction) and 3 units up (vertical direction). Its coordinates are (4, 3). The coordinates of the other points are as follows:

B: $(-3, 5)$;   *C*: $(-4, -3)$;   *D*: $(2, -4)$;
E: $(1, 5)$;   *F*: $(-2, 0)$;   *G*: $(0, 3)$.   ◀

DO EXERCISE 18.

EXERCISE SET 6.1

ANSWERS

a Use the bar graph in Example 1 to answer Exercises 1–3.

1. What reason was given most for dropping out?

2. What reason was given by about 20% for dropping out?

3. What percent in the survey dropped out because of grades?

Use the line graph in Example 2 to answer Exercises 4–6.

4. For which week was the DJIA closing the highest?

5. For which week was the DJIA closing about 2100?

6. About how many points was the increase in the DJIA between weeks 1 and 6?

Use the circle graph in Example 3 to answer Exercises 7 and 8.

7. What percent of the income is spent on medical expenses?

8. A family with $2000 monthly income would typically spend how much on clothing per month?

Use the following bar graph for Exercises 9–14. The graph shows the average daily expenses for lodging, food, and car rental for traveling executives in various cities.

9. What are the average daily expenses in New York?

10. What are the average daily expenses in Chicago?

11. Which city is least expensive?

12. Which city is most expensive?

13. How much more are the average daily expenses in Washington than in San Francisco?

14. How much less are the average daily expenses in Chicago than in Washington?

1. _____

2. _____

3. _____

4. _____

5. _____

6. _____

7. _____

8. _____

9. _____

10. _____

11. _____

12. _____

13. _____

14. _____

Use the following line graph for Exercises 15–20. The graph shows the estimated sales (in millions) for a company for several years.

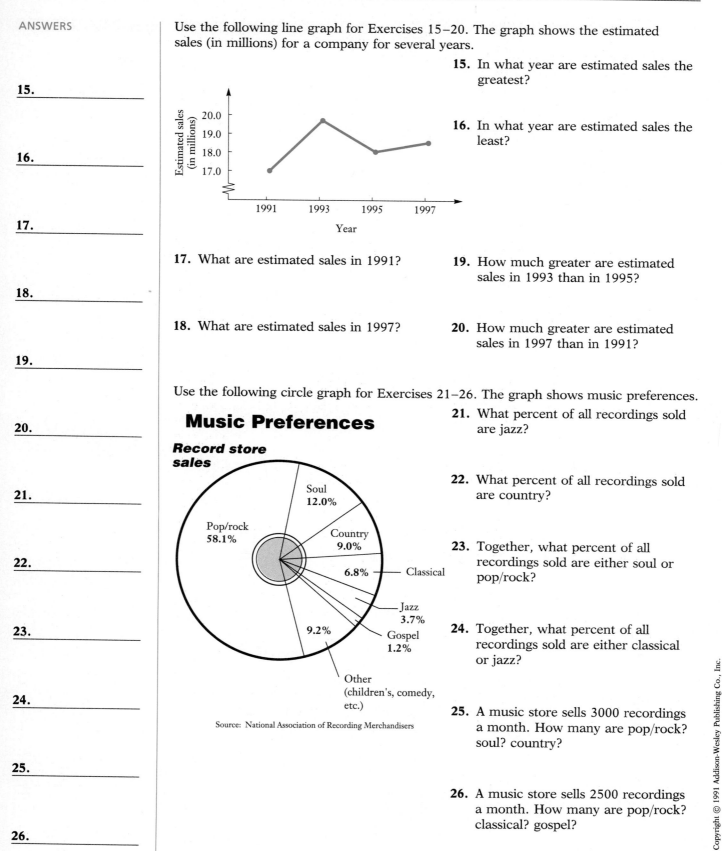

15. _____

16. _____

17. _____

18. _____

19. _____

20. _____

21. _____

22. _____

23. _____

24. _____

25. _____

26. _____

15. In what year are estimated sales the greatest?

16. In what year are estimated sales the least?

17. What are estimated sales in 1991?

18. What are estimated sales in 1997?

19. How much greater are estimated sales in 1993 than in 1995?

20. How much greater are estimated sales in 1997 than in 1991?

Use the following circle graph for Exercises 21–26. The graph shows music preferences.

Music Preferences

Record store sales

Soul 12.0%

Pop/rock 58.1%

Country 9.0%

6.8% — Classical

Jazz 3.7%

9.2%

Gospel 1.2%

Other (children's, comedy, etc.)

Source: National Association of Recording Merchandisers

21. What percent of all recordings sold are jazz?

22. What percent of all recordings sold are country?

23. Together, what percent of all recordings sold are either soul or pop/rock?

24. Together, what percent of all recordings sold are either classical or jazz?

25. A music store sells 3000 recordings a month. How many are pop/rock? soul? country?

26. A music store sells 2500 recordings a month. How many are pop/rock? classical? gospel?

b

27. Plot these points.

$(2, 5)$ $(-1, 3)$ $(3, -2)$ $(-2, -4)$

$(0, 4)$ $(0, -5)$ $(5, 0)$ $(-5, 0)$

28. Plot these points.

$(4, 4)$ $(-2, 4)$ $(5, -3)$ $(-5, -5)$

$(0, 4)$ $(0, -4)$ $(3, 0)$ $(-4, 0)$

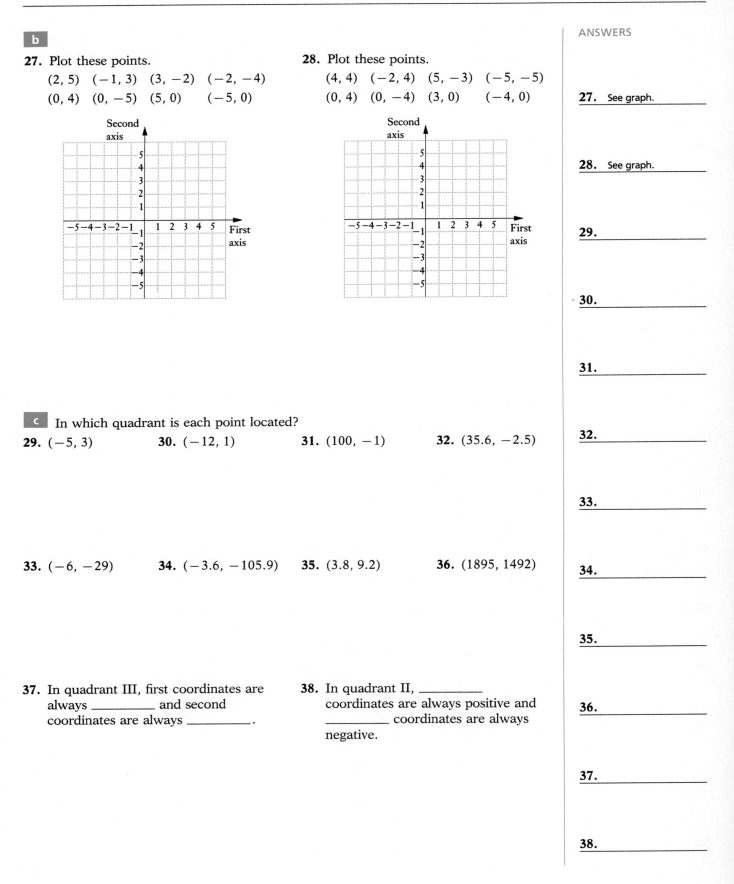

c In which quadrant is each point located?

29. $(-5, 3)$ **30.** $(-12, 1)$ **31.** $(100, -1)$ **32.** $(35.6, -2.5)$

33. $(-6, -29)$ **34.** $(-3.6, -105.9)$ **35.** $(3.8, 9.2)$ **36.** $(1895, 1492)$

37. In quadrant III, first coordinates are always _____ and second coordinates are always _____ .

38. In quadrant II, _____ coordinates are always positive and _____ coordinates are always negative.

ANSWERS

27. See graph.

28. See graph.

29. _____

30. _____

31. _____

32. _____

33. _____

34. _____

35. _____

36. _____

37. _____

38. _____

d

39. Find the coordinates of points A, B, C, D, and E.

40. Find the coordinates of points A, B, C, D, and E.

SKILL MAINTENANCE

41. A family making $19,600 a year will spend $5096 on food. At this rate, how much would a family making $20,500 spend on food?

42. Two cars leave town at the same time and drive in opposite directions. One travels at 54 mph and the other travels at 43 mph. In how many hours will they be 291 mi apart?

SYNTHESIS

In Exercises 43–46, tell in which quadrant(s) the point can be located.

43. The first coordinate is positive.

44. The second coordinate is negative.

45. The first and second coordinates are equal.

46. The first coordinate is the additive inverse of the second coordinate.

47. The points $(-1, 1)$, $(4, 1)$, and $(4, -5)$ are three vertices of a rectangle. Find the coordinates of the fourth vertex.

48. Three parallelograms share the vertices $(-2, -3)$, $(-1, 2)$, and $(4, -3)$. Find the fourth vertex of each parallelogram.

49. Graph eight points such that the sum of the coordinates in each pair is 6.

50. Graph eight points such that the first coordinate minus the second coordinate is 1.

51. Find the perimeter of a rectangle whose vertices have coordinates $(5, 3)$, $(5, -2)$, $(-3, -2)$, and $(-3, 3)$.

52. Find the area of a triangle whose vertices have coordinates $(0, 9)$, $(0, -4)$, and $(5, -4)$.

6.2 Graphing Linear Equations

A **linear equation** is equivalent to an equation of the type $Ax + By = C$. We now learn how to graph certain kinds of linear equations. We will see that the graphs of linear equations are straight lines.

a Solutions of Equations

An equation with two variables has *ordered pairs* of numbers for solutions. If not directed otherwise, we usually take the variables in alphabetical *order*. Then we get *ordered* pairs for solutions.

▶ **EXAMPLE 1** Determine whether $(-3, -5)$ is a solution of $y = 2x + 1$. We substitute:

$$y = 2x + 1$$

$$\begin{array}{c|c} -5 & 2(-3) + 1 \\ \hline & -6 + 1 \\ & -5 \end{array}$$

We substitute -3 for x and -5 for y (alphabetical order of variables).

Since the equation becomes true, $(-3, -5)$ is a solution. ◀

Similarly, in Example 1 we can show that $(3, 7)$ and $(0, 1)$ are solutions. In fact, there are more solutions than we can list: There is an infinite number of solutions.

▶ **EXAMPLE 2** Determine whether $(-2, 3)$ is a solution of $2t = 4s - 8$. We substitute:

$$2t = 4s - 8$$

$$\begin{array}{c|c} 2 \cdot 3 & 4(-2) - 8 \\ \hline 6 & -8 - 8 \\ & -16 \end{array}$$

We substitute -2 for s and 3 for t.

Since the equation becomes false, $(-2, 3)$ is not a solution. ◀

DO EXERCISES 1 AND 2.

b Graphing Equations of the Type $y = mx$ and $y = mx + b$

The equations considered in Examples 1 and 2 have an infinite number of solutions, meaning that we cannot list them all. Because of this, it is convenient to make a drawing that represents the solutions. Such a drawing is called a **graph**.

> To *graph* an equation means to make a drawing that represents its solutions.

The graphs of linear equations of the type $y = mx$ and $y = mx + b$ are straight lines. If an equation has a graph that is a straight line, we can graph it by plotting two or more points and then drawing a line through them.

OBJECTIVES

After finishing Section 6.2, you should be able to:

a Determine whether an ordered pair of numbers is a solution of an equation with two variables.

b Graph equations of the type $y = mx$ and $y = mx + b$.

FOR EXTRA HELP

Tape 12E   Tape 12B   MAC: 6
                          IBM: 6

1. Determine whether $(2, 3)$ is a solution of $y = 2x + 3$.

2. Determine whether $(-2, 4)$ is a solution of $4q - 3p = 22$.

ANSWERS ON PAGE A-6

Graph.

3. $y = 3x$

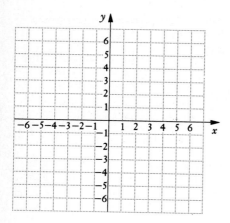

4. $y = \dfrac{1}{2}x$

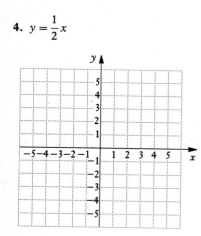

▶ **EXAMPLE 3**   Graph: $y = x$.

We will use alphabetical order. Thus the first (horizontal) axis will be the x-axis and the second (vertical) axis will be the y-axis. Next, we find some ordered pairs that are solutions of the equation, keeping the results in a table. We choose *any* number for x and then find y by substitution. In this case, it is easy. Here are a few:

Let $x = 0$.   Then $y = x = 0$.   We get a solution: the ordered pair (0, 0).

Let $x = 1$.   Then $y = x = 1$.   We get a solution: the ordered pair (1, 1).

Let $x = 5$.   Then $y = x = 5$.   We get a solution: the ordered pair (5, 5).

Let $x = -2$.   Then $y = x = -2$.   We get a solution: the ordered pair $(-2, -2)$.

Let $x = -4$.   Then $y = x = -4$.   We get a solution: the ordered pair $(-4, -4)$.

We gather our results in a table like the one shown below. Then we plot the points. We look for a pattern in the points plotted. It looks as if the points resemble a straight line. We draw the line with a ruler. Since the line is the graph of the equation $y = x$, we label the line $y = x$ on the graph paper.

x	y $y = x$	(x, y)
0	0	(0, 0)
1	1	(1, 1)
5	5	(5, 5)
-2	-2	$(-2, -2)$
-4	-4	$(-4, -4)$

(1) Choose x.
(2) Compute y.
(3) Form the pair (x, y).
(4) Plot the points.

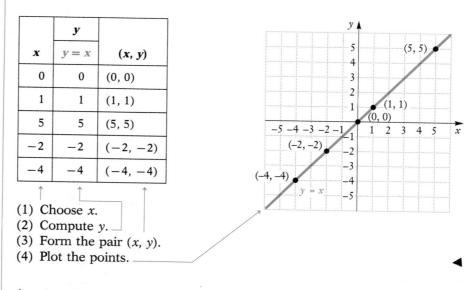

◀

▶ **EXAMPLE 4**   Graph: $y = 2x$.

We find some ordered pairs that are solutions. Since the graph is a line, we really need to find only two, but we will usually plot a third point as a check. We keep the results in a table. We choose *any* number for x and then determine y by substitution. Suppose we choose 3 for x. Then

$$y = 2x = 2 \cdot 3 = 6.$$

We get a solution: the ordered pair (3, 6). Suppose we choose 0 for x. Then

$$y = 2x = 2 \cdot 0 = 0.$$

We get a solution: the ordered pair (0, 0). For a third point, we make a negative choice for x. We now have enough points to plot the line, but if we wish we can compute more. If a number takes us off the graph paper, we generally do not use it. Continuing in this manner, we get a table like the one shown below. Since $y = 2x$, we get y by doubling x.

Now we plot these points. We draw the line, or graph, with a ruler and label it $y = 2x$.

x	y $y = 2x$	(x, y)
3	6	$(3, 6)$
1	2	$(1, 2)$
0	0	$(0, 0)$
-2	-4	$(-2, -4)$
-3	-6	$(-3, -6)$

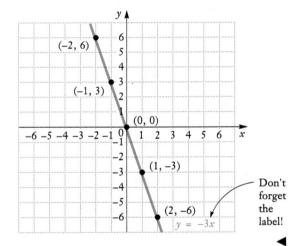

DO EXERCISES 3 AND 4 ON THE PRECEDING PAGE.

▶ **EXAMPLE 5** Graph: $y = -3x$.

We make a table of solutions. Then we plot points. We draw the line with a ruler and label it $y = -3x$.

x	y $y = -3x$	(x, y)
0	0	$(0, 0)$
1	-3	$(1, -3)$
-1	3	$(-1, 3)$
2	-6	$(2, -6)$
-2	6	$(-2, 6)$

Don't forget the label!

DO EXERCISES 5 AND 6.

▶ **EXAMPLE 6** Graph: $y = -\frac{5}{3}x$.

We make a table of solutions.

When $x = 0$, $y = -\frac{5}{3} \cdot 0 = 0$.
When $x = 3$, $y = -\frac{5}{3} \cdot 3 = -5$.
When $x = -3$, $y = -\frac{5}{3} \cdot (-3) = 5$.
When $x = 1$, $y = -\frac{5}{3} \cdot 1 = -\frac{5}{3}$.

Note that if we substitute multiples of 3, we can avoid fractions.

Next we plot the points and complete the graph by drawing a line through them.

Graph.

5. $y = -x$ (or $-1 \cdot x$)

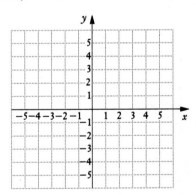

6. $y = -2x$

ANSWERS ON PAGE A-6

Graph.

7. $y = \dfrac{3}{4}x$

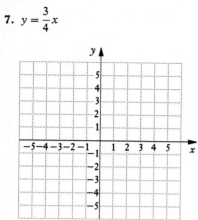

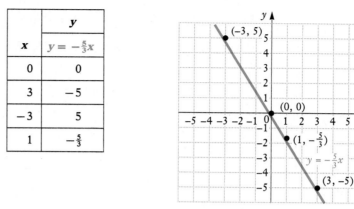

		y
x		$y = -\frac{5}{3}x$
0		0
3		-5
-3		5
1		$-\frac{5}{3}$

DO EXERCISES 7 AND 8.

Every equation $y = mx$ has a graph that is a straight line. It contains the origin, $(0, 0)$. What will happen if we add a number b on the right side to get an equation $y = mx + b$?

▶ **EXAMPLE 7**   Graph $y = x$ and $y = x + 2$ using the same set of axes. Compare.

We first make a table containing values for both equations.

x	y $y = x$	y $y = x + 2$
0	0	2
1	1	3
-1	-1	1
2	2	4
-2	-2	0
3	3	5

8. $y = -\dfrac{4}{5}x$

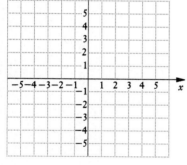

We then plot these points. We draw a dashed line for $y = x$ and a solid line for $y = x + 2$. We see that the graph of $y = x + 2$ can be obtained from the graph of $y = x$ by moving, or translating, the graph of $y = x$ up 2 units.

9. Graph $y = x + 3$ and compare it with the graph of $y = x$.

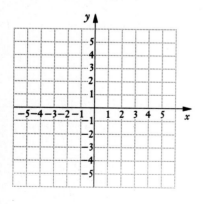

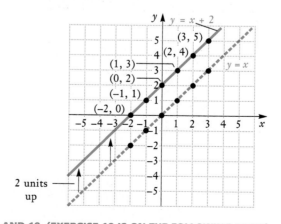

DO EXERCISES 9 AND 10. (EXERCISE 10 IS ON THE FOLLOWING PAGE.)

▶ **EXAMPLE 8** Graph $y = 2x$ and $y = 2x - 3$ using the same set of axes. Compare.

We first make a table containing values for both equations.

x	y y = 2x	y y = 2x − 3
0	0	−3
1	2	−1
2	4	1
−1	−2	−5

The graph of $y = 2x - 3$ looks just like the graph of $y = 2x$, but $y = 2x$ is moved, or translated, down 3 units.

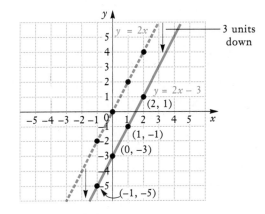

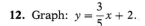

DO EXERCISE 11.

▶ **EXAMPLE 9** Graph: $y = \frac{2}{5}x + 4$.

We make a table of values. Using multiples of 5 avoids fractions.

When $x = 0$, $y = \frac{2}{5} \cdot 0 + 4 = 0 + 4 = 4$.

When $x = 5$, $y = \frac{2}{5} \cdot 5 + 4 = 2 + 4 = 6$.

When $x = -5$, $y = \frac{2}{5} \cdot (-5) + 4 = -2 + 4 = 2$.

Since two points determine a line, that is all we really need to graph a line, but we will usually plot a third point as a check.

x	y
0	4
5	6
−5	2

10. Graph $y = x - 1$ and compare it with the graph of $y = x$.

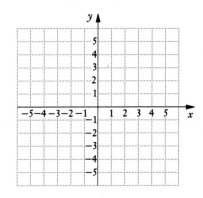

11. Graph $y = 2x + 3$ and compare it with the graph of $y = 2x$.

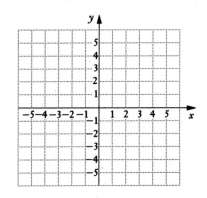

12. Graph: $y = \dfrac{3}{5}x + 2$.

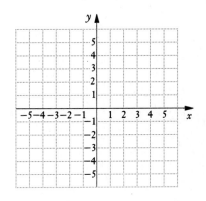

ANSWERS ON PAGE A-6

Graph.

13. $y = \dfrac{3}{5}x - 2$

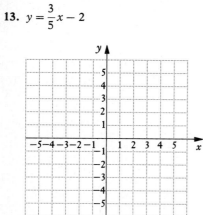

14. $y = -\dfrac{3}{5}x - 1$

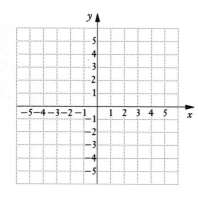

15. $y = -\dfrac{3}{5}x + 4$

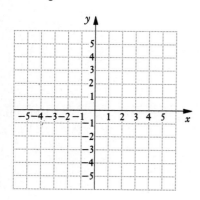

We draw the graph of $y = \frac{2}{5}x + 4$.

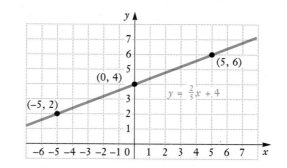

▶ **EXAMPLE 10**　　Graph: $y = -\frac{3}{4}x - 2$.

We first make a table of values.

When $x = 0$, $y = -\frac{3}{4} \cdot 0 - 2 = 0 - 2 = -2$.

When $x = 4$, $y = -\frac{3}{4} \cdot 4 - 2 = -3 - 2 = -5$.

When $x = -4$, $y = -\frac{3}{4}(-4) - 2 = 3 - 2 = 1$.

x	y
0	-2
4	-5
-4	1

We plot these points and draw a line through them. This line is a graph of the equation. We label the graph $y = -\frac{3}{4}x - 2$.

We plot this point for a check to see whether it is on the line.

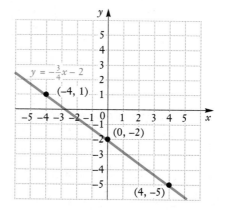

DO EXERCISES 12–15. (EXERCISE 12 IS ON THE PRECEDING PAGE.)

NAME   SECTION   DATE

EXERCISE SET 6.2

a Determine whether the given point is a solution of the equation.

1. $(2, 5);$ $\quad y = 3x - 1$

2. $(1, 7);$ $\quad y = 2x + 5$

3. $(2, -3);$ $\quad 3x - y = 4$

4. $(-1, 4);$ $\quad 2x + y = 6$

5. $(-2, -1);$ $2c + 2d = -7$

6. $(0, -4);$ $4p + 2q = -9$

1. _____

2. _____

3. _____

4. _____

5. _____

6. _____

b Graph.

7. $y = 4x$

8. $y = 2x$

9. $y = -2x$

10. $y = -4x$

11. $y = \dfrac{1}{3}x$

12. $y = \dfrac{1}{4}x$

13. $y = -\dfrac{3}{2}x$

14. $y = -\dfrac{5}{4}x$

15. $y = x + 1$

16. $y = -x + 1$

17. $y = 2x + 2$

18. $y = 3x - 2$

19. $y = \frac{1}{3}x - 1$

20. $y = \frac{1}{2}x + 1$

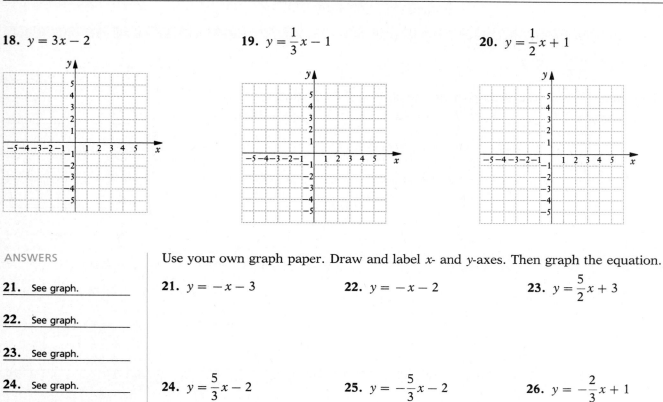

21. See graph.

22. See graph.

23. See graph.

24. See graph.

25. See graph.

26. See graph.

27. See graph.

28. See graph.

29. See graph.

30. See graph.

31. See graph.

32. See graph.

33. _____

34. _____

35. _____

36. _____

37. _____

38. _____

Use your own graph paper. Draw and label x- and y-axes. Then graph the equation.

21. $y = -x - 3$

22. $y = -x - 2$

23. $y = \frac{5}{2}x + 3$

24. $y = \frac{5}{3}x - 2$

25. $y = -\frac{5}{3}x - 2$

26. $y = -\frac{2}{3}x + 1$

27. $y = x$

28. $y = -x$

29. $y = 3 - 2x$

30. $y = 7 - 5x$

31. $y = \frac{4}{3} - \frac{1}{3}x$

32. $y = -\frac{1}{4}x - \frac{1}{2}$

SKILL MAINTENANCE

33. An airplane flew for 7 hr with a 5 km/h tailwind. The return flight against the wind took 8 hr. Find the speed of the plane in still air.

Solve.

34. $\dfrac{3}{x - 5} + \dfrac{1}{x + 5} = \dfrac{2}{x^2 - 25}$ **35.** $25t^2 - 49 = 0$ **36.** $x^2 - 4x = 0$

SYNTHESIS

37. Find all the whole-number solutions of $x + y = 6$.

38. Find three solutions of $y = |x|$.

6.3 More on Graphing Linear Equations

a Graphing Using Intercepts

We graphed equations of the type $y = mx$ and $y = mx + b$ in Section 6.2. We now consider equations of the type $Ax + By = C$, where $A \neq 0$ and $B \neq 0$. These equations can be graphed conveniently using **intercepts.** Look at the graph of $y - 2x = 4$ shown below. We could graph this equation by solving for y to get $y = 2x + 4$ and proceed as before, but we want to develop a faster method.

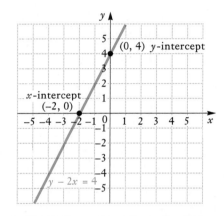

The y-intercept is $(0, 4)$. It occurs where the line crosses the y-axis and thus will always have 0 as the first coordinate. The x-intercept is $(-2, 0)$. It occurs where the line crosses the x-axis and thus will always have 0 as the second coordinate.

DO EXERCISE 1.

We find intercepts as follows.

> **The y-intercept is $(0, b)$. To find b, let $x = 0$ and solve the original equation for y.**
>
> **The x-intercept is $(a, 0)$. To find a, let $y = 0$ and solve the original equation for x.**

Now let's draw a graph using intercepts.

▶ **EXAMPLE 1** Graph: $4x + 3y = 12$.

To find the y-intercept, let $x = 0$. Then solve for y:

$$4 \cdot 0 + 3y = 12$$
$$3y = 12$$
$$y = 4.$$

Thus, $(0, 4)$ is the y-intercept. Note that this amounts to covering up the x-term and looking at the rest of the equation.

To find the x-intercept, let $y = 0$. Then solve for x:

$$4x + 3 \cdot 0 = 12$$
$$4x = 12$$
$$x = 3.$$

OBJECTIVES

After finishing Section 6.3, you should be able to:

a Find the intercepts of a linear equation, and graph using intercepts.

b Graph equations of the type $x = a$ or $y = b$.

FOR EXTRA HELP

Tape 12F   Tape 13A   MAC: 6
                        IBM: 6

1. Look at the graph shown below.

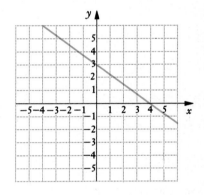

a) Find the coordinates of the x-intercept.

b) Find the coordinates of the y-intercept.

ANSWERS ON PAGE A-6

Graph using intercepts.

2. $2x + 3y = 6$

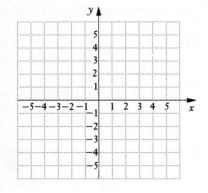

3. $3y - 4x = 12$

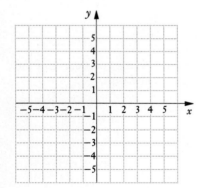

Thus, $(3, 0)$ is the x-intercept. Note that this amounts to covering up the y-term and looking at the rest of the equation.

We plot these points and draw the line, or graph.

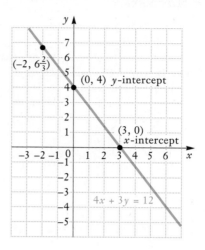

A third point should be used as a check. We substitute any convenient value for x and solve for y. In this case, we choose $x = -2$. Then

$$4(-2) + 3y = 12 \qquad \text{Substituting } -2 \text{ for } x$$
$$-8 + 3y = 12$$
$$3y = 12 + 8 = 20$$
$$y = \tfrac{20}{3}, \text{ or } 6\tfrac{2}{3}. \qquad \text{Solving for } y$$

It appears that the point $(-2, 6\tfrac{2}{3})$ is on the graph, though graphing fractional values can be inexact. The graph is probably correct. ◀

Graphs of equations of the type $y = mx$ pass through the origin. Thus the x-intercept and the y-intercept are the same, $(0, 0)$. In such cases, we must calculate another point in order to complete the graph. Another point would also have to be calculated if a check is desired.

DO EXERCISES 2 AND 3.

b **Equations Whose Graphs are Horizontal or Vertical Lines**

▶ **EXAMPLE 2** Graph: $y = 3$.

Consider $y = 3$. We can also think of this equation as $0 \cdot x + y = 3$. No matter what number we choose for x, we find that y is 3. We make up a table with all 3's in the y-column.

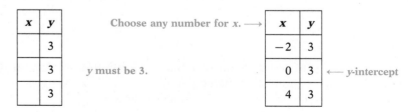

x	y			x	y	
	3	Choose any number for x. →		-2	3	
	3	y must be 3.		0	3	← y-intercept
	3			4	3	

Now when we plot the ordered pairs $(-2, 3)$, $(0, 3)$, and $(4, 3)$ and connect the points, we will obtain a horizontal line. Any ordered pair $(x, 3)$ is a solution. So the line is parallel to the x-axis with y-intercept $(0, 3)$.

Graph.

4. $x = 5$

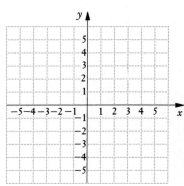

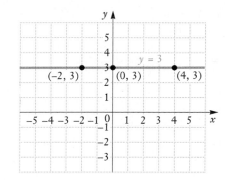

► **EXAMPLE 3** Graph: $x = -4$.

Consider $x = -4$. We can also think of this equation as $x + 0 \cdot y = -4$. We make up a table with all -4's in the x-column.

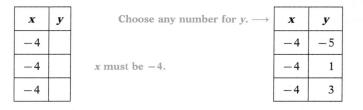

x	y
-4	
-4	
-4	

Choose any number for y. →

x must be -4.

x	y
-4	-5
-4	1
-4	3

When we plot the ordered pairs $(-4, -5)$, $(-4, 1)$, and $(-4, 3)$ and connect them, we will obtain a vertical line. Any ordered pair $(-4, y)$ is a solution. So the line is parallel to the y-axis with x-intercept $(-4, 0)$.

5. $y = -2$

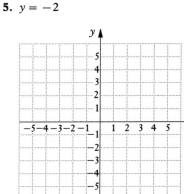

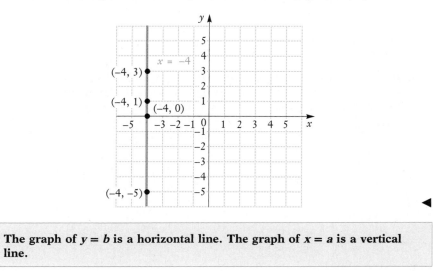

The graph of $y = b$ is a horizontal line. The graph of $x = a$ is a vertical line.

6. $x = 0$

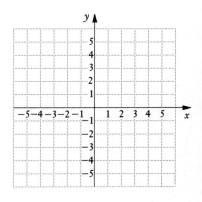

DO EXERCISES 4–7. (EXERCISE 7 IS ON THE FOLLOWING PAGE.)

ANSWERS ON PAGE A-7

7. Graph: $x = -3$.

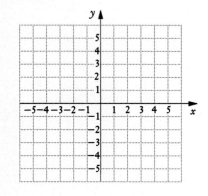

The following is a general procedure for graphing linear equations.

To Graph Linear Equations

1. **Is the equation of the type $x = a$ or $y = b$? If so, the graph will be a line parallel to an axis.**

   Examples.

2. **If the line is of the type $y = mx$, both intercepts are the origin, $(0, 0)$. Plot $(0, 0)$ and one other point. A third point can be calculated as a check.**

   Example.

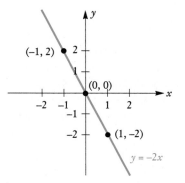

3. **If the equation is of the type $Ax + By = C$, but not of the type $x = a$, $y = b$, or $y = mx$, graph using intercepts. If the intercepts are too close together, choose another point farther from the origin.**

   Examples.

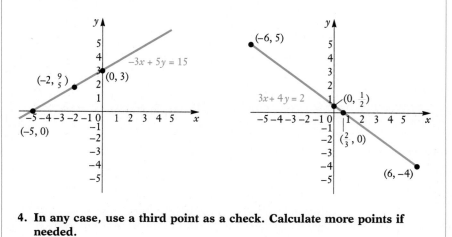

4. **In any case, use a third point as a check. Calculate more points if needed.**

NAME   SECTION   DATE

EXERCISE SET 6.3

a Find the intercepts. Then graph.

1. $x + 3y = 6$ **2.** $x + 2y = 8$ **3.** $-x + 2y = 4$ **4.** $-x + 3y = 9$

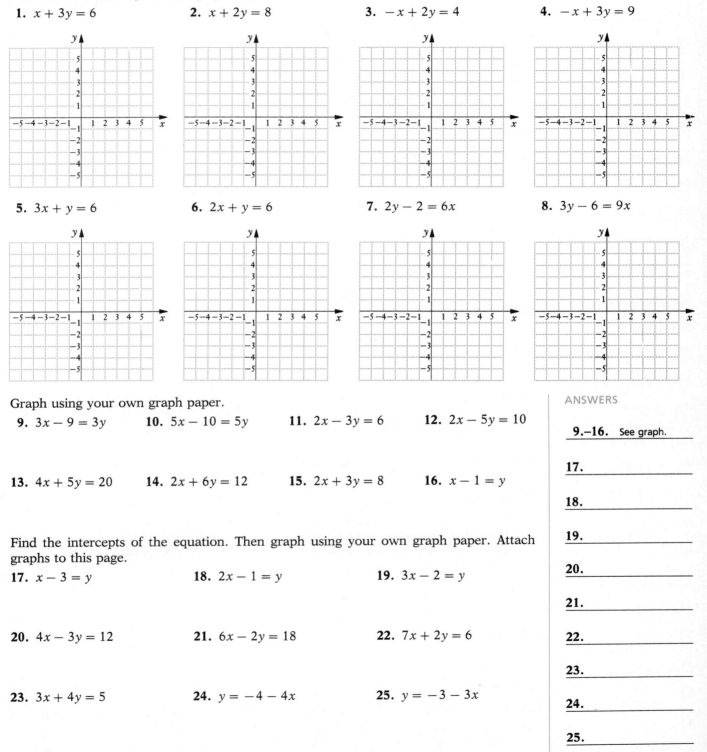

5. $3x + y = 6$ **6.** $2x + y = 6$ **7.** $2y - 2 = 6x$ **8.** $3y - 6 = 9x$

Graph using your own graph paper.

9. $3x - 9 = 3y$ **10.** $5x - 10 = 5y$ **11.** $2x - 3y = 6$ **12.** $2x - 5y = 10$

13. $4x + 5y = 20$ **14.** $2x + 6y = 12$ **15.** $2x + 3y = 8$ **16.** $x - 1 = y$

Find the intercepts of the equation. Then graph using your own graph paper. Attach graphs to this page.

17. $x - 3 = y$ **18.** $2x - 1 = y$ **19.** $3x - 2 = y$

20. $4x - 3y = 12$ **21.** $6x - 2y = 18$ **22.** $7x + 2y = 6$

23. $3x + 4y = 5$ **24.** $y = -4 - 4x$ **25.** $y = -3 - 3x$

ANSWERS

9.–16. See graph.

17. _____

18. _____

19. _____

20. _____

21. _____

22. _____

23. _____

24. _____

25. _____

26. $-3x = 6y - 2$

27. $-4x = 8y - 5$

28. $3 = 2x - 5y$

29. $y - 3x = 0$

30. $x + 2y = 0$

b Graph.

31. $x = -2$

32. $x = -1$

33. $y = 2$

34. $y = 4$

Graph using your own graph paper. Attach graphs to this page.

35. $x = 2$

36. $x = 3$

37. $y = 0$

38. $y = -1$

39. $x = \dfrac{3}{2}$

40. $x = -\dfrac{5}{2}$

41. $3y = -5$

42. $12y = 45$

43. $4x + 3 = 0$

44. $-3x + 12 = 0$

45. $48 - 3y = 0$

46. $63 + 7y = 0$

SKILL MAINTENANCE

47. Crew A can paint a house in 10 hr. Crew B can paint the same house in 12 hr. How many hours would it take both crews to paint the house if they worked together?

48. One car travels 20 km/h faster than another. While one car goes 250 km, the other goes 140 km. Find their speeds.

Solve.

49. $\dfrac{7}{1 + x} - 1 = \dfrac{5x}{x^2 + 3x + 2}$

50. $\dfrac{x}{x + 1} = 3 - \dfrac{33}{x^2 - 1}$

SYNTHESIS

51. Write an equation for the y-axis.

52. Write an equation for the x-axis.

53. Find the coordinates of the point of intersection of the graphs of the equations $x = -3$ and $y = 6$.

54. Find the value of m in $y = mx + 3$ so that the x-intercept of its graph will be (2, 0).

6.4 Slope and Equations of Lines

OBJECTIVES

After finishing Section 6.4, you should be able to:

a Given two points of a line, find the slope.

b Find the slope of a line from an equation.

c Given any equation, derive the equivalent slope–intercept equation and determine the slope and the *y*-intercept.

d Find an equation of a line given a point on the line and the slope, or given two points on the line.

FOR EXTRA HELP

Tape 13A   Tape 13A   MAC: 6
                                    IBM: 6

a Slope

The graphs of some linear equations slant upward from left to right. Others slant downward. Some are vertical and some are horizontal. Some slant more steeply than others. We now look for a way to describe such possibilities with numbers.

Consider a line with two points marked P and Q. As we move from P to Q, the y-coordinate changes from 1 to 3 and the x-coordinate changes from 2 to 6. The change in y is $3 - 1$, or 2. The change in x is $6 - 2$, or 4.

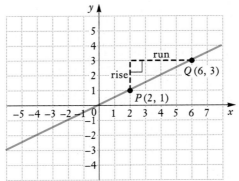

We call the change in y the **rise** and the change in x the **run**. The ratio rise/run is the same for any two points on a line. We call this ratio the **slope**. Slope describes the slant of a line. The slope of the line in the graph above is given by

$$\frac{\text{rise}}{\text{run}} = \frac{\text{the change in } y}{\text{the change in } x}, \text{ or } \frac{2}{4}, \text{ or } \frac{1}{2}.$$

The *slope* of a line containing points (x_1, y_1) and (x_2, y_2) is given by

$$m = \frac{\text{rise}}{\text{run}} = \frac{\text{the change in } y}{\text{the change in } x} = \frac{y_2 - y_1}{x_2 - x_1}.$$

▶ **EXAMPLE 1** Graph the line containing the points $(-4, 3)$ and $(2, -6)$ and find the slope.

The graph is shown below. From $(-4, 3)$ and $(2, -6)$, we see that the change in y, or the rise, is $-6 - 3$, or -9. The change in x, or the run, is $2 - (-4)$, or 6.

$$\text{Slope} = \frac{\text{rise}}{\text{run}} = \frac{\text{change in } y}{\text{change in } x}$$

$$= \frac{-6 - 3}{2 - (-4)}$$

$$= \frac{-9}{6} = -\frac{9}{6}, \text{ or } -\frac{3}{2}.$$

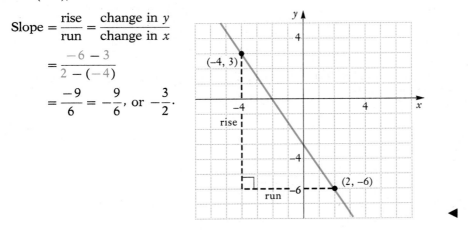

Graph the line containing the points and find the slope two different ways.

1. $(-2, 3)$ and $(3, 5)$

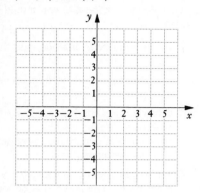

2. $(0, -3)$ and $(-3, 2)$

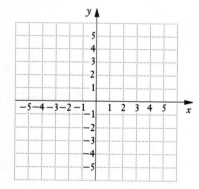

When we use the formula

$$m = \frac{y_2 - y_1}{x_2 - x_1},$$

we can subtract in two ways. We must remember, however, to subtract the y-coordinates in the same order that we subtract the x-coordinates. Let's do Example 1 again:

$$\text{Slope} = \frac{\text{change in } y}{\text{change in } x} = \frac{3 - (-6)}{-4 - 2} = \frac{9}{-6} = -\frac{3}{2}.$$

The slope of a line tells how it slants. A line with positive slope slants up from left to right. The larger the slope, the steeper the slant. A line with negative slope slants downward from left to right.

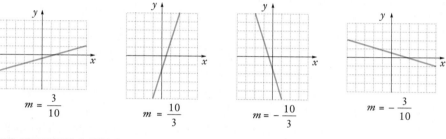

$$m = \frac{3}{10} \qquad m = \frac{10}{3} \qquad m = -\frac{10}{3} \qquad m = -\frac{3}{10}$$

DO EXERCISES 1 AND 2.

b   **Finding the Slope from an Equation**

What about the slope of a horizontal or a vertical line?

▶ **EXAMPLE 2**   Find the slope of the line $y = 5$.

Consider the points $(-3, 5)$ and $(4, 5)$, which are on the line. The change in $y = 5 - 5$, or 0. The change in $x = -3 - 4$, or -7. We have

$$m = \frac{5 - 5}{-3 - 4}$$

$$= \frac{0}{-7}$$

$$= 0.$$

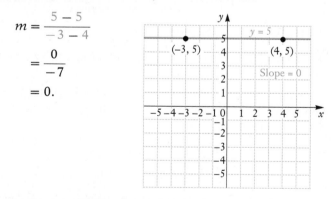

Any two points on a horizontal line have the same y-coordinate. Thus the change in y is 0.   ◀

▶ **EXAMPLE 3**   Find the slope of the line $x = -4$.

Consider the points $(-4, 3)$ and $(-4, -2)$, which are on the line. The change in $y = 3 - (-2)$, or 5. The change in $x = -4 - (-4)$, or 0. We have

$$m = \frac{3 - (-2)}{-4 - (-4)}$$

$$= \frac{5}{0}. \quad \text{Not defined}$$

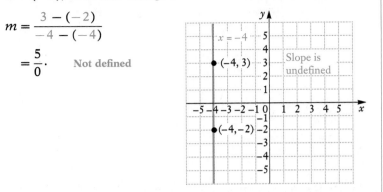

Since division by 0 is not defined, the slope of this line is not defined. The answer in this example is "The slope of this line is not defined."   ◀

A horizontal line has slope 0. The slope of a vertical line is not defined.

DO EXERCISES 3 AND 4.

It is possible to find the slope of a line from its equation. Let us consider the equation

$$y = 2x + 3.$$

We can find two points by choosing convenient values for x, say 0 and 1, and substituting to find the corresponding y-values. We find the two points on the line to be $(0, 3)$ and $(1, 5)$. The slope of the line is found using the definition of slope:

$$m = \frac{\text{change in } y}{\text{change in } x} = \frac{5 - 3}{1 - 0} = \frac{2}{1} = 2.$$

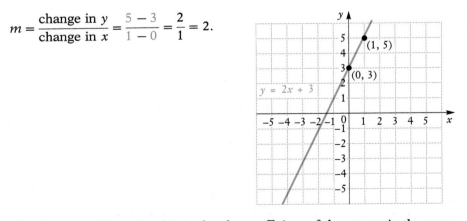

The slope is 2. Note that this is also the coefficient of the x-term in the equation $y = 2x + 3$.

The slope of the line $y = mx + b$ is m. To find the slope of a nonvertical line, solve the linear equation in x and y for y and get the resulting equation in the form $y = mx + b$. The coefficient of the x-term, m, is the slope of the line. The slope of $x = a$ is not defined.

Find the slope, if it exists, of the line.

3. $x = 7$

4. $y = -5$

ANSWERS ON PAGE A-7

Find the slope of the line.

5. $4x + 4y = 7$

▶ **EXAMPLE 4**   Find the slope of the line $2x + 3y = 7$.

We solve for y in two ways, first multiplying by $\frac{1}{3}$ or first dividing by 3:

$$2x + 3y = 7 \qquad\qquad\qquad 2x + 3y = 7$$
$$3y = -2x + 7 \qquad\qquad\qquad 3y = -2x + 7$$
$$y = \frac{1}{3}(-2x + 7) \qquad\qquad\qquad y = \frac{-2x + 7}{3}$$
$$y = -\frac{2}{3}x + \frac{7}{3}; \qquad\qquad\qquad y = -\frac{2}{3}x + \frac{7}{3}.$$

The slope is $-\frac{2}{3}$.   ◀

DO EXERCISES 5 AND 6.

---

c **The Slope–Intercept Equation of a Line**

In the equation $y = mx + b$, we know that m is the slope. What is the y-intercept? To find out, we let $x = 0$ and solve for y:

$$y = mx + b$$
$$y = m(0) + b$$
$$y = b.$$

Thus the y-intercept is $(0, b)$.

> **The Slope–Intercept Equation**
>
> **The equation $y = mx + b$ is called the *slope–intercept equation*. The slope is m and the y-intercept is $(0, b)$.**

6. $5x - 4y = 8$

▶ **EXAMPLE 5**   Find the slope and the y-intercept of $y = 3x - 4$.

Since the equation is already in the form $y = mx + b$, we simply read the slope and the y-intercept from the equation.

$$y = 3x - 4$$

The slope is 3.   The y-intercept is $(0, -4)$.   ◀

▶ **EXAMPLE 6**   Find the slope and the y-intercept of $2x - 3y = 8$.

We first solve for y:

$$2x - 3y = 8$$
$$-3y = -2x + 8 \qquad \text{This equation is not yet solved for } y.$$
$$y = -\frac{1}{3}(-2x + 8) \qquad \text{Multiplying by } -\frac{1}{3}$$
$$y = \frac{2}{3}x - \frac{8}{3}.$$

> CAUTION!   Only the coefficient of x is the slope.

The slope is $\frac{2}{3}$ and the y-intercept is $(0, -\frac{8}{3})$.   ◀

▶ **EXAMPLE 7**   A line has slope -2.4 and y-intercept 11. Find an equation of the line.

We use the slope–intercept equation and substitute -2.4 for m and 11 for b:

$$y = mx + b$$
$$y = -2.4x + 11. \qquad ◄$$

DO EXERCISES 7–12.

d The Point–Slope Equation of a Line

Suppose we know the slope of a line and a certain point on that line. We can use the slope–intercept equation to find an equation of the line.

▶ **EXAMPLE 8** Find the equation of the line with slope 3 that contains the point (4, 1).

Step 1. Using the point (4, 1), we substitute 4 for x and 1 for y in $y = mx + b$. We also substitute 3 for m, the slope. Then we solve for b:

$$y = mx + b$$
$$1 = 3 \cdot 4 + b \qquad \text{Substituting}$$
$$-11 = b. \qquad \text{Solving for } b, \text{ the } y\text{-intercept}$$

Step 2. We use the equation $y = mx + b$ and substitute 3 for m and -11 for b:

$$y = 3x - 11. \qquad ◄$$

DO EXERCISES 13 AND 14.

Let's redo Example 8 using another method. The new method leads us to another equation for a line: the *point–slope equation*. Consider a line with slope 3 and containing the point (4, 1) as shown. Suppose (x, y) is any point on this line. Using the definition of slope and the two points (4, 1) and (x, y), we get

$$m = \frac{\text{change in } y}{\text{change in } x}$$
$$= \frac{y - 1}{x - 4}.$$

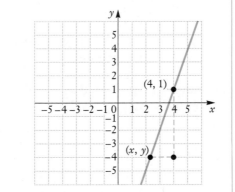

We know that the slope is 3, so

$$3 = \frac{y - 1}{x - 4}, \quad \text{or} \quad \frac{y - 1}{x - 4} = 3.$$

If we choose to solve for y, we obtain

$$y - 1 = 3x - 12$$
$$y = 3x - 11.$$

In most cases we will solve for y, but in some situations it may not be appropriate.

We can find a general equation for a line with slope m containing the point (x_1, y_1). Consider any other point (x, y) of the line.

Find the slope and the y-intercept.

7. $y = 5x$

8. $y = -\dfrac{3}{2}x - 6$

9. $3x + 4y = 15$

10. $2y = 4x - 17$

11. $-7x - 5y = 22$

12. A line has slope 3.5 and y-intercept $(0, -23)$. Find an equation of the line.

Find an equation of the line that contains the given point and has the given slope.

13. $(4, 2), \quad m = 5$

14. $(-2, 1), \ m = -3$

ANSWERS ON PAGE A-7

Find an equation of the line containing the given point and having the given slope.

15. $(3, 5)$, $m = 6$

16. $(1, 4)$, $m = -\dfrac{2}{3}$

Find an equation of the line containing the given points.

17. $(2, 4)$ and $(3, 5)$

18. $(-1, 2)$ and $(-3, -2)$

Slope has many real-world applications. For example, numbers like 2%, 3%, and 6% are often used to represent the *grade* of a road. Such a number is meant to tell how steep a road up a hill or mountain is. For example, a 3% grade means that for every horizontal distance of 100 ft, the road rises 3 ft. The concept of grade also occurs in cardiology when a person runs on a treadmill. A physician may change the steepness of the treadmill to measure its effect on heartbeat.

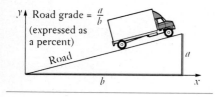

Road grade = $\dfrac{a}{b}$ (expressed as a percent)

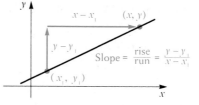

$$\text{Slope} = \frac{\text{rise}}{\text{run}} = \frac{y - y_1}{x - x_1}$$

From this drawing, we see that the slope of the line is

$$\frac{y - y_1}{x - x_1}.$$

The slope is also m. This gives us the equation

$$\frac{y - y_1}{x - x_1} = m.$$

Multiplying by $x - x_1$ gives us the following equation.

The Point–Slope Equation

A nonvertical line that contains a point (x_1, y_1) with slope m has an equation

$$y - y_1 = m(x - x_1).$$

▶ **EXAMPLE 9**   Find an equation of the line with slope 5 that contains the point $(-2, -3)$.

We substitute 5 for m, -2 for x_1, and -3 for y_1:

$$y - y_1 = m(x - x_1) \qquad \text{Using the point–slope equation}$$
$$y - (-3) = 5[x - (-2)]$$
$$y + 3 = 5(x + 2)$$
$$y + 3 = 5x + 10$$
$$y = 5x + 7. \qquad \blacktriangleleft$$

DO EXERCISES 15 AND 16.

We can also use the point–slope equation to find an equation of a line containing two given points.

▶ **EXAMPLE 10**   Find an equation of the line containing $(2, 3)$ and $(-6, 1)$.

First we find the slope:

$$m = \frac{3 - 1}{2 - (-6)} = \frac{2}{8}, \quad \text{or } \frac{1}{4}.$$

Either point can be used for (x_1, y_1), but we use $(2, 3)$:

$$y - y_1 = m(x - x_1) \qquad \text{Using the point–slope equation}$$
$$y - 3 = \tfrac{1}{4}(x - 2) \qquad \text{Substituting 2 for } x_1, 3 \text{ for } y_1, \text{ and } \tfrac{1}{4} \text{ for } m$$
$$y - 3 = \tfrac{1}{4}x - \tfrac{1}{2}$$
$$y = \tfrac{1}{4}x + \tfrac{5}{2}.$$

Try substituting the point $(-6, 1)$ for (x_1, y_1). You will find that the same equation will be obtained.   ◀

DO EXERCISES 17 AND 18.

EXERCISE SET 6.4

a   Find the slope, if it exists, of the line containing the pair of points.

1. $(3, 2)$ and $(-1, 2)$ **2.** $(4, 1)$ and $(-2, 3)$ **3.** $(-2, 4)$ and $(3, 0)$

4. $(-4, 2)$ and $(2, -3)$ **5.** $(4, 0)$ and $(5, 7)$ **6.** $(3, 0)$ and $(6, 2)$

7. $(-3, -2)$ and $(-5, -6)$ **8.** $(-2, -4)$ and $(-6, -7)$ **9.** $\left(-2, \dfrac{1}{2}\right)$ and $\left(-5, \dfrac{1}{2}\right)$

10. $(8, -3)$ and $(10, -3)$ **11.** $(9, -4)$ and $(9, -7)$ **12.** $(-10, 3)$ and $(-10, 4)$

b   Find the slope, if it exists, of the line.

13. $x = -8$ **14.** $x = -4$ **15.** $y = 2$ **16.** $y = 17$

17. $x = 9$ **18.** $y = -9$ **19.** $y = -4$ **20.** $x = 6$

Find the slope of the line.

21. $3x + 2y = 6$ **22.** $4x - y = 5$ **23.** $x + 4y = 8$

24. $x + 3y = 6$ **25.** $-2x + y = 4$ **26.** $-5x + y = 5$

c   Find the slope and the y-intercept of the line.

27. $y = -4x - 9$ **28.** $y = -3x - 5$ **29.** $y = 1.8x$

ANSWERS

1. _____
2. _____
3. _____
4. _____
5. _____
6. _____
7. _____
8. _____
9. _____
10. _____
11. _____
12. _____
13. _____
14. _____
15. _____
16. _____
17. _____
18. _____
19. _____
20. _____
21. _____
22. _____
23. _____
24. _____
25. _____
26. _____
27. _____
28. _____
29. _____

30. $y = -27.4x$ **31.** $-8x - 7y = 21$ **32.** $-2x - 9y = 13$

33. $9x = 3y + 5$ **34.** $4x = 9y + 7$ **35.** $-6x = 4y + 2$

36. $5x + 4y = 12$ **37.** $y = -17$ **38.** $y = 23$

d   Find an equation of the line containing the given point and having the given slope.

39. $(2, 5)$, $m = 5$ **40.** $(-3, 0)$, $m = -2$

41. $(2, 4)$, $m = \dfrac{3}{4}$ **42.** $\left(\dfrac{1}{2}, 2\right)$, $m = -1$

43. $(2, -6)$, $m = 1$ **44.** $(4, -2)$, $m = 6$

45. $(-3, 0)$, $m = -3$ **46.** $(0, 3)$, $m = -3$

Find an equation of the line that contains the given pair of points.

47. $(-6, 1)$, and $(2, 3)$ **48.** $(12, 16)$, and $(1, 5)$

49. $(0, 4)$, and $(4, 2)$ **50.** $(0, 0)$, and $(4, 2)$

51. $(3, 2)$, and $(1, 5)$ **52.** $(-4, 1)$, and $(-1, 4)$

53. $(-2, -4)$, and $(2, -1)$ **54.** $(-3, 5)$, and $(-1, -3)$

SKILL MAINTENANCE

Simplify.

55. $11 \cdot 6 \div 3 \cdot 2 \div 7$ **56.** $2^4 - 2^4 \div 2^2 - 2$

57. $[10 - 3(7 - 2)]$ **58.** $5^3 - 4^2 + 6(5 \cdot 7 + 4 \cdot 3)$

SYNTHESIS

59. Find an equation of the line that contains the point $(2, -3)$ and has the same slope as the line $3x - y + 4 = 0$.

60. Find an equation of the line that has the same y-intercept as the line $x - 3y = 6$ and contains the point $(5, -1)$.

61. Find an equation of the line with the same slope as $3x - 2y = 8$ and the same y-intercept as $2y + 3x = -4$.

62. Graph several equations that have the same slope. How are they related?

6.5   Parallel and Perpendicular Lines

When we graph a pair of linear equations, there are three possibilities:

1. The graphs are the same.
2. The graphs intersect at exactly one point.
3. The graphs are parallel (they do not intersect).

a   Parallel Lines

The graphs shown at the right
are of the linear equations

$$y = 2x + 5$$

and $$y = 2x - 3.$$

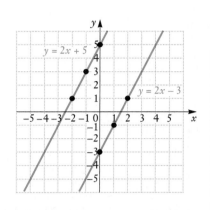

The slope of each line is 2. The y-intercepts are $(0, 5)$ and $(0, -3)$ and are different. The lines do not intersect and are parallel.

> **Parallel Lines**
>
> **Parallel nonvertical lines have the same slope and different y-intercepts.**
> **Parallel vertical lines have equations $x = p$ and $x = q$, where $p \neq q$.**
> **Parallel horizontal lines have equations $y = p$ and $y = q$, where $p \neq q$.**

▶ **EXAMPLE 1**   Determine whether the graphs of $y = -3x + 4$ and $6x + 2y = -10$ are parallel.

The graphs of these equations are shown below. By simply graphing, we may find it difficult to determine whether lines are parallel. Sometimes they may intersect only very far from the origin. We can use the preceding result about slopes, y-intercepts, and parallel lines to determine for certain whether lines are parallel.

We first solve each equation for y. The first equation is already solved for y.

a) $y = -3x + 4$

b) $6x + 2y = -10$

$$2y = -6x - 10$$

$$y = \frac{1}{2}(-6x - 10)$$

$$y = -3x - 5$$

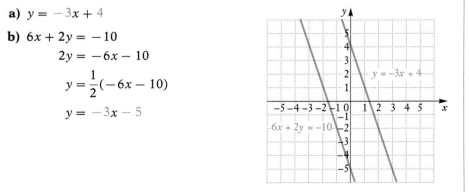

OBJECTIVES

After finishing Section 6.5, you should be able to:

a Determine whether the graphs of two linear equations are parallel.

b Determine whether the graphs of two linear equations are perpendicular.

FOR EXTRA HELP

Tape NC   Tape 13B   MAC: 6
                                   IBM: 6

Determine whether the graphs of the pair of equations are parallel.

1. $y - 3x = 1,$
$-2y = 3x + 2$

2. $3x - y = -5,$
$y - 3x = -2$

The slope of each line is -3. The y-intercepts are $(0, 4)$ and $(0, -5)$ and are different. The lines are parallel.   ◄

DO EXERCISES 1 AND 2.

b **Perpendicular Lines**

Perpendicular lines in a plane are lines that intersect at a right angle. The measure of a right angle is 90°. The lines whose graphs are shown at the right are perpendicular. You can check this partially by using a protractor or placing a rectangular piece of paper at the intersection.

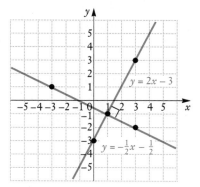

The slopes of the lines are 2 and $-\frac{1}{2}$. Note that $2(-\frac{1}{2}) = -1$. That is, the product of the slopes is -1.

> **Two nonvertical lines are perpendicular if the product of their slopes is -1. (If one line has slope m, the slope of the line perpendicular to it is $-1/m$. If one equation in a pair of perpendicular lines is vertical, then the other is horizontal. These equations are of the form $x = a$ and $y = b$.**

Determine whether the graphs of the pair of equations are perpendicular.

3. $y = -\dfrac{3}{4}x + 7,$

$y = \dfrac{4}{3}x - 9$

► **EXAMPLE 2** Determine whether the graphs of $3y = 9x + 3$ and $6y + 2x = 6$ are perpendicular.

The graphs are shown below, but they are not necessary in order to determine whether the lines are perpendicular.

We first solve each equation for y in order to determine the slopes:

a) $3y = 9x + 3$
$y = \frac{1}{3}(9x + 3)$
$y = 3x + 1;$

b) $6y + 2x = 6$
$6y = -2x + 6$
$y = \frac{1}{6}(-2x + 6)$
$y = -\frac{1}{3}x + 1.$

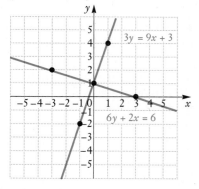

4. $4x - 5y = 8,$
$6x + 9y = -12$

The slopes are 3 and $-\frac{1}{3}$. The product of the slopes is

$$3(-\tfrac{1}{3}) = -1.$$

The lines are perpendicular.   ◄

DO EXERCISES 3 AND 4.

EXERCISE SET 6.5

a Determine whether the graphs of the equations are parallel lines.

1. $x + 4 = y,$
$\quad y - x = -3$

2. $3x - 4 = y,$
$\quad y - 3x = 8$

3. $y + 3 = 6x,$
$\quad -6x - y = 2$

4. $y = -4x + 2,$
$\quad -5 = -2y + 8x$

5. $y + 3.5 = 0.3125x,$
$\quad 5y = -32x + 23.5$

6. $y = 6.4x + 8.9,$
$\quad 5y - 32x = 5$

7. $y = 2x + 7,$
$\quad 5y + 10x = 20$

8. $y = -7x - 5,$
$\quad 2y = -7x - 10$

9. $3x - y = -9,$
$\quad 2y - 6x = -2$

10. $y - 6 = -6x,$
$\quad -2x + y = 5$

11. $x = 3,$
$\quad x = 4$

12. $y = -4,$
$\quad y = 5$

b Determine whether the graphs of the equations are perpendicular lines.

13. $y = -4x + 3,$
$\quad 4y + x = -1$

14. $y = -\dfrac{2}{3}x + 4,$
$\quad 3x + 2y = 1$

15. $x + y = 6,$
$\quad 4y - 4x = 12$

16. $2x - 5y = -3,$
$\quad 5x + 2y = 6$

ANSWERS

1. _____

2. _____

3. _____

4. _____

5. _____

6. _____

7. _____

8. _____

9. _____

10. _____

11. _____

12. _____

13. _____

14. _____

15. _____

16. _____

ANSWERS

17. _____

18. _____

19. _____

20. _____

21. _____

22. _____

23. _____

24. _____

25. _____

26. _____

27. _____

28. _____

29. _____

30. _____

31. _____

32. _____

33. _____

34. _____

35. _____

17. $y = -6.4x - 7,$
$64y - 5x = 32$

18. $y = -0.3125x + 11,$
$y - 3.2x = -14$

19. $y = -x + 8,$
$x - y = -1$

20. $2x + 6y = -3,$
$12y = 4x + 20$

21. $\dfrac{3}{8}x - \dfrac{y}{2} = 1,$

$\dfrac{4}{3}x - y + 1 = 0$

22. $\dfrac{1}{2}x + \dfrac{3}{4}y = 6,$

$-\dfrac{3}{2}x + y = 4$

SKILL MAINTENANCE

23. One car travels 10 km/h faster than another. While one car goes 130 km, the other goes 140 km. What is the speed of each car?

24. A train leaves a station and travels west at 70 km/h. Two hours later, a second train leaves on a parallel track and travels west at 90 km/h. When will it overtake the first train?

Solve.

25. $\dfrac{x^2}{x + 4} = \dfrac{16}{x + 4}$

26. $x^2 - 10x + 25 = 0$

SYNTHESIS

27. Find an equation of a line that contains the point $(0, 6)$ and is parallel to $y - 3x = 4$.

28. Find an equation of the line that contains $(-2, 4)$ and is parallel to $y = 2x - 3$.

29. Find an equation of the line that contains the point $(0, 2)$ and is perpendicular to the line $3y - x = 0$.

30. Find an equation of the line that contains the point $(1, 0)$ and is perpendicular to $2x + y = -4$.

31. Find an equation of the line that has x-intercept $(-2, 0)$ and is parallel to $4x - 8y = 12$.

32. Find the value of k so that $4y = kx - 6$ and $5x + 20y = 12$ are parallel.

33. Find the value of k so that $4y = kx - 6$ and $5x + 20y = 12$ are perpendicular.

The lines in each graph are perpendicular. Find an equation of each line.

34.

35.

6.6 Direct and Inverse Variation

a Equations of Direct Variation

A bicycle is traveling at 10 km/h. In 1 hr, it goes 10 km. In 2 hr, it goes 20 km. In 3 hr, it goes 30 km, and so on. We will use the number of hours as the first coordinate and the number of kilometers traveled as the second coordinate: (1, 10), (2, 20), (3, 30), (4, 40), and so on. Note that as the first number gets larger, so does the second. Note also that the ratio of distance to time for each of these ordered pairs is $\frac{10}{1}$, or 10.

Whenever a situation produces pairs of numbers in which the *ratio is constant*, we say that there is **direct variation.** Here the distance varies directly as the time:

$$\frac{d}{t} = 10 \ \text{(a constant)}, \quad \text{or} \quad d = 10t.$$

> **Direct Variation**
>
> If a situation translates to an equation described by $y = kx$, where k is a positive constant, $y = kx$ is called an *equation of direct variation*, and k is called the *variation constant.* We say that y varies directly as x.

The terminologies

   "*y* varies as *x*,"

   "*y* is directly proportional to *x*,"

and   "*y* is proportional to *x*"

also imply direct variation and are used in many situations. The constant k is often referred to as a **constant of proportionality.**

When there is direct variation $y = kx$, the variation constant can be found if one pair of values of x and y is known. Then other values can be found.

▶ **EXAMPLE 1**   Find an equation of variation where y varies directly as x and $y = 7$ and $x = 25$.

We substitute to find k:

$$y = kx$$
$$7 = k \cdot 25$$
$$\frac{7}{25} = k, \quad \text{or } k = 0.28.$$

Then the equation of variation is $y = 0.28x$. Note that the answer is an *equation*.   ◀

DO EXERCISES 1 AND 2.

OBJECTIVES

After finishing Section 6.6, you should be able to:

a Find an equation of direct variation given a pair of values of the variables.

b Solve problems involving direct variation.

c Find an equation of inverse variation given a pair of values of the variables.

d Solve problems involving inverse variation.

FOR EXTRA HELP

Tape 13B   Tape 13B   MAC: 6
                                    IBM: 6

Find an equation of variation where y varies directly as x and the following is true.

1. $y = 84$ when $x = 12$

2. $y = 50$ when $x = 80$

3. The cost C of operating a television varies directly as the number n of hours it is in operation. It costs $14.00 to operate a standard-size color TV continuously for 30 days. At this rate, how much would it cost to operate the TV for 1 day? for 1 hour?

4. The weight V of an object on Venus varies directly as its weight E on earth. A person weighing 165 lb on earth would weigh 145.2 lb on Venus. How much would a person weighing 198 lb on earth weigh on Venus?

b   Solving Problems with Direct Variation

▶ **EXAMPLE 2**   It is known that the karat rating K of a gold object varies directly as the actual percentage P of gold in the object. A 14-karat gold object is 58.25% gold. What is the percentage of gold in a 24-karat object?

1., 2. *Familiarize* and *Translate.*   The problem states that we have direct variation between the variables K and P. Thus an equation $K = kP$, $k > 0$, applies. As the percentage of gold increases, the karat rating increases. The letters K and k represent different quantities.

3. *Solve.*   The mathematical manipulation has two steps. First, we find the equation of variation by substituting known values for k. Second, we compute the percentage of gold in a 24-karat object.

a) First, we find an equation of variation:

$$K = kP$$
$$14 = k(0.5825) \qquad \text{Substituting 14 for } K \text{ and 58.25\%, or 0.5825, for } P$$
$$\frac{14}{0.5825} = k$$
$$24.03 \approx k. \qquad \text{Dividing and rounding to the nearest hundredth}$$

The equation of variation is $K = 24.03P$.

b) We then use the equation to find the percentage of gold in a 24-karat object:

$$K = 24.03P$$
$$24 = 24.03P \qquad \text{Substituting 24 for } K$$
$$\frac{24}{24.03} = P$$
$$0.999 \approx P$$
$$99.9\% \approx P.$$

4. *Check.*   The check might be done by repeating the computations. You might also do some reasoning about the answer. The karat rating increased from 14 to 24. Similarly, the percentage increased from 58.25% to 99.9%.

5. *State.*   A 24-karat object is 99.9% gold.   ◀

DO EXERCISES 3 AND 4.

Let us consider direct variation from the standpoint of a graph. The graph of $y = kx$, $k > 0$, always goes through the origin and rises from left to right. Note that as x increases, y increases; and as x decreases, y decreases. This is why the terminology "direct" is used. What one variable does, so does the other.

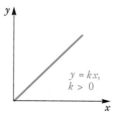

$y = kx$, $k > 0$

C Equations of Inverse Variation

A car is traveling a distance of 10 km. At a speed of 10 km/h, it will take 1 hr. At 20 km/h, it will take $\frac{1}{2}$ hr. At 30 km/h, it will take $\frac{1}{3}$ hr, and so on. This determines a set of pairs of numbers, all having the same product:

$$(10, 1), \quad (20, \tfrac{1}{2}), \quad (30, \tfrac{1}{3}), \quad (40, \tfrac{1}{4}), \quad \text{and so on.}$$

Note that as the first number gets larger, the second number gets smaller. Whenever a situation produces pairs of numbers whose *product is constant*, we say that there is **inverse variation.** Here the time varies inversely as the speed:

$$rt = 10 \text{ (a constant)}, \quad \text{or } t = \frac{10}{r}.$$

> **Inverse Variation**
>
> **If a situation translates to an equation described by $y = k/x$, where k is a positive constant, $y = k/x$ is called an *equation of inverse variation*. We say that y varies inversely as x.**

The terminology

"y is inversely proportional to x"

also implies inverse variation and is used in some situations.

▶ **EXAMPLE 3** Find an equation of variation where y varies inversely as x and $y = 145$ when $x = 0.8$.

We substitute to find k:

$$y = \frac{k}{x}$$
$$145 = \frac{k}{0.8}$$
$$(0.8)145 = k$$
$$116 = k.$$

The equation of variation is $y = \dfrac{116}{x}$. ◀

DO EXERCISES 5 AND 6.

The graph of $y = k/x$, $k > 0$, is shaped like the following figure for positive values of x. (You do not need to know how to graph such equations at this time.) Note that as x increases, y decreases; and as x decreases, y increases. This is why the terminology "inverse" is used. One variable does the opposite of what the other does.

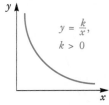

Find an equation of variation where y varies inversely as x and following is true.

5. $y = 105$ when $x = 0.6$

6. $y = 45$ when $x = 20$

ANSWERS ON PAGE A-7

7. In Example 4, how long would it take 10 people to do the job?

Solving Problems with Inverse Variation

Often in an applied situation we must decide which kind of variation, if any, might apply to the problem.

► **EXAMPLE 4**   It takes 4 hr for 20 people to wash and wax the floors in a building. How long would it then take 25 people to do the job?

1. *Familiarize.*   Think about the problem situation. What kind of variation would be used? It seems reasonable that the more people there are working on the job, the less time it will take to finish. (One might argue that too many people in a crowded area would be counterproductive, but we will disregard that possibility.) Thus inverse variation might apply. We let T = the time to do the job, in hours, and N = the number of people. Assuming inverse variation, we know that an equation $T = k/N$, $k > 0$, applies. As the number of people increases, the time it takes to do the job decreases.

2. *Translate.*   We write an equation of variation:

$$T = \frac{k}{N}.$$

Time varies inversely as the number of people.

3. *Solve.*   The mathematical manipulation has two steps. First, we find the equation of variation by substituting known values to find k. Second, we compute the amount of time it would take 25 people to do the job.

   a) First, we find an equation of variation:

   $$T = \frac{k}{N}$$

   $$4 = \frac{k}{20} \qquad \text{Substituting 4 for } T \text{ and 20 for } N$$

   $$20 \cdot 4 = k$$

   $$80 = k.$$

   The equation of variation is $T = \dfrac{80}{N}$.

   b) We then use the equation to find the amount of time that it takes 25 people to do the job:

   $$T = \frac{80}{N}$$

   $$T = \frac{80}{25} \qquad \text{Substituting 25 for } N$$

   $$T = 3.2.$$

4. *Check.*   The check might be done by repeating the computations. We might also analyze the results. The number of people increased from 20 to 25. Did the time decrease? It did, and this confirms what we expect with inverse variation.

5. *State.*   It should take 3.2 hr for 25 people to do the job.   ◄

DO EXERCISES 7 AND 8.

8. The time required to drive a fixed distance varies inversely as the speed r. It takes 5 hr at 60 km/h to drive a fixed distance. How long would it take at 40 km/h?

EXERCISE SET 6.6

a Find an equation of variation where y varies directly as x and the following are true.

1. $y = 28$ when $x = 7$ **2.** $y = 30$ when $x = 8$ **3.** $y = 0.7$ when $x = 0.4$

1. _____

4. $y = 0.8$ when $x = 0.5$ **5.** $y = 400$ when $x = 125$ **6.** $y = 630$ when $x = 175$

2. _____

3. _____

7. $y = 200$ when $x = 300$ **8.** $y = 500$ when $x = 60$

4. _____

5. _____

b Solve.

6. _____

9. A person's paycheck P varies directly as the number H of hours worked. For working 15 hr, the pay is $78.75. Find the pay for 35 hr of work.

10. The number of bolts B that a machine can make varies directly as the time it operates. It can make 6578 bolts in 2 hr. How many can it make in 5 hr?

7. _____

8. _____

11. The number of servings S of meat that can be obtained from a turkey varies directly as its weight W. From a turkey weighing 14 kg, one can get 40 servings of meat. How many servings can be obtained from an 8-kg turkey?

12. The number of servings S of meat that can be obtained from round steak varies directly as the weight W. From 9 kg of round steak, one can get 70 servings of meat. How many servings can one get from 12 kg of round steak?

9. _____

10. _____

11. _____

12. _____

ANSWERS

13. _____

14. _____

15. _____

16. _____

17. _____

18. _____

19. _____

20. _____

21. _____

22. _____

23. _____

24. _____

25. _____

26. _____

13. The weight M of an object on the moon varies directly as its weight E on earth. A person who weighs 171.6 lb on earth weighs 28.6 lb on the moon. How much would a 220-lb person weigh on the moon?

14. The weight M of an object on Mars varies directly as its weight E on earth. A person who weighs 209 lb on earth weighs 79.42 lb on Mars. How much would a 176-lb person weigh on Mars?

15. The number of kilograms W of water in a human body varies directly as the total body weight B. A person weighing 75 kg contains 54 kg of water. How many kilograms of water are in a person weighing 95 kg?

16. The amount C that a family spends on car expenses varies directly as its income I. A family making \$21,760 a year will spend \$3264 a year for car expenses. How much will a family making \$30,000 a year spend for car expenses?

c Find an equation of variation where y varies inversely as x and the following are true.

17. $y = 25$ when $x = 3$

18. $y = 45$ when $x = 2$

19. $y = 8$ when $x = 10$

20. $y = 7$ when $x = 10$

21. $y = 0.125$ when $x = 8$

22. $y = 6.25$ when $x = 0.16$

23. $y = 42$ when $x = 25$

24. $y = 42$ when $x = 50$

25. $y = 0.2$ when $x = 0.3$

26. $y = 0.4$ when $x = 0.6$

d Solve.

27. It takes 16 hr for 2 people to resurface a gym floor. How long will it take 6 people to do the job?
   a) What kind of variation might apply to this situation?
   b) Solve the problem.

28. It takes 4 hr for 9 cooks to prepare a school lunch. How long will it take 8 cooks to prepare the lunch?
   a) What kind of variation might apply to the situation?
   b) Solve the problem.

29. A production line produces 15 compact disc players every 8 hr. How many players can it produce in 37 hr?
   a) What kind of variation might apply to the situation?
   b) Solve the problem.

30. A person works for 15 hr and makes $93.75. How much will the person make by working 35 hr?
   a) What kind of variation might apply to this situation?
   b) Solve the problem.

31. The volume V of a gas varies inversely as the pressure P on it. The volume of a gas is 200 cubic centimeters (cm^3) under a pressure of 32 kg/cm^2. What will be its volume under a pressure of 20 kg/cm^2?

32. The current I in an electrical conductor varies inversely as the resistance R of the conductor. The current is 2 amperes when the resistance is 960 ohms. What is the current when the resistance is 540 ohms?

33. The time t required to empty a tank varies inversely as the rate r of pumping. A pump can empty a tank in 90 min at the rate of 1200 L/min. How long will it take the pump to empty the tank at 2000 L/min?

34. The height H of triangles of fixed area varies inversely as the base B. Suppose the height is 50 cm when the base is 40 cm. Find the height when the base is 8 cm. What is the fixed area?

35. The pitch P of a musical tone varies inversely as its wavelength W. One tone has a pitch of 660 vibrations per second and a wavelength of 1.6 ft. Find the wavelength of another tone that has a pitch of 440 vibrations per second.

36. The time t required to drive a fixed distance varies inversely as the speed r. It takes 5 hr at 55 mph to drive a fixed distance. How long would it take at 40 mph?

ANSWERS

27. a) _____

b) _____

28. a) _____

b) _____

29. a) _____

b) _____

30. a) _____

b) _____

31. _____

32. _____

33. _____

34. _____

35. _____

36. _____

Copyright © 1991 Addison-Wesley Publishing Co., Inc.

ANSWERS

37. _____

38. _____

39. _____

40. _____

41. _____

42. _____

43. _____

44. _____

45. _____

46. _____

47. _____

48. _____

49. _____

50. _____

51. _____

52. _____

53. _____

54. _____

55. _____

SYNTHESIS

Write an equation of direct variation for each situation in Exercises 37–40. If possible, give a value for k and graph the equation.

37. The perimeter P of an equilateral polygon varies directly as the length S of a side.

38. The circumference C of a circle varies directly as the radius r.

39. The number of bags B of peanuts sold at a baseball game varies directly as the number N of people in attendance.

40. The cost C of building a new house varies directly as the area A of the floor space of the house.

41. Show that if p varies directly as q, then q varies directly as p.

42. The area of a circle varies directly as the square of the length of the radius. What is the variation constant?

Write an equation of variation for each situation.

43. In a stream, the amount S of salt carried varies directly as the sixth power of the speed V of the stream.

44. The square of the pitch P of a vibrating string varies directly as the tension t on the string.

45. The volume V of a sphere varies directly as the cube of the radius r.

46. The power P in a windmill varies directly as the cube of the wind speed V.

Write an equation of inverse variation for each situation.

47. The cost C per person of chartering a fishing boat varies inversely as the number N of persons sharing the cost.

48. The number N of revolutions of a tire rolling over a given distance varies inversely as the circumference C of the tire.

49. The amount of current I flowing in an electrical circuit varies inversely with the resistance R of the circuit.

50. The density D of a given mass varies inversely as its volume V.

51. The intensity of illumination I from a light source varies inversely as the square of the distance d from the source.

Determine whether the given situation varies inversely.

52. The cost of mailing a letter in the United States and the distance it travels

53. A runner's speed in a race and the time it takes to run it

54. The number of plays to go 80 yd for a touchdown and the average gain per play

55. The weight of a turkey and the cooking time

6.7 Graphing Inequalities in Two Variables

A graph of an inequality is a drawing that represents its solutions. An inequality in one variable can be graphed on a number line. An inequality in two variables can be graphed on a coordinate plane.

a Solutions of Inequalities In Two Variables

The solutions of inequalities in two variables are ordered pairs.

▶ **EXAMPLE 1** Determine whether $(-3, 2)$ is a solution of $5x + 4y < 13$.

We use alphabetical order of the variables and replace x by -3 and y by 2.

$$\frac{5x + 4y < 13}{\begin{array}{c|c} 5(-3) + 4 \cdot 2 & 13 \\ -15 + 8 & \\ -7 & \text{TRUE} \end{array}}$$

Since $-7 < 13$ is true, $(-3, 2)$ is a solution. ◀

▶ **EXAMPLE 2** Determine whether $(6, 8)$ is a solution of $5x + 4y < 13$.

We use alphabetical order of the variables and replace x by 6 and y by 8.

$$\frac{5x + 4y < 13}{\begin{array}{c|c} 5(6) + 4(8) & 13 \\ 30 + 32 & \\ 62 & \text{FALSE} \end{array}}$$

Since $62 < 13$ is false, $(6, 8)$ is not a solution. ◀

DO EXERCISES 1 AND 2.

b Graphing Inequalities In Two Variables

▶ **EXAMPLE 3** Graph: $y > x$.

We first graph the line $y = x$ for comparison. Every solution of $y = x$ is an ordered pair like $(3, 3)$. The first and second coordinates are the same. The graph of $y = x$ is shown on the left below. We draw it dashed because these points are *not* solutions of $y > x$.

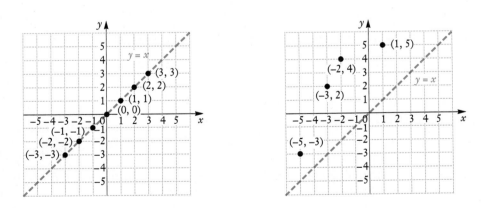

OBJECTIVES

After finishing Section 6.7, you should be able to:

a Determine whether an ordered pair of numbers is a solution of an inequality in two variables.

b Graph linear inequalities.

FOR EXTRA HELP

Tape 13C   Tape 14A   MAC: 6
                            IBM: 6

1. Determine whether $(4, 3)$ is a solution of $3x - 2y < 1$.

2. Determine whether $(2, -5)$ is a solution of $4x + 7y \geq 12$.

ANSWERS ON PAGE A-7

3. Graph: $y < x$.

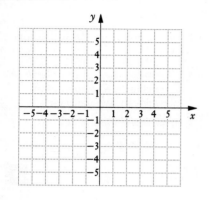

Now look at the graph on the right on the preceding page. Several pairs of numbers are plotted on the half-plane above the line $y = x$. Note that each of these ordered pairs is a solution of $y > x$. We can check a pair such as $(-2, 4)$ as follows:

$$\begin{array}{c|c} \multicolumn{2}{c}{y > x} \\ \hline 4 & -2 \end{array} \quad \text{TRUE}$$

It turns out that any point on the same side of $y = x$ as $(-2, 4)$ is also a solution. If we know that one point in a half-plane is a solution, then all points in that half-plane are solutions. The graph of $y > x$ is shown below. (Solutions will be indicated by color shading throughout.) We shade the half-plane above $y = x$.

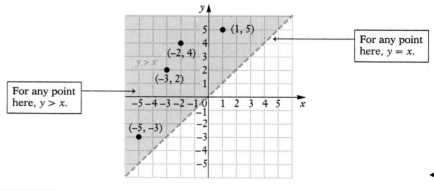

For any point here, $y = x$.

For any point here, $y > x$.

DO EXERCISE 3.

A **linear inequality** is one that we can get from a linear equation by changing the equals symbol to an inequality symbol. Every linear equation has a graph that is a straight line. The graph of a linear inequality is a half-plane, sometimes including the line along the edge.

To graph an inequality in two variables:

1. **Replace the inequality symbol with an equals sign and graph this related equation.**

2. **If the inequality symbol is $<$ or $>$, draw the line dashed. If the inequality symbol is $\leq$ or $\geq$, draw the line solid.**

3. **The graph consists of a half-plane, either above or below or left or right of the line, and, if the line is solid, the line as well. To determine which half-plane to shade, choose a point not on the line as a test point. Substitute to find whether that point is a solution of the inequality. If so, shade the half-plane containing that point. If not, shade the opposite half-plane.**

▶ **EXAMPLE 4** Graph: $6x - 2y < 10$.

1. We first graph the line $6x - 2y = 10$. The intercepts are $(0, -5)$ and $(\frac{5}{3}, 0)$. The point $(3, 4)$ is also on the graph. This line forms the boundary of the solutions of the inequality.

2. Since the inequality contains the $<$ symbol, points on the line are not solutions of the inequality, so we draw a dashed line.

3. To determine which half-plane to shade, we consider a test point *not* on the line. We try $(3, -2)$ and substitute:

$$\frac{6x - 2y < 10}{\begin{array}{c|c} 6(3) - 2(-2) & 10 \\ 18 + 4 & \\ 22 & \text{FALSE} \end{array}}$$

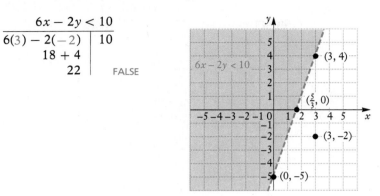

Since this inequality is false, the point $(3, -2)$ is *not* a solution; no point in the half-plane containing $(3, -2)$ is a solution. Thus the points in the opposite half-plane are solutions. The graph is shown above. ◀

DO EXERCISE 4.

▶ **EXAMPLE 5** Graph: $2x + 3y \le 6$.

First we graph the line $2x + 3y = 6$. The intercepts are $(0, 2)$ and $(3, 0)$. Since the inequality contains the $\le$ symbol, we draw the line solid to indicate that any pair on the line is a solution. Next, we pick a test point that does not belong to the line. We substitute to determine whether this point is a solution. The origin $(0, 0)$ is usually an easy one to use:

$$\frac{2x + 3y \le 6}{\begin{array}{c|c} 2 \cdot 0 + 3 \cdot 0 & 6 \\ 0 & \text{TRUE} \end{array}}$$

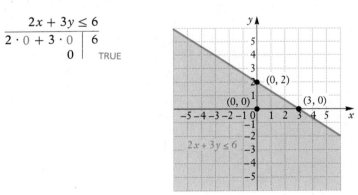

We see that $(0, 0)$ is a solution, so we shade the lower half-plane. Had the substitution given us a false inequality, we would have shaded the other half-plane. ◀

DO EXERCISES 5 AND 6.

▶ **EXAMPLE 6** Graph $x < 3$ on a plane.

There is a missing variable in this inequality. If we graph it on a line, its graph is as follows:

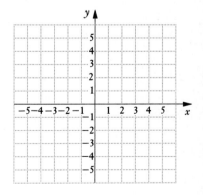

4. Graph: $2x + 4y < 8$.

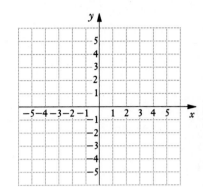

Graph.

5. $3x - 5y < 15$

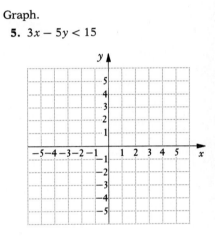

6. $2x + 3y \ge 12$

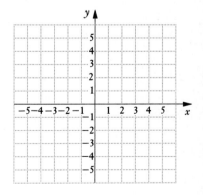

Graph.

7. $x > -3$

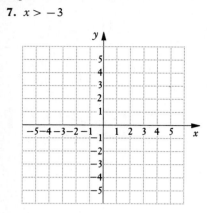

8. $y \le 4$

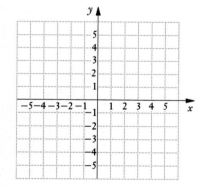

However, we can also write this inequality as $x + 0y < 3$ and consider graphing it on a plane. We use the same technique that we have used with the other examples. We first graph the related equation $x = 3$ on the plane and draw the graph with a dashed line.

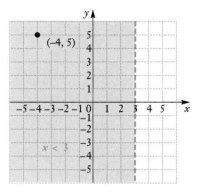

The rest of the graph is a half-plane either to the right or to the left of the line $x = 3$. To determine which, we consider a test point, $(-4, 5)$:

$$\frac{x + 0y < 3}{\begin{array}{c|c} -4 + 0(5) & 3 \\ -4 & \text{TRUE} \end{array}}$$

We see that $(-4, 5)$ is a solution, so all the pairs in the half-plane containing $(-4, 5)$ are solutions. We shade that half-plane.

We see from the graph that the solutions of $x < 3$ are all those ordered pairs whose first coordinates are less than 3. ◄

► **EXAMPLE 7** Graph $y \ge -4$ on a plane.

We first graph $y = -4$ using a solid line to indicate that all points on the line are solutions. We then use $(2, 3)$ as a test point and substitute:

$$\frac{0x + y \ge -4}{\begin{array}{c|c} 0(2) + 3 & -4 \\ 3 & \text{TRUE} \end{array}}$$

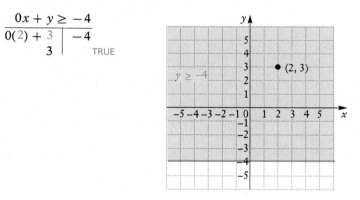

Since $(2, 3)$ is a solution, all points in the half-plane containing $(2, 3)$ are solutions. Note that this half-plane consists of all ordered pairs whose second coordinate is greater than or equal to -4. ◄

DO EXERCISES 7 AND 8.

NAME   SECTION   DATE

EXERCISE SET 6.7

a

1. Determine whether $(-3, -5)$ is a solution of
$$-x - 3y < 18.$$

2. Determine whether $(5, -3)$ is a solution of
$$-2x + 4y \le -2.$$

3. Determine whether $(\frac{1}{2}, -\frac{1}{4})$ is a solution of
$$7y - 9x > -3.$$

4. Determine whether $(-6, 5)$ is a solution of
$$x + 0 \cdot y < 3.$$

1. _____

2. _____

3. _____

4. _____

b Graph on a plane.

5. $x > 2y$

6. $x > 3y$

7. $y \le x - 3$

8. $y \le x - 5$

9. $y < x + 1$

10. $y < x + 4$

11. $y \ge x - 2$

12. $y \ge x - 1$

13. $y \le 2x - 1$

14. $y \le 3x + 2$

15. $x + y \le 3$

16. $x + y \le 4$

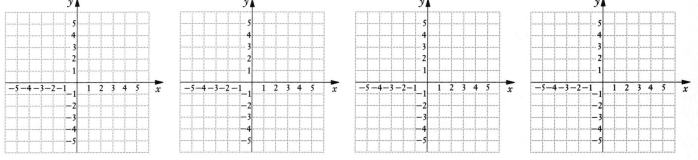

17. $x - y > 7$

18. $x - y > -2$

19. $2x + 3y \leq 12$

20. $5x + 4y \geq 20$

21. $y \geq 1 - 2x$

22. $y - 2x \leq -1$

23. $y + 4x > 0$

24. $y - x < 0$

25. $y \leq 3$

26. $y > -1$

27. $y \geq -5$

28. $y < 0$

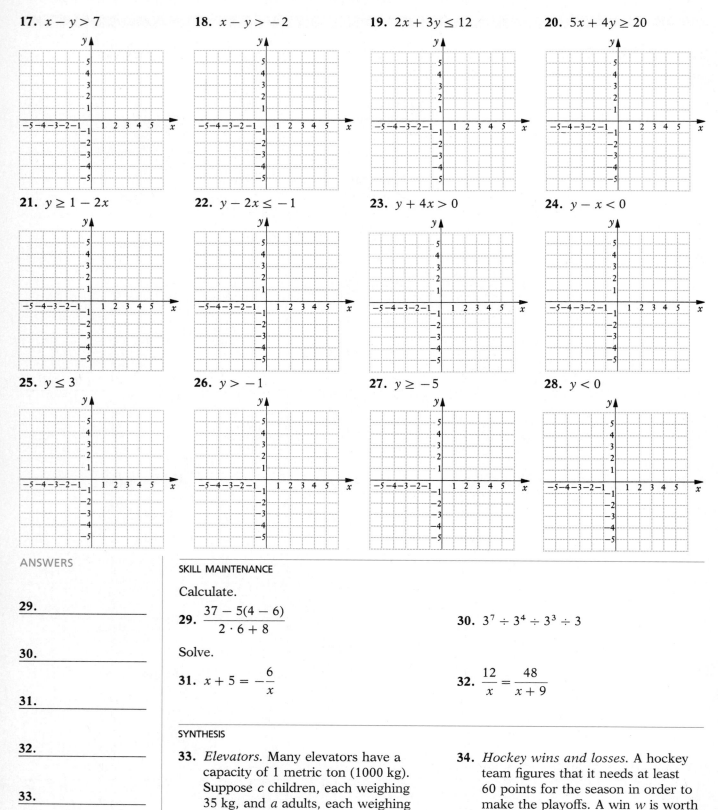

ANSWERS

29. _____

30. _____

31. _____

32. _____

33. _____

34. _____

SKILL MAINTENANCE

Calculate.

29. $\dfrac{37 - 5(4 - 6)}{2 \cdot 6 + 8}$

30. $3^7 \div 3^4 \div 3^3 \div 3$

Solve.

31. $x + 5 = -\dfrac{6}{x}$

32. $\dfrac{12}{x} = \dfrac{48}{x + 9}$

SYNTHESIS

33. *Elevators.* Many elevators have a capacity of 1 metric ton (1000 kg). Suppose c children, each weighing 35 kg, and a adults, each weighing 75 kg, are on an elevator. Find and graph an inequality that asserts that the elevator is overloaded.

34. *Hockey wins and losses.* A hockey team figures that it needs at least 60 points for the season in order to make the playoffs. A win w is worth 2 points and a tie t is worth 1 point. Find and graph an inequality that describes the situation.

SUMMARY AND REVIEW: CHAPTER 6

IMPORTANT PROPERTIES AND FORMULAS

$Slope = m = \dfrac{y_2 - y_1}{x_2 - x_1}$

Slope–Intercept Equation: $y = mx + b$

Point–Slope Equation: $y - y_1 = m(x - x_1)$

Parallel Lines: Slopes equal, y-intercepts different

Perpendicular Lines: Product of slopes $= -1$

REVIEW EXERCISES

The review sections and objectives to be tested in addition to the material in this chapter are [1.8d], [4.7b], [5.6a], and [5.7a, b].

1. This line graph shows the prime rate (the interest rate charged by banks to their best customers) in June for several years.

   a) What was the highest prime rate?

   b) Between what two consecutive years did the prime rate decrease the most?

Find the coordinates of each point.

2. A **3.** B **4.** C

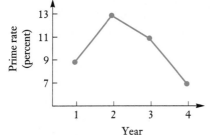

Plot these points using graph paper.

5. $(2, 5)$ **6.** $(0, -3)$ **7.** $(-4, -2)$

In which quadrant is the point located?

8. $(3, -8)$ **9.** $(-20, -14)$ **10.** $(4.9, 1.3)$

Determine whether the given point is a solution of the equation $2y - x = 10$.

11. $(2, -6)$ **12.** $(0, 5)$

Graph on a plane.

13. $y = 2x - 5$ **14.** $y = -\frac{3}{4}x$ **15.** $y = -x + 4$ **16.** $y = 3 - 4x$

17. $5x - 2y = 10$ **18.** $y = 3$ **19.** $4x + 3 = 0$ **20.** $x - 2y = 6$

Find the slope, if it exists, of the line containing the given pair of points.

21. $(6, 8)$ and $(-2, -4)$ **22.** $(5, 1)$ and $(-1, 1)$

23. $(-3, 0)$ and $(-3, 5)$ **24.** $(-8.3, 4.6)$ and $(-9.9, 1.4)$

Find the slope, if it exists, of the line.

25. $y = -6$

26. $x = 90$

27. $4x + 3y = -12$

Find the slope and the y-intercept of the line.

28. $y = -9x + 46$

29. $x + y = 9$

30. $3x - 5y = 4$

Find an equation of the line containing the given point and with the given slope.

31. $(1, 2), \quad m = 3$

32. $(-2, -5), \quad m = \frac{2}{3}$

33. $(0, -4), \quad m = -2$

Find an equation of the line that contains the given pair of points.

34. $(5, 7)$ and $(-1, 1)$

35. $(2, 0)$ and $(-4, -3)$

Determine whether the graphs of the equations are parallel, perpendicular, or neither.

36. $4x + y = 6,$
   $4x + y = 8$

37. $2x + y = 10,$
   $y = \frac{1}{2}x - 4$

38. $x + 4y = 8,$
   $x = -4y - 10$

39. $3x - y = 6,$
   $3x + y = 8$

Find an equation of variation where y varies directly as x and the following are true.

40. $y = 12$ when $x = 4$

41. $y = 4$ when $x = 8$

42. $y = 0.4$ when $x = 0.5$

Find an equation of variation where y varies inversely as x and the following are true.

43. $y = 5$ when $x = 6$

44. $y = 0.5$ when $x = 2$

45. $y = 1.3$ when $x = 0.5$

Solve.

46. A person's paycheck P varies directly as the number H of hours worked. The pay is $165.00 for working 20 hr. Find the pay for 30 hr of work.

47. It takes 5 hr for 2 washing machines to wash a fixed amount. How long would it take 10 washing machines? (The number of hours varies inversely as the number of washing machines.)

Determine whether the given point is a solution of the inequality $x - 2y > 1$.

48. $(0, 0)$

49. $(1, 3)$

50. $(4, -1)$

Graph on a plane.

51. $x < y$

52. $x + 2y \geq 4$

53. $x > -2$

SKILL MAINTENANCE

54. Judd can paint a shed alone in 5 hr. Mo can paint the same shed alone in 10 hr. How long would it take both of them working together to paint the shed?

55. Compute: $13 \cdot 6 \div 3 \cdot 26 \div 13$.

Solve.

56. $\dfrac{x^2}{x - 4} = \dfrac{16}{x - 4}$

57. $a^2 + 6a - 55 = 0$

❖ **THINKING IT THROUGH**

1. Briefly describe the concept of slope.

2. Graph $x < 1$ on a number line and on a plane, and explain the difference in the graphs.

NAME   SECTION   DATE

TEST: CHAPTER 6

Consider the bar graph shown here for Exercises 1–4.

1. What kind of degree was awarded most?

2. How many more bachelor's degrees than associate degrees were awarded?

3. How many more master's degrees than doctoral degrees were awarded?

4. In all, how many graduate degrees were awarded; that is, how many master's, doctoral, and professional degrees were awarded?

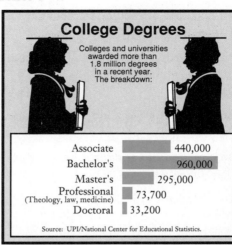

In which quadrant is the given point located?

5. $\left(-\frac{1}{2}, 7\right)$

6. $(-5, -6)$

Find the coordinates of the point.

7. A

8. B

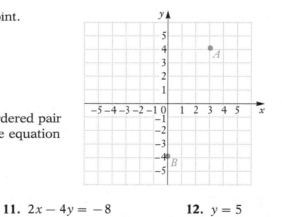

9. Determine whether the ordered pair $(2, -4)$ is a solution of the equation $y - 3x = -10$.

Graph.

10. $y = 2x - 1$ 　　　 **11.** $2x - 4y = -8$ 　　　 **12.** $y = 5$

13. $y = -\frac{3}{2}x$ 　　　 **14.** $2x + 8 = 0$

Find the slope, if it exists, of the line containing the given pair of points.

15. $(4, 7)$ and $(4, -1)$ 　　　 **16.** $(9, 2)$ and $(-3, -5)$

Find the slope, if it exists, of the given line.

17. $y = -7$ 　　　 **18.** $x = 6$

1. _____

2. _____

3. _____

4. _____

5. _____

6. _____

7. _____

8. _____

9. _____

10. _____

11. _____

12. _____

13. _____

14. _____

15. _____

16. _____

17. _____

18. _____

Find the slope and the y-intercept.

19. $y = 2x - \frac{1}{4}$ **20.** $-4x + 3y = -6$

Find an equation of the line that contains the given point and has the given slope.

21. $(3, 5)$, $m = 1$ **22.** $(-2, 0)$, $m = -3$

Find an equation of the line that contains the given pair of points.

23. $(1, 1)$ and $(2, -2)$ **24.** $(4, -1)$ and $(-4, -3)$

Determine whether the graphs of the equations are parallel, perpendicular, or neither.

25. $2x + y = 8$,   **26.** $2x + 5y = 2$,   **27.** $x + 2y = 8$,
     $2x + y = 4$ $y = 2x + 4$ $-2x + y = 8$

Find an equation of variation where y varies directly as x and the following are true.

28. $y = 6$ when $x = 3$ **29.** $y = 1.5$ when $x = 3$

Find an equation of variation where y varies inversely as x and the following are true.

30. $y = 6$ when $x = 2$ **31.** $y = \frac{1}{3}$ when $x = 3$

32. The distance d traveled by a train varies directly as the time t that it travels. The train travels 60 km in $\frac{1}{2}$ hr. How far will it travel in 2 hr?

33. It takes 3 hr for 2 cement mixers to mix a certain amount. The number of hours varies inversely as the number of cement mixers. How long would it take 5 cement mixers to do the job?

Determine whether the given point is a solution of the inequality $3y - 2x < -2$.

34. $(0, 0)$ **35.** $(-4, -10)$

Graph on a plane.

36. $y > x - 1$ **37.** $2x - y \leq 4$

SKILL MAINTENANCE

38. The speed of a freight train is 15 mph slower than the speed of a passenger train. The freight train travels 360 mi in the same time that it takes the passenger train to travel 420 mi. Find the speed of each train.

39. Compute: $\dfrac{3^2 - 2^3}{2^2 + 3 - 12 \div 2}$

Solve.

40. $3x^2 + 14x - 5 = 0$ **41.** $\dfrac{x + 1}{x + 3} = \dfrac{x + 2}{x + 4}$

SYNTHESIS

42. Find the area and the perimeter of a rectangle whose vertices are $(-3, 1)$, $(5, 1)$, $(5, 8)$, and $(-3, 8)$.

43. Find the slope–intercept equation of the line that contains the point $(-4, 1)$ and has the same slope as the line $2x - 3y = -6$.

CUMULATIVE REVIEW: CHAPTERS 1–6

1. Find the absolute value: $|3.5|$.

2. Identify the coefficient of each term of the polynomial
$$x^3 - 2x^2 + x - 1.$$

3. Identify the degree of each term and the degree of the polynomial
$$x^3 - 2x^2 + x - 1.$$

4. Classify this polynomial as a monomial, binomial, trinomial, or none of these:
$$x^3 - 2x^2 + x - 1.$$

5. Collect like terms: $x^2 - 3x^3 - 4x^2 + 5x^3 - 2$.

Simplify.

6. $\dfrac{1}{2}x - \left[\dfrac{3}{8}x - \left(\dfrac{2}{3} + \dfrac{1}{4}x\right) - \dfrac{1}{3}\right]$

7. $\left(\dfrac{2x^3}{3x^{-1}}\right)^{-2}$

8. $\dfrac{\dfrac{4}{x} - \dfrac{6}{x^2}}{\dfrac{5}{x} + \dfrac{7}{2x}}$

Perform the indicated operations. Simplify, if possible.

9. $(5xy^2 - 6x^2y^2 - 3xy^3) - (-4xy^3 + 7xy^2 - 2x^2y^2)$

10. $(4x^4 + 6x^3 - 6x^2 - 4) + (2x^5 + 2x^4 - 4x^3 - 4x^2 + 3x - 5)$

11. $\dfrac{2y + 4}{21} \cdot \dfrac{7}{y^2 + 4y + 4}$

12. $\dfrac{x^2 - 9}{x^2 + 8x + 15} \div \dfrac{x - 3}{2x + 10}$

13. $\dfrac{x^2}{x - 4} + \dfrac{16}{4 - x}$

14. $\dfrac{5x}{x^2 - 4} - \dfrac{-3}{2 - x}$

Multiply.

15. $(2.5a + 7.5)(0.4a - 1.2)$

16. $(2x^2 - 1)(x^3 + x - 3)$

17. $(2x^3 + 1)(2x^3 - 1)$

18. $(6x - 5)^2$

19. $4x(3x^3 + 4x^2 + x)$

20. $(2x^5 + 3)(3x^2 - 6)$

Solve.

21. $x - [x - (x - 1)] = 2$

22. $2x^2 + 7x = 4$

23. $x^2 + x - 20 = 0$

24. $3(x - 2) \leq 4(x + 5)$

25. $x(x - 4) = 0$

26. $x^2 = 10x$

27. $2x^2 = 800$

28. $t = ax + ay$ for a

29. $\dfrac{5x - 2}{4} - \dfrac{4x - 5}{3} = 1$

30. $\dfrac{2x}{x - 3} - \dfrac{6}{x} = \dfrac{18}{x^2 - 3x}$

Factor.

31. $-6 - 2x - 12y$ **32.** $x^2 - 10x + 24$ **33.** $2x^2 - 18$

34. $m^4 + 2m^3 - 3m - 6$ **35.** $16x^2 + 40x + 25$ **36.** $8x^2 + 10x + 3$

Solve.

37. The product of a number and 1 more than the number is 20. Find the number.

38. A person's salary varies directly as the number of hours worked. For working 9 hr, the salary is $117. Find the salary for working 6 hr.

39. Money is borrowed at 14% simple interest. After 1 year, $2850 pays off the loan. How much was originally borrowed?

40. One car travels 105 mi in the same time that a car traveling 10 mph slower travels 75 mi. Find the speed of each car.

41. If the sides of a square are increased by 2 ft, the sum of the areas of the two squares is 452 ft^2. Find the length of a side of the original square.

42. One number is 7 more than another number. The quotient of the larger divided by the smaller is $\frac{5}{4}$. Find the numbers.

Graph on a plane.

43. $y = \frac{1}{2}x$ **44.** $3x - 5y = 15$ **45.** $y = 1$

46. $y < -x - 2$ **47.** $x \le -3$

48. Find an equation of variation where y varies directly as x and $y = 8$ when $x = 12$.

49. Find an equation of variation where y varies inversely as x and $y = 20$ when $x = 0.5$.

Find the slope, if it exists, of the line containing the given pair of points.

50. $(-2, 6)$ and $(-2, -1)$ **51.** $(-4, 1)$ and $(3, -2)$

52. Find the slope and the y-intercept of $4x - 3y = 6$.

53. Find an equation for the line containing the point $(2, -3)$ and having slope $m = -4$.

54. Find an equation of the line containing the points $(-1, -3)$ and $(5, -2)$.

Determine whether the graphs of the equations are parallel, perpendicular, or neither.

55. $2x = 7 - 3y$,
$7 + 2x = 3y$

56. $x - y = 4$,
$y = x + 5$

SYNTHESIS

57. Compute: $(x + 7)(x - 4) - (x + 8)(x - 5)$.

58. Multiply: $[4y^3 - (y^2 - 3)][4y^3 + (y^2 - 3)]$.

59. Factor: $2a^{32} - 13{,}122b^{40}$.

60. Solve: $(x - 4)(x + 7)(x - 12) = 0$.

61. Find an equation of the line that contains the point $(-3, -2)$ and is parallel to the line $2x - 3y = -12$.

62. Find the meaningful replacements:

$$\frac{\dfrac{1}{x} + x}{2 + \dfrac{1}{x - 3}}.$$

INTRODUCTION We now consider how two graphs of linear equations might intersect. Such a point is a solution of what is called a *system of equations*. Many problems involve two facts about two quantities and are easier to solve by translating to a system of two equations in two variables. Systems of equations have extensive applications in many fields such as sociology, psychology, business, education, engineering, and science.

The review sections to be tested in addition to the material in this chapter are 3.1, 5.1, 5.5, and 6.3. ❖

Systems of Equations

7

AN APPLICATION

Denny's® is a national restaurant firm. The ad shown on p. 457 once appeared on the tables as a special. Determine the price of one item from the *A* side of the menu and the price of one item from the *B* side of the menu.

THE MATHEMATICS

The ad gives us the following pair of equations:

$$\left. \begin{array}{l} a + b = \$5.49, \\ a + 2b = \$6.99, \end{array} \right\} \leftarrow \begin{array}{l} \text{This is a } \textit{system} \\ \textit{of equations.} \end{array}$$

where a = the price of one item from the *A* side of the menu and b = the price of one item from the *B* side of the menu.

Supplementary Angles:	Two angles are supplementary if the sum of their measures is 180°.
Complementary Angles:	Two angles are complementary if the sum of their measures is 90°.
Perimeter of a Rectangle:	$P = 2l + 2w$
Motion Formula:	$d = rt$
Sum of the Angle Measures of a Triangle = 180°	
Simple-Interest Formula:	$I = Prt$

PRETEST: CHAPTER 7

1. Determine whether the ordered pair $(-1, 1)$ is a solution of the system of equations
$$2x + y = -1,$$
$$3x - 2y = -5.$$

2. Solve this system by graphing.
$$2x = y + 1,$$
$$2x - y = 5$$

Solve by the substitution method.

3. $x + y = 7,$
$\quad x = 2y + 1$

4. $2x - 3y = 7,$
$\quad x + y = 1$

Solve by the elimination method.

5. $2x - y = 1,$
$\quad 2x + y = 2$

6. $2x - 3y = -4,$
$\quad 3x - 4y = -7$

7. $\dfrac{3}{5}x - \dfrac{1}{4}y = 4,$
$\quad \dfrac{1}{5}x + \dfrac{3}{4}y = 8$

8. Find two numbers whose sum is 74 and whose difference is 26.

9. Two angles are complementary. One angle is 15° more than twice the other. Find the angles. (Complementary angles are angles whose sum is 90°.)

10. A train leaves a station and travels north at 96 mph. Two hours later, a second train leaves on a parallel track and travels north at 120 mph. When will it overtake the first train?

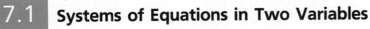

7.1 Systems of Equations in Two Variables

a Systems of Equations and Solutions

Many problems can be solved more easily by translating to two equations in two variables. The following is a system of equations:

$$x + y = 8,$$
$$2x - y = 1.$$

> A *solution* of a system of two equations is an ordered pair that makes both equations true.

Consider the system shown above. Look at the graph below. **Recall that a graph of an equation is a drawing that represents its solution set. Each point on the graph corresponds to a solution of that equation.** Which points (ordered pairs) are solutions of *both* equations?

The point P with coordinates (3, 5) is a drawing of the set of common solutions.

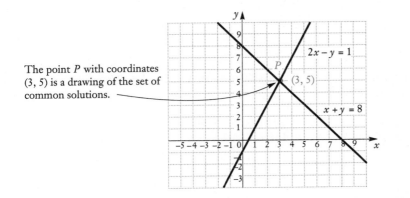

The graph shows that there is only one. It is the point *P* where the graphs cross. This point looks as if its coordinates are (3, 5). We check to see if (3, 5) is a solution of *both* equations. We substitute 3 for *x* and 5 for *y*:

$$\frac{x + y = 8}{3 + 5 \;\bigg|\; 8}$$
$$8 \;\bigg|\; \text{TRUE}$$

$$\frac{2x - y = 1}{2 \cdot 3 - 5 \;\bigg|\; 1}$$
$$6 - 5 \;\bigg|\;$$
$$1 \;\bigg|\; \text{TRUE}$$

There is just one solution of the system of equations. It is (3, 5). In other words, $x = 3$ and $y = 5$.

OBJECTIVES

After finishing Section 7.1, you should be able to:

a Determine whether an ordered pair is a solution of a system of equations.

b Solve systems of two linear equations in two variables by graphing.

FOR EXTRA HELP

Tape 13D   Tape 14A   MAC: 7
IBM: 7

Determine whether the given ordered pair is a solution of the system of equations.

1. $(2, -3);$ $x = 2y + 8,$
$\qquad 2x + y = 1$

▶ **EXAMPLE 1**   Determine whether $(1, 2)$ is a solution of the system

$$y = x + 1,$$
$$2x + y = 4.$$

We check by substituting alphabetically 1 for x and 2 for y:

$$\begin{array}{c|c} y = x + 1 \\ \hline 2 & 1 + 1 \\ & 2 \quad \text{TRUE} \end{array} \qquad \begin{array}{c|c} 2x + y = 4 \\ \hline 2 \cdot 1 + 2 & 4 \\ 2 + 2 & \\ & 4 \quad \text{TRUE} \end{array}$$

This checks, so $(1, 2)$ is a solution of the system.   ◀

▶ **EXAMPLE 2**   Determine whether $(-3, 2)$ is a solution of the system

$$p + q = -1,$$
$$q + 3p = 4.$$

We check by substituting alphabetically -3 for p and 2 for q:

$$\begin{array}{c|c} p + q = -1 \\ \hline -3 + 2 & -1 \\ -1 & \text{TRUE} \end{array} \qquad \begin{array}{c|c} q + 3p = 4 \\ \hline 2 + 3(-3) & 4 \\ 2 - 9 & \\ -7 & \text{FALSE} \end{array}$$

The point $(-3, 2)$ is not a solution of $q + 3p = 4$. Thus it is not a solution of the system.   ◀

Example 2 illustrates that an ordered pair may be a solution of one equation but not *both*. If that is the case, it is *not* a solution of the system.

DO EXERCISES 1 AND 2.

2. $(20, 40);$ $a = \dfrac{1}{2}b,$
$\qquad b - a = 60$

b   Graphing Systems of Equations

Recall that the **graph** of an equation is a drawing that represents its solution set. If the graph of an equation is a line, then every point on the line corresponds to an ordered pair that is a solution of the equation. If we graph a **system** of two linear equations, we graph both equations and find the coordinates of the points of intersection, if any exist.

▶ **EXAMPLE 3**   Solve this system of equations by graphing:

$$x + y = 6,$$
$$x = y + 2.$$

We graph the equations using any of the methods studied in Chapter 6. Point P with coordinates $(4, 2)$ looks as if it is the solution.

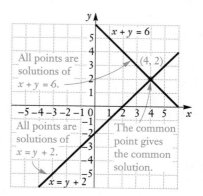

Graphing is not perfectly accurate, so solving by graphing may give only approximate solutions. We check the pair as follows.

Check:

$$\frac{x + y = 6}{4 + 2 \mid 6}$$
$$\qquad 6 \mid \text{TRUE}$$

$$\frac{x = y + 2}{4 \mid 2 + 2}$$
$$\qquad 4 \qquad \text{TRUE}$$

The solution is (4, 2). ◀

DO EXERCISE 3.

Sometimes the equations in a system have graphs that are parallel lines.

▶ **EXAMPLE 4** Solve this system of equations by graphing:

$$y = 3x + 4,$$
$$y = 3x - 3.$$

We graph the equations, again using any of the methods studied in Chapter 6. The lines have the same slope, 3, and different intercepts, (0, 4) and (0, −3), so they are parallel.

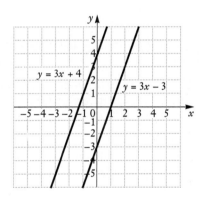

There is no point at which they cross, so the system has no solution. The solution set is the empty set, denoted ∅, or { }. ◀

DO EXERCISE 4.

3. Solve this system by graphing:
$$2x + y = 1,$$
$$x = 2y + 8.$$

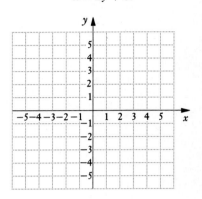

4. Solve this system by graphing:
$$y + 4 = x,$$
$$x - y = -2.$$

5. Solve this system by graphing:

$$2x + y = 4,$$
$$-6x - 3y = -12.$$

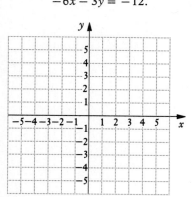

Sometimes the equations in a system have the same graph.

▶ **EXAMPLE 5**   Solve this system by graphing:

$$2x + 3y = 6,$$
$$-8x - 12y = -24.$$

We graph the equations and see that the graphs are the same. Thus any solution of one of the equations is a solution of the other. Each equation has an infinite number of solutions, some of which are indicated on the graph.

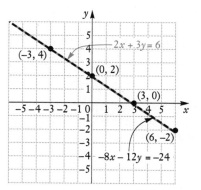

We check one such solution, (0, 2), which is the *y*-intercept of each equation.

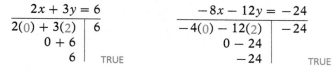

On your own, check that $(-3, 4)$ is also a solution of the system. If (0, 2) and $(-3, 4)$ are solutions, then all points on the line containing them are solutions. The system has an infinite number of solutions.   ◀

DO EXERCISE 5.

When we graph a system of two equations in two variables, one of the following three things can happen.

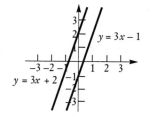

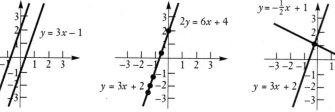

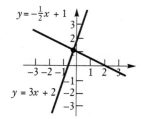

No solution.
Graphs are parallel.

Infinitely
many solutions.
Equations have
the same graph.

One solution.
Graphs intersect.

ANSWER ON PAGE A-7

EXERCISE SET 7.1

a Determine whether the given ordered pair is a solution of the system of equations. Use alphabetical order of the variables.

1. $(3, 2)$; $2x + 3y = 12$,
            $x - 4y = -5$

2. $(1, 5)$; $5x - 2y = -5$,
            $3x - 7y = -32$

3. $(3, 2)$; $3t - 2s = 0$,
            $t + 2s = 15$

4. $(2, -2)$; $b + 2a = 2$,
            $b - a = -4$

5. $(15, 20)$; $3x - 2y = 5$,
            $6x - 5y = -10$

6. $(-1, -3)$; $3r + s = -6$,
            $2r = 1 + s$

7. $(-1, 1)$; $x = -1$,
            $x - y = -2$

8. $(-3, 4)$; $2x = -y - 2$,
            $y = -4$

9. $(12, 3)$; $y = \dfrac{1}{4}x$,
            $3x - y = 33$

10. $(-3, 1)$; $y = -\dfrac{1}{3}x$,
             $3y = -5x - 12$

b Solve the system of equations by graphing.

11. $x + y = 3$,
         $x - y = 1$

12. $x - y = 2$,
         $x + y = 6$

13. $8x - y = 29$,
         $2x + y = 11$

14. $4x - y = 10$,
         $3x + 5y = 19$

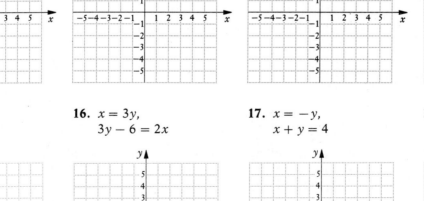

15. $u = v$,
         $4u = 2v - 6$

16. $x = 3y$,
         $3y - 6 = 2x$

17. $x = -y$,
         $x + y = 4$

18. $-3x = 5 - y$,
          $2y = 6x + 10$

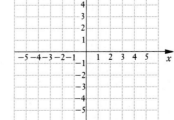

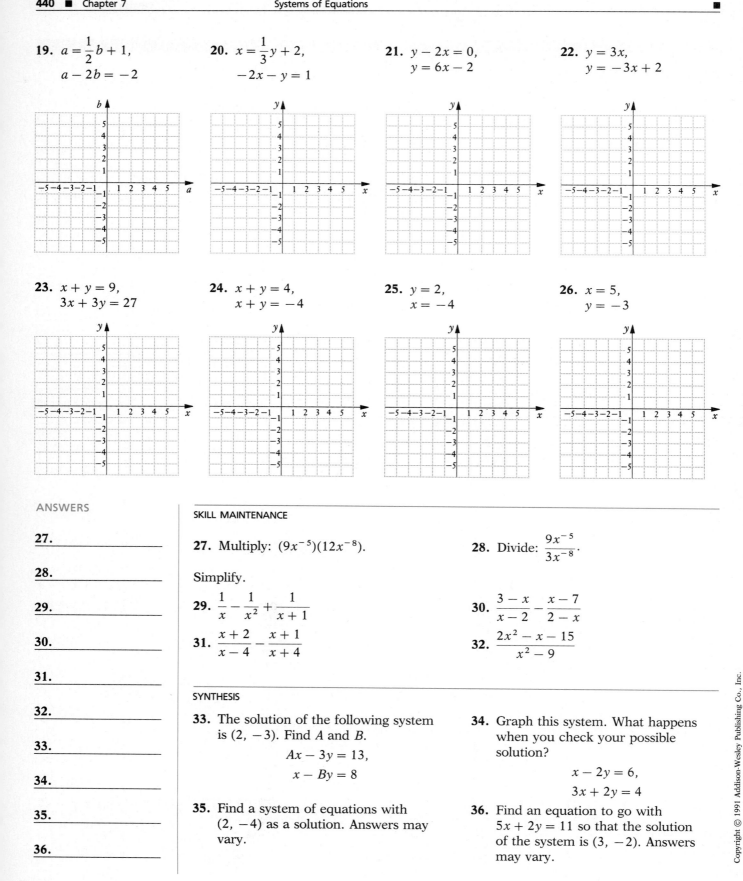

19. $a = \frac{1}{2}b + 1,$
　　$a - 2b = -2$

20. $x = \frac{1}{3}y + 2,$
　　$-2x - y = 1$

21. $y - 2x = 0,$
　　$y = 6x - 2$

22. $y = 3x,$
　　$y = -3x + 2$

23. $x + y = 9,$
　　$3x + 3y = 27$

24. $x + y = 4,$
　　$x + y = -4$

25. $y = 2,$
　　$x = -4$

26. $x = 5,$
　　$y = -3$

ANSWERS

27. _____

28. _____

29. _____

30. _____

31. _____

32. _____

33. _____

34. _____

35. _____

36. _____

SKILL MAINTENANCE

27. Multiply: $(9x^{-5})(12x^{-8})$.

28. Divide: $\dfrac{9x^{-5}}{3x^{-8}}$.

Simplify.

29. $\dfrac{1}{x} - \dfrac{1}{x^2} + \dfrac{1}{x + 1}$

30. $\dfrac{3 - x}{x - 2} - \dfrac{x - 7}{2 - x}$

31. $\dfrac{x + 2}{x - 4} - \dfrac{x + 1}{x + 4}$

32. $\dfrac{2x^2 - x - 15}{x^2 - 9}$

SYNTHESIS

33. The solution of the following system is $(2, -3)$. Find A and B.
$$Ax - 3y = 13,$$
$$x - By = 8$$

34. Graph this system. What happens when you check your possible solution?
$$x - 2y = 6,$$
$$3x + 2y = 4$$

35. Find a system of equations with $(2, -4)$ as a solution. Answers may vary.

36. Find an equation to go with $5x + 2y = 11$ so that the solution of the system is $(3, -2)$. Answers may vary.

7.2 The Substitution Method

Consider the following system of equations:

$$3x + 7y = 5,$$
$$6x - 7y = 1.$$

Suppose we try to solve this system graphically. We obtain the following graph.

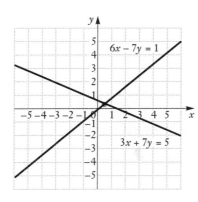

What is the solution? It is rather difficult to tell exactly. It would appear that fractions are involved. It turns out that the solution is $(\frac{2}{3}, \frac{3}{7})$. We need techniques involving algebra to determine the solution exactly. Graphing helps us picture the solution of a system of equations, but solving by graphing, though practical in many applied situations, is not always fast or accurate in cases where solutions are not integers. We now learn other methods using algebra. Because they use algebra, they are called **algebraic.**

a Solving by the Substitution Method

One nongraphical method for solving systems is known as the **substitution method.** In Example 1, we use the substitution method to solve a system we graphed in Example 3 of Section 7.1.

▶ **EXAMPLE 1**   Solve the system:

$$x + y = 6, \quad \textbf{(1)}$$
$$x = y + 2. \quad \textbf{(2)}$$

We graphed this system in Example 3 of Section 7.1. Equation (2) says that x and $y + 2$ name the same thing. Thus in Equation (1), we can substitute $y + 2$ for x:

$$x + y = 6 \qquad \text{Equation (1)}$$
$$(y + 2) + y = 6. \qquad \text{Substituting } y + 2 \text{ for } x$$

This last equation has only one variable. We solve it:

$$y + 2 + y = 6 \qquad \text{Removing parentheses}$$
$$2y + 2 = 6 \qquad \text{Collecting like terms}$$
$$2y = 4 \qquad \text{Subtracting 2 on both sides}$$
$$y = 2. \qquad \text{Dividing by 2}$$

OBJECTIVES

After finishing Section 7.2, you should be able to:

a Solve a system of two equations in two variables by the substitution method when one of the equations has a variable alone on one side.

b Solve a system of two equations in two variables by the substitution method when neither equation has a variable alone on one side.

c Solve problems by translating to a system of two equations and then solving using the substitution method.

FOR EXTRA HELP

Tape 13D   Tape 14B   MAC: 7
                                IBM: 7

1. Solve by the substitution method. Do not graph.

$$x + y = 5,$$
$$x = y + 1$$

We have found the y-value of the solution. To find the x-value, we return to the original pair of equations. Substituting into either equation will give us the x-value. We choose Equation (1):

$$x + y = 6 \qquad \text{Equation (1)}$$
$$x + 2 = 6 \qquad \text{Substituting 2 for } y$$
$$x = 4. \qquad \text{Subtracting 2}$$

The ordered pair (4, 2) may be a solution. We check.

Check:

$$\frac{x + y = 6}{4 + 2 \;\bigl|\; 6}$$
$$6 \;\bigl|\; \text{TRUE}$$

$$\frac{x = y + 2}{4 \;\bigl|\; 2 + 2}$$
$$4 \;\bigl|\; \text{TRUE}$$

Since (4, 2) checks, we have the solution. We could also express the answer as $x = 4$, $y = 2$.   ◀

Note in Example 1 that substituting 2 for y in the second equation will also give us the x-value of the solution: $x = y + 2 = 2 + 2 = 4$. Note also that we are using alphabetical order in listing the coordinates in an ordered pair. That is, since x precedes y, we have 4 before 2 in the pair (4, 2).

DO EXERCISE 1.

2. Solve by the substitution method:

$$a - b = 4,$$
$$b = 2 - a.$$

▶ **EXAMPLE 2**   Solve:

$$s = 13 - 3t, \qquad (1)$$
$$s + t = 5. \qquad (2)$$

We substitute $13 - 3t$ for s in Equation (2):

$$s + t = 5 \qquad \text{Equation (2)}$$
$$(13 - 3t) + t = 5. \qquad \text{Substituting } 13 - 3t \text{ for } s$$

> Remember to use parentheses when you substitute. Then remove them carefully.

Now we solve for t:

$$13 - 2t = 5 \qquad \text{Collecting like terms}$$
$$-2t = -8 \qquad \text{Subtracting 13}$$
$$t = \frac{-8}{-2}, \text{ or } 4. \qquad \text{Dividing by } -2$$

Next we substitute 4 for t in Equation (2) of the original system:

$$s + t = 5 \qquad \text{Equation (2)}$$
$$s + 4 = 5 \qquad \text{Substituting 4 for } t$$
$$s = 1. \qquad \text{Subtracting 4}$$

The pair (1, 4) checks and is the solution.   ◀

DO EXERCISE 2.

b Solving for the Variable First

Sometimes neither equation of a pair has a variable alone on one side. Then we solve one equation for one of the variables and proceed as before, substituting into the *other* equation. If possible, we solve in either equation for a variable that has a coefficient of 1.

▶ **EXAMPLE 3** Solve:

$$x - 2y = 6, \qquad (1)$$
$$3x + 2y = 4. \qquad (2)$$

We solve one equation for one variable. Since the coefficient of x is 1 in Equation (1), it is easier to solve that equation for x:

$$x - 2y = 6 \qquad \text{Equation (1)}$$
$$x = 6 + 2y. \qquad \text{Adding } 2y \qquad (3)$$

We substitute $6 + 2y$ for x in Equation (2) of the original pair and solve for y:

$$3x + 2y = 4 \qquad \text{Equation (2)}$$
$$3(6 + 2y) + 2y = 4 \qquad \text{Substituting } 6 + 2y \text{ for } x$$
$$18 + 6y + 2y = 4 \qquad \text{Removing parentheses}$$
$$18 + 8y = 4 \qquad \text{Collecting like terms}$$
$$8y = -14 \qquad \text{Subtracting 18}$$
$$y = \frac{-14}{8}, \text{ or } -\frac{7}{4}. \qquad \text{Dividing by 8}$$

To find x, we go back to either of the original Equations (1) or (2) or to Equation (3), which we solved for x. It is generally easier to use an equation like Equation (3) where we have solved for a specific variable. We substitute $-\frac{7}{4}$ for y in Equation (3) and compute x:

$$x = 6 + 2y = 6 + 2\left(-\frac{7}{4}\right) = 6 - \frac{7}{2} = \frac{5}{2}.$$

We check the ordered pair $(\frac{5}{2}, -\frac{7}{4})$.

Check:

$$\begin{array}{c|c} x - 2y = 6 \\ \hline \dfrac{5}{2} - 2\left(-\dfrac{7}{4}\right) & 6 \\ \dfrac{5}{2} + \dfrac{7}{2} \\ \dfrac{12}{2} \\ 6 & \text{TRUE} \end{array} \qquad \begin{array}{c|c} 3x + 2y = 4 \\ \hline 3 \cdot \dfrac{5}{2} + 2\left(-\dfrac{7}{4}\right) & 4 \\ \dfrac{15}{2} - \dfrac{7}{2} \\ \dfrac{8}{2} \\ 4 & \text{TRUE} \end{array}$$

Since $(\frac{5}{2}, -\frac{7}{4})$ checks, it is the solution. ◀

This solution would have been difficult to find graphically because it involves fractions.

DO EXERCISE 3.

3. Solve:

$$x - 2y = 8,$$
$$2x + y = 8.$$

ANSWER ON PAGE A-7

4. The perimeter of a rectangle is 76 cm. The length is 17 cm more than the width. Find the length and the width.

ANSWER ON PAGE A-7

C   **Solving Problems**

Now let us use the substitution method to solve a problem.

▶ **EXAMPLE 4**   The state of Colorado is a rectangle whose perimeter is 1300 mi. The length is 110 mi more than the width. Find the length and the width.

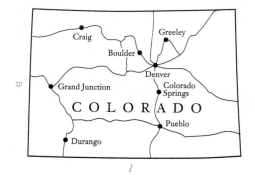

1. *Familiarize.*   We make a drawing and label it. We have called the length l and the width w.

2. *Translate.*   The perimeter of the rectangle is $2l + 2w$. We translate the first statement.

The perimeter is 1300 mi.

$$2l + 2w \quad = \quad 1300$$

We translate the second statement.

The length is 110 mi greater than the width.

$$l \quad = \quad 110 + w$$

We have translated to a system of equations:

$$2l + 2w = 1300, \qquad \textbf{(1)}$$
$$l = 110 + w. \qquad \textbf{(2)}$$

3. *Solve.*   We solve the system. We substitute $110 + w$ for l in the first equation and solve:

$$
\begin{aligned}
2(110 + w) + 2w &= 1300 &&\text{Substituting } 110 + w \text{ for } l \text{ in Equation (1)} \\
220 + 2w + 2w &= 1300 &&\text{Removing parentheses} \\
220 + 4w &= 1300 &&\text{Collecting like terms} \\
4w &= 1080 &&\text{Subtracting 220} \\
w &= \frac{1080}{4}, \text{ or } 270. &&\text{Dividing by 4}
\end{aligned}
$$

We go back to the original equations and substitute 270 for w. We use Equation (2):

$$
\begin{aligned}
l &= 110 + w &&\text{Equation (2)} \\
l &= 110 + 270 &&\text{Substituting 270 for } w \\
l &= 380.
\end{aligned}
$$

4. *Check.*   A possible solution is a length of 380 mi and a width of 270 mi. The perimeter would be $2(380) + 2(270)$, or $760 + 540$, or 1300. Also, the length is 110 mi greater than the width. These check.

5. *State.*   The length is 370 mi, and the width is 110 mi.   ◀

This problem illustrates that many problems that can be solved by translating to *one* equation in *one* variable are actually easier to solve by translating to *two* equations in *two* variables.

DO EXERCISE 4.

EXERCISE SET 7.2

a Solve by the substitution method.

1. $x + y = 4,$
$y = 2x + 1$

2. $x + y = 10,$
$y = x + 8$

3. $y = x + 1,$
$2x + y = 4$

4. $y = x - 6,$
$x + y = -2$

5. $y = 2x - 5,$
$3y - x = 5$

6. $y = 2x + 1,$
$x + y = -2$

7. $x = -2y,$
$x + 4y = 2$

8. $r = -3s,$
$r + 4s = 10$

b Solve by the substitution method. First, solve one equation for one variable.

9. $s + t = -4,$
$s - t = 2$

10. $x - y = 6,$
$x + y = -2$

11. $y - 2x = -6,$
$2y - x = 5$

12. $x - y = 5,$
$x + 2y = 7$

13. $2x + 3y = -2,$
$2x - y = 9$

14. $x + 2y = 10,$
$3x + 4y = 8$

15. $x - y = -3,$
$2x + 3y = -6$

16. $3b + 2a = 2,$
$-2b + a = 8$

17. $r - 2s = 0,$
$4r - 3s = 15$

18. $y - 2x = 0,$
$3x + 7y = 17$

c Solve.

19. The sum of two numbers is 27. One number is three more than the other. Find the numbers.

20. The sum of two numbers is 36. One number is two more than the other. Find the numbers.

21. Find two numbers whose sum is 58 and whose difference is 16.

22. Find two numbers whose sum is 66 and whose difference is 8.

ANSWERS

1. _____

2. _____

3. _____

4. _____

5. _____

6. _____

7. _____

8. _____

9. _____

10. _____

11. _____

12. _____

13. _____

14. _____

15. _____

16. _____

17. _____

18. _____

19. _____

20. _____

21. _____

22. _____

Copyright © 1991 Addison-Wesley Publishing Co., Inc.

ANSWERS

23. _____

24. _____

25. _____

26. _____

27. _____

28. _____

29.   See graph. _____

30.   See graph. _____

31.   See graph. _____

32.   See graph. _____

33. _____

34. _____

35. _____

36. _____

37. _____

38. _____

39. _____

23. The difference between two numbers is 16. Three times the larger number is seven times the smaller. What are the numbers?

24. The difference between two numbers is 18. Twice the smaller number plus three times the larger is 74. What are the numbers?

25. The state of Wyoming is a rectangle whose perimeter is 1280 mi. The width is 90 mi less than the length. Find the length and the width.

26. The perimeter of a standard-sized rectangular rug is 42 ft. The length is 3 ft more than the width. Find the length and the width.

27. The perimeter of a rectangle is 400 m. The length is 3 m more than twice the width. Find the length and the width.

28. The perimeter of a rectangle is 876 cm. The length is 1 cm less than three times the width. Find the length and the width.

SKILL MAINTENANCE

Graph.

29. $2x - 3y = 6$

30. $2x + 3y = 6$

31. $2x - 3 = 0$

32. $y = 2x - 5$

SYNTHESIS

Solve by the substitution method.

33. ▦ $y - 2.35x = -5.97,$
$2.14y - x = 4.88$

34. $\frac{1}{4}(a - b) = 2,$
$\frac{1}{6}(a + b) = 1$

35. $\frac{x}{2} + \frac{3y}{2} = 2,$
$\frac{x}{5} - \frac{y}{2} = 3$

36. A rectangle has a perimeter of P ft. The width is 5 ft less than the length. Find the length in terms of P.

37. The perimeter of a football field (excluding the end zones) is $306\frac{2}{3}$ yd. The length is $46\frac{2}{3}$ yd longer than the width. Find the length and the width.

38. Consider this system of equations:
$$3y + 3x = 14,$$
$$y = -x + 4.$$
Try to solve by the substitution method. Can you explain your results?

39. Consider this system of equations:
$$y = x + 5,$$
$$-3x + 3y = 15.$$
Try to solve by the substitution method. Can you explain your results?

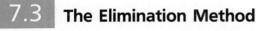

7.3 The Elimination Method

a Solving by the Elimination Method

The **elimination method** for solving systems of equations makes use of the *addition principle*. Some systems are much easier to solve using this method. Trying to solve the system in Example 1 by substitution would necessitate the use of fractions and extra steps. Instead we use the elimination method.

▶ **EXAMPLE 1** Solve:

$$2x + 3y = 13, \quad (1)$$
$$4x - 3y = 17. \quad (2)$$

The key to the advantage of the elimination method for solving this system involves the $3y$ in one equation and the $-3y$ in the other. The terms are opposites. If we add the terms on the sides of the equations, these terms will add to 0, and in effect, the variable y will be eliminated.

We will use the addition principle for equations. According to Equation (2), $4x - 3y$ and 17 are the same number. Thus we can use a vertical form and add $4x - 3y$ to the left side of Equation (1) and 17 to the right side:

$$2x + 3y = 13 \quad (1)$$
$$\underline{4x - 3y = 17} \quad (2)$$
$$6x + 0y = 30. \quad \text{Adding}$$

We have "eliminated" one variable. This is why we call this the **elimination method.** We now have an equation with just one variable that can be solved for x:

$$6x = 30$$
$$x = 5.$$

Next we substitute 5 for x in either of the original equations:

$$2x + 3y = 13 \quad \text{Equation (1)}$$
$$2(5) + 3y = 13 \quad \text{Substituting 5 for } x$$
$$10 + 3y = 13$$
$$3y = 3$$
$$y = 1. \quad \text{Solving for } y$$

We check the ordered pair (5, 1).

$$\textit{Check:} \quad \begin{array}{c|c} 2x + 3y = 13 \\ \hline 2(5) + 3(1) & 13 \\ 10 + 3 & \\ 13 & \text{TRUE} \end{array} \qquad \begin{array}{c|c} 4x - 3y = 17 \\ \hline 4(5) - 3(1) & 17 \\ 20 - 3 & \\ 17 & \text{TRUE} \end{array}$$

Since (5, 1) checks, it is the solution. ◀

DO EXERCISES 1 AND 2.

b Using the Multiplication Principle First

The elimination method allows us to eliminate a variable. We may need to multiply by certain numbers first, however, so that terms become opposites.

Solve using the elimination method.

1. $x + y = 5,$
    $2x - y = 4$

2. $3x - 3y = 6,$
    $3x + 3y = 0$

3. Solve. Multiply one equation by -1 first.

$$5x + 3y = 17,$$
$$5x - 2y = -3$$

▶ **EXAMPLE 2**   Solve:

$$2x + 3y = 8, \quad \textbf{(1)}$$
$$x + 3y = 7. \quad \textbf{(2)}$$

If we add, we will not eliminate a variable. However, if the $3y$ were $-3y$ in one equation, we could eliminate y. We multiply on both sides of Equation (2) by -1 and then add, using a vertical form.

$$
\begin{array}{ll}
2x + 3y = 8 & \text{Equation (1)} \\
\underline{-x - 3y = -7} & \text{Multiplying Equation (2) by } -1 \\
x = 1. & \text{Adding}
\end{array}
$$

Now we substitute 1 for x in one of the original equations:

$$
\begin{array}{ll}
x + 3y = 7 & \text{Equation (2)} \\
1 + 3y = 7 & \text{Substituting 1 for } x \\
3y = 6 & \\
y = 2. & \text{Solving for } y
\end{array}
$$

We can check the ordered pair $(1, 2)$.

Check:

$$
\begin{array}{c|c}
2x + 3y = 8 & \\
\hline
2 \cdot 1 + 3 \cdot 2 & 8 \\
2 + 6 & \\
8 & \text{TRUE}
\end{array}
\qquad
\begin{array}{c|c}
x + 3y = 7 & \\
\hline
1 + 3 \cdot 2 & 7 \\
1 + 6 & \\
7 & \text{TRUE}
\end{array}
$$

Since $(1, 2)$ checks, it is the solution. ◀

DO EXERCISE 3.

In Example 2, we used the multiplication principle, multiplying by -1. We often need to multiply by something other than -1.

▶ **EXAMPLE 3**   Solve:

$$3x + 6y = -6, \quad \textbf{(1)}$$
$$5x - 2y = 14. \quad \textbf{(2)}$$

Looking at the terms with variables, we see that if $-2y$ were $-6y$, we would have terms that are opposites. We can achieve this by multiplying on both sides of Equation (2) by 3. Then we add:

$$
\begin{array}{ll}
3x + 6y = -6 & \text{Equation (1)} \\
\underline{15x - 6y = 42} & \text{Multiplying Equation (2) by 3} \\
18x = 36 & \text{Adding} \\
x = 2. & \text{Solving for } x
\end{array}
$$

We go back to Equation (1) and substitute 2 for x:

$$
\begin{array}{ll}
3 \cdot 2 + 6y = -6 & \text{Substituting} \\
6 + 6y = -6 & \\
6y = -12 & \\
y = -2. & \text{Solving for } y
\end{array}
$$

We check the ordered pair $(2, -2)$.

Check:

$$3x + 6y = -6$$

$3 \cdot 2 + 6 \cdot (-2)$	-6
$6 + (-12)$	
-6	TRUE

$$5x - 2y = 14$$

$5 \cdot 2 - 2 \cdot (-2)$	14
$10 - (-4)$	
14	TRUE

4. Solve:

$$4a + 7b = 11,$$
$$2a + 3b = 5.$$

Since $(2, -2)$ checks, it is the solution.   ◀

> CAUTION!   Solving a *system* of equations in two variables requires finding an ordered *pair* of numbers. Once you have solved for one variable, don't forget the other.

DO EXERCISE 4.

We have used the elimination method in Examples 1–3. Part of the strategy in doing so is making a decision about which variable to eliminate. So long as the algebra has been carried out correctly, the solution can be found by eliminating *either* variable. We multiply so that terms involving the variable to be eliminated are opposites. It is helpful to first get each equation in a form equivalent to $Ax + By = C$.

▶ **EXAMPLE 4**   Solve:

$$3y + 1 + 2x = 0,$$
$$5x = 7 - 4y.$$

We first get each equation into a form equivalent to $Ax + By = C$:

$$2x + 3y = -1, \qquad \text{Subtracting 1 on both sides and rearranging terms}$$
$$5x + 4y = 7. \qquad \text{Adding } 4y \text{ on both sides}$$

We then use the multiplication principle with both equations:

$$2x + 3y = -1, \qquad (1)$$
$$5x + 4y = 7. \qquad (2)$$

We decide to eliminate the x-term. We do this by multiplying on both sides of Equation (1) by 5 and on both sides of Equation (2) by -2:

$10x + 15y = -5$	Multiplying on both sides of Equation (1) by 5
$-10x - 8y = -14$	Multiplying on both sides of Equation (2) by -2
$7y = -19$	Adding

$$y = \frac{-19}{7}, \text{ or } -\frac{19}{7}. \qquad \text{Dividing by 7}$$

We substitute $-\frac{19}{7}$ for y in one of the original equations:

$$2x + 3y = -1 \qquad \text{Equation (1)}$$
$$2x + 3(-\tfrac{19}{7}) = -1 \qquad \text{Substituting } -\tfrac{19}{7} \text{ for } y$$
$$2x - \tfrac{57}{7} = -1$$
$$2x = -1 + \tfrac{57}{7}$$
$$2x = -\tfrac{7}{7} + \tfrac{57}{7}$$
$$2x = \tfrac{50}{7}$$
$$x = \tfrac{50}{7} \cdot \tfrac{1}{2}, \text{ or } \tfrac{25}{7}. \qquad \text{Solving for } x$$

We check the ordered pair $(\tfrac{25}{7}, -\tfrac{19}{7})$.

ANSWER ON PAGE A-7

5. Solve:

$$3x = 5 + 2y,$$
$$2x + 3y - 1 = 0.$$

Check:

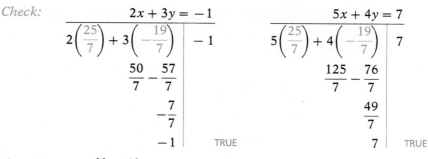

The solution is $(\frac{25}{7}, -\frac{19}{7})$.   ◀

DO EXERCISE 5.

Let us consider a system with no solution and see what happens when we apply the elimination method.

▶ **EXAMPLE 5**   Solve:

$$y - 3x = 2, \qquad (1)$$
$$y - 3x = 1. \qquad (2)$$

We multiply by -1 on both sides of Equation (2) and then add:

$$\begin{array}{ll} y - 3x = 2 & \\ \underline{-y + 3x = -1} & \text{Multiplying by } -1 \\ 0 = 1. & \text{Adding} \end{array}$$

We obtain a *false* equation, $0 = 1$, so there is *no solution*. The slope–intercept forms of these equations are

$$y = 3x + 2,$$
$$y = 3x + 1.$$

The slopes are the same and the y-intercepts are different. Thus the lines are parallel. They do not intersect.   ◀

DO EXERCISE 6.

6. Solve:

$$2x + y = 15,$$
$$4x + 2y = 23.$$

Sometimes there is an infinite number of solutions. Let's look at a system that we graphed in Example 5 of Section 7.1.

▶ **EXAMPLE 6**   Solve:

$$2x + 3y = 6,$$
$$-8x - 12y = -24.$$

We multiply on both sides of Equation (1) by 4 and then add the two equations:

$$\begin{array}{ll} 8x + 12y = 24 & \text{Multiplying by 4} \\ \underline{-8x - 12y = -24} & \\ 0 = 0. & \text{Adding} \end{array}$$

7. Solve:

$$5x - 2y = 3,$$
$$-15x + 6y = -9$$

We have eliminated both variables, and what remains is an equation easily seen to be true. If this happens when we use the elimination method, we have an infinite number of solutions.   ◀

DO EXERCISE 7.

When decimals or fractions appear, we first multiply to clear of them. Then we proceed as before.

► **EXAMPLE 7**   Solve:

$$\frac{1}{3}x + \frac{1}{2}y = -\frac{1}{6}, \quad \text{(1)}$$

$$\frac{1}{2}x + \frac{2}{5}y = \frac{7}{10}. \quad \text{(2)}$$

The number 6 is a multiple of all the denominators of Equation (1). The number 10 is a multiple of all the denominators of Equation (2). We multiply on both sides of Equation (1) by 6 and on both sides of Equation (2) by 10:

$$6\left(\frac{1}{3}x + \frac{1}{2}y\right) = 6\left(-\frac{1}{6}\right) \qquad\qquad 10\left(\frac{1}{2}x + \frac{2}{5}y\right) = 10\left(\frac{7}{10}\right)$$

$$6 \cdot \frac{1}{3}x + 6 \cdot \frac{1}{2}y = -1 \qquad\qquad 10 \cdot \frac{1}{2}x + 10 \cdot \frac{2}{5}y = 7$$

$$2x + 3y = -1; \qquad\qquad\qquad 5x + 4y = 7.$$

The resulting system is

$$2x + 3y = -1,$$
$$5x + 4y = 7.$$

As we saw in Example 4, the solution of this system is $\left(\frac{25}{9}, -\frac{19}{7}\right)$.   ◄

DO EXERCISE 8.

The following is a summary that compares the graphical, substitution, and elimination methods for solving systems of equations.

Method	Strengths	Weaknesses
Graphical	Can "see" solution.	Inexact when solution involves numbers that are not integers or are very large and off the graph.
Substitution	Works well when solutions are not integers. Easy to use when a variable is alone on one side.	Introduces extensive computations with fractions for more complicated systems where coefficients are not 1 or −1. Cannot "see" the solution.
Elimination	Works well when solutions are not integers, when coefficients are not 1 or −1, and when coefficients involve decimals or fractions	Cannot "see" the solution.

When deciding which method to use, consider the preceding chart and directions from your instructor. The situation is like having a piece of wood to cut and three saws with which to cut it. The saw you use depends on the type of wood, the type of cut you are making, and how you want the wood to turn out.

8. Solve:

$$\frac{1}{2}x + \frac{3}{10}y = \frac{1}{5},$$

$$\frac{3}{5}x + \quad y = -\frac{2}{5}.$$

ANSWER ON PAGE A-8

9. Acme Rent-A-Car rents a car at a daily rate of $41.95 plus 43 cents per mile. Speedo Rentzit rents a car for $44.95 plus 39 cents per mile. For what mileage is the cost the same?

c   Solving Problems

We now use the elimination method to solve a problem.

▶ **EXAMPLE 8**   At one time, Budget Rent-A-Car rented compact cars at a daily rate of $43.95 plus 40 cents per mile. Thrifty Rent-A-Car rented compact cars at a daily rate of $42.95 plus 42 cents per mile. For what mileage is the cost the same?

1. *Familiarize.*   To become familiar with the problem, we make a guess. Suppose a person rents a compact car from each rental agency and drives it 100 miles. The total cost at Budget is $43.95 + \$0.40(100) =$43.95 + $40.00, or $83.95. The total cost at Thrifty is $42.95 + \$0.42(100) =$42.95 + $42.00, or $84.95. Note that we converted all of our money units to dollars. The resulting costs are very nearly the same, so our guess is close. We can, of course, refine our guess. Instead, we will use algebra to solve the problem. We let $M =$ the number of miles driven and $C =$ the total cost of the car rental.

2. *Translate.*   We translate the first statement, using $0.40 for 40 cents. It helps to reword the problem before translating.

Rewording: $43.95 plus 40 cents times the number of miles driven is cost.

Translating: $43.95   + $0.40   · M = C

We translate the second statement, but again it helps to reword it first.

Rewording: $42.95 plus 42 cents times the number of miles driven is cost.

Translating: $42.95   + $0.42   · M = C

We have now translated to a system of equations:

$$43.95 + 0.40M = C,$$
$$42.95 + 0.42M = C.$$

3. *Solve.*   We solve the system of equations. We clear the system of decimals by multiplying on both sides by 100. Then we multiply the second equation by -1 and add.

$$
\begin{array}{r}
4395 + 40M = 100C \\
-4295 - 42M = -100C \\
\hline
100 - 2M = 0 \\
100 = 2M \\
50 = M
\end{array}
$$

4. *Check.*   For 50 mi, the cost of the Budget car is $43.95 + 0.40(50)$, or $43.95 + 20$, or $63.95, and the cost of the Thrifty car is $42.95 + 0.42(50)$, or $42.95 + 21$, or $63.95. Thus the costs are the same when the mileage is 50.

5. *State.*   When the cars are driven 50 miles, the costs will be the same.

◀

DO EXERCISE 9.

ANSWER ON PAGE A-8

EXERCISE SET 7.3

ANSWERS

a   Solve using the elimination method.

1. $x + y = 10,$
$\quad x - y = 8$

2. $x - y = 7,$
$\quad x + y = 3$

3. $\quad x + \ y = 8,$
$\quad -x + 2y = 7$

4. $\quad x + \ y = 6,$
$\quad -x + 3y = -2$

5. $3x - y = 9,$
$\quad 2x + y = 6$

6. $4x - y = 1,$
$\quad 3x + y = 13$

7. $\quad 4a + 3b = 7,$
$\quad -4a + \ b = 5$

8. $7c + 5d = 18,$
$\quad c - 5d = -2$

9. $8x - 5y = -9,$
$\quad 3x + 5y = -2$

10. $\quad 3a - 3b = -15,$
$\quad -3a - 3b = -3$

11. $\quad 4x - 5y = 7,$
$\quad -4x + 5y = 7$

12. $\quad 2x + 3y = 4,$
$\quad -2x - 3y = -4$

1. _____

2. _____

3. _____

4. _____

5. _____

6. _____

7. _____

8. _____

9. _____

10. _____

11. _____

12. _____

b Solve using the multiplication principle first. Then add.

13. $-x - y = 8,$
$\quad 2x - y = -1$

14. $x + y = -7,$
$\quad 3x + y = -9$

15. $x + 3y = 19,$
$\quad x - y = -1$

16. $3x - y = 8,$
$\quad x + 2y = 5$

17. $x + y = 5,$
$\quad 5x - 3y = 17$

18. $x - y = 7,$
$\quad 4x - 5y = 25$

19. $2w - 3z = -1,$
$\quad 3w + 4z = 24$

20. $7p + 5q = 2,$
$\quad 8p - 9q = 17$

21. $2a + 3b = -1,$
$\quad 3a + 5b = -2$

22. $3x - 4y = 16,$
$\quad 5x + 6y = 14$

23. $x = 3y,$
$\quad 5x + 14 = y$

24. $5a = 2b,$
$\quad 2a + 11 = 3b$

25. $3x - 2y = 10,$
$\quad 5x + 3y = 4$

26. $2p + 5q = 9,$
$\quad 3p - 2q = 4$

27. $3x = 8y + 11,$
$\quad x + 6y - 8 = 0$

28. $m = 32 + n,$
$\quad 3m = 8n + 6$

29. $3x - 2y = 10,$
$\quad -6x + 4y = -20$

30. $2x + y = 13,$
$\quad 4x + 2y = 23$

ANSWERS

13. _____

14. _____

15. _____

16. _____

17. _____

18. _____

19. _____

20. _____

21. _____

22. _____

23. _____

24. _____

25. _____

26. _____

27. _____

28. _____

29. _____

30. _____

31. $0.06x + 0.05y = 0.07,$
$\quad 0.4x - 0.3y = 1.1$

32. $1.8x - 2y = 0.9,$
$\quad 0.04x + 0.18y = 0.15$

33. $\dfrac{1}{3}x + \dfrac{3}{2}y = \dfrac{5}{4},$
$\quad \dfrac{3}{4}x - \dfrac{5}{6}y = \dfrac{3}{8}$

34. $x - \dfrac{3}{2}y = 13,$
$\quad \dfrac{3}{2}x - y = 17$

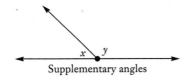 **Solve.**

35. At one time, Avis Rent-A-Car rented an intermediate-size car at a daily rate of $53.95 plus 30 cents per mile. Another company rents an intermediate-size car for $54.95 plus 20 cents per mile. For what mileage is the cost the same?

36. Budget Rent-A-Car rented a basic car at a daily rate of $45.95 plus 40 cents per mile. Another company rents a basic car for $46.95 plus 20 cents per mile. For what mileage is the cost the same?

37. Two angles are supplementary. One is 30° more than two times the other. Find the angles. (Supplementary angles are angles whose sum is 180°.)

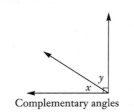

Supplementary angles

38. Two angles are supplementary. One is 8° less than three times the other. Find the angles.

39. Two angles are complementary. Their difference is 34°. Find the angles. (Complementary angles are angles whose sum is 90°.)

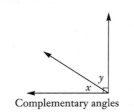

Complementary angles

40. Two angles are complementary. One angle is 42° more than one half the other. Find the angles.

31. _____

32. _____

33. _____

34. _____

35. _____

36. _____

37. _____

38. _____

39. _____

40. _____

ANSWERS

41. _____

42. _____

43. _____

44. _____

45. _____

46. _____

47. _____

48. _____

49. _____

50. _____

51. _____

52. _____

53. _____

54. _____

41. In a vineyard, a vintner uses 820 hectares to plant Chardonnay and Riesling grapes. The vintner knows that the profits will be greatest by planting 140 hectares more of Chardonnay than of Riesling. How many hectares of each grape should be planted?

42. The Hayburner Horse Farm allots 650 hectares to plant hay and oats. The owners know that their needs are best met if they plant 180 hectares more of hay than of oats. How many hectares of each should they plant?

SKILL MAINTENANCE

43. Simplify: $(a^2 b^{-3})(a^5 b^{-6})$.

44. Simplify: $\dfrac{a^2 b^{-3}}{a^5 b^{-6}}$.

45. Simplify: $\dfrac{x^2 - 5x + 6}{x^2 - 4}$.

46. Subtract: $\dfrac{x + 7}{x^2 - 1} - \dfrac{3}{x + 1}$.

SYNTHESIS

47. Several ancient Chinese books included problems that can be solved by translating to systems of equations. *Arithmetical Rules in Nine Sections* is a book of 246 problems compiled by a Chinese mathematician, Chang Tsang, who died in 152 B.C. One of the problems is: Suppose there are a number of rabbits and pheasants confined in a cage. In all, there are 35 heads and 94 feet. How many rabbits and how many pheasants are there? Solve the problem.

48. Patrick's age is 20% of his father's age. Twenty years from now, Patrick's age will be 52% of his father's age. How old are Patrick and his father now?

49. If 5 is added to a man's age and the total is divided by 5, the result will be his daughter's age. Five years ago, the man's age was eight times his daughter's age. Find their present ages.

50. When the base of a triangle is increased by 2 ft and the height is decreased by 1 ft, the height becomes one third of the base, and the area becomes 24 ft^2. Find the original dimensions of the triangle.

Solve.

51. $3(x - y) = 9,$
$\quad x + y = 7$

52. $2(x - y) = 3 + x,$
$\quad x = 3y + 4$

53. $\quad 2(5a - 5b) = 10,$
$-5(6a + 2b) = 10$

54. $\dfrac{x}{3} + \dfrac{y}{2} = 1\dfrac{1}{3},$
$\quad x + 0.05y = 4$

7.4 More on Solving Problems

OBJECTIVE

After finishing Section 7.4, you should be able to:

a Solve problems by translating them to systems of two equations in two variables.

FOR EXTRA HELP

Tape 14C   Tape 15A   MAC: 7
                            IBM: 7

a We continue solving problems using the five steps for problem solving and our methods for solving systems of equations.

▶ **EXAMPLE 1** *Denny's Restaurants.* Denny's® is a national restaurant firm. The ad shown here once appeared on the tables as a special.

Determine the price of one item from the *A* side of the menu and the price of one item from the *B* side of the menu.

1, 2. *Familiarize* and *Translate.* The ad gives us the system of equations at the outset:

$$a + b = \$5.49, \quad \textbf{(1)}$$
$$a + 2b = \$6.99, \quad \textbf{(2)}$$

where a = the price of one item from the *A* side of the menu and b = the price of one item from the *B* side of the menu.

3. *Solve.* We solve the system of equations. Which method should we use? As we discussed in Section 7.3, any method can be used. Each has its advantages and disadvantages. We decide to proceed with the elimination method, because we see that if we multiply Equation (1) by −1 and then add, the *a*-terms can be eliminated:

$$
\begin{array}{ll}
-a - b = -5.49 & \text{Multiplying by } -1 \\
\underline{a + 2b = 6.99} & \\
\ b = 1.50. & \text{Adding}
\end{array}
$$

We substitute 1.50 for b in Equation (1) and solve for a:

$$a + b = 5.49$$
$$a + 1.50 = 5.49$$
$$a = 3.99.$$

1. Suppose, in a later promotion, that Denny's changed the menu as follows:

$$A + \ B = \$6.59,$$
$$A + 2B = \$9.95.$$

Determine the price of one item from the A side of the menu and the price of one item from the B side.

4. *Check.* The sum of the two prices is $\$3.99 + \$1.50$, or $\$5.49$. The A price plus twice the B price is $\$3.99 + 2(\$1.50) = \$3.99 + \$3.00$, or $\$6.99$. The prices check. Sometimes a "common sense" check is appropriate. If you look at the foods on the B side of the menu, it does not seem reasonable that such items would sell for $\$1.50$ each. Note that the menu does not say that you can buy a B item alone for $\$1.50$. That price is a bonus for buying an A item. Mathematically, the prices given do stand, though it is doubtful that you can buy a B item by itself.

5. *State.* The price of one A item is $\$3.99$. The price of one B item is $\$1.50$.

◀

DO EXERCISE 1.

▶ **EXAMPLE 2**   Howie is 21 years older than Judy. In six years, Howie will be twice as old as Judy. How old are they now?

1. *Familiarize.*   Let us consider some conditions of the problem. We let $H =$ Howie's age now and $J =$ Judy's age now. Everyone ages together. As one person gets 1 year older, so does the other. How do the ages relate in 6 years? In 6 years, Judy will be $J + 6$ and Howie will be $H + 6$. We make a table to organize our information.

	Howie	**Judy**
Age now	H	J	$\longrightarrow H = 21 + J$
Age in 6 years	$H + 6$	$J + 6$	$\longrightarrow H + 6 = 2(J + 6)$

2. *Translate.*   From the present ages, we get the following rewording and translation.

Howie's age is 21 more than Judy's age.   **Rewording**

$$H \quad = 21 \quad + \qquad J \qquad$$ **Translating**

From their ages in 6 years, we get the following rewording and translation.

Howie's age in six years will be twice Judy's age in six years.   **Rewording**

$$H + 6 \qquad = \quad 2 \cdot \qquad (J + 6)$$ **Translating**

The problem has been translated to the following system of equations.

$$H = 21 + J, \qquad \textbf{(1)}$$
$$H + 6 = 2(J + 6). \qquad \textbf{(2)}$$

3. *Solve.*   We solve the system of equations. This time we use the substitution method since there is a variable alone on one side. We substitute $21 + J$ for H in Equation (2):

$$H + 6 = 2(J + 6)$$
$$(21 + J) + 6 = 2(J + 6)$$
$$J + 27 = 2J + 12$$
$$15 = J.$$

We find H by substituting 15 for J in the first equation:

$$H = 21 + J$$
$$H = 21 + 15$$
$$H = 36.$$

4. *Check.* Howie's age is 36, which is 21 more than 15, Judy's age. In 6 years, when Howie will be 42 and Judy 21, Howie's age will be twice Judy's age.

5. *State.* Howie is now 36 and Judy is 15. ◄

DO EXERCISE 2.

▶ **EXAMPLE 3** There were 411 people at a movie. Admission was $7.00 for adults and $3.75 for children. The receipts were $2678.75. The box office manager loses the records of how many adults and how many children attended. Can you use algebra to help her? How many adults and how many children did attend?

1. *Familiarize.* There are many ways to familiarize ourselves with a problem situation. This time, let us make a guess and do some calculations. The total number of people at the movie was 411, so we choose numbers that total 411. Let's try

240 adults and
171 children.

How much money was taken in? The problem says that adults paid $7.00 each, so the total amount of money collected from the adults was

240($7), or $1680.

The children paid $3.75 each, so the total amount of money collected from the children was

171($3.75), or $641.25.

This makes the total receipts $1680 + $641.25, or $2321.25.

Our guess is not the answer to the problem because the total taken in, according to the problem, was $2678.75. If we were to continue guessing, we would need to add more adults and fewer children, since our first guess was too low. The steps we have used to see if our guesses are correct help us to understand the actual steps involved in solving the problem.

Let us list the information in a table. That usually helps in the familiarization process. We let a = the number of adults and c = the number of children.

	Adults	Children	Total
Admission	$7.00	$3.75	
Number attending	a	c	411	$\longrightarrow a + c = 411$
Money taken in	$7.00a$	$3.75c$	$2678.75	$\longrightarrow 7.00a + 3.75c = 2678.75$

2. *Translate.* The total number of people attending was 411, so

$$a + c = 411.$$

2. Jackie is 26 years older than Jack. In 5 years, Jackie will be twice as old as Jack. How old are they now?

Complete the following table to aid with the familiarization.

	Jackie	Jack
Age now	J	K
Age in 5 years		

ANSWER ON PAGE A-8

3. There were 166 paid admissions to a game. The price was $2.10 each for adults and $0.75 each for children. The amount taken in was $293.25. How many adults and how many children attended?

Complete the following table to aid with the familiarization.

	Adults	Children	Total	
Paid		$0.75		
Number attending	x	y		→ $x + y =$ ( )
Money taken in			$293.25	→ $2.10x +$ ( ) $= 293.25$

The amount taken in from the adults was $7.00a$, and the amount taken in from the children was $3.75c$. These amounts are in dollars. The total was $2678.75, so we have

$$7.00a + 3.75c = 2678.75.$$

We can multiply on both sides by 100 to clear of decimals. Thus we have a translation to a system of equations:

$$a + c = 411, \qquad \textbf{(1)}$$
$$700a + 375c = 267{,}875. \qquad \textbf{(2)}$$

3. *Solve.* We solve the system of equations. We use the elimination method since the equations are both in the form $Ax + By = C$. (A case can certainly be made for using the substitution method since we can solve for one of the variables quite easily in the first equation. Very often a decision is just a matter of choice.) We multiply on both sides of Equation (1) by -375 and then add:

$$
\begin{array}{ll}
-375a - 375c = -154{,}125 & \textbf{Multiplying by} -375 \\
\underline{700a + 375c = 267{,}875} & \\
325a = 113{,}750 & \textbf{Adding} \\
a = \dfrac{113{,}750}{325} & \textbf{Dividing by 325} \\
a = 350. &
\end{array}
$$

We go back to Equation (1) and substitute 350 for a:

$$a + c = 411$$
$$350 + c = 411$$
$$c = 61.$$

4. *Check.* We leave the check to the student. It is similar to what we did in the familiarization step.

5. *State.* 350 adults and 61 children attended.   ◀

DO EXERCISE 3.

▶ **EXAMPLE 4** A chemist has one solution that is 80% acid (the rest is water) and another solution that is 30% acid. What is needed is 200 liters (L) of a solution that is 62% acid. The chemist will prepare it by mixing the two solutions. How much of each should be used?

1. *Familiarize.* We can draw a picture of the situation. The chemist uses x liters of the first solution and y liters of the second solution.

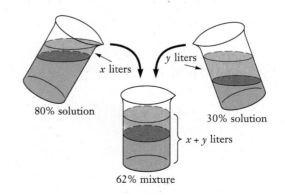

We can also arrange the information in a table.

Type of solution	First	Second	Mixture
Amount of solution	x	y	200 liters	→ $x + y = 200$
Percent of acid	80%	30%	62%	
Amount of acid in solution	$0.8x$	$0.3y$	0.62×200, or 124 liters	→ $0.8x + 0.3y = 124$

2. *Translate.* The chemist uses x liters of the first solution and y liters of the second. Since the total is to be 200 liters, we have

$$\text{Total amount of solution:} \quad x + y = 200.$$

The amount of acid in the new mixture is to be 62% of 200 liters, or 124 liters. The amounts of acid from the two solutions are $80\%x$ and $30\%y$. Thus,

$$\text{Total amount of acid:} \quad 80\%x + 30\%y = 124$$

or

$$0.8x + 0.3y = 124.$$

We clear of decimals by multiplying on both sides by 10:

$$10(0.8x + 0.3y) = 10 \cdot 124$$
$$8x + 3y = 1240.$$

Thus we have a translation to a system of equations:

$$x + y = 200, \qquad \textbf{(1)}$$
$$8x + 3y = 1240. \qquad \textbf{(2)}$$

3. *Solve.* We solve the system. We use the elimination method, again because equations are in the form $Ax + By = C$ and a multiplication in one equation will allow us to eliminate a variable, but substitution would also work. We multiply on both sides of Equation (1) by -3 and then add:

$$
\begin{array}{rl}
-3x - 3y = -600 & \text{Multiplying by } -3 \\
\underline{8x + 3y = 1240} & \\
5x = 640 & \text{Adding} \\
x = \dfrac{640}{5} & \text{Dividing by 5} \\
x = 128. &
\end{array}
$$

We go back to Equation (1) and substitute 128 for x:

$$x + y = 200$$
$$128 + y = 200$$
$$y = 72.$$

The solution is $x = 128$ and $y = 72$.

4. *Check.* The sum of 128 and 72 is 200. Also, 80% of 128 is 102.4 and 30% of 72 is 21.6. These add up to 124.

5. *State.* The chemist should use 128 liters of the 80%-acid solution and 72 liters of the 30%-acid solution. ◄

4. One solution is 50% alcohol and a second is 70% alcohol. How much of each should be mixed to make 30 L of a solution that is 55% alcohol?

Complete the following table to aid in the familiarization.

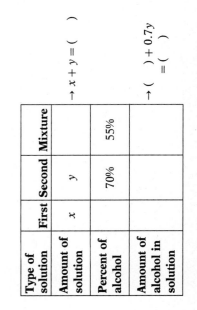

Type of solution	First	Second	Mixture	
Amount of solution	x	y		↑ $x + y = (\quad)$
Percent of alcohol		70%	55%	
Amount of alcohol in solution				↑ $(\quad) + 0.7y = (\quad)$

ANSWER ON PAGE A-8

▶ **EXAMPLE 5**   A grocer wishes to mix some nuts worth 45 cents per pound and some worth 80 cents per pound to make 350 lb of a mixture worth 65 cents per pound. How much of each should be used?

1. *Familiarize.*   Arranging the information in a table will help. We let $x =$ the amount of 45-cents nuts and $y =$ the amount of 80-cents nuts.

Type of nuts	Inexpensive nuts	Expensive nuts	Mixture	
Cost of nuts	45 cents	80 cents	65 cents	
Amount (in pounds)	x	y	350	$\longrightarrow x + y = 350$
Mixture	$45x$	$80y$	65 cents · (350), or 22,750 cents	$\longrightarrow 45x + 80y$ $= 22{,}750$

Note the similarity of this problem with Example 3. Here we consider nuts instead of tickets.

2. *Translate.*   We translate as follows. From the second row of the table, we find that

$$\text{Total amount of nuts:}\quad x + y = 350.$$

Our second equation will come from the costs. The value of the inexpensive nuts, in cents, is $45x$ (x lb at 45 cents per pound). The value of the expensive nuts is $80y$, and the value of the mixture is 65×350, or 22,750 cents. Thus we have

$$\text{Total cost of mixture:}\quad 45x + 80y = 22{,}750.$$

Remember the problem-solving tip about dimension symbols. In this last equation, all expressions stand for cents. We could have expressed them all in dollars, but we do not want some in cents and some in dollars. Thus we have a translation to a system of equations:

$$x + y = 350,$$
$$45x + 80y = 22{,}750.$$

3. *Solve.*   We solve the system using the elimination method again. We multiply on both sides of Equation (1) by -45 and then add:

$$
\begin{array}{rl}
-45x - 45y = -15{,}750 & \textbf{Multiplying by } -45 \\
\underline{45x + 80y = 22{,}750} & \\
35y = 7{,}000 & \textbf{Adding} \\
y = \dfrac{7{,}000}{35} & \\
y = 200. &
\end{array}
$$

We go back to Equation (1) and substitute 200 for y:

$$x + y = 350$$
$$x + 200 = 350$$
$$x = 150.$$

4. *Check.* We consider $x = 150$ lb and $y = 200$ lb. The sum is 350 lb. The value of the nuts is $45(150) + 80(200)$, or 22,750 cents. These values check.

5. *State.* The grocer should mix 150 lb of the 45-cents nuts with 200 lb of the 80-cents nuts. ◄

DO EXERCISE 5.

▶ **EXAMPLE 6** A student has some nickels and dimes. The value of the coins is $1.65. There are 12 more nickels than dimes. How many of each kind of coin are there?

1. *Familiarize.* We let d = the number of dimes and n = the number of nickels.

2. *Translate.* We have one equation at once:

$$d + 12 = n.$$

The value of the nickels, in cents, is $5n$, since each is worth 5¢. The value of the dimes, in cents, is $10d$, since each is worth 10¢. The total value is given as $1.65. Since we have the values of the nickels and dimes in *cents*, we must use *cents* for the total value. This is 165. This gives us another equation:

$$10d + 5n = 165.$$

Thus we have a system of equations:

$$d + 12 = n, \qquad \textbf{(1)}$$
$$10d + 5n = 165. \qquad \textbf{(2)}$$

3. *Solve.* Since we have n alone on one side of one equation, we use the substitution method. We substitute $d + 12$ for n in Equation (2):

$$10d + 5n = 165$$
$$10d + 5(d + 12) = 165 \qquad \text{Substituting } d + 12 \text{ for } n$$
$$10d + 5d + 60 = 165 \qquad \text{Removing parentheses}$$
$$15d + 60 = 165 \qquad \text{Collecting like terms}$$
$$15d = 105 \qquad \text{Subtracting 60}$$
$$d = \frac{105}{15}, \text{ or } 7. \qquad \text{Dividing by 15}$$

We substitute 7 for d in either of the original equations to find n. We use Equation (1):

$$d + 12 = n$$
$$7 + 12 = n$$
$$19 = n.$$

4. *Check.* We have 7 dimes and 19 nickels. There are 12 more nickels than dimes. The value of the coins is $7(0.10) + 19(0.05)$, which is $1.65. This checks.

5. *State.* The student has 7 dimes and 19 nickels. ◄

DO EXERCISE 6.

5. Grass seed A is worth $1.00 per pound and seed B is worth $1.35 per pound. How much of each should be mixed to make 50 lb of a mixture worth $1.14 per pound?

Complete the following table to aid in the familiarization.

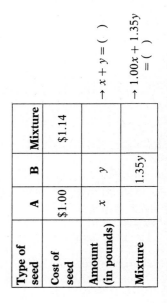

Type of seed	A	B	Mixture
Cost of seed	$1.00		$1.14
Amount (in pounds)	x	y	$\rightarrow x + y = (\ )$
Mixture		$1.35y$	$\rightarrow 1.00x + 1.35y = (\ )$

6. On a table are 20 coins, quarters and dimes. Their value is $3.05. How many of each kind of coin are there?

You should look back over Examples 3–6. The problems are quite similar in their structure. Compare them and try to see the similarities. The problems in Examples 3–6 are often called *mixture problems*. In each case a situation is considered in two different ways. These problems provide a pattern, or model, for many related problems.

> **Problem-Solving Tip**
>
> **When solving problems, see if they are patterned or modeled after other problems that you have studied.**

❖ SIDELIGHTS

Study Tips: Extra Tips on Problem Solving

We will often present some tips and guidelines to enhance your learning abilities. The following tips are focused on problem solving. They summarize some points already considered and propose some new tips.

The following are the five steps for problem solving.

1. **Familiarize** yourself with the problem situation.
2. **Translate** the problem to an equation. As you study more mathematics, you will find that the translation may be to some other kind of mathematical language, such as an inequality.
3. **Solve** the equation. If the translation is to some other kind of mathematical language, you will carry out some other kind of mathematical manipulation.
4. **Check** the answer in the original equation. This does not mean to check in the translated equation. It means to go back to the original worded problem.
5. **State** the answer to the problem clearly.

For Step 4 on checking, some further comment is appropriate. *You may find that although you were able to translate to an equation and solve the equation, none of the solutions of the equation are solutions of the original problem.* To see how this can happen, consider the following problem.

EXAMPLE The sum of two even consecutive integers is 537. Find the integers.

1. *Familiarize.* Suppose we let x = the first number. Then $x + 2$ = the second number.
2. *Translate.* The problem can be translated to the following equation:

$$x + (x + 2) = 537.$$

3. *Solve.* We solve the equation as follows:

$$2x + 2 = 537$$
$$2x = 535$$
$$x = \frac{535}{2}, \quad \text{or } 267.5.$$

4. *Check.* Then $x + 2 = 269.5$. However, not only are the numbers 267.5 and 269.5 not even, they are also not integers.
5. *State.* The problem has no solution. ◄

The following are some additional tips for problem solving.

- *To be good at problem solving, do lots of problems.* The situation is similar to what happens when we learn to play tennis. At first, we are not successful. But the more we practice and work at the game, the more successful we become. For problem solving, do more than just two or three odd-numbered problems assigned—do them all, and if you have time, do the even-numbered problems. Then find another book on the same subject and do problems in that book.

- *Look for patterns when solving problems.* By using the preceding tip and doing lots of problems, you will eventually see patterns in similar kinds of problems. For example, there is a pattern in the way we solve problems involving consecutive integers.

- *When translating to an equation, or some other mathematical language, consider the dimensions of the variables and constants in the equation.* The variables that represent length should be all in the same unit, those that represent money should be all in dollars, or all in cents, and so on.

EXERCISE SET 7.4

a Solve.

1. A firm sells cars and trucks. There is room on its lot for 510 vehicles. From experience they know that profits will be greatest if there are 190 more cars than trucks on the lot. How many of each vehicle should the firm have for the greatest profit?

2. A family went camping at a park 45 km from town. They drove 23 km more than they walked to get to the campsite. How far did they walk?

3. Sammy is twice as old as his daughter. In four years, Sammy's age will be three times what his daughter's age was six years ago. How old are they now?

4. Ann is eighteen years older than her son. She was three times as old one year ago. How old are they now?

5. Marge is twice as old as Consuelo. The sum of their ages seven years ago was 13. How old are they now?

6. Andy is four times as old as Wendy. In twelve years, Wendy's age will be half of Andy's. How old are they now?

7. A collection of dimes and quarters is worth $15.25. There are 103 coins in all. How many of each are there?

8. A collection of quarters and nickels is worth $1.25. There are 13 coins in all. How many of each are there?

9. A collection of nickels and dimes is worth $25. There are three times as many nickels as dimes. How many of each are there?

10. A collection of nickels and dimes is worth $2.90. There are nineteen more nickels than dimes. How many of each are there?

1. _____
2. _____
3. _____
4. _____
5. _____
6. _____
7. _____
8. _____
9. _____
10. _____

11. There were 429 people at a play. Admission was $1 each for adults and 75 cents each for children. The receipts were $372.50. How many adults and how many children attended?

12. The attendance at a school concert was 578. Admission was $2 each for adults and $1.50 each for children. The receipts were $985. How many adults and how many children attended?

11. _____

13. There were 200 tickets sold for a women's basketball game. Tickets for students were $0.50 each and for adults were $0.75 each. The total amount of money collected was $132.50. How many of each type of ticket were sold?

14. There were 203 tickets sold for a volleyball game. For activity-card holders, the price was $1.25 each and for noncard holders the price was $2 each. The total amount of money collected was $310. How many of each type of ticket were sold?

12. _____

13. _____

15. Solution A is 50% acid and solution B is 80% acid. How many of each should be used to make 100 L of a solution that is 68% acid? (*Hint:* 68% of what is acid?) Complete the following to aid in the familiarization.

16. Solution A is 30% alcohol and solution B is 75% alcohol. How much of each should be used to make 100 L of a solution that is 50% alcohol?

14. _____

Type of solution	A	B	Mixture	
Amount of solution	x	y	liters	$\longrightarrow x + y = ($ $)$
Percent of acid	50%		68%	
Amount of acid in solution		$0.8y$	0.68×100, or liters	$\longrightarrow 0.5x + ($ $) = ($ $)$

15. _____

16. _____

17. A solution containing 30% insecticide is to be mixed with a solution containing 50% insecticide to make 200 L of a solution containing 42% insecticide. How much of each solution should be used?

18. A solution containing 28% fungicide is to be mixed with a solution containing 40% fungicide to make 300 L of a solution containing 36% fungicide. How much of each solution should be used?

17. _____

18. _____

19. The Nuthouse has 10 kg of mixed cashews and pecans worth $8.40 per kilogram. Cashews alone sell for $8.00 per kilogram and pecans sell for $9.00 per kilogram. How many kilograms of each are in the mixture?

20. A coffee shop mixes Brazilian coffee worth $5 per kilogram with Turkish coffee worth $8 per kilogram. The mixture is to sell for $7 per kilogram. How much of each type of coffee should be used to make a 300-kg mixture? Complete the following table to aid in the familiarization.

Type of coffee	Brazilian	Turkish	Mixture
Cost of coffee	$5		$7
Amount (in kilograms)	x	y	300
Mixture		$8y$	$7(300)$, or $2100

$\longrightarrow x + y = (\quad)$

$\longrightarrow 5x + (\quad) = 2100$

21. Grass seed A is worth $2.50 per pound and seed B is worth $1.75 per pound. How much of each would you use to make 75 lb of a mixture worth $2.14 per pound?

22. A grocer wishes to mix some nuts worth 63 cents per pound and some worth 95 cents per pound to make 480 lb of a mixture worth 86 cents per pound. How much of each should be used?

23. You are taking a test in which items of type A are worth 10 points and items of type B are worth 15 points. It takes 3 min for each item of type A and 6 min for each item of type B. The total time allowed is 60 min and you can do exactly 16 questions. Your score is 180 points by using the entire 60 min. How many questions of each type did you answer correctly?

24. The goldsmith has two alloys that are different purities of gold. The first is three-fourths pure gold and the second is five-twelfths pure gold. How many ounces of each should be melted and mixed to obtain a 6-oz mixture that is two-thirds pure gold?

25. A merchant has two kinds of paint. If 9 gal of the inexpensive paint is mixed with 7 gal of the expensive paint, the mixture will be worth $19.70 per gallon. If 3 gal of the inexpensive paint is mixed with 5 gal of the expensive paint, the mixture will be worth $19.825 per gallon. What is the price per gallon of each type of paint?

26. A printer knows that a page of print contains 1300 words if large type is used and 1850 words if small type is used. A document containing 18,526 words fills exactly 12 pages. How many pages are in the large types? in the small type?

19. _____

20. _____

21. _____

22. _____

23. _____

24. _____

25. _____

26. _____

SYNTHESIS

27. A total of $27,000 is invested, part of it at 12% and part of it at 13%. The total yield after one year is $3385. How much was invested at each rate?

28. A student earned $288 on investments. If $1100 was invested at one yearly rate and $1800 at a rate that was 1.5% higher, find the two rates of interest.

29. A two-digit number is six times the sum of its digits. The tens digit is one more than the units digit. Find the number.

30. The sum of the digits of a two-digit number is 12. When the digits are reversed, the number is decreased by 18. Find the original number.

31. A farmer has 100 L of milk that is 4.6% butterfat. How much skim milk (no butterfat) should be mixed with it to make milk that is 3.2% butterfat?

32. A tank contains 8000 L of a solution that is 40% acid. How much water should be added to make a solution that is 30% acid?

33. An automobile radiator contains 16 L of antifreeze and water. This mixture is 30% antifreeze. How much of this mixture should be drained and replaced with pure antifreeze so that the mixture will be 50% antifreeze?

34. An employer has a daily payroll of $325 when employing some workers at $20 per day and others at $25 per day. When the number of $20 workers is increased by 50% and the number of $25 workers is decreased by $\frac{1}{5}$, the new daily payroll is $400. Find how many were originally employed at each rate.

35. In a two-digit number, the sum of the units digit and the number is 43 more than five times the tens digit. The sum of the digits is 11. Find the number.

36. The sum of the digits of a three-digit number is 9. If the digits are reversed, the number increases by 495. The sum of the tens and hundreds digits is half the units digit. Find the number.

37. Together, a bat, ball, and glove cost $99.00. The bat costs $9.95 more than the ball, and the glove costs $65.45 more than the bat. How much does each cost?

38. In Lewis Carroll's ''Through the Looking Glass,'' Tweedledum says to Tweedledee, ''The sum of your weight and twice mine is 361 pounds.'' Then Tweedledee says to Tweedledum, ''Contrariwise, the sum of your weight and twice mine is 362 pounds.'' Find the weight of Tweedledum and Tweedledee.

27. _____

28. _____

29. _____

30. _____

31. _____

32. _____

33. _____

34. _____

35. _____

36. _____

37. _____

38. _____

7.5   Motion Problems

a   We have studied problems involving motion in Chapter 5. Here we solve certain motion problems whose solutions can be found using systems of equations. Recall the motion formula.

> **The Motion Formula**
>
> $$\text{Distance = Rate (or speed)} \cdot \text{Time}$$
> $$d = rt$$

We have five steps for problem solving. The following tips are also helpful when solving motion problems.

> **Tips for Solving Motion Problems**
>
> 1. **Draw a diagram using an arrow or arrows to represent distance and the direction of each object in motion.**
> 2. **Organize the information in a chart.**
> 3. **Look for as many things as you can that are the same so that you can write equations.**

▶ **EXAMPLE 1**   A train leaves Podunk traveling east at 35 kilometers per hour (km/h). An hour later, another train leaves Podunk on a parallel track at 40 km/h. How far from Podunk will the trains meet?

1. *Familiarize.*   We first make a drawing.

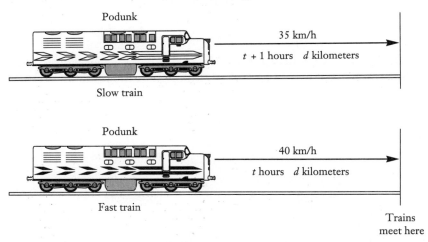

From the drawing, we see that the distances are the same. Let's call the distance d. We don't know the times. Let $t =$ the time for the faster train. Then the time for the slower train $= t + 1$, since it left 1 hr earlier. We can organize the information in a chart.

$$d \quad = \quad r \quad \cdot \quad t$$

	Distance	Speed	Time	
Slow train	d	35	$t + 1$	$d = 35(t + 1)$
Fast train	d	40	t	$\longrightarrow d = 40t$

OBJECTIVE

After finishing Section 7.5, you should be able to:

a Solve motion problems using the formula $d = rt$.

FOR EXTRA HELP

Tape 14D   Tape 15A   MAC: 7
                                IBM: 7

1. A car leaves Hereford traveling north at 56 km/h. Another car leaves Hereford one hour later traveling north at 84 km/h. How far from Hereford will the second car overtake the first? (*Hint:* The cars travel the same distance.)

2. *Translate.* In motion problems, we look for things that are the same so that we can write equations. From each row of the chart, we get an equation, $d = rt$. Thus we have two equations:

$$d = 35(t + 1), \qquad \textbf{(1)}$$
$$d = 40t. \qquad \textbf{(2)}$$

3. *Solve.* Since we have a variable alone on one side, we solve the system using the substitution method:

$$35(t + 1) = 40t \qquad \text{Using the substitution method (substituting } 35(t+1) \text{ for } d \text{ in Equation 2)}$$
$$35t + 35 = 40t \qquad \text{Removing parentheses}$$
$$35 = 5t \qquad \text{Subtracting } 35t$$
$$\frac{35}{5} = t \qquad \text{Dividing by 5}$$
$$7 = t.$$

The problem asks us to find how far from Podunk the trains meet. Thus we need to find d. We can do this by substituting 7 for t in the equation $d = 40t$:

$$d = 40(7)$$
$$= 280.$$

4. *Check.* If the time is 7 hr, then the distance that the slow train travels is $35(7 + 1)$, or 280 km. The fast train travels $40(7)$, or 280 km. Since the distances are the same, we know how far from Podunk the trains will meet.

5. *State.* The trains meet 280 km from Podunk.   ◀

DO EXERCISE 1.

▶ **EXAMPLE 2**   A motorboat took 3 hr to make a downstream trip with a 6-km/h current. The return trip against the same current took 5 hr. Find the speed of the boat in still water.

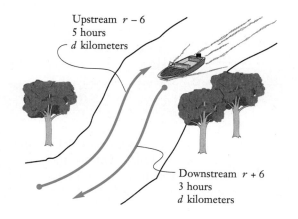

Upstream $r - 6$
5 hours
d kilometers

Downstream $r + 6$
3 hours
d kilometers

1. *Familiarize.* We first make a drawing. From the drawing, we see that the distances are the same. Let's call the distance d. Let $r =$ the speed of the boat in still water. Then, when the boat is traveling downstream, its speed is $r + 6$ (the current helps the boat along). When it is traveling upstream, its speed is $r - 6$ (the current holds the boat back). We can organize the information in a chart. In this case, the distances are the same, so we use the formula $d = rt$.

	Distance	Speed	Time	
Downstream	d	$r + 6$	3	$\longrightarrow d = (r + 6)3$
Upstream	d	$r - 6$	5	$\longrightarrow d = (r - 6)5$

$$d \quad = \quad r \quad \cdot \quad t$$

2. *Translate.* From each row of the chart, we get an equation, $d = rt$:

$$d = (r + 6)3, \quad \textbf{(1)}$$
$$d = (r - 6)5. \quad \textbf{(2)}$$

3. *Solve.* Since there is a variable alone on one side of an equation, we solve the system using substitution:

$(r + 6)3 = (r - 6)5$ Substituting $(r + 6)3$ for d in the second equation

$3r + 18 = 5r - 30$ Removing parentheses

$-2r + 18 = -30$ Subtracting $5r$

$-2r = -48$ Subtracting 18

$r = \dfrac{-48}{-2}$, or 24.   Dividing by -2

4. *Check.* When $r = 24$, $r + 6 = 30$, and $30 \cdot 3 = 90$, the distance downstream. When $r = 24$, $r - 6 = 18$, and $18 \cdot 5 = 90$, the distance upstream. In both cases we get the same distance. Now in this type of problem a problem-solving tip to keep in mind is ''Have I found what the problem asked for?'' We could solve for a certain variable but still have not answered the question of the original problem. For example, we might have found speed when the problem wanted distance. In this problem, we want the speed of the boat in still water, and that is r.

5. *State.* The speed in still water is 24 km/h. ◄

More Tips for Solving Motion Problems

1. **Translating to a system of equations eases the solution of many motion problems.**

2. **When checking, be sure that you have solved for what the problem asked for.**

DO EXERCISE 2.

2. An airplane flew for 5 hr with a 25-km/h tail wind. The return flight against the same wind took 6 hr. Find the speed of the airplane in still air. (*Hint:* The distance is the same both ways. The speeds are $r + 25$ and $r - 25$, where r is the speed in still air.)

ANSWER ON PAGE A-8

3. Two cars leave town at the same time traveling in opposite directions. One travels at 48 mph and the other at 60 mph. How far apart will they be 3 hr later? (*Hint:* The times are the same. Be *sure* to make a drawing.)

4. Two cars leave town at the same time traveling in the same direction. One travels at 35 mph and the other at 40 mph. In how many hours will they be 15 mi apart? (*Hint:* The times are the same. Be *sure* to make a drawing.)

As we saw in Chapter 5, there are motion problems that can be solved with just one equation. The following is another such problem.

▶ **EXAMPLE 3**   Two cars leave town at the same time going in opposite directions. One of them travels at 60 mph and the other at 30 mph. In how many hours will they be 150 mi apart?

1. *Familiarize.*   We first make a drawing.

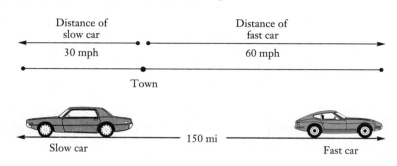

From the wording of the problem and the drawing, we see that the distances may *not* be the same. But the times the cars travel are the same, so we can just use t for time. We can organize the information in a chart.

$$d = r \cdot t$$

	Distance	Speed	Time
Fast car	Distance of fast car	60	t
Slow car	Distance of slow car	30	t

2. *Translate.*   From the drawing we see that

(Distance of fast car) + (Distance of slow car) = 150.

Then using $d = rt$ in each row of the table, we get

$$60t + 30t = 150.$$

3. *Solve.*   We solve the equation:

$$60t + 30t = 150$$
$$90t = 150 \qquad \text{Collecting like terms}$$
$$t = \frac{150}{90}, \text{ or } \frac{5}{3}, \text{ or } 1\frac{2}{3} \text{ hours.} \qquad \text{Dividing by 90}$$

4. *Check.*   When $t = \frac{5}{3}$ hr,

$$(\text{Distance of fast car}) + (\text{Distance of slow car}) = 60\left(\frac{5}{3}\right) + 30\left(\frac{5}{3}\right)$$
$$= 100 + 50, \text{ or } 150 \text{ mi.}$$

Thus the time of $\frac{5}{3}$ hr, or $1\frac{2}{3}$ hr, checks.

5. *State.*   In $1\frac{2}{3}$ hr, the cars will be 150 mi apart.   ◄

DO EXERCISES 3 AND 4.

EXERCISE SET 7.5

ANSWERS

a Solve.

1. A truck and a car leave a service station at the same time and travel in the same direction. The truck travels at 55 mph and the car at 40 mph. They can maintain CB radio contact within a range of 10 mi. When will they lose contact? Complete the following table to aid the translation.

2. Two cars leave town at the same time going in the same direction. One travels at 30 mph and the other travels at 46 mph. In how many hours will they be 72 mi apart?

1. _____

$$d = r \cdot t$$

	Distance	Speed	Time
Truck	Distance of truck	55	
Car	Distance of car		t

2. _____

3. A train leaves a station and travels east at 72 km/h. Three hours later, a second train leaves on a parallel track and travels east at 120 km/h. When will it overtake the first train? Complete the following table to aid the translation.

$$d = r \cdot t$$

	Distance	Speed	Time	
Slow train	d		$t + 3$	$\longrightarrow d = 72(\quad)$
Fast train	d	120		$\longrightarrow d = (\quad)t$

3. _____

4. A private airplane leaves an airport and flies due south at 192 km/h. Two hours later, a jet leaves the same airport and flies due south at 960 km/h. When will the jet overtake the plane?

4. _____

5. A canoeist paddled for 4 hr with a 6-km/h current to reach a campsite. The return trip against the same current took 10 hr. Find the speed of the canoe in still water. Complete the following table to aid the translation.

$$d = r \cdot t$$

	Distance	Speed	Time	
Downstream	d	$r + 6$		$\longrightarrow d = (\quad)4$
Upstream	d		10	$\longrightarrow \quad = (r - 6)10$

5. _____

6. An airplane flew for 4 hr with a 20-km/h tailwind. The return flight against the same wind took 5 hr. Find the speed of the plane in still air.

6. _____

ANSWERS

7. _____

8. _____

9. _____

10. _____

11. _____

12. _____

13. _____

14. _____

15. _____

16. _____

17. _____

18. _____

19. _____

20. _____

7. It takes a passenger train 2 hr less time than it takes a freight train to make the trip from Central City to Clear Creek. The passenger train averages 96 km/h, while the freight train averages 64 km/h. How far is it from Central City to Clear Creek?

8. It takes a small jet 4 hr less time than it takes a propeller-driven plane to travel from Glen Rock to Oakville. The jet averages 637 km/h, while the propeller plane averages 273 km/h. How far is it from Glen Rock to Oakville?

9. An airplane took 2 hr to fly 600 km against a headwind. The return trip with the wind took $1\frac{2}{3}$ hr. Find the speed of the plane in still air.

10. It took 3 hr to row a boat 18 km against the current. The return trip with the current took $1\frac{1}{2}$ hr. Find the speed of the rowboat in still water.

11. Two cars leave different towns at the same time traveling toward each other. The towns are 880 mi apart. One travels at 55 mph and the other travels at 48 mph. In how many hours will they meet?

12. Two airplanes start at the same time and fly toward each other from points 1000 km apart at rates of 420 km/h and 330 km/h. When will they meet?

13. A motorcycle breaks down and the rider has to walk the rest of the way to work. The motorcycle was being driven at 45 mph, and the rider walks at a speed of 6 mph. The distance from home to work is 25 mi, and the total time for the trip was 2 hr. How far did the motorcycle go before it broke down?

14. A student walks and jogs to college each day. The student averages 5 km/h walking and 9 km/h jogging. The distance from home to college is 8 km, and the student makes the trip in 1 hr. How far does the student jog?

SYNTHESIS

15. An airplane flew for 4.23 hr with a 25.5-km/h tailwind. The return flight against the same wind took 4.97 hr. Find the speed of the plane in still air.

16. An airplane took $2\frac{1}{2}$ hr to fly 625 mi with the wind. It took 4 hr and 10 min to make the return trip against the same wind. Find the wind speed and the speed of the plane in still air.

17. To deliver a package, a messenger must travel at a speed of 60 mph on land and then use a motorboat whose speed is 20 mph in still water. While delivering the package, the messenger goes by land to a dock and then travels on a river against a current of 4 mph. The messenger reaches the destination in 4.5 hr and then returns to the starting point in 3.5 hr. How far did the messenger travel by land and how far by water?

18. Against a headwind, Gary computes his flight time for a trip of 2900 mi at 5 hr. The flight would take 4 hr and 50 min if the headwind were half as much. Find the headwind and the plane's air speed.

19. A car travels from one town to another at a speed of 32 mph. If it had gone 4 mph faster, it could have made the trip in $\frac{1}{2}$ hr less time. How far apart are the towns?

20. Charles Lindbergh flew the Spirit of St. Louis in 1927 from New York to Paris at an average speed of 107.4 mph. Eleven years later, Howard Hughes flew the same route, averaged 217.1 mph, and took 16 hr and 57 min less time. Find the length of their route.

SUMMARY AND REVIEW: CHAPTER 7

> ## IMPORTANT PROPERTIES AND FORMULAS
>
> *Motion Formula:* $d = rt$

REVIEW EXERCISES

The review sections and objectives to be tested in addition to the material in this chapter are [3.1d, e], [5.1c], [5.5c], and [6.3a].

Determine whether the given ordered pair is a solution of the system of equations.

1. $(6, -1)$; $x - y = 3,$
$2x + 5y = 6$

2. $(2, -3)$; $2x + y = 1,$
$x - y = 5$

3. $(-2, 1)$; $x + 3y = 1,$
$2x - y = -5$

4. $(-4, -1)$; $x - y = 3,$
$x + y = -5$

Solve the system of equations by graphing.

5. $x + y = 4,$
$x - y = 8$

6. $x + 3y = 12,$
$2x - 4y = 4$

7. $y = 5 - x,$
$3x - 4y = -20$

8. $3x - 2y = -4,$
$2y - 3x = -2$

Solve using the substitution method.

9. $y = 5 - x,$
$3x - 4y = -20$

10. $x + 2y = 6,$
$2x + 3y = 8$

11. $3x + y = 1,$
$x - 2y = 5$

12. $x + y = 6,$
$y = 3 - 2x$

13. $s + t = 5,$
$s = 13 - 3t$

14. $x - y = 4,$
$y = 2 - x$

Solve using the elimination method.

15. $x + y = 4,$
$2x - y = 5$

16. $x + 2y = 9,$
$3x - 2y = -5$

17. $x - y = 8,$
$2x + y = 7$

18. $\frac{2}{3}x + y = -\frac{5}{3},$
$x - \frac{1}{3}y = -\frac{13}{3}$

19. $2x + 3y = 8,$
$5x + 2y = -2$

20. $5x - 2y = 2,$
$3x - 7y = 36$

21. $-x - y = -5,$
$2x - y = 4$

22. $6x + 2y = 4,$
$10x + 7y = -8$

23. $-6x - 2y = 5,$
$12x + 4y = -10$

Solve.

24. The sum of two numbers is 8. Their difference is 12. Find the numbers.

25. The sum of two numbers is 27. One half of the first number plus one third of the second number is 11. Find the numbers.

26. The perimeter of a rectangle is 96 cm. The length is 27 cm more than the width. Find the length and the width.

27. An airplane flew for 4 hr with a 15-km/h tailwind. The return flight against the wind took 5 hr. Find the speed of the airplane in still air.

28. There were 508 people at an organ recital. Orchestra seats cost $5.00 per person and balcony seats cost $3.00. The total receipts were $2118. Find the number of orchestra seats and the number of balcony seats sold.

29. Solution A is 30% alcohol, and solution B is 60% alcohol. How much of each is needed to make 80 L of a solution that is 45% alcohol?

30. Jeff is three times as old as his son. In nine years, Jeff will be twice as old as his son. How old is each now?

SKILL MAINTENANCE

Simplify.

31. $t^{-5} \cdot t^{13}$

32. $\dfrac{t^{-5}}{t^{13}}$

33. Subtract: $\dfrac{x}{x^2 - 9} - \dfrac{x - 1}{x^2 - 5x + 6}$.

34. Graph: $2y - x = 6$.

SYNTHESIS

35. The solution of the following system is (6, 2). Find C and D.

$$2x - Dy = 6,$$
$$Cx + 4y = 14$$

36. Solve:

$$3(x - y) = 4 + x,$$
$$x = 5y + 2.$$

37. For a two-digit number, the sum of the units digit and the tens digit is 6. When the digits are reversed, the new number is eighteen more than the original number. Find the original number.

38. A stablehand agreed to work for one year. At the end of that time, she was to receive $240 and one horse. After 7 months she quit the job, but still received the horse and $100. What was the value of the horse?

❖ **THINKING IT THROUGH**

1. Briefly compare the strengths and the weaknesses of the graphical, substitution, and elimination methods.

2. List a system of equations with no solution. Answers may vary.

3. List a system of equations with infinitely many solutions. Answers may vary.

4. Explain the advantages of using a system of equations to solve certain kinds of problems.

TEST: CHAPTER 7

1. Determine whether the given ordered pair is a solution of the system of equations.

$$(-2, -1); \quad x = 4 + 2y,$$
$$2y - 3x = 4$$

2. Solve this system by graphing:

$$x - y = 3,$$
$$x - 2y = 4.$$

Solve using the substitution method.

3. $y = 6 - x,$
$2x - 3y = 22$

4. $x + 2y = 5,$
$x + y = 2$

5. $y = 5x - 2,$
$y - 2 = 5x$

Solve using the elimination method.

6. $x - y = 6,$
$3x + y = -2$

7. $\dfrac{1}{2}x - \dfrac{1}{3}y = 8,$

$\dfrac{2}{3}x + \dfrac{1}{2}y = 5$

8. $4x + 5y = 5,$
$6x + 7y = 7$

9. $2x + 3y = 13,$
$3x - 5y = 10$

1. _____

2. _____

3. _____

4. _____

5. _____

6. _____

7. _____

8. _____

9. _____

ANSWERS

10. _____

11. _____

12. _____

13. _____

14. _____

15. _____

16. _____

17. _____

18. _____

10. The difference of two numbers is 12. One fourth of the larger number plus one half of the smaller is 9. Find the numbers.

11. A motorboat traveled for 2 hr with an 8-km/h current. The return trip against the same current took 3 hr. Find the speed of the motorboat in still water.

12. Solution A is 25% acid, and solution B is 40% acid. How much of each is needed to make 60 L of a solution that is 30% acid?

SKILL MAINTENANCE

13. Subtract: $\dfrac{1}{x^2 - 16} - \dfrac{x - 4}{x^2 - 3x - 4}$.

Simplify.

15. $(2x^{-2}y^7)(5x^6y^{-9})$

14. Graph: $3x - 4y = -12$.

16. $\dfrac{a^4b^2}{a^{-6}b^8}$

SYNTHESIS

17. Find the numbers C and D such that $(-2, 3)$ is a solution of the system

$$Cx - 4y = 7,$$
$$3x + Dy = 8.$$

18. You are in line at a ticket window. There are two more people ahead of you than there are behind you. In the entire line there are three times as many people as there are behind you. How many are ahead of you in line?

CUMULATIVE REVIEW: CHAPTERS 1–7

Compute and simplify.

1. $-2[1.4 - (-0.8 - 1.2)]$

2. $(1.3 \times 10^8)(2.4 \times 10^{-10})$

3. $\left(-\dfrac{1}{6}\right) \div \left(\dfrac{2}{9}\right)$

4. $\dfrac{2^{12}2^{-7}}{2^8}$

Simplify.

5. $\dfrac{x^2 - 9}{2x^2 - 7x + 3}$

6. $\dfrac{t^2 - 16}{(t + 4)^2}$

7. $\dfrac{x - \dfrac{x}{x + 2}}{\dfrac{2}{x} - \dfrac{1}{x + 2}}$

Perform the indicated operations and simplify.

8. $(1 - 3x^2)(2 - 4x^2)$

9. $(2a^2b - 5ab^2)^2$

10. $(3x^2 + 4y)(3x^2 - 4y)$

11. $-2x^2(x - 2x^2 + 3x^3)$

12. $(1 + 2x)(4x^2 - 2x + 1)$

13. $\left(8 - \dfrac{1}{3}x\right)\left(8 + \dfrac{1}{3}x\right)$

14. $(-8y^2 - y + 2) - (y^3 - 6y^2 + y - 5)$

15. $(2x^3 - 3x^2 - x - 1) \div (2x - 1)$

16. $\dfrac{7}{5x - 25} + \dfrac{x + 7}{5 - x}$

17. $\dfrac{2x - 1}{x - 2} - \dfrac{2x}{2 - x}$

18. $\dfrac{y^2 + y}{y^2 + y - 2} \cdot \dfrac{y + 2}{y^2 - 1}$

19. $\dfrac{7x + 7}{x^2 - 2x} \div \dfrac{14}{3x - 6}$

Factor completely.

20. $6x^5 - 36x^3 + 9x^2$

21. $16y^4 - 81$

22. $3x^2 + 10x - 8$

23. $4x^4 - 12x^2y + 9y^2$

24. $3m^3 + 6m^2 - 45m$

25. $x^3 + x^2 - x - 1$

Solve.

26. $3x - 4(x + 1) = 5$

27. $x(2x - 5) = 0$

28. $5x + 3 \geq 6(x - 4) + 7$

29. $1.5x - 2.3x = 0.4(x - 0.9)$

30. $2x^2 = 338$

31. $3x^2 + 15 = 14x$

32. $\dfrac{2}{x} - \dfrac{3}{x - 2} = \dfrac{1}{x}$

33. $1 + \dfrac{3}{x} + \dfrac{x}{x + 1} = \dfrac{1}{x^2 + x}$

34. $y = 2x - 9,$
$2x + 3y = -3$

35. $6x + 3y = -6,$
$-2x + 5y = 14$

36. $2x = y - 2,$
$3y - 6x = 6$

37. $\dfrac{1}{x} - \dfrac{1}{y} = \dfrac{1}{xy}$, for x

Solve.

38. The vice-president of a sorority has $100 to spend on promotional buttons. There is a set-up fee of $18 and a cost of 35¢ per button. How many buttons can she purchase?

39. It takes David 15 hr to put a roof on a house. It takes Loren 9 hr to put a roof on the same type of house. How long would it take if they worked together?

40. The length of one leg of a right triangle is 8 m. The length of the hypotenuse is 4 m longer than the length of the other leg. Find the lengths of the hypotenuse and the other leg.

41. To determine the number of fish in a lake, a conservationist catches 85 fish, tags them, and throws them back into the lake. Later, 60 fish are caught, 25 of which are tagged. How many fish are in the lake?

42. The height of a triangle is 3 cm less than the base. The area is 27 cm². Find the height and the base.

43. The height h of a parallelogram of fixed area varies inversely as the base b. Suppose that the height is 24 ft when the base is 15 ft. Find the height when the base is 5 ft. What is the variation constant?

44. Two cars leave town at the same time going in the same direction. One travels 50 mph and the other travels 55 mph. In how many hours will they be 50 miles apart?

45. Solution A is 20% alcohol, and solution B is 60% alcohol. How much of each should be used to make 10 L of a solution that is 50% alcohol?

46. Find an equation of variation where y varies directly as x and $y = 2.4$ when $x = 12$.

47. Find the slope of the line containing the points $(2, 3)$ and $(-1, 3)$.

48. Find the slope and the y-intercept of the line $2x + 3y = 6$.

49. Find an equation of the line that contains the points $(-5, 6)$ and $(2, -4)$.

50. Find an equation of the line containing the point $(0, -3)$ and having the slope $m = 6$.

Graph on a plane.

51. $y = -2$ **52.** $2x + 5y = 10$ **53.** $y \le 5x$ **54.** $5x - 1 < 24$

Solve by graphing.

55. $x = 5 + y,$
$\quad x - y = 1$

56. $3x - y = 4,$
$\quad x + 3y = -2$

SYNTHESIS

57. The solution of the following system is $(-5, 2)$. Find A and B.
$$3x - Ay = -7,$$
$$Bx + 4y = 15$$

58. Solve: $x^2 + 2 < 0$.

59. Simplify:
$$\frac{x-5}{x+3} - \frac{x^2 - 6x + 5}{x^2 + x - 2} \div \frac{x^2 + 4x + 3}{x^2 + 3x + 2}.$$

60. Find the value of k so that $y - kx = 4$ and $10x - 3y = -12$ are perpendicular.

INTRODUCTION The formula below illustrates the use of another kind of expression in problem solving. It is called a *radical expression* and involves a square root. We say that 3 is a square root of 9 because $3^2 = 9$. In this chapter, we study manipulations of radical expressions in addition, subtraction, multiplication, division, and simplifying. Finally, we consider another equation-solving principle and apply it to problem solving.

The review sections to be tested in addition to the material in this chapter are 5.2, 6.6, and 7.3. ❖

Radical Expressions and Equations

8

AN APPLICATION

How can we use the length of the skid marks of a car to estimate its speed before the brakes were applied?

THE MATHEMATICS

The formula

$r = 2\sqrt{5L}$ ← This is a *radical expression*.

can be used to approximate the speed r, in miles per hour, of a car that has left a skid mark of length L, in feet.

❖ POINTS TO REMEMBER: CHAPTER 8

Pythagorean Theorem: In a right triangle, the sum of the squares of the legs is equal to the square of the hypotenuse: $a^2 + b^2 = c^2$.

Product Rule of Exponents: $a^m a^n = a^{m+n}$

Quotient Rule of Exponents: $\dfrac{a^m}{a^n} = a^{m-n}$

Power Rule of Exponents: $(a^m)^n = a^{mn}$

Raising a Product to a Power: $(ab)^n = a^n b^n$

Raising a Quotient to a Power: $\left(\dfrac{a}{b}\right)^n = \dfrac{a^n}{b^n}$

PRETEST: CHAPTER 8

1. Find the square roots of 49.

2. Identify the radicand in $\sqrt{3t}$.

3. Determine whether 8 is a meaningful replacement in $\sqrt{3x - 40}$.

4. Determine the meaningful replacements in $\sqrt{x - 2}$.

5. Approximate $\sqrt{47}$ to three decimal places.

6. Solve: $\sqrt{2x + 1} = 3$.

In the remaining questions on this Pretest, assume that *all* expressions under radicals represent positive numbers.
Simplify.

7. $\sqrt{4x^2}$

8. $4\sqrt{18} - 2\sqrt{8} + \sqrt{32}$

9. $(2 - \sqrt{3})^2$

10. $(2 - \sqrt{3})(2 + \sqrt{3})$

Multiply and simplify.

11. $\sqrt{6}\sqrt{10}$

12. $(2\sqrt{6} - 1)^2$

Divide and simplify.

13. $\dfrac{\sqrt{15}}{\sqrt{3}}$

14. $\sqrt{\dfrac{24a^7}{3a^3}}$

15. In a right triangle, $a = 5$ and $b = 8$. Find c. Give an exact answer and an approximation to three decimal places.

16. How long is a guy wire reaching from the top of a 12-m pole to a point 7 m from the base of the pole?

17. Rationalize the denominator:
$$\frac{\sqrt{5}}{\sqrt{x}}.$$

18. Rationalize the denominator:
$$\frac{8}{6 + \sqrt{5}}.$$

8.1 Introduction to Square Roots and Radical Expressions

a Square Roots

When we raise a number to the second power, we have squared the number. Sometimes we may need to find the number that was squared. We call this process finding a square root of a number.

> The number c is a *square root* of a if $c^2 = a$.

Every positive number has two square roots. For example, the square roots of 25 are 5 and -5 because $5^2 = 25$ and $(-5)^2 = 25$. The positive square root is also called the **principal square root**. The symbol $\sqrt{}$ is called a **radical** symbol. The radical symbol refers only to the principal root. Thus, $\sqrt{25} = 5$. To name the negative square root of a number, we use $-\sqrt{}$. The number 0 has only one square root, 0.

▶ **EXAMPLE 1** Find the square roots of 81.

The square roots are 9 and -9. ◀

▶ **EXAMPLE 2** Find $\sqrt{225}$.

The symbol $\sqrt{225}$ represents the principal square root. There are two square roots, 15 and -15. We want the positive square root since this is what $\sqrt{}$ represents. Thus, $\sqrt{225} = 15$. ◀

Table 2 at the back of the book contains a list of squares and square roots. It would be most helpful to memorize the squares of whole numbers from 0 to 25.

▶ **EXAMPLE 3** Find $-\sqrt{64}$.

The symbol $\sqrt{64}$ represents the positive square root. Then $-\sqrt{64}$ represents the negative square root. That is, $\sqrt{64} = 8$, so $-\sqrt{64} = -8$. ◀

DO EXERCISES 1–10.

b Approximating Square Roots

We often need to use rational numbers to approximate square roots that are irrational. Such approximations can be found using a calculator with a square root key $\boxed{\sqrt{}}$. They can also be found using Table 2 at the back of the book.

▶ **EXAMPLE 4** Use your calculator or Table 2 to approximate $\sqrt{10}$. Round to three decimal places.

Before finding square roots on a calculator, you will need to consult the instruction manual. Calculators vary in their methods of operation.

$$\sqrt{10} \approx 3.162277660 \quad \text{Using a calculator with a 10-digit readout}$$

Different calculators give different numbers of digits in their readouts. This may cause some variance in answers. We round to the third decimal place. Then $\sqrt{10} \approx 3.162$. This can also be found in Table 2. ◀

DO EXERCISES 11 AND 12.

Find the square roots.

1. 36   2. 64

3. 225   4. 100

Find the following.

5. $\sqrt{16}$ 6. $\sqrt{49}$

7. $\sqrt{100}$ 8. $\sqrt{441}$

9. $-\sqrt{49}$ 10. $-\sqrt{169}$

Use a calculator or Table 2. Approximate to three decimal places.

11. $\sqrt{30}$ 12. $-\sqrt{98}$

13. In the situation of Example 5, find the number of spaces needed when the average number of arrivals in peak hours is **(a)** 64; **(b)** 83.

Identify the radicand.

14. $\sqrt{45 + x}$

15. $\sqrt{\dfrac{x}{x + 2}}$

---

c   Applications of Square Roots

We now consider an application involving a formula with a radical expression.

▶ **EXAMPLE 5**   *Parking lot arrival spaces.* A parking lot has attendants to park cars, and it uses spaces for cars to be left before they are taken to permanent parking stalls. The number N of such spaces needed is approximated by the formula

$$N = 2.5\sqrt{A},$$

where A is the average number of arrivals in peak hours. Find the number of spaces needed when the average number of arrivals in peak hours is 77.

We substitute 77 into the formula. We use a calculator or Table 2 to find an approximation:

$$N = 2.5\sqrt{77} \approx 2.5(8.775) = 21.938 \approx 22.$$

Note that we round up to 22 spaces because 21.938 spaces would give us part of a space, which we could not use. To ensure that we have enough, we need 22. ◀

> *Calculator note.* In most situations when using a calculator for a calculation like that in Example 5, we find the approximation to some number of decimal places, say 10, and then multiply by 2.5 and round. Thus, on a calculator, we might find
>
> $$N = 2.5\sqrt{77} \approx 2.5(8.774964387) = 21.93741097.$$
>
> Note that this gives a variance in the third decimal place. If your instructor is allowing you to use a calculator for approximation, you should be aware of possible variance in answers. You may get answers different than those given at the back of the text. Answers to the exercises have been found by rounding at the end.

DO EXERCISE 13.

d   Radicands and Radical Expressions

When an expression is written under a radical, we have a **radical expression.** Here are some examples:

$$\sqrt{14}, \qquad \sqrt{x}, \qquad \sqrt{x^2 + 4}, \qquad \sqrt{\dfrac{x^2 - 5}{2}}.$$

The expression written under the radical is called the **radicand.**

▶ **EXAMPLES**   Identify the radicand in each expression.

6. $\sqrt{x}$ The radicand is x.

7. $\sqrt{y^2 - 5}$ The radicand is $y^2 - 5$.

8. $\sqrt{\dfrac{a - b}{a + b}}$ The radicand is $\dfrac{a - b}{a + b}$. ◀

DO EXERCISES 14 AND 15.

e Expressions That Are Meaningful as Real Numbers

The square of any nonzero number is always positive. For example, $8^2 = 64$ and $(-11)^2 = 121$. There are no real numbers that can be squared to get negative numbers.

> **Radical expressions with negative radicands do not represent real numbers.**

Thus the following expressions do not represent real numbers (they are meaningless as real numbers):

$$\sqrt{-100}, \qquad \sqrt{-49}, \qquad -\sqrt{-3}.$$

Later in your study of mathematics, you may encounter a number system called the **complex numbers** in which negative numbers have square roots.

▶ **EXAMPLE 9** Determine whether 6 is a meaningful replacement in $\sqrt{1 - y}$.

   If we replace y by 6, we get $\sqrt{1 - 6} = \sqrt{-5}$, which has no meaning as a real number because the radicand is negative. ◀

▶ **EXAMPLE 10** Determine whether 7 is a meaningful replacement in $\sqrt{3 + 2x}$.

   If we replace x by 7, we get $\sqrt{3 + 2(7)} = \sqrt{17}$, which has meaning as a real number because the radicand is nonnegative. ◀

DO EXERCISES 16–21.

▶ **EXAMPLES** Determine the meaningful replacements in each expression.

11. $\sqrt{x}$ Any number greater than or equal to 0 is meaningful.

12. $\sqrt{x + 2}$ We solve the inequality $x + 2 \geq 0$. Any number greater than or equal to -2 is meaningful.

13. $\sqrt{x^2}$ Squares of numbers are never negative. All replacements are meaningful.

14. $\sqrt{x^2 + 1}$ Since x^2 is never negative, $x^2 + 1$ is never negative. All real-number replacements are meaningful. ◀

DO EXERCISES 22–25.

f Perfect-Square Radicands

The expression $\sqrt{x^2}$, with a perfect-square radicand, can be troublesome. If x represents a nonnegative number, $\sqrt{x^2}$ simplifies to x. If x represents a negative number, $\sqrt{x^2}$ simplifies to $-x$ (the opposite of x). That is because $\sqrt{a}$ denotes the *principal* square root of a.

   Suppose $x = 3$. Then $\sqrt{x^2} = \sqrt{3^2}$, which is $\sqrt{9}$, or 3. Suppose $x = -3$. Then $\sqrt{x^2} = \sqrt{(-3)^2}$, which is $\sqrt{9}$, or 3, the *absolute value* of -3. In either case, when replacements for x are considered to be any real number, it follows that $\sqrt{x^2} = |x|$.

Is the expression meaningless as a real number? Write "yes" or "no."

16. $-\sqrt{25}$ **17.** $\sqrt{-25}$

18. $-\sqrt{-36}$ **19.** $-\sqrt{36}$

20. Determine whether 8 is a meaningful replacement in $\sqrt{x}$.

21. Determine whether 10 is a meaningful replacement in $\sqrt{4 - x}$.

Determine the meaningful replacements.

22. $\sqrt{a}$ **23.** $\sqrt{x - 3}$

24. $\sqrt{2x - 5}$ **25.** $\sqrt{x^2 + 3}$

ANSWERS ON PAGE A-8

Simplify. Assume that expressions under radicals represent any real number.

26. $\sqrt{(xy)^2}$

27. $\sqrt{x^2 y^2}$

28. $\sqrt{(x-1)^2}$

29. $\sqrt{x^2 + 8x + 16}$

Simplify. Assume that expressions under radicals represent nonnegative real numbers.

30. $\sqrt{(xy)^2}$

31. $\sqrt{x^2 y^2}$

32. $\sqrt{(x-1)^2}$

33. $\sqrt{x^2 + 8x + 16}$

34. $\sqrt{25y^2}$

35. $\sqrt{\frac{1}{4} t^2}$

> For any real number A,
> $$\sqrt{A^2} = |A|.$$
> (That is, for any real number A, the principal square root of A squared is the absolute value of A.)

▶ **EXAMPLES**   Simplify. Assume that expressions under radicals represent any real number.

15. $\sqrt{(3x)^2} = |3x|$ Absolute-value notation is necessary.

16. $\sqrt{a^2 b^2} = \sqrt{(ab)^2} = |ab|$

17. $\sqrt{x^2 + 2x + 1} = \sqrt{(x+1)^2} = |x+1|$ ◀

DO EXERCISES 26–29.

Fortunately, in most uses of radicals, it can be assumed that expressions under radicals are nonnegative or positive. Indeed, many computers are programmed to consider only nonnegative radicands. Suppose that $x \geq 0$. Then

$$\sqrt{x^2} = |x| = x,$$

since x is nonnegative.

> For any nonnegative real number A,
> $$\sqrt{A^2} = A.$$
> (That is, for any nonnegative real number A, the principal square root of A squared is A.)

▶ **EXAMPLES**   Simplify. Assume that expressions under radicals represent nonnegative real numbers.

18. $\sqrt{(3x)^2} = 3x$ Since $3x$ is assumed to be nonnegative

19. $\sqrt{a^2 b^2} = \sqrt{(ab)^2} = ab$ Since ab is assumed to be nonnegative

20. $\sqrt{x^2 + 2x + 1} = \sqrt{(x+1)^2} = x + 1$ Since $x + 1$ is assumed to be nonnegative   ◀

DO EXERCISES 30–35.

> Henceforth, in this text we will assume that all expressions under radicals represent nonnegative real numbers.

We make this assumption in order to eliminate some confusion and because it is valid in many applications. As you study further in mathematics, however, you will frequently have to make a determination about expressions under radicals being nonnegative or positive. This will often be necessary in calculus.

NAME   SECTION   DATE

EXERCISE SET 8.1

a Find the square roots.

1. 1   **2.** 4   **3.** 16   **4.** 9

5. 100   **6.** 121   **7.** 169   **8.** 144

Simplify.

9. $\sqrt{4}$ **10.** $\sqrt{1}$ **11.** $-\sqrt{9}$ **12.** $-\sqrt{25}$ **13.** $-\sqrt{64}$

14. $-\sqrt{81}$ **15.** $-\sqrt{225}$ **16.** $\sqrt{400}$ **17.** $\sqrt{361}$ **18.** $\sqrt{441}$

b Use the calculator or Table 2 to approximate these square roots. Round to three decimal places.

19. $\sqrt{5}$ **20.** $\sqrt{6}$ **21.** $\sqrt{17}$ **22.** $\sqrt{19}$ **23.** $\sqrt{93}$ **24.** $\sqrt{43}$

c Solve. Use the formula $N = 2.5\sqrt{A}$ of Example 5.

25. Find the number of spaces needed when the average number of arrivals is **(a)** 25; **(b)** 89.

26. Find the number of spaces needed when the average number of arrivals is **(a)** 62; **(b)** 100.

d Identify the radicand.

27. $\sqrt{a-4}$ **28.** $\sqrt{t+3}$ **29.** $5\sqrt{t^2+1}$

30. $8\sqrt{x^2+5}$ **31.** $x^2y\sqrt{\dfrac{3}{x+2}}$ **32.** $ab^2\sqrt{\dfrac{a}{a-b}}$

e Determine whether the expression is meaningful as a real number. Write "yes" or "no."

33. $\sqrt{-16}$ **34.** $\sqrt{-81}$ **35.** $-\sqrt{81}$ **36.** $-\sqrt{64}$

1. _____
2. _____
3. _____
4. _____
5. _____
6. _____
7. _____
8. _____
9. _____
10. _____
11. _____
12. _____
13. _____
14. _____
15. _____
16. _____
17. _____
18. _____
19. _____
20. _____
21. _____
22. _____
23. _____
24. _____
25. a) _____
b) _____
26. a) _____
b) _____
27. _____
28. _____
29. _____
30. _____
31. _____
32. _____
33. _____
34. _____
35. _____
36. _____

Determine whether the given number is a meaningful replacement in the given radical expression.

37. 4; $\sqrt{y}$ **38.** -8; $\sqrt{m}$ **39.** -11; $\sqrt{t-5}$ **40.** -11; $\sqrt{2-x}$

Determine the meaningful replacements.

41. $\sqrt{5x}$ **42.** $\sqrt{3y}$ **43.** $\sqrt{t-5}$ **44.** $\sqrt{y-8}$ **45.** $\sqrt{y+8}$

46. $\sqrt{m-18}$ **47.** $\sqrt{2y-7}$ **48.** $\sqrt{3x+8}$ **49.** $\sqrt{t^2+5}$ **50.** $\sqrt{y^2+1}$

f   Simplify. Remember that we have assumed that expressions under radicals represent nonnegative real numbers.

51. $\sqrt{t^2}$ **52.** $\sqrt{x^2}$ **53.** $\sqrt{9x^2}$ **54.** $\sqrt{4a^2}$

55. $\sqrt{(ab)^2}$ **56.** $\sqrt{(6y)^2}$ **57.** $\sqrt{(34d)^2}$ **58.** $\sqrt{(53b)^2}$

59. $\sqrt{(x+3)^2}$ **60.** $\sqrt{(x-7)^2}$ **61.** $\sqrt{a^2-10a+25}$ **62.** $\sqrt{x^2+2x+1}$

63. $\sqrt{4x^2-20x+25}$ **64.** $\sqrt{9p^2+12p+4}$

SKILL MAINTENANCE

65. The amount F that a family spends on food varies directly as its income I. A family making \$19,600 a year will spend \$5096 on food. At this rate, how much would a family making \$20,500 spend on food?

Divide and simplify.

66. $\dfrac{x-3}{x+4} \div \dfrac{x^2-9}{x+4}$ **67.** $\dfrac{x^2-x-2}{x-1} \div \dfrac{x-2}{x^2-1}$

SYNTHESIS

68. Simplify: $\sqrt{\sqrt{16}}$.   **69.** Simplify: $\sqrt{3^2+4^2}$.

70. Between what two consecutive integers is $-\sqrt{33}$?

▦ Use a calculator to approximate these square roots. Round to three decimal places.

71. $\sqrt{12.8}$ **72.** $\sqrt{930}$ **73.** $\sqrt{1043.89}$

Solve.

74. $\sqrt{x^2}=6$ **75.** $\sqrt{y^2}=-7$ **76.** $t^2=49$

77. Suppose the area of a square is 3. Find the length of a side.

8.2 Multiplying and Simplifying with Radical Expressions

a Simplifying by Factoring

To see how to multiply with radical notation, consider the following.

a) $\sqrt{9} \cdot \sqrt{4} = 3 \cdot 2 = 6$ This is a product of square roots.

b) $\sqrt{9 \cdot 4} = \sqrt{36} = 6$ This is the square root of a product.

Note that

$$\sqrt{9} \cdot \sqrt{4} = \sqrt{9 \cdot 4}.$$

DO EXERCISE 1.

We can multiply radical expressions by multiplying the radicands.

> **The Product Rule for Radicals**
>
> **For any nonnegative radicands A and B,**
> $$\sqrt{A} \cdot \sqrt{B} = \sqrt{A \cdot B}.$$
> **(The product of square roots, provided they exist, is the square root of the product of the radicands.)**

▶ **EXAMPLES** Multiply.

1. $\sqrt{5}\sqrt{7} = \sqrt{5 \cdot 7} = \sqrt{35}$

2. $\sqrt{8}\sqrt{8} = \sqrt{8 \cdot 8} = \sqrt{64} = 8$

3. $\sqrt{\dfrac{2}{3}}\sqrt{\dfrac{4}{5}} = \sqrt{\dfrac{2}{3} \cdot \dfrac{4}{5}} = \sqrt{\dfrac{8}{15}}$

4. $\sqrt{2x}\sqrt{3x-1} = \sqrt{2x(3x-1)}$
$= \sqrt{6x^2 - 2x}$

DO EXERCISES 2–5.

To factor radical expressions, we can use the product rule for radicals in reverse. That is,

$$\sqrt{AB} = \sqrt{A}\sqrt{B}.$$

In some cases, we can simplify after factoring.

> **A radical expression is simplified when its radicand has no factors that are perfect squares.**

When simplifying a radical expression, we first determine whether a radicand is a perfect square. Then we determine whether it has perfect-square factors. The radicand is then factored and the radical expression simplified using the preceding rule.

Compare the following:

$$\sqrt{50} = \sqrt{10 \cdot 5} = \sqrt{10}\sqrt{5};$$
$$\sqrt{50} = \sqrt{25 \cdot 2} = \sqrt{25}\sqrt{2} = 5\sqrt{2}.$$

In the second case, the radicand has the perfect-square factor 25. If you do not recognize perfect-square factors, try factoring the radicand into its prime

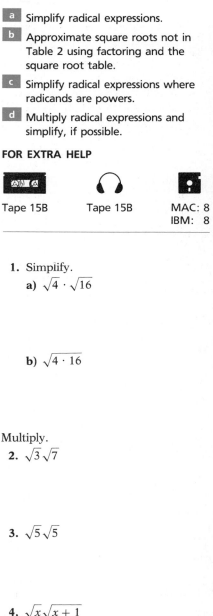

1. Simplify.
a) $\sqrt{4} \cdot \sqrt{16}$

b) $\sqrt{4 \cdot 16}$

Multiply.
2. $\sqrt{3}\sqrt{7}$

3. $\sqrt{5}\sqrt{5}$

4. $\sqrt{x}\sqrt{x+1}$

5. $\sqrt{x+1}\sqrt{x-1}$

Simplify by factoring.

6. $\sqrt{32}$

7. $\sqrt{x^2 + 14x + 49}$

8. $\sqrt{25x^2}$

9. $\sqrt{36m^2}$

10. $\sqrt{76}$

11. $\sqrt{x^2 - 8x + 16}$

12. $\sqrt{64t^2}$

13. $\sqrt{100a^2}$

factors. For example,

$$\sqrt{50} = \sqrt{2 \cdot \underbrace{5 \cdot 5}} = 5\sqrt{2}.$$

$\uparrow$
Perfect square (a pair of the same numbers)

Radical expressions in which the radicand has no perfect-square factors, such as $5\sqrt{2}$, are considered to be in simplest form.

▶ **EXAMPLES**

5. $\sqrt{18} = \sqrt{9 \cdot 2}$ Identifying a perfect-square factor and factoring the radicand. The factor 9 is a perfect square.

     $= \sqrt{9} \cdot \sqrt{2}$ Factoring into a product of radicals

     $= 3\sqrt{2}$ $3\sqrt{2}$ means $3 \cdot \sqrt{2}$

     $\uparrow$ The radicand has no factors that are perfect squares.

6. $\sqrt{48t} = \sqrt{16 \cdot 3t}$ Identifying a perfect-square factor and factoring the radicand. The factor 16 is a perfect square.

     $= \sqrt{16}\sqrt{3t}$ Factoring into a product of radicals

     $= 4\sqrt{3t}$ Taking a square root

7. $\sqrt{20t^2} = \sqrt{4 \cdot t^2 \cdot 5}$ Identifying perfect-square factors and factoring the radicand. The factors 4 and t^2 are perfect squares.

     $= \sqrt{4}\sqrt{t^2}\sqrt{5}$ Factoring into a product of several radicals

     $= 2t\sqrt{5}$ Taking square roots. No absolute-value signs are necessary since we have assumed that expressions under radicals are nonnegative.

8. $\sqrt{x^2 - 6x + 9} = \sqrt{(x-3)^2} = x - 3$ No absolute-value signs are necessary since we have assumed that expressions under radicals are nonnegative.

9. $\sqrt{36x^2} = \sqrt{36}\sqrt{x^2} = 6x$, or $\sqrt{36x^2} = \sqrt{(6x)^2} = 6x$

10. $\sqrt{3x^2 + 6x + 3} = \sqrt{3(x^2 + 2x + 1)}$ Factoring the radicand

     $= \sqrt{3}\sqrt{x^2 + 2x + 1}$ Factoring into a product of radicals

     $= \sqrt{3}\sqrt{(x+1)^2}$

     $= \sqrt{3}(x+1)$ Taking the square root   ◀

DO EXERCISES 6–13.

b   **Approximating Square Roots**

Some numbers might be too large to find in a table of square roots. For example, Table 2 goes only to 100. We may still be able to find approximate square roots for other numbers, however. We do this by first looking for the largest perfect-square factor, if there is one. If there is none, we use any factorization for which all factors appear in Table 2.

▶ **EXAMPLES**   Approximate the square roots. Use Table 2.

11. $\sqrt{160} = \sqrt{16 \cdot 10}$ Factoring the radicand. (Make one factor a perfect square, if you can.)

     $= \sqrt{16}\sqrt{10}$ Factoring the radical expression

     $= 4\sqrt{10} \approx 4(3.162) = 12.648$ From Table 2, $\sqrt{10} \approx 3.162$.

12. $\sqrt{341} = \sqrt{11 \cdot 31}$ Factoring into a product where each factor is in Table 2. (There is no perfect-square factor.)

$= \sqrt{11}\sqrt{31}$

$\approx 3.317 \times 5.568$ From Table 2

≈ 18.469 Rounded to three decimal places   ◄

If the approximations in Examples 11 and 12 were done on a calculator, the simplifying would not be necessary and the square roots would be found directly, but there is variance in the answers. For example, on a calculator with a 10-digit readout, we would get

$$\sqrt{160} \approx 12.64911064 \quad \text{and} \quad \sqrt{341} \approx 18.46618531.$$

DO EXERCISES 14 AND 15.

C   Simplifying Square Roots of Powers

To take the square root of an even power such as x^{10}, we note that $x^{10} = (x^5)^2$. Then

$$\sqrt{x^{10}} = \sqrt{(x^5)^2} = x^5.$$

We can find the answer by taking half the exponent. That is,

$$\sqrt{x^{10}} = x^5. \longleftarrow \tfrac{1}{2}(10) = 5$$

► **EXAMPLES**   Simplify.

13. $\sqrt{x^6} = \sqrt{(x^3)^2} = x^3 \longleftarrow \tfrac{1}{2}(6) = 3$

14. $\sqrt{x^8} = x^4$

15. $\sqrt{t^{22}} = t^{11}$ ◄

DO EXERCISES 16–18.

If an odd power occurs, we express the power in terms of the largest even power. Then we simplify the even power as in Examples 13–15.

► **EXAMPLE 16**   Simplify by factoring: $\sqrt{x^9}$.

$$\sqrt{x^9} = \sqrt{x^8 \cdot x}$$
$$= \sqrt{x^8}\sqrt{x}$$
$$= x^4\sqrt{x}$$ ◄

Note in Example 16 that $\sqrt{x^9} \neq x^3$.

► **EXAMPLE 17**   Simplify by factoring: $\sqrt{32x^{15}}$.

$$\sqrt{32x^{15}} = \sqrt{16x^{14}(2x)}$$ The largest even power is 14. Then we factor the radicand.

$$= \sqrt{16}\sqrt{x^{14}}\sqrt{2x}$$ Factoring into a product of radicals

$$= 4x^7\sqrt{2x}$$ Simplifying   ◄

DO EXERCISES 19 AND 20.

Approximate the square roots using Table 2. Round to three decimal places.

14. $\sqrt{275}$

15. $\sqrt{102}$

Simplify.

16. $\sqrt{t^4}$

17. $\sqrt{t^{20}}$

18. $\sqrt{q^{34}}$

Simplify by factoring.

19. $\sqrt{x^7}$

20. $\sqrt{24x^{11}}$

ANSWERS ON PAGE A-8

Multiply and simplify.

21. $\sqrt{3}\sqrt{6}$

22. $\sqrt{2}\sqrt{50}$

Multiply and simplify.

23. $\sqrt{2x^3}\sqrt{8x^3y^4}$

24. $\sqrt{10xy^2}\sqrt{5x^2y^3}$

d Multiplying and Simplifying

Sometimes we can simplify after multiplying. We leave the radicand in factored form and factor further to determine perfect-square factors. Then we simplify the perfect-square factors.

▶ **EXAMPLE 18**   Multiply and then simplify by factoring: $\sqrt{2}\sqrt{14}$.

$$\begin{aligned}
\sqrt{2}\sqrt{14} &= \sqrt{2 \cdot 14} && \text{Multiplying}\\
&= \sqrt{2 \cdot 2 \cdot 7} && \text{Factoring}\\
&= \sqrt{2 \cdot 2}\sqrt{7} && \text{Looking for perfect-square factors; pairs of factors}\\
&= 2\sqrt{7}
\end{aligned}$$
◀

DO EXERCISES 21 AND 22.

▶ **EXAMPLE 19**   Multiply and then simplify by factoring: $\sqrt{3x^2}\sqrt{9x^3}$.

$$\begin{aligned}
\sqrt{3x^2}\sqrt{9x^3} &= \sqrt{3x^2 \cdot 9x^3} && \text{Multiplying}\\
&= \sqrt{9 \cdot x^2 \cdot x^2 \cdot 3 \cdot x} && \begin{array}{l}\text{Looking for perfect-square factors or}\\ \text{largest even powers}\end{array}\\
& && \text{Perfect squares are listed first.}\\
&= \sqrt{9}\sqrt{x^2}\sqrt{x^2}\sqrt{3x}\\
&= 3 \cdot x \cdot x \cdot \sqrt{3x}\\
&= 3x^2\sqrt{3x}
\end{aligned}$$
◀

DO EXERCISES 23 AND 24.

We know that $\sqrt{AB} = \sqrt{A}\sqrt{B}$. That is, the square root of a product is the product of the square roots. What about the square root of a sum? That is, is the square root of a sum equal to the sum of the square roots? To check, consider $\sqrt{A + B}$ and $\sqrt{A} + \sqrt{B}$ when $A = 16$ and $B = 9$:

$$\sqrt{A + B} = \sqrt{16 + 9} = \sqrt{25} = 5;$$

and

$$\sqrt{A} + \sqrt{B} = \sqrt{16} + \sqrt{9} = 4 + 3 = 7.$$

Thus we see the following.

CAUTION!   The square root of a sum is not the sum of the square roots.

$$\sqrt{A + B} \neq \sqrt{A} + \sqrt{B}$$

EXERCISE SET 8.2

a Simplify by factoring.

1. $\sqrt{12}$　　　**2.** $\sqrt{8}$　　　**3.** $\sqrt{75}$　　　**4.** $\sqrt{50}$　　　**5.** $\sqrt{20}$

6. $\sqrt{45}$　　　**7.** $\sqrt{200}$　　　**8.** $\sqrt{300}$　　　**9.** $\sqrt{9x}$　　　**10.** $\sqrt{4y}$

11. $\sqrt{48x}$　　　**12.** $\sqrt{40m}$　　　**13.** $\sqrt{16a}$　　　**14.** $\sqrt{49b}$　　　**15.** $\sqrt{64y^2}$

16. $\sqrt{9x^2}$　　　**17.** $\sqrt{13x^2}$　　　**18.** $\sqrt{29t^2}$　　　**19.** $\sqrt{8t^2}$　　　**20.** $\sqrt{125a^2}$

21. $\sqrt{180}$　　　**22.** $\sqrt{448}$　　　**23.** $\sqrt{288y}$　　　**24.** $\sqrt{363p}$　　　**25.** $\sqrt{20x^2}$

26. $\sqrt{28x^2}$　　　　　**27.** $\sqrt{8x^2 + 8x + 2}$　　　　　**28.** $\sqrt{27x^2 - 36x + 12}$

29. $\sqrt{36y + 12y^2 + y^3}$　　　**30.** $\sqrt{x - 2x^2 + x^3}$

ANSWERS

1. _____
2. _____
3. _____
4. _____
5. _____
6. _____
7. _____
8. _____
9. _____
10. _____
11. _____
12. _____
13. _____
14. _____
15. _____
16. _____
17. _____
18. _____
19. _____
20. _____
21. _____
22. _____
23. _____
24. _____
25. _____
26. _____
27. _____
28. _____
29. _____
30. _____

b Approximate the square roots using Table 2. Round to three decimal places.

31. $\sqrt{125}$ **32.** $\sqrt{180}$ **33.** $\sqrt{360}$ **34.** $\sqrt{105}$

35. $\sqrt{300}$ **36.** $\sqrt{143}$ **37.** $\sqrt{122}$ **38.** $\sqrt{2000}$

Speed of a skidding car. How do police determine the speed of a car after an accident? The formula

$$r = 2\sqrt{5L}$$

can be used to approximate the speed r, in miles per hour, of a car that has left a skid mark of length L, in feet.

39. What was the speed of a car that left skid marks of 20 ft? of 150 ft?

40. What was the speed of a car that left skid marks of 30 ft? of 70 ft?

c Simplify by factoring.

41. $\sqrt{x^6}$ **42.** $\sqrt{x^{10}}$ **43.** $\sqrt{x^{12}}$ **44.** $\sqrt{x^{16}}$

45. $\sqrt{x^5}$ **46.** $\sqrt{x^3}$ **47.** $\sqrt{t^{19}}$ **48.** $\sqrt{p^{17}}$

49. $\sqrt{(y-2)^8}$ **50.** $\sqrt{(x+3)^6}$ **51.** $\sqrt{4(x+5)^{10}}$ **52.** $\sqrt{16(a-7)^4}$

53. $\sqrt{36m^3}$ **54.** $\sqrt{250y^3}$ **55.** $\sqrt{8a^5}$ **56.** $\sqrt{12b^7}$

57. $\sqrt{104p^{17}}$ **58.** $\sqrt{284m^{23}}$ **59.** $\sqrt{448x^6y^3}$ **60.** $\sqrt{243x^5y^4}$

d Multiply and then simplify by factoring, if possible.

61. $\sqrt{3}\sqrt{18}$ **62.** $\sqrt{5}\sqrt{10}$ **63.** $\sqrt{15}\sqrt{6}$

64. $\sqrt{3}\sqrt{27}$ **65.** $\sqrt{18}\sqrt{14x}$ **66.** $\sqrt{12}\sqrt{18x}$

67. $\sqrt{3x}\sqrt{12y}$ **68.** $\sqrt{7x}\sqrt{21y}$ **69.** $\sqrt{10}\sqrt{10}$

70. $\sqrt{11}\sqrt{11x}$ **71.** $\sqrt{5b}\sqrt{15b}$ **72.** $\sqrt{6a}\sqrt{18a}$

73. $\sqrt{2t}\sqrt{2t}$ **74.** $\sqrt{3a}\sqrt{3a}$ **75.** $\sqrt{ab}\sqrt{ac}$

76. $\sqrt{xy}\sqrt{xz}$ **77.** $\sqrt{2x^2y}\sqrt{4xy^2}$ **78.** $\sqrt{15mn^2}\sqrt{5m^2n}$

ANSWERS

53. _____

54. _____

55. _____

56. _____

57. _____

58. _____

59. _____

60. _____

61. _____

62. _____

63. _____

64. _____

65. _____

66. _____

67. _____

68. _____

69. _____

70. _____

71. _____

72. _____

73. _____

74. _____

75. _____

76. _____

77. _____

78. _____

79. $\sqrt{18}\sqrt{18}$ **80.** $\sqrt{16}\sqrt{16}$ **81.** $\sqrt{5}\sqrt{2x-1}$

82. $\sqrt{3}\sqrt{4x+2}$ **83.** $\sqrt{x+2}\sqrt{x+2}$ **84.** $\sqrt{x-3}\sqrt{x-3}$

85. $\sqrt{18x^2y^3}\sqrt{6xy^4}$ **86.** $\sqrt{12x^3y^2}\sqrt{8xy}$

87. $\sqrt{50ab}\sqrt{10a^2b^4}$ **88.** $\sqrt{10xy^2}\sqrt{5x^2y^3}$

SKILL MAINTENANCE

Solve.

89. $x - y = 7,$
$\quad\;\, x + y = 9$

90. $3x + 5y = 6,$
$\quad\;\, 5x + 3y = 4$

91. The perimeter of a rectangle is 642 ft. The length is 15 ft greater than the width. Find the area of the rectangle.

SYNTHESIS

Factor.

92. $\sqrt{3x-3}$ **93.** $\sqrt{x^2-x-2}$ **94.** $\sqrt{x^2-4}$

95. $\sqrt{2x^2-5x-12}$ **96.** $\sqrt{x^3-2x^2}$ **97.** $\sqrt{a^2-b^2}$

Simplify.

98. $\sqrt{0.01}$ **99.** $\sqrt{0.25}$ **100.** $\sqrt{x^8}$ **101.** $\sqrt{9a^6}$

102. Find $\sqrt{49}$, $\sqrt{490}$, $\sqrt{4900}$, $\sqrt{49{,}000}$, $\sqrt{490{,}000}$. What pattern do you see?

Use the proper symbol ($>$, $<$, or $=$) between each pair of values to make a true sentence.

103. $15\quad 4\sqrt{14}$ **104.** $15\sqrt{2}\quad\sqrt{450}$ **105.** $16\quad\sqrt{15}\sqrt{17}$

106. $3\sqrt{11}\quad 7\sqrt{2}$ **107.** $5\sqrt{7}\quad 4\sqrt{11}$ **108.** $8\quad\sqrt{15}+\sqrt{17}$

Multiply and then simplify by factoring.

109. $(\sqrt{2y})(\sqrt{3})(\sqrt{8y})$ **110.** $\sqrt{a}(\sqrt{a^3}-5)$

111. $\sqrt{27(x+1)}\sqrt{12y(x+1)^2}$ **112.** $\sqrt{18(x-2)}\sqrt{20(x-2)^3}$

113. $\sqrt{x}\sqrt{2x}\sqrt{10x^5}$ **114.** $\sqrt{2^{109}}\sqrt{x^{306}}\sqrt{x^{11}}$

Simplify.

115. $\sqrt{x^{8n}}$ **116.** $\sqrt{0.04x^{4n}}$

117. Determine whether it is true that $\sqrt{A}-\sqrt{B}=\sqrt{A-B}$.

8.3 Quotients Involving Square Roots

a Dividing Radical Expressions

Consider the expressions

$$\frac{\sqrt{25}}{\sqrt{16}} \quad \text{and} \quad \sqrt{\frac{25}{16}}.$$

Let us evaluate them separately:

a) $\dfrac{\sqrt{25}}{\sqrt{16}} = \dfrac{5}{4}$ since $\sqrt{25} = 5$ and $\sqrt{16} = 4$;

b) $\sqrt{\dfrac{25}{16}} = \dfrac{5}{4}$ because $\dfrac{5}{4} \cdot \dfrac{5}{4} = \dfrac{25}{16}$.

We see that both expressions represent the same number. This suggests that the quotient of two square roots is the square root of the quotient of the radicands.

> **The Quotient Rules for Radicals**
>
> **For any nonnegative number A and any positive number B,**
> $$\frac{\sqrt{A}}{\sqrt{B}} = \sqrt{\frac{A}{B}}.$$
>
> **(The quotient of two square roots, provided they exist, is the square root of the quotients of the radicands.)**

▶ **EXAMPLES**   Divide and simplify.

1. $\dfrac{\sqrt{27}}{\sqrt{3}} = \sqrt{\dfrac{27}{3}} = \sqrt{9} = 3$

2. $\dfrac{\sqrt{30a^5}}{\sqrt{6a^2}} = \sqrt{\dfrac{30a^5}{6a^2}} = \sqrt{5a^3} = \sqrt{a^2 \cdot 5a} = \sqrt{a^2} \cdot \sqrt{5a} = a\sqrt{5a}$ ◀

DO EXERCISES 1–3.

b Roots of Quotients

To find the square root of a quotient, we can reverse the quotient rule for radicals. We can take the square root of a quotient by taking the square roots of the numerator and the denominator separately.

> **For any nonnegative number A and any positive number B,**
> $$\sqrt{\frac{A}{B}} = \frac{\sqrt{A}}{\sqrt{B}}.$$
>
> **(We can take the square roots of the numerator and the denominator separately.)**

OBJECTIVES

After finishing Section 8.3, you should be able to:

a Divide radical expressions with fractional radicands.

b Simplify square roots of quotients.

c Rationalize the denominator of a radical expression.

d Approximate radical expressions involving division.

FOR EXTRA HELP

Tape 15C   Tape 16A   MAC: 8
                              IBM: 8

Divide and simplify.

1. $\dfrac{\sqrt{48}}{\sqrt{3}}$

2. $\dfrac{\sqrt{75}}{\sqrt{3}}$

3. $\dfrac{\sqrt{42x^5}}{\sqrt{7x^2}}$

ANSWERS ON PAGE A-8

Simplify.

4. $\sqrt{\dfrac{16}{9}}$

5. $\sqrt{\dfrac{1}{25}}$

6. $\sqrt{\dfrac{36}{x^2}}$

Simplify.

7. $\sqrt{\dfrac{18}{32}}$

8. $\sqrt{\dfrac{2250}{2560}}$

9. $\sqrt{\dfrac{75x}{3x^7}}$

▶ **EXAMPLES**   Simplify by taking the square roots of the numerator and the denominator separately.

3. $\sqrt{\dfrac{25}{9}} = \dfrac{\sqrt{25}}{\sqrt{9}} = \dfrac{5}{3}$ Taking the square roots of the numerator and the denominator

4. $\sqrt{\dfrac{1}{16}} = \dfrac{\sqrt{1}}{\sqrt{16}} = \dfrac{1}{4}$ Taking the square roots of the numerator and the denominator

5. $\sqrt{\dfrac{49}{t^2}} = \dfrac{\sqrt{49}}{\sqrt{t^2}} = \dfrac{7}{t}$ ◀

DO EXERCISES 4–6.

We are assuming that expressions for numerators are nonnegative and expressions for denominators are positive. Thus we need not be concerned about absolute-value signs or zero denominators.

Sometimes a rational expression can be simplified to one that has a perfect-square numerator and a perfect-square denominator.

▶ **EXAMPLES**   Simplify.

6. $\sqrt{\dfrac{18}{50}} = \sqrt{\dfrac{9 \cdot 2}{25 \cdot 2}} = \sqrt{\dfrac{9}{25} \cdot \dfrac{2}{2}} = \sqrt{\dfrac{9}{25} \cdot 1} = \sqrt{\dfrac{9}{25}} = \dfrac{\sqrt{9}}{\sqrt{25}} = \dfrac{3}{5}$

7. $\sqrt{\dfrac{2560}{2890}} = \sqrt{\dfrac{256 \cdot 10}{289 \cdot 10}} = \sqrt{\dfrac{256}{289} \cdot \dfrac{10}{10}} = \sqrt{\dfrac{256}{289} \cdot 1} = \sqrt{\dfrac{256}{289}} = \dfrac{\sqrt{256}}{\sqrt{289}} = \dfrac{16}{17}$

8. $\dfrac{\sqrt{48x^3}}{\sqrt{3x^7}} = \sqrt{\dfrac{48x^3}{3x^7}} = \sqrt{\dfrac{16}{x^4}} = \dfrac{\sqrt{16}}{\sqrt{x^4}} = \dfrac{4}{x^2}$ ◀

DO EXERCISES 7–9.

C   **Rationalizing Denominators**

Sometimes in mathematics it is useful to find an equivalent expression without a radical in the denominator. This provides a standard notation for expressing results. The procedure for finding such an expression is called **rationalizing the denominator.** We carry this out by multiplying by 1 in either of two ways. One way is to multiply by 1 under the radical to make the denominator a perfect square. Another way is to multiply by 1 outside the radical to make the denominator a perfect square.

▶ **EXAMPLE 9** Rationalize the denominator: $\sqrt{\dfrac{2}{3}}$.

Method 1. We multiply by 1, choosing $\frac{3}{3}$ for 1. This makes the denominator a perfect square:

$$\sqrt{\frac{2}{3}} = \sqrt{\frac{2}{3} \cdot \frac{3}{3}} \qquad \text{Multiplying by 1}$$
$$= \sqrt{\frac{6}{9}}$$
$$= \frac{\sqrt{6}}{\sqrt{9}}$$
$$= \frac{\sqrt{6}}{3}.$$

Method 2. We can also rationalize by first taking the square roots of the numerator and the denominator. Then we multiply by 1, using $\sqrt{3}/\sqrt{3}$:

$$\sqrt{\frac{2}{3}} = \frac{\sqrt{2}}{\sqrt{3}} = \frac{\sqrt{2}}{\sqrt{3}} \cdot \frac{\sqrt{3}}{\sqrt{3}} = \frac{\sqrt{2} \cdot \sqrt{3}}{\sqrt{3} \cdot \sqrt{3}} = \frac{\sqrt{6}}{\sqrt{9}} = \frac{\sqrt{6}}{3}. \qquad ◀$$

DO EXERCISE 10.

We can always multiply by 1 to make a denominator a perfect square. Then we can take the square root of the denominator.

▶ **EXAMPLE 10** Rationalize the denominator: $\sqrt{\dfrac{5}{18}}$.

The denominator 18 is not a perfect square. Factoring, we get $18 = 3 \cdot 3 \cdot 2$. If we had another factor of 2, however, we would have a perfect square, 36. Thus we multiply by 1, choosing $\frac{2}{2}$. This makes the denominator a perfect square.

$$\sqrt{\frac{5}{18}} = \sqrt{\frac{5}{18} \cdot \frac{2}{2}} = \sqrt{\frac{10}{36}} = \frac{\sqrt{10}}{\sqrt{36}} = \frac{\sqrt{10}}{6} \qquad ◀$$

▶ **EXAMPLE 11** Rationalize the denominator: $\dfrac{8}{\sqrt{7}}$.

This time we obtain an expression without a radical in the denominator by multiplying by 1, choosing $\sqrt{7}/\sqrt{7}$:

$$\frac{8}{\sqrt{7}} = \frac{8}{\sqrt{7}} \cdot \frac{\sqrt{7}}{\sqrt{7}} = \frac{8\sqrt{7}}{\sqrt{49}} = \frac{8\sqrt{7}}{7}. \qquad ◀$$

DO EXERCISES 11 AND 12.

10. Rationalize the denominator:
$$\sqrt{\frac{3}{5}}.$$

Rationalize the denominator.

11. $\sqrt{\dfrac{5}{8}}$

(*Hint:* Multiply the radicand by $\frac{2}{2}$.)

12. $\dfrac{2}{\sqrt{3}}$

ANSWERS ON PAGE A-8

Rationalize the denominator.

13. $\dfrac{\sqrt{5}}{\sqrt{7}}$

14. $\dfrac{\sqrt{3}}{\sqrt{t}}$

15. $\dfrac{\sqrt{64y^2}}{\sqrt{7}}$

Approximate to three decimal places.

16. $\sqrt{\dfrac{2}{3}}$

17. $\dfrac{\sqrt{14}}{\sqrt{11}}$

▶ **EXAMPLE 12**   Rationalize the denominator: $\dfrac{\sqrt{3}}{\sqrt{2}}$.

We look at the denominator. It is $\sqrt{2}$. We multiply by 1, choosing $\sqrt{2}/\sqrt{2}$:

$$\dfrac{\sqrt{3}}{\sqrt{2}} = \dfrac{\sqrt{3}}{\sqrt{2}} \cdot \dfrac{\sqrt{2}}{\sqrt{2}} = \dfrac{\sqrt{3} \cdot \sqrt{2}}{\sqrt{2} \cdot \sqrt{2}} = \dfrac{\sqrt{6}}{2}, \quad \text{or} \quad \dfrac{1}{2}\sqrt{6}. \qquad ◀$$

▶ **EXAMPLES**   Rationalize the denominator.

13. $\dfrac{\sqrt{5}}{\sqrt{x}} = \dfrac{\sqrt{5}}{\sqrt{x}} \cdot \dfrac{\sqrt{x}}{\sqrt{x}}$ Multiplying by 1

$\qquad = \dfrac{\sqrt{5}\sqrt{x}}{\sqrt{x}\sqrt{x}}$

$\qquad = \dfrac{\sqrt{5x}}{x}$

14. $\dfrac{\sqrt{49a^5}}{\sqrt{12}} = \dfrac{\sqrt{49a^5}}{\sqrt{12}} \cdot \dfrac{\sqrt{3}}{\sqrt{3}} = \dfrac{\sqrt{49a^5}\sqrt{3}}{\sqrt{12}\sqrt{3}}$

$\qquad = \dfrac{\sqrt{49a^4 \cdot 3a}}{\sqrt{36}} = \dfrac{7a^2\sqrt{3a}}{6} \qquad ◀$

DO EXERCISES 13–15.

d **Approximating Expressions with Square Roots**

We can use a calculator or Table 2 to approximate square roots of quotients. There are at least two ways to do it.

▶ **EXAMPLE 15**   Approximate $\sqrt{\frac{3}{5}}$ to three decimal places.

Method 1. Suppose we are using a calculator. We divide and then approximate the square root:

$$\sqrt{\dfrac{3}{5}} = \sqrt{0.6} \approx 0.774596669 \approx 0.775.$$

Method 2. Suppose we are using a table like Table 2. We first rationalize the denominator and then use Table 2 to approximate the square root in the numerator. Then we divide:

$$\sqrt{\dfrac{3}{5}} = \sqrt{\dfrac{3}{5} \cdot \dfrac{5}{5}} = \sqrt{\dfrac{15}{25}} = \dfrac{\sqrt{15}}{\sqrt{25}} = \dfrac{\sqrt{15}}{5} \approx \dfrac{3.873}{5} \approx 0.775. \qquad \text{Rounding to three decimal places} \quad ◀$$

DO EXERCISES 16 AND 17.

NAME   SECTION   DATE

EXERCISE SET 8.3

a Divide and simplify.

1. $\dfrac{\sqrt{18}}{\sqrt{2}}$

2. $\dfrac{\sqrt{20}}{\sqrt{5}}$

3. $\dfrac{\sqrt{60}}{\sqrt{15}}$

4. $\dfrac{\sqrt{108}}{\sqrt{3}}$

5. $\dfrac{\sqrt{75}}{\sqrt{15}}$

6. $\dfrac{\sqrt{18}}{\sqrt{3}}$

7. $\dfrac{\sqrt{3}}{\sqrt{75}}$

8. $\dfrac{\sqrt{3}}{\sqrt{48}}$

9. $\dfrac{\sqrt{12}}{\sqrt{75}}$

10. $\dfrac{\sqrt{18}}{\sqrt{32}}$

11. $\dfrac{\sqrt{8x}}{\sqrt{2x}}$

12. $\dfrac{\sqrt{18b}}{\sqrt{2b}}$

13. $\dfrac{\sqrt{63y^3}}{\sqrt{7y}}$

14. $\dfrac{\sqrt{48x^3}}{\sqrt{3x}}$

b Simplify.

15. $\sqrt{\dfrac{9}{49}}$

16. $\sqrt{\dfrac{16}{25}}$

17. $\sqrt{\dfrac{1}{36}}$

18. $\sqrt{\dfrac{1}{4}}$

19. $-\sqrt{\dfrac{16}{81}}$

1. _____

2. _____

3. _____

4. _____

5. _____

6. _____

7. _____

8. _____

9. _____

10. _____

11. _____

12. _____

13. _____

14. _____

15. _____

16. _____

17. _____

18. _____

19. _____

20. $-\sqrt{\dfrac{25}{49}}$ **21.** $\sqrt{\dfrac{64}{289}}$ **22.** $\sqrt{\dfrac{81}{361}}$ **23.** $\sqrt{\dfrac{1690}{1960}}$ **24.** $\sqrt{\dfrac{1440}{6250}}$

25. $\sqrt{\dfrac{36}{a^2}}$ **26.** $\sqrt{\dfrac{25}{x^2}}$ **27.** $\sqrt{\dfrac{9a^2}{625}}$ **28.** $\sqrt{\dfrac{x^2y^2}{256}}$

C   Rationalize the denominator.

29. $\sqrt{\dfrac{2}{5}}$ **30.** $\sqrt{\dfrac{2}{7}}$ **31.** $\sqrt{\dfrac{3}{8}}$ **32.** $\sqrt{\dfrac{7}{8}}$ **33.** $\sqrt{\dfrac{7}{12}}$

34. $\sqrt{\dfrac{1}{12}}$ **35.** $\sqrt{\dfrac{1}{18}}$ **36.** $\sqrt{\dfrac{5}{18}}$ **37.** $\dfrac{3}{\sqrt{5}}$ **38.** $\dfrac{4}{\sqrt{3}}$

39. $\sqrt{\dfrac{8}{3}}$ **40.** $\sqrt{\dfrac{12}{5}}$ **41.** $\sqrt{\dfrac{3}{x}}$ **42.** $\sqrt{\dfrac{2}{x}}$ **43.** $\sqrt{\dfrac{x}{y}}$

44. $\sqrt{\dfrac{a}{b}}$ **45.** $\sqrt{\dfrac{x^2}{18}}$ **46.** $\sqrt{\dfrac{x^2}{20}}$ **47.** $\dfrac{\sqrt{7}}{\sqrt{3}}$ **48.** $\dfrac{\sqrt{2}}{\sqrt{5}}$

49. $\dfrac{\sqrt{9}}{\sqrt{8}}$ **50.** $\dfrac{\sqrt{4}}{\sqrt{27}}$ **51.** $\dfrac{\sqrt{2}}{\sqrt{5}}$ **52.** $\dfrac{\sqrt{3}}{\sqrt{2}}$ **53.** $\dfrac{2}{\sqrt{2}}$

54. $\dfrac{3}{\sqrt{3}}$ **55.** $\dfrac{\sqrt{5}}{\sqrt{11}}$ **56.** $\dfrac{\sqrt{7}}{\sqrt{27}}$ **57.** $\dfrac{\sqrt{7}}{\sqrt{12}}$ **58.** $\dfrac{\sqrt{5}}{\sqrt{18}}$

59. $\dfrac{\sqrt{48}}{\sqrt{32}}$ **60.** $\dfrac{\sqrt{56}}{\sqrt{40}}$ **61.** $\dfrac{\sqrt{450}}{\sqrt{18}}$ **62.** $\dfrac{\sqrt{224}}{\sqrt{14}}$ **63.** $\dfrac{\sqrt{3}}{\sqrt{x}}$

64. $\dfrac{\sqrt{2}}{\sqrt{y}}$ **65.** $\dfrac{4y}{\sqrt{3}}$ **66.** $\dfrac{8x}{\sqrt{5}}$ **67.** $\dfrac{\sqrt{a^{3}}}{\sqrt{8}}$ **68.** $\dfrac{\sqrt{x^{3}}}{\sqrt{27}}$

69. $\dfrac{\sqrt{56}}{\sqrt{12x}}$ **70.** $\dfrac{\sqrt{45}}{\sqrt{8a}}$ **71.** $\dfrac{\sqrt{27c}}{\sqrt{32c^{3}}}$ **72.** $\dfrac{\sqrt{7x^{3}}}{\sqrt{12x}}$ **73.** $\dfrac{\sqrt{y^{5}}}{\sqrt{xy^{2}}}$

74. $\dfrac{\sqrt{x^{3}}}{\sqrt{xy}}$ **75.** $\dfrac{\sqrt{16a^{4}b^{6}}}{\sqrt{128a^{6}b^{6}}}$ **76.** $\dfrac{\sqrt{45mn^{2}}}{\sqrt{32m}}$

ANSWERS

49. _____

50. _____

51. _____

52. _____

53. _____

54. _____

55. _____

56. _____

57. _____

58. _____

59. _____

60. _____

61. _____

62. _____

63. _____

64. _____

65. _____

66. _____

67. _____

68. _____

69. _____

70. _____

71. _____

72. _____

73. _____

74. _____

75. _____

76. _____

d Approximate to three decimal places.

77. $\sqrt{\dfrac{1}{3}}$ **78.** $\sqrt{\dfrac{3}{2}}$ **79.** $\sqrt{\dfrac{7}{8}}$ **80.** $\sqrt{\dfrac{3}{8}}$ **81.** $\sqrt{\dfrac{1}{12}}$ **82.** $\sqrt{\dfrac{5}{12}}$

83. $\sqrt{\dfrac{1}{2}}$ **84.** $\sqrt{\dfrac{1}{7}}$ **85.** $\dfrac{17}{\sqrt{20}}$ **86.** $\dfrac{28}{\sqrt{13}}$ **87.** $\dfrac{\sqrt{13}}{\sqrt{18}}$ **88.** $\dfrac{\sqrt{11}}{\sqrt{18}}$

SKILL MAINTENANCE

Solve.

89. $x = y + 2,$
$x + y = 6$

90. $2x - 3y = 7,$
$2x + 3y = 9$

91. $2x - 3y = 7,$
$2x - 3y = 9$

92. $2x - 3y = 7,$
$-4x + 6y = -14$

SYNTHESIS

The period T of a pendulum is the time it takes to move from one side to the other and back. A formula for the period is

$$T = 2\pi\sqrt{\frac{L}{32}},$$

where T is in seconds and L is in feet. Use 3.14 for π.

93. Find the periods of pendulums of lengths 2 ft, 8 ft, 64 ft, and 100 ft.

94. Find the period of a pendulum of length $\frac{2}{3}$ in.

95. The pendulum of a grandfather clock is $(32/\pi^2)$ ft long. How long does it take to swing from one side to the other?

96. The pendulum of a grandfather clock is $(45/\pi^2)$ ft long. How long does it take to swing from one side to the other?

Rationalize the denominator.

97. $\sqrt{\dfrac{5}{1600}}$ **98.** $\sqrt{\dfrac{3}{1000}}$ **99.** $\sqrt{\dfrac{1}{5x^3}}$ **100.** $\sqrt{\dfrac{3x^2y}{a^2x^5}}$

101. $\sqrt{\dfrac{3a}{b}}$ **102.** $\sqrt{\dfrac{1}{5zw^2}}$ **103.** $\sqrt{0.007}$ **104.** $\sqrt{0.012}$

Simplify.

105. $\sqrt{\dfrac{1}{x^2} - \dfrac{2}{xy} + \dfrac{1}{y^2}}$ **106.** $\sqrt{2 - \dfrac{4}{z^2} + \dfrac{2}{z^4}}$

8.4 Addition, Subtraction, and More Multiplication

a Addition and Subtraction

We can add any two real numbers. The sum of 5 and $\sqrt{2}$ can be expressed as

$$5 + \sqrt{2}.$$

We cannot simplify this unless we use rational approximations. However, when we have *like radicals*, a sum can be simplified using the distributive laws and collecting like terms. **Like radicals** have the same radicands.

▶ **EXAMPLE 1** Add: $3\sqrt{5} + 4\sqrt{5}$.

Suppose we were considering $3x + 4x$. Recall that to add, we use the distributive laws as follows:

$$3x + 4x = (3 + 4)x = 7x.$$

The situation is similar in this example, but we let $x = \sqrt{5}$:

$$3\sqrt{5} + 4\sqrt{5} = (3 + 4)\sqrt{5} \qquad \text{Using the distributive law to factor out } \sqrt{5}$$
$$= 7\sqrt{5}. \qquad\qquad\qquad ◀$$

To add or subtract as we did in Example 1, the radicands must be the same. Sometimes after simplifying the radical terms, we discover that we have like radicals.

▶ **EXAMPLES** Add or subtract. Simplify, if possible, by collecting like radical terms.

2. $5\sqrt{2} - \sqrt{18} = 5\sqrt{2} - \sqrt{9 \cdot 2} \qquad \text{Factoring 18}$
$$= 5\sqrt{2} - \sqrt{9}\sqrt{2}$$
$$= 5\sqrt{2} - 3\sqrt{2}$$
$$= (5 - 3)\sqrt{2} \qquad \text{Using the distributive law to factor out the common factor, } \sqrt{2}$$
$$= 2\sqrt{2}$$

3. $\sqrt{4x^3} + 7\sqrt{x} = \sqrt{4x^2 \cdot x} + 7\sqrt{x}$
$$= 2x\sqrt{x} + 7\sqrt{x}$$
$$= (2x + 7)\sqrt{x} \qquad \text{Using the distributive law to factor out } \sqrt{x}$$

Don't forget the parentheses!

4. $\sqrt{x^3 - x^2} + \sqrt{4x - 4} = \sqrt{x^2(x - 1)} + \sqrt{4(x - 1)} \qquad \text{Factoring radicands}$
$$= \sqrt{x^2}\sqrt{x - 1} + \sqrt{4}\sqrt{x - 1}$$
$$= x\sqrt{x - 1} + 2\sqrt{x - 1}$$
$$= (x + 2)\sqrt{x - 1} \qquad \text{Using the distributive law to factor out the common factor, } \sqrt{x - 1}$$

Don't forget the parentheses! ◀

DO EXERCISES 1–5.

OBJECTIVES

After finishing Section 8.4, you should be able to:

a Add or subtract with radical notation, using the distributive law to simplify.

b Multiply expressions involving radicals, where some of the expressions contain more than one term.

c Rationalize denominators having two terms.

FOR EXTRA HELP

Tape 15D   Tape 16A   MAC: 8
                              IBM: 8

Add or subtract and simplify by collecting like radical terms, if possible.

1. $3\sqrt{2} + 9\sqrt{2}$

2. $8\sqrt{5} - 3\sqrt{5}$

3. $2\sqrt{10} - 7\sqrt{40}$

4. $\sqrt{24} + \sqrt{54}$

5. $\sqrt{9x + 9} - \sqrt{4x + 4}$

ANSWERS ON PAGE A-8

Add or subtract.

6. $\sqrt{2} + \sqrt{\dfrac{1}{2}}$

Sometimes rationalizing denominators enables us to combine like radicals.

► **EXAMPLE 5**   Add: $\sqrt{3} + \sqrt{\dfrac{1}{3}}$.

$$\sqrt{3} + \sqrt{\dfrac{1}{3}} = \sqrt{3} + \sqrt{\dfrac{1}{3} \cdot \dfrac{3}{3}} \qquad \text{Multiplying by 1 in order to rationalize denominators}$$

$$= \sqrt{3} + \sqrt{\dfrac{3}{9}}$$

$$= \sqrt{3} + \dfrac{\sqrt{3}}{\sqrt{9}}$$

$$= \sqrt{3} + \dfrac{\sqrt{3}}{3}$$

$$= 1 \cdot \sqrt{3} + \dfrac{1}{3}\sqrt{3}$$

$$= \left(1 + \dfrac{1}{3}\right)\sqrt{3} \qquad \text{Factoring out the common factor, } \sqrt{3}$$

$$= \dfrac{4}{3}\sqrt{3} \qquad\qquad\qquad\qquad\qquad ◄$$

DO EXERCISES 6 AND 7.

7. $\sqrt{\dfrac{5}{3}} + \sqrt{\dfrac{3}{5}}$

b   Multiplication

Now let us multiply where some of the expressions may contain more than one term. To do this, we use procedures already studied in this chapter as well as the distributive law and special products for multiplying with polynomials.

► **EXAMPLE 6**   Multiply: $\sqrt{2}(\sqrt{3} + \sqrt{7})$.

$$\sqrt{2}(\sqrt{3} + \sqrt{7}) = \sqrt{2}\sqrt{3} + \sqrt{2}\sqrt{7} \qquad \text{Multiplying using a distributive law}$$

$$= \sqrt{6} + \sqrt{14} \qquad \text{Using the rule for multiplying with radicals} ◄$$

► **EXAMPLE 7**   Multiply: $(2 + \sqrt{3})(5 - 4\sqrt{3})$.

$$(2 + \sqrt{3})(5 - 4\sqrt{3}) = 2 \cdot 5 - 2 \cdot 4\sqrt{3} + \sqrt{3} \cdot 5 - \sqrt{3} \cdot 4\sqrt{3} \qquad \text{Using FOIL}$$

$$= 10 - 8\sqrt{3} + 5\sqrt{3} - 4 \cdot 3$$

$$= 10 - 12 - 3\sqrt{3}$$

$$= -2 - 3\sqrt{3} \qquad\qquad\qquad\qquad ◄$$

▶ **EXAMPLE 8**   Multiply: $(\sqrt{3} - \sqrt{x})(\sqrt{3} + \sqrt{x})$.

$$(\sqrt{3} - \sqrt{x})(\sqrt{3} + \sqrt{x}) = (\sqrt{3})^2 - (\sqrt{x})^2 \qquad \text{Using } (A - B)(A + B) = A^2 - B^2$$
$$= 3 - x \qquad\qquad ◀$$

▶ **EXAMPLE 9**   Multiply: $(3 - \sqrt{p})^2$.

$$(3 - \sqrt{p})^2 = 3^2 - 2 \cdot 3 \cdot \sqrt{p} + (\sqrt{p})^2 \qquad \text{Using } (A - B)^2 = A^2 - 2AB + B^2$$
$$= 9 - 6\sqrt{6} + p \qquad\qquad ◀$$

▶ **EXAMPLE 10**   Multiply: $(2 - \sqrt{5})(2 + \sqrt{5})$.

$$(2 - \sqrt{5})(2 + \sqrt{5}) = 2^2 - (\sqrt{5})^2 \qquad \text{Using } (A - B)(A + B) = A^2 - B^2$$
$$= 4 - 5$$
$$= -1 \qquad\qquad ◀$$

DO EXERCISES 8–12.

[c] More Rationalizing Denominators

Note in Example 10 that the result has no radicals. This will happen whenever we multiply expressions such as $\sqrt{a} - \sqrt{b}$ and $\sqrt{a} + \sqrt{b}$, where a and b are rational numbers. We see this in the following:

$$(\sqrt{a} + \sqrt{b})(\sqrt{a} - \sqrt{b}) = (\sqrt{a})^2 - (\sqrt{b})^2 = a - b.$$

Expressions such as $\sqrt{3} - \sqrt{5}$ and $\sqrt{3} + \sqrt{5}$ are known as **conjugates;** so too are $2 + \sqrt{5}$ and $2 - \sqrt{5}$. We can use conjugates to rationalize a denominator that involves a sum or difference of two terms, where one or both are radicals. To do so, we multiply by 1 using the conjugate in the numerator and the denominator.

▶ **EXAMPLE 11**   Rationalize the denominator: $\dfrac{3}{2 + \sqrt{5}}$.

We multiply by 1 using the conjugate of $2 + \sqrt{5}$, which is $2 - \sqrt{5}$, as the numerator and the denominator:

$$\frac{3}{2 + \sqrt{5}} = \frac{3}{2 + \sqrt{5}} \cdot \frac{2 - \sqrt{5}}{2 - \sqrt{5}} \qquad \text{Multiplying by 1}$$

$$= \frac{3(2 - \sqrt{5})}{(2 + \sqrt{5})(2 - \sqrt{5})} \qquad \text{Multiplying}$$

$$= \frac{6 - 3\sqrt{5}}{2^2 - (\sqrt{5})^2}$$

$$= \frac{6 - 3\sqrt{5}}{4 - 5}$$

$$= \frac{6 - 3\sqrt{5}}{-1}$$

$$= -6 + 3\sqrt{5}, \text{ or } 3\sqrt{5} - 6 \qquad\qquad ◀$$

Multiply.

8. $\sqrt{3}(\sqrt{5} + \sqrt{2})$

9. $(1 - \sqrt{2})(4 + 3\sqrt{5})$

10. $(\sqrt{2} + \sqrt{a})(\sqrt{2} - \sqrt{a})$

11. $(5 + \sqrt{x})^2$

12. $(3 - \sqrt{7})(3 + \sqrt{7})$

ANSWERS ON PAGE A-8

Rationalize the denominator.

13. $\dfrac{6}{7 + \sqrt{5}}$

14. $\dfrac{\sqrt{5} + \sqrt{2}}{\sqrt{5} - \sqrt{2}}$

▶ **EXAMPLE 12**　Rationalize the denominator: $\dfrac{\sqrt{3} + \sqrt{5}}{\sqrt{3} - \sqrt{5}}$.

We multiply by 1 using the conjugate of $\sqrt{3} - \sqrt{5}$, which is $\sqrt{3} + \sqrt{5}$, as the numerator and the denominator:

$$\frac{\sqrt{3} + \sqrt{5}}{\sqrt{3} - \sqrt{5}} = \frac{\sqrt{3} + \sqrt{5}}{\sqrt{3} - \sqrt{5}} \cdot \frac{\sqrt{3} + \sqrt{5}}{\sqrt{3} + \sqrt{5}} \qquad \text{Multiplying by 1}$$

$$= \frac{(\sqrt{3} + \sqrt{5})^2}{(\sqrt{3} - \sqrt{5})(\sqrt{3} + \sqrt{5})}$$

$$= \frac{(\sqrt{3})^2 + 2\sqrt{3}\sqrt{5} + (\sqrt{5})^2}{(\sqrt{3})^2 - (\sqrt{5})^2}$$

$$= \frac{3 + 2\sqrt{15} + 5}{3 - 5}$$

$$= \frac{8 + 2\sqrt{15}}{-2}$$

$$= \frac{2(4 + \sqrt{15})}{2(-1)} \qquad \text{Factoring in order to simplify}$$

$$= \frac{2}{2} \cdot \frac{4 + \sqrt{15}}{-1}$$

$$= \frac{4 + \sqrt{15}}{-1} = -4 - \sqrt{15}. \qquad ◀$$

ANSWERS ON PAGE A-8　　　　　　**DO EXERCISES 13 AND 14.**

❖　SIDELIGHTS

▦ Wind Chill Temperature

Calculators are often used to approximate square roots. For example, using a calculator, we can approximate $\sqrt{73}$:

$$\sqrt{73} \approx 8.544003745.$$

Different calculators may give different numbers of digits in their readouts.

　　We can use approximations of square roots to consider an application involving the effect of wind on the feeling of cold in the winter. In cold weather, we feel colder when there is wind than when there is not. The **wind chill temperature** is what the temperature would have to be with no wind in order to give the same chilling effect. A formula for finding the wind chill temperature, T_w, is

$$T_w = 91.4 - \frac{(10.45 + 6.68\sqrt{v} - 0.447v)(457 - 5T)}{110},$$

where T is the actual temperature given by a thermometer, in degrees Fahrenheit, and v is the wind speed, in miles per hour.

EXERCISES

▦ Use a calculator to find the wind chill temperature in each case. You can find the square roots from Table 2. Round to the nearest degree.

1. $T = 30°\text{F}$, $v = 25$ mph

2. $T = 10°\text{F}$, $v = 25$ mph

3. $T = 20°\text{F}$, $v = 20$ mph

4. $T = 20°\text{F}$, $v = 40$ mph

5. $T = -10°\text{F}$, $v = 30$ mph

6. $T = -30°\text{F}$, $v = 30$ mph

EXERCISE SET 8.4

a Add or subtract. Simplify by collecting like radical terms, if possible.

1. $3\sqrt{2} + 4\sqrt{2}$ **2.** $8\sqrt{3} + 3\sqrt{3}$ **3.** $7\sqrt{5} - 3\sqrt{5}$

4. $8\sqrt{2} - 5\sqrt{2}$ **5.** $6\sqrt{x} + 7\sqrt{x}$ **6.** $9\sqrt{y} + 3\sqrt{y}$

7. $9\sqrt{x} - 11\sqrt{x}$ **8.** $6\sqrt{a} - 14\sqrt{a}$ **9.** $5\sqrt{8} + 15\sqrt{2}$

10. $3\sqrt{12} + 2\sqrt{3}$ **11.** $\sqrt{27} - 2\sqrt{3}$ **12.** $7\sqrt{50} - 3\sqrt{2}$

13. $\sqrt{45} - \sqrt{20}$ **14.** $\sqrt{27} - \sqrt{12}$ **15.** $\sqrt{72} + \sqrt{98}$

16. $\sqrt{45} + \sqrt{80}$ **17.** $2\sqrt{12} + \sqrt{27} - \sqrt{48}$ **18.** $9\sqrt{8} - \sqrt{72} + \sqrt{98}$

19. $3\sqrt{18} - 2\sqrt{32} - 5\sqrt{50}$ **20.** $\sqrt{18} - 3\sqrt{8} + \sqrt{50}$

21. $2\sqrt{27} - 3\sqrt{48} + 3\sqrt{12}$ **22.** $3\sqrt{48} - 2\sqrt{27} - 3\sqrt{12}$

23. $\sqrt{4x} + \sqrt{81x^3}$ **24.** $\sqrt{12x^2} + \sqrt{27}$

25. $\sqrt{27} - \sqrt{12x^2}$ **26.** $\sqrt{81x^3} - \sqrt{4x}$

27. $\sqrt{8x + 8} + \sqrt{2x + 2}$ **28.** $\sqrt{12x + 12} + \sqrt{3x + 3}$

29. $\sqrt{x^5 - x^2} + \sqrt{9x^3 - 9}$ **30.** $\sqrt{16x - 16} + \sqrt{25x^3 - 25x^2}$

31. $3x\sqrt{y^3x} - x\sqrt{yx^3} + y\sqrt{y^3x}$ **32.** $4a\sqrt{a^2b} + a\sqrt{a^2b^3} - 5\sqrt{b^3}$

33. $\sqrt{3} - \sqrt{\dfrac{1}{3}}$ **34.** $\sqrt{2} - \sqrt{\dfrac{1}{2}}$ **35.** $5\sqrt{2} + 3\sqrt{\dfrac{1}{2}}$ **36.** $4\sqrt{3} + 2\sqrt{\dfrac{1}{3}}$

ANSWERS

1. ___
2. ___
3. ___
4. ___
5. ___
6. ___
7. ___
8. ___
9. ___
10. ___
11. ___
12. ___
13. ___
14. ___
15. ___
16. ___
17. ___
18. ___
19. ___
20. ___
21. ___
22. ___
23. ___
24. ___
25. ___
26. ___
27. ___
28. ___
29. ___
30. ___
31. ___
32. ___
33. ___
34. ___
35. ___
36. ___

37. $\sqrt{\dfrac{2}{3}} - \sqrt{\dfrac{1}{6}}$ **38.** $\sqrt{\dfrac{1}{2}} - \sqrt{\dfrac{1}{8}}$ **39.** $\sqrt{\dfrac{1}{12}} - \sqrt{\dfrac{1}{27}}$ **40.** $\sqrt{\dfrac{5}{6}} - \sqrt{\dfrac{6}{5}}$

b Multiply.

41. $(\sqrt{5} + 7)(\sqrt{5} - 7)$ **42.** $(1 + \sqrt{5})(1 - \sqrt{5})$

43. $(\sqrt{6} - \sqrt{3})(\sqrt{6} + \sqrt{3})$ **44.** $(\sqrt{2} + \sqrt{6})(\sqrt{2} - \sqrt{6})$

45. $(3\sqrt{5} - 2)(\sqrt{5} + 1)$ **46.** $(\sqrt{5} - 2\sqrt{2})(\sqrt{10} - 1)$

47. $(\sqrt{x} - \sqrt{y})^2$ **48.** $(\sqrt{w} + 11)^2$

c Rationalize the denominator.

49. $\dfrac{2}{\sqrt{3} - \sqrt{5}}$ **50.** $\dfrac{5}{3 + \sqrt{7}}$ **51.** $\dfrac{\sqrt{3} - \sqrt{2}}{\sqrt{3} + \sqrt{2}}$ **52.** $\dfrac{2 - \sqrt{7}}{\sqrt{3} - \sqrt{2}}$

53. $\dfrac{4}{\sqrt{10} + 1}$ **54.** $\dfrac{6}{\sqrt{11} - 3}$ **55.** $\dfrac{1 - \sqrt{7}}{3 + \sqrt{7}}$ **56.** $\dfrac{2 + \sqrt{8}}{1 - \sqrt{5}}$

SKILL MAINTENANCE

57. The time t it takes a bus to travel a fixed distance varies inversely as its speed r. At a speed of 40 mph, it takes $\frac{1}{2}$ hr to travel a fixed distance. How long will it take to travel the same distance at 60 mph? Describe the variation constant.

58. Solution A is 3% alcohol, and solution B is 6% alcohol. A service station attendant wants to mix the two to get 80 gal of a solution that is 5.4% alcohol. How many gallons of each should the attendant use?

SYNTHESIS

59. Three students were asked to simplify $\sqrt{10} + \sqrt{50}$. Their answers were $\sqrt{10}(1 + \sqrt{5})$, $\sqrt{10} + 5\sqrt{2}$, and $\sqrt{2}(5 + \sqrt{5})$. Which, if any, are correct?

Add or subtract.

60. $\frac{3}{5}\sqrt{24} + \frac{2}{5}\sqrt{150} - \sqrt{96}$ **61.** $\frac{1}{3}\sqrt{27} + \sqrt{8} + \sqrt{300} - \sqrt{18} - \sqrt{162}$

62. Evaluate $\sqrt{a^2 + b^2}$ and $\sqrt{a^2} + \sqrt{b^2}$ for $a = 2$ and $b = 3$.

63. On the basis of Exercise 62, determine whether $\sqrt{a^2 + b^2}$ and $\sqrt{a^2} + \sqrt{b^2}$ are equivalent.

Determine whether each of the following is true. Show why or why not.

64. $(\sqrt{x + 2})^2 = x + 2$ **65.** $(3\sqrt{x + 2})^2 = 9(x + 2)$

8.5 Radical Equations

a Solving Radical Equations

The following are examples of *radical equations:*

$$\sqrt{2x} - 4 = 7, \qquad \sqrt{x+1} = \sqrt{2x-5}.$$

A **radical equation** has variables in one or more radicands. To solve radical equations, we first convert them to equations without radicals. We do this by squaring both sides of the equation, using the following principle.

> **The Principle of Squaring**
>
> **If an equation $a = b$ is true, then the equation $a^2 = b^2$ is true.**

To solve radical equations, we first try to get a radical by itself. That is, we try to isolate the radical. Then we use the principle of squaring. This allows us to eliminate one radical.

▶ **EXAMPLE 1** Solve: $\sqrt{2x} - 4 = 7$.

$$\sqrt{2x} - 4 = 7$$
$$\sqrt{2x} = 11 \qquad \text{Adding 4 to isolate the radical}$$
$$(\sqrt{2x})^2 = 11^2 \qquad \text{Squaring both sides}$$
$$2x = 121$$
$$x = \frac{121}{2}$$

Check:

$$\begin{array}{c|c} \sqrt{2x} - 4 = 7 \\ \hline \sqrt{2 \cdot \dfrac{121}{2}} - 4 & 7 \\ \sqrt{121} - 4 & \\ 11 - 4 & \\ 7 & \text{TRUE} \end{array}$$

The solution is $\frac{121}{2}$. ◀

DO EXERCISE 1.

▶ **EXAMPLE 2** Solve: $2\sqrt{x+2} = \sqrt{x+10}$.

Each is already isolated. We proceed with the principle of squaring.

$$(2\sqrt{x+2})^2 = (\sqrt{x+10})^2 \qquad \text{Squaring both sides}$$
$$2^2(\sqrt{x+2})^2 = x + 10 \qquad \text{Raising the product to the power 2 on the left;}$$
$$\qquad\qquad\qquad\qquad\qquad \text{simplifying on the right}$$
$$4(x+2) = x + 10$$
$$4x + 8 = x + 10 \qquad \text{Removing parentheses}$$
$$3x = 2 \qquad \text{Subtracting } x \text{ and 8}$$
$$x = \frac{2}{3} \qquad \text{Dividing by 3}$$

OBJECTIVES

After finishing Section 8.5, you should be able to:

a Solve radical equations with one or more radical terms isolated, using the principle of squaring once.

b Solve radical expressions with two or more radical terms using the principle of squaring twice.

c Solve applied problems using radical equations.

FOR EXTRA HELP

Tape 15E   Tape 16B   MAC: 8
                                     IBM: 8

1. Solve: $\sqrt{3x} - 5 = 3$.

ANSWER ON PAGE A-8

Solve.

2. $\sqrt{3x + 1} = \sqrt{2x + 3}$

3. $3\sqrt{x + 1} = \sqrt{x + 12}$

4. Solve: $x - 1 = \sqrt{x + 5}$.

Check:

$$2\sqrt{x + 2} = \sqrt{x + 10}$$

$$\begin{array}{c|c} 2\sqrt{\dfrac{2}{3} + 2} & \sqrt{\dfrac{2}{3} + 10} \\[2mm] 2\sqrt{\dfrac{8}{3}} & \sqrt{\dfrac{32}{3}} \\[2mm] 4\sqrt{\dfrac{2}{3}} & 4\sqrt{\dfrac{2}{3}} \quad \text{TRUE} \end{array}$$

The number $\frac{2}{3}$ checks. The solution is $\frac{2}{3}$.   ◀

DO EXERCISES 2 AND 3.

It is important to check when using the principle of squaring. This principle may not produce equivalent equations. When we square both sides of an equation, the new equation may have solutions that the first one does not. For example, the equation

$$x = 1 \qquad \textbf{(1)}$$

has just one solution, the number 1. When we square both sides, we get

$$x^2 = 1, \qquad \textbf{(2)}$$

which has two solutions, 1 and -1. Thus the equations $x = 1$ and $x^2 = 1$ do not have the same solutions and thus are not equivalent. Whereas it is true that any solution of Equation (1) is a solution of Equation (2), it is not true that any solution of Equation (2) is a solution of Equation (1).

> **When the principle of squaring is used to solve an equation, solutions of an equation found by squaring *must* be checked in the original equation!**

Sometimes we may need to apply the principle of zero products after squaring. (See Section 4.7.)

▶ **EXAMPLE 3**   Solve: $x - 5 = \sqrt{x + 7}$.

$$\begin{aligned} x - 5 &= \sqrt{x + 7} \\ (x - 5)^2 &= (\sqrt{x + 7})^2 \qquad \text{Using the principle of squaring} \\ x^2 - 10x + 25 &= x + 7 \\ x^2 - 11x + 18 &= 0 \\ (x - 9)(x - 2) &= 0 \qquad \text{Factoring} \\ x - 9 = 0 \;\; &\text{or} \;\; x - 2 = 0 \qquad \text{Using the principle of zero products} \\ x = 9 \;\; &\text{or} \qquad\quad x = 2 \end{aligned}$$

Check:

$$\begin{array}{c|c} x - 5 = \sqrt{x + 7} & \\ \hline 9 - 5 & \sqrt{9 + 7} \\ 4 & 4 \qquad \text{TRUE} \end{array} \qquad \begin{array}{c|c} x - 5 = \sqrt{x + 7} & \\ \hline 2 - 5 & \sqrt{2 + 7} \\ -3 & 3 \qquad \text{FALSE} \end{array}$$

The number 9 checks, but 2 does not. Thus the solution is 9.   ◀

DO EXERCISE 4.

▶ **EXAMPLE 4**   Solve: $3 + \sqrt{27 - 3x} = x$.

In this case, we must first isolate the radical.

$$3 + \sqrt{27 - 3x} = x$$
$$\sqrt{27 - 3x} = x - 3 \qquad \text{Subtracting 3 to isolate the radical}$$
$$(\sqrt{27 - 3x})^2 = (x - 3)^2 \qquad \text{Using the principle of squaring}$$
$$27 - 3x = x^2 - 6x + 9$$
$$0 = x^2 - 3x - 18 \qquad \text{We can have 0 on the left.}$$
$$0 = (x - 6)(x + 3) \qquad \text{Factoring}$$
$$x - 6 = 0 \quad \text{or} \quad x + 3 = 0 \qquad \text{Using the principle of zero products}$$
$$x = 6 \quad \text{or} \qquad x = -3$$

Check:

$3 + \sqrt{27 - 3x} = x$	
$3 + \sqrt{27 - 3 \cdot 6}$	6
$3 + \sqrt{9}$	
$3 + 3$	
6	TRUE

$3 + \sqrt{27 - 3x} = x$	
$3 + \sqrt{27 - 3 \cdot (-3)}$	-3
$3 + \sqrt{27 + 9}$	
$3 + \sqrt{36}$	
$3 + 6$	
9	FALSE

The number 6 checks, but -3 does not. The solution is 6.   ◀

DO EXERCISE 5.

Suppose that in Example 4 we did not isolate the radical before squaring. Then we get an expression on the left side of the equation in which we have *not* eliminated the radical.

$$(3 + \sqrt{27 - 3x})^2 = (x)^2$$
$$3^2 + 2 \cdot 3 \cdot \sqrt{27 - 3x} + (\sqrt{27 - 3x})^2 = x^2$$
$$9 + 6\sqrt{27 - 3x} + (27 - 3x) = x^2$$

In fact, we have ended up with a more complicated expression than the one we squared.

b   Using the Principle of Squaring More Than Once

Sometimes when we have two radical terms, we may need to apply the principle of squaring a second time.

▶ **EXAMPLE 5**   Solve: $\sqrt{x} - 1 = \sqrt{x - 5}$.

$$\sqrt{x} - 1 = \sqrt{x - 5}$$
$$(\sqrt{x} - 1)^2 = (\sqrt{x - 5})^2 \qquad \text{Using the principle of squaring}$$
$$(\sqrt{x})^2 - 2 \cdot \sqrt{x} \cdot 1 + 1^2 = x - 5 \qquad \text{Using } (A + B)^2 = A^2 + 2AB = B^2 \text{ on the left side}$$
$$x - 2\sqrt{x} + 1 = x - 5 \qquad \text{Simplifying}$$
$$-2\sqrt{x} = -6 \qquad \text{Isolating the radical}$$
$$\sqrt{x} = 3$$
$$(\sqrt{x})^2 = 3^2 \qquad \text{Using the principle of squaring}$$
$$x = 9$$

The check is left to the student. The number 9 checks and is the solution.   ◀

5. Solve: $1 + \sqrt{1 - x} = x$.

ANSWER ON PAGE A-8

6. Solve: $\sqrt{x} - 1 = \sqrt{x - 3}$.

7. How far can you see to the horizon through an airplane window at a height of 8000 m?

8. How far can a sailor see to the horizon from the top of a 20-m mast?

9. A sailor can see 91 km to the horizon from the top of a mast. How high is the mast?

The following is a procedure for solving radical equations.

> **To solve radical equations:**
> 1. **Isolate one of the radical terms.**
> 2. **Use the principle of squaring.**
> 3. **If a radical term remains, perform steps (1) and (2) again.**
> 4. **Solve the equation and check possible solutions.**

DO EXERCISE 6.

c   Applications

How far can you see from a given height? There is a formula for this. At a height of h meters, you can see V kilometers to the horizon. These numbers are related as follows:

$$V = 3.5\sqrt{h}. \quad \textbf{(1)}$$

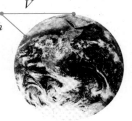

Earth

▶ **EXAMPLE 6**   How far to the horizon can you see through an airplane window at a height, or altitude, of 9000 m?

We substitute 9000 for h in Equation (1) and find an approximation.

Method 1.   We use a calculator and approximate $\sqrt{9000}$ directly:

$$V = 3.5\sqrt{9000} \approx 3.5(94.868) = 332.038.$$

Method 2.   We simplify and then approximate:

$$V = 3.5\sqrt{9000} = 3.5\sqrt{900 \cdot 10} = 3.5 \times 30 \times \sqrt{10}$$
$$\approx 3.5 \times 30 \times 3.162 \approx 332.010 \text{ km}.$$

You can see about 332 km at a height of 9000 m.   ◀

DO EXERCISES 7 AND 8.

▶ **EXAMPLE 7**   A person can see 50.4 km to the horizon from the top of a cliff. What is the altitude of the eyes of the person?

We substitute 50.4 for V in Equation (1) and solve:

$$50.4 = 3.5\sqrt{h}$$
$$\frac{50.4}{3.5} = \sqrt{h}$$
$$14.4 = \sqrt{h}$$
$$(14.4)^2 = (\sqrt{h})^2$$
$$207.36 = h.$$

The altitude of the eyes of the person is about 207 m.   ◀

DO EXERCISE 9.

NAME   SECTION   DATE

EXERCISE SET 8.5

a Solve.

1. $\sqrt{x} = 5$

2. $\sqrt{x} = 7$

3. $\sqrt{x} = 6.2$

4. $\sqrt{x} = 4.3$

5. $\sqrt{x + 3} = 20$

6. $\sqrt{x + 4} = 11$

7. $\sqrt{2x + 4} = 25$

8. $\sqrt{2x + 1} = 13$

9. $3 + \sqrt{x - 1} = 5$

10. $4 + \sqrt{y - 3} = 11$

11. $6 - 2\sqrt{3n} = 0$

12. $8 - 4\sqrt{5n} = 0$

13. $\sqrt{5x - 7} = \sqrt{x + 10}$

14. $\sqrt{4x - 5} = \sqrt{x + 9}$

15. $\sqrt{x} = -7$

16. $\sqrt{x} = -5$

17. $\sqrt{2y + 6} = \sqrt{2y - 5}$

18. $2\sqrt{3x - 2} = \sqrt{2x - 3}$

19. $x - 7 = \sqrt{x - 5}$

20. $\sqrt{x + 7} = x - 5$

21. $\sqrt{x + 18} = x - 2$

22. $x - 9 = \sqrt{x - 3}$

23. $2\sqrt{x - 1} = x - 1$

24. $x + 4 = 4\sqrt{x + 1}$

25. $\sqrt{5x + 21} = x + 3$

26. $\sqrt{27 - 3x} = x - 3$

27. $x = 1 + 6\sqrt{x - 9}$

28. $\sqrt{2x - 1} + 2 = x$

29. $\sqrt{x^2 + 6} - x + 3 = 0$

1. _____

2. _____

3. _____

4. _____

5. _____

6. _____

7. _____

8. _____

9. _____

10. _____

11. _____

12. _____

13. _____

14. _____

15. _____

16. _____

17. _____

18. _____

19. _____

20. _____

21. _____

22. _____

23. _____

24. _____

25. _____

26. _____

27. _____

28. _____

29. _____

Copyright © 1991 Addison-Wesley Publishing Co., Inc.

ANSWERS

30. _____

31. _____

32. _____

33. _____

34. _____

35. _____

36. _____

37. _____

38. _____

39. _____

40. _____

41. _____

42. _____

43. _____

44. _____

45. _____

46. _____

47. _____

48. _____

49. _____

50. _____

51. _____

52. _____

53. _____

54. _____

55. _____

56. _____

30. $\sqrt{x^2 + 5} - x + 2 = 0$

31. $\sqrt{(p + 6)(p + 1)} - 2 = p + 1$

32. $\sqrt{(4x + 5)(x + 4)} = 2x + 5$

33. $\sqrt{2 - x} = \sqrt{3x - 7}$

34. $\sqrt{4x - 10} = \sqrt{2 - x}$

b Solve. Use the principle of squaring twice.

35. $\sqrt{x + 9} = 1 + \sqrt{x}$

36. $\sqrt{x - 5} = 5 - \sqrt{x}$

37. $\sqrt{3x + 1} = 1 - \sqrt{x + 4}$

38. $\sqrt{y + 8} - \sqrt{y} = 2$

c Solve.

Use $V = 3.5\sqrt{h}$ for Exercises 39–42.

39. How far can you see to the horizon through an airplane window at a height of 9800 m?

40. How far can a sailor see to the horizon from the top of a 24-m mast?

41. A person can see 371 km to the horizon from an airplane window. How high is the airplane?

42. A sailor can see 99.4 km to the horizon from the top of a mast. How high is the mast?

The formula $r = 2\sqrt{5L}$ can be used to approximate the speed r, in miles per hour, of a car that has left a skid mark of length L, in feet.

43. How far will a car skid at 50 mph? at 70 mph?

44. How far will a car skid at 60 mph? at 100 mph?

45. Find the number such that twice its square root is 14.

46. Find a number such that the square root of four more than five times the number is 8.

SKILL MAINTENANCE

Divide and simplify.

47. $\dfrac{x^2 - 49}{x + 8} \div \dfrac{x^2 - 14x + 49}{x^2 + 15x + 56}$

48. $\dfrac{x - 2}{x - 3} \div \dfrac{x - 4}{x - 5}$

49. Two angles are supplementary. One angle is 3° less than twice the other. Find the measures of the angles.

50. Two angles are complementary. The sum of the measure of the first angle and half the second is 64°. Find the measures of the angles.

SYNTHESIS

Solve.

51. $\sqrt{5x^2 + 5} = 5$

52. $\sqrt{x} = -x$

53. $4 + \sqrt{19 - x} = 6 + \sqrt{4 - x}$

54. $x = (x - 2)\sqrt{x}$

55. $\sqrt{x + 3} = \dfrac{8}{\sqrt{x - 9}}$

56. $\dfrac{12}{\sqrt{5x + 6}} = \sqrt{2x + 5}$

8.6 Right Triangles and Applications

a Right Triangles

A **right triangle** is a triangle with a 90° angle, as shown in the figure below. The small square in the corner indicates the 90° angle.

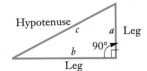

In a right triangle, the longest side is called the **hypotenuse.** It is also the side opposite the right angle. The other two sides are called **legs.** We generally use the letters a and b for the lengths of the legs and c for the length of the hypotenuse. They are related as follows.

The Pythagorean Theorem

In any right triangle, if a and b are the lengths of the legs and c is the length of the hypotenuse, then

$$a^2 + b^2 = c^2.$$

The equation $a^2 + b^2 = c^2$ is called the *Pythagorean equation.*

The Pythagorean theorem is named after the ancient Greek mathematician Pythagoras (569?–500? B.C.). It is uncertain who actually proved this result the first time. The proof can be found in most geometry books.

If we know the lengths of any two sides of a right triangle, we can find the length of the third side.

▶ **EXAMPLE 1** Find the length of the hypotenuse of this right triangle. Give an exact answer and an approximation to three decimal places.

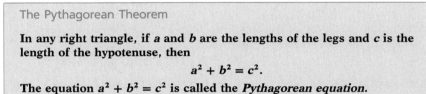

$$4^2 + 5^2 = c^2 \quad \text{Substituting in the Pythagorean equation}$$
$$16 + 25 = c^2$$
$$41 = c^2$$
$$c = \sqrt{41}$$
$$c \approx 6.403 \quad \text{Using a calculator or Table 2}$$

◀

▶ **EXAMPLE 2** Find the length of the leg of this right triangle. Give an exact answer and an approximation to three decimal places.

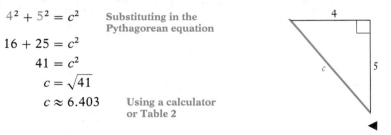

$$10^2 + b^2 = 12^2 \quad \text{Substituting in the Pythagorean equation}$$
$$100 + b^2 = 144$$
$$b^2 = 144 - 100$$
$$b^2 = 44$$
$$b^2 = \sqrt{44}$$
$$b \approx 6.633 \quad \text{Using a calculator or Table 2}$$

◀

DO EXERCISES 1 AND 2.

OBJECTIVES

After finishing Section 8.6, you should be able to:

a Given the lengths of any two sides of a right triangle, find the length of the third side.

b Solve applied problems involving right triangles.

FOR EXTRA HELP

Tape 16A   Tape 16B   MAC: 8
                                     IBM: 8

1. Find the length of the hypotenuse of this right triangle. Give an exact answer and an approximation to three decimal places.

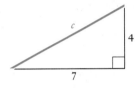

2. Find the length of the leg of this right triangle. Give an exact answer and an approximation to three decimal places.

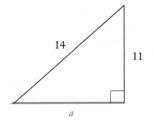

ANSWERS ON PAGE A-8

Find the length of the leg of the right triangle. Give an exact answer and an approximation to three decimal places.

3.

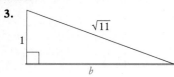

4.

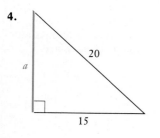

5. How long is a guy wire reaching from the top of a 15-ft pole to a point on the ground 10 ft from the pole? Give an exact answer and an approximation to three decimal places.

▶ **EXAMPLE 3**   Find the length of the leg of this right triangle. Give an exact answer and an approximation to three decimal places.

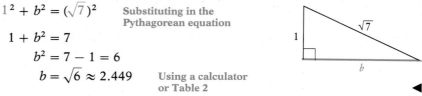

$1^2 + b^2 = (\sqrt{7})^2$ **Substituting in the Pythagorean equation**

$1 + b^2 = 7$

$b^2 = 7 - 1 = 6$

$b = \sqrt{6} \approx 2.449$ **Using a calculator or Table 2**

◀

▶ **EXAMPLE 4**   Find the length of the leg of this right triangle. Give an exact answer and an approximation to three decimal places.

$a^2 + 10^2 = 15^2$

$a^2 + 100 = 225$

$a^2 = 225 - 100$

$a^2 = 125$

$a = \sqrt{125} \approx 11.180$ **Using a calculator**

◀

In Example 4, if you use Table 2 to find an approximation, you will need to simplify before finding an approximation:

$$\sqrt{125} = \sqrt{25 \cdot 5} = 5\sqrt{5} \approx 5(2.236) = 11.180.$$

A possible variance in answers can occur depending on the procedure used.

DO EXERCISES 3 AND 4.

b   Applications

▶ **EXAMPLE 5**   A slow-pitch softball diamond is actually a square 65 ft on a side. How far is it from home plate to second base? (This can be helpful information when lining up the bases.) Give an exact answer and an approximation to three decimal places.

a) We first make a drawing. We note that the first and second base lines, together with a line from home to second, form a right triangle. We label the unknown distance d.

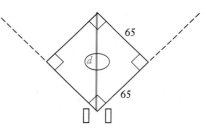

b) We know that $65^2 + 65^2 = d^2$. We solve this equation:

$4225 + 4225 = d^2$

$8450 = d^2.$

Exact answer: $\sqrt{8450} = d$

Approximation: $91.924 \approx d$

If you use Table 2 to find an approximation, you will need to simplify before finding an approximation in the table:

$$d = \sqrt{8450} = \sqrt{25 \cdot 169 \cdot 2} = \sqrt{25}\sqrt{169}\sqrt{2}$$
$$\approx 5(13)(1.414) = 91.910.$$

Note that we get a variance in the last two decimal places. ◀

NAME   SECTION   DATE

EXERCISE SET 8.6

a Find the length of the third side of the right triangle. Give an exact answer and an approximation to three decimal places.

1.

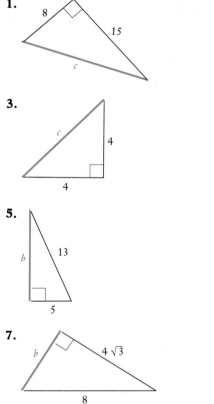

2.

3.

4.

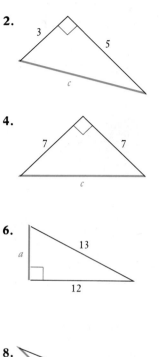

5.

6.

7.

8.

In a right triangle, find the length of the side not given. Give an exact answer and an approximation to three decimal places.

9. $a = 10, \quad b = 24$

10. $a = 5, \quad b = 12$

11. $a = 9, \quad c = 15$

12. $a = 18, \quad c = 30$

13. $b = 1, \quad c = \sqrt{5}$

14. $b = 1, \quad c = \sqrt{2}$

15. $a = 1, \quad c = \sqrt{3}$

16. $a = \sqrt{3}, \quad b = \sqrt{5}$

17. $c = 10, \quad b = 5\sqrt{3}$

18. $a = 5, \quad b = 5$

b Solve. Don't forget to make a drawing. Give an exact answer and an approximation to three decimal places.

19. A 10-m ladder is leaning against a building. The bottom of the ladder is 5 m from the building. How high is the top of the ladder?

20. Find the length of a diagonal of a square whose sides are 3 cm long.

1. _____

2. _____

3. _____

4. _____

5. _____

6. _____

7. _____

8. _____

9. _____

10. _____

11. _____

12. _____

13. _____

14. _____

15. _____

16. _____

17. _____

18. _____

19. _____

20. _____

ANSWERS

21.

22.

23.

24.

25.

26.

27.

28.

29.

30.

31.

32.

33.

21. How long is a guy wire reaching from the top of a 12-ft pole to a point 8 ft from the pole?

22. How long must a wire be to reach from the top of a 13-m telephone pole to a point on the ground 9 m from the foot of the pole?

23. A surveyor had poles located at points P, Q, and R. The distances that the surveyor was able to measure are marked on the drawing. What is the approximate distance from P to R?

24. An airplane is flying at an altitude of 4100 ft. The slanted distance directly to the airport is 15,100 ft. How far is the airplane horizontally from the airport?

SKILL MAINTENANCE

Solve.

25. $5x + 7 = 8y,$
$3x = 8y - 4$

26. $5x + y = 17,$
$-5x + 2y = 10$

SYNTHESIS

27. The length and the width of a rectangle are given by consecutive integers. The area of the rectangle is 90 cm². Find the length of the diagonal of the rectangle.

28. Two cars leave a service station at the same time. One car travels east at a speed of 50 mph, and the other travels south at a speed of 60 mph. After one-half hour, how far apart are they?

Find x.

29.

30.

31.

An **equilateral triangle** is shown at the right.

32. Find an expression for its height h in terms of a.

33. Find an expression for its area A in terms of a.

SUMMARY AND REVIEW: CHAPTER 8

IMPORTANT PROPERTIES AND FORMULAS

Products: $\sqrt{A}\sqrt{B} = \sqrt{AB}$

Quotients: $\dfrac{\sqrt{A}}{\sqrt{B}} = \sqrt{\dfrac{A}{B}}$

Principle of Squaring: If an equation $a = b$ is true, then the equation $a^2 = b^2$ is true.

Pythagorean Equation: $a^2 + b^2 = c^2$, where a and b are the lengths of the legs of a right triangle and c is the length of the hypotenuse.

REVIEW EXERCISES

The review sections and objectives to be tested in addition to the material in this chapter are [5.2b], [7.3a, b], [6.6b], and [7.3c].

Find the square roots.

1. 64

2. 400

Simplify.

3. $\sqrt{36}$

4. $-\sqrt{169}$

Approximate the square roots to three decimal places.

5. $\sqrt{3}$

6. $\sqrt{99}$

7. $\sqrt{108}$

8. $\sqrt{320}$

9. $\sqrt{\dfrac{1}{8}}$

10. $\sqrt{\dfrac{11}{20}}$

Identify the radicand.

11. $\sqrt{x^2 + 4}$

12. $\sqrt{5ab^3}$

Determine whether the expression is meaningless. Write "yes" or "no."

13. $\sqrt{-22}$

14. $-\sqrt{49}$

15. $\sqrt{-36}$

16. $\sqrt{-100}$

17. Determine whether the given number is a meaningful replacement in the given radical expression:

$$-3; \quad \sqrt{2x}.$$

18. Determine the meaningful replacements in

$$\sqrt{2y - 20}.$$

Simplify.

19. $\sqrt{m^2}$

20. $\sqrt{(x-4)^2}$

Multiply.

21. $\sqrt{3}\sqrt{7}$

22. $\sqrt{x-3}\sqrt{x+3}$

Simplify by factoring.

23. $-\sqrt{48}$

24. $\sqrt{32t^2}$

25. $\sqrt{x^2 + 16x + 64}$

26. $\sqrt{t^2 - 49}$

27. $\sqrt{x^8}$

28. $\sqrt{m^{15}}$

Multiply and simplify.

29. $\sqrt{6}\sqrt{10}$

30. $\sqrt{5x}\sqrt{8x}$

31. $\sqrt{5x}\sqrt{10xy^2}$

32. $\sqrt{20a^3b}\sqrt{5a^2b^2}$

Simplify.

33. $\sqrt{\dfrac{25}{64}}$

34. $\sqrt{\dfrac{20}{45}}$

35. $\sqrt{\dfrac{49}{t^2}}$

Rationalize the denominator.

36. $\sqrt{\dfrac{1}{2}}$

37. $\sqrt{\dfrac{1}{8}}$

38. $\sqrt{\dfrac{5}{y}}$

39. $\dfrac{2}{\sqrt{3}}$

Divide and simplify.

40. $\dfrac{\sqrt{27}}{\sqrt{45}}$

41. $\dfrac{\sqrt{45x^2y}}{\sqrt{54y}}$

42. Rationalize the denominator: $\dfrac{4}{2 + \sqrt{3}}$.

Simplify.

43. $10\sqrt{5} + 3\sqrt{5}$

44. $\sqrt{80} - \sqrt{45}$

45. $3\sqrt{2} - 5\sqrt{\dfrac{1}{2}}$

46. $(2 + \sqrt{3})^2$

47. $(2 + \sqrt{3})(2 - \sqrt{3})$

In a right triangle, find the length of the side not given.

48. $a = 15, \quad c = 25$

49. $a = 1, \quad b = \sqrt{2}$

50. Find the length of the diagonal of a square whose sides are 7 m long.

Solve.

51. $\sqrt{x - 3} = 7$

52. $\sqrt{5x + 3} = \sqrt{2x - 1}$

53. $\sqrt{x} = \sqrt{x - 5} + 1$

54. $1 + x = \sqrt{1 + 5x}$

Solve.

55. The formula $r = 2\sqrt{5L}$ can be used to approximate the speed r, in miles per hour, of a car that has left a skid mark of length L, in feet. How far will a car skid at 90 mph?

SKILL MAINTENANCE

56. A person's paycheck P varies directly as the number H of hours worked. For working 15 hr, the pay is $168.75. Find the pay for 40 hr of work.

57. There were 12,000 people at a rock concert. Admission was $7.00 at the door and $6.50 if bought in advance. Total receipts were $81,165. How many people bought their tickets in advance?

58. Solve:

$$2x - 3y = 4,$$
$$3x + 4y = 2.$$

59. Divide and simplify:

$$\dfrac{x^2 - 10x + 25}{x^2 + 14x + 49} \div \dfrac{x^2 - 25}{x^2 - 49}.$$

SYNTHESIS

60. Simplify: $\sqrt{\sqrt{\sqrt{256}}}$.

61. Solve $A = \sqrt{a^2 + b^2}$ for b.

❖ **THINKING IT THROUGH**

1. Explain why the following is incorrect:

$$\sqrt{\dfrac{9 + 100}{25}} = \dfrac{3 + 10}{5}.$$

Determine whether each of the following is true or false. Explain your answer.

2. $\sqrt{5x^2} = x\sqrt{5}$

3. $\sqrt{b^2 - 4} = b - 2$

4. The solution of $\sqrt{11 - 2x} = -3$ is 1.

TEST: CHAPTER 8

1. Find the square roots of 81.

Simplify.

2. $\sqrt{64}$

3. $-\sqrt{25}$

Approximate the expression involving square roots to three decimal places.

4. $\sqrt{116}$

5. $\sqrt{87}$

6. $\dfrac{3}{\sqrt{3}}$

7. Identify the radicand in $\sqrt{4 - y^3}$.

Determine whether the expression is meaningless. Write "yes" or "no."

8. $\sqrt{24}$

9. $\sqrt{-23}$

10. Determine whether 6 is a meaningful replacement in $\sqrt{3 - 4x}$.

11. Determine the meaningful replacements in $\sqrt{8 - x}$.

Simplify.

12. $\sqrt{a^2}$

13. $\sqrt{36y^2}$

Multiply.

14. $\sqrt{5}\,\sqrt{6}$

15. $\sqrt{x - 8}\,\sqrt{x + 8}$

Simplify by factoring.

16. $\sqrt{27}$

17. $\sqrt{25x - 25}$

18. $\sqrt{t^5}$

Multiply and simplify.

19. $\sqrt{5}\,\sqrt{10}$

20. $\sqrt{3ab}\,\sqrt{6ab^3}$

ANSWERS

1. _____

2. _____

3. _____

4. _____

5. _____

6. _____

7. _____

8. _____

9. _____

10. _____

11. _____

12. _____

13. _____

14. _____

15. _____

16. _____

17. _____

18. _____

19. _____

20. _____

Simplify.

21. $\sqrt{\dfrac{27}{12}}$

22. $\sqrt{\dfrac{144}{a^2}}$

Rationalize the denominator.

23. $\sqrt{\dfrac{2}{5}}$

24. $\sqrt{\dfrac{2x}{y}}$

Divide and simplify.

25. $\dfrac{\sqrt{27}}{\sqrt{32}}$

26. $\dfrac{\sqrt{35x}}{\sqrt{80xy^2}}$

Add or subtract.

27. $3\sqrt{18} - 5\sqrt{18}$

28. $\sqrt{5} + \sqrt{\dfrac{1}{5}}$

Simplify.

29. $(4 - \sqrt{5})^2$

30. $(4 - \sqrt{5})(4 + \sqrt{5})$

31. Rationalize the denominator: $\dfrac{10}{4 - \sqrt{5}}$.

32. In a right triangle, $a = 8$ and $b = 4$. Find c.

Solve.

33. $\sqrt{3x} + 2 = 14$

34. $\sqrt{6x + 13} = x + 3$

35. $\sqrt{1 - x} + 1 = \sqrt{6 - x}$

36. A person can see 247.49 km to the horizon from an airplane window. How high is the airplane? Use the formula $V = 3.5\sqrt{h}$.

SKILL MAINTENANCE

37. The perimeter of a rectangle is 118 yd. The width is 18 yd less than the length. Find the area of the rectangle.

38. The number of switches N that a production line can make varies directly as the time it operates. It can make 7240 switches in 6 hr. How many can it make in 13 hr?

39. Solve:
$$-6x + 5y = 10,$$
$$5x + 6y = 12.$$

40. Divide and simplify:
$$\dfrac{x^2 - 11x + 30}{x^2 - 12x + 35} \div \dfrac{x^2 - 36}{x^2 - 14x + 49}.$$

SYNTHESIS

Simplify.

41. $\sqrt{\sqrt{\sqrt{625}}}$

42. $\sqrt{y^{16n}}$

CUMULATIVE REVIEW: CHAPTERS 1–8

1. Evaluate $x^3 - x^2 + x - 1$ when $x = -2$.

2. Collect like terms: $2x^3 - 7 + \dfrac{3}{7}x^2 - 6x^3 - \dfrac{4}{7}x^2 + 5$.

Find the meaningful replacements.

3. $\sqrt{3x - 5}$

4. $\dfrac{x - 6}{2x + 1}$

Simplify.

5. $(2 + \sqrt{3})(2 - \sqrt{3})$

6. $-\sqrt{196}$

7. $\sqrt{3}\,\sqrt{75}$

8. $(1 - \sqrt{2})^2$

9. $\dfrac{\sqrt{162}}{\sqrt{125}}$

10. $2\sqrt{45} + 3\sqrt{20}$

Perform the indicated operations and simplify.

11. $(3x^4 - 2y^5)(3x^4 + 2y^5)$

12. $(x^2 + 4)^2$

13. $\left(2x + \dfrac{1}{4}\right)\left(4x - \dfrac{1}{2}\right)$

14. $\dfrac{x}{2x - 1} - \dfrac{3x + 2}{1 - 2x}$

15. $(3x^2 - 2x^3) - (x^3 - 2x^2 + 5) + (3x^2 - 5x + 5)$

16. $\dfrac{2x + 2}{3x - 9} \cdot \dfrac{x^2 - 8x + 15}{x^2 - 1}$

17. $\dfrac{2x^2 - 2}{2x^2 + 7x + 3} \div \dfrac{4x - 4}{2x^2 - 5x - 3}$

18. $(3x^3 - 2x^2 + x - 5) \div (x - 2)$

Simplify.

19. $\sqrt{2x^2 - 4x + 2}$

20. $x^{-9} \cdot x^{-3}$

21. $\sqrt{\dfrac{50}{2x^8}}$

22. $\dfrac{x - \dfrac{1}{x}}{1 - \dfrac{x - 1}{2x}}$

Factor completely.

23. $3 - 12x^8$

24. $12t - 4t^2 - 48t^4$

25. $6x^2 - 28x + 16$

26. $4x^3 + 4x^2 - x - 1$

27. $16x^4 - 56x^2 + 49$

28. $x^2 + 3x - 180$

Solve.

29. $x^2 = -17x$

30. $-3x < 30 + 2x$

31. $\dfrac{1}{x} + \dfrac{2}{3} = \dfrac{1}{4}$

32. $x^2 - 30 = x$

33. $-4(x + 5) \geq 2(x + 5) - 3$

34. $2x^2 = 162$

35. $\sqrt{2x - 1} + 5 = 14$

36. $\sqrt{4x} + 1 = \sqrt{x} + 4$

37. $\dfrac{1}{4}x + \dfrac{2}{3}x = \dfrac{2}{3} - \dfrac{3}{4}x$

38. $\dfrac{x}{x - 1} - \dfrac{x}{x + 1} = \dfrac{1}{2x - 2}$

39. $x = y + 3,$
$3y - 4x = -13$

40. $2x - 3y = 30,$
$5y - 2x = -46$

41. $\dfrac{E}{r} = \dfrac{R + r}{R}$, for R

42. $4A = pr + pq$, for p

Graph on a plane.

43. $3y - 3x > -6$

44. $x = 5$

45. $2x - 6y = 12$

46. Solve by graphing:

$$x + 2y = 9,$$
$$2x + y = 6.$$

47. Find an equation of the line containing the points $(1, -2)$ and $(5, 9)$.

48. Find the slope and the y-intercept of the line $5x - 3y = 9$.

Determine whether the lines are parallel, perpendicular, or neither.

49. $2x - 3y = 5,$
$3x + 2y = 5$

50. $x - y = 3,$
$2y - 5 = 2x$

Solve.

51. The second angle of a triangle is twice as large as the first. The third angle is 48° less than the sum of the other two angles. Find the measures of the angles.

52. The cost of 6 hamburgers and 4 milkshakes is $11.40. Three hamburgers and 1 milkshake cost $4.80. Find the cost of a hamburger and the cost of a milkshake.

53. An 8-m ladder is leaning against a building. The bottom of the ladder is 4 m from the building. How high is the top of the ladder?

54. The amount C that a family spends on housing varies directly as its income I. A family making $25,000 a year will spend $6250 a year for housing. How much will a family making $30,000 a year spend for housing?

55. A sample of 150 resistors contained 12 defective resistors. How many defective resistors would you expect among 250 resistors?

56. The length of a rectangle is 3 m greater than the width. The area of the rectangle is 180 m². Find the length and the width.

57. A collection of dimes and quarters is worth $19.00. There are 115 coins in all. How many of each are there?

58. The winner of an election won by a margin of 2 to 1, with 238 votes. How many voted in the election?

59. Money is invested in an account at 10.5% simple interest. At the end of one year, there is $2873 in the account. How much was originally invested?

60. A person traveled 600 mi in one direction. The return trip took 2 hr longer at a speed that was 10 mph less. Find the speed going.

SYNTHESIS

Write a true sentence using $<$ or $>$.

61. $-4 \quad -3$

62. $|-4| \quad |-3|$

63. Find the meaningful replacements: $\dfrac{x}{\sqrt{x - 5}}$.

64. A tank contains 200 L of a 30%-salt solution. How much pure water should be added to make a solution that is 12% salt?

65. Solve: $\sqrt{x} + 1 = y,$
$\sqrt{x} + \sqrt{y} = 5.$

INTRODUCTION   A *quadratic equation* contains a polynomial of second degree. In this chapter we first learn to solve quadratic equations by factoring. Because certain quadratic equations are difficult to solve by factoring, we learn to use the *quadratic formula*, which is a "recipe" for finding solutions of quadratic equations.

We apply our skills for solving quadratic equations to problem solving. Then we graph quadratic equations.

The review sections to be tested in addition to the material in this chapter are 6.6, 8.2, 8.4, and 8.6. ❖

Quadratic Equations

9

AN APPLICATION

The Sears Tower in Chicago is 1451 ft tall. How long would it take an object to fall from the top to the ground?

THE MATHEMATICS

Let $t =$ the time required for the object to fall from the top to the ground. We determine t by solving the equation

$$1451 = 16t^2.$$

This is a *quadratic equation*.

Every positive real number has two square roots.
Factoring Skills: Chapter 4
Skills at Manipulating Square-Root Symbolism: Chapter 8
Pythagorean Theorem: $a^2 + b^2 = c^2$
Motion Formula: $d = rt$

PRETEST: CHAPTER 9

Solve.

1. $x^2 + 9 = 6x$

2. $x^2 - 7 = 0$

3. $3x^2 + 3x - 1 = 0$

4. $5y^2 - 3y = 0$

5. $\dfrac{3}{3x + 2} - \dfrac{2}{3x + 4} = 1$

6. $(x + 4)^2 = 5$

7. Solve $x^2 - 2x - 5 = 0$ by completing the square. Show your work.

8. Solve for n: $A = n^2 - pn$.

9. The length of a rectangle is three times the width. The area is 48 cm². Find the length and the width.

10. Find the x-intercepts: $y = 2x^2 + x - 4$.

11. The current in a stream moves at a speed of 2 km/h. A boat travels 24 km upstream and 24 km downstream in a total time of 5 hr. What is the speed of the boat in still water?

12. Graph: $y = 4 - x^2$.

9.1 Introduction to Quadratic Equations

a Standard Form

The following are **quadratic equations.** They contain polynomials of second degree.

$$x^2 + 7x - 5 = 0, \qquad 3t^2 - \tfrac{1}{2}t = 9, \qquad 5y^2 = -6y, \qquad 5m^2 = 15.$$

The quadratic equation

$$4x^2 + 7x - 5 = 0$$

is said to be in **standard form.** The quadratic equation

$$4x^2 = 5 - 7x$$

is equivalent to the preceding equation, but it is *not* in standard form.

> A quadratic equation of the type $ax^2 + bx + c = 0$, where a, b, and c are real-number constants and $a > 0$, is called the *standard form of a quadratic equation.*

Often a quadratic equation is defined so that $a \neq 0$. We use $a > 0$ to ease the proof of the quadratic formula, which we consider later, and to ease solving by factoring, which we review in this section. Suppose we are studying an equation like $-3x^2 + 8x - 2 = 0$. It is not in standard form. We can find an equivalent equation that is in standard form by multiplying on both sides by -1:

$$-1(-3x^2 + 8x - 2) = -1(0)$$
$$3x^2 - 8x + 2 = 0.$$

▶ **EXAMPLES**   Write in standard form and determine a, b, and c.

1. $4x^2 + 7x - 5 = 0$ The equation is already in standard form.

$a = 4; \quad b = 7; \quad c = -5$

2. $3x^2 - 0.5x = 9$

$3x^2 - 0.5x - 9 = 0$ Subtracting 9. This is standard form.

$a = 3; \quad b = -0.5; \quad c = -9$

3. $-4y^2 = 5y$

$-4y^2 - 5y = 0$ Subtracting $5y$

Not positive!

$4y^2 + 5y = 0$ Multiplying by -1. This is standard form.

$a = 4; \quad b = 5; \quad c = 0$ ◀

DO EXERCISES 1–3.

OBJECTIVES

After finishing Section 9.1, you should be able to:

a Write a quadratic equation in standard form $ax^2 + bx + c = 0$, $a > 0$, and determine the coefficients a, b, and c.

b Solve quadratic equations of the type $ax^2 + bx = 0$, where $a \neq 0$ and $b \neq 0$, by factoring.

c Solve quadratic equations of the type $ax^2 + bx + c = 0$, where $a \neq 0$, $b \neq 0$, and $c \neq 0$, by factoring.

d Solve problems involving quadratic equations.

FOR EXTRA HELP

Tape 16B   Tape 17A   MAC: 9
                                IBM: 9

Write in standard form and determine a, b, and c.

1. $x^2 = 7x$

2. $3 - x^2 = 9x$

3. $3x + 5x^2 = x^2 - 4 + x$

ANSWERS ON PAGE A-8

Solve.

4. $3x^2 + 5x = 0$

b   Solving Quadratic Equations of the Type $ax^2 + bx = 0$

Sometimes we can use factoring and the principle of zero products to solve quadratic equations. We are actually reviewing methods that we introduced in Section 4.7.

When c is 0 and $b \neq 0$, we can always factor and use the principle of zero products.

▶ **EXAMPLE 4**   Solve: $7x^2 + 2x = 0$.

$$7x^2 + 2x = 0$$
$$x(7x + 2) = 0 \qquad \text{Factoring}$$

$x = 0 \quad$ or $\quad 7x + 2 = 0 \qquad$ Using the principle of zero products

$x = 0 \quad$ or $\qquad 7x = -2$

$x = 0 \quad$ or $\qquad x = -\frac{2}{7}$

Check: For 0:

$$\frac{7x^2 + 2x = 0}{7 \cdot 0^2 + 2 \cdot 0 \;\big|\; 0}$$
$$0 \;\big|\; \text{TRUE}$$

For $-\frac{2}{7}$:

$$\frac{7x^2 + 2x = 0}{7(-\frac{2}{7})^2 + 2(-\frac{2}{7}) \;\big|\; 0}$$
$$7(\frac{4}{49}) - \frac{4}{7}$$
$$\frac{4}{7} - \frac{4}{7}$$
$$0 \;\big|\; \text{TRUE}$$

The solutions are 0 and $-\frac{2}{7}$. ◀

You may be tempted to divide each term in an equation like the one in Example 4 by x. This method would yield the equation

$$7x + 2 = 0,$$

whose only solution is $-\frac{2}{7}$. In effect, since 0 is also a solution of the original equation, we have divided by 0. The error of such division causes the loss of one of the solutions.

5. $10x^2 - 6x = 0$

▶ **EXAMPLE 5**   Solve: $20x^2 - 15x = 0$.

$$20x^2 - 15x = 0$$
$$5x(4x - 3) = 0 \qquad \text{Factoring}$$

$5x = 0 \quad$ or $\quad 4x - 3 = 0 \qquad$ Using the principle of zero products

$x = 0 \quad$ or $\qquad 4x = 3$

$x = 0 \quad$ or $\qquad x = \frac{3}{4}$

The solutions are 0 and $\frac{3}{4}$. ◀

A quadratic equation of the type $ax^2 + bx = 0$, where $a \neq 0$ and $b \neq 0$, will always have 0 as one solution and a nonzero number as the other solution.

DO EXERCISES 4 AND 5.

c **Solving Quadratic Equations of the Type** $ax^2 + bx + c = 0$

When neither b nor c is 0, we can sometimes solve by factoring.

▶ **EXAMPLE 6** Solve: $5x^2 - 8x + 3 = 0$.

$$5x^2 - 8x + 3 = 0$$
$$(5x - 3)(x - 1) = 0 \qquad \text{Factoring}$$
$$5x - 3 = 0 \quad \text{or} \quad x - 1 = 0 \qquad \text{Using the principle of zero products}$$
$$5x = 3 \quad \text{or} \qquad x = 1$$
$$x = \tfrac{3}{5} \quad \text{or} \qquad x = 1$$

The solutions are $\tfrac{3}{5}$ and 1. ◀

▶ **EXAMPLE 7** Solve: $(y - 3)(y - 2) = 6(y - 3)$.

We write the equation in standard form and then try to factor:

$$y^2 - 5y + 6 = 6y - 18 \qquad \text{Multiplying}$$
$$y^2 - 11y + 24 = 0 \qquad \text{Standard form}$$
$$(y - 8)(y - 3) = 0$$
$$y - 8 = 0 \quad \text{or} \quad y - 3 = 0$$
$$y = 8 \quad \text{or} \qquad y = 3.$$

The solutions are 8 and 3. ◀

DO EXERCISES 6 AND 7.

Recall that to solve a rational equation, we multiply on both sides by the LCM of all the denominators. We may obtain a quadratic equation after a few steps. When that happens, we know how to finish solving, but we must remember to check possible solutions because a replacement may result in division by 0.

▶ **EXAMPLE 8** Solve: $\dfrac{3}{x - 1} + \dfrac{5}{x + 1} = 2$.

We multiply by the LCM, which is $(x - 1)(x + 1)$:

$$(x - 1)(x + 1) \cdot \left(\frac{3}{x - 1} + \frac{5}{x + 1} \right) = 2 \cdot (x - 1)(x + 1).$$

We use the distributive law on the left:

$$(x - 1)(x + 1) \cdot \frac{3}{x - 1} + (x - 1)(x + 1) \cdot \frac{5}{x + 1} = 2(x - 1)(x + 1)$$
$$3(x + 1) + 5(x - 1) = 2(x - 1)(x + 1)$$
$$3x + 3 + 5x - 5 = 2(x^2 - 1)$$
$$8x - 2 = 2x^2 - 2$$
$$0 = 2x^2 - 8x$$
$$0 = 2x(x - 4) \qquad \text{Factoring}$$
$$2x = 0 \quad \text{or} \quad x - 4 = 0$$
$$x = 0 \quad \text{or} \qquad x = 4.$$

The check is left to the student. Since both numbers check, the solutions are 0 and 4. ◀

Solve.

6. $3x^2 + x - 2 = 0$

7. $(x - 1)(x + 1) = 5(x - 1)$

ANSWERS ON PAGE A-8

8. Solve:

$$\frac{20}{x+5} - \frac{1}{x-4} = 1.$$

9. Use $d = \dfrac{n^2 - 3n}{2}$.

a) A heptagon has 7 sides. How many diagonals does it have?

b) A polygon has 44 diagonals. How many sides does it have?

DO EXERCISE 8.

d   **Solving Problems**

▶ **EXAMPLE 9**   The number of diagonals d of a polygon of n sides is given by the formula

$$d = \frac{n^2 - 3n}{2}.$$

If a polygon has 27 diagonals, how many sides does it have?

1. *Familiarize.*   We can make a drawing to familiarize ourselves with the problem. We draw an octagon (8 sides) and count the diagonals and see that there are 20. Let us check this in the formula. We evaluate the formula for $n = 8$:

$$d = \frac{8^2 - 3(8)}{2} = \frac{64 - 24}{2} = \frac{40}{2} = 20.$$

2. *Translate.*   We know that the number of diagonals is 27. We substitute 27 for d:

$$27 = \frac{n^2 - 3n}{2}.$$

This gives us a translation.

3. *Solve.*   We solve the equation for n, reversing the equation first for convenience:

$$\frac{n^2 - 3n}{2} = 27$$

$$n^2 - 3n = 54 \qquad \text{Multiplying by 2 to clear of fractions}$$

$$n^2 - 3n - 54 = 0$$

$$(n - 9)(n + 6) = 0$$

$$n - 9 = 0 \quad \text{or} \quad n + 6 = 0$$

$$n = 9 \quad \text{or} \qquad n = -6.$$

4. *Check.*   Since the number of sides cannot be negative, -6 cannot be a solution. We leave it to the student to show that 9 checks by substitution.

5. *State.*   The polygon has 9 sides (it is a nonagon).   ◀

DO EXERCISE 9.

NAME   SECTION   DATE

EXERCISE SET 9.1

a Write standard form and determine a, b, and c.

1. $x^2 - 3x + 2 = 0$

2. $x^2 - 8x - 5 = 0$

3. $7x^2 = 4x - 3$

4. $9x^2 = x + 5$

5. $5 = -2x^2 + 3x$

6. $2x - 1 = 3x^2 + 7$

b Solve.

7. $x^2 + 7x = 0$

8. $x^2 + 5x = 0$

9. $3x^2 + 6x = 0$

10. $4x^2 + 8x = 0$

11. $5x^2 = 2x$

12. $7x = 3x^2$

13. $4x^2 + 4x = 0$

14. $2x^2 - 2x = 0$

15. $0 = 10x^2 - 30x$

16. $0 = 10x^2 - 50x$

17. $11x = 55x^2$

18. $33x^2 = -11x$

19. $14t^2 = 3t$

20. $8m = 17m^2$

21. $5y^2 - 3y^2 = 72y + 9y$

22. $63p - 16p^2 = 17p + 58p^2$

c Solve.

23. $x^2 + 8x - 48 = 0$

24. $x^2 - 16x + 48 = 0$

25. $5 + 6x + x^2 = 0$

26. $x^2 + 6 + 7x = 0$

27. $18 = 7p + p^2$

28. $t^2 + 4t = 21$

29. $-15 = -8y + y^2$

30. $m^2 + 14 = 9m$

31. $x^2 + 6x + 9 = 0$

32. $x^2 + 10x + 25 = 0$

33. $r^2 = 8r - 16$

34. $x^2 + 1 = 2x$

35. $6x^2 + x - 2 = 0$

36. $2x^2 - 13x + 15 = 0$

37. $15b - 9b^2 = 4$

38. $3a^2 = 10a + 8$

ANSWERS

1. _____
2. _____
3. _____
4. _____
5. _____
6. _____
7. _____
8. _____
9. _____
10. _____
11. _____
12. _____
13. _____
14. _____
15. _____
16. _____
17. _____
18. _____
19. _____
20. _____
21. _____
22. _____
23. _____
24. _____
25. _____
26. _____
27. _____
28. _____
29. _____
30. _____
31. _____
32. _____
33. _____
34. _____
35. _____
36. _____
37. _____
38. _____

ANSWERS
39. _____
40. _____
41. _____
42. _____
43. _____
44. _____
45. _____
46. _____
47. _____
48. _____
49. _____
50. _____
51. _____
52. _____
53. _____
54. _____
55. _____
56. _____
57. _____
58. _____
59. _____
60. _____
61. _____
62. _____
63. _____
64. _____
65. _____
66. _____
67. _____
68. _____
69. _____
70. _____
71. _____
72. _____
73. _____
74. _____
75. _____
76. _____
77. _____
78. _____

39. $6x^2 - 4x = 10$

40. $3x^2 - 7x = 20$

41. $12w^2 - 5w = 2$

42. $2t^2 + 12t = -10$

43. $t(t - 5) = 14$

44. $6z^2 + z - 1 = 0$

45. $t(9 + t) = 4(2t + 5)$

46. $3y^2 + 8y = 12y + 15$

47. $16(p - 1) = p(p + 8)$

48. $(2x - 3)(x + 1) = 4(2x - 3)$

49. $(x - 2)(x + 2) = x + 2$

50. $(t - 1)(t + 3) = t - 1$

Solve.

51. $\dfrac{8}{x + 2} + \dfrac{8}{x - 2} = 3$

52. $\dfrac{24}{x - 2} + \dfrac{24}{x + 2} = 5$

53. $\dfrac{1}{x} + \dfrac{1}{x + 6} = \dfrac{1}{4}$

54. $\dfrac{1}{x} + \dfrac{1}{x + 9} = \dfrac{1}{20}$

55. $1 + \dfrac{12}{x^2 - 4} = \dfrac{3}{x - 2}$

56. $\dfrac{5}{t - 3} - \dfrac{30}{t^2 - 9} = 1$

57. $\dfrac{r}{r - 1} + \dfrac{2}{r^2 - 1} = \dfrac{8}{r + 1}$

58. $\dfrac{x + 2}{x^2 - 2} = \dfrac{2}{3 - x}$

59. $\dfrac{4 - x}{x - 4} + \dfrac{x + 3}{x - 3} = 0$

60. $\dfrac{x - 1}{1 - x} = -\dfrac{x + 8}{x - 8}$

d Solve.

61. A hexagon is a figure with 6 sides. How many diagonals does a hexagon have?

62. A decagon is a figure with 10 sides. How many diagonals does a decagon have?

63. A polygon has 14 diagonals. How many sides does it have?

64. A polygon has 9 diagonals. How many sides does it have?

SKILL MAINTENANCE

Simplify.

65. $\sqrt{20}$

66. $\sqrt{\dfrac{2890}{2560}}$

67. $\sqrt{\dfrac{3240}{2560}}$

68. $\sqrt{88}$

SYNTHESIS

Solve.

69. $4m^2 - (m + 1)^2 = 0$

70. $x^2 + \sqrt{3}\,x = 0$

71. $\sqrt{5}\,x^2 - x = 0$

72. $\sqrt{7}\,x^2 + \sqrt{3}\,x = 0$

73. $\dfrac{5}{y + 4} - \dfrac{3}{y - 2} = 4$

74. $\dfrac{2z + 11}{2z + 8} = \dfrac{3z - 1}{z - 1}$

75. Solve for x: $ax^2 + bx = 0$.

76. ▤ Solve: $0.0025x^2 + 70{,}400x = 0$.

Solve.

77. $z - 10\sqrt{z} + 9 = 0$ (Let $x = \sqrt{z}$.)

78. $(x - 2)^2 + 3(x - 2) = 4$

9.2 Solving Quadratic Equations by Completing the Square

a Solving Quadratic Equations of the Type $ax^2 = p$

For equations of the type $ax^2 = p$, we solve for x^2 and apply the *principle of square roots*, which states that a positive number has two square roots. The number 0 has one square root, 0.

> **The Principle of Square Roots**
>
> The equation $x^2 = k$ has two real solutions when $k > 0$. The solutions are $\sqrt{k}$ and $-\sqrt{k}$.
>
> The equation $x^2 = 0$ has 0 as its only solution.
>
> The equation $x^2 = k$ has no real-number solution when $k < 0$.

► **EXAMPLE 1** Solve: $x^2 = 3$.

$$x^2 = 3$$
$$x = \sqrt{3} \quad \text{or} \quad x = -\sqrt{3}. \qquad \text{Using the principle of square roots}$$

Check: For $\sqrt{3}$:   For $-\sqrt{3}$:

$$\begin{array}{c|c} x^2 = 3 & \\ \hline (\sqrt{3})^2 & 3 \\ 3 & \text{TRUE} \end{array} \qquad \begin{array}{c|c} x^2 = 3 & \\ \hline (-\sqrt{3})^2 & 3 \\ 3 & \text{TRUE} \end{array}$$

The solutions are $\sqrt{3}$ and $-\sqrt{3}$. ◄

DO EXERCISE 1.

► **EXAMPLE 2** Solve: $\frac{1}{3}x^2 = 0$.

$$\frac{1}{3}x^2 = 0$$
$$x^2 = 0 \qquad \text{Multiplying by 3}$$
$$x = 0 \qquad \text{Using the principle of square roots}$$

The solution is 0. ◄

DO EXERCISE 2.

► **EXAMPLE 3** Solve: $-3x^2 + 7 = 0$.

$$-3x^2 + 7 = 0$$
$$-3x^2 = -7 \qquad \text{Subtracting 7}$$
$$x^2 = \frac{-7}{-3} \qquad \text{Dividing by } -3$$
$$x^2 = \frac{7}{3}$$
$$x = \sqrt{\frac{7}{3}} \quad \text{or} \quad x = -\sqrt{\frac{7}{3}} \qquad \text{Using the principle of square roots}$$
$$x = \sqrt{\frac{7}{3} \cdot \frac{3}{3}} \quad \text{or} \quad x = -\sqrt{\frac{7}{3} \cdot \frac{3}{3}} \qquad \text{Rationalizing the denominators}$$
$$x = \frac{\sqrt{21}}{3} \quad \text{or} \quad x = -\frac{\sqrt{21}}{3}$$

OBJECTIVES

After finishing Section 9.2, you should be able to:

a Solve equations of the type $ax^2 = p$.

b Solve equations of the type $(x + k)^2 = p$.

c Solve quadratic equations by completing the square.

d Solve certain problems involving quadratic equations of the type $ax^2 = p$.

FOR EXTRA HELP

Tape NC   Tape 17A   MAC: 9
                             IBM: 9

1. Solve: $x^2 = 5$.

2. Solve: $2x^2 = 0$.

ANSWERS ON PAGE A-8

3. Solve: $2x^2 - 3 = 0$.

Solve.

4. $(x - 3)^2 = 16$

5. $(x + 3)^2 = 10$

6. $(x - 1)^2 = 5$

7. Solve: $x^2 - 6x + 9 = 64$.

Check: For $\dfrac{\sqrt{21}}{3}$:   For $-\dfrac{\sqrt{21}}{3}$:

$$
\begin{array}{r|l}
-3x^2 + 7 = 0 & \\
\hline
-3\left(\dfrac{\sqrt{21}}{3}\right)^2 + 7 & 0 \\
-3 \cdot \dfrac{21}{9} + 7 & \\
-7 + 7 & \\
0 & \text{TRUE}
\end{array}
\qquad
\begin{array}{r|l}
-3x^2 + 7 = 0 & \\
\hline
-3\left(-\dfrac{\sqrt{21}}{3}\right)^2 + 7 & 0 \\
-3 \cdot \dfrac{21}{9} + 7 & \\
-7 + 7 & \\
0 & \text{TRUE}
\end{array}
$$

The solutions are $\dfrac{\sqrt{21}}{3}$ and $-\dfrac{\sqrt{21}}{3}$.   ◄

DO EXERCISE 3.

b   **Solving Quadratic Equations of the Type $(x + k)^2 = p$**

The equation $(x - 5)^2 = 9$ can be solved by using the principle of square roots. We will see that other equations can be made to look like this one.

In an equation of the type $(x + k)^2 = p$, we have the square of a binomial equal to a constant. We can use the principle of square roots to solve such an equation.

►   **EXAMPLE 4**   Solve: $(x - 5)^2 = 9$.

$$(x - 5)^2 = 9$$
$$x - 5 = 3 \quad \text{or} \quad x - 5 = -3 \qquad \text{Using the principle of square roots}$$
$$x = 8 \quad \text{or} \qquad\quad x = 2$$

The solutions are 8 and 2.   ◄

►   **EXAMPLE 5**   Solve: $(x + 2)^2 = 7$.

$$(x + 2)^2 = 7$$
$$x + 2 = \sqrt{7} \qquad \text{or} \quad x + 2 = -\sqrt{7} \qquad \text{Using the principle of square roots}$$
$$x = -2 + \sqrt{7} \quad \text{or} \qquad\quad x = -2 - \sqrt{7}$$

The solutions are $-2 + \sqrt{7}$ and $-2 - \sqrt{7}$, or simply $-2 \pm \sqrt{7}$ (read "-2 plus or minus $\sqrt{7}$").   ◄

DO EXERCISES 4–6.

In Examples 4 and 5, the left sides of the equations are squares of binomials. If we can express an equation in such a form, we can proceed as we did in those examples.

►   **EXAMPLE 6**   Solve: $x^2 + 8x + 16 = 49$.

$$x^2 + 8x + 16 = 49 \qquad \text{The left side is the square of a binomial.}$$
$$(x + 4)^2 = 49$$
$$x + 4 = 7 \quad \text{or} \quad x + 4 = -7 \qquad \text{Using the principle of square roots}$$
$$x = 3 \quad \text{or} \qquad\quad x = -11$$

The solutions are 3 and -11.   ◄

DO EXERCISE 7.

c Completing the Square

We have seen that a quadratic equation like $(x - 5)^2 = 9$ can be solved by using the principle of square roots. We also noted that an equation like $x^2 + 8x + 16 = 49$ can be solved in the same manner because the expression on the left side is the square of a binomial, $(x + 4)^2$. This second procedure is the basis for a method of solving quadratic equations called **completing the square.** It can be used to solve any quadratic equation.

Suppose we have the following quadratic equation:

$$x^2 + 10x = 4.$$

If we could add to both sides of the equation a constant that would make the expression on the left the square of a binomial, we could then solve the equation using the principle of square roots.

How can we determine what to add to $x^2 + 10x$ in order to construct the square of a binomial? We want to find a number a such that the following equation is satisfied:

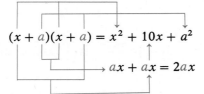

$$(x + a)(x + a) = x^2 + 10x + a^2$$
$$ax + ax = 2ax$$

Thus, a is such that $2ax = 10x$. Solving for a, we get

$$a = \frac{10x}{2x} = \frac{10}{2} = 5;$$

that is, a is half of the coefficient of x in $x^2 + 10x$. Since $a^2 = (\frac{10}{2})^2 = 5^2 = 25$, we add 25 to our original expression:

$$x^2 + 10x + 25 \text{ is the square of } x + 5;$$

that is, $$x^2 + 10x + 25 = (x + 5)^2.$$

> To *complete the square* of an expression like $x^2 + bx$, we take half the coefficient of x and square. Then we add that number, which is $(b/2)^2$.

Returning to solve our original equation, we first add 25 on both sides to complete the square. Then we solve as follows:

$$x^2 + 10x \qquad = 4$$
$$x^2 + 10x + 25 = 4 + 25 \qquad \text{Adding 25: } (\tfrac{10}{2})^2 = 5^2 = 25$$
$$(x + 5)^2 = 29$$
$$x + 5 = \sqrt{29} \qquad \text{or} \quad x + 5 = -\sqrt{29} \qquad \text{Using the principle of square roots}$$
$$x = -5 + \sqrt{29} \quad \text{or} \qquad x = -5 - \sqrt{29}.$$

The solutions are $-5 \pm \sqrt{29}$.

We have seen that a quadratic equation $(x + k)^2 = p$ can be solved by using the principle of square roots. Any equation can be put in this form by completing the square. Then we can solve as before.

Solve.

8. $x^2 - 6x + 8 = 0$

9. $x^2 + 8x - 20 = 0$

10. Solve: $x^2 + 6x - 1 = 0$.

11. Solve: $x^2 - 3x - 10 = 0$.

▶ **EXAMPLE 7**   Solve: $x^2 + 6x + 8 = 0$.

We have

$$x^2 + 6x + 8 = 0$$
$$x^2 + 6x \qquad = -8. \qquad \text{Subtracting 8}$$

We take half of 6 and square it, to get 9. Then we add 9 on *both* sides of the equation. This makes the left side the square of a binomial. We have now completed the square.

$$x^2 + 6x + 9 = -8 + 9 \qquad \text{Adding 9}$$
$$(x + 3)^2 = 1$$
$$x + 3 = 1 \qquad \text{or} \quad x + 3 = -1 \qquad \text{Using the principle of square roots}$$
$$x = -2 \quad \text{or} \qquad x = -4$$

The solutions are -2 and -4.   ◀

This method of solving is called *completing the square*.

DO EXERCISES 8 AND 9.

▶ **EXAMPLE 8**   Solve $x^2 - 4x - 7 = 0$ by completing the square.

$$x^2 - 4x - 7 = 0$$
$$x^2 - 4x \qquad = 7 \qquad \text{Adding 7}$$
$$x^2 - 4x + 4 = 7 + 4 \qquad \text{Adding 4: } (\tfrac{-4}{2})^2 = (-2)^2 = 4$$
$$(x - 2)^2 = 11$$
$$x - 2 = \sqrt{11} \qquad \text{or} \quad x - 2 = -\sqrt{11} \qquad \text{Using the principle of square roots}$$
$$x = 2 + \sqrt{11} \quad \text{or} \qquad x = 2 - \sqrt{11}$$

The solutions are $2 \pm \sqrt{11}$.   ◀

DO EXERCISE 10.

Example 7, as well as the following example, can be solved more easily by factoring. We solved it by completing the square only to illustrate that completing the square can be used to solve any quadratic equation.

▶ **EXAMPLE 9**   Solve $x^2 + 3x - 10 = 0$ by completing the square.

We have

$$x^2 + 3x - 10 = 0$$
$$x^2 + 3x \qquad = 10$$
$$x^2 + 3x + \tfrac{9}{4} = 10 + \tfrac{9}{4} \qquad \text{Adding } \tfrac{9}{4}: (\tfrac{3}{2})^2 = \tfrac{9}{4}$$
$$(x + \tfrac{3}{2})^2 = \tfrac{40}{4} + \tfrac{9}{4} = \tfrac{49}{4}$$
$$x + \tfrac{3}{2} = \tfrac{7}{2} \quad \text{or} \quad x + \tfrac{3}{2} = -\tfrac{7}{2} \qquad \text{Using the principle of square roots}$$
$$x = \tfrac{4}{2} \quad \text{or} \qquad x = -\tfrac{10}{2}$$
$$x = 2 \quad \text{or} \qquad x = -5.$$

The solutions are 2 and -5.   ◀

DO EXERCISE 11.

ANSWERS ON PAGE A-8

When the coefficient of x^2 is not 1, we can make it 1, as shown in the following example.

▶ **EXAMPLE 10** Solve $2x^2 = 3x + 1$ by completing the square.

We first obtain standard form. Then we multiply on both sides by $\frac{1}{2}$ to make the x^2-coefficient 1.

$$2x^2 = 3x + 1$$

$$2x^2 - 3x - 1 = 0 \qquad \text{Finding standard form}$$

$$\frac{1}{2}(2x^2 - 3x - 1) = \frac{1}{2} \cdot 0 \qquad \text{Multiplying by } \frac{1}{2} \text{ to make the } x^2\text{-coefficient 1}$$

$$x^2 - \frac{3}{2}x - \frac{1}{2} = 0$$

$$x^2 - \frac{3}{2}x = \frac{1}{2} \qquad \text{Adding } \frac{1}{2}$$

$$x^2 - \frac{3}{2}x + \frac{9}{16} = \frac{1}{2} + \frac{9}{16} \qquad \text{Adding } \frac{9}{16}: \left[\frac{1}{2}\left(-\frac{3}{2}\right)\right]^2 = \left[-\frac{3}{4}\right]^2 = \frac{9}{16}$$

$$\left(x - \frac{3}{4}\right)^2 = \frac{8}{16} + \frac{9}{16} \qquad \text{Finding a common denominator}$$

$$\left(x - \frac{3}{4}\right)^2 = \frac{17}{16}$$

$$x - \frac{3}{4} = \frac{\sqrt{17}}{4} \qquad \text{or} \quad x - \frac{3}{4} = -\frac{\sqrt{17}}{4} \qquad \text{Using the principle of square roots}$$

$$x = \frac{3}{4} + \frac{\sqrt{17}}{4} \quad \text{or} \qquad x = \frac{3}{4} - \frac{\sqrt{17}}{4}$$

The solutions are $\dfrac{3 \pm \sqrt{17}}{4}$. ◀

Solving by Completing the Square

To solve a quadratic equation $ax^2 + bx + c = 0$ by completing the square:

1. **If $a \neq 1$, multiply by $1/a$ so that the x^2-coefficient is 1.**

2. **If the x^2-coefficient is 1, add so that the equation is in the form**

   $$x^2 + bx = -c, \quad \text{or } x^2 + \frac{b}{a}x = -\frac{c}{a} \text{ if step (1) has been applied.}$$

3. **Take half of the x-coefficient and square it. Add the result on both sides of the equation.**

4. **Express the side with the variables as the square of a binomial.**

5. **Use the principle of square roots and complete the solution.**

DO EXERCISE 12.

12. Solve: $2x^2 + 3x - 3 = 0$.

13. The Texas Building in Houston is 1002 ft tall. How long would it take an object to fall to the ground from the top?

ANSWER ON PAGE A-8

---

d   **Applications**

▶ **EXAMPLE 11**   The Sears Tower in Chicago is 1451 ft tall. How long would it take an object to fall to the ground from the top?

1. *Familiarize.* If we did not know anything about this problem, we might consider looking up a formula in a mathematics or physics book. A formula that fits this situation is

$$s = 16t^2,$$

where s is the distance, in feet, traveled by a body falling freely from rest in t seconds. This formula is actually an approximation in that it does not account for air resistance. In this problem, we know the distance s to be 1451. We want to determine the time t for the object to reach the ground.

$s = 16t^2$

2. *Translate.* We know that the distance is 1451 and that we need to solve for t. We substitute 1451 for s:

$$1451 = 16t^2.$$

This gives us a translation.

3. *Solve.* We solve the equation:

$$1451 = 16t^2$$

$$\frac{1451}{16} = t^2 \qquad \text{Solving for } t^2$$

$$90.6875 = t^2 \qquad \text{Dividing}$$

$$\sqrt{90.6875} = t \quad \text{or} \quad -\sqrt{90.6875} = t \qquad \text{Using the principle of square roots}$$

$$9.5 \approx t \quad \text{or} \qquad -9.5 \approx t. \qquad \text{Using a calculator to find the square root and rounding to the nearest tenth}$$

4. *Check.* The number -9.5 cannot be a solution because time cannot be negative in this situation. We substitute 9.5 in the original equation:

$$s = 16(9.5)^2 = 16(90.25) = 1444.$$

This is close. Remember that we approximated a solution. Thus we have a check.

5. *State.* It takes about 9.5 sec for the object to fall to the ground from the top of the Sears Tower.   ◄

DO EXERCISE 13.

EXERCISE SET 9.2

a Solve.

1. $x^2 = 121$ **2.** $x^2 = 10$ **3.** $5x^2 = 35$ **4.** $3x^2 = 30$

5. $5x^2 = 3$ **6.** $2x^2 = 5$ **7.** $4x^2 - 25 = 0$ **8.** $9x^2 - 4 = 0$

9. $3x^2 - 49 = 0$ **10.** $5x^2 - 16 = 0$ **11.** $4y^2 - 3 = 9$ **12.** $49y^2 - 16 = 0$

13. $25y^2 - 36 = 0$ **14.** $5x^2 - 100 = 0$

b Solve.

15. $(x - 2)^2 = 49$ **16.** $(x + 1)^2 = 6$ **17.** $(x + 3)^2 = 21$

18. $(x - 3)^2 = 6$ **19.** $(x + 13)^2 = 8$ **20.** $(x - 13)^2 = 64$

21. $(x - 7)^2 = 12$ **22.** $(x + 1)^2 = 14$ **23.** $(x + 9)^2 = 34$

24. $(t + 2)^2 = 25$ **25.** $(x + \frac{3}{2})^2 = \frac{7}{2}$ **26.** $(y - \frac{3}{4})^2 = \frac{17}{16}$

27. $x^2 - 6x + 9 = 64$ **28.** $x^2 - 10x + 25 = 100$

29. $y^2 + 14y + 49 = 4$ **30.** $p^2 + 8p + 16 = 1$

c Solve by completing the square. Show your work.

31. $x^2 - 6x - 16 = 0$ **32.** $x^2 + 8x + 15 = 0$ **33.** $x^2 + 22x + 21 = 0$

34. $x^2 + 14x - 15 = 0$ **35.** $x^2 - 2x - 5 = 0$ **36.** $x^2 - 4x - 11 = 0$

ANSWERS

1. _____
2. _____
3. _____
4. _____
5. _____
6. _____
7. _____
8. _____
9. _____
10. _____
11. _____
12. _____
13. _____
14. _____
15. _____
16. _____
17. _____
18. _____
19. _____
20. _____
21. _____
22. _____
23. _____
24. _____
25. _____
26. _____
27. _____
28. _____
29. _____
30. _____
31. _____
32. _____
33. _____
34. _____
35. _____
36. _____

Copyright © 1991 Addison-Wesley Publishing Co., Inc.

ANSWERS

37. _____

38. _____

39. _____

40. _____

41. _____

42. _____

43. _____

44. _____

45. _____

46. _____

47. _____

48. _____

49. _____

50. _____

51. _____

52. _____

53. _____

54. _____

55. _____

56. _____

57. _____

58. _____

59. _____

60. _____

61. _____

62. _____

63. _____

64. _____

65. _____

66. _____

67. _____

68. _____

69. _____

70. _____

71. _____

37. $x^2 - 22x + 102 = 0$ **38.** $x^2 - 18x + 74 = 0$ **39.** $x^2 + 10x - 4 = 0$

40. $x^2 - 10x - 4 = 0$ **41.** $x^2 - 7x - 2 = 0$ **42.** $x^2 + 7x - 2 = 0$

43. $x^2 + 3x - 28 = 0$ **44.** $x^2 - 3x - 28 = 0$ **45.** $x^2 + \frac{3}{2}x - \frac{1}{2} = 0$

46. $x^2 - \frac{3}{2}x - 2 = 0$ **47.** $2x^2 + 3x - 17 = 0$ **48.** $2x^2 - 3x - 1 = 0$

49. $3x^2 + 4x - 1 = 0$ **50.** $3x^2 - 4x - 3 = 0$ **51.** $2x^2 = 9x + 5$

52. $2x^2 = 5x + 12$ **53.** $4x^2 + 12x = 7$ **54.** $6x^2 + 11x = 10$

d   Solve.

55. The height of the World Trade Center in New York is 1377 ft (excluding TV towers and antennas). How long would it take an object to fall to the ground from the top?

56. A body falls 2496 ft. How many seconds does this take?

57. The world record for free-fall by a woman to the ground, without a parachute, into a cushioned landing area is 175 ft and is held by Kitty O'Neill. Approximately how long did the fall take?

58. The world record for free-fall to the ground, without a parachute, by a man is 311 ft and is held by Dar Robinson. Approximately how long did the fall take?

SKILL MAINTENANCE

59. Find an equation of variation where y varies inversely as x, and $y = 235$ when $x = 0.6$.

60. The time T to do a certain job varies inversely as the number N of people working. It takes 5 hr for 24 people to wash and wax the floors in a building. How long would it take 36 people to do the job?

SYNTHESIS

Find b such that the trinomial is a square.

61. $x^2 + bx + 36$ **62.** $x^2 + bx + 55$ **63.** $x^2 + bx + 128$

64. $4x^2 + bx + 16$ **65.** $x^2 + bx + c$ **66.** $ax^2 + bx + c$

Solve.

67. ▦ $4.82x^2 = 12{,}000$ **68.** $\dfrac{x}{4} = \dfrac{9}{x}$ **69.** $1 = \dfrac{1}{3}x^2$

70. $\dfrac{x}{9} = \dfrac{36}{4x}$ **71.** $\dfrac{4}{m^2 - 7} = 1$

9.3 The Quadratic Formula

We learn to complete the square to enhance our ability to graph certain second-degree equations and to prove a general formula that can be used to solve quadratic equations.

a Solving Using the Quadratic Formula

Each time you solve by completing the square, you continually do nearly the same thing. When we repeat the same kind of computation many times, we look for a formula so we can speed up our work. Consider

$$ax^2 + bx + c = 0, \quad a > 0.$$

Let's solve by completing the square. As we carry out the steps, compare them with Example 10 in the preceding section.

$$x^2 + \frac{b}{a}x + \frac{c}{a} = 0 \qquad \text{Multiplying by } \frac{1}{a}$$

$$x^2 + \frac{b}{a}x \qquad = -\frac{c}{a} \qquad \text{Adding } -\frac{c}{a}$$

Half of $\frac{b}{a}$ is $\frac{b}{2a}$. The square is $\frac{b^2}{4a^2}$. Thus we add $\frac{b^2}{4a^2}$ on both sides.

$$x^2 + \frac{b}{a}x + \frac{b^2}{4a^2} = -\frac{c}{a} + \frac{b^2}{4a^2} \qquad \text{Adding } \frac{b^2}{4a^2}$$

$$\left(x + \frac{b}{2a}\right)^2 = -\frac{4ac}{4a^2} + \frac{b^2}{4a^2} \qquad \begin{array}{l}\text{Factoring the left side and finding a}\\\text{common denominator on the right}\end{array}$$

$$\left(x + \frac{b}{2a}\right)^2 = \frac{b^2 - 4ac}{4a^2}$$

$$x + \frac{b}{2a} = \sqrt{\frac{b^2 - 4ac}{4a^2}} \quad \text{or} \quad x + \frac{b}{2a} = -\sqrt{\frac{b^2 - 4ac}{4a^2}} \qquad \begin{array}{l}\text{Using the principle}\\\text{of square roots}\end{array}$$

Since $a > 0$, $\sqrt{4a^2} = 2a$, so we can simplify as follows:

$$x + \frac{b}{2a} = \frac{\sqrt{b^2 - 4ac}}{2a} \quad \text{or} \quad x + \frac{b}{2a} = -\frac{\sqrt{b^2 - 4ac}}{2a}.$$

Thus,

$$x = -\frac{b}{2a} + \frac{\sqrt{b^2 - 4ac}}{2a} \quad \text{or} \quad x = -\frac{b}{2a} + \frac{\sqrt{b^2 - 4ac}}{2a},$$

so

$$x = -\frac{b}{2a} \pm \frac{\sqrt{b^2 - 4ac}}{2a},$$

or

$$x = \frac{-b \pm \sqrt{b^2 - 4ac}}{2a}.$$

We now have the following.

The Quadratic Formula

The solutions of $ax^2 + bx + c = 0$ are given by

$$x = \frac{-b \pm \sqrt{b^2 - 4ac}}{2a}.$$

OBJECTIVES

After finishing Section 9.3, you should be able to:

a Solve quadratic equations using the quadratic formula.

b Find approximate solutions of quadratic equations using a calculator or a square-root table.

FOR EXTRA HELP

Tape 16C   Tape 17B   MAC: 9
                                IBM: 9

1. Solve using the quadratic formula:
$$2x^2 = 4 - 7x.$$

Note that the formula also holds when $a < 0$. A similar proof would show this, but we will not consider it here.

▶ **EXAMPLE 1**   Solve $5x^2 - 8x = -3$ using the quadratic formula.

We first find standard form and determine a, b, and c:

$$5x^2 - 8x + 3 = 0,$$
$$a = 5, \quad b = -8, \quad c = 3,$$

We then use the quadratic formula:

$$x = \frac{-b \pm \sqrt{b^2 - 4ac}}{2a}$$

$$x = \frac{-(-8) \pm \sqrt{(-8)^2 - 4 \cdot 5 \cdot 3}}{2 \cdot 5} \qquad \text{Substituting}$$

> Be sure to write the fraction bar all the way across.

$$x = \frac{8 \pm \sqrt{64 - 60}}{10}$$

$$x = \frac{8 \pm \sqrt{4}}{10}$$

$$x = \frac{8 \pm 2}{10}$$

$$x = \frac{8 + 2}{10} \quad \text{or} \quad x = \frac{8 - 2}{10}$$

$$x = \frac{10}{10} \quad \text{or} \quad x = \frac{6}{10}$$

$$x = 1 \quad \text{or} \quad x = \frac{3}{5}.$$

The solutions are 1 and $\frac{3}{5}$.   ◀

DO EXERCISE 1.

It would have been easier to solve the equation in Example 1 by factoring. We used the quadratic formula only to illustrate that it can be used to solve any quadratic equation. The following is a general procedure for solving a quadratic equation.

To solve a quadratic equation:

1. **Check to see if it is in the form $ax^2 = p$ or $(x + k)^2 = p$. If it is, use the principle of square roots as in Section 9.2.**

2. **If it is not in the form of (1), write it in standard form, $ax^2 + bx + c = 0$ with a and b nonzero.**

3. **Then try factoring.**

4. **If it is not possible to factor or if factoring seems difficult, use the quadratic formula.**

The solutions of a quadratic equation can always be found using the quadratic formula. They cannot always be found by factoring. When $b^2 - 4ac \geq 0$, the equation has real-number solutions. When $b^2 - 4ac < 0$, the equation has no real-number solutions.

ANSWER ON PAGE A-9

The expression $b^2 - 4ac$ is called the **discriminant.** The square root of the discriminant is part of the quadratic formula.

When using the quadratic formula, it is wise to compute the discriminant first. If it is negative, there are no real-number solutions because we are taking the square root of a negative number. If it is a perfect square, you can solve by factoring if you wish.

▶ **EXAMPLE 2**   Solve $x^2 + 3x - 10 = 0$ using the quadratic formula.

The equation is in standard form. So we determine a, b, and c:

$$x^2 + 3x - 10 = 0,$$
$$a = 1, \quad b = 3, \quad c = -10.$$

We compute the discriminant:

$$b^2 - 4ac = 3^2 - 4 \cdot 1 \cdot (-10) = 9 + 40 = 49.$$

The discriminant is positive and is also a perfect square, so we could use factoring to solve. But for purposes of illustration, we will use the quadratic formula (try factoring on your own):

$$x = \frac{-3 \pm \sqrt{49}}{2(1)} = \frac{-3 \pm 7}{2}.$$

Thus,

$$x = \frac{-3 + 7}{2} = \frac{4}{2} = 2 \quad \text{or} \quad x = \frac{-3 - 7}{2} = \frac{-10}{2} = -5.$$

The solutions are 2 and -5.   ◀

DO EXERCISE 2.

▶ **EXAMPLE 3**   Solve $x^2 = 4x + 7$ using the quadratic formula. Compare with Example 8 in Section 9.2.

We first find standard form and determine a, b, and c:

$$x^2 - 4x - 7 = 0,$$
$$a = 1, \quad b = -4, \quad c = -7.$$

We then compute the discriminant:

$$b^2 - 4ac = (-4)^2 - 4 \cdot (1) \cdot (-7) = 16 + 28 = 44.$$

The discriminant is positive, so there are real-number solutions. They are given by

$$x = \frac{-(-4) \pm \sqrt{44}}{2(1)} \qquad \text{Substituting into the quadratic formula}$$

$$= \frac{4 \pm \sqrt{44}}{2} = \frac{4 \pm \sqrt{4 \cdot 11}}{2} = \frac{4 \pm \sqrt{4}\sqrt{11}}{2}$$

$$= \frac{4 \pm 2\sqrt{11}}{2} = \frac{2 \cdot 2 \pm 2\sqrt{11}}{2 \cdot 1} \qquad \text{Factoring out 2 in the numerator and the denominator}$$

$$= \frac{2(2 \pm \sqrt{11})}{2 \cdot 1} = \frac{2}{2} \cdot \frac{2 \pm \sqrt{11}}{1} = 2 \pm \sqrt{11}.$$

The solutions are $2 + \sqrt{11}$ and $2 - \sqrt{11}$, or $2 \pm \sqrt{11}$.   ◀

DO EXERCISE 3.

2. Solve using the quadratic formula:
$$x^2 - 3x - 10 = 0.$$

3. Solve using the quadratic formula:
$$x^2 + 4x = 7.$$

ANSWERS ON PAGE A-9

4. Solve using the quadratic formula:
$$x^2 = x - 1.$$

▶ **EXAMPLE 4**   Solve $x^2 + x = -1$ using the quadratic formula.

We first find standard form and determine a, b, and c:
$$x^2 + x + 1 = 0,$$
$$a = 1, \quad b = 1, \quad c = 1.$$

We then compute the discriminant:
$$b^2 - 4ac = 1^2 - 4 \cdot 1 \cdot 1 = 1 - 4 = -3.$$

Since the discriminant is negative, there are no real-number solutions because square roots of negative numbers do not exist as real numbers. ◀

DO EXERCISE 4.

▶ **EXAMPLE 5**   Solve $3x^2 = 7 - 2x$ using the quadratic formula.

We first find standard form and determine a, b, and c:
$$3x^2 + 2x - 7 = 0,$$
$$a = 3, \quad b = 2, \quad c = -7.$$

5. Solve using the quadratic formula:
$$5x^2 - 8x = 3.$$

We then compute the discriminant:
$$b^2 - 4ac = 2^2 - 4 \cdot 3 \cdot (-7) = 4 + 84 = 88.$$

This is positive, so there are real-number solutions. They are given by

$$x = \frac{-2 \pm \sqrt{88}}{2(3)} \quad \text{Substituting into the quadratic formula}$$

$$= \frac{-2 \pm \sqrt{4 \cdot 22}}{6} = \frac{-2 \pm 2\sqrt{22}}{6}$$

$$= \frac{2(-1 \pm \sqrt{22})}{2 \cdot 3} \quad \text{Factoring out 2 in the numerator and the denominator}$$

$$= \frac{-1 \pm \sqrt{22}}{3}.$$

The solutions are $\dfrac{-1 + \sqrt{22}}{3}$ and $\dfrac{-1 - \sqrt{22}}{3}$, or $\dfrac{-1 \pm \sqrt{22}}{3}$. ◀

DO EXERCISE 5.

6. Approximate the solutions to the equation in Margin Exercise 5. Round to the nearest tenth.

b   **Approximate Solutions**

A calculator or Table 2 can be used to approximate solutions.

▶ **EXAMPLE 6**   Use a calculator or Table 2 to approximate to the nearest tenth the solutions to the equation in Example 5.

Using a calculator or Table 2, we see that $\sqrt{22} \approx 4.690$. Thus we have

$$\frac{-1 + \sqrt{22}}{3} \approx \frac{-1 + 4.690}{3} \qquad \text{or} \qquad \frac{-1 - \sqrt{22}}{3} \approx \frac{-1 - 4.690}{3}$$

$$= \frac{3.69}{3} \qquad\qquad \text{or} \qquad\qquad = \frac{-5.69}{3}$$

$$\approx 1.2 \quad \text{to the} \qquad \text{or} \qquad \approx -1.9 \quad \text{to the}$$
$$\text{nearest tenth} \qquad\qquad\qquad\qquad \text{nearest tenth.}$$

The approximate solutions are 1.2 and -1.9. ◀

DO EXERCISE 6.

NAME   SECTION   DATE

EXERCISE SET 9.3

a Solve. Try factoring first. If factoring is not possible or is difficult, use the quadratic formula.

1. $x^2 - 4x = 21$ **2.** $x^2 + 7x = 18$ **3.** $x^2 = 6x - 9$

4. $x^2 = 8x - 16$ **5.** $3y^2 - 2y - 8 = 0$ **6.** $3y^2 - 7y + 4 = 0$

7. $4x^2 + 12x = 7$ **8.** $4x^2 + 4x = 15$ **9.** $x^2 - 9 = 0$

10. $x^2 - 4 = 0$ **11.** $x^2 - 2x - 2 = 0$ **12.** $x^2 - 4x - 7 = 0$

13. $y^2 - 10y + 22 = 0$ **14.** $y^2 + 6y - 1 = 0$ **15.** $x^2 + 4x + 4 = 7$

16. $x^2 - 2x + 1 = 5$ **17.** $3x^2 + 8x + 2 = 0$ **18.** $3x^2 - 4x - 2 = 0$

19. $2x^2 - 5x = 1$ **20.** $2x^2 + 2x = 3$ **21.** $4y^2 - 4y - 1 = 0$

22. $4y^2 + 4y - 1 = 0$ **23.** $2t^2 + 6t + 5 = 0$ **24.** $4y^2 + 3y + 2 = 0$

25. $3x^2 = 5x + 4$ **26.** $2x^2 + 3x = 1$ **27.** $2y^2 - 6y = 10$

ANSWERS

1. _____
2. _____
3. _____
4. _____
5. _____
6. _____
7. _____
8. _____
9. _____
10. _____
11. _____
12. _____
13. _____
14. _____
15. _____
16. _____
17. _____
18. _____
19. _____
20. _____
21. _____
22. _____
23. _____
24. _____
25. _____
26. _____
27. _____

28. $5m^2 = 3 + 11m$

29. $\dfrac{x^2}{x-4} - \dfrac{7}{x-4} = 0$

30. $\dfrac{x^2}{x+3} - \dfrac{5}{x+3} = 0$

31. $x + 2 = \dfrac{3}{x+2}$

32. $x - 3 = \dfrac{5}{x-3}$

33. $\dfrac{1}{x} + \dfrac{1}{x+6} = \dfrac{1}{5}$

34. $\dfrac{1}{x} + \dfrac{1}{x+1} = \dfrac{1}{3}$

b   Solve using the quadratic formula. Use a calculator or Table 2 to approximate the solutions to the nearest tenth.

35. $x^2 - 4x - 7 = 0$

36. $x^2 + 2x - 2 = 0$

37. $y^2 - 6y - 1 = 0$

38. $y^2 + 10y + 22 = 0$

39. $4x^2 + 4x = 1$

40. $4x^2 = 4x + 1$

41. $3x^2 + 4x - 2 = 0$

42. $3x^2 - 8x + 2 = 0$

SKILL MAINTENANCE

43. Multiply and simplify: $\sqrt{3x^2}\sqrt{9x^3}$.

44. Subtract: $\sqrt{54} - \sqrt{24}$.

45. Simplify: $\sqrt{80}$.

46. Rationalize the denominator: $\sqrt{\tfrac{7}{3}}$.

SYNTHESIS

Solve.

47. $5x + x(x - 7) = 0$

48. $x(3x + 7) - 3x = 0$

49. $3 - x(x - 3) = 4$

50. $x(5x - 7) = 1$

51. $(y + 4)(y + 3) = 15$

52. $(y + 5)(y - 1) = 27$

53. $x^2 + (x + 2)^2 = 7$

54. $x^2 + (x + 1)^2 = 5$

9.4 Formulas

a To solve a formula for a given letter, we try to get the letter alone on one side.

▶ **EXAMPLE 1** Solve for h: $V = 3.5\sqrt{h}$ (the distance to the horizon).

This is a radical equation. Recall that we first isolate the radical. Then we use the principle of squaring.

$$\frac{V}{3.5} = \sqrt{h} \qquad \text{Isolating the radical}$$

$$\left(\frac{V}{3.5}\right)^2 = (\sqrt{h})^2 \qquad \text{Using the principle of squaring (Section 8.5)}$$

$$\frac{V^2}{12.25} = h \qquad \text{Simplifying} \qquad ◀$$

▶ **EXAMPLE 2** Solve for g: $T = 2\pi\sqrt{\dfrac{L}{g}}$ (the period of a pendulum).

$$\frac{T}{2\pi} = \sqrt{\frac{L}{g}} \qquad \text{Isolating the radical}$$

$$\left(\frac{T}{2\pi}\right)^2 = \left(\sqrt{\frac{L}{g}}\right)^2 \qquad \text{Using the principle of squaring}$$

$$\frac{T^2}{4\pi^2} = \frac{L}{g}$$

$$gT^2 = 4\pi^2 L \qquad \text{Multiplying by } 4\pi^2 g \text{ to clear of fractions}$$

$$g = \frac{4\pi^2 L}{T^2} \qquad \text{Dividing by } T^2 \text{ to get } g \text{ alone} \qquad ◀$$

DO EXERCISES 1–3.

In most formulas, the letters represent nonnegative numbers, so we need not use absolute values when taking square roots.

▶ **EXAMPLE 3** *Torricelli's theorem.* The speed v of a liquid leaving a tank from an orifice is related to the height h of the top of the liquid above the orifice by the formula

$$h = \frac{v^2}{2g}.$$

Solve for v.

Since v^2 appears by itself and there is no expression involving v, we first solve for v^2. Then we use the principle of square roots, taking only the nonnegative square root because v is nonnegative.

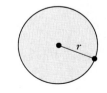

$$2gh = v^2 \qquad \text{Multiplying by } 2g \text{ to clear of fractions}$$

$$\sqrt{2gh} = v \qquad \text{Using the principle of square roots.}$$
$$\text{Assume that } v \text{ is nonnegative.} \qquad ◀$$

DO EXERCISE 4.

OBJECTIVE

After finishing Section 9.4, you should be able to:

a Solve a formula for a given letter.

FOR EXTRA HELP

Tape 17A   Tape 17B   MAC: 9
                                    IBM: 9

1. Solve for L: $r = 2\sqrt{5L}$ (the speed of a skidding car).

2. Solve for L: $T = 2\pi\sqrt{\dfrac{L}{g}}.$

3. Solve for m: $c = \sqrt{\dfrac{E}{m}}.$

4. Solve for r: $A = \pi r^2$ (the area of a circle).

ANSWERS ON PAGE A-9

5. Solve for d: $C = P(d - 1)^2$.

▶ **EXAMPLE 4**   Solve for r: $A = P(1 + r)^2$ (a compound-interest formula).

$$A = P(1 + r)^2$$

$$\frac{A}{P} = (1 + r)^2 \qquad \text{Dividing by } P$$

$$\sqrt{\frac{A}{P}} = 1 + r \qquad \begin{array}{l}\text{Using the principle of square roots.}\\ \text{Assume that } 1 + r \text{ is nonnegative.}\end{array}$$

$$-1 + \sqrt{\frac{A}{P}} = r \qquad \text{Subtracting 1 to get } r \text{ alone}$$ ◀

DO EXERCISE 5.

Sometimes we must use the quadratic formula to solve a formula for a certain letter.

6. Solve for n: $N = n^2 - n$.

▶ **EXAMPLE 5**   Solve for n: $d = \dfrac{n^2 - 3n}{2}$, where d is the number of diagonals of a polygon.

This time there is a term involving n as well as an n^2-term. Thus we must use the quadratic formula.

$$d = \frac{n^2 - 3n}{2}$$

$$n^2 - 3n = 2d \qquad \text{Multiplying by 2 to clear of fractions}$$

$$n^2 - 3n - 2d = 0 \qquad \text{Finding standard form}$$

$$a = 1, \quad b = -3, \quad c = -2d \qquad \text{The letter } d \text{ represents a constant.}$$

$$n = \frac{-b \pm \sqrt{b^2 - 4ac}}{2a} \qquad \text{Quadratic formula}$$

$$n = \frac{-(-3) \pm \sqrt{(-3)^2 - 4 \cdot 1 \cdot (-2d)}}{2 \cdot 1} \qquad \begin{array}{l}\text{Substituting into the}\\ \text{quadratic formula}\end{array}$$

$$n = \frac{3 \pm \sqrt{9 + 8d}}{2}$$ ◀

DO EXERCISE 6.

7. Solve for t: $h = vt + 8t^2$.

▶ **EXAMPLE 6**   Solve for t: $S = gt + 16t^2$.

$$S = gt + 16t^2$$

$$16t^2 + gt - S = 0 \qquad \text{Finding standard form}$$

$$a = 16, \quad b = g, \quad c = -S$$

$$t = \frac{-b \pm \sqrt{b^2 - 4ac}}{2a}$$

$$t = \frac{-g \pm \sqrt{g^2 - 4 \cdot 16 \cdot (-S)}}{2 \cdot 16} \qquad \begin{array}{l}\text{Substituting into the}\\ \text{quadratic formula}\end{array}$$

$$t = \frac{-g \pm \sqrt{g^2 + 64S}}{32}$$ ◀

DO EXERCISE 7.

NAME   SECTION   DATE

EXERCISE SET 9.4

a Solve for the indicated letter.

1. $N = 2.5\sqrt{A}$, for A

2. $T = 2\pi\sqrt{\dfrac{L}{32}}$, for L

3. $Q = \sqrt{\dfrac{aT}{c}}$, for T

4. $v = \sqrt{\dfrac{2gE}{m}}$, for E

5. $E = mc^2$, for c

6. $S = 4\pi r^2$, for r

7. $Q = ad^2 - cd$, for d

8. $P = kA^2 + mA$, for A

9. $c^2 = a^2 + b^2$, for a

10. $c = \sqrt{a^2 + b^2}$, for b

11. $s = 16t^2$, for t

12. $V = \pi r^2 h$, for r

13. $A = \pi r^2 + 2\pi rh$, for r

14. $A = 2\pi r^2 + 2\pi rh$, for r

15. $A = \dfrac{\pi r^2 S}{360}$, for r

16. $H = \dfrac{D^2 N}{2.5}$, for D

17. $c = \sqrt{a^2 + b^2}$, for a

18. $c^2 = a^2 + b^2$, for b

19. $h = \dfrac{a}{2}\sqrt{3}$, for a
(The height of an equilateral triangle with sides of length a)

20. $d = s\sqrt{2}$, for s
(The hypotenuse of an isosceles right triangle with s the length of the legs)

1. _____

2. _____

3. _____

4. _____

5. _____

6. _____

7. _____

8. _____

9. _____

10. _____

11. _____

12. _____

13. _____

14. _____

15. _____

16. _____

17. _____

18. _____

19. _____

20. _____

21. _____

22. _____

23. _____

24. _____

25. _____

26. _____

27. _____

28. _____

29. _____

30. _____

31. _____

32. _____

33. a) _____

b) _____

34. _____

35. _____

36. _____

21. $n = aT^2 - 4T + m$, for T

22. $y = ax^2 + bx + c$, for x

23. $v = 2\sqrt{\dfrac{2kT}{\pi m}}$, for T

24. $E = \dfrac{1}{2}mv^2 + mgy$, for v

25. $c = \sqrt{\dfrac{E}{m}}$, for E

26. $3x^2 = d^2$, for x

27. $N = \dfrac{n^2 - n}{2}$, for n

28. $M = \dfrac{m}{\sqrt{1 - \left(\dfrac{v}{c}\right)^2}}$, for c

---

SKILL MAINTENANCE

In a right triangle, find the length of the side not given. Given an exact answer and an approximation to three decimal places.

29. $a = 4$, $b = 7$ **30.** $b = 11$, $c = 14$ **31.** $a = 4$, $b = 5$ **32.** $a = 10$, $c = 12$

SYNTHESIS

33. The circumference C of a circle is given by $C = 2\pi r$.

    a) Solve $C = 2\pi r$ for r.

    b) The area is given by $A = \pi r^2$. Express the area in terms of the circumference C.

34. In reference to Exercise 33, express the circumference C in terms of the area A.

35. Solve $3ax^2 - x - 3ax + 1 = 0$ for x.

36. Solve $h = 16t^2 + vt + s$ for t.

9.5 Solving Problems

a Using Quadratic Equations to Solve Problems

▶ **EXAMPLE 1** The area of a rectangle is 76 in². The length is 7 in. longer than three times the width. Find the dimensions of the rectangle.

1. *Familiarize.* We first make a drawing and label it with both known and unknown information. We let w = the width of the rectangle. The length of the rectangle is 7 in. longer than three times the width. Thus the length is $3w + 7$.

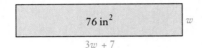

2. *Translate.* Recall that area is length × width. Thus we have two expressions for the area of the rectangle: $(3w + 7)(w)$ and 76. This gives us a translation:

$$(3w + 7)(w) = 76.$$

3. *Solve.* We solve the equation:

$$3w^2 + 7w = 76$$
$$3w^2 + 7w - 76 = 0$$
$$(3w + 19)(w - 4) = 0 \quad \text{Factoring (the quadratic formula could also be used)}$$
$$3w + 19 = 0 \quad \text{or} \quad w - 4 = 0 \quad \text{Using the principle of zero products}$$
$$3w = -19 \quad \text{or} \quad w = 4$$
$$w = -\tfrac{19}{3} \quad \text{or} \quad w = 4.$$

4. *Check.* We check in the original problem. We know that $-\frac{19}{3}$ is not a solution because width cannot be negative. When $w = 4$, $3w + 7 = 19$, and the area is 4(19), or 76. This checks.

5. *State.* The width of the rectangle is 4 in., and the length is 19 in. ◀

DO EXERCISE 1.

▶ **EXAMPLE 2** The hypotenuse of a right triangle is 6 m long. One leg is 1 m longer than the other. Find the lengths of the legs. Round to the nearest tenth.

1. *Familiarize.* We first make a drawing and the area is 4(19), or 76. This checks. one leg. Then $s + 1$ = the length of the other leg.

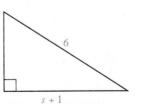

2. *Translate.* To translate, we use the Pythagorean equation:

$$s^2 + (s + 1)^2 = 6^2.$$

OBJECTIVE

After finishing Section 9.5, you should be able to:

a Solve problems using quadratic equations.

FOR EXTRA HELP

Tape 17B   Tape 18A   MAC: 9
                            IBM: 9

1. The area of a rectangle is 68 in². The length is 1 in. longer than three times the width. Find the dimensions of the rectangle.

ANSWER ON PAGE A-9

2. The hypotenuse of a right triangle is 4 cm long. One leg is 1 cm longer than the other. Find the lengths of the legs. Round to the nearest tenth.

3. *Solve.* We solve the equation:

$$s^2 + (s + 1)^2 = 6^2$$
$$s^2 + s^2 + 2s + 1 = 36$$
$$2s^2 + 2s - 35 = 0.$$

Since we cannot factor, we use the quadratic formula:

$$a = 2, \quad b = 2, \quad c = -35$$

$$s = \frac{-b \pm \sqrt{b^2 - 4ac}}{2a}$$

$$= \frac{-2 \pm \sqrt{2^2 - 4 \cdot 2(-35)}}{2 \cdot 2}$$

$$= \frac{-2 \pm \sqrt{4 + 280}}{4} = \frac{-2 \pm \sqrt{284}}{4}$$

$$= \frac{-2 \pm \sqrt{4 \cdot 71}}{4}$$

$$= \frac{-2 \pm 2 \cdot \sqrt{71}}{2 \cdot 2}$$

$$= \frac{2(-1 \pm \sqrt{71})}{2 \cdot 2}$$

$$= \frac{2}{2} \cdot \frac{-1 \pm \sqrt{71}}{2}$$

$$= \frac{-1 \pm \sqrt{71}}{2}.$$

Using a calculator or Table 2, we get an approximation: $\sqrt{71} \approx 8.426$. Thus,

$$\frac{-1 + 8.426}{2} \approx 3.7 \quad \text{or} \quad \frac{-1 - 8.426}{2} \approx -4.7.$$

4. *Check.* Since the length of a leg cannot be negative, -4.7 does not check. But 3.7 does check. If the smaller leg is 3.7, the other leg is 4.7. Then

$$(3.7)^2 + (4.7)^2 = 13.69 + 22.09 = 35.78.$$

Using a calculator, we get $\sqrt{35.78} \approx 5.96 \approx 6$. Note that our check is not exact because we are using an approximation.

5. *State.* One leg is about 3.7 m long, and the other is about 4.7 m long.

◀

DO EXERCISE 2.

▶ **EXAMPLE 3** The current in a stream moves at a speed of 2 km/h. A boat travels 24 km upstream and 24 km downstream in a total time of 5 hr. What is the speed of the boat in still water?

1. *Familiarize.* We first make a drawing. The distances are the same. Let r = the speed of the boat in still water. Then when the boat is traveling upstream, its speed is $r - 2$. When it is traveling downstream, its speed is $r + 2$. We let t_1 represent the time it takes the boat to go upstream and t_2 represent the time it takes to go downstream. We summarize in a table.

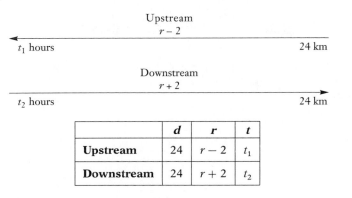

Upstream
$r - 2$
t_1 hours   24 km

Downstream
$r + 2$
t_2 hours   24 km

	d	**r**	**t**
Upstream	24	$r - 2$	t_1
Downstream	24	$r + 2$	t_2

2. *Translate.* Recall the basic formula for motion: $d = rt$. From it we can obtain an equation for time: $t = d/r$. Total time consists of the time to go upstream, t_1, plus the time to go downstream, t_2. Using $t = d/r$ and the rows of the table, we have

$$t_1 = \frac{24}{r - 2} \quad \text{and} \quad t_2 = \frac{24}{r + 2}.$$

Since the total time is 5 hr, $t_1 + t_2 = 5$, and we have

$$\frac{24}{r - 2} + \frac{24}{r + 2} = 5.$$

3. *Solve.* We solve the equation. We multiply on both sides by the LCM, which is $(r - 2)(r + 2)$:

$$(r - 2)(r + 2) \cdot \left[\frac{24}{r - 2} + \frac{24}{r + 2} \right] = (r - 2)(r + 2)5 \quad \text{\textbf{Multiplying by the LCM}}$$

$$(r - 2)(r + 2) \cdot \frac{24}{r - 2} + (r - 2)(r + 2) \cdot \frac{24}{r + 2} = (r^2 - 4)5$$

$$24(r + 2) + 24(r - 2) = 5r^2 - 20$$

$$24r + 48 + 24r - 48 = 5r^2 - 20$$

$$-5r^2 + 48r + 20 = 0$$

$$5r^2 - 48r - 20 = 0 \quad \text{\textbf{Multiplying by} } -1$$

$$(5r + 2)(r - 10) = 0 \quad \text{\textbf{Factoring}}$$

$$5r + 2 = 0 \quad \text{or} \quad r - 10 = 0 \quad \text{\textbf{Using the principle of zero products}}$$

$$5r = -2 \quad \text{or} \quad r = 10$$

$$r = -\tfrac{2}{5} \quad \text{or} \quad r = 10.$$

4. *Check.* Since speed cannot be negative, $-\frac{2}{5}$ cannot be a solution. But suppose the speed of the boat in still water is 10 km/h. The speed upstream is then $10 - 2$, or 8 km/h. The speed downstream is $10 + 2$, or 12 km/h. The time upstream, using $t = d/r$, is 24/8, or 3 hr. The time downstream is 24/12, or 2 hr. The total time is 5 hr. This checks.

5. *State.* The speed of the boat in still water is 10 km/h. ◀

DO EXERCISE 3.

3. The speed of a boat in still water is 12 km/h. The boat travels 45 km upstream and 45 km downstream in a total time of 8 hr. What is the speed of the stream? (*Hint:* Let $s =$ the speed of the stream. Then $12 - s$ is the speed upstream and $12 + s$ is the speed downstream. Note also that $12 - s$ cannot be negative, because the boat must be going faster than the current if it is moving forward.)

ANSWER ON PAGE A-9

❖ SIDELIGHTS

Handling Dimension Symbols

In many applications, we add, subtract, multiply and divide quantities having units, or dimensions, such as ft, km, sec, hr, etc. For example, to find average speed, we divide total distance by total time. What results is notation very much like a rational expression.

EXAMPLE 1 A car travels 150 km in 2 hr. What is its average speed?

$$\text{Speed} = \frac{150 \text{ km}}{2 \text{ hr}}, \text{ or } 75 \frac{\text{km}}{\text{hr}}$$

(The standard abbreviation for km/hr is km/h, but it does not suit our present discussion well.)

The symbol km/hr makes it look as if we are dividing kilometers by hours. It may be argued that we can divide only numbers. Nevertheless, we treat dimension symbols, such as km, ft, and hr, as if they were numerals or variables, obtaining correct results mechanically.

EXAMPLE 2 Compare

$$\frac{150x}{2y} = \frac{150}{2} \cdot \frac{x}{y} = 75\frac{x}{y}$$

with

$$\frac{150 \text{ km}}{2 \text{ hr}} = \frac{150}{2} \frac{\text{km}}{\text{hr}} = 75\frac{\text{km}}{\text{hr}}.$$

EXAMPLE 3 Compare

$$3x + 2x = (3 + 2)x = 5x$$

with

$$3 \text{ ft} + 2 \text{ ft} = (3 + 2) \text{ ft} = 5 \text{ ft}.$$

EXAMPLE 4 Compare

$$5x \cdot 3x = 15x^2$$

with

$$5 \text{ ft} \cdot 3 \text{ ft} = 15 \text{ ft}^2 \text{ (square feet)}.$$

EXAMPLE 5 Compare

$$5x \cdot 8y = 40xy$$

with

$$5 \text{ men} \cdot 8 \text{ hours} = 40 \text{ man-hours}.$$

If 5 men work 8 hours, the total amount of labor is 40 man-hours, which is the same as 4 men working 10 hours.

EXAMPLE 6 Compare

$$\frac{300x \cdot 240y}{15t} = 4800\frac{xy}{t}$$

with

$$\frac{300 \text{ kW} \cdot 240 \text{ hr}}{15 \text{ da}} = 4800\frac{\text{kW-hr}}{\text{da}}.$$

If an electrical device uses 300 kilowatts for 240 hours over a period of 15 days, its rate of usage of energy is 4800 kilowatt-hours per day. The standard abbreviation for kilowatt-hours is kWh.

These "multiplications" and "divisions" can have humorous interpretations. For example,

$$2 \text{ barns} \cdot 4 \text{ dances} = 8 \text{ barn-dances},$$

$$2 \text{ dances} \cdot 4 \text{ dances} = 8 \text{ dances}^2 \text{ (8 square dances)},$$

and

$$\text{Ice} \cdot \text{Ice} \cdot \text{Ice} = \text{Ice}^3 \text{ (Ice cubed)}.$$

However, the fact that such amusing examples exist causes us no trouble, since they do not come up in practice.

EXERCISES

Add these measures.

1. 45 ft + 23 ft
2. 55 km/hr + 27 km/hr
3. 17 g + 28 g
4. 3.4 lb + 5.2 lb

Find average speeds, given total distance and total time.

5. 90 mi, 6 hr
6. 640 km, 20 hr
7. 9.9 m, 3 sec
8. 76 ft, 4 min

Perform these calculations.

9. $\dfrac{3 \text{ in.} \cdot 8 \text{ lb}}{6 \text{ sec}}$
10. $\dfrac{60 \text{ men} \cdot 8 \text{ hr}}{20 \text{ da}}$

11. $36 \text{ ft} \cdot \dfrac{1 \text{ yd}}{3 \text{ ft}}$
12. $55\dfrac{\text{mi}}{\text{hr}} \cdot 4 \text{ hr}$

13. $5 \text{ ft}^3 + 11 \text{ ft}^3$
14. $\dfrac{3 \text{ lb}}{14 \text{ ft}} \cdot \dfrac{7 \text{ lb}}{6 \text{ ft}}$

15. Divide $4850 by 5 days.
16. Divide $25.60 by 8 hr.

NAME   SECTION   DATE

EXERCISE SET 9.5

a Solve.

1. The hypotenuse of a right triangle is 25 ft long. One leg is 17 ft longer than the other. Find the lengths of the legs.

2. The hypotenuse of a right triangle is 26 yd long. One leg is 14 yd longer than the other. Find the lengths of the legs.

1. _____

2. _____

3. The length of a rectangle is 2 cm greater than the width. The area is 80 cm². Find the length and the width.

4. The length of a rectangle is 3 m greater than the width. The area is 70 m². Find the length and the width.

3. _____

4. _____

5. The width of a rectangle is 4 cm less than the length. The area is 320 cm². Find the length and the width.

6. The width of a rectangle is 3 cm less than the length. The area is 340 cm². Find the length and the width.

5. _____

6. _____

7. The length of a rectangle is twice the width. The area is 50 m². Find the length and the width.

8. The length of a rectangle is twice the width. The area is 32 cm². Find the length and the width.

7. _____

8. _____

ANSWERS

Find the approximate answers for Exercises 9–14. Round to the nearest tenth.

9. The hypotenuse of a right triangle is 8 m long. One leg is 2 m longer than the other. Find the lengths of the legs.

10. The hypotenuse of a right triangle is 5 cm long. One leg is 2 cm longer than the other. Find the lengths of the legs.

9. _____

10. _____

11. The length of a rectangle is 2 in. greater than the width. The area is 20 in². Find the length and the width.

12. The length of a rectangle is 3 ft greater than the width. The area is 15 ft². Find the length and the width.

11. _____

13. The length of a rectangle is twice the width. The area is 10 m². Find the length and the width.

14. The length of a rectangle is twice the width. The area is 20 cm². Find the length and the width.

12. _____

13. _____

15. A picture frame measures 20 cm by 12 cm. There is 84 cm² of picture showing. The frame is of uniform thickness. Find the thickness of the frame.

16. A picture frame measures 18 cm by 14 cm. There is 192 cm² of picture showing. The frame is of uniform thickness. Find the thickness of the frame.

14. _____

15. _____

16. _____

17. The current in a stream moves at a speed of 3 km/h. A boat travels 40 km upstream and 40 km downstream in a total time of 14 hr. What is the speed of the boat in still water? Complete the following table to help with the familiarization.

	d	r	t
Upstream		$r - 3$	t_1
Downstream	40		t_2

Upstream
$r - 3$

←————————————————————————
t_1 hours   40 km

Downstream
$r + 3$

————————————————————————→
t_2 hours   40 km

18. The current in a stream moves at a speed of 3 km/h. A boat travels 45 km upstream and 45 km downstream in a total time of 8 hr. What is the speed of the boat in still water?

19. The current in a stream moves at a speed of 4 mph. A boat travels 4 mi upstream and 12 mi downstream in a total time of 2 hr. What is the speed of the boat in still water?

20. The current in a stream moves at a speed of 4 mph. A boat travels 5 mi upstream and 13 mi downstream in a total time of 2 hr. What is the speed of the boat in still water?

21. The speed of a boat in still water is 10 km/h. The boat travels 12 km upstream and 28 km downstream in a total time of 4 hr. What is the speed of the stream?

22. The speed of a boat in still water is 8 km/h. The boat travels 60 km upstream and 60 km downstream in a total time of 16 hr. What is the speed of the stream?

ANSWERS

17. _____

18. _____

19. _____

20. _____

21. _____

22. _____

23. An airplane flies 738 mi against the wind and 1062 mi with the wind in a total time of 9 hr. The speed of the airplane in still air is 200 mph. What is the speed of the wind?

24. An airplane flies 520 km against the wind and 680 km with the wind in a total time of 4 hr. The speed of the airplane in still air is 300 km/h. What is the speed of the wind?

25. The speed of a boat in still water is 9 km/h. The boat travels 80 km upstream and 80 km downstream in a total time of 18 hr. What is the speed of the stream?

26. The speed of a boat in still water is 10 km/h. The boat travels 48 km upstream and 48 km downstream in a total time of 10 hr. What is the speed of the stream?

SYNTHESIS

27. Find the area of a square for which the diagonal is one unit longer than the length of the sides.

28. Two consecutive integers have squares that differ by 25. Find the integers.

29. Find r in this figure. Round to the nearest hundredth.

30. A 20-ft pole is struck by lightning and, while not completely broken, falls over and touches the ground 10 ft from the bottom of the pole. How high up did the pole break?

31. What should the diameter d of a pizza be so that it has the same area as two 10-in. pizzas? Do you get more to eat with a 13-in. pizza or with two 10-in. pizzas?

32. Find the side of a square whose diagonal is 3 cm longer than a side.

9.6 Graphs of Quadratic Equations

In this section, we will graph equations of the form

$$y = ax^2 + bx + c, \quad a \neq 0.$$

The polynomial on the right side of the equation is of second degree, or **quadratic.** Examples of the types of equations we are going to graph are

$$y = x^2, \qquad y = x^2 + 2x - 3, \qquad y = -2x^2 + 3.$$

OBJECTIVES

After finishing Section 9.6, you should be able to:

a   Graph quadratic equations.

b   Find the *x*-intercepts of a quadratic equation.

FOR EXTRA HELP

Tape 17C   Tape 18A   MAC: 9
                                IBM: 9

a   Graphing Quadratic Equations of the Type $y = ax^2 + bx + c$

Graphs of quadratic equations of the type $y = ax^2 + bx + c$ (where $a \neq 0$) are always cup-shaped. They have a **line of symmetry** like the dashed lines shown in the figures below. If we fold on this line, the two halves will match exactly. The curve goes on forever. The top or bottom point where the curve changes is called the **vertex.** The second coordinate is either the largest value of y or the smallest value of y. The vertex is also thought of as a turning point. Graphs of quadratic equations are called **parabolas.**

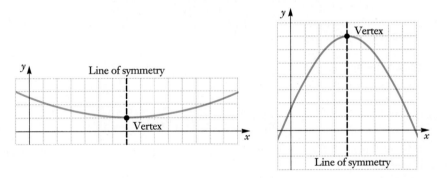

To graph a quadratic equation, we begin by choosing some numbers for x and computing the corresponding values of y.

▶ **EXAMPLE 1**   Graph: $y = x^2$.

We choose numbers for x and find the corresponding values for y. Then we plot the ordered pairs (x, y) resulting from the computations and connect them with a smooth curve.

For $x = -3, y = x^2 = (-3)^2 = 9$.
For $x = -2, y = x^2 = (-2)^2 = 4$.
For $x = -1, y = x^2 = (-1)^2 = 1$.
For $x = 0, y = x^2 = (0)^2 = 0$.
For $x = 1, y = x^2 = (1)^2 = 1$.
For $x = 2, y = x^2 = (2)^2 = 4$.
For $x = 3, y = x^2 = (3)^2 = 9$.

x	y	(x, y)
-3	9	$(-3, 9)$
-2	4	$(-2, 4)$
-1	1	$(-1, 1)$
0	0	$(0, 0)$
1	1	$(1, 1)$
2	4	$(2, 4)$
3	9	$(3, 9)$

◀

Graph. List the ordered pair for the vertex.

1. $y = x^2 - 3$

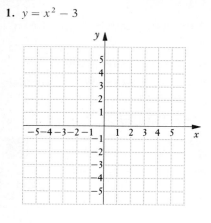

2. $y = -3x^2 + 6x$

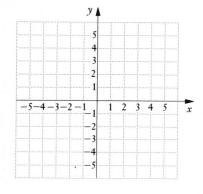

3. $y = x^2 - 4x + 4$

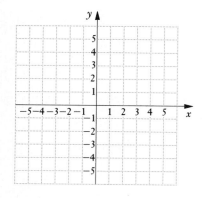

   In Example 1, the vertex is the point (0, 0). The second coordinate of the vertex, 0, is the smallest y-value. The y-axis is the line of symmetry. Parabolas whose equations are $y = ax^2$ always have the origin (0, 0) as the vertex and the y-axis as the line of symmetry.

   How do we graph a general equation? There are many methods, some of which you will study in your next mathematics course. Our goal here is to give you a basic graphing technique that is fairly easy to apply. A key in the graphing is knowing the vertex. By graphing it and then choosing x-values on both sides of the vertex, we can compute more points and complete the graph.

> **Finding the Vertex**
>
> **For a parabola given by the quadratic equation $y = ax^2 + bx + c$:**
>
> **1. The x-coordinate of the vertex is $-\dfrac{b}{2a}$.**
>
> **2. The second coordinate of the vertex is found by substituting the x-coordinate into the equation and computing y.**

   The proof that the vertex can be found in this way can be shown by completing the square in a manner similar to the proof of the quadratic formula, but it will not be considered here.

▶ **EXAMPLE 2**   Graph: $y = -2x^2 + 3$.

   We first find the vertex. The x-coordinate of the vertex is

$$-\frac{b}{2a} = -\frac{0}{2(-2)} = 0.$$

   We substitute 0 for x into the equation to find the second coordinate of the vertex:

$$y = -2x^2 + 3 = -2(0)^2 + 3 = 3.$$

The vertex is (0, 3). The line of symmetry is $x = 0$, which is the y-axis. We choose some x-values on both sides of the vertex and graph the parabola.

For $x = 1$, $y = -2x^2 + 3 = -2(1)^2 + 3 = -2 + 3 = 1.$
For $x = -1$, $y = -2x^2 + 3 = -2(-1)^2 + 3 = -2 + 3 = 1.$
For $x = 2$, $y = -2x^2 + 3 = -2(2)^2 + 3 = -8 + 3 = -5.$
For $x = -2$, $y = -2x^2 + 3 = -2(-2)^2 + 3 = -8 + 3 = -5.$

x	y
0	3
1	1
-1	1
2	-5
-2	-5

← This is the vertex.

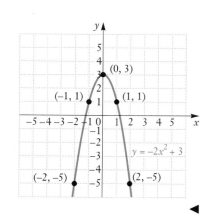

$y = -2x^2 + 3$

There are two other tips you might use when graphing quadratic equations. The first involves the coefficient of x^2. Note that a in $y = ax^2 + bx + c$ tells us whether the graph opens up or down. When a is positive, as in Example 1, the graph opens up; when a is negative, as in Example 2, the graph opens down. It is also helpful to plot the y-intercept. It occurs when $x = 0$.

Tips for Graphing Quadratic Equations

1. **Graphs of quadratic equations $y = ax^2 + bx + c$ are all parabolas. They are *smooth* cup-shaped symmetric curves, with no sharp points or kinks in them.**
2. **The graph of $y = ax^2 + bx + c$ opens up if $a > 0$. It opens down if $a < 0$.**
3. **Find the y-intercept. It occurs when $x = 0$, and it is easy to compute.**

▶ **EXAMPLE 3** Graph: $y = x^2 + 2x - 3$.

We first find the vertex. The x-coordinate of the vertex is

$$-\frac{b}{2a} = -\frac{2}{2(1)} = -1.$$

We substitute -1 for x into the equation to find the second coordinate of the vertex:

$$y = x^2 + 2x - 3 = (-1)^2 + 2(-1) - 3 = 1 - 2 - 3 = -4.$$

The vertex is $(-1, -4)$. The line of symmetry is $x = -1$.

We choose some x-values on both sides of the vertex and graph the parabola. Since the coefficient of x^2 is 1, which is positive, we know that the graph opens up. Be sure to find y when $x = 0$. This gives the y-intercept.

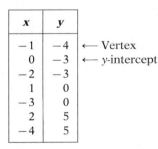

x	y	
-1	-4	← Vertex
0	-3	← y-intercept
-2	-3	
1	0	
-3	0	
2	5	
-4	5	

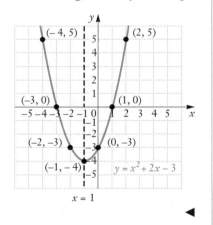

◀

DO EXERCISES 1–3 ON THE PRECEDING PAGE.

Find the x-intercepts.

4. $y = x^2 - 3$

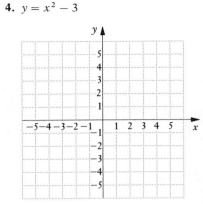

5. $y = x^2 + 6x + 8$

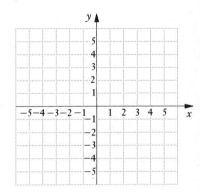

6. $y = -2x^2 - 4x + 1$

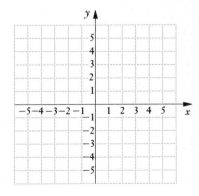

ANSWERS ON PAGE A-9

7. $y = x^2 + 3$

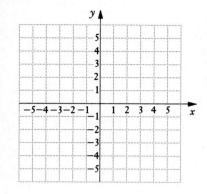

The *x*-intercepts of $y = ax^2 + bx + c$ occur at those values of *x* for which $y = 0$. Thus the first coordinates of the *x*-intercepts are solutions of the equation

$$0 = ax^2 + bx + c.$$

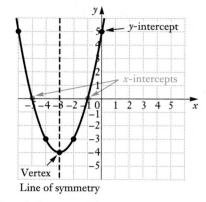

▶ **EXAMPLE 4**   Find the *x*-intercepts of $y = x^2 - 4x + 1$.

We solve the equation

$$x^2 - 4x + 1 = 0.$$

Factoring is not convenient, so we use the quadratic formula.

$$a = 1, \quad b = -4, \quad c = 1$$

$$x = \frac{-b \pm \sqrt{b^2 - 4ac}}{2a}$$

$$= \frac{-(-4) \pm \sqrt{(-4)^2 - 4(1)(1)}}{2(1)}$$

$$= \frac{4 \pm \sqrt{16 - 4}}{2}$$

$$= \frac{4 \pm \sqrt{12}}{2} = \frac{4 \pm \sqrt{4 \cdot 3}}{2}$$

$$= \frac{4 \pm 2\sqrt{3}}{2} = \frac{2 \cdot 2 \pm 2\sqrt{3}}{2 \cdot 1}$$

$$= \frac{2}{2} \cdot \frac{2 \pm \sqrt{3}}{1} = 2 \pm \sqrt{3}.$$

The *x*-intercepts are $(2 - \sqrt{3}, 0)$ and $(2 + \sqrt{3}, 0)$.   ◀

The discriminant, $b^2 - 4ac$, tells how many real-number solutions the equation $0 = ax^2 + bx + c$ has, so it also tells how many *x*-intercepts there are.

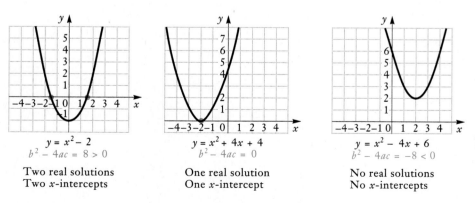

$y = x^2 - 2$
$b^2 - 4ac = 8 > 0$

Two real solutions
Two *x*-intercepts

$y = x^2 + 4x + 4$
$b^2 - 4ac = 0$

One real solution
One *x*-intercept

$y = x^2 - 4x + 6$
$b^2 - 4ac = -8 < 0$

No real solutions
No *x*-intercepts

DO EXERCISES 4–7. (EXERCISES 4–6 ARE ON THE PRECEDING PAGE.)

NAME SECTION DATE

EXERCISE SET 9.6

a Graph the quadratic equation. List the ordered pair for the vertex.

1. $y = x^2 + 1$

2. $y = 2x^2$

3. $y = -1 \cdot x^2$

4. $y = x^2 - 1$

5. $y = -x^2 + 2x$

6. $y = x^2 + x - 6$

7. $y = 5 - x - x^2$

8. $y = x^2 + 2x + 1$

9. $y = x^2 - 2x + 1$

10. $y = -\dfrac{1}{2}x^2$

11. $y = -x^2 + 2x + 3$

12. $y = -x^2 - 2x + 3$

13. $y = -2x^2 - 4x + 1$

14. $y = 2x^2 + 4x - 1$

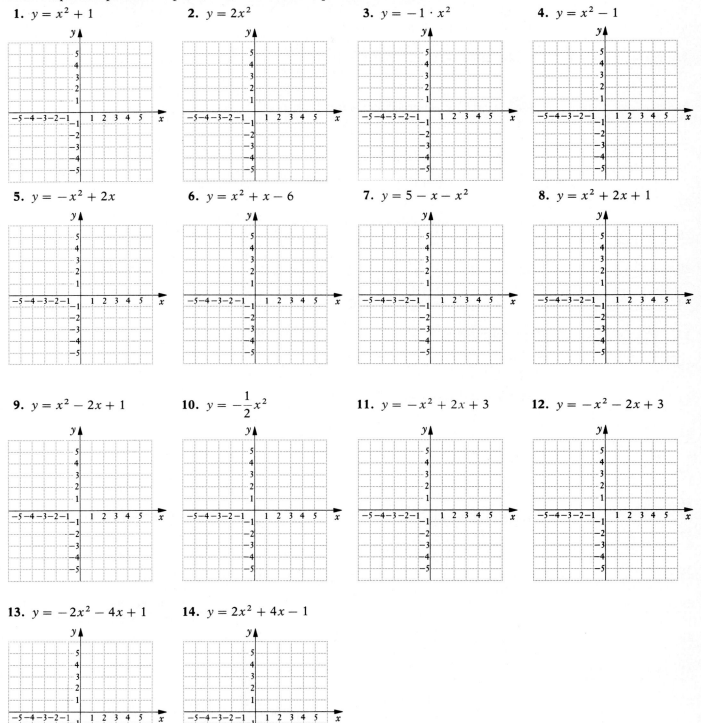

Graph the quadratic equation. Use your own graph paper.

15. $y = \dfrac{1}{4}x^2$　　　　**16.** $y = -0.1x^2$　　　　**17.** $y = 3 - x^2$　　　　**18.** $y = x^2 + 3$

19. $y = -x^2 + x - 1$　　　**20.** $y = x^2 + 2x$　　　**21.** $y = -2x^2$　　　**22.** $y = -x^2 - 1$

23. $y = x^2 - x - 6$　　　**24.** $y = 8 + x - x^2$

ANSWERS

25. _____

26. _____

27. _____

28. _____

29. _____

30. _____

31. _____

32. _____

33. _____

34. _____

35. _____

36. _____

37. _____

38. _____

39. _____

40. _____

41. a) _____

b) _____

c) _____

b　Find the x-intercepts exactly.

25. $y = x^2 - 5$　　　　**26.** $y = x^2 - 3$　　　　**27.** $y = x^2 + 2x$

28. $y = x^2 - 2x$　　　　**29.** $y = 8 - x - x^2$　　　**30.** $y = 8 + x - x^2$

31. $y = x^2 + 10x + 25$　　**32.** $y = x^2 - 8x + 16$　　**33.** $y = -x^2 - 4x + 1$

34. $y = x^2 + 4x - 1$　　　**35.** $y = x^2 + 5$　　　　**36.** $y = x^2 + 3$

SKILL MAINTENANCE

Add.

37. $\sqrt{x^3 - x^2} + \sqrt{4x - 4}$　　　　　　　　**38.** $\sqrt{8} + \sqrt{50} + \sqrt{98} + \sqrt{128}$

Multiply and simplify.

39. $\sqrt{2}\,\sqrt{14}$　　　　　　　　　　　　**40.** $\sqrt{3x^2}\,\sqrt{9x^3}$

SYNTHESIS

41. *Height of a projectile.* The height H, in feet, of a projectile with an initial velocity of 96 ft/sec is given by the equation

$$H = -16t^2 + 96t,$$

where $t =$ time, in seconds. Use the graph of this function, shown here, or any equation-solving technique to answer the following questions.

a) How many seconds after launch is the projectile 128 ft above ground?

b) When does the projectile reach its maximum height?

c) How many seconds after launch does the projectile return to the ground?

SUMMARY AND REVIEW: CHAPTER 9

IMPORTANT PROPERTIES AND FORMULAS

Standard Form: $ax^2 + bx + c = 0,\ a > 0$

Principle of Square Roots: The equation $x^2 = k$, where $k > 0$, has two solutions, $\sqrt{k}$ and $-\sqrt{k}$. The solution of $x^2 = 0$ is 0.

Quadratic Formula: $x = \dfrac{-b \pm \sqrt{b^2 - 4ac}}{2a}$

Discriminant: $b^2 - 4ac$

The x-coordinate of the vertex of a parabola $= -\dfrac{b}{2a}$.

REVIEW EXERCISES

The review sections and objectives to be tested in addition to the material in this chapter are [6.6c, d], [8.6b], [8.2d], and [8.4a].

Solve.

1. $8x^2 = 24$

2. $5x^2 - 8x + 3 = 0$

3. $x^2 - 2x - 10 = 0$

4. $3y^2 + 5y = 2$

5. $(x + 8)^2 = 13$

6. $9x^2 = 0$

7. $5t^2 - 7t = 0$

8. $9x^2 - 6x - 9 = 0$

9. $x^2 + 6x = 9$

10. $1 + 4x^2 = 8x$

11. $6 + 3y = y^2$

12. $3m = 4 + 5m^2$

13. $3x^2 = 4x$

14. $40 = 5y^2$

15. $\dfrac{15}{x} - \dfrac{15}{x + 2} = 2$

16. $x + \dfrac{1}{x} = 2$

Solve by completing the square. Show your work.

17. $3x^2 - 2x - 5 = 0$

18. $x^2 - 5x + 2 = 0$

Approximate the solutions to the nearest tenth.

19. $x^2 - 5x + 2 = 0$

20. $4y^2 + 8y + 1 = 0$

21. Solve for T: $V = \dfrac{1}{2}\sqrt{1 + \dfrac{T}{L}}$.

Graph the quadratic equation.

22. $y = 2 - x^2$

23. $y = x^2 - 4x - 2$

Find the x-intercepts.

24. $y = 2 - x^2$

25. $y = x^2 - 4x - 2$

Solve.

26. The hypotenuse of a right triangle is 5 m long. One leg is 3 m longer than the other. Find the lengths of the legs. Round to the nearest tenth.

27. The length of a rectangle is 3 m greater than the width. The area is 70 m². Find the length and the width.

28. The current in a stream moves at a speed of 2 km/h. A boat travels 56 km upstream and 64 km downstream in a total time of 4 hr. What is the speed of the boat in still water?

SKILL MAINTENANCE

Multiply and simplify.

29. $\sqrt{18a}\sqrt{2}$

30. $\sqrt{12xy^2}\sqrt{5xy}$

31. Find an equation of variation where y varies inversely as x and $y = 16$ when $x = 0.0625$.

32. The sides of a rectangle are 1 and $\sqrt{2}$. Find the length of a diagonal.

Add or subtract.

33. $5\sqrt{11} + 7\sqrt{11}$

34. $2\sqrt{90} - \sqrt{40}$

SYNTHESIS

35. Two consecutive integers have squares that differ by 63. Find the integers.

36. Find b such that the trinomial $x^2 + bx + 49$ is a square.

37. Solve: $x - 4\sqrt{x} - 5 = 0$.

38. A square with sides of length s has the same area as a circle with radius of 5 in. Find s.

❖ THINKING IT THROUGH

1. Briefly explain the connection between the number of real-number solutions of a quadratic equation and its x-intercepts.

2. List a quadratic equation with exactly one real-number solution.

3. Solve the following system of equations graphically:

$$y = x^2 - 4x + 1,$$
$$y = 1 - x.$$

4. List the names and give an example of as many types of equations as you can that you have learned to solve in this text.

5. List a quadratic equation with no real-number solutions.

TEST: CHAPTER 9

ANSWERS

Solve.

1. $7x^2 = 35$

2. $7x^2 + 8x = 0$

3. $48 = t^2 + 2t$

1. _____

2. _____

4. $3y^2 - 5y = 2$

5. $(x - 8)^2 = 13$

6. $x^2 = x + 3$

3. _____

4. _____

7. $m^2 - 3m = 7$

8. $10 = 4x + x^2$

9. $3x^2 - 7x + 1 = 0$

5. _____

6. _____

7. _____

10. $x - \dfrac{2}{x} = 1$

11. $\dfrac{4}{x} - \dfrac{4}{x + 2} = 1$

8. _____

9. _____

10. _____

11. _____

12. Solve $x^2 - 4x - 10 = 0$ by completing the square. Show your work.

13. Approximate the solutions to $x^2 - 4x - 10 = 0$ to the nearest tenth.

12. _____

14. Solve for n: $d = an^2 + bn$.

13. _____

14. _____

ANSWERS

Graph.

15. $y = 4 - x^2$

16. $y = -x^2 + x + 5$

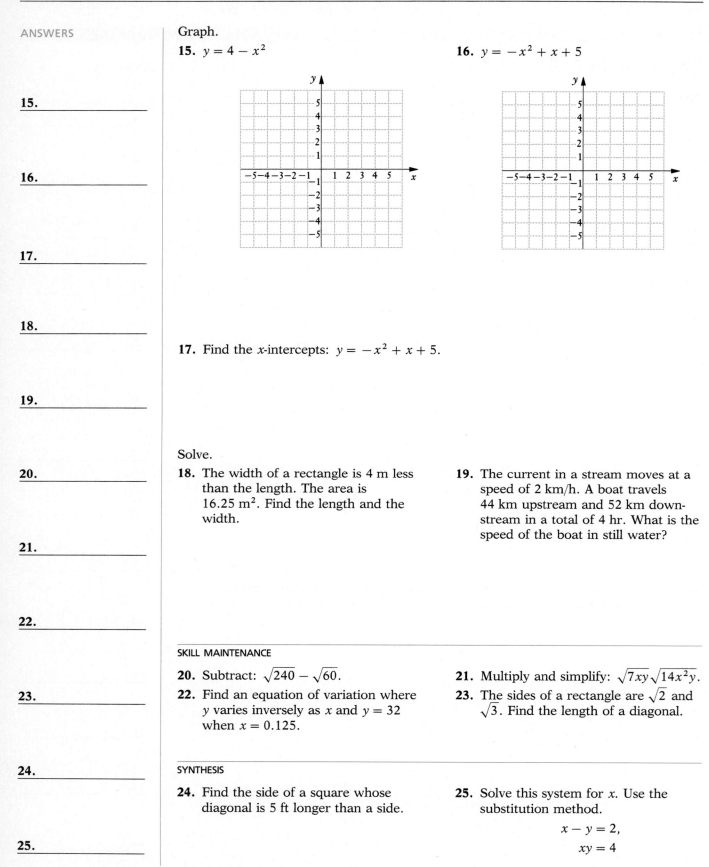

15. _____

16. _____

17. _____

17. Find the x-intercepts: $y = -x^2 + x + 5$.

18. _____

Solve.

19. _____

18. The width of a rectangle is 4 m less than the length. The area is 16.25 m². Find the length and the width.

19. The current in a stream moves at a speed of 2 km/h. A boat travels 44 km upstream and 52 km downstream in a total of 4 hr. What is the speed of the boat in still water?

20. _____

21. _____

22. _____

SKILL MAINTENANCE

23. _____

20. Subtract: $\sqrt{240} - \sqrt{60}$.

21. Multiply and simplify: $\sqrt{7xy}\sqrt{14x^2y}$.

22. Find an equation of variation where y varies inversely as x and $y = 32$ when $x = 0.125$.

23. The sides of a rectangle are $\sqrt{2}$ and $\sqrt{3}$. Find the length of a diagonal.

24. _____

SYNTHESIS

24. Find the side of a square whose diagonal is 5 ft longer than a side.

25. Solve this system for x. Use the substitution method.

$$x - y = 2,$$
$$xy = 4$$

25. _____

CUMULATIVE REVIEW: CHAPTERS 1–9

1. What is the meaning of x^3?

2. Evaluate $(x - 3)^2 + 5$ when $x = 10$.

3. Find decimal notation: $-\dfrac{3}{11}$.

4. Find the LCM of 15 and 48.

5. Find the absolute value: $|-7|$.

Compute and simplify.

6. $-6 + 12 + (-4) + 7$　　　**7.** $2.8 - (-12.2)$　　　**8.** $-\dfrac{3}{8} \div \dfrac{5}{2}$　　　　**9.** $13 \cdot 6 \div 3 \cdot 2 \div 13$

10. Remove parentheses and simplify: $4m + 9 - (6m + 13)$.

Solve.

11. $3x = -24$

12. $3x + 7 = 2x - 5$

13. $3(y - 1) - 2(y + 2) = 0$

14. $x^2 - 8x + 15 = 0$

15. $y - x = 1,$
$y = 3 - x$

16. $x + y = 17,$
$x - y = 7$

17. $4x - 3y = 3,$
$3x - 2y = 4$

18. $x^2 - x - 6 = 0$

19. $x^2 + 3x = 5$

20. $3 - x = \sqrt{x^2 - 3}$

21. $5 - 9x \le 19 + 5x$

22. $-\dfrac{7}{8}x + 7 = \dfrac{3}{8}x - 3$

23. $0.6x - 1.8 = 1.2x$

24. $-3x > 24$

25. $23 - 19y - 3y \ge -12$

26. $3y^2 = 30$

27. $(x - 3)^2 = 6$

28. $\dfrac{6x - 2}{2x - 1} = \dfrac{9x}{3x + 1}$

29. $\dfrac{2x}{x + 1} = 2 - \dfrac{5}{2x}$

30. $\dfrac{2x}{x + 3} + \dfrac{6}{x} + 7 = \dfrac{18}{x^2 + 3x}$

31. $\sqrt{x + 9} = \sqrt{2x - 3}$

Solve the formula for the given letter.

32. $A = \dfrac{4b}{t}$, for b

33. $\dfrac{1}{t} = \dfrac{1}{m} - \dfrac{1}{n}$, for m

34. $r = \sqrt{\dfrac{A}{\pi}}$, for A

35. $y = ax^2 - bx$, for x

Simplify.

36. $x^{-6} \cdot x^2$

37. $\dfrac{y^3}{y^{-4}}$

38. $(2y^6)^2$

39. Collect like terms and arrange in descending order: $2x - 3 + 5x^3 - 2x^3 + 7x^3 + x$.

Compute and simplify.

40. $(4x^3 + 3x^2 - 5) + (3x^3 - 5x^2 + 4x - 12)$

41. $(6x^2 - 4x + 1) - (-2x^2 + 7)$

42. $-2y^2(4y^2 - 3y + 1)$

43. $(2t - 3)(3t^2 - 4t + 2)$

44. $\left(t - \dfrac{1}{4}\right)\left(t + \dfrac{1}{4}\right)$

45. $(3m - 2)^2$

46. $(15x^2y^3 + 10xy^2 + 5) - (5xy^2 - x^2y^2 - 2)$

47. $(x^2 - 0.2y)(x^2 + 0.2y)$

48. $(3p + 4q^2)^2$

49. $\dfrac{4}{2x - 6} \cdot \dfrac{x - 3}{x + 3}$

50. $\dfrac{3a^4}{a^2 - 1} \div \dfrac{2a^3}{a^2 - 2a + 1}$

51. $\dfrac{3}{3x - 1} + \dfrac{4}{5x}$

52. $\dfrac{2}{x^2 - 16} - \dfrac{x - 3}{x^2 - 9x + 20}$

Factor.

53. $8x^2 - 4x$

54. $25x^2 - 4$

55. $6y^2 - 5y - 6$

56. $m^2 - 8m + 16$

57. $x^3 - 8x^2 - 5x + 40$

58. $3a^4 + 6a^2 - 72$

59. $16x^4 - 1$

60. $49a^2b^2 - 4$

61. $9x^2 + 30xy + 25y^2$

62. $2ac - 6ab - 3db + dc$

63. $15x^2 + 14xy - 8y^2$

Simplify.

64. $\dfrac{\dfrac{3}{x}+\dfrac{1}{2x}}{\dfrac{1}{3x}-\dfrac{3}{4x}}$

65. $\sqrt{49}$

66. $-\sqrt{625}$

67. $\sqrt{64x^2}$

68. Multiply: $\sqrt{a+b}\sqrt{a-b}$.

69. Multiply and simplify: $\sqrt{32ab}\sqrt{6a^4b^2}$.

Simplify.

70. $\sqrt{150}$

71. $\sqrt{243x^3y^2}$

72. $\sqrt{\dfrac{100}{81}}$

73. $\sqrt{\dfrac{64}{x^2}}$

74. $4\sqrt{12}+2\sqrt{12}$

75. Divide and simplify: $\dfrac{\sqrt{72}}{\sqrt{45}}$.

76. In a right triangle, $a=9$ and $c=41$. Find b.

Graph on a plane.

77. $y=\dfrac{1}{3}x-2$

78. $2x+3y=-6$

79. $y=-3$

80. $4x-3y>12$

81. $y=x^2+2x+1$

82. $x\ge-3$

83. Solve $9x^2-12x-2=0$ by completing the square. Show your work.

84. Approximate the solutions of $4x^2=4x+1$ to the nearest tenth.

Solve.

85. What percent of 52 is 13?

86. 12 is 20% of what?

87. The speed of a boat in still water is 8 km/h. It travels 60 km upstream and 60 km downstream in a total time of 16 hr. What is the speed of the stream?

88. The length of a rectangle is 7 m more than the width. The length of a diagonal is 13 m. Find the length.

89. Three-fifths of the automobiles entering the city each morning will be parked in city parking lots. There are 3654 such parking spaces filled each morning. How many cars enter the city each morning?

90. A candy shop mixes nuts worth $1.10 per pound with another variety worth $0.80 per pound to make 42 lb of a mixture worth $0.90 per pound. How many pounds of each kind of nuts should be used?

91. In checking records, a contractor finds that crew A can pave a certain length of highway in 8 hr. Crew B can do the same job in 10 hr. How long would they take if they worked together?

92. A student's paycheck varies directly as the number of hours worked. The pay was $242.52 for 43 hr of work. What would the pay be for 80 hr of work? Explain the meaning of the variation constant.

93. Determine whether the graphs of the following equations are parallel, perpendicular, or neither.
$$y - x = 4,$$
$$3y + x = 8$$

94. Find the slope and the y-intercept:
$$-6x + 3y = -24.$$

95. Find the slope of the line containing the points $(-5, -6)$ and $(-4, 9)$.

Find an equation of variation where:

96. y varies directly as x and $y = 100$ when $x = 10$.

97. y varies inversely as x and $y = 100$ when $x = 10$.

SYNTHESIS

98. Solve: $|x| = 12$.

99. Simplify: $\sqrt{\sqrt{\sqrt{81}}}$.

100. Find b such that the trinomial $x^2 - bx + 225$ is a square.

101. Find x.

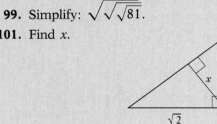

Determine whether the pair of expressions is equivalent.

102. $x^2 - 9$, $(x - 3)(x + 3)$

103. $\dfrac{x + 3}{3}$, x

104. $(x + 5)^2$, $x^2 + 25$

105. $\sqrt{x^2 + 16}$, $x + 4$

106. $\sqrt{x^2}$, $|x|$

FINAL EXAMINATION

ANSWERS

1. Evaluate $x^3 + 5$ when $x = -10$.

2. Find the LCM of 16 and 24.

1. _____

2. _____

3. Find the absolute value of $|-9|$.

3. _____

4. _____

Compute and simplify.

4. $-6.3 + (-8.4) + 5$

5. $-8 - (-3)$

6. $\dfrac{3}{11} \cdot \left(-\dfrac{22}{7}\right)$

5. _____

6. _____

7. _____

7. Remove parentheses and simplify:
$$4y - 5(9 - 3y).$$

8. Simplify:
$$2^3 - 14 \cdot 10 + (3 + 4)^3.$$

8. _____

9. _____

Solve.

9. $x + 8 = 13.6$

10. $4x = -28$

11. $5x + 3 = 2x - 27$

10. _____

11. _____

12. $5(x - 3) - 2(x + 3) = 0$

13. $x^2 - 2x - 24 = 0$

14. $y = x - 7,$
$2x + y = 5$

12. _____

13. _____

14. _____

15. $5x - 3y = -1,$
$4x + 2y = 30$

16. $\dfrac{1}{x} - 2 = 8x$

17. $\sqrt{x^2 - 11} = x - 1$

15. _____

16. _____

17. _____

Copyright © 1991 Addison-Wesley Publishing Co., Inc.

ANSWERS

18. _____

19. _____

20. _____

21. _____

22. _____

23. _____

24. _____

25. _____

26. _____

27. _____

28. _____

29. _____

30. _____

31. _____

32. _____

33. _____

34. _____

35. _____

36. _____

18. $x^2 = 7 - 3x$

19. $2 - 3x \le 12 - 7x$

Solve the formula for the given letter.

20. $A = \dfrac{Bw + 1}{w}$, for w

21. $K = MT + 2$, for M

Simplify.

22. $\dfrac{x^8}{x^{-2}}$

23. $(x^{-5})^2$

24. $x^{-5} \cdot x^{-7}$

25. Collect like terms and arrange in descending order:
$$2y^3 - 3 + 4y^3 - 3y^2 + 12 - y.$$

Compute and simplify.

26. $(2x^2 - 6x + 3) - (4x^2 + 2x - 4)$

27. $-3t^2(2t^4 + 4t^2 + 1)$

28. $(4x - 1)(x^2 - 5x + 2)$

29. $(x - 8)(x + 8)$

30. $(2m - 7)^2$

31. $(3ab^2 + 2c)^2$

32. $(3x^2 - 2y)(3x^2 + 4y)$

33. $\dfrac{x}{x^2 - 9} \cdot \dfrac{x - 3}{x^3}$

34. $\dfrac{3x^5}{4x - 4} \div \dfrac{x}{x^2 - 2x + 1}$

35. $\dfrac{2}{3x - 1} + \dfrac{1}{4x}$

36. $\dfrac{3}{x - 3} - \dfrac{x - 1}{x^2 - 2x - 3}$

Factor.

37. $3x^3 - 15x$

38. $16x^2 - 25$

39. $6x^2 - 13x + 6$

40. $x^2 - 10x + 25$

41. $2ax + 6bx - ay - 3by$

42. $x^8 - 81y^4$

Simplify.

43. $\sqrt{72}$

44. $\dfrac{\sqrt{54}}{\sqrt{45}}$

45. $2\sqrt{8} + 3\sqrt{18}$

46. $\sqrt{24a^2 b}\sqrt{a^3 b^2}$

Graph on a plane.

47. $3x + 2y = -4$

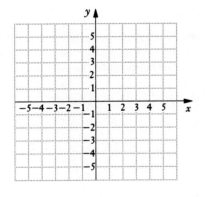

48. $x = -2$

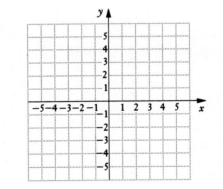

49. $3x - 2y < 6$

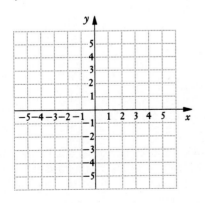

50. $y = x^2 - 2x + 1$

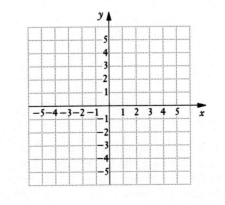

37. _____

38. _____

39. _____

40. _____

41. _____

42. _____

43. _____

44. _____

45. _____

46. _____

47. _____

48. _____

49. _____

50. _____

Solve.

51. The sum of the squares of two consecutive odd integers is 74. Find the integers.

52. Solution A is 75% alcohol and solution B is 50% alcohol. How much of each is needed to make 60 L of a solution that is $66\frac{2}{3}\%$ alcohol?

51. _____

52. _____

53. An airplane flew for 6 hr with a 10-km/h tailwind. The return flight against the same wind took 8 hr. Find the speed of the plane in still air.

54. The width of a rectangle is 3 m less than the length. The area is 88 m². Find the length and the width.

53. _____

54. _____

55. Find the slope of the line containing the points $(-2, 3)$ and $(4, -5)$.

56. Determine whether the graphs of the following equations are parallel, perpendicular, or neither.

$$y = 2x + 7,$$
$$2y + x = 6$$

55. _____

56. _____

Find an equation of variation where:

57. y varies directly as x and $y = 200$ when $x = 25$.

58. y varies inversely as x and $y = 200$ when $x = 25$.

57. _____

58. _____

SYNTHESIS

59. A side of a square is five less than a side of an equilateral triangle. The perimeter of the square is the same as the perimeter of the triangle. Find the length of a side of the square and the length of a side of the triangle.

60. Find c such that the trinomial $x^2 - 24x + c$ is a square.

59. _____

60. _____

Tables

TABLE 1

FRACTIONAL AND DECIMAL EQUIVALENTS

Fractional Notation	Decimal Notation	Percent Notation
$\dfrac{1}{10}$	0.1	10%
$\dfrac{1}{8}$	0.125	12.5%, or $12\dfrac{1}{2}$%
$\dfrac{1}{6}$	$0.16\overline{6}$	$16.6\overline{6}$%, or $16\dfrac{2}{3}$%
$\dfrac{1}{5}$	0.2	20%
$\dfrac{1}{4}$	0.25	25%
$\dfrac{3}{10}$	0.3	30%
$\dfrac{1}{3}$	$0.333\overline{3}$	$33.3\overline{3}$%, or $33\dfrac{1}{3}$%
$\dfrac{3}{8}$	0.375	37.5%, or $37\dfrac{1}{2}$%
$\dfrac{2}{5}$	0.4	40%
$\dfrac{1}{2}$	0.5	50%
$\dfrac{3}{5}$	0.6	60%
$\dfrac{5}{8}$	0.625	62.5%, or $62\dfrac{1}{2}$%
$\dfrac{2}{3}$	$0.666\overline{6}$	$66.6\overline{6}$%, or $66\dfrac{2}{3}$%
$\dfrac{7}{10}$	0.7	70%
$\dfrac{3}{4}$	0.75	75%
$\dfrac{4}{5}$	0.8	80%
$\dfrac{5}{6}$	$0.83\overline{3}$	$83.3\overline{3}$%, or $83\dfrac{1}{3}$%
$\dfrac{7}{8}$	0.875	87.5%, or $87\dfrac{1}{2}$%
$\dfrac{9}{10}$	0.9	90%
$\dfrac{1}{1}$	1	100%

TABLE 2

SQUARES AND SQUARE ROOTS

N	$\sqrt{N}$	N^2	N	$\sqrt{N}$	N^2	N	$\sqrt{N}$	N^2	N	$\sqrt{N}$	N^2
2	1.414	4	27	5.196	729	52	7.211	2704	77	8.775	5929
3	1.732	9	28	5.292	784	53	7.280	2809	78	8.832	6084
4	2	16	29	5.385	841	54	7.348	2916	79	8.888	6241
5	2.236	25	30	5.477	900	55	7.416	3025	80	8.944	6400
6	2.449	36	31	5.568	961	56	7.483	3136	81	9	6561
7	2.646	49	32	5.657	1024	57	7.550	3249	82	9.055	6724
8	2.828	64	33	5.745	1089	58	7.616	3364	83	9.110	6889
9	3	81	34	5.831	1156	59	7.681	3481	84	9.165	7056
10	3.162	100	35	5.916	1225	60	7.746	3600	85	9.220	7225
11	3.317	121	36	6	1296	61	7.810	3721	86	9.274	7396
12	3.464	144	37	6.083	1369	62	7.874	3844	87	9.327	7569
13	3.606	169	38	6.164	1444	63	7.937	3969	88	9.381	7744
14	3.742	196	39	6.245	1521	64	8	4096	89	9.434	7921
15	3.873	225	40	6.325	1600	65	8.062	4225	90	9.487	8100
16	4	256	41	6.403	1681	66	8.124	4356	91	9.539	8281
17	4.123	289	42	6.481	1764	67	8.185	4489	92	9.592	8464
18	4.243	324	43	6.557	1849	68	8.246	4624	93	9.644	8649
19	4.359	361	44	6.633	1936	69	8.307	4761	94	9.695	8836
20	4.472	400	45	6.708	2025	70	8.367	4900	95	9.747	9025
21	4.583	441	46	6.782	2116	71	8.426	5041	96	9.798	9216
22	4.690	484	47	6.856	2209	72	8.485	5184	97	9.849	9409
23	4.796	529	48	6.928	2304	73	8.544	5329	98	9.899	9604
24	4.899	576	49	7	2401	74	8.602	5476	99	9.950	9801
25	5	625	50	7.071	2500	75	8.660	5625	100	10	10,000
26	5.099	676	51	7.141	2601	76	8.718	5776			

Answers

CHAPTER R

Margin Exercises, Section R.1, pp. 3-6

1. $1 \cdot 9$, or $3 \cdot 3$; 1, 3, 9 **2.** $1 \cdot 16$, $2 \cdot 8$, or $4 \cdot 4$ (other answers are possible); 1, 2, 4, 8, 16 **3.** $1 \cdot 18$; $2 \cdot 9$; $3 \cdot 6$; $2 \cdot 3 \cdot 3$ **4.** $1 \cdot 20$; $2 \cdot 10$; $4 \cdot 5$; $2 \cdot 2 \cdot 5$ **5.** 13 **6.** $2 \cdot 2 \cdot 2 \cdot 2 \cdot 3$ **7.** $2 \cdot 5 \cdot 5$ **8.** $2 \cdot 5 \cdot 7 \cdot 11$ **9.** 15, 30, 45, 60, ... **10.** 40 **11.** 360 **12.** 2520 **13.** 18 **14.** 24 **15.** 36 **16.** 210

Margin Exercises, Section R.2, pp. 9-16

1. $\frac{8}{12}$ **2.** $\frac{21}{35}$ **3.** $\frac{14}{16}, \frac{21}{24}, \frac{28}{32}$; answers may vary **4.** $\frac{2}{3}$ **5.** $\frac{19}{9}$ **6.** $\frac{8}{7}$ **7.** $\frac{1}{2}$ **8.** 4 **9.** $\frac{5}{2}$ **10.** $\frac{35}{16}$ **11.** $\frac{7}{5}$ **12.** 2 **13.** $\frac{23}{15}$ **14.** $\frac{7}{12}$ **15.** $\frac{2}{15}$ **16.** $\frac{7}{36}$ **17.** $\frac{11}{4}$ **18.** $\frac{7}{15}$ **19.** $\frac{1}{5}$ **20.** 3 **21.** $\frac{21}{20}$ **22.** $\frac{8}{21}$ **23.** $\frac{5}{6}$ **24.** $\frac{8}{15}$ **25.** $\frac{1}{64}$ **26.** 81

Margin Exercises, Section R.3, pp. 19-24

1. $\frac{568}{1000}$ **2.** $\frac{23}{10}$ **3.** $\frac{8904}{100}$ **4.** 4.131 **5.** 0.4131 **6.** 5.73 **7.** 284.455 **8.** 268.63 **9.** 27.676 **10.** 64.683 **11.** 99.59 **12.** 239.883 **13.** 5.868 **14.** 0.5868 **15.** 51.53808 **16.** 48.9 **17.** 15.82 **18.** 1.28 **19.** 17.95 **20.** 856 **21.** 0.85 **22.** 0.625 **23.** $0.\overline{6}$ **24.** $7.\overline{63}$ **25.** 2.8 **26.** 13.9 **27.** 7.0 **28.** 7.83 **29.** 34.68 **30.** 0.03 **31.** 0.943 **32.** 8.004 **33.** 43.112 **34.** 37.401 **35.** 7459.355 **36.** 7459.35 **37.** 7459.4 **38.** 7459 **39.** 7460

Margin Exercises, Section R.4, pp. 27-28

1. 0.462 **2.** 1 **3.** $\frac{67}{100}$ **4.** $\frac{456}{1000}$ **5.** $\frac{1}{400}$ **6.** 677% **7.** 99.44% **8.** 25% **9.** 37.5% **10.** $66.\overline{6}\%$, or $66\frac{2}{3}\%$

Margin Exercises, Section R.5, pp. 31-34

1. 5^3 **2.** 6^5 **3.** $(1.08)^2$ **4.** 10,000 **5.** 512 **6.** 1.331

7. 13   **8.** 1000   **9.** 250   **10.** 178   **11.** 48   **12.** 3.55625
13. 4

CHAPTER 1

Margin Exercises, Section 1.1, pp. 43-46

1. $2,866 - 2,128 = x$ **2.** 64   **3.** 28   **4.** 60   **5.** 192 ft^2
6. 3.375 hr   **7.** 25   **8.** 16   **9.** $x - 12$ **10.** $y + 12$, or
$12 + y$ **11.** $m - 4$ **12.** $\frac{1}{2}p$ **13.** $6 + 8x$, or $8x + 6$
14. $a - b$ **15.** $59\%x$, or $0.59x$ **16.** $xy - 200$ **17.** $p + q$

Margin Exercises, Section 1.2, pp. 49-56

1. 8, −5   **2.** 134, −76   **3.** −10, 148   **4.** −137, 289

5. $-\frac{7}{2}$ 0

6. 1.4, 0

7. $-\frac{11}{4}$ 0

8. −0.375   **9.** $-0.\overline{54}$ **10.** $1.\overline{3}$ **11.** <   **12.** <   **13.** >   **14.** >
15. >   **16.** <   **17.** <   **18.** >   **19.** $7 > -5$ **20.** $4 < x$
21. False   **22.** True   **23.** True   **24.** 8   **25.** 0   **26.** 9   **27.** $\frac{2}{3}$
28. 5.6

Margin Exercises, Section 1.3, pp. 59-62

1. −8   **2.** −3   **3.** −8   **4.** 4   **5.** 0   **6.** −2   **7.** −11   **8.** −12
9. 2   **10.** −4   **11.** −2   **12.** 0   **13.** −22   **14.** 3   **15.** 0.53
16. 2.3   **17.** −7.7   **18.** −6.2   **19.** $-\frac{2}{9}$ **20.** $-\frac{19}{20}$ **21.** −58
22. −56   **23.** −14   **24.** −12   **25.** 4   **26.** −8.7   **27.** 7.74
28. $\frac{8}{9}$ **29.** 0   **30.** −12   **31.** −14, 14   **32.** −1, 1   **33.** 19,
−19   **34.** 1.6, −1.6   **35.** $-\frac{2}{3}, \frac{2}{3}$ **36.** $\frac{9}{8}, -\frac{9}{8}$ **37.** 4   **38.** 13.4
39. 0   **40.** $-\frac{1}{4}$

Margin Exercises, Section 1.4, pp. 65-68

1. −10   **2.** 3   **3.** −5   **4.** −2   **5.** −11   **6.** 4   **7.** −2   **8.** −6
9. −16   **10.** 7.1   **11.** 3   **12.** 0   **13.** $\frac{3}{2}$ **14.** −8   **15.** 7
16. −3   **17.** −23.3   **18.** 0   **19.** −9   **20.** 17   **21.** 12.7
22. $17 profit   **23.** 50° C

Margin Exercises, Section 1.5, pp. 73-76

1. $2 \cdot 10 = 20$; $1 \cdot 10 = 10$; $0 \cdot 10 = 0$; $-1 \cdot 10 = -10$;
$-2 \cdot 10 = -20$; $-3 \cdot 10 = -30$ **2.** −18   **3.** −100   **4.** −80
5. $-\frac{5}{9}$ **6.** −30.033   **7.** $-\frac{7}{10}$ **8.** $1 \cdot (-10) = -10$;
$0 \cdot (-10) = 0$; $-1 \cdot (-10) = 10$; $-2 \cdot (-10) = 20$;

$-3 \cdot (-10) = 30$ **9.** 12   **10.** 32   **11.** 35   **12.** $\frac{20}{63}$ **13.** $\frac{2}{3}$
14. 13.455   **15.** −30   **16.** 30   **17.** 0   **18.** $-\frac{8}{3}$ **19.** −30
20. −30.75   **21.** $-\frac{5}{3}$ **22.** 120   **23.** −120   **24.** 6   **25.** 4, −4
26. 9, −9   **27.** 48, 48

Margin Exercises, Section 1.6, pp. 79-82

1. −2   **2.** 5   **3.** −3   **4.** 8   **5.** −6   **6.** $-\frac{30}{7}$ **7.** Undefined
8. 0   **9.** $\frac{3}{2}$ **10.** $-\frac{4}{5}$ **11.** $-\frac{1}{3}$ **12.** −5   **13.** $\frac{1}{5.78}$ **14.** $\frac{2}{3}$
15. *First row:* $\frac{2}{3}, -\frac{2}{3}, \frac{3}{2}$; *second row:* $-\frac{5}{4}, \frac{5}{4}, -\frac{4}{5}$; *third row:*
0, 0, undefined; *fourth row:* 1, −1, 1; *fifth row:*
−4.5, 4.5, $-\frac{1}{4.5}$ **16.** $\frac{4}{7} \cdot \left(-\frac{5}{3}\right)$ **17.** $5 \cdot \left(-\frac{1}{8}\right)$ **18.** $(a - b) \cdot \left(\frac{1}{7}\right)$
19. $-23 \cdot a$ **20.** $-5 \cdot \left(\frac{1}{7}\right)$ **21.** $-\frac{20}{21}$ **22.** $-\frac{12}{5}$ **23.** $\frac{16}{7}$ **24.** −7
25. $\frac{5}{-6}, -\frac{5}{6}$ **26.** $\frac{-8}{7}, \frac{8}{-7}$ **27.** $\frac{-10}{3}, -\frac{10}{3}$

Margin Exercises, Section 1.7, pp. 85-92

1.

	$x + x$	$2x$
$x = 3$	6	6
$x = -6$	−12	−12
$x = 4.8$	9.6	9.6

2.

	$x + 3x$	$5x$
$x = 2$	8	10
$x = -6$	−24	−30
$x = 4.8$	19.2	24

3. $\frac{3y}{4y}$ **4.** $\frac{3}{4}$ **5.** $-\frac{4}{3}$ **6.** 1; 1   **7.** −10; −10   **8.** $9 + x$ **9.** qp
10. $t + xy$, or $yx + t$ **11.** 19; 19   **12.** 150; 150
13. $(a + b) + 2$ **14.** $(3v)w$ **15.** $(4t)u$, $(tu)4$, $t(4u)$;
answers may vary   **16.** $(2 + r) + s$, $(r + s) + 2$, $s + (r + 2)$;
answers may vary   **17.** (a) 63; (b) 63   **18.** (a) 80; (b) 80
19. (a) 28; (b) 28   **20.** (a) 8; (b) 8   **21.** (a) −4; (b) −4
22. (a) −25; (b) −25   **23.** $5x, -4y, 3$ **24.** $-4y, -2x, 3z$
25. $3x - 15$ **26.** $5x - 5y + 20$ **27.** $-2x + 6$
28. $-5x + 10y - 20z$ **29.** $6(x - 2)$ **30.** $3(x - 2y + 3)$
31. $b(x + y - z)$ **32.** $2(8a - 18b + 21)$
33. $\frac{1}{8}(3x - 5y + 7)$ **34.** $-4(3x - 8y + 4z)$ **35.** $3x$ **36.** $6x$
37. $-8x$ **38.** $0.59x$ **39.** $3x + 3y$ **40.** $-4x - 5y - 7$

Margin Exercises, Section 1.8, pp. 97-100

1. $-x - 2$ **2.** $-5x - 2y - 8$ **3.** $-6 + t$ **4.** $-x + y$

5. $4a - 3t + 10$　**6.** $-18 + m + 2n - 4z$　**7.** $2x - 9$
8. $3y + 2$　**9.** $2x - 7$　**10.** $3y + 3$　**11.** $-2a + 8b - 3c$
12. $-9x - 8y$　**13.** $-16a + 18$　**14.** $-26a + 41b - 48c$
15. 2　**16.** 18　**17.** 6　**18.** 17　**19.** $5x - y - 8$　**20.** -1237
21. 381

CHAPTER 2

Margin Exercises, Section 2.1, pp. 111-114

1. False　**2.** True　**3.** Neither true nor false　**4.** Yes
5. No　**6.** No　**7.** -5　**8.** 13.2　**9.** -6.5　**10.** -2　**11.** $\frac{31}{8}$

Margin Exercises, Section 2.2, pp. 117-120

1. $-\frac{4}{5}$　**2.** $-\frac{7}{4}$　**3.** 8　**4.** -18　**5.** 10　**6.** 28　**7.** 7800　**8.** -3

Margin Exercises, Section 2.3, pp. 123-128

1. 5　**2.** 4　**3.** 4　**4.** 39　**5.** $-\frac{3}{2}$　**6.** -4.3　**7.** -3　**8.** 800
9. 1　**10.** 2　**11.** 2　**12.** $\frac{17}{2}$　**13.** $\frac{8}{3}$　**14.** -4.3　**15.** 2　**16.** 3
17. -2　**18.** $-\frac{1}{2}$

Margin Exercises, Section 2.4, pp. 133-138

1. 28 in., 30 in.　**2.** 5　**3.** 228, 229　**4.** 240.9 mi
5. Width: 9 ft; length: 12 ft　**6.** $30°, 90°, 60°$

Margin Exercises, Section 2.5, pp. 143-146

1. 32%　**2.** 25%　**3.** 225　**4.** 50　**5.** 11.04
6. $111,416 \text{ mi}^2$　**7.** \$8400　**8.** \$658

Margin Exercises, Section 2.6, pp. 149-150

1. 2.8 mi　**2.** $I = \frac{E}{R}$　**3.** $D = \frac{C}{\pi}$　**4.** $c = 4A - a - b - d$
5. $I = \frac{9R}{E}$

Margin Exercises, Section 2.7, pp. 153-158

1. (a) No; (b) no; (c) no; (d) yes; (e) no　**2.** (a) Yes;
(b) yes; (c) yes; (d) no; (e) yes

3. $x < 4$

4. $y \geq -2$

5. $-2 \leq x < 4$

6. $\{x \mid x > 2\}$;

7. $\{x \mid x \leq 3\}$;

8. $\{x \mid x < -3\}$;

9. $\left\{x \mid x \geq \frac{2}{15}\right\}$　**10.** $\{y \mid y \leq -3\}$

11. $\{x \mid x < 8\}$;

12. $\{y \mid y \geq 32\}$;

13. $\{x \mid x \geq -6\}$　**14.** $\left\{y \mid y < -\frac{13}{5}\right\}$　**15.** $\left\{x \mid x > -\frac{1}{4}\right\}$
16. $\left\{y \mid y \geq \frac{19}{9}\right\}$　**17.** $\left\{y \mid y \geq \frac{19}{9}\right\}$　**18.** $\{x \mid x \geq -2\}$

Margin Exercises, Section 2.8, pp. 163-164

1. $x \leq 8$　**2.** $y > -2$　**3.** $s \leq 180$　**4.** $p \geq \$5800$
5. $2x - 32 > 5$　**6.** $\{x \mid x \geq 84\}$　**7.** $\{C \mid C < 1063°\}$

CHAPTER 3

Margin Exercises, Section 3.1, pp. 175-180

1. $5 \cdot 5 \cdot 5 \cdot 5$　**2.** $x \cdot x \cdot x \cdot x \cdot x$　**3.** $3t \cdot 3t$　**4.** $3 \cdot t \cdot t$　**5.** 6
6. 1　**7.** 8.4　**8.** 1　**9.** 125　**10.** 3215.36 cm^2　**11.** 119
12. $3; -3$　**13.** (a) 144; (b) 36 (c) No　**14.** 3^{10}　**15.** x^{10}
16. p^{24}　**17.** x^5　**18.** $a^9 b^8$　**19.** 4^3　**20.** y^4　**21.** p^9
22. $a^4 b^2$　**23.** $\frac{1}{4^3} = \frac{1}{64}$　**24.** $\frac{1}{5^2} = \frac{1}{25}$　**25.** $\frac{1}{2^4} = \frac{1}{16}$　**26.** $\frac{1}{(-2)^3} = -\frac{1}{8}$
27. $\frac{4}{p^3}$　**28.** x^2　**29.** 5^2　**30.** $\frac{1}{x^7}$　**31.** $\frac{1}{7^5}$　**32.** b　**33.** t^6

Margin Exercises, Section 3.2, pp. 185-190

1. 3^{20}　**2.** $\frac{1}{x^{12}}$　**3.** y^{15}　**4.** $\frac{1}{x^{32}}$　**5.** $\frac{16x^{20}}{y^{12}}$　**6.** $\frac{25x^{10}}{y^{12}z^6}$　**7.** x^{74}　**8.** $\frac{27z^{24}}{y^6 x^{15}}$
9. $\frac{x^{12}}{25}$　**10.** $\frac{8t^{15}}{w^{12}}$　**11.** $\frac{9}{x^8}$　**12.** 5.17×10^{-4}　**13.** 5.23×10^8
14. $689,300,000,000$　**15.** 0.0000567　**16.** 5.6×10^{-15}
17. 7.462×10^{-13}　**18.** 2.0×10^3　**19.** 5.5×10^2
20. $3.\overline{3} \times 10^{-2}$　**21.** 2.3725×10^9 gal

Margin Exercises, Section 3.3, pp. 193-198

1. $4x^2 - 3x + \frac{5}{4}$; $15y^3$; $-7x^3 + 1.1$; answers may vary
2. -19　**3.** -104　**4.** -13　**5.** 8　**6.** 132　**7.** 360 ft
8. 7.55 parts per million　**9.** $-9x^3 + (-4x^5)$
10. $-2y^3 + 3y^7 + (-7y)$　**11.** $3x^2, 6x, \frac{1}{2}$　**12.** $-4y^5, 7y^2,$
$-3y, -2$　**13.** $4x^3$ and $-x^3$　**14.** $4t^4$ and $-7t^4$; $-9t^3$ and
$10t^3$　**15.** $2, -7, -8.5, 10, -4$　**16.** $8x^2$　**17.** $2x^3 + 7$
18. $-\frac{1}{4}x^5 + 2x^2$　**19.** $-4x^3$　**20.** $5x^3$　**21.** $25 - 3x^5$　**22.** $6x$

23. $4x^3 + 4$ **24.** $-\frac{1}{4}x^3 + 4x^2 + 7$ **25.** $3x^2 + x^3 + 9$

26. $6x^7 + 3x^5 - 2x^4 + 4x^3 + 5x^2 + x$ **27.** $7x^5 - 5x^4 + 2x^3 +$ $4x^2 - 3$ **28.** $14t^7 - 10t^5 + 7t^2 - 14$ **29.** $-2x^2 - 3x + 2$

30. $10x^4 - 8x - \frac{1}{2}$ **31.** 4, 2, 1, 0; 4   **32.** x **33.** $x^3, x^2, x,$ x^0 **34.** x^2, x **35.** x^3 **36.** Monomial   **37.** None of these **38.** Binomial   **39.** Trinomial

Margin Exercises, Section 3.4, pp. 203-206

1. $x^2 + 7x + 3$ **2.** $-4x^5 + 7x^4 + x^3 + 2x^2 + 4$

3. $24x^4 + 5x^3 + x^2 + 1$ **4.** $2x^3 + \frac{10}{3}$ **5.** $2x^2 - 3x - 1$

6. $8x^3 - 2x^2 - 8x + \frac{5}{2}$ **7.** $-8x^4 + 4x^3 + 12x^2 + 5x - 8$

8. $-x^3 + x^2 + 3x + 3$ **9.** $-(12x^4 - 3x^2 + 4x);$ $-12x^4 + 3x^2 - 4x$ **10.** $-(-4x^4 + 3x^2 - 4x); \ 4x^4 - 3x^2 + 4x$

11. $-\left(-13x^6 + 2x^4 - 3x^2 + x - \frac{5}{13}\right); \ 13x^6 - 2x^4 + 3x^2 - x + \frac{5}{13}$

12. $-(-7y^3 + 2y^2 - y + 3); \ 7y^3 - 2y^2 + y - 3$

13. $-4x^3 + 6x - 3$ **14.** $-5x^4 - 3x^2 - 7x + 5$

15. $-14x^{10} + \frac{1}{2}x^5 - 5x^3 + x^2 - 3x$ **16.** $2x^3 + 2x + 8$

17. $x^2 - 6x - 2$ **18.** $-8x^4 - 5x^3 + 8x^2 - 1$

19. $x^3 - x^2 - \frac{4}{3}x - 0.9$ **20.** $2x^3 - 5x^2 - 2x - 5$

21. $-x^5 - 2x^3 + 3x^2 - 2x + 2$ **22.** $\frac{7}{2}x^2$ **23.** $\pi x^2 - x^2,$ or $(\pi - 1)x^2$

Margin Exercises, Section 3.5, pp. 211-214

1. $-15x$ **2.** $-x^2$ **3.** x^2 **4.** $-x^5$ **5.** $12x^7$ **6.** $-8y^{11}$ **7.** $7y^5$ **8.** 0   **9.** $8x^2 + 16x$ **10.** $-15t^3 + 6t^2$ **11.** $5x^6 + 25x^5 -$ $30x^4 + 40x^3$ **12.** $x^3 + 13x + 40$ **13.** $x^2 + x - 20$ **14.** $5x^2 - 17x - 12$ **15.** $6x^2 - 19x + 15$ **16.** $x^4 + 3x^3 +$ $x^2 + 15x - 20$ **17.** $6y^5 - 20y^3 + 15y^2 + 14y - 35$ **18.** $3x^3 + 13x^2 - 6x + 20$ **19.** $20x^4 - 16x^3 + 32x^2 -$ $32x - 16$ **20.** $6x^4 - x^3 - 18x^2 - x + 10$

Margin Exercises, Section 3.6, pp. 217-222

1. $x^2 + 7x + 12$ **2.** $x^2 - 2x - 15$ **3.** $2x^2 + 9x + 4$

4. $2x^3 - 4x^2 - 3x + 6$ **5.** $12x^5 + 6x^2 + 10x^3 + 5$ **6.** $y^6 - 49$

7. $-2x^7 + x^5 + x^3$ **8.** $t^2 + 8t + 15$ **9.** $x^2 - \frac{16}{25}$

10. $x^5 + 0.5x^3 - 0.5x^2 - 0.25$ **11.** $8 + 2x^2 - 15x^4$

12. $30x^5 - 3x^4 - 6x^3$ **13.** $x^2 - 25$ **14.** $4x^2 - 9$ **15.** $x^2 - 4$

16. $x^2 - 49$ **17.** $36 - 16y^2$ **18.** $4x^6 - 1$

19. $x^2 + 16x + 64$ **20.** $x^2 - 10x + 25$ **21.** $x^2 + 4x + 4$

22. $a^2 - 8a + 16$ **23.** $4x^2 + 20x + 25$

24. $16x^4 - 24x^3 + 9x^2$ **25.** $49 + 14y + y^2$

26. $9x^4 - 30x^2 + 25$ **27.** $x^2 + 11x + 30$ **28.** $t^2 - 16$

29. $-8x^5 + 20x^4 + 40x^2$ **30.** $81x^4 + 18x^2 + 1$

31. $4a^2 + 6a - 40$ **32.** $25x^2 + 5x + \frac{1}{4}$ **33.** $4x^2 - 2x + \frac{1}{4}$

34. $x^3 - 3x^2 + 6x - 8$

Margin Exercises, Section 3.7, pp. 227-230

1. -7940 **2.** -176 **3.** 433.32 ft^2 **4.** $-3, 3, -2, 1, 2$ **5.** 3, 7, 1, 1, 0; 7   **6.** $2x^2y + 3xy$ **7.** $5pq + 4$ **8.** $-4x^3 + 2x^2 -$ $4x + 2$ **9.** $14x^3y + 7x^2y - 3xy - 2y$

10. $-5p^2q^4 + 2p^2q^2 + 3p^2q + 6pq^2 + 3q + 5$

11. $-8s^4t + 6s^3t^2 + 2s^2t^3 - s^2t^2$ **12.** $-9p^4q + 9p^3q^2 -$ $4p^2q^3 - 9q^4$ **13.** $x^5y^5 + 2x^4y^2 + 3x^3y^3 + 6x^2$

14. $p^5q - 4p^3q^3 + 3pq^3 + 6q^4$ **15.** $3x^3y + 6x^2y^3 +$ $2x^3 + 4x^2y^2$ **16.** $2x^2 - 11xy + 15y^2$

17. $16x^2 + 40xy + 25y^2$ **18.** $9x^4 - 12x^3y^2 + 4x^2y^4$

19. $4x^2y^4 - 9x^2$ **20.** $16y^2 - 9x^2y^4$ **21.** $9y^2 + 24y +$ $16 - 9x^2$ **22.** $4a^2 - 25b^2 - 10bc - c^2$

Margin Exercises, Section 3.8, pp. 235-238

1. $x^2 + 3x + 2$ **2.** $2x^2 + x - \frac{2}{3}$ **3.** $4x^2 - \frac{3}{2}x + \frac{1}{2}$

4. $2x^2y^4 - 3xy^2 + 5y$ **5.** $x - 2$ **6.** $x + 4$ **7.** $x + 4, \text{R} -2,$ or $x + 4 + \frac{-2}{x+3}$ **8.** $x^2 + x + 1$

CHAPTER 4

Margin Exercises, Section 4.1, pp. 249-252

1. (a) $12x^2$; (b) $(3x)(4x), (2x)(6x),$ answers may vary

2. (a) $16x^3$; (b) $(2x)(8x^2), (4x)(4x^2),$ answers may vary

3. $(8x)(x^3); (4x^2)(2x^2); (2x^3)(4x);$ answers may vary

4. $(7x)(3x); (-7x)(-3x); (21x)(x);$ answers may vary

5. $(6x^4)(x); (-2x^3)(-3x^2); (3x^3)(2x^2);$ answers may vary

6. (a) $3x + 6$; (b) $3(x + 2)$ **7.** (a) $2x^3 + 10x^2 + 8x;$

(b) $2x(x^2 + 5x + 4)$ **8.** $x(x + 3)$ **9.** $y^2(3y^4 - 5y + 2)$

10. $3x^2(3x^2 - 5x + 1)$ **11.** $\frac{1}{4}(3t^3 + 5t^2 + 7t + 1)$

12. $7x^3(5x^4 - 7x^3 + 2x^2 - 9)$ **13.** $(x^2 + 3)(x + 7)$

14. $(x^2 + 2)(a + b)$ **15.** $(x^2 + 3)(x + 7)$

16. $(2t^2 + 3)(4t + 1)$ **17.** $(3m^3 + 2)(m^2 - 5)$

18. $(2x^2 - 3)(2x - 3)$ **19.** Not factorable by grouping

Margin Exercises, Section 4.2, pp. 255-258

1. $(x + 4)(x + 3)$ **2.** $(x + 9)(x + 4)$ **3.** $(x - 5)(x - 3)$ **4.** $(t - 5)(t - 4)$ **5.** $x(x + 6)(x - 2)$ **6.** $(y - 6)(y + 2)$ **7.** $(t^2 + 7)(t^2 - 2)$ **8.** $p(p - q - 3q^2)$ **9.** Not factorable **10.** $(x + 4)^2$

Margin Exercises, Section 4.3, pp. 261-264

1. $(2x + 5)(x - 3)$ **2.** $(4x + 1)(3x - 5)$ **3.** $(3x - 4)(x - 5)$ **4.** $2(5x - 4)(2x - 3)$ **5.** $(2x + 1)(3x + 2)$

6. $(2a - b)(3a - b)$ **7.** $3(2x + 3y)(x + y)$

Margin Exercises, Section 4.4, pp. 267-268

1. $(2x + 1)(3x + 2)$ **2.** $(4x + 1)(3x - 5)$
3. $3(2x + 3)(x + 1)$ **4.** $2(5x - 4)(2x - 3)$

Margin Exercises, Section 4.5, pp. 271-276

1. Yes **2.** No **3.** No **4.** Yes **5.** No **6.** Yes **7.** No
8. Yes **9.** $(x + 1)^2$ **10.** $(x - 1)^2$ **11.** $(t + 2)^2$
12. $(5x - 7)^2$ **13.** $(7 - 4y)^2$ **14.** $3(4m + 5)^2$ **15.** $(p^2 + 9)^2$
16. $z^3(2z - 5)^2$ **17.** $(3a + 5b)^2$ **18.** Yes **19.** No **20.** No
21. No **22.** Yes **23.** Yes **24.** Yes **25.** $(x + 3)(x - 3)$
26. $4(4 + t)(4 - t)$ **27.** $(a + 5b)(a - 5b)$
28. $x^4(8 + 5x)(8 - 5x)$ **29.** $5(1 + 2t^3)(1 - 2t^3)$
30. $(9x^2 + 1)(3x + 1)(3x - 1)$ **31.** $(7p^2 + 5q^3)(7p^2 - 5q^3)$

Margin Exercises, Section 4.6, pp. 281-284

1. $3(m^2 + 1)(m + 1)(m - 1)$ **2.** $(x^3 + 4)^2$
3. $2x^2(x + 1)(x + 3)$ **4.** $(3x^2 - 2)(x + 4)$
5. $8x(x - 5)(x + 5)$ **6.** $x^2y(x^2y + 2x + 3)$
7. $2p^4q^2(5p^2 + 2pq + q^2)$ **8.** $(a - b)(2x + 5 + y^2)$
9. $(a + b)(x^2 + y)$ **10.** $(x^2 + y^2)^2$ **11.** $(xy + 1)(xy + 4)$
12. $(p^2 + 9q^2)(p + 3q)(p - 3q)$

Margin Exercises, Section 4.7, pp. 289-292

1. $3, -4$ **2.** $7, 3$ **3.** $-\frac{1}{4}, \frac{2}{3}$ **4.** $0, \frac{17}{3}$ **5.** $-2, 3$ **6.** $7, -4$
7. 3 **8.** $0, 4$ **9.** $\frac{4}{3}, -\frac{4}{3}$ **10.** $3, -3$

Margin Exercises, Section 4.8, pp. 295-298

1. $5, -5$ **2.** $7, 8$ **3.** $-4, 5$ **4.** Length: 5 cm; width: 3 cm
5. **(a)** 342 **(b)** 9 **6.** 22 and 23 **7.** 3 m, 4 m

CHAPTER 5

Margin Exercises, Section 5.1, pp. 311-316

1. $\{x \mid x \neq 3\}$ **2.** $\{x \mid x \neq -8 \text{ and } x \neq 3\}$ **3.** $\frac{x(2x+1)}{x(3x-2)}$
4. $\frac{(x+1)(x+2)}{(x-2)(x+2)}$ **5.** $\frac{-1(x-8)}{-1(x-y)}$ **6.** 5 **7.** $\frac{x}{4}$ **8.** $\frac{2x+1}{3x+2}$ **9.** $\frac{x+1}{2x+1}$
10. $x + 2$ **11.** $\frac{y+2}{4}$ **12.** -1 **13.** -1 **14.** $\frac{a-2}{a-3}$ **15.** $\frac{x-5}{2}$

Margin Exercises, Section 5.2, pp. 321-322

1. $\frac{2}{7}$ **2.** $\frac{2x^3-1}{x^2+5}$ **3.** $\frac{1}{x-5}$ **4.** $x^2 - 3$ **5.** $\frac{6}{35}$ **6.** $\frac{(x-3)(x-2)}{(x+5)(x+5)}$ **7.** $\frac{x-3}{x+2}$
8. $\frac{(x-3)(x-2)}{x+2}$ **9.** $\frac{y+1}{y-1}$

Margin Exercises, Section 5.3, pp. 325-326

1. 144 **2.** 12 **3.** 10 **4.** 120 **5.** $\frac{35}{144}$ **6.** $\frac{1}{4}$ **7.** $\frac{11}{10}$ **8.** $\frac{9}{40}$

9. $60x^3y^2$ **10.** $(y + 1)^2(y + 4)$ **11.** $7(t^2 + 16)(t - 2)$
12. $3x(x + 1)^2(x - 1)$

Margin Exercises, Section 5.4, pp. 329-332

1. $\frac{7}{9}$ **2.** $\frac{3+x}{x-2}$ **3.** $\frac{6x+4}{x-1}$ **4.** $\frac{x-5}{4}$ **5.** $\frac{x-1}{x-3}$ **6.** $\frac{10x^2+9x}{48}$ **7.** $\frac{9x+10}{48x^2}$
8. $\frac{4x^2-x+3}{x(x-1)(x+1)^2}$ **9.** $\frac{2x^2+16x+5}{(x+3)(x+8)}$ **10.** $\frac{8x+88}{(x+16)(x+1)(x+8)}$ **11.** $\frac{-2x-11}{3(x+4)(x-4)}$

Margin Exercises, Section 5.5, pp. 337-340

1. $\frac{4}{11}$ **2.** $\frac{x^2+2x+1}{2x+1}$ **3.** $\frac{3x-1}{3}$ **4.** $\frac{4x-3}{x-2}$ **5.** $\frac{-x-7}{15x}$ **6.** $\frac{x^2-48}{(x+7)(x+8)(x+6)}$
7. $\frac{-8y-28}{(y+4)(y-4)}$ **8.** $\frac{x-13}{(x+3)(x-3)}$ **9.** $\frac{6x^2-2x-2}{3x(x+1)}$

Margin Exercises, Section 5.6, pp. 345-348

1. $\frac{33}{2}$ **2.** 3 **3.** $\frac{3}{2}$ **4.** $-\frac{1}{8}$ **5.** 1 **6.** 2 **7.** 4

Margin Exercises, Section 5.7, pp. 351-356

1. -3 **2.** 40 km/h, 50 km/h **3.** $\frac{24}{7}$, or $3\frac{3}{7}$ hr **4.** 58 km/L
5. 0.280 **6.** 124 km/h **7.** 2.4 fish/yd^2 **8.** 100 oz **9.** 42
10. 2074

Margin Exercises, Section 5.8, pp. 361-362

1. $M = \frac{Fd^2}{km}$ **2.** $b_1 = \frac{2A - hb_2}{h}$ **3.** $f = \frac{pq}{p+q}$ **4.** $a = 2Qb + b$

Margin Exercises, Section 5.9, pp. 365-368

1. $\frac{136}{5}$ **2.** $\frac{7x^2}{3(2-x^2)}$ **3.** $\frac{x}{x-1}$ **4.** $\frac{136}{5}$ **5.** $\frac{7x^2}{3(2-x^2)}$ **6.** $\frac{x}{x-1}$

CHAPTER 6

Margin Exercises, Section 6.1, pp. 379-382

1. **(a)** 40%; **(b)** Grades **2.** **(a)** The first week; **(b)** The
third and fifth weeks **3.** **(a)** \$200; **(b)** 28%
4.-11.

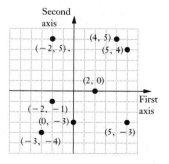

12. Both are negative numbers **13.** First positive; second negative **14.** I **15.** III **16.** IV **17.** II **18.** *A:* $(-5, 1)$; *B:* $(-3, 2)$; *C:* $(0, 4)$; *D:* $(3, 3)$; *E:* $(1, 0)$; *F:* $(0, -3)$; *G:* $(-5, -4)$

Margin Exercises, Section 6.2, pp. 387-392

1. No   **2.** Yes

3.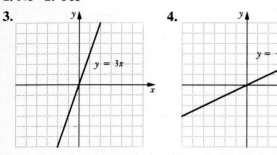

4.

5.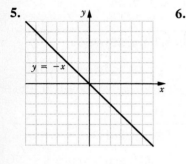

6.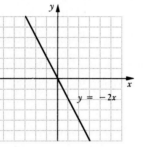

7.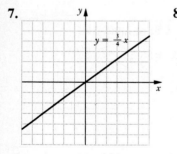

8.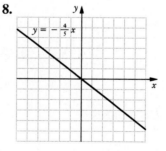

9. $y = x + 3$ looks like $y = x$ moved <u>up</u> 3 units.   **10.** $y = x - 1$ looks like $y = x$ moved <u>down</u> 1 unit.

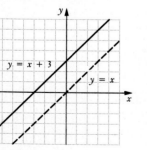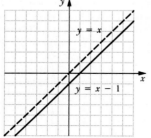

11. $y = 2x + 3$ looks like $y = 2x$ moved <u>up</u> 3 units.

12.

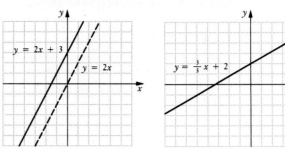

13.

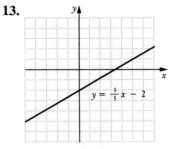

14.

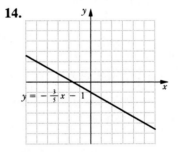

15.

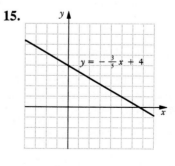

Margin Exercises, Section 6.3, pp. 395-398

1. (a) $(4, 0)$; **(b)** $(0, 3)$

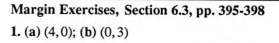
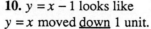
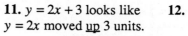

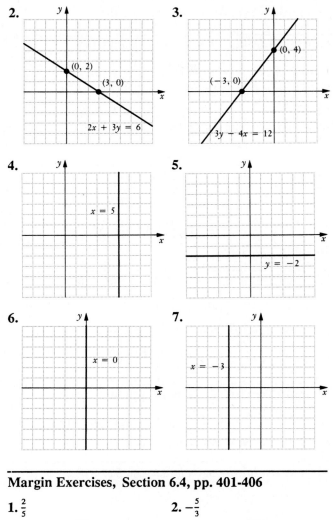

2. $2x + 3y = 6$ (0, 2), (3, 0)

3. $3y - 4x = 12$ (0, 4), (−3, 0)

4. $x = 5$

5. $y = -2$

6. $x = 0$

7. $x = -3$

Margin Exercises, Section 6.4, pp. 401-406

1. $\frac{2}{5}$ **2.** $-\frac{5}{3}$

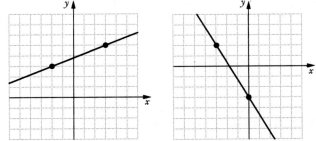

3. Not defined **4.** 0 **5.** −1 **6.** $\frac{5}{4}$ **7.** 5, (0,0) **8.** $-\frac{3}{2}$, (0,−6) **9.** $-\frac{3}{4}$, $\left(0, \frac{15}{4}\right)$ **10.** 2, $\left(0, -\frac{17}{2}\right)$ **11.** $-\frac{7}{5}$, $\left(0, -\frac{22}{5}\right)$ **12.** $y = 3.5x - 23$ **13.** $y = 5x - 18$ **14.** $y = -3x - 5$ **15.** $y = 6x - 13$ **16.** $y = -\frac{2}{3}x + \frac{14}{3}$ **17.** $y = x + 2$ **18.** $y = 2x + 4$

Margin Exercises, Section 6.5, pp. 409-410

1. No **2.** Yes **3.** Yes **4.** No

Margin Exercises, Section 6.6, pp. 413-416

1. $y = 7x$ **2.** $y = \frac{5}{8}x$ **3.** \$0.4667; \$0.0194 **4.** 174.24 lb **5.** $y = \frac{63}{x}$ **6.** $y = \frac{900}{x}$ **7.** 8 hr **8.** $7\frac{1}{2}$ hr

Margin Exercises, Section 6.7, pp. 421-424

1. No **2.** No

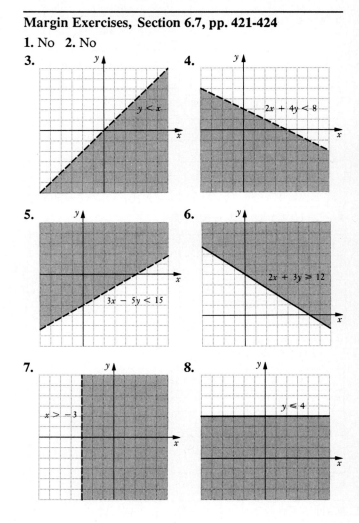

3. $y < x$

4. $2x + 4y < 8$

5. $3x - 5y < 15$

6. $2x + 3y \geqslant 12$

7. $x > -3$

8. $y \leqslant 4$

CHAPTER 7

Margin Exercises, Section 7.1, pp. 435-438

1. Yes **2.** No **3.** (2, −3) **4.** No solution. The lines are parallel. **5.** Infinite number of solutions. Same line.

Margin Exercises, Section 7.2, pp. 441-444

1. (3,2) **2.** (3,−1) **3.** $\left(\frac{24}{5}, -\frac{8}{5}\right)$ **4.** Length: 27.5 cm; width: 10.5 cm

Margin Exercises, Section 7.3, pp. 447-452

1. (3,2) **2.** (1,−1) **3.** (1,4) **4.** (1,1) **5.** $\left(\frac{17}{13}, -\frac{7}{13}\right)$ **6.** No

solution   **7.** Infinitely many solutions   **8.** $(1, -1)$
9. 75 miles

Margin Exercises, Section 7.4, pp. 457-464

1. A: \$3.23; B: \$3.36

2. Jackie is 47; Jack is 21

	Jackie	Jack
Age now	J	K
Age in 5 yrs	$J + 5$	$K + 5$

3. 125 adults and 41 children

Adults	Children	Totals	
\$2.10	\$0.75		
x	y	166	$\longrightarrow x + y = 166$
\$2.10x	\$0.75y	\$293.25	$\longrightarrow 2.10x + 0.75y = 293.25$

4. 22.5 L of 50%, 7.5 L of 70%

First	Second	Mixture	
x	y	30 L	$\longrightarrow \quad x + y = 30$
50%	70%	55%	
0.5x	0.7y	0.55×30, or 16.5 L	$\longrightarrow \quad 0.5x + 0.7y = 16.5$

5. 30 lb of A, 20 lb of B

A	B	Mixture	
\$1.00	\$1.35	\$1.14	
x	y	50	$\longrightarrow \quad x + y = 50$
1.00x	1.35y	\$1.14(50), or 57	$\longrightarrow \quad 1.00x + 1.35y = 57$

6. 7 quarters; 13 dimes

Margin Exercises, Section 7.5, pp. 469-472

1. 168 km   **2.** 275 km/h   **3.** 324 mi   **4.** 3 hr

CHAPTER 8

Margin Exercises, Section 8.1, pp. 483-486

1. 6, -6 **2.** 8, -8 **3.** 15, -15 **4.** 10, -10 **5.** 4   **6.** 7
7. 10   **8.** 21   **9.** -7 **10.** -13 **11.** 5.477   **12.** -9.899
13. (a) 20; (b) 23   **14.** $45 + x$ **15.** $\frac{x}{x+2}$ **16.** No   **17.** Yes
18. Yes   **19.** No   **20.** Yes   **21.** No   **22.** $\{a \mid a \geq 0\}$
23. $\{x \mid x \geq 3\}$ **24.** $\left\{x \mid x \geq \frac{5}{2}\right\}$ **25.** All real numbers
26. $|xy|$ **27.** $|xy|$ **28.** $|x-1|$ **29.** $|x+4|$ **30.** xy

31. xy **32.** $x - 1$ **33.** $x + 4$ **34.** $5y$ **35.** $\frac{1}{2}t$

Margin Exercises, Section 8.2, pp. 489-492

1. (a) 8; (b) 8   **2.** $\sqrt{21}$ **3.** 5   **4.** $\sqrt{x^2 + x}$ **5.** $\sqrt{x^2 - 1}$
6. $4\sqrt{2}$ **7.** $x + 7$ **8.** $5x$ **9.** $6m$ **10.** $2\sqrt{19}$ **11.** $x - 4$
12. $8t$ **13.** $10a$ **14.** 16.585   **15.** 10.097   **16.** t^2 **17.** t^{10}
18. q^{17} **19.** $x^3\sqrt{x}$ **20.** $2x^5\sqrt{6x}$ **21.** $3\sqrt{2}$ **22.** 10
23. $4x^3y^2$ **24.** $5xy^2\sqrt{2xy}$

Margin Exercises, Section 8.3, pp. 497-500

1. 4   **2.** 5   **3.** $x\sqrt{6x}$ **4.** $\frac{4}{3}$ **5.** $\frac{1}{5}$ **6.** $\frac{6}{x}$ **7.** $\frac{3}{4}$ **8.** $\frac{15}{16}$ **9.** $\frac{5}{x^3}$
10. $\frac{\sqrt{15}}{5}$ **11.** $\frac{\sqrt{10}}{4}$ **12.** $\frac{2\sqrt{3}}{3}$ **13.** $\frac{\sqrt{35}}{7}$ **14.** $\frac{\sqrt{3t}}{t}$ **15.** $\frac{8y\sqrt{7}}{7}$
16. 0.816   **17.** 1.128

Margin Exercises, Section 8.4, pp. 505-508

1. $12\sqrt{2}$ **2.** $5\sqrt{5}$ **3.** $-12\sqrt{10}$ **4.** $5\sqrt{6}$ **5.** $\sqrt{x+1}$ **6.** $\frac{3}{2}\sqrt{2}$
7. $\frac{8\sqrt{15}}{15}$ **8.** $\sqrt{15} + \sqrt{6}$ **9.** $4 + 3\sqrt{5} - 4\sqrt{2} - 3\sqrt{10}$ **10.** $2 - a$
11. $5 + 10\sqrt{x} + x$ **12.** 2   **13.** $\frac{21 - 3\sqrt{5}}{22}$ **14.** $\frac{7 + 2\sqrt{10}}{3}$

Margin Exercises, Section 8.5, pp. 511-514

1. $\frac{64}{3}$ **2.** 2   **3.** $\frac{3}{8}$ **4.** 4   **5.** 1   **6.** 4   **7.** Approx. 313 km
8. Approx. 16 km   **9.** 676 m

Margin Exercises, Section 8.6, pp. 517-518

1. $c = \sqrt{65} \approx 8.062$ **2.** $a = \sqrt{75} \approx 8.660$
3. $b = \sqrt{10} \approx 3.162$ **4.** $a = \sqrt{175} \approx 13.229$
5. $\sqrt{325} \approx 18.028$ ft

CHAPTER 9

Margin Exercises, Section 9.1, pp. 529-532

1. $x^2 - 7x = 0$; $a = 1, b = -7, c = 0$ **2.** $x^2 + 9x - 3 = 0$;
$a = 1, b = 9, c = -3$ **3.** $4x^2 + 2x + 4 = 0$; $a = 4, b = 2$,
$c = 4$ **4.** $0, -\frac{5}{3}$ **5.** $0, \frac{3}{5}$ **6.** $\frac{2}{3}, -1$ **7.** 4, 1   **8.** 13, 5   **9.** (a)
14; (b) 11

Margin Exercises, Section 9.2, pp. 535-540

1. $\sqrt{5}, -\sqrt{5}$ **2.** 0   **3.** $\frac{\sqrt{6}}{2}, -\frac{\sqrt{6}}{2}$ **4.** 7, -1 **5.** $-3 \pm \sqrt{10}$
6. $1 \pm \sqrt{5}$ **7.** $-5, 11$ **8.** 2, 4   **9.** 2, -10 **10.** $-3 \pm \sqrt{10}$
11. 5, -2 **12.** $\frac{-3 \pm \sqrt{33}}{4}$ **13.** About 7.9 sec

Margin Exercises, Section 9.3, pp. 543-546

1. $\frac{1}{2}, -4$ **2.** $5, -2$ **3.** $-2 \pm \sqrt{11}$ **4.** No real-number solution **5.** $\frac{4 \pm \sqrt{31}}{5}$ **6.** $-0.3, 1.9$

Margin Exercises, Section 9.4, pp. 549-550

1. $L = \frac{r^2}{20}$ **2.** $L = \frac{T^2 g}{4\pi^2}$ **3.** $m = \frac{E}{c^2}$ **4.** $r = \sqrt{\frac{A}{\pi}}$ **5.** $d = \sqrt{\frac{C}{P}} + 1$
6. $n = \frac{1 \pm \sqrt{1 + 4N}}{2}$ **7.** $t = \frac{-v \pm \sqrt{v^2 + 32h}}{16}$

Margin Exercises, Section 9.5, pp. 553-556

1. Length: $\frac{1 + \sqrt{817}}{2} \approx 14.8$ in.; width: $\frac{-1 + \sqrt{817}}{6} \approx 4.6$ in.
2. 2.3 cm; 3.3 cm **3.** 3 km/h

Margin Exercises, Section 9.6, pp. 561-564

1. $(0, -3)$ **2.** $(1, 3)$

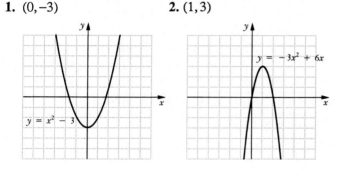

3. $(2, 0)$

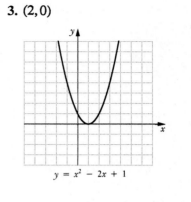

4. $(\sqrt{3}, 0)$; $(-\sqrt{3}, 0)$ **5.** $(-4, 0)$; $(-2, 0)$

6. $\left(\frac{-2 - \sqrt{6}}{2}, 0\right)$; $\left(\frac{-2 + \sqrt{6}}{2}, 0\right)$ **7.** None

EXERCISE SET ANSWERS

Book Diagnostic Test, p. xxiii

1. [R.2c] $\frac{8}{15}$ **2.** [R.2c] $\frac{11}{12}$ **3.** [R.3b] 26 **4.** [R.3b]
10.983 **5.** [1.4a] 1.18 **6.** [1.5a] -11.7 **7.** [1.4b]
$-\$9.93$ **8.** [1.7c] $39a - 84$ **9.** [2.3c] -5
10. [2.7e] $\left\{ x \mid x \geq \frac{9}{23} \right\}$ **11.** [2.4a] 9 in. and 27 in.

12. [2.5a] \$1500 **13.** [3.1e] $\frac{x^4}{y}$ **14.** [3.1d, 3.2b]

$-32x^{11}$ **15.** [3.4c] $-x^2 + 3x + 4$ **16.** [3.6b] $4x^4 - 9$
17. [4.5d] $2(x + 9)(x - 9)$ **18.** [4.3a, 4.4a]
$(5x + 1)(x - 3)$ **19.** [4.7b] $-5, 2$ **20.** [4.8a] 8 m;
17 m **21.** [5.2b] $\frac{x(x - 4)}{2(x + 5)(x - 3)}$ **22.** [5.4a] $\frac{2 - x}{x(x + 1)(x + 2)}$
23. [5.6a] 4 **24.** [5.7a] 55 mi/hr; 40 mi/hr

25. [6.2b]

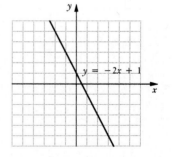

26. [6.4c] $m = -\frac{2}{3}$; y-intercept $= \left(0, \frac{8}{3}\right)$ **27.** [6.4d]
$y = -3x + 11$

28. [6.7b]

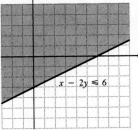

29. [7.2b] $(8, -3)$ **30. [7.3b]** $\left(\frac{23}{8}, -\frac{3}{16}\right)$ **31. [7.4a]**
25 liters of each   **32. [7.5a]** 10 hr   **33. [8.2d]**
$2xy^2\sqrt{3x}$ **34. [8.3a]** $\frac{x}{3y}$ **35. [8.4c]** $6 + 3\sqrt{3}$
36. [8.5a] $\frac{77}{2}$ **37. [9.1c]** $-1, \frac{1}{3}$ **38. [9.3a]** No real
solutions   **39. [9.5a]** 30 m; 16 m   **40. [9.6a]**

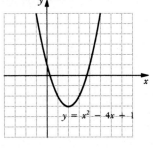

CHAPTER R

Pretest: Chapter R, p. 2

1. [R.1a] $2 \cdot 2 \cdot 2 \cdot 31$ **2. [R.1b]** 168   **3. [R.2a]** $\frac{10}{15}$
4. [R.2a] $\frac{44}{48}$ **5. [R.2b]** $\frac{23}{64}$ **6. [R.2b]** $\frac{2}{3}$ **7. [R.2c]** $\frac{11}{10}$
8. [R.2c] $\frac{2}{21}$ **9. [R.2c]** $\frac{1}{2}$ **10. [R.2c]** $\frac{5}{21}$ **11. [R.3a]** $\frac{3217}{100}$
12. [R.3a] 0.0789   **13. [R.3b]** 134.0362   **14. [R.3b]**
212.05   **15. [R.3b]** 350.5824   **16. [R.3b]** 12.4
17. [R.3a] $1.\overline{4}$ **18. [R.3c]** 345.84   **19. [R.3c]** 345.8
20. [R.5b] 8   **21. [R.5b]** 1.21   **22. [R.5c]** 18   **23. [R.4a]**
0.116   **24. [R.4b]** $\frac{87}{100}$ **25. [R.4d]** 87.5%   **26. [R.5a]** 5^4

Exercise Set R.1, p. 7

1. $1 \cdot 24; 2 \cdot 12; 3 \cdot 8; 4 \cdot 6; 1, 2, 3, 4, 6, 8, 12, 24$
3. $1 \cdot 81; 3 \cdot 27; 9 \cdot 9; 1, 3, 9, 27, 81$ **5.** $2 \cdot 7$ **7.** $3 \cdot 11$
9. $3 \cdot 3$ **11.** $7 \cdot 7$ **13.** $3 \cdot 3 \cdot 2$ **15.** $2 \cdot 2 \cdot 2 \cdot 5$
17. $2 \cdot 3 \cdot 3 \cdot 5$ **19.** $2 \cdot 3 \cdot 5 \cdot 7$ **21.** $7 \cdot 13$ **23.** $7 \cdot 17$
25. 3; 7; 21   **27.** $2 \cdot 2 \cdot 5; 2 \cdot 3 \cdot 5; 60$ **29.** 3; $3 \cdot 5$; 15
31. $2 \cdot 3 \cdot 5; 2 \cdot 2 \cdot 2 \cdot 5; 120$ **33.** 13; 23; 299   **35.** $2 \cdot 3 \cdot 3$;
$2 \cdot 3 \cdot 5; 90$ **37.** $2 \cdot 3 \cdot 5; 2 \cdot 2 \cdot 3 \cdot 3; 180$ **39.** $2 \cdot 2 \cdot 2 \cdot 3$;
$2 \cdot 3 \cdot 5; 120$ **41.** 17; 29; 493   **43.** $2 \cdot 2 \cdot 3; 2 \cdot 2 \cdot 7; 84$
45. 2; 3; 5; 30   **47.** $2 \cdot 2 \cdot 2 \cdot 3; 2 \cdot 2 \cdot 3; 2 \cdot 2 \cdot 3; 72$
49. 5; $2 \cdot 2 \cdot 3; 3 \cdot 5; 60$ **51.** $2 \cdot 3; 2 \cdot 2 \cdot 3; 2 \cdot 3 \cdot 3; 36$
53. (a) No; not a multiple of 8; **(b)** No; not a multiple of
8; **(c)** No; not a multiple of 8 or 12; **(d)** Yes; it is a

multiple both of 8 and 12 and is the smallest such multiple   **55.** 70,200   **57.** Every 420 years

Exercise Set R.2, p. 17

1. $\frac{40}{48}$ **3.** $\frac{600}{700}$ **5.** $\frac{78}{120}$ **7.** $\frac{21}{24}$ **9.** $\frac{5x}{15}$ **11.** $\frac{2}{5}$ **13.** $\frac{7}{2}$ **15.** $\frac{1}{7}$ **17.** 8
19. $\frac{1}{4}$ **21.** 5   **23.** $\frac{17}{21}$ **25.** $\frac{13}{7}$ **27.** $\frac{4}{3}$ **29.** $\frac{1}{8}$ **31.** $\frac{51}{8}$ **33.** 1
35. $\frac{7}{6}$ **37.** $\frac{5}{6}$ **39.** $\frac{1}{2}$ **41.** $\frac{5}{18}$ **43.** $\frac{31}{60}$ **45.** $\frac{35}{18}$ **47.** $\frac{10}{3}$ **49.** $\frac{1}{2}$
51. $\frac{5}{36}$ **53.** 500   **55.** $\frac{3}{40}$ **57.** $2 \cdot 2 \cdot 2 \cdot 2 \cdot 2 \cdot 2 \cdot 3$ **59.** $\frac{2}{3}$
61. 3   **63.** 1

Exercise Set R.3, p. 25

1. $\frac{49}{10}$ **3.** $\frac{59}{100}$ **5.** $\frac{20,007}{10,000}$ **7.** $\frac{78,898}{10}$ **9.** 0.1   **11.** 0.0001
13. 307.9   **15.** 9.999   **17.** 0.0039   **19.** 0.00001
21. 444.94   **23.** 390.617   **25.** 155.724   **27.** 63.79
29. 32.234   **31.** 26.835   **33.** 47.91   **35.** 1.9193
37. 9.968   **39.** 0.4368   **41.** 179.5   **43.** 0.1894
45. 1.40756   **47.** 3.60558   **49.** 2.3   **51.** 5.2   **53.** 0.023
55. 18.75   **57.** 660   **59.** 0.53125   **61.** $1.41\overline{6}$ **63.** $0.8\overline{3}$
65. $0.\overline{81}$ **67.** 317.19; 317.2; 317; 320; 300   **69.** 840.15;
840.2; 840; 840; 800   **71.** $20.49; $20   **73.** $4.72; $5
75. $18   **77.** $567   **79.** 1.3529; 1.353; 1.35; 1.4; 1

Exercise Set R.4, p. 29

1. 0.76   **3.** 0.547   **5.** 1   **7.** 0.0061   **9.** 2.4   **11.** 0.0325
13. $\frac{20}{100}$ **15.** $\frac{786}{1000}$ **17.** $\frac{125}{1000}$, or $\frac{25}{200}$ **19.** $\frac{42}{100,000}$ **21.** $\frac{250}{100}$
23. $\frac{347}{10,000}$ **25.** 454%   **27.** 99.8%   **29.** 200%   **31.** 7.2%
33. 920%   **35.** 0.68%   **37.** 12.5%   **39.** 68%   **41.** 17%
43. 70%   **45.** 60%   **47.** $66\frac{2}{3}$%, or $66.\overline{6}$%   **49.** 175%
51. 75%   **53.** 26%   **55.** 90%   **57.** 1200%   **59.** 345%
61. 2.5%

Sidelight: Factors and Sums, p. 34

First row: 48, 90, 432, 63; second row: 7, 18, 36, 14, 12,
6, 21, 11; third row: 9, 2, 2, 10, 8, 10, 21; fourth row: 29,
19, 42

Exercise Set R.5, p. 35

1. 3^4 **3.** 10^5 **5.** 1^5 **7.** 25   **9.** 59,049   **11.** 100   **13.** 1
15. 5.29   **17.** 0.008   **19.** 416.16   **21.** $\frac{9}{64}$ **23.** 125
25. 1225.043   **27.** 19   **29.** 86   **31.** 7   **33.** 5   **35.** 12
37. 324   **39.** 100   **41.** 512   **43.** 22   **45.** 1   **47.** 4
49. 1500   **51.** 96   **53.** 24   **55.** 90   **57.** 8   **59.** 1   **61.** $\frac{22}{45}$
63. $\frac{19}{66}$ **65.** 31.25%   **67.** $2 \cdot 2 \cdot 2 \cdot 2 \cdot 3$ **69.** 10^2 **71.** 5^2
73. $3 = \frac{5+5}{5} + \frac{5}{5}$; $4 = \frac{5+5+5+5}{5}$; $5 = \frac{5(5+5)}{5} - 5$; $6 = \frac{5}{5} + \frac{5 \cdot 5}{5}$;
$7 = \frac{5}{5} + \frac{5}{5} + 5$; $8 = 5 + \frac{5+5+5}{5}$; $9 = \frac{5 \cdot 5 - 5}{5} + 5$; $10 = \frac{5 \cdot 5 + 5 \cdot 5}{5}$

Summary and Review: Chapter R, p. 37

1. [R.1a] $2 \cdot 2 \cdot 23$ **2.** [R.1a] $2 \cdot 2 \cdot 2 \cdot 5 \cdot 5 \cdot 7$ **3.** [R.1b]
416 **4.** [R.1b] 90 **5.** [R.2a] $\frac{12}{30}$ **6.** [R.2a] $\frac{96}{184}$ **7.** [R.2a] $\frac{40}{64}$
8. [R.2a] $\frac{91}{84}$ **9.** [R.2b] $\frac{5}{12}$ **10.** [R.2b] $\frac{51}{91}$ **11.** [R.2c] $\frac{31}{36}$
12. [R.2c] $\frac{1}{4}$ **13.** [R.2c] $\frac{3}{5}$ **14.** [R.2c] $\frac{72}{25}$ **15.** [R.3a] $\frac{1797}{100}$
16. [R.3a] 0.2337 **17.** [R.3b] 2442.905 **18.** [R.3b]
86.0298 **19.** [R.3b] 9.342 **20.** [R.3b] 133.264
21. [R.3b] 430.8 **22.** [R.3b] 110.483 **23.** [R.3b] 55.6
24. [R.3b] 0.45 **25.** [R.3a] $1.58\overline{3}$ **26.** [R.3c] 34.1
27. [R.5a] 6^3 **28.** [R.5b] 1.1236 **29.** [R.5c] 119
30. [R.5c] 29 **31.** [R.5c] 7 **32.** [R.5c] $\frac{103}{17}$ **33.** [R.4a]
0.047 **34.** [R.4b] $\frac{60}{100}$ **35.** [R.4c] 88.6% **36.** [R.4d]
62.5% **37.** [R.4d] 116% **38.** [R.4d] 0.0000006%

Test: Chapter R, p. 39

1. [R.1a] $2 \cdot 2 \cdot 3 \cdot 5 \cdot 5$ **2.** [R.1b] 120 **3.** [R.2a] $\frac{21}{49}$
4. [R.2a] $\frac{33}{48}$ **5.** [R.2b] $\frac{2}{3}$ **6.** [R.2b] $\frac{37}{61}$ **7.** [R.2c] $\frac{5}{36}$
8. [R.2c] $\frac{11}{40}$ **9.** [R.3a] $\frac{678}{100}$ **10.** [R.3a] 1.895 **11.** [R.3b]
99.0187 **12.** [R.3b] 1796.58 **13.** [R.3b] 435.47728
14. [R.3b] 1.6 **15.** [R.3a] $2.\overline{09}$ **16.** [R.3c] 234.7
17. [R.3c] 234.728 **18.** [R.5b] 625 **19.** [R.5b] 1.44
20. [R.5c] 207 **21.** [R.4a] 0.007 **22.** [R.4b] $\frac{91}{100}$
23. [R.4d] 44% **24.** [R.2b] $\frac{33}{100}$

CHAPTER 1

Pretest: Chapter 1, p. 42

1. [1.1a] $\frac{5}{16}$ **2.** [1.1b] $78\%x$, or $0.78x$ **3.** [1.1a] 360 ft^2
4. [1.3b] 12 **5.** [1.2d] > **6.** [1.2d] > **7.** [1.2d] >
8. [1.2d] < **9.** [1.2e] 12 **10.** [1.2e] 2.3 **11.** [1.2e] 0
12. [1.3b] -5.4 **13.** [1.3b] $\frac{2}{3}$ **14.** [1.3a] -17 **15.** [1.4a]
38.6 **16.** [1.4a] $-\frac{17}{15}$ **17.** [1.3a] -5 **18.** [1.5a] 63
19. [1.5a] $-\frac{5}{12}$ **20.** [1.6c] -98 **21.** [1.6a] 8 **22.** [1.4a]
24 **23.** [1.8d] 26 **24.** [1.7c] $9z - 18$ **25.** [1.7c]
$-4a - 2b + 10c$ **26.** [1.7d] $4(x - 3)$ **27.** [1.7d]
$3(2y - 3z - 6)$ **28.** [1.8b] $-y - 13$ **29.** [1.8c] $y + 18$
30. [1.2d] $12 < x$

Exercise Set 1.1, p. 47

1. 23, 28, 41 **3.** 100.1 cm^2 **5.** 220 mi **7.** 42 **9.** 3
11. 1 **13.** 6 **15.** 2 **17.** $b + 6$, or $6 + b$ **19.** $c - 9$
21. $6 + q$, or $q + 6$ **23.** $b + a$, or $a + b$ **25.** $y - x$
27. $x + w$, or $w + x$ **29.** $n - m$ **31.** $r + s$, or $s + r$
33. $2x$ **35.** $5t$ **37.** $97\%x$, or $0.97x$ **39.** $d - \$29.95$
41. $2 \cdot 3 \cdot 3 \cdot 3$ **43.** 18 **45.** $x + 3y$ **47.** $2x - 3$

Exercise Set 1.2, p. 57

1. 18, -2 **3.** 750, -125 **5.** 20, -150, 300
7. **9.**
11. $x < -6$ **13.** $y \geq -10$ **15.** -0.375 **17.** $1.\overline{6}$ **19.** $1.1\overline{6}$
21. $0.\overline{6}$ **23.** -0.5 **25.** 0.1 **27.** > **29.** < **31.** < **33.** <
35. > **37.** < **39.** > **41.** < **43.** < **45.** < **47.** True
49. False **51.** 3 **53.** 10 **55.** 0 **57.** 24 **59.** $\frac{2}{3}$ **61.** 0
63. $-\frac{5}{6}, -\frac{3}{4}, -\frac{2}{3}, \frac{1}{6}, \frac{3}{8}, \frac{1}{2}$

Exercise Set 1.3, p. 63

1. -7 **3.** -4 **5.** 0 **7.** -8 **9.** -7 **11.** -27 **13.** 0
15. -42 **17.** 0 **19.** 0 **21.** 3 **23.** -9 **25.** 7 **27.** 0
29. 45 **31.** -1.8 **33.** -8.1 **35.** $-\frac{1}{5}$ **37.** $-\frac{8}{7}$ **39.** $-\frac{3}{8}$
41. $-\frac{29}{35}$ **43.** $\frac{7}{16}$ **45.** 39 **47.** 50 **49.** -1409 **51.** -24
53. 26.9 **55.** -9 **57.** $\frac{14}{3}$ **59.** -65 **61.** $\frac{5}{3}$ **63.** 14 **65.** -10
67. All positive **69.** Negative

Exercise Set 1.4, p. 69

1. -4 **3.** -7 **5.** -6 **7.** 0 **9.** -4 **11.** -7 **13.** -6 **15.** 0
17. 0 **19.** 14 **21.** 11 **23.** -14 **25.** 5 **27.** -7 **29.** -1
31. 18 **33.** -5 **35.** -3 **37.** -21 **39.** 5 **41.** -8 **43.** 12
45. -23 **47.** -68 **49.** -73 **51.** 116 **53.** 0 **55.** $-\frac{1}{4}$
57. $\frac{1}{12}$ **59.** $-\frac{17}{12}$ **61.** $\frac{1}{8}$ **63.** 19.9 **65.** -9 **67.** -0.01
69. -193 **71.** 500 **73.** -2.8 **75.** -3.53 **77.** $-\frac{1}{2}$ **79.** $\frac{6}{7}$
81. $-\frac{41}{30}$ **83.** $-\frac{1}{156}$ **85.** 37 **87.** -62 **89.** -139 **91.** 6
93. 107 **95.** 219 **97.** $-\$330.54$ **99.** $y + \$215.50$
101. 116 m **103.** 125 **105.** $-309,882$ **107.** False;
$3 - 0 \neq 0 - 3$ **109.** True **111.** True

Exercise Set 1.5, p. 77

1. -16 **3.** -42 **5.** -24 **7.** -72 **9.** 16 **11.** 42 **13.** -120
15. -238 **17.** 1200 **19.** 98 **21.** -72 **23.** -12.4 **25.** 24
27. 21.7 **29.** $-\frac{2}{5}$ **31.** $\frac{1}{12}$ **33.** -17.01 **35.** $-\frac{5}{12}$ **37.** 420
39. $\frac{2}{7}$ **41.** -60 **43.** 150 **45.** $-\frac{2}{45}$ **47.** 1911 **49.** 50.4
51. $\frac{10}{189}$ **53.** -960 **55.** 17.64 **57.** $-\frac{5}{784}$ **59.** 0 **61.** -720
63. $-30,240$ **65.** 441, -147 **67.** 20, 20 **69.** 72
71. 1944 **73.** -17 **75. (a)** One must be negative and
one must be positive; **(b)** either or both must be zero;
(c) both must be negative or both must be positive

Exercise Set 1.6, p. 83

1. -6 **3.** -13 **5.** -2 **7.** 4 **9.** -8 **11.** 2 **13.** -12
15. -8 **17.** Undefined **19.** $-\frac{88}{9}$ **21.** $\frac{7}{15}$ **23.** $-\frac{13}{47}$ **25.** $\frac{1}{13}$

27. $\frac{1}{4.3}$ **29.** -7.1 **31.** $\frac{q}{p}$ **33.** $4y$ **35.** $\frac{3b}{2a}$ **37.** $3 \cdot \left(\frac{1}{19}\right)$
39. $6 \cdot \left(-\frac{1}{13}\right)$ **41.** $13.9 \cdot \left(-\frac{1}{1.5}\right)$ **43.** $x \cdot y$
45. $(3x+4) \cdot \left(\frac{1}{5}\right)$ **47.** $(5a-b)\left(\frac{1}{5a+b}\right)$ **49.** $-\frac{9}{8}$ **51.** $\frac{5}{3}$
53. $\frac{9}{14}$ **55.** $\frac{9}{64}$ **57.** -2 **59.** $\frac{11}{13}$ **61.** -16.2 **63.** Undefined
65. $\frac{22}{39}$ **67.** 33 **69.** $-\frac{1}{10.5}$, or $-0.\overline{095238}$ **71.** No real
numbers   **73.** Negative   **75.** Positive   **77.** Negative

---

Exercise Set 1.7, p. 93

1. $\frac{2x}{5x}$ **3.** $\frac{10y}{15y}$ **5.** $-\frac{3}{2}$ **7.** $-\frac{7}{6}$ **9.** $8+y$ **11.** nm **13.** $xy+9$,
or $9+yx$ **15.** $c+ab$, or $ba+c$ **17.** $(a+b)+2$
19. $8(xy)$ **21.** $a+(b+3)$ **23.** $(3a)b$ **25.** $2+(b+a)$,
$(2+a)+b$, $(b+2)+a$; answers may vary
27. $(5+w)+v$, $(v+5)+w$, $(w+v)+5$; answers may
vary   **29.** $(3x)y$, $y(x \cdot 3)$, $3(yx)$; answers may vary
31. $a(7b)$, $b(7a)$, $(7b)a$; answers may vary   **33.** $2b+10$
35. $7+7t$ **37.** $30x+12$ **39.** $7x+28+42y$ **41.** 7
43. 12 **45.** -12.71 **47.** $7x-14$ **49.** $-7y+14$
51. $45x+54y-72$ **53.** $-4x+12y+8z$
55. $-3.72x+9.92y-3.41$ **57.** $4x, 3z$ **59.** $7x, 8y, -9z$
61. $2(x+2)$ **63.** $5(6+y)$ **65.** $7(2x+3y)$
67. $5(x+2+3y)$ **69.** $8(x-3)$ **71.** $4(8-y)$
73. $2(4x+5y-11)$ **75.** $a(x-1)$ **77.** $a(x-y-z)$
79. $6(3x-2y+1)$ **81.** $19a$ **83.** $9a$ **85.** $8x+9z$
87. $7x+15y^2$ **89.** $-19a+88$ **91.** $4t+6y-4$ **93.** b
95. $\frac{13}{4}y$ **97.** $8x$ **99.** $5n$ **101.** $-16y$ **103.** $17a-12b-1$
105. $4x+2y$ **107.** $7x+y$ **109.** $0.8x+0.5y$
111. $\frac{3}{5}x+\frac{3}{5}y$ **113.** $\frac{89}{48}$ **115.** 144 **117.** Not equivalent
119. Equivalent; commutative law of addition
121. $q(1+r+rs+rst)$

Exercise Set 1.8, p. 101

1. $-2x-7$ **3.** $-5x+8$ **5.** $-4a+3b-7c$
7. $-6x+8y-5$ **9.** $-3x+5y+6$ **11.** $8x+6y+43$
13. $5x-3$ **15.** $-3a+9$ **17.** $5x-6$ **19.** $-19x+2y$
21. $9y-25z$ **23.** $-7x+10y$ **25.** $37a-23b+35c$
27. 7 **29.** -40 **31.** 19 **33.** $12x+30$ **35.** $3x+30$
37. $9x-18$ **39.** $-4x-64$ **41.** -7 **43.** -7 **45.** -16
47. -334 **49.** 14 **51.** 1880 **53.** 12 **55.** 8 **57.** -86
59. 37 **61.** -1 **63.** -10 **65.** 25 **67.** -7988 **69.** -3000
71. 60 **73.** 1 **75.** 10 **77.** $-\frac{13}{45}$ **79.** $-\frac{23}{18}$ **81.** -118
83. $6y-(-2x+3a-c)$ **85.** $6m-(-3n+5m-4b)$
87. $-2x-f$ **89.** True. $(-a)(-b)=(-1 \cdot a)(-1 \cdot b)=$
$(-1)(-1)(a)(b)=1 \cdot (a \cdot b)=ab$

Summary and Review: Chapter 1, p. 105

1. [1.1a] 4 **2.** [1.1b] $19\%x$, or $0.19x$ **3.** [1.2a] $-45, 72$
4. [1.2e] 38

5. [1.2b]

6. [1.2b]

7. [1.2d] $<$ **8.** [1.2d] $>$ **9.** [1.2d] $>$ **10.** [1.2d] $<$
11. [1.3b] -3.8 **12.** [1.3b] $\frac{3}{4}$ **13.** [1.6b] $\frac{8}{3}$ **14.** [1.6b] $-\frac{1}{7}$
15. [1.3b] 34 **16.** [1.3b] 5 **17.** [1.3a] -3 **18.** [1.3a] -4
19. [1.3a] -5 **20.** [1.4a] 4 **21.** [1.4a] $-\frac{7}{5}$ **22.** [1.4a] -7.9
23. [1.5a] 54 **24.** [1.5a] -9.18 **25.** [1.5a] $-\frac{2}{7}$ **26.** [1.5a]
-210 **27.** [1.6a] -7 **28.** [1.6c] -3 **29.** [1.6c] $\frac{3}{4}$
30. [1.8d] 71.6 **31.** [1.8d] 62 **32.** [1.4b] 8-yd gain
33. [1.4b] $-\$130$ **34.** [1.7c] $15x-35$ **35.** [1.7c]
$-8x+10$ **36.** [1.7c] $4x+15$ **37.** [1.7c] $-24+48x$
38. [1.7d] $2(x-7)$ **39.** [1.7d] $6(x-1)$ **40.** [1.7d]
$5(x+2)$ **41.** [1.7d] $3(4-x)$ **42.** [1.7e] $7a-3b$
43. [1.7e] $-2x+5y$ **44.** [1.7e] $5x-y$ **45.** [1.7e]
$-a+8b$ **46.** [1.8b] $-3a+9$ **47.** [1.8b] $-2b+21$
48. [1.8c] 6 **49.** [1.8c] $12y-34$ **50.** [1.8c] $5x+24$
51. [1.8c] $-15x+25$ **52.** [1.2d] True   **53.** [1.2d] False
54. [1.2d] $x>-3$ **55.** [R.2c] $\frac{55}{42}$ **56.** [R.5c] $\frac{109}{18}$
57. [R.1a] $2 \cdot 2 \cdot 2 \cdot 3 \cdot 3 \cdot 3 \cdot 3$ **58.** [R.4d] 62.5%
59. [R.4a] 0.0567 **60.** [R.1b] 270 **61.** [1.8d] $-\frac{5}{8}$
62. [1.8d] -2.1

Test: Chapter 1, p. 107

1. [1.1a] 6 **2.** [1.1b] $x-9$ **3.** [1.1a] $240\ \text{ft}^2$ **4.** [1.2d] $<$
5. [1.2d] $>$ **6.** [1.2d] $>$ **7.** [1.2d] $<$ **8.** [1.2e] 7 **9.** [1.2e]
$\frac{9}{4}$ **10.** [1.2e] 2.7 **11.** [1.3b] $-\frac{2}{3}$ **12.** [1.3b] 1.4
13. [1.3b] 8 **14.** [1.6b] $-\frac{1}{2}$ **15.** [1.6b] $\frac{7}{4}$ **16.** [1.4a] 7.8
17. [1.3a] -8 **18.** [1.3a] $\frac{7}{40}$ **19.** [1.4a] 10 **20.** [1.4a]
-2.5 **21.** [1.4a] $\frac{7}{8}$ **22.** [1.5a] -48 **23.** [1.5a] $\frac{3}{16}$
24. [1.6a] -9 **25.** [1.6c] $\frac{3}{4}$ **26.** [1.6c] -9.728 **27.** [1.8d]
-173 **28.** [1.4b] $\$148$ **29.** [1.7c] $18-3x$ **30.** [1.7c]
$-5y+5$ **31.** [1.7d] $2(6-11x)$ **32.** [1.7d] $7(x+3+2y)$
33. [1.4a] 12 **34.** [1.8b] $2x+7$ **35.** [1.8b] $9a-12b-7$
36. [1.8c] $68y-8$ **37.** [1.8d] -4 **38.** [1.8d] 448
39. [1.2d] $-2 \geq x$ **40.** [R.5b] 1.728 **41.** [R.4d] 12.5%
42. [R.1a] $2 \cdot 2 \cdot 2 \cdot 5 \cdot 7$ **43.** [R.1b] 240 **44.** [1.8d] 15
45. [1.8c] $4a$

CHAPTER 2

Pretest: Chapter 2, p. 110

1. [2.2a] -7 **2.** [2.3b] -1 **3.** [2.3a] 2 **4.** [2.1b] 8
5. [2.3c] -5 **6.** [2.3a] $\frac{135}{32}$ **7.** [2.3c] 1 **8.** [2.7d]

$\{x \mid x \geq -6\}$ **9.** [2.7c] $\{y \mid y > -4\}$ **10.** [2.7e]
$\{a \mid a > -1\}$ **11.** [2.7e] $\{x \mid x \geq 3\}$ **12.** [2.7d]
$\left\{ y \mid y < -\frac{9}{4} \right\}$ **13.** [2.6a] $G = \frac{P}{3K}$ **14.** [2.6a] $a = \frac{Ab+b}{3}$
15. [2.4a] Width: 34 m; length: 39 m **16.** [2.5a] $650
17. [2.4a] 81, 82, 83 **18.** [2.8b] Numbers less than 17
19. [2.7b] **20.** [2.7b]

Exercise Set 2.1, p. 115

1. Yes **3.** No **5.** No **7.** Yes **9.** No **11.** No **13.** 4
15. −20 **17.** −14 **19.** −18 **21.** 15 **23.** −14 **25.** 2
27. 20 **29.** −6 **31.** $\frac{7}{3}$ **33.** $-\frac{7}{4}$ **35.** $\frac{41}{24}$ **37.** $-\frac{1}{20}$ **39.** 5.1
41. 12.4 **43.** −5 **45.** $1\frac{5}{6}$ **47.** $-\frac{10}{21}$ **49.** −11 **51.** $-\frac{5}{12}$
53. 342.246 **55.** $-\frac{26}{15}$ **57.** −10 **59.** All real numbers
61. $-\frac{5}{17}$ **63.** 13, −13

Exercise Set 2.2, p. 121

1. 6 **3.** 9 **5.** 12 **7.** −40 **9.** 1 **11.** −7 **13.** −6 **15.** 6
17. −63 **19.** 36 **21.** −21 **23.** $-\frac{3}{5}$ **25.** $-\frac{3}{2}$ **27.** $\frac{9}{2}$ **29.** 7
31. −7 **33.** 8 **35.** 15.9 **37.** $7x$ **39.** $x - 4$ **41.** −8655
43. No solution **45.** No solution **47.** $x = \frac{b}{3a}$ **49.** $x = \frac{4b}{a}$

Exercise Set 2.3, p. 129

1. 5 **3.** 8 **5.** 10 **7.** 14 **9.** −8 **11.** −8 **13.** −7 **15.** 15
17. 6 **19.** 4 **21.** 6 **23.** −3 **25.** 1 **27.** −20 **29.** 6 **31.** 7
33. 2 **35.** 5 **37.** 2 **39.** 10 **41.** 4 **43.** 0 **45.** −1 **47.** $-\frac{4}{3}$
49. $\frac{2}{5}$ **51.** −2 **53.** −4 **55.** $\frac{4}{5}$ **57.** $-\frac{28}{27}$ **59.** 6 **61.** 2 **63.** 6
65. 8 **67.** 1 **69.** 17 **71.** $-\frac{5}{3}$ **73.** −3 **75.** 2 **77.** 6 **79.** $\frac{4}{7}$
81. 8 **83.** $\frac{11}{18}$ **85.** $-\frac{51}{31}$ **87.** 2 **89.** −6.5 **91.** <
93. 4.4233464 **95.** $-\frac{7}{2}$ **97.** −2 **99.** 0 **101.** 10

Exercise Set 2.4, p. 139

1. 52 **3.** 56 **5.** 29 **7.** $1.99 **9.** 19 **11.** −10 **13.** 40
15. 20 m, 40 m, 120 m **17.** 36, 37 **19.** 56, 58 **21.** 35,
36, 37 **23.** 61, 63, 65 **25.** Width: 100 ft; length: 160 ft;
area: 16,000 ft² **27.** Length: 27.9 cm; width: 21.6 cm
29. 22.5° **31.** 450.5 mi **33.** 28°, 84°, 68° **35.** 2020
37. 20 **39.** 19 **41.** 120 **43.** 5 half dollars, 10 quarters,
20 dimes, 60 nickels

Exercise Set 2.5, p. 147

1. 25% **3.** 24% **5.** 150 **7.** 2.5 **9.** 546 **11.** 125%
13. 0.8 **15.** 5% **17.** 3.75% **19.** 27 **21.** Approx.
86.36% **23.** $480,000 **25.** $800 **27.** 36 cm³; 436 cm³

29. $6540 **31.** $16 **33.** 30 **35.** $9.17, not $9.10
37. Approximately 3.7% **39.** Both are equal. In A, x is
increased to $x + 0.25x = 1.25x$, then decreased to
$1.25x - 0.25(1.25x) = 0.9375x$. In B, x is decreased to
$x - 0.25x = 0.75x$, then increased to
$0.75x + 0.25(0.75x) = 0.9375x$.

Exercise Set 2.6, p. 151

1. $b = \frac{A}{h}$ **3.** $r = \frac{d}{t}$ **5.** $P = \frac{I}{rt}$ **7.** $a = \frac{F}{m}$ **9.** $w = \frac{P-2l}{2}$
11. $r^2 = \frac{A}{\pi}$ **13.** $b = \frac{2A}{h}$ **15.** $m = \frac{E}{c^2}$ **17.** $d = 2Q - c$
19. $b = 3A - a - c$ **21.** $t = \frac{3k}{v}$ **23.** $y = \frac{C-Ax}{B}$
25. $b = \frac{2A - ah}{h}$; $h = \frac{2A}{a+b}$ **27.** $a = \frac{Q}{3+5c}$ **29.** $D^2 = \frac{2.5H}{N}$
31. $S = \frac{360A}{\pi r^2}$ **33.** $t = \frac{R - 3.85}{-0.0075}$ **35.** 0.92 **37.** −13.2 **39.** A
quadruples **41.** A increases by $2h$ units

Exercise Set 2.7, p. 159

1. (a) Yes; (b) yes; (c) no; (d) yes; (e) yes **3.** (a) No;
(b) no; (c) yes; (d) yes; (e) no

5.

7.

9.

11.

13.

15. $\{x \mid x > -5\}$;

17. $\{x \mid x \leq -18\}$;

19. $\{y \mid y > -5\}$ **21.** $\{x \mid x > 2\}$ **23.** $\{x \mid x \leq -3\}$
25. $\{x \mid x < 4\}$ **27.** $\{c \mid c > 14\}$ **29.** $\left\{ y \mid y \leq \frac{1}{4} \right\}$
31. $\left\{ x \mid x > \frac{7}{12} \right\}$
33. $\{x \mid x < 7\}$;
35. $\{x \mid x < 3\}$;
37. $\left\{ y \mid y \geq -\frac{2}{5} \right\}$ **39.** $\{x \mid x \geq -6\}$ **41.** $\{y \mid y \leq 4\}$
43. $\left\{ x \mid x > \frac{17}{3} \right\}$ **45.** $\left\{ y \mid y < -\frac{1}{14} \right\}$ **47.** $\left\{ x \mid x \leq \frac{3}{10} \right\}$
49. $\{x \mid x < 8\}$ **51.** $\{x \mid x \leq 6\}$ **53.** $\{x \mid x < -3\}$
55. $\{x \mid x > -3\}$ **57.** $\{x \mid x \leq 7\}$ **59.** $\{x \mid x > -10\}$
61. $\{y \mid y < 2\}$ **63.** $\{y \mid y \geq 3\}$ **65.** $\{y \mid y > -2\}$

67. $\{x \mid x > -4\}$ **69.** $\{x \mid x \leq 9\}$ **71.** $\{y \mid y \leq -3\}$
73. $\{y \mid y < 6\}$ **75.** $\{m \mid m \geq 6\}$ **77.** $\left\{t \mid t < -\frac{5}{3}\right\}$
79. $\{r \mid r > -3\}$ **81.** $\{y \mid y > 5\}$ **83.** $\{x \mid x \geq 8\}$
85. $\left\{x \mid x \geq -\frac{57}{34}\right\}$ **87.** $\{a \mid a < 1\}$ **89.** -74 **91.** $-\frac{5}{8}$

93. True   **95.**

$$|x| < 3$$
$-3 \quad 0 \quad 3$

97. All real numbers   **99.** $x \geq 6$

Exercise Set 2.8, p. 165

1. $x > 4$ **3.** $y \leq -6$ **5.** $n \geq 1200$ **7.** $a \leq 500$
9. $2 + 3x < 13$ **11.** $\{x \mid x \geq 97\}$ **13.** $\{Y \mid Y \geq 1935\}$
15. $\{x \mid x \leq 341.4\}$ **17.** $\{x \mid x > 5\}$ **19.** $\left\{L \mid L < \frac{43}{2} \text{ cm}\right\}$
21. $\{b \mid b > 6 \text{ cm}\}$ **23.** $\{x \mid x \geq 20\}$ **25.** \$49.02
27. $\{l \mid l \geq 64 \text{ km}\}$ **29.** $\{l \mid 0 < l \leq 8 \text{ cm}\}$

Summary and Review: Chapter 2, p. 167

1. [2.1b] -22 **2.** [2.2a] 7 **3.** [2.2a] -192 **4.** [2.1b] 1
5. [2.2a] $-\frac{7}{3}$ **6.** [2.1b] 25 **7.** [2.1b] -8 **8.** [2.2a] $-\frac{15}{64}$
9. [2.1b] 9.99 **10.** [2.1b] $\frac{1}{2}$ **11.** [2.3b] -5 **12.** [2.3b] $-\frac{1}{3}$
13. [2.3a] 4 **14.** [2.3b] 3 **15.** [2.3b] 4 **16.** [2.3b] 16
17. [2.3c] 6 **18.** [2.3c] -3 **19.** [2.3c] 12 **20.** [2.3c] 4
21. [2.7a] Yes   **22.** [2.7a] No   **23.** [2.7a] Yes   **24.** [2.7c]
$\left\{y \mid y \geq -\frac{1}{2}\right\}$ **25.** [2.7d] $\{x \mid x \geq 7\}$ **26.** [2.7e]
$\{y \mid y > 2\}$ **27.** [2.7e] $\{y \mid y \leq -4\}$ **28.** [2.7e]
$\{x \mid x < -11\}$ **29.** [2.7d] $\{y \mid y > -7\}$ **30.** [2.7e]
$\{x \mid x > -6\}$ **31.** [2.7e] $\left\{x \mid x > -\frac{9}{11}\right\}$ **32.** [2.7d]
$\{y \mid y \leq 7\}$ **33.** [2.7d] $\left\{x \mid x \geq -\frac{1}{12}\right\}$

34. [2.7b, e]

$$x < 3$$
$0 \quad 3$

35. [2.7b]

$$-2 < x \leq 5$$
$-2 \quad 0 \quad 5$

36. [2.7b]

$$y > 0$$
0

37. [2.6a] $d = \frac{C}{\pi}$ **38.** [2.6a] $B = \frac{3V}{h}$ **39.** [2.6a] $a = 2A - b$
40. [2.4a] \$591   **41.** [2.4a] 27 **42.** [2.4a] 3 m, 5 m
43. [2.4a] 9 **44.** [2.4a] 57, 59   **45.** [2.4a] Width: 11 cm;
length: 17 cm   **46.** [2.5a] \$220   **47.** [2.5a] \$26,087
48. [2.4a] $35°, 85°, 60°$ **49.** [2.8b] 86 **50.** [2.8b]
$\{w \mid w > 17 \text{ cm}\}$ **51.** [R.3a] $1.41\overline{6}$ **52.** [R.3b] 2.3
53. [1.3a] -45 **54.** [1.8b] $-43x + 8y$ **55.** [2.4a]
Amazon: 6437 km; Nile: 6671 km   **56.** [2.5a] \$14,150
57. [1.2e; 2.3a] 23, -23 **58.** [1.2e; 2.2a] 20, -20
59. [2.6a] $a = \frac{y-3}{2-b}$

Test: Chapter 2, p. 169

1. [2.1b] 8 **2.** [2.1b] 26 **3.** [2.2a] -6 **4.** [2.2a] 49
5. [2.3b] -12 **6.** [2.3a] 2 **7.** [2.1b] -8 **8.** [2.1b] $-\frac{7}{20}$
9. [2.3c] 7 **10.** [2.3c] -5 **11.** [2.3b] 2.5 **12.** [2.7c]
$\{x \mid x \leq -4\}$ **13.** [2.7e] $\{x \mid x > -13\}$ **14.** [2.7d]
$\{x \mid x \leq 5\}$ **15.** [2.7d] $\{y \mid y \leq -13\}$ **16.** [2.7d]
$\{y \mid y \geq 8\}$ **17.** [2.7d] $\left\{x \mid x \leq -\frac{1}{20}\right\}$ **18.** [2.7e]
$\{x \mid x < -6\}$ **19.** [2.7e] $\{x \mid x \leq -1\}$

20. [2.7b]

$$y \leq 9$$
$0 \quad 9$

21. [2.7b, e]

$$x < 1$$
$0 \quad 1$

22. [2.7b]

$$-2 \leq x \leq 2$$
$-2 \quad 0 \quad 2$

23. [2.4a] Width: 7 cm; length: 11 cm   **24.** [2.4a] 6
25. [2.4a] 81, 83, 85   **26.** [2.5a] \$750   **27.** [2.6a] $r = \frac{A}{2\pi h}$
28. [2.6a] $l = \frac{2w - P}{-2}$ **29.** [2.8b] $\{x \mid x > 6\}$ **30.** [2.8b]
$\{l \mid l \geq 174 \text{ yd}\}$ **31.** [1.3a] $-\frac{2}{9}$ **32.** [1.1a] $\frac{8}{3}$ **33.** [1.1b]
$73\%p$, or $0.73p$ **34.** [1.8b] $-18x + 37y$ **35.** [2.6a]
$d = \frac{ca-1}{c}$ **36.** [1.2e; 2.3a] 15, -15 **37.** [2.4a] 60

Cumulative Review: Chapters 1-2, p. 171

1. [1.1a] $\frac{3}{2}$ **2.** [1.1a] $\frac{15}{4}$ **3.** [1.1a] 0 **4.** [1.1b] $2w - 4$
5. [1.2d] $>$ **6.** [1.2d] $>$ **7.** [1.2d] $<$ **8.** [1.3b; 1.6b] $-\frac{2}{5}, \frac{5}{2}$
9. [1.2e] 3 **10.** [1.2e] $\frac{3}{4}$ **11.** [1.2e] 0 **12.** [1.3a] -4.4
13. [1.4a] $-\frac{5}{2}$ **14.** [1.5a] $\frac{5}{6}$ **15.** [1.5a] -105 **16.** [1.6a] -9
17. [1.6c] -3 **18.** [1.6c] $\frac{32}{125}$ **19.** [1.7c] $15x + 25y + 10z$
20. [1.7c] $-12x - 8$ **21.** [1.7c] $-12y + 24x$ **22.** [1.7d]
$2(32 + 9x + 12y)$ **23.** [1.7d] $8(2y - 7)$ **24.** [1.7d]
$5(a - 3b + 5)$ **25.** [1.7e] $15b + 22y$ **26.** [1.7e]
$4 + 9y + 6z$ **27.** [1.7e] $1 - 3a - 9d$ **28.** [1.7e]
$-2.6x - 5.2y$ **29.** [1.8b] $-1 + 3x$ **30.** [1.8b] $-2x - y$
31. [1.8b] $-7x + 6$ **32.** [1.8b] $8x$ **33.** [1.8c] $5x - 13$
34. [2.1b] 4.5 **35.** [2.2a] $\frac{4}{25}$ **36.** [2.1b] 10.9 **37.** [2.1b]
$3\frac{5}{6}$ **38.** [2.2a] -48 **39.** [2.2a] 12 **40.** [2.2a] -6.2
41. [2.3a] -3 **42.** [2.3b] $-\frac{12}{5}$ **43.** [2.3b] 8 **44.** [2.3c] 7
45. [2.3b] $-\frac{4}{5}$ **46.** [2.3b] $-\frac{10}{3}$ **47.** [2.7e] $\{x \mid x < 2\}$
48. [2.7e] $\{y \mid y \geq 4\}$ **49.** [2.7e] $\{y \mid y < -3\}$
50. [2.6a] $h = \frac{2A}{b+c}$ **51.** [2.6a] $q = p - 2Q$ **52.** [2.4a] 154
53. [2.4a] \$45   **54.** [2.5a] \$1500   **55.** [2.4a] 50 m, 53 m,

40 m   **56.** [2.8a, b] $\{x \mid x \geq 78\}$ **57.** [2.5a] \$7.80
58. [2.5a] \$25,000   **59.** [1.2e; 2.3a] 4, −4   **60.** [2.3c] All
real numbers   **61.** [2.3c] No solution   **62.** [2.3b] 3
63. [2.3c] All real numbers   **64.** [2.6a] $Q = \frac{2-pm}{p}$

CHAPTER 3

Pretest: Chapter 3, p. 174

1. [3.1d] x^2 **2.** [3.1e] $\frac{1}{x^7}$ **3.** [3.2b] $\frac{16x^4}{y^6}$ **4.** [3.1f] $\frac{1}{p^3}$

5. [3.2c] 3.47×10^{-4} **6.** [3.2c] 3,400,000   **7.** [3.3g] 3, 2,
1, 0; 3   **8.** [3.7c] $-3a^3b - 2a^2b^2 + ab^3 + 12b^3 + 9$
9. [3.4a] $11x^2 + 4x - 11$ **10.** [3.4c] $-x^2 - 18x + 27$
11. [3.5b] $15x^4 - 20x^3 + 5x^2$ **12.** [3.6c] $x^2 + 10x + 25$
13. [3.6b] $x^2 - 25$ **14.** [3.6a] $4x^6 + 19x^3 - 30$ **15.** [3.7f]
$4x^2 - 12xy + 9y^2$ **16.** [3.8b] $x^2 + x + 3$, R 8; or
$x^2 + x + 3 + \frac{8}{x-2}$

Exercise Set 3.1, p. 181

1. $2 \cdot 2 \cdot 2 \cdot 2$ **3.** $(1.4)(1.4)(1.4)(1.4)(1.4)$ **5.** $(7p)(7p)$
7. $(19k)(19k)(19k)(19k)$ **9.** 1   **11.** a **13.** 1   **15.** ab
17. 27   **19.** 19   **21.** 256   **23.** 93   **25.** 3629.84 ft^2
27. $\frac{1}{3^2} = \frac{1}{9}$ **29.** $\frac{1}{10^4} = \frac{1}{10,000}$ **31.** $\frac{1}{7^3} = \frac{1}{343}$ **33.** $\frac{1}{a^3}$ **35.** y^4
37. z^n **39.** 4^{-3} **41.** x^{-3} **43.** 2^7 **45.** 8^{14} **47.** x^7 **49.** 9^{38}
51. $(3y)^{12}$ **53.** $(7y)^{17}$ **55.** 3^3 **57.** $\frac{1}{x}$ **59.** x^7 **61.** $\frac{1}{x^{13}}$

63. 1 · **65.** 7^3 **67.** 8^6 **69.** y^4 **71.** $\frac{1}{16^6}$ **73.** $\frac{1}{m^6}$ **75.** $\frac{1}{(8x)^4}$

77. 1   **79.** x^2 **81.** x^9 **83.** $\frac{1}{z^4}$ **85.** x^3 **87.** 1   **89.** 64, $\frac{1}{64}$,
64, −64, 64   **91.** 64%t, or 0.64t **93.** 64   **95.** No;
$(5y)^0 = 1$ and $5y^0 = 5$ **97.** a^{2k} **99.** 2   **101.** No; for
example, $(3+4)^2 = 49$, but $3^2 + 4^2 = 25$ **103.** <   **105.** <
107. Let $a = 3$; then $(a+3)^2 = 36$ and $a^2 + 3^2 = 18$
109. Let $y = 2$; then $\frac{y^6}{y^3} = 8$ and $y^2 = 4$

Exercise Set 3.2, p. 191

1. 2^6 **3.** $\frac{1}{5^6}$ **5.** x^{12} **7.** $16x^6$ **9.** $\frac{1}{x^{12}y^{15}}$ **11.** $x^{24}y^8$ **13.** $\frac{9x^6}{y^{16}z^6}$

15. $\frac{a^8}{b^{12}}$ **17.** $\frac{y^6}{4}$ **19.** $\frac{8}{y^6}$ **21.** $\frac{x^6y^3}{z^3}$ **23.** $\frac{c^2d^6}{a^4b^2}$ **25.** 7.8×10^{10}

27. 9.07×10^{17} **29.** 3.74×10^{-6} **31.** 1.8×10^{-8} **33.** 10^{11}
35. 784,000,000   **37.** 0.0000000008764
39. 100,000,000   **41.** 0.0001   **43.** 6×10^9
45. 3.38×10^4 **47.** 8.1477×10^{-13} **49.** 2.5×10^{13}
51. 5.0×10^{-4} **53.** 3.0×10^{-21} **55.** 9.125×10^7
57. $6.\overline{6} \times 10^{-2}$ **59.** 2.478125×10^{-1} **61.** 2^{13} **63.** $\frac{1}{5}$

65. 3^{11} **67.** a^n **69.** False   **71.** False

Exercise Set 3.3, p. 199

1. −18   **3.** 19   **5.** −12   **7.** 2   **9.** 4   **11.** 11
13. Approximately 449   **15.** 1024 ft   **17.** \$18,750
19. \$155,000   **21.** 2, −3x, x^2 **23.** $6x^2$ and $-3x^2$ **25.** $2x^4$
and $-3x^4$; $5x$ and $-7x$ **27.** −3, 6   **29.** 5, 3, 3   **31.** −7, 6,
3, 7   **33.** −5, 6, −3, 8, −2   **35.** −3x **37.** −8x
39. $11x^3 + 4$ **41.** $x^3 - x$ **43.** $4b^5$ **45.** $\frac{3}{4}x^5 - 2x - 42$
47. x^4 **49.** $\frac{15}{16}x^3 - \frac{7}{6}x^2$ **51.** $x^5 + 6x^3 + 2x^2 + x + 1$
53. $15y^9 + 7y^8 + 5y^3 - y^2 + y$ **55.** $x^6 + x^4$
57. $13x^3 - 9x + 8$ **59.** $-5x^2 + 9x$ **61.** $12x^4 - 2x + \frac{1}{4}$
63. 1, 0; 1   **65.** 2, 1, 0; 2   **67.** 3, 2, 1, 0; 3   **69.** 2, 1, 6, 4; 6

71.

Term	Coefficient	Degree of Term	Degree of Polynomial
$6x^3$	6	3	
$-3x^2$	−3	2	
$8x$	8	1	4
-2	−2	0	
$-7x^4$	−7	4	

73. x^2, x **75.** x^3, x^2, x^0 **77.** None missing
79. Trinomial   **81.** None of these   **83.** Binomial
85. Monomial   **87.** 27   **89.** $5x^9 + 4x^8 + x^2 + 5x$
91. $4x^5 - 3x^3 + x^2 - 7x$; answers may vary
93. $x^3 - 2x^2 - 6x + 3$

Exercise Set 3.4, p. 207

1. $-x + 5$ **3.** $x^2 - 5x - 1$ **5.** $2x^2$ **7.** $5x^2 + 3x - 30$
9. $-2.2x^3 - 0.2x^2 - 3.8x + 23$ **11.** $12x^2 + 6$
13. $9x^8 + 8x^7 - 3x^4 + 2x^2 - 2x + 5$ **15.** $-\frac{1}{2}x^4 + \frac{2}{3}x^3 + x^2$
17. $0.01x^5 + x^4 - 0.2x^3 + 0.2x + 0.06$ **19.** $-3x^4 + 3x^2 + 4x$
21. $1.05x^4 + 0.36x^3 + 14.22x^2 + x + 0.97$ **23.** $-(-5x)$, $5x$
25. $-(-x^2 + 10x - 2)$, $x^2 - 10x + 2$ **27.** $-(12x^4 - 3x^3 + 3)$,
$-12x^4 + 3x^3 - 3$ **29.** $-3x + 7$ **31.** $-4x^2 + 3x - 2$
33. $4x^4 - 6x^2 - \frac{3}{4}x + 8$ **35.** $7x - 1$ **37.** $-x^2 - 7x + 5$
39. −18   **41.** $6x^4 + 3x^3 - 4x^2 + 3x - 4$
43. $4.6x^3 + 9.2x^2 - 3.8x - 23$ **45.** $\frac{3}{4}x^3 - \frac{1}{2}x$
47. $0.06x^3 - 0.05x^2 + 0.01x + 1$ **49.** $3x + 6$
51. $11x^4 + 12x^3 - 9x^2 - 8x - 9$ **53.** $x^4 - x^3 + x^2 - x$
55. $5x^2 + 4x$ **57.** $\frac{23}{2}a + 10$ **59.** $20 + 5(m-4) +$
$4(m-5) + (m-5)(m-4)$; m^2 **61.** $m^2 - 28$
63. $144 - 4x^2$ **65.** $12y^2 - 23y + 21$
67. $-3y^4 - y^3 + 5y - 2$

Sidelight: Factors and Sums, p. 214

First row: 90, −432, −63; second row:
7, −18, −36, −14, 12, −6, −21, −11; third row:
9, −2, −2, 10, −8, −8, −8, −10, 21; fourth row; −19, −6

Exercise Set 3.5, p. 215

1. $42x^2$ **3.** x^4 **5.** $28x^8$ **7.** $-0.02x^{10}$ **9.** $\frac{1}{15}x^4$ **11.** 0

13. $-24x^{11}$ **15.** $-3x^2 + 15x$ **17.** $-3x^2 + 3x$ **19.** $x^5 + x^2$

21. $6x^3 - 18x^2 + 3x$ **23.** $-6x^4 - 6x^3$ **25.** $18y^6 + 24y^5$

27. $x^2 + 9x + 18$ **29.** $x^2 + 3x - 10$ **31.** $x^2 - 7x + 12$

33. $x^2 - 9$ **35.** $25 - 15x + 2x^2$ **37.** $4x^2 + 20x + 25$

39. $x^2 - \frac{21}{10}x - 1$ **41.** $x^3 - 1$ **43.** $4x^3 + 14x^2 + 8x + 1$

45. $3y^4 - 6y^3 - 7y^2 + 18y - 6$ **47.** $x^6 + 2x^5 - x^3$

49. $-10x^5 - 9x^4 + 7x^3 + 2x^2 - x$ **51.** $x^4 - x^2 - 2x - 1$

53. $6t^4 + t^3 - 16t^2 - 7t + 4$ **55.** $x^9 - x^5 + 2x^3 - x$

57. $x^4 - 1$ **59.** $-\frac{3}{4}$ **61.** $78t^2 + 40t$ **63.** $A = \frac{1}{2}b^2 + 2b$

65. 0

Exercise Set 3.6, p. 223

1. $x^3 + x^2 + 3x + 3$ **3.** $x^4 + x^3 + 2x + 2$ **5.** $y^2 - y - 6$

7. $9x^2 + 15x + 6$ **9.** $5x^2 + 4x - 12$ **11.** $9t^2 - 1$

13. $4x^2 - 6x + 2$ **15.** $p^2 - \frac{1}{16}$ **17.** $x^2 - 0.01$

19. $2x^3 + 2x^2 + 6x + 6$ **21.** $-2x^2 - 11x + 6$

23. $a^2 + 14a + 49$ **25.** $1 - x - 6x^2$ **27.** $x^5 + 3x^3 - x^2 - 3$

29. $3x^6 - 2x^4 - 6x^2 + 4$ **31.** $6x^7 + 18x^5 + 4x^2 + 12$

33. $8x^6 + 65x^3 + 8$ **35.** $4x^3 - 12x^2 + 3x - 9$

37. $4y^6 + 4y^5 + y^4 + y^3$ **39.** $x^2 - 16$ **41.** $4x^2 - 1$

43. $25m^2 - 4$ **45.** $4x^4 - 9$ **47.** $9x^8 - 16$ **49.** $x^{12} - x^4$

51. $x^8 - 9x^2$ **53.** $x^{24} - 9$ **55.** $4y^{16} - 9$ **57.** $x^2 + 4x + 4$

59. $9x^4 + 6x^2 + 1$ **61.** $a^2 - a + \frac{1}{4}$ **63.** $9 + 6x + x^2$

65. $x^4 + 2x^2 + 1$ **67.** $4 - 12x^4 + 9x^8$ **69.** $25 + 60t^2 + 36t^4$

71. $9 - 12x^3 + 4x^6$ **73.** $4x^3 + 24x^2 - 12x$

75. $4x^4 - 2x^2 + \frac{1}{4}$ **77.** $9p^2 - 1$ **79.** $15t^5 - 3t^4 + 3t^3$

81. $36x^8 + 48x^4 + 16$ **83.** $12x^3 + 8x^2 + 15x + 10$

85. $64 - 96x^4 + 36x^8$ **87.** $t^3 - 1$ **89.** $25;\ 49$ **91.** $56;\ 16$
93. Lamps: 500 watts; air conditioner: 2000 watts; television: 50 watts **95.** $8y^3 + 72y^2 + 160y$
97. $-2x^4 + x^3 + 5x^2 - x - 2$ **99.** -7
101. $V = w^3 + 3w^2 + 2w$ **103.** $V = h^3 - 3h^2 + 2h$
105. $F^2 - (F - 17)(F - 7);\ 24F - 119$
107. $4567.0564x^2 + 435.891x + 10.400625$
109. $10,000 - 49 = 9951$ **111.** $5a^2 + 12a - 9$ **113. (a)**
$A^2 + AB$; **(b)** $AB + B^2$; **(c)** $A^2 - B^2$;
(d) $(A + B)(A - B) = A^2 - B^2$

Exercise Set 3.7, p. 231

1. -1 **3.** -7 **5.** $\$11,664$ **7.** $\$12,597.12$ **9.** $141.3\ \text{in}^2$
11. Coefficients: 1, −2, 3, −5; degrees: 4, 2, 2, 0; 4
13. Coefficients: 17, −3, −7; degrees: 5, 5, 0; 5
15. $-a - 2b$ **17.** $3x^2y - 2xy^2 + x^2$ **19.** $8u^2v - 5uv^2$
21. $20au + 10av$ **23.** $x^2 - 4xy + 3y^2$ **25.** $3r + 7$
27. $-x^2 - 8xy - y^2$ **29.** $2ab$ **31.** $-2a + 10b - 5c + 8d$
33. $6z^2 + 7zu - 3u^2$ **35.** $a^4b^2 - 7a^2b + 10$
37. $a^4 + a^3 - a^2y - ay + a + y - 1$ **39.** $a^6 - b^2c^2$
41. $y^6x + y^4x + y^4 + 2y^2 + 1$ **43.** $12x^2y^2 + 2xy - 2$
45. $12 - c^2d^2 - c^4d^4$ **47.** $m^3 + m^2n - mn^2 - n^3$
49. $x^9y^9 - x^6y^6 + x^5y^5 - x^2y^2$ **51.** $x^2 + 2xh + h^2$
53. $r^6t^4 - 8r^3t^2 + 16$ **55.** $p^8 + 2m^2n^2p^4 + m^4n^4$
57. $4a^6 - 2a^3b^3 + \frac{1}{4}b^6$ **59.** $3a^3 - 12a^2b + 12ab^2$
61. $4a^2 - b^2$ **63.** $c^4 - d^2$ **65.** $a^2b^2 - c^2d^4$
67. $x^2 + 2xy + y^2 - 9$ **69.** $x^2 - y^2 - 2yz - z^2$
71. $a^2 - b^2 - 2bc - c^2$ **73.** $4xy - 4y^2$ **75.** $2xy + \pi x^2$
77. $2.4747\ \text{L}$ **79.** $A^3 + 3A^2B + 3AB^2 + B^3$

Exercise Set 3.8, p. 239

1. $3x^4 - \frac{1}{2}x^3 + \frac{1}{8}x^2 - 2$ **3.** $1 - 2u - u^4$ **5.** $5t^2 + 8t - 2$

7. $-4x^4 + 4x^2 + 1$ **9.** $6x^2 - 10x + \frac{3}{2}$ **11.** $9x^2 - \frac{5}{2}x + 1$

13. $6x^2 + 13x + 4$ **15.** $3rs + r - 2s$ **17.** $x + 2$

19. $x - 5 + \frac{-50}{x-5}$ **21.** $x - 2 + \frac{-2}{x+6}$ **23.** $x - 3$

25. $x^4 - x^3 + x^2 - x + 1$ **27.** $2x^2 - 7x + 4$ **29.** $x^3 - 6$

31. $x^3 + 2x^2 + 4x + 8$ **33.** $t^2 + 1$ **35.** 6.8

37. $25,543.75\ \text{ft}^2$ **39.** $x^2 + 5$ **41.** $a + 3 + \frac{5}{5a^2 - 7a - 2}$

43. $2x^2 + x - 3$ **45.** $a^5 + a^4b + a^3b^2 + a^2b^3 + ab^4 + b^5$
47. -5 **49.** 1

Summary and Review: Chapter 3, p. 241

1. [3.1d] $\frac{1}{7^2}$ **2.** [3.1d] y^{11} **3.** [3.1d] $(3x)^{14}$ **4.** [3.1d] t^8

5. [3.1e] 4^3 **6.** [3.1e] $\frac{1}{a^3}$ **7.** [3.1e] 1 **8.** [3.2a, b] $9t^8$

9. [3.1d; 3.2a, b] $36x^8$ **10.** [3.2b] $\frac{y^3}{8x^3}$ **11.** [3.1f] t^{-5}

12. [3.1f] $\frac{1}{y^4}$ **13.** [3.2c] 3.28×10^{-5} **14.** [3.2c] $8,300,000$

15. [3.2d] 2.09×10^4 **16.** [3.2d] 5.12×10^{-5} **17.** [3.2e]
6.205×10^{10} **18.** [3.3a] 10 **19.** [3.3b] $-4y^5,\ 7y^2,\ -3y,$
-2 **20.** [3.3h] x^2, x^0 **21.** [3.3g] 3, 2, 1, 0; 3 **22.** [3.3i]
Binomial **23.** [3.3i] None of these **24.** [3.3i] Monomial
25. [3.3f] $-2x^2 - 3x + 2$ **26.** [3.3f] $10x^4 - 7x^2 - x - \frac{1}{2}$
27. [3.4a] $x^5 - 2x^4 + 6x^3 + 3x^2 - 9$ **28.** [3.4a]

$2x^5 - 6x^4 + 2x^3 - 2x^2 + 2$ **29.** [3.4c] $2x^2 - 4x - 6$
30. [3.4c] $x^5 - 3x^3 - x^2 + 8$ **31.** [3.4d] $P = 4w + 8$;
$A = w^2 + 4w$ **32.** [3.6a] $x^2 + \frac{7}{6}x + \frac{1}{3}$ **33.** [3.6c]
$49x^2 + 14x + 1$ **34.** [3.5d] $12x^3 - 23x^2 + 13x - 2$
35. [3.6b] $9x^4 - 16$ **36.** [3.5b] $15x^7 - 40x^6 + 50x^5 + 10x^4$
37. [3.6a] $x^2 - 3x - 28$ **38.** [3.6c] $9y^4 - 12y^3 + 4y^2$
39. [3.6a] $2t^4 - 11t^2 - 21$ **40.** [3.7a] 49 **41.** [3.7b] Coefficients: 1, –7, 9, –8; degrees: 6, 2, 2, 0; 6 **42.** [3.7c]
$-y + 9w - 5$ **43.** [3.7c]
$m^6 - 2m^2n + 2m^2n^2 + 8n^2m - 6m^3$ **44.** [3.7d] $-9xy - 2y^2$
45. [3.7e] $11x^3y^2 - 8x^2y - 6x^2 - 6x + 6$ **46.** [3.7f] $p^3 - q^3$
47. [3.6c] $9a^8 - 2a^4b^3 + \frac{1}{9}b^6$ **48.** [3.8a] $5x^2 - \frac{1}{2}x + 3$
49. [3.8b] $3x^2 - 7x + 4 + \frac{1}{2x+3}$ **50.** [1.7d] $25(t - 2 + 4m)$
51. [2.3b] $\frac{9}{4}$ **52.** [1.4a] -11.2 **53.** [2.4a] Width:
125.5 m; length: 144.5 m **54.** [3.1d; 3.2a, b; 3.3e] $-28x^8$
55. [2.3b; 3.6a] $\frac{94}{13}$

Test: Chapter 3, p. 243

1. [3.1d] 6^{-5} **2.** [3.1d] x^9 **3.** [3.1d] $(4a)^{11}$ **4.** [3.1e] 3^3
5. [3.1e] $\frac{1}{x^5}$ **6.** [3.1b, e] 1 **7.** [3.2a] x^6 **8.** [3.2a, b] $-27y^6$
9. [3.2a, b] $16a^{12}b^4$ **10.** [3.2b] $\frac{a^3b^3}{c^3}$ **11.** [3.1d; 3.2a, b]
$-216x^{21}$ **12.** [3.1d; 3.2a, b] $-24x^{21}$ **13.** [3.1d; 3.2a, b]
$162x^{10}$ **14.** [3.1d; 3.2a, b] $324x^{10}$ **15.** [3.1f] $\frac{1}{5^3}$
16. [3.1f] y^{-8} **17.** [3.2c] 3.9×10^9 **18.** [3.2c]
0.00000005 **19.** [3.2d] 1.75×10^{17} **20.** [3.2d]
1.296×10^{22} **21.** [3.2e] Approximately 2.55×10^2
22. [3.3a] -43 **23.** [3.3d] $\frac{1}{3}$, -1, 7 **24.** [3.3g] 3, 0, 1, 6; 6
25. [3.3i] Binomial **26.** [3.3e] $5a^2 - 6$ **27.** [3.3e]
$\frac{7}{4}y^2 - 4y$ **28.** [3.3e, f] $x^5 + 2x^3 + 4x^2 - 8x + 3$ **29.** [3.4a]
$4x^5 + x^4 + 2x^3 - 8x^2 + 2x - 7$ **30.** [3.4a] $5x^4 + 5x^2 + x + 5$
31. [3.4c] $-4x^4 + x^3 - 8x - 3$ **32.** [3.4c]
$-x^5 + 0.7x^3 - 0.8x^2 - 21$ **33.** [3.5b] $-12x^4 + 9x^3 + 15x^2$
34. [3.6c] $x^2 - \frac{2}{3}x + \frac{1}{9}$ **35.** [3.6b] $9x^2 - 100$ **36.** [3.6a]
$3b^2 - 4b - 15$ **37.** [3.6a] $x^{14} - 4x^8 + 4x^6 - 16$ **38.** [3.6a]
$48 + 34y - 5y^2$ **39.** [3.5d] $6x^3 - 7x^2 - 11x - 3$ **40.** [3.6c]
$25t^2 + 20t + 4$ **41.** [3.7c] $-5x^3y - y^3 + xy^3 - x^2y^2 + 19$
42. [3.7e] $8a^2b^2 + 6ab - 4b^3 + 6ab^2 + ab^3$ **43.** [3.7f]
$9x^{10} - 16y^{10}$ **44.** [3.8a] $4x^2 + 3x - 5$ **45.** [3.8b]
$2x^2 - 4x - 2 + \frac{17}{3x+2}$ **46.** [2.3b] 13 **47.** [1.7d]
$16(4t - 2m + 1)$ **48.** [1.4a] $\frac{23}{20}$ **49.** [2.4a] 100°, 25°, 55°
50. [3.6a] $V = l^3 - 3l^2 + 2l$ **51.** [2.3b; 3.6a] $-\frac{61}{12}$

Cumulative Review: Chapters 1-3, p. 245

1. [1.1a] $\frac{5}{2}$ **2.** [3.3a] -4 **3.** [3.7a] -14 **4.** [1.2e] 4
5. [1.6b] $\frac{1}{5}$ **6.** [1.3a] $-\frac{11}{60}$ **7.** [1.4a] 4.2 **8.** [1.5a] 7.28
9. [1.6c] $-\frac{5}{12}$ **10.** [3.2d] 2.2×10^{22} **11.** [3.2d] 4×10^{-5}
12. [1.7a] -3 **13.** [1.8b] $-2y - 7$ **14.** [1.8c] $5x + 11$
15. [1.8d] -2 **16.** [3.4a] $2x^5 - 2x^4 + 3x^3 + 2$ **17.** [3.7d]
$3x^2 + xy - 2y^2$ **18.** [3.4c] $x^3 + 5x^2 - x - 7$ **19.** [3.4c]
$-\frac{1}{3}x^2 - \frac{3}{4}x$ **20.** [1.7c] $12x - 15y + 21$ **21.** [3.5a] $6x^8$
22. [3.5b] $2x^5 - 4x^4 + 8x^3 - 10x^2$ **23.** [3.5d]
$p^3q^2 + 2p^2q^2 - p^2q^3 + pq^3$ **25.** [3.6a] $6x^2 + 13x + 6$
26. [3.6c] $9x^4 + 6x^2 + 1$ **27.** [3.6b] $t^2 - \frac{1}{4}$ **28.** [3.6b]
$4y^4 - 25$ **29.** [3.6a] $4x^6 + 6x^4 - 6x^2 - 9$ **30.** [3.6c]
$t^2 - 4t^3 + 4t^4$ **31.** [3.7f] $15p^2 - pq - 2q^2$ **32.** [3.8a]
$6x^2 + 2x - 3$ **33.** [3.8b] $3x^2 - 2x - 7$ **34.** [2.1b] -1.2
35. [2.2a] -21 **36.** [2.3a] 9 **37.** [2.2a] $-\frac{20}{3}$ **38.** [2.3b] 2
39. [2.1b] $\frac{13}{8}$ **40.** [2.3c] $-\frac{17}{21}$ **41.** [2.3b] -17 **42.** [2.3b] 2
43. [2.7e] $\{x \mid x < 16\}$ **44.** [2.7e] $\left\{x \mid x \le -\frac{11}{8}\right\}$
45. [2.6a] $h = \frac{A - \pi r^2}{2\pi r}$ **46.** [3.4d] $\pi r^2 - 18$ **47.** [2.4a]
18 and 19 **48.** [2.4a] 20 ft, 24 ft **49.** [2.4a] 10^0
50. [2.4a] -45 **51.** [2.5a] $3.50 **52.** [3.1d] y^4
53. [3.1e] $\frac{1}{x}$ **54.** [3.2a, b] $-\frac{27x^9}{y^6}$ **55.** [3.1d, e] x^3
56. [3.3d] $\frac{2}{3}$, 4, –6 **57.** [3.3g] 4, 2, 1, 0; 4 **58.** [3.3i]
Binomial **59.** [3.3i] Trinomial **60.** [3.4d] $4x - 4$
61. [3.1d; 3.2a, b; 3.4a] $12x^5 - 15x^4 - 27x^3 + 4x^2$
62. [3.4a; 3.6c] $5x^2 - 2x + 10$ **63.** [3.4a; 3.8b]
$4x^2 - 2x + 7$ **64.** [2.3b; 3.6c] $\frac{11}{7}$ **65.** [2.3b; 3.8b] 1
66. [1.2e; 2.3a] -5, 5 **67.** [2.3b; 3.6c] No solution
68. [2.3b; 3.6a; 3.8b] All real numbers

CHAPTER 4

Pretest: Chapter 4, p. 248

1. [4.1a] $4(-5x^6)$, $(-2x^3)(10x^3)$, $x^2(-20x^4)$; answers may
vary **2.** [4.5b] $2(x + 1)^2$ **3.** [4.2a] $(x + 4)(x + 2)$
4. [4.6a] $4a(2a^4 + a^2 - 5)$ **5.** [4.3a, 4.4a] $(5x + 2)(x - 3)$
6. [4.5d] $(9 + z^2)(3 + z)(3 - z)$ **7.** [4.6a] $(y^3 - 2)^2$
8. [4.1c] $(x^2 + 4)(3x + 2)$ **9.** [4.2a] $(p - 6)(p + 5)$
10. [4.7b] 0, 5 **11.** [4.7a] 4, $\frac{3}{5}$ **12.** [4.7b] $\frac{2}{3}$, –4
13. [4.8a] 6, –1 **14.** [4.8a] Base: 8 cm; height: 11 cm
15. [4.6a] $(x^2y + 8)(x^2y - 8)$ **16.** [4.6a]
$(2p - q)(p + 4q)$

Exercise Set 4.1, p. 253

1. $6x^2 \cdot x$; $3x^2 \cdot 2x$; $(-3x^2)(-2x)$; answers may vary

3. $(-9x^4) \cdot x$; $(-3x^2)(3x^3)$; $(-3x)(3x^4)$; answers may vary
5. $(8x^2)(3x^2)$; $(-8x^2)(-3x^2)$; $(4x^3)(6x)$; answers may vary
7. $x(x-4)$ **9.** $2x(x+3)$ **11.** $x^2(x+6)$ **13.** $8x^2(x^2-3)$
15. $2(x^2+x-4)$ **17.** $17xy(x^4y^2+2x^2y+3)$
19. $x^2(6x^2-10x+3)$ **21.** $x^2y^2(x^3y^3+x^2y+xy-1)$
23. $2x^3(x^4-x^3-32x^2+2)$ **25.** $0.8x(2x^3-3x^2+4x+8)$
27. $\frac{1}{3}x^3(5x^3+4x^2+x+1)$ **29.** $(x^2+2)(x+3)$
31. $(x^2+2)(x+3)$ **33.** $(2x^2+1)(x+3)$
35. $(4x^2+3)(2x-3)$ **37.** $(4x^2+1)(3x-4)$
39. $(x^2-3)(x+8)$ **41.** $(2x^2-9)(x-4)$
43. $\{x \mid x > -24\}$ **45.** 27   **47.** $y^2+12y+35$ **49.** y^2-49
51. $(2x^3+3)(2x^2+3)$ **53.** $(x^7+1)(x^5+1)$ **55.** Not factorable by grouping

Exercise Set 4.2, p. 259

1. $(x+3)(x+5)$ **3.** $(x+3)(x+4)$ **5.** $(x-3)^2$
7. $(x+2)(x+7)$ **9.** $(b+1)(b+4)$ **11.** $\left(x+\frac{1}{3}\right)^2$
13. $(d-2)(d-5)$ **15.** $(y-1)(y-10)$ **17.** $(x-6)(x+7)$
19. $(x-9)(x+2)$ **21.** $x(x-8)(x+2)$ **23.** $(y-9)(y+5)$
25. $(x-11)(x+9)$ **27.** $(c^2+8)(c^2-7)$
29. $(a^2+7)(a^2-5)$ **31.** Not factorable   **33.** Not factorable   **35.** $(x+10)^2$ **37.** $(x-25)(x+4)$
39. $(x-24)(x+3)$ **41.** $(x-9)(x-16)$
43. $(a+12)(a-11)$ **45.** $(x-15)(x-8)$
47. $(12+x)(9-x)$, or $-(x+12)(x-9)$
49. $(y-0.4)(y+0.2)$ **51.** $(p+5q)(p-2q)$
53. $(m+4n)(m+n)$ **55.** $(s+3t)(s-5t)$
57. $16x^3-48x^2+8x$ **59.** $49w^2+84w+36$ **61.** 15, −15, 27, −27, 51, −51   **63.** $\left(x+\frac{1}{4}\right)\left(x-\frac{3}{4}\right)$ **65.** $(x+5)\left(x-\frac{5}{7}\right)$
67. $(b^n+5)(b^n+2)$ **69.** $2x^2(4-\pi)$

Exercise Set 4.3, p. 265

1. $(2x+1)(x-4)$ **3.** $(5x-9)(x+2)$ **5.** $(3x+1)(2x+7)$
7. $(3x+1)(x+1)$ **9.** $(2x-3)(2x+5)$
11. $(2x+1)(x-1)$ **13.** $(3x-2)(3x+8)$
15. $(3x+1)(x-2)$ **17.** $(3x+4)(4x+5)$
19. $(7x-1)(2x+3)$ **21.** $(3x+2)(3x+4)$
23. $(3x-7)(3x-7)$ **25.** $(24x-1)(x+2)$
27. $(5x-11)(7x+4)$ **29.** $2(5-x)(2+x)$
31. $4(3x-2)(x+3)$ **33.** $6(5x-9)(x+1)$
35. $2(3x+5)(x-1)$ **37.** $(3x-1)(x-1)$
39. $4(3x+2)(x-3)$ **41.** $(2x+1)(x-1)$
43. $(3x+2)(3x-8)$ **45.** $5(3x+1)(x-2)$
47. $x(3x+4)(4x+5)$ **49.** $x^2(7x-1)(2x+3)$
51. $3x(8x-1)(7x-1)$ **53.** $(5x^2-3)(3x^2-2)$
55. $(5t+8)^2$ **57.** $2x(3x+5)(x-1)$ **59.** Not factorable
61. Not factorable   **63.** $(4m-5n)(3m+4n)$
65. $(2a+3b)(3a-5b)$ **67.** $(3a+2b)(3a+4b)$
69. $(5p+2q)(7p+4q)$ **71.** $6(3x-4y)(x+y)$

73. $q = \frac{A+7}{p}$ **75.** $(2x^n+1)(10x^n+3)$
77. $(x^{3a}-1)(3x^{3a}+1)$

Exercise Set 4.4, p. 269

1. $(y+1)(y+4)$ **3.** $(x-1)(x-4)$ **5.** $(2x+3)(3x+2)$
7. $(x-4)(3x-4)$ **9.** $(5x+3)(7x-8)$
11. $(2x-3)(2x+3)$ **13.** $(2x^2+5)(x^2+3)$
15. $(2x+1)(x-4)$ **17.** $(5x-9)(x+2)$
19. $(2x+7)(3x+1)$ **21.** $(3x+1)(x+1)$
23. $(2x-3)(2x+5)$ **25.** $(2x+1)(x-1)$
27. $(3x-2)(3x+8)$ **29.** $(3x+1)(x-2)$
31. $(3x+4)(4x+5)$ **33.** $(7x-1)(2x+3)$
35. $(3x+2)(3x+4)$ **37.** $(3x-7)^2$ **39.** $(24x-1)(x+2)$
41. $x^3(5x-11)(7x+4)$ **43.** $6x(5-x)(2+x)$
45. $\left\{x \mid x > \frac{26}{7}\right\}$ **47.** $(3x^5-2)^2$ **49.** $(4x^5+1)^2$

Exercise Set 4.5, p. 277

1. Yes   **3.** No   **5.** No   **7.** No   **9.** $(x-7)^2$ **11.** $(x+8)^2$
13. $(x-1)^2$ **15.** $(x+2)^2$ **17.** $(y^2+3)^2$ **19.** $(4y-7)^2$
21. $2(x-1)^2$ **23.** $x(x-9)^2$ **25.** $5(2x+5)^2$ **27.** $(7-3x)^2$
29. $5(y^2+1)^2$ **31.** $(1+2x^2)^2$ **33.** $(2p+3q)^2$
35. $(a-7b)^2$ **37.** $(8m+n)^2$ **39.** $(4s-5t)^2$ **41.** Yes
43. No   **45.** No   **47.** Yes   **49.** $(y+2)(y-2)$
51. $(p+3)(p-3)$ **53.** $(t+7)(t-7)$ **55.** $(a+b)(a-b)$
57. $(5t+m)(5t-m)$ **59.** $(10+k)(10-k)$
61. $(4a+3)(4a-3)$ **63.** $(2x+5y)(2x-5y)$
65. $2(2x+7)(2x-7)$ **67.** $x(6+7x)(6-7x)$
69. $(7a^2+9)(7a^2-9)$ **71.** $(x^2+1)(x+1)(x-1)$
73. $4(x^2+4)(x+2)(x-2)$
75. $(y^4+1)(y^2+1)(1+y)(1-y)$
77. $(x^6+4)(x^3+2)(x^3-2)$ **79.** $\left(y+\frac{1}{4}\right)\left(y-\frac{1}{4}\right)$
81. $\left(5+\frac{1}{7}x\right)\left(5-\frac{1}{7}x\right)$ **83.** $(4m^2+t^2)(2m+t)(2m-t)$
85. −11   **87.** $-\frac{5}{6}$ **89.** Not factorable   **91.** $(x+11)^2$
93. $2x(3x+1)^2$ **95.** $(x^4+2^4)(x^2+2^2)(x+2)(x-2)$
97. $3x^3(x+2)(x-2)$ **99.** $2x\left(3x+\frac{2}{5}\right)\left(3x-\frac{2}{5}\right)$
101. $p(0.7+p)(0.7-p)$ **103.** $(0.8x+1.1)(0.8x-1.1)$
105. $x(x+6)$ **107.** $\left(x+\frac{1}{x}\right)\left(x-\frac{1}{x}\right)$
109. $(9+b^{2k})(3+b^k)(3-b^k)$ **111.** $(3b^n+2)^2$
113. $(y+4)^2$ **115.** 9

Exercise Set 4.6, p. 285

1. $2(x+8)(x-8)$ **3.** $(a-5)^2$ **5.** $(2x-3)(x-4)$
7. $x(x+12)^2$ **9.** $(x+2)(x-2)(x+3)$
11. $6(2x+3)(2x-3)$ **13.** $4x(5x+9)(x-2)$ **15.** Not factorable   **17.** $x(x^2+7)(x-3)$ **19.** $x^3(x-7)^2$

21. $2(2-x)(5+x)$, or $-2(x-2)(x+5)$　**23.** Not factorable　**25.** $4(x^2+4)(x+2)(x-2)$
27. $(1+y^4)(1+y^2)(1+y)(1-y)$　**29.** $x^3(x-3)(x-1)$
31. $\left(6a-\frac{5}{4}\right)^2$　**33.** $12n^2(1+2n)$　**35.** $9xy(xy-4)$
37. $2\pi r(h+r)$　**39.** $(a+b)(2x+1)$
41. $(x+1)(x-1-y)$　**43.** $(n+p)(n+2)$
45. $(2x+z)(x-2)$　**47.** $(x-y)^2$　**49.** $(3c+d)^2$
51. $(7m^2-8n)^2$　**53.** $(y^2+5z^2)^2$　**55.** $\left(\frac{1}{2}a+\frac{1}{3}b\right)^2$
57. $(a+b)(a-2b)$　**59.** $(m+20n)(m-18n)$
61. $(mn-8)(mn+4)$　**63.** $a^3(ab+5)(ab-2)$
65. $a^3(a-b)(a+5b)$　**67.** $(x^3-y)(x^3+2y)$
69. $(x-y)(x+y)$　**71.** $7(p^2+q^2)(p+q)(p-q)$
73. $(9a^2+b^2)(3a+b)(3a-b)$
75. $(w+2)(w-2)(w-7)$　**77.** $-\frac{14}{11}$　**79.** $X=\frac{A+7}{a+b}$
81. $(a+1)^2(a-1)^2$　**83.** $(3.5x-1)^2$　**85.** $(5x+4)(x+1.8)$
87. $(y+3)(y-3)(y-2)$　**89.** $(a^2+1)(a+4)$
91. $(x+3)(x-3)(x^2+2)$　**93.** $(x+2)(x-2)(x-1)$
95. $(y-1)^3$　**97.** $(y+4+x)^2$
99. $(x^4+2^4)(x^2+2^2)(x+2)(x-2)$

Exercise Set 4.7, p. 293

1. $-8,-6$　**3.** $3,-5$　**5.** $-12,11$　**7.** $0,-5$　**9.** $0,-10$
11. $-\frac{5}{2},-4$　**13.** $-\frac{1}{5},3$　**15.** $4,\frac{1}{4}$　**17.** $0,\frac{2}{3}$　**19.** $0,18$
21. $-\frac{1}{10},\frac{1}{27}$　**23.** $\frac{1}{3},20$　**25.** $0,\frac{2}{3},\frac{1}{2}$　**27.** $-1,-5$　**29.** $-9,2$
31. $3,5$　**33.** $0,8$　**35.** $0,-19$　**37.** $4,-4$　**39.** $\frac{2}{3},-\frac{2}{3}$
41. -3　**43.** 4　**45.** $0,\frac{6}{5}$　**47.** $\frac{5}{3},-1$　**49.** $\frac{2}{3},-\frac{1}{4}$　**51.** $7,-2$
53. $\frac{9}{8},-\frac{9}{8}$　**55.** $-3,1$　**57.** $\frac{4}{5},\frac{3}{2}$　**59.** $(a+b)^2$　**61.** -16
63. $4,-5$　**65.** $9,-3$　**67.** $\frac{1}{8},-\frac{1}{8}$　**69.** $4,-4$　**71. (a)**
$x^2-x-12=0$; **(b)** $x^2+7x+12=0$; **(c)** $4x^2-4x+1=0$;
(d) $x^2-25=0$; **(e)** $40x^3-14x^2+x=0$

Exercise Set 4.8, p. 299

1. $-\frac{3}{4},1$　**3.** $4,2$　**5.** 14 and 15　**7.** 12 and 14; -12 and -14　**9.** 15 and 17; -15 and -17　**11.** Length: 12 m; width: 8 m　**13.** 5　**15.** Height: 4 cm; base: 14 cm
17. 6 km　**19.** 5 and 7　**21.** 506　**23.** 12　**25.** 780　**27.** 20
29. Hypotenuse: 17 ft; leg: 15 ft　**31.** 5 ft　**33.** 37
35. 30 cm by 15 cm　**37.** 100 cm²; 225 cm²

Summary and Review: Chapter 4, p. 303

1. [4.1a] $-10x\cdot x$, $-5x\cdot 2x$, $(5x)(-2x)$; answers may vary
2. [4.1a] $6x\cdot 6x^4$, $4x^2\cdot 9x^3$, $2x^4\cdot 18x$; answers may vary
3. [4.5d] $5(1+2x^3)(1-2x^3)$　**4.** [4.1b] $x(x-3)$　**5.** [4.5d]

$(3x+2)(3x-2)$　**6.** [4.2a] $(x+6)(x-2)$　**7.** [4.5b]
$(x+7)^2$　**8.** [4.1b] $3x(2x^2+4x+1)$　**9.** [4.1c]
$(x^2+3)(x+1)$　**10.** [4.3a; 4.4a] $(3x-1)(2x-1)$
11. [4.5d] $(x^2+9)(x+3)(x-3)$　**12.** [4.3a; 4.4a]
$3x(3x-5)(x+3)$　**13.** [4.5d] $2(x+5)(x-5)$　**14.** [4.1c]
$(x^3-2)(x+4)$　**15.** [4.5d] $(4x^2+1)(2x+1)(2x-1)$
16. [4.1b] $4x^4(2x^2-8x+1)$　**17.** [4.5b] $3(2x+5)^2$
18. [4.5c] Not factorable　**19.** [4.2a] $x(x-6)(x+5)$
20. [4.5d] $(2x+5)(2x-5)$　**21.** [4.5b] $(3x-5)^2$
22. [4.3a; 4.4a] $2(3x+4)(x-6)$　**23.** [4.5b] $(x-3)^2$
24. [4.3a; 4.4a] $(2x+1)(x-4)$　**25.** [4.5b] $2(3x-1)^2$
26. [4.5d] $3(x+3)(x-3)$　**27.** [4.2a] $(x-5)(x-3)$
28. [4.5b] $(5x-2)^2$　**29.** [4.7a] $1,-3$　**30.** [4.7b] $-7,5$
31. [4.7b] $-4,3$　**32.** [4.7b] $\frac{2}{3},1$　**33.** [4.7b] $\frac{3}{2},-4$
34. [4.7b] $8,-2$　**35.** [4.8a] 3 and -2　**36.** [4.8a] -18 and -16; 16 and 18　**37.** [4.8a] -19 and -17; 17 and 19
38. [4.8a] $\frac{5}{2}$ and -2　**39.** [4.5b] $(7b^5-2a^4)^2$　**40.** [4.2a]
$(xy+4)(xy-3)$　**41.** [4.5b] $3(2a+7b)^2$　**42.** [4.1c]
$(m+t)(m+5)$　**43.** [4.5d] $32(x^2-2y^2z^2)(x^2+2y^2z^2)$
44. [1.6c] $\frac{8}{35}$　**45.** [2.7e] $\left\{x\;\middle|\;x\le\frac{4}{3}\right\}$　**46.** [2.6a] $b=\frac{A-a}{2}$
47. [3.6d] $4a^2+12a+9$　**48.** [3.6d] $4a^2-9$　**49.** [3.6d]
$10a^2-a-21$　**50.** [4.8a] 2.5 cm　**51.** [4.8a] 0, 2
52. [4.8a] Length: 12; width: 6　**53.** [4.7b] No solution
54. [4.7a] $2,-3,\frac{5}{2}$　**55.** [4.7a] $-2000,-\frac{5000}{97}$　**56.** [4.6a] a, i; b, k; c, g; d, h; e, j; f, *l*

Test: Chapter 4, p. 305

1. [4.1a] $(4x)(x^2)$; $(2x^2)(2x)$; $(-2x)(-2x^2)$; answers may vary　**2.** [4.2a] $(x-5)(x-2)$　**3.** [4.5b] $(x-5)^2$　**4.** [4.1b]
$2y^2(2y^2-4y+3)$　**5.** [4.1c] $(x^2+2)(x+1)$　**6.** [4.1b]
$x(x-5)$　**7.** [4.2a] $x(x+3)(x-1)$　**8.** [4.3a; 4.4a]
$2(5x-6)(x+4)$　**9.** [4.5d] $(2x+3)(2x-3)$　**10.** [4.2a]
$(x-4)(x+3)$　**11.** [4.3a; 4.4a] $3m(2m+1)(m+1)$
12. [4.5d] $3(w+5)(w-5)$　**13.** [4.5b] $5(3x+2)^2$
14. [4.5d] $3(x^2+4)(x+2)(x-2)$　**15.** [4.5b] $(7x-6)^2$
16. [4.3a; 4.4a] $(5x-1)(x-5)$　**17.** [4.1c] $(x^3-3)(x+2)$
18. [4.5d] $5(4+x^2)(2+x)(2-x)$　**19.** [4.3a; 4.4a]
$(2x+3)(2x-5)$　**20.** [4.3a; 4.4a] $3t(2t+5)(t-1)$
21. [4.7b] $5,-4$　**22.** [4.7b] $\frac{3}{2},-5$　**23.** [4.7b] $7,-4$
24. [4.8a] $8,-3$　**25.** [4.8a] Length: 10 m; width: 4 m
26. [4.2a] $3(m+2n)(m-5n)$　**27.** [1.6c] $-\frac{10}{11}$
28. [2.7e; 2.8a] $\left\{x\;\middle|\;x<\frac{19}{3}\right\}$　**29.** [2.6a] $T=\frac{I}{PR}$　**30.** [3.6d]
$25x^4-70x^2+49$　**31.** [4.8a] Length: 15; width: 3
32. [4.2a] $(a-4)(a+8)$　**33.** [4.7a] (c)　**34.** [3.6b; 4.5d]
(d)

Cumulative Review: Chapters 1-4, p. 307

1. [1.2d] <   **2.** [1.2d] >   **3.** [1.4a] 0.35   **4.** [1.6c] −1.57
5. [1.5a] $-\frac{1}{14}$ **6.** [1.6c] $-\frac{6}{5}$ **7.** [1.8c] $4x+1$ **8.** [1.8d] −8
9. [3.2a, b] $\frac{8x^6}{y^3}$ **10.** [3.1d, e] $-\frac{1}{6x^3}$ **11.** [3.4a]
$x^4-3x^3-3x^2-4$ **12.** [3.7e] $2x^2y^2-x^2y-xy$ **13.** [3.8b]
$x^2+3x+2+\frac{3}{x-1}$ **14.** [3.6c] $4t^2-12t+9$ **15.** [3.6b]
x^4-9 **16.** [3.6a] $6x^2+4x-16$ **17.** [3.5b]
$2x^4+6x^3+8x^2$ **18.** [3.5d] $4y^3+4y^2+5y-4$ **19.** [3.6b]
$x^2-\frac{4}{9}$ **20.** [4.2a] $(x+4)(x-2)$ **21.** [4.5d]
$(2x+5)(2x-5)$ **22.** [4.1c] $(3x-4)(x^2+1)$ **23.** [4.5b]
$(x-13)^2$ **24.** [4.5d] $3(5x+6y)(5x-6y)$
25. [4.3a; 4.4a] $(3x+7)(2x-9)$ **26.** [4.2a]
$(x^2-3)(x^2+1)$ **27.** [4.6a] $2(y-1)(y+1)(2y-3)$
28. [4.3a; 4.4a] $(3p-q)(2p+q)$ **29.** [4.3a; 4.4a]
$2x(5x+1)(x+5)$ **30.** [4.5b] $x(7x-3)^2$ **31.** [4.3a; 4.4a]
Cannot be factored   **32.** [4.1b] $3x(25x^2+9)$ **33.** [4.5d]
$3(x^4+4y^4)(x^2+2y^2)(x^2-2y^2)$ **34.** [4.2a]
$14(x+2)(x+1)$ **35.** [4.6a] $(x^3+1)(x+1)(x-1)$
36. [2.3b] 15   **37.** [2.7e] $\{y \mid y < 6\}$ **38.** [4.7a] 15, $-\frac{1}{4}$
39. [4.7a] 0, −37   **40.** [4.7b] 5, −5, −1   **41.** [4.7b] 6, −6
42. [4.7b] $\frac{1}{3}$ **43.** [4.7b] −10, −7   **44.** [4.7b] 0, $\frac{3}{2}$
45. [2.3a] 0.2   **46.** [4.7b] −4, 5   **47.** [2.7e] $\{x \mid x \le 20\}$
48. [2.3c] All real numbers   **49.** [2.6a] $m = \frac{y-b}{x}$
50. [2.4a] 50, 52   **51.** [4.8a] −20, −18 and 18, 20
52. [4.8a] 6 ft, 3 ft   **53.** [2.4a] 150 m by 350 m
54. [2.5a] \$6500   **55.** [4.8a] 17 m   **56.** [2.4a] 30 m,
60 m, 10 m   **57.** [2.5a] \$29.00   **58.** [4.8a] 18 cm, 16 cm
59. [2.7e; 3.6a] $\left\{ x \mid x \ge -\frac{13}{3} \right\}$ **60.** [2.3b] 22   **61.** [4.7b]
−6, 4   **62.** [4.6a] $(x-2)(x+1)(x-3)$ **63.** [4.6a]
$(2a+3b+3)(2a-3b-5)$ **64.** [4.5a] 25   **65.** [4.8a] 2 cm

CHAPTER 5

Pretest: Chapter 5, p. 310

1. [5.3c] $(x+2)(x+3)^2$ **2.** [5.4a] $\frac{-b-1}{b^2-4}$, or $\frac{b+1}{4-b^2}$ **3.** [5.5a]
$\frac{1}{y-2}$ **4.** [5.4a] $\frac{7a+6}{a(a+2)}$ **5.** [5.5b] $\frac{2x}{x+1}$ **6.** [5.1d] $\frac{2(x-3)}{x-2}$
7. [5.9a] $\frac{y+x}{y-x}$ **8.** [5.6a] −5   **9.** [5.6a] 0   **10.** [5.8a]
$M = \frac{3R}{a-b}$ **11.** [5.7b] 10.5 hr   **12.** [5.7a] $\frac{30}{11}$ hr   **13.** [5.7a]
60 mph, 80 mph

Exercise Set 5.1, p. 317

1. $\{x \mid x \ne 0\}$ **3.** $\{a \mid a \ne 8\}$ **5.** $\left\{ y \mid y \ne -\frac{5}{2} \right\}$
7. $\{x \mid x \ne 7 \text{ and } x \ne -4\}$ **9.** $\{m \mid m \ne 5 \text{ and } m \ne -5\}$

11. $\frac{(3a)(5a^2)}{(3a)(2c)}$ **13.** $\frac{2x(x-1)}{2x(x+4)}$ **15.** $\frac{-1(3-x)}{-1(4-x)}$ **17.** $\frac{(y+6)(y-7)}{(y+6)(y+2)}$ **19.** $\frac{x^2}{4}$
21. $\frac{8p^2q}{3}$ **23.** $\frac{x-3}{x}$ **25.** $\frac{m+1}{2m+3}$ **27.** $\frac{a-3}{a+2}$ **29.** $\frac{a-3}{a-4}$ **31.** $\frac{x+5}{x-5}$
33. $a+1$ **35.** $\frac{x^2+1}{x+1}$ **37.** $\frac{3}{2}$ **39.** $\frac{6}{t-3}$ **41.** $\frac{t+2}{2(t-4)}$ **43.** $\frac{t-2}{t+2}$
45. −1   **47.** −1   **49.** −6   **51.** $-a-1$ **53.** $\frac{56x}{3}$ **55.** $\frac{2}{dc^2}$
57. $\frac{x+2}{x-2}$ **59.** $\frac{(a+3)(a-3)}{a(a+4)}$ **61.** $\frac{2a}{a-2}$ **63.** $\frac{(x+2)(x-2)}{(x+1)(x-1)}$ **65.** $\frac{t-2}{t-1}$
67. $\frac{5(a+6)}{a-1}$ **69.** 18 and 20; −18 and −20
71. $(2y^2+1)(y-5)$ **73.** $x+2y$ **75.** $\frac{(t-9)^2(t-1)}{(t^2+9)(t+1)}$ **77.** $\frac{x-y}{x-5y}$

Exercise Set 5.2, p. 323

1. $\frac{x}{4}$ **3.** $\frac{1}{x^2-y^2}$ **5.** $\frac{x^2-4x+7}{x^2+2x-5}$ **7.** $\frac{3}{10}$ **9.** $\frac{1}{4}$ **11.** $\frac{y^2}{x}$ **13.** $\frac{(a+2)(a+3)}{(a-3)(a-1)}$
15. $\frac{(x-1)^2}{x}$ **17.** $\frac{1}{2}$ **19.** $\frac{15}{8}$ **21.** $\frac{15}{4}$ **23.** $\frac{a-5}{3(a-1)}$ **25.** $\frac{(x+2)^2}{x}$ **27.** $\frac{3}{2}$
29. $\frac{c+1}{c-1}$ **31.** $\frac{y-3}{2y-1}$ **33.** $\frac{x+1}{x-1}$ **35.** 4   **37.** $-\frac{1}{b^2}$ **39.** $\frac{x(x^2+1)}{3(x+y-1)}$

Exercise Set 5.3, p. 327

1. 108   **3.** 72   **5.** 126   **7.** 360   **9.** 420   **11.** $\frac{65}{72}$ **13.** $\frac{29}{120}$
15. $\frac{151}{180}$ **17.** $12x^3$ **19.** $18x^2y^2$ **21.** $6(y-3)$
23. $t(t+2)(t-2)$ **25.** $(x+2)(x-2)(x+3)$
27. $t(t+2)^2(t-4)$ **29.** $(a+1)(a-1)^2$
31. $(m-3)(m-2)^2$ **33.** $(2+3x)(2-3x)$
35. $10v(v+4)(v+3)$ **37.** $18x^3(x-2)^2(x+1)$
39. $6x^3(x+2)^2(x-2)$ **41.** $(x-3)^2$ **43.** $(x+3)(x-3)$
45. 1440   **47.** 24 min

Exercise Set 5.4, p. 333

1. 1   **3.** $\frac{6}{3+x}$ **5.** $\frac{2x+3}{x-5}$ **7.** $\frac{1}{4}$ **9.** $-\frac{1}{t}$ **11.** $\frac{-x+7}{x-6}$ **13.** $y+3$
15. $\frac{2b-14}{b^2-16}$ **17.** $-\frac{1}{y+z}$ **19.** $\frac{5x+2}{x-5}$ **21.** −1   **23.** $\frac{-x^2+9x-14}{(x-3)(x+3)}$
25. $\frac{2x+5}{x^2}$ **27.** $\frac{41}{24r}$ **29.** $\frac{4x+6y}{x^2y^2}$ **31.** $\frac{4+3t}{18t^3}$ **33.** $\frac{x^2+4xy+y^2}{x^2y^2}$
35. $\frac{6x}{(x-2)(x+2)}$ **37.** $\frac{11x+2}{3x(x+1)}$ **39.** $\frac{x^2+6x}{(x+4)(x-4)}$ **41.** $\frac{6}{z+4}$ **43.** $\frac{3x-1}{(x-1)^2}$
45. $\frac{11a}{10(a-2)}$ **47.** $\frac{2x^2+8x+16}{x(x+4)}$ **49.** $\frac{x^2+5x+1}{(x+1)^2(x+4)}$ **51.** $\frac{2x^2-4x+34}{(x-5)(x+3)}$
53. $\frac{3a+2}{(a+1)(a-1)}$ **55.** $\frac{2x+6y}{(x+y)(x-y)}$ **57.** $\frac{y^2+10y+11}{(y+7)(y-7)}$ **59.** $\frac{3x^2+19x-20}{(x+3)(x-2)^2}$
61. x^2-1 **63.** $\frac{1}{8x^{12}y^9}$ **65.** Perimeter: $\frac{16y+28}{15}$; area: $\frac{y^2+2y-8}{15}$
67. $\frac{(z+6)(2z-3)}{(z-2)(z+2)}$ **69.** $\frac{11z^4-22z^2+6}{(z^2+2)(z^2-2)(2z^2-3)}$

Exercise Set 5.5, p. 341

1. $\frac{1}{2}$ **3.** 1   **5.** $\frac{4}{x-1}$ **7.** $\frac{8}{3}$ **9.** $\frac{13}{a}$ **11.** $\frac{4x-5}{4}$ **13.** $\frac{x-2}{x-7}$ **15.** $\frac{2x-16}{x^2-16}$

17. $\frac{2x-4}{x-9}$ **19.** $\frac{-9}{2x-3}$ **21.** $\frac{-x-4}{6}$ **23.** $\frac{7z-12}{12z}$ **25.** $\frac{4x^2-13xt+9t^2}{3x^2t^2}$

27. $\frac{2x-40}{(x+5)(x-5)}$ **29.** $\frac{3-5t}{2t(t-1)}$ **31.** $\frac{2s-st-s^2}{(t+s)(t-s)}$ **33.** $\frac{y-19}{4y}$

35. $\frac{-2a^2}{(x+a)(x-a)}$ **37.** $\frac{-13x-20}{(x+4)(x-4)}$ **39.** $\frac{1}{2}$ **41.** $\frac{x-3}{(x+3)(x+1)}$ **43.** $\frac{18x+5}{x-1}$

45. 0 **47.** $\frac{20}{2y-1}$ **49.** $\frac{2}{y(y-1)}$ **51.** $\frac{z-3}{2z-1}$ **53.** $\frac{2}{x+y}$ **55.** x^5

57. $\frac{b^{20}}{a^8}$ **59.** $\frac{1-3x}{(2x-3)(x+1)}$

Exercise Set 5.6, p. 349

1. $\frac{47}{2}$ **3.** -6 **5.** $\frac{24}{7}$ **7.** $-4, -1$ **9.** $4, -4$ **11.** 3 **13.** $\frac{14}{3}$

15. 10 **17.** 5 **19.** $\frac{5}{2}$ **21.** -1 **23.** $\frac{17}{2}$ **25.** No solution

27. -5 **29.** $\frac{5}{3}$ **31.** $\frac{1}{2}$ **33.** No solution **35.** -13 **37.** $\frac{1}{a^6b^{15}}$

39. $\frac{16x^4}{t^8}$ **41.** 7 **43.** $-\frac{1}{30}$ **45.** $2, -2$ **47.** 4

Exercise Set 5.7, p. 357

1. $\frac{20}{9}$ **3.** 20 and 15

5. 30 km/h, 70 km/h

Speed	Time
r	t
$r+40$	t

$150 = r(t)$

$350 = (r+40)t$

7. Passenger: 80 km/h; freight: 66 km/h

Speed	Time
$r-14$	t
r	t

$330 = (r-14)t$

$400 = r(t)$

9. 20 mph **11.** $2\frac{2}{9}$ hr **13.** $5\frac{1}{7}$ hr **15.** 9 **17.** 2.3 km/h

19. 582 **21.** 702 km **23.** 1.92 g **25.** 287 **27. (a)**
1.92 tons; **(b)** 28.8 lb **29.** $\frac{36}{68}$ **31.** $\frac{3}{4}$ **33.** 2 mph

35. $\frac{A}{B}=\frac{C}{D}; \frac{A}{C}=\frac{B}{D}; \frac{D}{B}=\frac{C}{A}; \frac{D}{C}=\frac{B}{A}$ **37.** $9\frac{3}{13}$ days **39.** 45 mph

Exercise Set 5.8, p. 363

1. $r=\frac{S}{2\pi h}$ **3.** $b=\frac{2A}{h}$ **5.** $n=\frac{S+360}{180}$ **7.** $b=\frac{3V-kB-4kM}{k}$

9. $r=\frac{S-a}{S-l}$ **11.** $h=\frac{2A}{b_1+b_2}$ **13.** $B=\frac{A}{AQ+1}$ **15.** $p=\frac{qf}{q-f}$

17. $A=P(1+r)$ **19.** $R=\frac{r_1r_2}{r_1+r_2}$ **21.** $D=\frac{BC}{A}$

23. $h_2=\frac{p(h_1-q)}{q}$ **25.** $a=\frac{b}{K-C}$ **27.** $23x^4+50x^3+$
$23x^2-163x+41$ **29.** $3(2y^2-1)(5y^2+4)$ **31.** $T=\frac{FP}{u+EF}$
33. $-40°$

Exercise Set 5.9, p. 369

1. $\frac{25}{4}$ **3.** $\frac{1}{3}$ **5.** -6 **7.** $\frac{1+3x}{1-5x}$ **9.** $\frac{5}{3y^2}$ **11.** 8 **13.** $x-8$ **15.** $\frac{y}{y-1}$

17. $-\frac{1}{a}$ **19.** $\frac{x+y}{x}$ **21.** $\frac{x-2}{x-3}$ **23.** $4x^4+3x^3+2x-7$

25. $\frac{(x-1)(3x-2)}{5x-3}$ **27.** $-\frac{ac}{bd}$ **29.** $\frac{5x+3}{3x+2}$

Summary and Review: Chapter 5, p. 371

1. [5.1a] 0 **2.** [5.1a] 6 **3.** [5.1a] 6, -6 **4.** [5.1a] -6, 5
5. [5.1a] -2 **6.** [5.1a] 0, 3, 5 **7.** [5.1c] $\frac{x-2}{x+1}$ **8.** [5.1c] $\frac{7x+3}{x-3}$
9. [5.1c] $\frac{y-5}{y+5}$ **10.** [5.1d] $\frac{a-6}{5}$ **11.** [5.1d] $\frac{6}{2t-1}$ **12.** [5.2b]
$-20t$ **13.** [5.2b] $\frac{2x^2-2x}{x+1}$ **14.** [5.3c] $30x^2y^2$ **15.** [5.3c]
$4(a-2)$ **16.** [5.3c] $(y-2)(y+2)(y+1)$ **17.** [5.4a]
$\frac{-3x+18}{x+7}$ **18.** [5.4a] -1 **19.** [5.5a] $\frac{4}{x-4}$ **20.** [5.5a] $\frac{x+5}{2x}$
21. [5.5a] $\frac{2x+3}{x-2}$ **22.** [5.4a] $\frac{2a}{a-1}$ **23.** [5.4a] $d+c$
24. [5.5a] $\frac{-x^2+x+26}{(x-5)(x+5)(x+1)}$ **25.** [5.5b] $\frac{2(x-2)}{x+2}$ **26.** [5.9a] $\frac{z}{1-z}$
27. [5.9a] $c-d$ **28.** [5.6a] 8 **29.** [5.6a] 3, -5 **30.** [5.7a]
$5\frac{1}{7}$ hr **31.** [5.7a] 240 km/h, 280 km/h **32.** [5.7a] -2
33. [5.7b] 160 **34.** [5.7a] 95 km/h, 175 km/h **35.** [5.8a]
$s=\frac{rt}{r-t}$ **36.** [5.8a] $C=\frac{5}{9}(F-32)$, or $C=\frac{5F-160}{9}$
37. [4.1c] $(5x^2-3)(x+4)$ **38.** [3.2a, b] $\frac{1}{125x^9y^6}$ **39.** [3.4c]
$-2x^3+3x^2+12x-18$ **40.** [4.8a] Length: 5 cm; width:
3 cm; perimeter: 16 cm **41.** [5.1c; 5.2b] $\frac{5(a+3)^2}{a}$
42. [5.5a] $\frac{10a}{(a-b)(b-c)}$ **43. (a)** [5.4a] $\frac{10x+6}{(x-3)(x+3)}$; **(b)** [5.6a] $\frac{1}{10}$;
(c) In (a) the LCM is used to find an equivalent expression for each rational expression with the LCM as the least common denominator. In (b) the LCM is used to clear fractions.

Test: Chapter 5, p. 373

1. [5.1a] 0 **2.** [5.1a] -8 **3.** [5.1a] 7, -7 **4.** [5.1a] 1, 2
5. [5.1a] 1 **6.** [5.1a] 0, -3, -5 **7.** [5.1c] $\frac{3x+7}{x+3}$ **8.** [5.1d]
$\frac{a+5}{2}$ **9.** [5.2b] $\frac{(5x+1)(x+1)}{3x(x+2)}$ **10.** [5.3c] $(y-3)(y+3)(y+7)$
11. [5.4a] $\frac{23-3x}{x^3}$ **12.** [5.5a] $\frac{8-2t}{t^2+1}$ **13.** [5.4a] $\frac{-3}{x-3}$ **14.** [5.5a]
$\frac{2x-5}{x-3}$ **15.** [5.4a] $\frac{8t-3}{t(t-1)}$ **16.** [5.5a] $\frac{-x^2-7x-15}{(x+4)(x-4)(x+1)}$ **17.** [5.5b]
$\frac{x^2+2x-7}{(x-1)^2(x+1)}$ **18.** [5.9a] $\frac{3y+1}{y}$ **19.** [5.6a] 12 **20.** [5.6a] 5, -3
21. [5.7a] 4 **22.** [5.7b] 16 **23.** [5.7a] 45 km/h, 65 km/h
24. [5.8a] $t=\frac{g}{M-L}$ **25.** [4.5c] $(4a+7)(4a-7)$
26. [3.2a, b] $\frac{y^{12}}{81x^8}$ **27.** [3.4c] $13x^2-29x+76$ **28.** [4.8a]

21 and 22, or −22 and −21   **29.** [5.7a] Team *A:* 4 hr;
team *B:* 10 hr   **30.** [5.9a] $\frac{3a+2}{2a+1}$

Cumulative Review: Chapters 1-5, p. 375

1. [1.1a] $-\frac{9}{5}$ **2.** [3.1c] 12   **3.** [1.8c] $2x+6$
4. [3.1d; 3.2a, b] $\frac{27x^7}{4}$ **5.** [3.1e] $\frac{4x^{10}}{3}$ **6.** [5.1c] $\frac{2(t-3)}{2t-1}$
7. [5.9a] $\frac{(x+2)^2}{x^2}$ **8.** [5.1c] $\frac{a+4}{a-4}$ **9.** [1.3a] $\frac{17}{42}$ **10.** [5.4a]
$\frac{x^2+4xy+y^2}{x^2y^2}$ **11.** [5.4a] $\frac{3z-2}{z^2-1}$ **12.** [3.4a] $2x^4+8x^3-2x+9$
13. [1.4a] 2.33   **14.** [3.7e] $-xy$ **15.** [5.5a] $\frac{1}{x-3}$ **16.** [5.5a]
$\frac{2x^2-14x-16}{(x+4)(x-5)^2}$ **17.** [1.5a] -1.3 **18.** [3.5b] $6x^4+12x^3-15x^2$
19. [3.6b] $9t^2-\frac{1}{4}$ **20.** [3.6c] $4p^2-4pq+q^2$ **21.** [3.6a]
$3x^2-7x-20$ **22.** [3.6b] $4x^4-1$ **23.** [5.1d] $\frac{2(t-1)}{t}$
24. [5.1d] $-\frac{2(a+1)}{a}$ **25.** [3.8b] $3x^2-4x+5$ **26.** [1.6c] $-\frac{9}{20}$
27. [5.2b] $\frac{x-2}{2x(x-3)}$ **28.** [5.2b] $-\frac{12}{x}$ **29.** [4.6a]
$(2x+3)(2x-3)(x+3)$ **30.** [4.2a] $(x+8)(x-1)$
31. [4.3a; 4.4a] $(3x+1)(x-5)$ **32.** [4.5b] $(4y+5x)^2$
33. [4.2a] $3x(x+5)(x+3)$ **34.** [4.5d] $2(x+1)(x-1)$
35. [4.5b] $(x-14)^2$ **36.** [4.1b, c] $2(y^2+3)(2y+5)$
37. [2.3c] -7 **38.** [4.7a] $0, -\frac{4}{3}$ **39.** [4.7b] $0, 8$ **40.** [4.7b]
4   **41.** [2.7e] $\{x \mid x \ge -9\}$ **42.** [4.7b] $3, -3$ **43.** [2.3b]
$-\frac{11}{7}$ **44.** [5.6a] $3, -3$ **45.** [5.6a] $-\frac{1}{11}$ **46.** [4.7a] $0, \frac{1}{10}$
47. [5.6a] No solution   **48.** [5.6a] $\frac{5}{7}$ **49.** [5.8a] $z = \frac{xy}{x+y}$
50. [2.6a] $N = \frac{DT}{3}$ **51.** [2.4a] 32, 33, 34   **52.** [5.7a]
12 km/h, 10 km/h   **53.** [5.7a] $\frac{30}{11}$ hr   **54.** [2.4a] 33 and 34
55. [4.8a] 16 and 17   **56.** [5.7b] $1\frac{2}{3}$ cups   **57.** [4.8a] 7
58. [4.8a] 9 and 11   **59.** [4.7b] $4, -\frac{2}{3}$ **60.** [4.7b; 5.6a] 2,
-1 **61.** [4.7b; 5.6a; 5.9a] 1, -4 **62.** [3.1d, e; 4.7b] 5, -5
63. [5.2a; 5.4a] $\frac{-x^2-x+6}{2x^2+x+5}$ **64.** [3.2d] 5×10^7

CHAPTER 6

Pretest: Chapter 6, p. 378

1. [6.2b]   **2.** [6.3b]

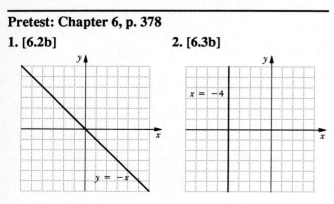

---

3. [6.3a]   **4.** [6.2b]

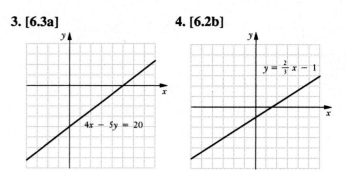

5. [6.1c] III   **6.** [6.2a] No   **7.** [6.4b] 4   **8.** [6.4b] 0
9. [6.4c] Slope: $\frac{1}{3}$; y-intercept: $\left(0, -\frac{7}{3}\right)$ **10.** [6.4a] Not
defined   **11.** [6.4d] $y = x - 4$ **12.** [6.4d] $y = 4x + 7$
13. [6.6a] $y = \frac{5}{2}x$ **14.** [6.6c] $y = \frac{40}{x}$

15. [6.7b]   **16.** [6.7b]

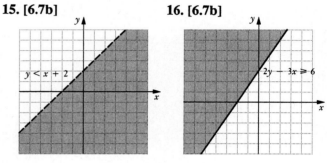

17. [6.5a] Parallel   **18.** [6.5b] Perpendicular   **19.** [6.5b]
Perpendicular   **20.** [6.7a] Yes

Exercise Set 6.1, p. 383

1. Boredom   **3.** Aproximately 5%   **5.** The second week
7. 12%   **9.** $270   **11.** San Francisco   **13.** $50   **15.** 1993
17. $17.0 million   **19.** $1.5 million   **21.** 3.7%
23. 70.1%   **25.** 1743; 360; 270

27.

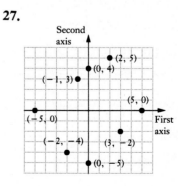

29. II   **31.** IV   **33.** III   **35.** I   **37.** Negative; negative
39. *A:* $(3, 3)$; *B:* $(0, -4)$; *C:* $(-5, 0)$; *D:* $(-1, -1)$; *E:* $(2, 0)$
41. $5330   **43.** I, IV   **45.** I, III   **47.** $(-1, -5)$

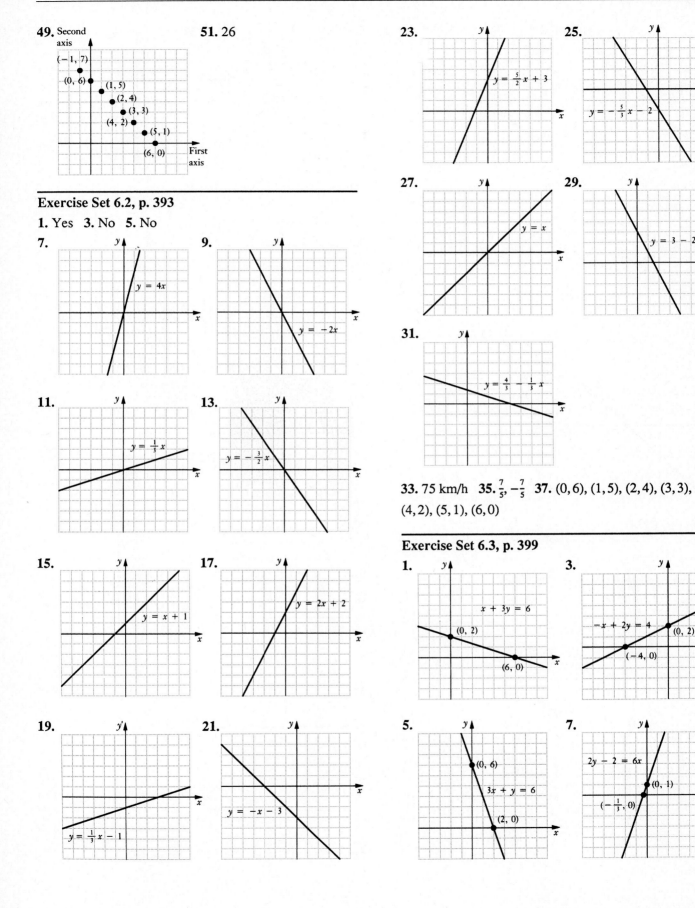

49.

51. 26

Exercise Set 6.2, p. 393

1. Yes   **3.** No   **5.** No

7. $y = 4x$

9. $y = -2x$

11. $y = \frac{1}{3}x$

13. $y = -\frac{3}{2}x$

15. $y = x + 1$

17. $y = 2x + 2$

19. $y = \frac{1}{3}x - 1$

21. $y = -x - 3$

23. $y = \frac{5}{2}x + 3$

25. $y = -\frac{5}{3}x - 2$

27. $y = x$

29. $y = 3 - 2x$

31. $y = \frac{4}{3} - \frac{1}{3}x$

33. 75 km/h   **35.** $\frac{7}{5}, -\frac{7}{5}$ **37.** $(0,6), (1,5), (2,4), (3,3),$ $(4,2), (5,1), (6,0)$

Exercise Set 6.3, p. 399

1. $x + 3y = 6$; $(0, 2)$, $(6, 0)$

3. $-x + 2y = 4$; $(0, 2)$, $(-4, 0)$

5. $3x + y = 6$; $(0, 6)$, $(2, 0)$

7. $2y - 2 = 6x$; $(0, 1)$, $(-\frac{1}{3}, 0)$

9.

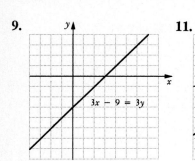

$3x - 9 = 3y$

11.

$2x - 3y = 6$

29.

$y - 3x = 0$ $(0, 0)$

31.

$x = -2$

13.

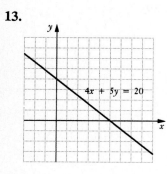

$4x + 5y = 20$

15.

$2x + 3y = 8$

33.

$y = 2$

35.

$x = 2$

17.

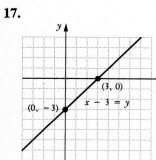

$(3, 0)$ $(0, -3)$ $x - 3 = y$

19.

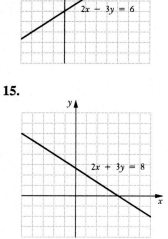

$3x - 2 = y$ $\left(\frac{2}{3}, 0\right)$ $(0, -2)$

37.

$y = 0$

39.

$x = \frac{3}{2}$

21.

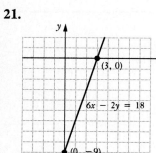

$(3, 0)$ $6x - 2y = 18$ $(0, -9)$

23.

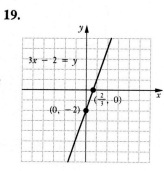

$3x + 4y = 5$ $\left(0, \frac{5}{4}\right)$ $\left(\frac{5}{3}, 0\right)$

41.

$3y = -5$

43.

$4x + 3 = 0$

25.

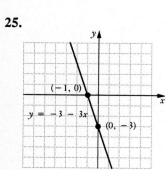

$(-1, 0)$ $y = -3 - 3x$ $(0, -3)$

27.

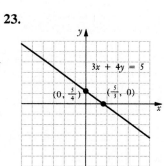

$-4x = 8y - 5$ $\left(\frac{5}{4}, 0\right)$ $\left(0, \frac{5}{8}\right)$

45.

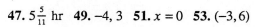

$48 - 3y = 0$

47. $5\frac{5}{11}$ hr   **49.** $-4, 3$ **51.** $x = 0$ **53.** $(-3, 6)$

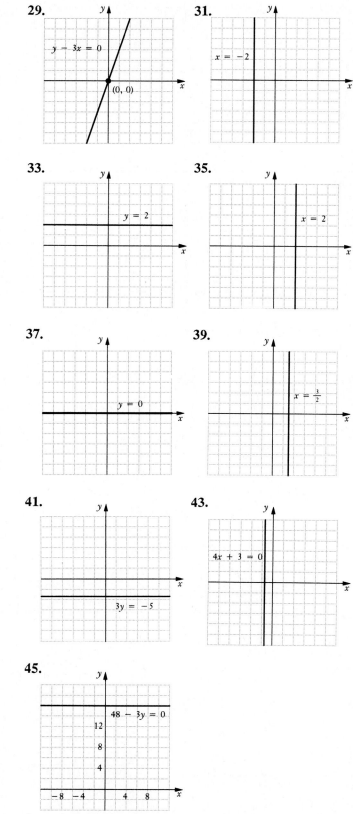

Exercise Set 6.4, p. 407

1. 0 **3.** $-\frac{4}{5}$ **5.** 7 **7.** 2 **9.** 0 **11.** Not defined **13.** Not defined **15.** 0 **17.** Not defined **19.** 0 **21.** $-\frac{3}{2}$ **23.** $-\frac{1}{4}$ **25.** 2 **27.** -4, $(0,-9)$ **29.** 1.8, $(0,0)$ **31.** $-\frac{8}{7}$, $(0,-3)$ **33.** 3, $\left(0,-\frac{5}{3}\right)$ **35.** $-\frac{3}{2}$, $\left(0,-\frac{1}{2}\right)$ **37.** 0, $(0,-17)$ **39.** $y = 5x - 5$ **41.** $y = \frac{3}{4}x + \frac{5}{2}$ **43.** $y = x - 8$ **45.** $y = -3x - 9$ **47.** $y = \frac{1}{4}x + \frac{5}{2}$ **49.** $y = -\frac{1}{2}x + 4$ **51.** $y = -\frac{3}{2}x + \frac{13}{2}$ **53.** $y = \frac{3}{4}x - \frac{5}{2}$ **55.** $\frac{44}{7}$ **57.** -5 **59.** $y = 3x - 9$ **61.** $y = \frac{3}{2}x - 2$

Exercise Set 6.5, p. 411

1. Yes **3.** No **5.** No **7.** No **9.** Yes **11.** Yes **13.** No **15.** Yes **17.** No **19.** Yes **21.** No **23.** 130 km/h; 140 km/h **25.** 4 **27.** $y = 3x + 6$ **29.** $y = -3x + 2$ **31.** $y = \frac{1}{2}x + 1$ **33.** $k = 16$ **35.** $A: y = \frac{4}{3}x - \frac{7}{3}, B:$ $y = -\frac{3}{4}x - \frac{1}{4}$

Exercise Set 6.6, p. 417

1. $y = 4x$ **3.** $y = 1.75x$ **5.** $y = 3.2x$ **7.** $y = \frac{2}{3}x$ **9.** \$183.75 **11.** $22\frac{6}{7}$ **13.** $36.\overline{6}$ lb **15.** 68.4 kg **17.** $y = \frac{75}{x}$ **19.** $y = \frac{80}{x}$ **21.** $y = \frac{1}{x}$ **23.** $y = \frac{1050}{x}$ **25.** $y = \frac{0.06}{x}$ **27. (a)** Inverse; **(b)** $5\frac{1}{3}$ hr **29. (a)** Direct; **(b)** $69\frac{3}{8}$ **31.** 320 cm³ **33.** 54 min **35.** 2.4 ft **37.** $P = nS$, $k = n$, where n is the number of sides of the polygon **39.** $B = kN$ **41.** If p varies directly as q, then $p = mq$. Then $q = \frac{1}{m}p$. Let $k = \frac{1}{m}$. Then $q = kp$, so q varies directly as p. **43.** $S = kV^6$ **45.** $V = kr^3$ **47.** $C = \frac{k}{N}$ **49.** $I = \frac{k}{R}$ **51.** $I = \frac{k}{d^2}$ **53.** Yes **55.** No

Exercise Set 6.7, p. 425

1. No **3.** No

5.

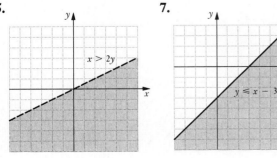

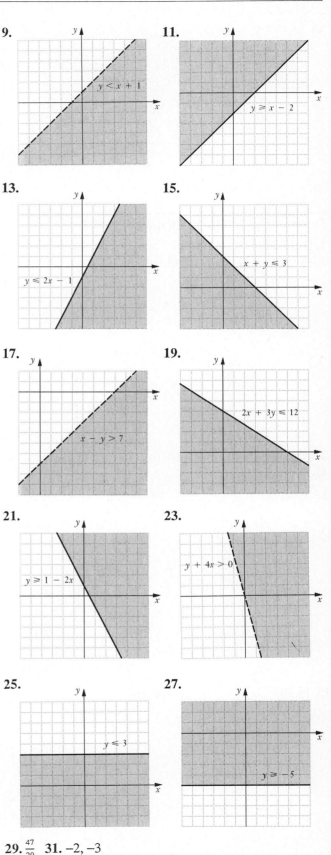

29. $\frac{47}{20}$ **31.** $-2, -3$

33. $35c + 75a > 1000$

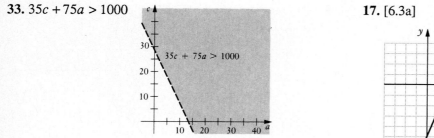

Summary and Review: Chapter 6, p. 427

1. [6.1a] (a) 13%; (b) Years 3 and 4   **2.** [6.1d] $(-5, -1)$
3. [6.1d] $(-2, 5)$ **4.** [6.1d] $(3, 0)$

5.-7. [6.1b]

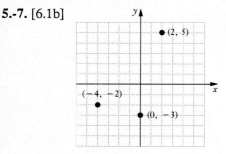

8. [6.1c] IV   **9.** [6.1c] III   **10.** [6.1c] I   **11.** [6.2a] No
12. [6.2a] Yes

13. [6.2b]   **14.** [6.2b]

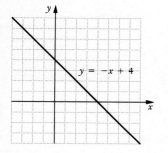

15. [6.2b]   **16.** [6.2b]

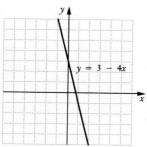

17. [6.3a]   **18.** [6.3b]

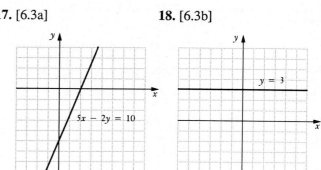

19. [6.3b]   **20.** [6.3a]

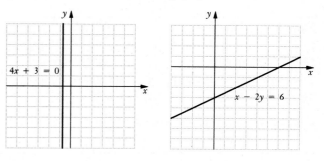

21. [6.4a] $\frac{3}{2}$ **22.** [6.4a] 0   **23.** [6.4a] Not defined
24. [6.4a] 2   **25.** [6.4b] 0   **26.** [6.4b] Not defined
27. [6.4b] $-\frac{4}{3}$ **28.** [6.4c] $-9, (0, 46)$ **29.** [6.4c] $-1, (0, 9)$
30. [6.4c] $\frac{3}{5}, \left(0, -\frac{4}{5}\right)$ **31.** [6.4d] $y = 3x - 1$ **32.** [6.4d]
$y = \frac{2}{3}x - \frac{11}{3}$ **33.** [6.4d] $y = -2x - 4$ **34.** [6.4d] $y = x + 2$
35. [6.4d] $y = \frac{1}{2}x - 1$ **36.** [6.5a] Parallel   **37.** [6.5b]
Perpendicular   **38.** [6.5a] Parallel   **39.** [6.5a, b] Neither
40. [6.6a] $y = 3x$ **41.** [6.6a] $y = \frac{1}{2}x$ **42.** [6.6a] $y = \frac{4}{5}x$
43. [6.6c] $y = \frac{30}{x}$ **44.** [6.6c] $y = \frac{1}{x}$ **45.** [6.6c] $y = \frac{0.65}{x}$
46. [6.6b] $247.50   **47.** [6.6d] 1 hr   **48.** [6.7a] No
49. [6.7a] No   **50.** [6.7a] Yes

51. [6.7b]   **52.** [6.7b]

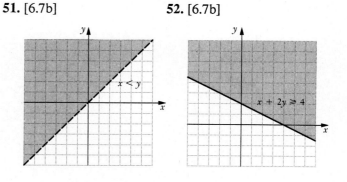

53. [6.7b]

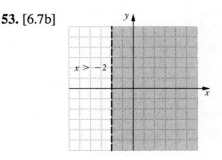

54. [5.7a] $3\frac{1}{3}$ hr **55.** [1.8d] 52 **56.** [5.6a] −4 **57.** [4.7b] 5, −11 **58.** [6.2a] −1 **59.** [6.2a] 19

Test: Chapter 6, p. 429

1. [6.1a] Bachelor's **2.** [6.1a] 520,000 **3.** [6.1a] 261,800 **4.** [6.1a] 401,900 **5.** [6.1c] II **6.** [6.1c] III **7.** [6.1d] (3, 4) **8.** [6.1d] (0, −4) **9.** [6.2a] Yes

10. [6.2b]   **11.** [6.3a]

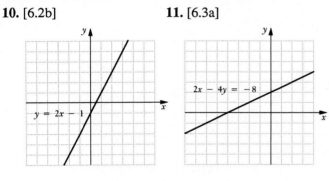

12. [6.3b]   **13.** [6.2b]

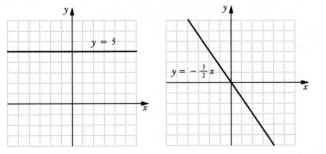

14. [6.3b]

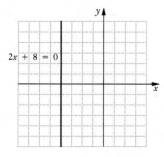

15. [6.4a] Not defined **16.** [6.4a] $\frac{7}{12}$ **17.** [6.4b] 0
18. [6.4b] Not defined **19.** [6.4c] 2, $\left(0, -\frac{1}{4}\right)$ **20.** [6.4c] $\frac{4}{3}$, (0, −2) **21.** [6.4d] $y = x + 2$ **22.** [6.4d] $y = -3x - 6$
23. [6.4d] $y = -3x + 4$ **24.** [6.4d] $y = \frac{1}{4}x - 2$ **25.** [6.5a] Parallel **26.** [6.5a, b] Neither **27.** [6.5b] Perpendicular
28. [6.6a] $y = 2x$ **29.** [6.6a] $y = 0.5x$ **30.** [6.6c] $y = \frac{12}{x}$
31. [6.6c] $y = \frac{1}{x}$ **32.** [6.6b] 240 km **33.** [6.6d] $1\frac{1}{5}$ hr
34. [6.7a] No **35.** [6.7a] Yes
36. [6.7b]   **37.** [6.7b]

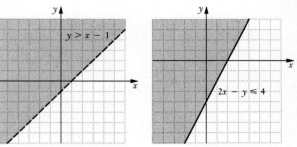

38. [5.7a] Freight: 90 mph; passenger: 105 mph
39. [1.8d] 1 **40.** [4.7b] $\frac{1}{3}$, −5 **41.** [5.6a] No solution
42. [6.1b] Area: 56; perimeter: 30 **43.** [6.4c, d] $y = \frac{2}{3}x + \frac{11}{3}$

Cumulative Review: Chapters 1-6, p. 431

1. [1.2e] 3.5 **2.** [3.3d] 1, −2, 1, −1 **3.** [3.3g] 3, 2, 1, 0; 3
4. [3.3i] None of these **5.** [3.3e] $2x^3 - 3x^2 - 2$ **6.** [1.8c] $\frac{3}{8}x + 1$ **7.** [3.1e; 3.2a, b] $\frac{9}{4x^8}$ **8.** [5.9a] $\frac{8x-12}{17x}$ **9.** [3.7e] $-2xy^2 - 4x^2y^2 + xy^3$ **10.** [3.4a] $2x^5 + 6x^4 + 2x^3 - 10x^2 + 3x - 9$ **11.** [5.1d] $\frac{2}{3(y+2)}$ **12.** [5.2b] 2 **13.** [5.4a] $x + 4$ **14.** [5.5a] $\frac{2x-6}{(x+2)(x-2)}$ **15.** [3.6a] $a^2 - 9$ **16.** [3.5d] $2x^5 + x^3 - 6x^2 - x + 3$ **17.** [3.6b] $4x^6 - 1$ **18.** [3.6c] $36x^2 - 60x + 25$ **19.** [3.5b] $12x^4 + 16x^3 + 4x^2$ **20.** [3.6a] $6x^7 - 12x^5 + 9x^2 - 18$ **21.** [2.3c] 3 **22.** [4.7b] $\frac{1}{2}$, −4
23. [4.7b] −5, 4 **24.** [2.7e] $\{x \mid x \geq -26\}$ **25.** [4.7a] 0, 4
26. [4.7b] 0, 10 **27.** [4.7b] 20, −20 **28.** [2.6a] $a = \frac{t}{x+y}$
29. [5.6a] 2 **30.** [5.6a] No solution **31.** [1.7d] $-2(3 + x + 6y)$ **32.** [4.2a] $(x - 4)(x - 6)$ **33.** [4.5c] $2(x + 3)(x - 3)$ **34.** [4.1c] $(m^3 - 3)(m + 2)$ **35.** [4.5b] $(4x + 5)^2$ **36.** [4.3a; 4.4a] $(2x + 1)(4x + 3)$ **37.** [4.8a] 4 or −5 **38.** [6.6b] $78 **39.** [2.5a] $2500 **40.** [5.7a] 35 mph, 25 mph **41.** [4.8a] 14 ft **42.** [5.7a] 35, 28

43. [6.2b]   **44.** [6.3a]

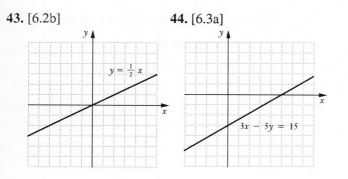

45. [6.3b]   **46.** [6.7b]

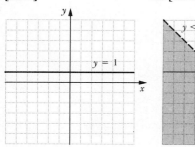

47. [6.7b]

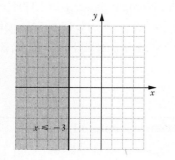

48. [6.6a] $y = \frac{2}{3}x$ **49.** [6.6c] $y = \frac{10}{x}$ **50.** [6.4a] Not defined   **51.** [6.4a] $-\frac{3}{7}$ **52.** [6.4c] $m = \frac{4}{3}$; y-intercept: $(0, -2)$ **53.** [6.4d] $y = -4x + 5$ **54.** [6.4d] $y = \frac{1}{6}x - \frac{17}{6}$ **55.** [6.5a, b] Neither   **56.** [6.5a] Parallel   **57.** [3.4c; 3.6a] 12   **58.** [3.6b, c] $16y^6 - y^4 + 6y^2 - 9$ **59.** [4.5d] $2(a^{16} + 81b^{20})(a^8 + 9b^{10})(a^4 + 3b^5)(a^4 - 3b^5)$ **60.** [4.7a] $4, -7, 12$ **61.** [6.4d; 6.5a] $y = \frac{2}{3}x$ **62.** [5.1a; 5.9a] All real numbers except 0, 3, and $\frac{5}{2}$

CHAPTER 7

Pretest: Chapter 7, p. 434

1. [7.1a] Yes   **2.** [7.1b] No solution. The lines are parallel.   **3.** [7.2a] $(5, 2)$ **4.** [7.2b] $(2, -1)$ **5.** [7.3a] $\left(\frac{3}{4}, \frac{1}{2}\right)$ **6.** [7.3b] $(-5, -2)$ **7.** [7.3b] $(10, 8)$ **8.** [7.2c] 50 and 24   **9.** [7.3c] $25°$ and $65°$ **10.** [7.5a] 8 hrs after the second train leaves

Exercise Set 7.1, p. 439

1. Yes   **3.** No   **5.** Yes   **7.** Yes   **9.** Yes   **11.** $(2, 1)$ **13.** $(4, 3)$ **15.** $(-3, -3)$ **17.** No solution. The lines are parallel.   **19.** $(2, 2)$ **21.** $\left(\frac{1}{2}, 1\right)$ **23.** Infinite number of solutions.   **25.** $(-4, 2)$ **27.** $\frac{108}{x^{13}}$ **29.** $\frac{2x^2 - 1}{x^2(x+1)}$ **31.** $\frac{9x + 12}{(x-4)(x+4)}$ **33.** $A = 2, B = 2$ **35.** $2x + y = 0, y - x = -6$

Exercise Set 7.2, p. 445

1. $(1, 3)$ **3.** $(1, 2)$ **5.** $(4, 3)$ **7.** $(-2, 1)$ **9.** $(-1, -3)$ **11.** $\left(\frac{17}{3}, \frac{16}{3}\right)$ **13.** $\left(\frac{25}{8}, -\frac{11}{4}\right)$ **15.** $(-3, 0)$ **17.** $(6, 3)$ **19.** 12 and 15   **21.** 21 and 37   **23.** 28 and 12   **25.** Length: 365 mi; width: 275 mi   **27.** Length: $134\frac{1}{3}$ m; width: $65\frac{2}{3}$ m

29.   **31.**

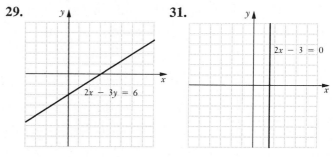

33. $(4.3821792, 4.3281211)$ **35.** $(10, -2)$ **37.** Length: 100 yd; width: $53\frac{1}{3}$ yd   **39.** You get $0 = 0$, a statement that is true for any values of x and y. There are infinitely many solutions; the lines coincide.

Exercise Set 7.3, p. 453

1. $(9, 1)$ **3.** $(3, 5)$ **5.** $(3, 0)$ **7.** $\left(-\frac{1}{2}, 3\right)$ **9.** $\left(-1, \frac{1}{5}\right)$ **11.** No solution   **13.** $(-3, -5)$ **15.** $(4, 5)$ **17.** $(4, 1)$ **19.** $(4, 3)$ **21.** $(1, -1)$ **23.** $(-3, -1)$ **25.** $(2, -2)$ **27.** $\left(5, \frac{1}{2}\right)$ **29.** Infinitely many solutions   **31.** $(2, -1)$ **33.** $\left(\frac{231}{202}, \frac{117}{202}\right)$ **35.** 10 miles   **37.** $50°$ and $130°$ **39.** $62°$ and $28°$ **41.** 480 hectares Chardonnay; 340 hectares Riesling   **43.** $\frac{a^7}{b^9}$ **45.** $\frac{x-3}{x+2}$ **47.** 12 rabbits and 23 pheasants   **49.** The father is 45; the daughter 10   **51.** $(5, 2)$ **53.** $(0, -1)$

Exercise Set 7.4, p. 465

1. 350 cars; 160 trucks   **3.** Sammy is 44; his daughter is 22   **5.** Marge is 18; Consuelo is 9   **7.** 70 dimes; 33 quarters   **9.** 300 nickels; 100 dimes   **11.** 203 adults; 226 children   **13.** 130 adults; 70 students   **15.** 40 L of A; 60 L of B

A	B	Mixture	
x	y	100 liters	$\longrightarrow$ $x + y = 100$
50%	80%	68%	
$0.5x$	$0.8y$	0.68×100, or 68 liters	$\longrightarrow$ $0.5x + 0.8y = 68$

17. 80 L of 30%; 120 L of 50% **19.** 6 kg of cashews; 4 kg of pecans **21.** 39 lb of A; 36 lb of B **23.** 12 of A; 4 of B **25.** Inexpensive: $19.408; expensive: $20.075 **27.** $12,500 at 12%; $14,500 at 13% **29.** 54 **31.** 43.75 L **33.** $4\frac{4}{7}$ L **35.** 74 **37.** Bat: $14.50; ball: $4.55; glove: $79.95

Exercise Set 7.5, p. 473

1. After $\frac{2}{3}$ hr, or 40 min

Speed	Time
55	t
40	t

3. $7\frac{1}{2}$ hr after the first train leaves, or $4\frac{1}{2}$ hr after the second train leaves

Speed	Time	
72	$t + 3$	$\longrightarrow$ $d = 72(t + 3)$
120	t	$\longrightarrow$ $d = 120t$

5. 14 km/h

Speed	Time	
$r + 6$	4	$\longrightarrow$ $d = (r + 6)4$
$r - 6$	10	$\longrightarrow$ $d = (r - 6)10$

7. 384 km **9.** 330 km/h **11.** 8.54 hr **13.** 15 mi **15.** 317.02702 km/h **17.** 180 mi by land; 96 mi by water **19.** 144 mi

---

Summary and Review: Chapter 7, p. 475

1. [7.1a] No **2.** [7.1a] Yes **3.** [7.1a] Yes **4.** [7.1a] No **5.** [7.1b] $(6, -2)$ **6.** [7.1b] $(6, 2)$ **7.** [7.1b] $(0, 5)$ **8.** [7.1b] No solution. The lines are parallel. **9.** [7.2a] $(0, 5)$ **10.** [7.2b] $(-2, 4)$ **11.** [7.2b] $(1, -2)$ **12.** [7.2a] $(-3, 9)$ **13.** [7.2a] $(1, 4)$ **14.** [7.2a] $(3, -1)$ **15.** [7.3a] $(3, 1)$ **16.** [7.3a] $(1, 4)$ **17.** [7.3a] $(5, -3)$ **18.** [7.3b] $(-4, 1)$ **19.** [7.3b] $(-2, 4)$ **20.** [7.3b] $(-2, -6)$ **21.** [7.3b] $(3, 2)$ **22.** [7.3b] $(2, -4)$ **23.** [7.3b] Infinitely many solutions **24.** [7.2c] 10 and -2 **25.** [7.2c] 12 and 15 **26.** [7.2c] Length: 37.5 cm; width: 10.5 cm **27.** [7.5a] 135 km/h **28.** [7.4a] 297 orchestra, 211 balcony **29.** [7.4a] 40 L of each **30.** [7.4a] Jeff: 27; son: 9

31. [3.1d] t^8 **32.** [3.1e] $\frac{1}{t^{18}}$ **33.** [5.5a] $\frac{-4x + 3}{(x - 2)(x - 3)(x + 3)}$

34. [6.3a]

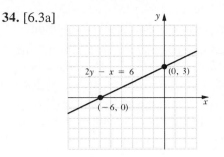

35. [7.1a] $C = 1, D = 3$ **36.** [1.7c; 7.2a] $(2, 0)$ **37.** [7.4a] 24 **38.** [7.4a] $96

Test: Chapter 7, p. 477

1. [7.1a] No **2.** [7.1b] $(2, -1)$ **3.** [7.2a] $(8, -2)$ **4.** [7.2b] $(-1, 3)$ **5.** [7.2a] No solution **6.** [7.3a] $(1, -5)$ **7.** [7.3a,b] $(12, -6)$ **8.** [7.3a,b] $(0, 1)$ **9.** [7.3a,b] $(5, 1)$ **10.** [7.2c] 20, 8 **11.** [7.5a] 40 km/h **12.** [7.4a] 40 L of A; 20 L of B **13.** [5.5a] $\frac{-x^2 + x + 17}{(x - 4)(x + 4)(x + 1)}$

14. [6.3a]

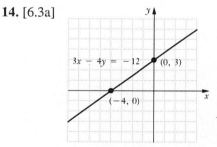

15. [3.1d] $\frac{10x^4}{y^2}$ **16.** [3.1e] $\frac{a^{10}}{b^6}$ **17.** [7.1a] $C = -\frac{19}{2}; D = \frac{14}{3}$ **18.** [7.4a] 5

Cumulative Review: Chapters 1-7, p. 479

1. [1.8d] -6.8 **2.** [3.2d] 3.12×10^{-2} **3.** [1.6c] $-\frac{3}{4}$

4. [3.1d, e] $\frac{1}{8}$ **5.** [5.1c] $\frac{x + 3}{2x - 1}$ **6.** [5.1c] $\frac{t - 4}{t + 4}$ **7.** [5.9a] $\frac{x^2(x + 1)}{x + 4}$

8. [3.6a] $2 - 10x^2 + 12x^4$ **9.** [3.6c] $4a^4b^2 - 20a^3b^3 + 25a^2b^4$ **10.** [3.6b] $9x^4 - 16y^2$ **11.** [3.5b] $-2x^3 + 4x^4 - 6x^5$ **12.** [3.5d] $8x^3 + 1$ **13.** [3.6b] $64 - \frac{1}{9}x^2$ **14.** [3.4c] $-y^3 - 2y^2 - 2y + 7$ **15.** [3.8b] $x^2 - x - 1 + \frac{-2}{2x - 1}$ **16.** [5.4a] $\frac{-5x - 28}{5x - 25}$ **17.** [5.5a] $\frac{4x - 1}{x - 2}$ **18.** [5.1d] $\frac{y}{(y - 1)^2}$ **19.** [5.2b] $\frac{3(x + 1)}{2x}$ **20.** [4.1b] $3x^2(2x^3 - 12x + 3)$ **21.** [4.5d] $(4y^2 + 9)(2y + 3)(2y - 3)$ **22.** [4.3a; 4.4a] $(3x - 2)(x + 4)$ **23.** [4.5b] $(2x^2 - 3y)^2$ **24.** [4.1b; 4.2a] $3m(m + 5)(m - 3)$ **25.** [4.6a]

$(x+1)^2(x-1)$ **26.** [2.3c] -9 **27.** [4.7a] $0, \frac{5}{2}$ **28.** [2.7e]
$\{x \mid x \le 20\}$ **29.** [2.3c] 0.3 **30.** [4.7b] $13, -13$
31. [4.7b] $\frac{5}{3}, 3$ **32.** [5.6a] -1 **33.** [5.6a] No solution
34. [7.2a] $(3,-3)$ **35.** [7.3b] $(-2,2)$ **36.** [7.3b] Infinitely
many solutions   **37.** [5.8a] $x = y - 1$ **38.** [2.4a] 234
39. [5.7a] $5\frac{5}{8}$ hr   **40.** [4.8a] 6 m, 10 m   **41.** [5.7b] 204
42. [4.8a] 6 cm, 9 cm   **43.** [6.6d] 72 ft; 360 **44.** [7.5a]
10 hr   **45.** [7.4a] 2.5 liters of A, 7.5 liters of B **46.** [6.6a]
$y = 0.2x$ **47.** [6.4a] 0 **48.** [6.4c] $m = -\frac{2}{3}, b = 2$
49. [6.4d] $y = -\frac{10}{7}x - \frac{8}{7}$ **50.** [6.4d] $y = 6x - 3$
51. [6.3b]   **52.** [6.3a]

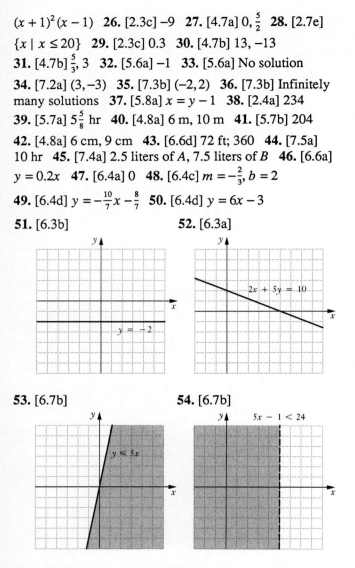

53. [6.7b]   **54.** [6.7b]

55. [7.1b] No solution. The lines are parallel.   **56.** [7.1b]
$(1,-1)$ **57.** [7.1a] $A = -4$, $B = -\frac{7}{5}$ **58.** [2.7e] No solution
59. [1.8 d; 5.2b; 5.5a] 0 **60.** [6.5b] $-\frac{3}{10}$

CHAPTER 8

Pretest: Chapter 8, p. 482

1. [8.1a] $7, -7$ **2.** [8.1d] $3t$ **3.** [8.1e] No   **4.** [8.1e]
$\{x \mid x \ge 2\}$ **5.** [8.1b] 6.856 **6.** [8.5a] 4 **7.** [8.1f] $2x$
8. [8.4a] $12\sqrt{2}$ **9.** [8.4b] $7 - 4\sqrt{3}$ **10.** [8.4b] 1
11. [8.2d] $2\sqrt{15}$ **12.** [8.4b] $25 - 4\sqrt{6}$ **13.** [8.3a] $\sqrt{5}$
14. [8.3b] $2a^2\sqrt{2}$ **15.** [8.6a] $c = \sqrt{89} \approx 9.434$
16. [8.6b] $\sqrt{193} \approx 13.892$ m   **17.** [8.3c] $\frac{\sqrt{5x}}{x}$ **18.** [8.4c]
$\frac{48 - 8\sqrt{5}}{31}$

Exercise Set 8.1, p. 487

1. $1, -1$ **3.** $4, -4$ **5.** $10, -10$ **7.** $13, -13$ **9.** 2 **11.** -3
13. -8 **15.** -15 **17.** 19 **19.** 2.236 **21.** 4.123
23. 9.644 **25.** (a) 13; (b) 24 **27.** $a - 4$ **29.** $t^2 + 1$
31. $\frac{3}{x+2}$ **33.** No   **35.** Yes   **37.** Yes   **39.** No
41. $\{x \mid x \ge 0\}$ **43.** $\{t \mid t \ge 5\}$ **45.** $\{y \mid y \ge -8\}$
47. $\left\{y \mid y \ge \frac{7}{2}\right\}$ **49.** All real numbers   **51.** t **53.** $3x$
55. ab **57.** $34d$ **59.** $x + 3$ **61.** $a - 5$ **63.** $2x - 5$
65. $\$5330$ **67.** $(x+1)^2$ **69.** 5 **71.** 3.578 **73.** 32.309
75. No solution   **77.** $\sqrt{3}$

Exercise Set 8.2, p. 493

1. $2\sqrt{3}$ **3.** $5\sqrt{3}$ **5.** $2\sqrt{5}$ **7.** $10\sqrt{2}$ **9.** $3\sqrt{x}$ **11.** $4\sqrt{3x}$
13. $4\sqrt{a}$ **15.** $8y$ **17.** $x\sqrt{13}$ **19.** $2t\sqrt{2}$ **21.** $6\sqrt{5}$
23. $12\sqrt{2y}$ **25.** $2x\sqrt{5}$ **27.** $\sqrt{2}(2x+1)$ **29.** $\sqrt{y}(6+y)$
31. 11.180 **33.** 18.972 **35.** 17.320 **37.** 11.043
39. 20 mph; 54.8 mph   **41.** x^3 **43.** x^6 **45.** $x^2\sqrt{x}$
47. $t^9\sqrt{t}$ **49.** $(y-2)^4$ **51.** $2(x+5)^5$ **53.** $6m\sqrt{m}$
55. $2a^2\sqrt{2a}$ **57.** $2p^8\sqrt{26p}$ **59.** $8x^3y\sqrt{7y}$ **61.** $3\sqrt{6}$
63. $3\sqrt{10}$ **65.** $6\sqrt{7x}$ **67.** $6\sqrt{xy}$ **69.** 10 **71.** $5b\sqrt{3}$
73. $2t$ **75.** $a\sqrt{bc}$ **77.** $2xy\sqrt{2xy}$ **79.** 18 **81.** $\sqrt{10x-5}$
83. $x + 2$ **85.** $6xy^3\sqrt{3xy}$ **87.** $10ab^2\sqrt{5ab}$ **89.** $(8,1)$
91. $25,704$ ft^2 **93.** $\sqrt{x-2}\sqrt{x+1}$ **95.** $\sqrt{2x+3}\sqrt{x-4}$
97. $\sqrt{a+b}\sqrt{a-b}$ **99.** 0.5 **101.** $3a^3$ **103.** $>$ **105.** $>$
107. $<$ **109.** $4y\sqrt{3}$ **111.** $18(x+1)\sqrt{y(x+1)}$
113. $2x^3\sqrt{5x}$ **115.** x^{4n} **117.** False. For example,
$\sqrt{25} - \sqrt{9} = 5 - 3 = 2$; but $\sqrt{25-9} = \sqrt{16} = 4$

Exercise Set 8.3, p. 501

1. 3 **3.** 2 **5.** $\sqrt{5}$ **7.** $\frac{1}{5}$ **9.** $\frac{2}{5}$ **11.** 2 **13.** $3y$ **15.** $\frac{3}{7}$ **17.** $\frac{1}{6}$
19. $-\frac{4}{9}$ **21.** $\frac{8}{17}$ **23.** $\frac{13}{14}$ **25.** $\frac{6}{a}$ **27.** $\frac{3a}{25}$ **29.** $\frac{\sqrt{10}}{5}$ **31.** $\frac{\sqrt{6}}{4}$
33. $\frac{\sqrt{21}}{6}$ **35.** $\frac{\sqrt{2}}{6}$ **37.** $\frac{3\sqrt{5}}{5}$ **39.** $\frac{2\sqrt{6}}{3}$ **41.** $\frac{\sqrt{3x}}{x}$ **43.** $\frac{\sqrt{xy}}{y}$ **45.** $\frac{x\sqrt{2}}{6}$
47. $\frac{\sqrt{21}}{3}$ **49.** $\frac{3\sqrt{2}}{4}$ **51.** $\frac{\sqrt{10}}{5}$ **53.** $\sqrt{2}$ **55.** $\frac{\sqrt{55}}{11}$ **57.** $\frac{\sqrt{21}}{6}$ **59.** $\frac{\sqrt{6}}{2}$
61. 5 **63.** $\frac{\sqrt{3x}}{x}$ **65.** $\frac{4y\sqrt{3}}{3}$ **67.** $\frac{a\sqrt{2a}}{4}$ **69.** $\frac{\sqrt{42x}}{3x}$ **71.** $\frac{3\sqrt{6}}{8c}$ **73.** $\frac{y\sqrt{xy}}{x}$
75. $\frac{\sqrt{2}}{4a}$ **77.** 0.577 **79.** 0.935 **81.** 0.289 **83.** 0.707
85. 3.801 **87.** 0.850 **89.** $(4,2)$ **91.** No solution
93. 1.57 sec; 3.14 sec; 8.88 sec; 11.10 sec   **95.** 1 sec
97. $\frac{\sqrt{5}}{40}$ **99.** $\frac{\sqrt{5x}}{5x^2}$ **101.** $\frac{\sqrt{3ab}}{b}$ **103.** $\frac{\sqrt{70}}{100}$ **105.** $\frac{y-x}{xy}$

Sidelight: Wind Chill Temperature, p. 508

1. $0°$ F   **2.** $-29°$ F   **3.** $-10°$ F   **4.** $-22°$ F   **5.** $-64°$ F
6. $-94°$ F

Exercise Set 8.4, p. 509

1. $7\sqrt{2}$ 3. $4\sqrt{5}$ 5. $13\sqrt{x}$ 7. $-2\sqrt{x}$ 9. $25\sqrt{2}$ 11. $\sqrt{3}$
13. $\sqrt{5}$ 15. $13\sqrt{2}$ 17. $3\sqrt{3}$ 19. $-24\sqrt{2}$ 21. 0
23. $(2+9x)\sqrt{x}$ 25. $(3-2x)\sqrt{3}$ 27. $3\sqrt{2x+2}$
29. $(x+3)\sqrt{x^3-1}$ 31. $(3xy-x^2+y^2)\sqrt{xy}$ 33. $\frac{2\sqrt{3}}{3}$
35. $\frac{13\sqrt{2}}{2}$ 37. $\frac{\sqrt{6}}{6}$ 39. $\frac{\sqrt{3}}{18}$ 41. -44 43. 3 45. $13+\sqrt{5}$
47. $x-2\sqrt{xy}+y$ 49. $-\sqrt{3}-\sqrt{5}$ 51. $5-2\sqrt{6}$ 53. $\frac{4\sqrt{10}-4}{9}$
55. $5-2\sqrt{7}$ 57. $\frac{1}{3}$ hr; the variation constant is the fixed
distance   59. All   61. $11\sqrt{3}-10\sqrt{2}$ 63. No   65. True;
$(3\sqrt{x+2})^2=(3\sqrt{x+2})(3\sqrt{x+2})=(3\cdot3)(\sqrt{x+2}\cdot\sqrt{x+2})=9(x+2)$.

Exercise Set 8.5, p. 515

1. 25 3. 38.44 5. 397 7. $\frac{621}{2}$ 9. 5 11. 3 13. $\frac{17}{4}$
15. No solution   17. No solution   19. 9 21. 7 23. $1, 5$
25. 3 27. $13, 25$ 29. No solution   31. 3 33. No
solution   35. 16 37. No solution   39. Approximately
346 km   41. $11,236$ m   43. 125 ft, 245 ft   45. 49
47. $\frac{(x+7)^2}{x-7}$ 49. $61°, 119°$ 51. $2, -2$ 53. $-\frac{57}{16}$ 55. 13

Exercise Set 8.6, p. 519

1. $c=17$ 3. $c=\sqrt{32}\approx5.657$ 5. $b=12$ 7. $b=4$
9. $c=26$ 11. $b=12$ 13. $a=2$ 15. $b=\sqrt{2}\approx1.414$
17. $a=5$ 19. $\sqrt{75}\approx8.660$ m   21. $\sqrt{208}\approx14.422$ ft
23. Approximately 43 yd   25. $\left(-\frac{3}{2},-\frac{1}{16}\right)$
27. $\sqrt{181}\approx13.454$ cm   29. $12-2\sqrt{6}\approx7.101$ 31. 6
33. $A=\frac{a^2\sqrt{3}}{4}$

Summary and Review: Chapter 8, p. 521

1. [8.1a] $8, -8$ 2. [8.1a] $20, -20$ 3. [8.1a] 6 4. [8.1a]
-13 5. [8.1b] 1.732 6. [8.1b] 9.950 7. [8.1b] 10.392
8. [8.1b] 17.889 9 [8.3d] 0.354 10. [8.3d] 0.742
11. [8.1d] x^2+4 12. [8.1d] $5ab^3$ 13. [8.1e] Yes
14. [8.1e] No   15. [8.1e] Yes   16. [8.1e] Yes   17. [8.1e]
No   18. [8.1e] $\{y \mid y\geq10\}$ 19. [8.1f] m 20. [8.1f]
$x-4$ 21. [8.2d] $\sqrt{21}$ 22. [8.2d] $\sqrt{x^2-9}$ 23. [8.2a]
$-4\sqrt{3}$ 24. [8.2a] $4t\sqrt{2}$ 25. [8.2a] $x+8$ 26. [8.2a]
$\sqrt{t-7}\sqrt{t+7}$ 27. [8.2c] x^4 28. [8.2c] $m^7\sqrt{m}$ 29. [8.2d]
$2\sqrt{15}$ 30. [8.2d] $2x\sqrt{10}$ 31. [8.2d] $5xy\sqrt{2}$ 32. [8.2d]
$10a^2b\sqrt{ab}$ 33. [8.3b] $\frac{5}{8}$ 34. [8.3b] $\frac{2}{3}$ 35. [8.3b] $\frac{7}{t}$
36. [8.3c] $\frac{\sqrt{2}}{2}$ 37. [8.3c] $\frac{\sqrt{2}}{4}$ 38. [8.3c] $\frac{\sqrt{5y}}{y}$ 39. [8.3c] $\frac{2\sqrt{3}}{3}$
40. [8.3a] $\frac{\sqrt{15}}{5}$ 41. [8.3a] $\frac{x\sqrt{30}}{6}$ 42. [8.4c] $8-4\sqrt{3}$
43. [8.4a] $13\sqrt{5}$ 44. [8.4a] $\sqrt{5}$ 45. [8.4a] $\frac{\sqrt{2}}{2}$ 46. [8.4b]

$7+4\sqrt{3}$ 47. [8.4b] 1 48. [8.6a] 20 49. [8.6a]
$\sqrt{3}\approx1.732$ 50. [8.6b] $\sqrt{98}\approx9.899$ 51. [8.5a] 52
52. [8.5a] No solution   53. [8.5b] 9 54. [8.5a] $0, 3$
55. [8.5c] 405 ft   56. [6.6b] $450   57. [7.3c] 5670
58. [7.3b] $\left(\frac{22}{17},-\frac{8}{17}\right)$ 59. [5.2b] $\frac{(x-5)(x-7)}{(x+7)(x+5)}$ 60. [8.1a] 2
61. [8.5a] $b=\sqrt{A^2-a^2}$

Test: Chapter 8, p. 567

1. [8.1a] $9, -9$ 2. [8.1a] 8 3. [8.1a] -5 4. [8.1b] 10.770
5. [8.1b] 9.327 6. [8.3d] 1.732 7. [8.1d] $4-y^3$
8. [8.1e] No   9. [8.1e] Yes   10. [8.1e] No   11. [8.1e]
$\{x \mid x\leq8\}$ 12. [8.1f] a 13. [8.1f] $6y$ 14. [8.2d] $\sqrt{30}$
15. [8.2d] $\sqrt{x^2-64}$ 16. [8.2a] $3\sqrt{3}$ 17. [8.2a] $5\sqrt{x-1}$
18. [8.2c] $t^2\sqrt{t}$ 19. [8.2d] $5\sqrt{2}$ 20. [8.2d] $3ab^2\sqrt{2}$
21. [8.3b] $\frac{3}{2}$ 22. [8.3b] $\frac{12}{a}$ 23. [8.3c] $\frac{\sqrt{10}}{5}$ 24. [8.3c] $\frac{\sqrt{2xy}}{y}$
25. [8.3a,c] $\frac{3\sqrt{6}}{8}$ 26. [8.3a,c] $\frac{\sqrt{7}}{4y}$ 27. [8.4a] $-6\sqrt{2}$
28. [8.4a] $\frac{6\sqrt{5}}{5}$ 29. [8.4b] $21-8\sqrt{5}$ 30. [8.4b] 11
31. [8.4c] $\frac{40+10\sqrt{5}}{11}$ 32. [8.6a] $c=\sqrt{80}\approx8.944$ 33. [8.5a]
48 34. [8.5a] $2, -2$ 35. [8.5b] -3 36. [8.5c] About
5000 m   37. [7.3c] 789.25 yd^2 38. [6.6b] $15,686\frac{2}{3}$
39. [7.3b] $(0, 2)$ 40. [5.2b] $\frac{x-7}{x+6}$ 41. [8.1a] $\sqrt{5}$ 42. [8.2c]
y^{8n}

Cumulative Review: Chapters 1-8, p. 525

1. [3.3a] -15 2. [3.3e] $-4x^3-\frac{1}{7}x^2-2$ 3. [8.1e]
$\left\{x \mid x\geq\frac{5}{3}\right\}$ 4. [5.1a] $\left\{x \mid x\neq-\frac{1}{2}\right\}$ 5. [8.4b] 1 6. [8.1a]
-14 7. [8.2d] 15 8. [8.4b] $3-2\sqrt{2}$ 9. [8.3a, c] $\frac{9\sqrt{10}}{25}$
10. [8.4a] $12\sqrt{5}$ 11. [3.7f] $9x^8-4y^{10}$ 12. [3.6c]
x^4+8x^2+16 13. [3.6a] $8x^2-\frac{1}{8}$ 14. [5.5a] $\frac{4x+2}{2x-1}$
15. [3.4a, c] $-3x^3+8x^2-5x$ 16. [5.1d] $\frac{2(x-5)}{3(x-1)}$ 17. [5.2b]
$\frac{(x+1)(x-3)}{2(x+3)}$ 18. [3.8b] $3x^2+4x+9+\frac{13}{x-2}$ 19. [8.2a]
$\sqrt{2}(x-1)$ 20. [3.1d] x^{-12} 21. [8.3b] $\frac{5}{x^4}$ 22. [5.9a]
$2(x-1)$ 23. [4.5d] $3(1+2x^4)(1-2x^4)$ 24. [4.1b]
$4t(3-t-12t^3)$ 25. [4.3a; 4.4a] $2(3x-2)(x-4)$
26. [4.6a] $(2x+1)(2x-1)(x+1)$ 27. [4.5b] $(4x^2-7)^2$
28. [4.2a] $(x+15)(x-12)$ 29. [4.7b] $0, -17$ 30. [2.7e]
$\{x \mid x>-6\}$ 31. [5.6a] $-\frac{12}{5}$ 32. [4.7b] $-5, 6$ 33. [2.7e]
$\left\{x \mid x\leq-\frac{9}{2}\right\}$ 34. [4.7b] $9, -9$ 35. [8.5a] 41 36. [8.5b]
9 37. [2.3b] $\frac{2}{5}$ 38. [5.6a] $\frac{1}{3}$ 39. [7.2a] $(4, 1)$

40. [7.3a] $(3, -8)$ **41.** [5.8a] $R = \frac{r^2}{E-r}$ **42.** [2.6a] $p = \frac{4A}{r+q}$

43. [6.7b]   **44.** [6.3b]

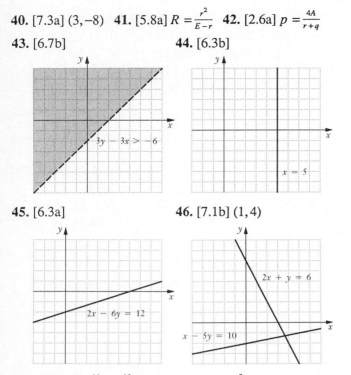

45. [6.3a]   **46.** [7.1b] $(1, 4)$

47. [6.4d] $y = \frac{11}{4}x - \frac{19}{4}$ **48.** [6.4c] Slope: $\frac{5}{3}$; y-intercept: $(0, -3)$ **49.** [6.5b] Perpendicular   **50.** [6.5a] Parallel **51.** [2.4a] $38°, 76°, 66°$ **52.** [7.4a] Hamburger: $1.30; milkshake: $0.90   **53.** [8.6b] $4\sqrt{3} \approx 6.9$ m   **54.** [6.6b] $7500   **55.** [5.7b] 20   **56.** [4.8a] 15 m, 12 m   **57.** [7.4a] 65 dimes, 50 quarters   **58.** [5.7b] 357   **59.** [2.5a] $2600 **60.** [5.7a] 60 mph   **61.** [1.2d] $<$ **62.** [1.2d, e] $>$ **63.** [5.1a; 8.1e] $\{x \mid x > 5\}$ **64.** [7.4a] 300 L **65.** [7.3b; 8.5a] $(9, 4)$

CHAPTER 9

Pretest: Chapter 9, p. 528

1. [9.1c] 3   **2.** [9.2a] $\sqrt{7}, -\sqrt{7}$ **3.** [9.3a] $\frac{-3 \pm \sqrt{21}}{6}$ **4.** [9.1b] $0, \frac{3}{5}$ **5.** [9.1b] $0, -\frac{5}{3}$ **6.** [9.2b] $-4 \pm \sqrt{5}$ **7.** [9.2c] $1 \pm \sqrt{6}$

8. [9.4a] $n = \frac{p \pm \sqrt{p^2 + 4A}}{2}$ **9.** [9.5a] Width: 4 cm; length: 12 cm   **10.** [9.6b] $\left(\frac{-1+\sqrt{33}}{4}, 0\right), \left(\frac{-1-\sqrt{33}}{4}, 0\right)$ **11.** [9.5a] 10 km/h **12.** [9.6a]

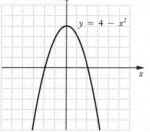

Exercise Set 9.1, p. 533

1. $a = 1, b = -3, c = 2$ **3.** $7x^2 - 4x + 3 = 0; a = 7, b = -4, c = 3$ **5.** $2x^2 - 3x + 5 = 0; a = 2, b = -3, c = 5$ **7.** $0, -7$ **9.** $0, -2$ **11.** $0, \frac{2}{5}$ **13.** $0, -1$ **15.** $0, 3$ **17.** $0, \frac{1}{5}$ **19.** $0, \frac{3}{14}$ **21.** $0, \frac{81}{2}$ **23.** $-12, 4$ **25.** $-5, -1$ **27.** $-9, 2$ **29.** $5, 3$ **31.** -3 **33.** 4 **35.** $-\frac{2}{3}, \frac{1}{2}$ **37.** $\frac{1}{3}, \frac{4}{3}$ **39.** $\frac{5}{3}, -1$ **41.** $-\frac{1}{4}, \frac{2}{3}$ **43.** $-2, 7$ **45.** $4, -5$ **47.** 4 **49.** $3, -2$ **51.** $6, -\frac{2}{3}$ **53.** $6,$ -4 **55.** 1 **57.** $5, 2$ **59.** No solution   **61.** 9 **63.** 7 **65.** $2\sqrt{5}$ **67.** $\frac{9}{8}$ **69.** $-\frac{1}{3}, 1$ **71.** $0, \frac{\sqrt{5}}{5}$ **73.** $-\frac{5}{2}, 1$ **75.** $0, -\frac{b}{a}$ **77.** $81, 1$

Exercise Set 9.2, p. 541

1. $11, -11$ **3.** $\sqrt{7}, -\sqrt{7}$ **5.** $\frac{\sqrt{15}}{5}, -\frac{\sqrt{15}}{5}$ **7.** $\frac{5}{2}, -\frac{5}{2}$ **9.** $\frac{7\sqrt{3}}{3}, -\frac{7\sqrt{3}}{3}$ **11.** $\sqrt{3}, -\sqrt{3}$ **13.** $\frac{6}{5}, -\frac{6}{5}$ **15.** $-5, 9$ **17.** $-3 \pm \sqrt{21}$ **19.** $-13 \pm 2\sqrt{2}$ **21.** $7 \pm 2\sqrt{3}$ **23.** $-9 \pm \sqrt{34}$ **25.** $\frac{-3 \pm \sqrt{14}}{2}$ **27.** $11, -5$ **29.** $-5, -9$ **31.** $-2, 8$ **33.** $-21, -1$ **35.** $1 \pm \sqrt{6}$ **37.** $11 \pm \sqrt{19}$ **39.** $-5 \pm \sqrt{29}$ **41.** $\frac{7 \pm \sqrt{57}}{2}$ **43.** $-7, 4$ **45.** $\frac{-3 \pm \sqrt{17}}{4}$ **47.** $\frac{-3 \pm \sqrt{145}}{4}$ **49.** $\frac{-2 \pm \sqrt{7}}{3}$ **51.** $-\frac{1}{2}, 5$ **53.** $-\frac{7}{2}, \frac{1}{2}$ **55.** About 9.3 sec   **57.** About 3.3 sec **59.** $y = \frac{141}{x}$ **61.** $12, -12$ **63.** $16\sqrt{2}, -16\sqrt{2}$ **65.** $2\sqrt{c}$, $-2\sqrt{c}$ **67.** Approximately $50, -50$ **69.** $\sqrt{3}, -\sqrt{3}$ **71.** $\sqrt{11}, -\sqrt{11}$

Exercise Set 9.3, p. 547

1. $-3, 7$ **3.** 3 **5.** $-\frac{4}{3}, 2$ **7.** $-\frac{7}{2}, \frac{1}{2}$ **9.** $-3, 3$ **11.** $1 \pm \sqrt{3}$ **13.** $5 \pm \sqrt{3}$ **15.** $-2 \pm \sqrt{7}$ **17.** $\frac{-4 \pm \sqrt{10}}{3}$ **19.** $\frac{5 \pm \sqrt{33}}{4}$ **21.** $\frac{1 \pm \sqrt{2}}{2}$ **23.** No real-number solutions   **25.** $\frac{5 \pm \sqrt{73}}{6}$ **27.** $\frac{3 \pm \sqrt{29}}{2}$ **29.** $\sqrt{7}, -\sqrt{7}$ **31.** $-2 \pm \sqrt{3}$ **33.** $2 \pm \sqrt{34}$ **35.** $-1.3, 5.3$ **37.** $-0.2, 6.2$ **39.** $-1.2, 0.2$ **41.** $-1.7, 0.4$ **43.** $3x^2\sqrt{3x}$ **45.** $4\sqrt{5}$ **47.** $0, 2$ **49.** $\frac{3 \pm \sqrt{5}}{2}$ **51.** $\frac{-7 \pm \sqrt{61}}{2}$ **53.** $\frac{-2 \pm \sqrt{10}}{2}$

Exercise Set 9.4, p. 551

1. $A = \frac{N^2}{6.25}$ **3.** $T = \frac{cQ^2}{a}$ **5.** $c = \sqrt{\frac{E}{m}}$ **7.** $d = \frac{c \pm \sqrt{c^2 + 4aQ}}{2a}$ **9.** $a = \sqrt{c^2 - b^2}$ **11.** $t = \frac{\sqrt{s}}{4}$ **13.** $r = \frac{-\pi h \pm \sqrt{\pi^2 h^2 + \pi A}}{\pi}$ **15.** $r = 6\sqrt{\frac{10A}{\pi S}}$ **17.** $a = \sqrt{c^2 - b^2}$ **19.** $a = \frac{2h\sqrt{3}}{3}$ **21.** $T = \frac{2 \pm \sqrt{4 - a(m-n)}}{a}$ **23.** $T = \frac{v^2 \pi m}{8k}$ **25.** $E = mc^2$ **27.** $n = \frac{1 \pm \sqrt{1 + 8N}}{2}$ **29.** $c = \sqrt{65} \approx 8.062$

31. $c = \sqrt{41} \approx 6.403$ **33. (a)** $r = \frac{C}{2\pi}$; **(b)** $A = \frac{C^2}{4\pi}$ **35.** $\frac{1}{3a}$, 1

Sidelight: Handling Dimension Symbols, p. 556

1. 68 ft **2.** 82 km/h **3.** 45 g **4.** 8.6 lb **5.** 15 mi/h
6. 32 km/h **7.** 3.3 m/sec **8.** 19 ft/min **9.** $4 \frac{\text{in.-lb}}{\text{sec}}$ **10.** 24 $\frac{\text{man-hr}}{\text{day}}$ **11.** 12 yd **12.** 220 mi **13.** 16 ft^3 **14.** $\frac{1}{4} \frac{\text{lb}^2}{\text{ft}^2}$ **15.** $\frac{\$970}{\text{day}}$
16. $\frac{\$3.20}{\text{hr}}$

Exercise Set 9.5, p. 555

1. 7 ft; 24 ft **3.** Width: 8 cm; length: 10 cm **5.** Length: 20 cm; width: 16 cm **7.** Length: 10 m; width: 5 m
9. 4.6 m; 6.6 m **11.** Length: 5.6 in.; width: 3.6 in.
13. Length: 4.4 m; width: 2.2 m **15.** 3 cm
17. 7 km/h

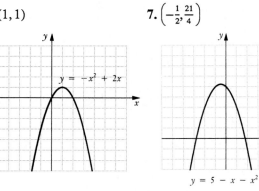

19. 8 mph **21.** 4 km/h **23.** 36 mph **25.** 1 km/h
27. $3 + 2\sqrt{2} \approx 5.828$ **29.** $1 + \sqrt{2} \approx 2.41$
31. $d = 10\sqrt{2} \approx 14.14$ in.; two 10-in. pizzas

Exercise Set 9.6, p. 565

1. $(0, 1)$ **3.** $(0, 0)$

5. $(1, 1)$ **7.** $\left(-\frac{1}{2}, \frac{21}{4}\right)$

9. $(1, 0)$ **11.** $(1, 4)$

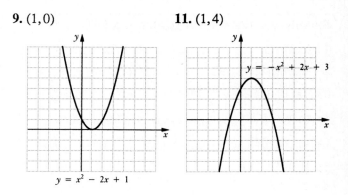

13. $(-1, 3)$ **15.**

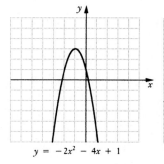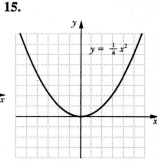

17.   **19.**

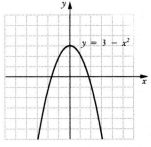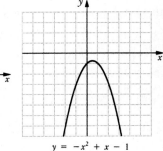

21.   **23.**

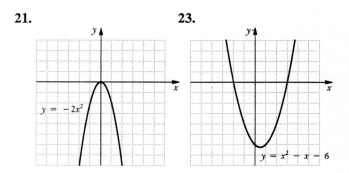

25. $(-\sqrt{5}, 0)$; $(\sqrt{5}, 0)$ **27.** $(-2, 0)$; $(0, 0)$ **29.** $\left(\frac{-1-\sqrt{33}}{2}, 0\right)$; $\left(\frac{-1+\sqrt{33}}{2}, 0\right)$ **31.** $(-5, 0)$ **33.** $(-2-\sqrt{5}, 0)$; $(-2+\sqrt{5}, 0)$
35. None **37.** $(x+2)\sqrt{x-1}$ **39.** $2\sqrt{7}$ **41. (a)** After 2 sec, after 4 sec; **(b)** after 3 sec; **(c)** after 6 sec

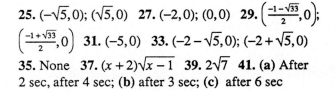

Summary and Review: Chapter 9, p. 567

1. [9.2a] $\sqrt{3}, -\sqrt{3}$ 2. [9.1c] $\frac{3}{5}, 1$ 3. [9.3a] $1 \pm \sqrt{11}$
4. [9.1c] $\frac{1}{3}, -2$ 5. [9.2b] $-8 \pm \sqrt{13}$ 6. [9.2a] 0 7. [9.1b]
$0, \frac{7}{5}$ 8. [9.3a] $\frac{1 \pm \sqrt{10}}{3}$ 9. [9.3a] $-3 \pm 3\sqrt{2}$ 10. [9.3a] $\frac{2 \pm \sqrt{3}}{2}$
11. [9.3a] $\frac{3 \pm \sqrt{33}}{2}$ 12. [9.3a] No real-number solution
13. [9.1b] $0, \frac{4}{3}$ 14. [9.2a] $-2\sqrt{2}, 2\sqrt{2}$ 15. [9.1c] $3, -5$
16. [9.1c] 1 17. [9.2c] $\frac{5}{3}, -1$ 18. [9.2c] $\frac{5 \pm \sqrt{17}}{2}$ 19. [9.3b]
$4.6, 0.4$ 20. [9.3b] $-1.9, -0.1$ 21. [9.4a] $T = L(4V^2 - 1)$
22. [9.6a] 23. [9.6a]

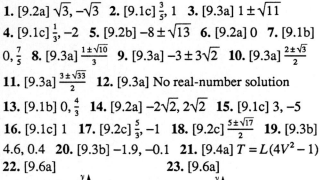

24. [9.6b] $(-\sqrt{2}, 0)$; $(\sqrt{2}, 0)$ 25. [9.6b] $(2 - \sqrt{6}, 0)$;
$(2 + \sqrt{6}, 0)$ 26. [9.5a] 1.7 m, 4.7 m 27. [9.5a] Length:
10 m; width: 7 m 28. [9.5a] 30 km/h 29. [8.2d] $6\sqrt{a}$
30. [8.2d] $2xy\sqrt{15y}$ 31. [6.6c] $y = \frac{1}{x}$ 32. [8.6b] $\sqrt{3}$
33. [8.4a] $12\sqrt{11}$ 34. [8.4a] $4\sqrt{10}$ 35. [9.5a] 31 and 32;
-32 and -31 36. [9.2c] $14, -14$ 37. [9.1c; 8.5a] 25
38. [9.5a] $s = 5\sqrt{\pi}$

Test: Chapter 9, p. 569

1. [9.2a] $\sqrt{5}, -\sqrt{5}$ 2. [9.1b] $0, -\frac{8}{7}$ 3. [9.1c] $-8, 6$
4. [9.1c] $-\frac{1}{3}, 2$ 5. [9.2b] $8 \pm \sqrt{13}$ 6. [9.3a] $\frac{1 \pm \sqrt{13}}{2}$
7. [9.3a] $\frac{3 \pm \sqrt{37}}{2}$ 8. [9.3a] $-2 \pm \sqrt{14}$ 9. [9.3a] $\frac{7 \pm \sqrt{37}}{6}$
10. [9.1c] $2, -1$ 11. [9.1c] $2, -4$ 12. [9.2c] $2 \pm \sqrt{14}$
13. [9.3b] $5.7, -1.7$ 14. [9.4a] $n = \frac{-b \pm \sqrt{b^2 + 4ad}}{2a}$
15. [9.6a] 16. [9.6a]

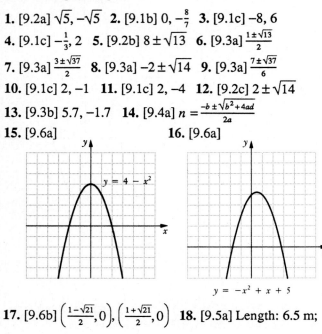

17. [9.6b] $\left(\frac{1 - \sqrt{21}}{2}, 0\right), \left(\frac{1 + \sqrt{21}}{2}, 0\right)$ 18. [9.5a] Length: 6.5 m;

width: 2.5 m 19. [9.5a] 24 km/h 20. [8.4a] $2\sqrt{15}$
21. [8.2d] $7xy\sqrt{2x}$ 22. [6.6c] $y = \frac{4}{x}$ 23. [8.6b] $\sqrt{5}$
24. [9.5a] $5 + 5\sqrt{2}$ 25. [7.2b; 9.3a] $1 \pm \sqrt{5}$

Cumulative Review: Chapters 1-9, p. 571

1. [3.1a] $x \cdot x \cdot x$ 2. [3.1c] 54 3. [1.2c] $-0.\overline{27}$ 4. [5.3a]
240 5. [1.2e] 7 6. [1.3a] 9 7. [1.4a] 15 8. [1.6c] $-\frac{3}{20}$
9. [1.8d] 4 10. [1.8b] $-2m - 4$ 11. [2.2a] -8 12. [2.3b]
-12 13. [2.3c] 7 14. [4.7b] 3, 5 15. [7.2a] $(1, 2)$
16. [7.3a] $(12, 5)$ 17. [7.3b] $(6, 7)$ 18. [4.7b] 3, -2
19. [9.3a] $\frac{-3 \pm \sqrt{29}}{2}$ 20. [8.5a] 2 21. [2.7e] $\{x \mid x \geq -1\}$
22. [2.3b] 8 23. [2.3b] -3 24. [2.7d] $\{x \mid x < -8\}$
25. [2.7e] $\left\{ y \mid y \leq \frac{35}{22} \right\}$ 26. [9.2a] $\sqrt{10}, -\sqrt{10}$ 27. [9.2b]
$3 \pm \sqrt{6}$ 28. [5.6a] $\frac{2}{9}$ 29. [5.6a] -5 30. [5.6a; 9.1b] No
solution 31. [8.5a] 12 32. [2.6a] $b = \frac{At}{4}$ 33. [5.8a]
$m = \frac{tn}{t+n}$ 34. [9.4a] $A = \pi r^2$ 35. [9.4a] $x = \frac{b \pm \sqrt{b^2 + 4ay}}{2a}$
36. [3.1d] $\frac{1}{x^4}$ 37. [3.1e] y^7 38. [3.2a, b] $4y^{12}$ 39. [3.3f]
$10x^3 + 3x - 3$ 40. [3.4a] $7x^3 - 2x^2 + 4x - 17$ 41. [3.4c]
$8x^2 - 4x - 6$ 42. [3.5b] $-8y^4 + 6y^3 - 2y^2$ 43. [3.5d]
$6t^3 - 17t^2 + 16t - 6$ 44. [3.6b] $t^2 - \frac{1}{16}$ 45. [3.6c]
$9m^2 - 12m + 4$ 46. [3.7e] $15x^2y^3 + x^2y^2 + 5xy^2 + 7$
47. [3.7f] $x^4 - 0.04y^2$ 48. [3.7f] $9p^2 + 24pq^2 + 16q^4$
49. [5.1d] $\frac{2}{x+3}$ 50. [5.2b] $\frac{3a(a-1)}{2(a+1)}$ 51. [5.4a] $\frac{27x-4}{5x(3x-1)}$
52. [5.5a] $\frac{-x^2+x+2}{(x+4)(x-4)(x-5)}$ 53. [4.1b] $4x(2x-1)$ 54. [4.5d]
$(5x-2)(5x+2)$ 55. [4.3a; 4.4a] $(3y+2)(2y-3)$
56. [4.5b] $(m-4)^2$ 57. [4.1c] $(x^2-5)(x-8)$ 58. [4.6a]
$3(a^2+6)(a+2)(a-2)$ 59. [4.5d]
$(4x^2+1)(2x+1)(2x-1)$ 60. [4.5d] $(7ab+2)(7ab-2)$
61. [4.5b] $(3x+5y)^2$ 62. [4.1c] $(2a+d)(c-3b)$
63. [4.3a; 4.4a] $(5x-2)(3x+4)$ 64. [5.9a] $-\frac{42}{5}$
65. [8.1a] 7 66. [8.1a] -25 67. [8.1f] $8x$ 68. [8.2d]
$\sqrt{a^2-b^2}$ 69. [8.2d] $8a^2b\sqrt{3ab}$ 70. [8.2a] $5\sqrt{6}$
71. [8.2c] $9xy\sqrt{3x}$ 72. [8.3b] $\frac{10}{9}$ 73. [8.3b] $\frac{8}{x}$ 74. [8.4a]
$12\sqrt{3}$ 75. [8.3a, c] $\frac{2\sqrt{10}}{5}$ 76. [8.6a] 40

77. [6.2b] 78. [6.3a]

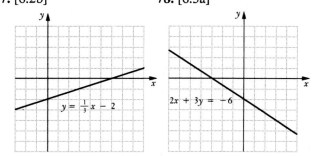

79. [6.3b]　　　　　　　　　　**80.** [6.7b]

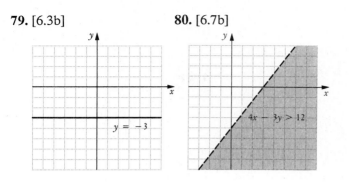

81. [9.6a]　　　　　　　　　　**82.** [6.7b]

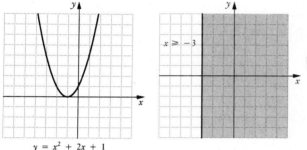

83. [9.2c] $\frac{2\pm\sqrt{6}}{3}$　**84.** [9.3b] −0.2, 1.2　**85.** [2.5a] 25%
86. [2.5a] 60　**87.** [9.5a] 2 km/h　**88.** [4.8a] 12 m
89. [2.4a] 6090　**90.** [7.4a] 14 lb of $1.10; 28 lb of $0.80
91. [5.7a] $4\frac{4}{9}$ hr　**92.** [6.6b] $451.20; variation constant
is the amount earned per hour　**93.** [6.5a, b] Neither
94. [6.4c] Slope: 2; y-intercept: $(0, -8)$　**95.** [6.4a] 15
96. [6.6a] $y = 10x$　**97.** [6.6c] $y = \frac{1000}{x}$
98. [1.2e] 12, −12　**99.** [8.1a] $\sqrt{3}$　**100.** [9.2c] −30, 30
101. [8.6a] $\frac{\sqrt{6}}{3}$　**102.** [4.5d] Yes　**103.** [5.1c] No
104. [3.6c] No　**105.** [4.5a; 8.2a] No　**106.** [8.1f] Yes

Final Examination: Chapters 1-9, p. 575

1. [3.1c] −995　**2.** [5.3a] 48　**3.** [1.2e] 9　**4.** [1.3a] −9.7
5. [1.4a] −5　**6.** [1.5a] $-\frac{6}{7}$　**7.** [1.8b] $19y - 45$　**8.** [1.8d]
211　**9.** [2.1b] 5.6　**10.** [2.2a] −7　**11.** [2.3b] −10
12. [2.3c] 7　**13.** [4.7b] 6, −4　**14.** [7.2a] $(4, -3)$
15. [7.3b] $(4, 7)$　**16.** [9.1c] $\frac{1}{4}, -\frac{1}{2}$　**17.** [8.5a] 6　**18.** [9.3a]
$\frac{-3\pm\sqrt{37}}{2}$　**19.** [2.7e] $\left\{ x \mid x \leq \frac{5}{2} \right\}$　**20.** [5.8a] $w = \frac{1}{A-B}$

21. [2.6a] $M = \frac{K-2}{T}$　**22.** [3.1e] x^{10}　**23.** [3.2a] $\frac{1}{x^{10}}$
24. [3.1d] $\frac{1}{x^{12}}$　**25.** [3.3f] $6y^3 - 3y^2 - y + 9$　**26.** [3.4c]
$-2x^2 - 8x + 7$　**27.** [3.5b] $-6t^6 - 12t^4 - 3t^2$　**28.** [3.5d]
$4x^3 - 21x^2 + 13x - 2$　**29.** [3.6b] $x^2 - 64$　**30.** [3.6c]
$4m^2 - 28m + 49$　**31.** [3.7f] $9a^2b^4 + 12ab^2c + 4c^2$
32. [3.7f] $9x^4 + 6x^2y - 8y^2$　**33.** [5.1d] $\frac{1}{x^2(x+3)}$　**34.** [5.2b]
$\frac{3x^4(x-1)}{4}$　**35.** [5.4a] $\frac{11x-1}{4x(3x-1)}$　**36.** [5.5a] $\frac{2x+4}{(x-3)(x+1)}$　**37.** [4.1b]
$3x(x^2 - 5)$　**38.** [4.5d] $(4x + 5)(4x - 5)$　**39.** [4.3a; 4.4a]
$(3x - 2)(2x - 3)$　**40.** [4.5b] $(x - 5)^2$　**41.** [4.1c]
$(2x - y)(a + 3b)$　**42.** [4.5d] $(x^4 + 9y^2)(x^2 - 3y)(x^2 + 3y)$
43. [8.2a] $6\sqrt{2}$　**44.** [8.3a, c] $\frac{\sqrt{30}}{5}$　**45.** [8.4a] $13\sqrt{2}$
46. [8.2d] $2a^2b\sqrt{6ab}$

47. [6.3a]　　　　　　　　　　**48.** [6.3b]

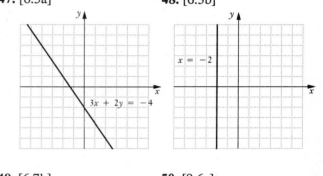

49. [6.7b]　　　　　　　　　　**50.** [9.6a]

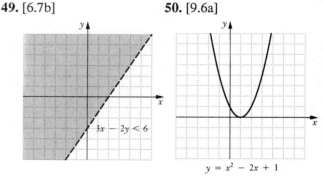

51. [4.8a] 5, 7; −7, −5　**52.** [7.4a] 40 L of A; 20 L of B
53. [7.5a] 70 km/h　**54.** [9.5a] Length: 11 m; width: 8 m
55. [6.4a] $-\frac{4}{3}$　**56.** [6.5a,b] Perpendicular　**57.** [6.6a]
$y = 8x$　**58.** [6.6c] $y = \frac{5000}{x}$　**59.** [2.4a] Side of square: 15;
side of triangle: 20　**60.** [9.2c] 144

Index

A

Absolute value, 55, 56
  and radicals, 486
Addition
  associative law of, 88
  commutative law of, 87
  with decimal notation, 20
  of exponents, 177
  of fractional expressions, 13, 326
  on number line, 59
  of polynomials, 203, 229
  of radical expressions, 505
  of rational expressions, 329, 340
  of real numbers, 59, 60
Addition principle
  for equations, 112
  for inequalities, 154
Additive identity, 85
Additive inverses, 61, 62
  and changing the sign, 62, 97, 204
  and multiplying by −1, 97
  of polynomials, 204, 229
  in subtraction, 65
  of sums, 97

Algebraic expressions, 43, 44
  evaluating, 44, 176
  LCM, 326
  translating to, 45
Angles
  complementary, 455
  sum of measures in a triangle, 138
  supplementary, 455
Applications, *see* Applied problems;
      Problems
Applied problems, 67, 133–138,
      143–146, 163, 164, 189, 190,
      194, 206, 295–298, 351–356,
      414, 416, 444, 452, 457–464,
      508, 514, 518, 532, 540,
      553–555. *See also* Problems.
Approximately equal to ($\approx$), 54
Approximating
  solutions of quadratic equations, 546
  square roots, 483, 490, 500
Area
  of a circle, 176
  of a parallelogram, 47
  of a rectangle, 44

Area (*continued*)
   of a right circular cylinder, 227
   of a square, 182
   of a trapezoid, 361
   of a triangle, 47
Arithmetic, numbers of, 9
Ascending order, 196
Associative laws, 88
Axes of graphs, 381

B

Bar graph, 379
Base, 31, 175
Binomials, 198
   difference of squares, 273
   product of, 212
      FOIL method, 217
      sum and difference of expressions,
         219
      squares of, 219
Braces, 32, 99
Brackets, 32, 99

C

Calculator, approximating square
      roots on, 483, 490, 500
Canceling, 11, 314
Careers and mathematics, 68, 316
Carry out, 133
Changing the sign, 62, 97, 204, 229
Checking
   in problem-solving process, 133
   quotients, 235, 236, 237
   solutions of equations, 113, 290,
      348, 512
Circle
   area of, 176
   circumference of, 552
Circle graph, 380
Circumference, 552
Clearing of decimals, 126
Clearing of fractions, 126, 345, 361
Coefficients, 118, 195, 228
   leading, 255
Collecting like terms, 92, 195, 228
   in equation solving, 124
Combining like terms, *see* Collecting
      like terms
Common denominators, 13, 325.
      See also Least common
      denominator.
Common factor, 91, 250
Common multiples, 5. *See also* Least
      common multiple.
Commutative laws, 87

Complementary angles, 455
Completing the square, 537
Complex fractional expressions, 365
Complex number system, 485
Complex rational expressions, 365
Composite number, 4
Compound interest, 231
Conjugate, 507
Consecutive integers, 135
Constant, 43
   of proportionality, 413
   variation, 413
Coordinates, 381
   finding, 382
Cost, total, 199
Cross products, 355
Cylinder, right circular, surface area,
      227

D

Decimal notation, 19
   addition with, 20
   converting to fractional notation, 19
   converting to percent notation, 28
   converting to scientific notation, 187
   division with, 22
   for irrational numbers, 53
   multiplication with, 21
   for rational numbers, 52, 53
   repeating 23, 53
   rounding, 24
   subtraction with, 21
   terminating, 52, 53
Decimals, clearing of, 126
Degrees of polynomials and terms,
      197, 228
Denominator, 9
   additive inverse of, 329, 332, 337,
      339
   least common, 13, 325
   rationalizing, 498, 507
Descending order, 196
Diagonals, number of, 532
Difference, *see* Subtraction
Differences of squares, 273
Dimension symbols, 556
Direct variation, 413
Discriminant, 545
   and x-intercepts, 564
Distance to the horizon, 514
Distributive law, 90
Division
   with decimal notation, 22
   using exponents, 178

of fractional expressions, 15
of integers, 79
of polynomials, 235
of rational expressions, 321
of real numbers, 79–82
and reciprocals, 15, 81
using scientific notation, 188
by zero, 79

E

Elimination method, solving systems
      of equations, 447
Empty set, 437
Equations, 111. *See also* Formulas.
  of direct variation, 413
  equivalent, 112
  false, 111
  fractional, 345
  graphs of, *see* Graphing
  of inverse variation, 415
  linear, 387. *See also* Graphing.
  containing parentheses, 127
  point–slope, 405
  Pythagorean, 517
  quadratic, 289, 529
  radical, 511
  rational, 345
  related, 422
  reversing, 120
  slope–intercept, 404
  solutions, 111, 387
  solving, *see* Solving equations
  systems of, 435
  translating to, 133
  true, 111
  of variation, 413, 415
Equilateral triangle, 520
Equivalent equations, 112
Equivalent expressions, 9, 85, 227, 312
  for one, 10
Equivalent inequalities, 155
Evaluating expressions, 44, 176
Evaluating polynomials, 193
Even integers, consecutive, 135
Exponential notation, 31, 175. *See*
      also Exponents.
Exponents, 31, 175
  dividing using, 178
  evaluating expressions with, 31, 176
  multiplying using, 177
  negative, 179
  one as, 176
  raising a power to a power, 185
  raising a product to a power, 186

raising a quotient to a power, 186
  rules for, 180, 190
  zero as, 176
Expressions
  algebraic, 43, 44
  equivalent, 9, 85, 227, 312
  evaluating, 31, 44
  exponential, 31
  fractional, *see* Rational expressions
  radical, 484
  rational, *see* Rational expressions
  simplifying, *see* Simplifying
  terms of, 90
  value of, 44

F

Factor, 3. *See also* Factoring;
      Factorizations.
Factor tree, 14
Factoring, 3, 91
  common factor, 91, 250
  completely, 275
  finding LCM by, 5, 326
  numbers, 3
  polynomials, 249
    with a common factor, 250
    differences of squares, 274
    general strategy, 281
    by grouping, 252
    monomials, 249
    trinomial squares, 272
    trinomials, 255, 261, 267
  radical expressions, 489
  solving equations by, 290, 530, 531
Factorizations, 3, 4. *See also* Factoring.
Factors, 3, 250. *See also* Factoring.
  and sums, 34, 214
Falling object, distance traveled, 199,
    540
False equation, 111
False inequality, 153
Familiarize, 133
First coordinate, 381
Five-step process for problem solving,
    133
FOIL method, 217
  and factoring, 255, 261
Formulas, 149. *See also* Equations;
    Problems.
  compound-interest, 231
  motion, 352, 469
  quadratic, 543
  solving for given letter, 149, 150,
    361, 549

Fraction bar
  as a division symbol, 44
  as a grouping symbol, 33
Fractional equations, 345
Fractional expressions, 311. *See also*
    Fractional notation; Rational
    expressions.
  addition of, 13
  complex, 365
  division of, 15
  multiplication of, 12
  multiplying by one, 10
  subtraction of, 13
Fractional notation, 9. *See also*
    Fractional expressions; Rational
    expressions.
  converting to decimal notation, 20
  converting to percent notation, 28
  simplifying, 10
Fractions, *see* Fractional expressions;
    Fractional notation; Rational
    expressions
Fractions, clearing of, 126, 345, 361

G

Games played, 194, 297
Grade, 406
Graph. *See also* Graphing.
  bar, 379
  circle, 380
  line, 380
Graphing
  equations, 387, 398
    of direct variation, 414
    horizontal line, 396, 397
    using intercepts, 395
    of inverse variation, 415
    quadratic, 561
    slope, 401
    vertical line, 397
    x-intercept, 395, 564
    $y = mx$, 388
    $y = mx + b$, 390
    y-intercept, 395
  inequalities, 153, 421
    linear, 422
    in one variable, 153
    in two variables, 421, 422
  numbers, 51, 54
  points on a plane, 381
  systems of equations, 436
Gravitational force, 361
Greater than ($>$), 54
Greater than or equal to ($\geq$), 55

Grouping
  in addition, 88
  factoring by, 252, 267
  in multiplication, 88
  symbols, 32, 33, 99

H

Half-plane, 422
Handshakes, number possible, 301
Height of a projectile, 566
Horizon, distance to, 514
Horizontal lines, 396, 397
  slope, 402, 403
Hypotenuse, 298, 517

I

Identity property
  of 1, 9, 85
  of 0, 9, 60, 85
Improper symbols, 13
Inequalities, 55, 153
  addition principle for, 154
  equivalent, 155
  false, 153
  graphs of, 153, 421
    linear, 422
    in one variable, 153
    in two variables, 421, 422
  linear, 422
  multiplication principle for, 156
  problem solving using, 163
  related equation, 422
  solution set, 153
  solutions of, 153, 421
  solving, *see* Solving inequalities
  translating to, 163
  true, 153
Integers, 49
  consecutive, 135
  division of, 79
  negative, 49
  positive, 49
Intercepts, 395, 564
  graphs using, 395
Interest
  compound, 231
  simple, 47
Inverse variation, 415
Inverses
  additive, 61, 62, 97
    and subtraction, 65
    of sums, 97
  multiplicative, 14, 80
Irrational numbers, 52
  graphing, 54

L

LCD, *see* Least common denominator
LCM, *see* Least common multiple
Leading coefficient, 255
Least common denominator, 13, 325, 330, 338
Least common multiple, 5, 6, 325
  of an algebraic expression, 326
  and clearing of fractions, 126, 345
Legs of a right triangle, 298, 517
Less than ($<$), 54
Less than or equal to ($\leq$), 55
Like radicals, 505
Like terms, 92, 195, 228
Line. *See also* Lines.
  horizontal, 396, 397
  point–slope equation, 405
  slope, 401
  slope–intercept equation, 404
  vertical, 397
Line graph, 380
Line of symmetry, 561
Linear equations, 387
Linear inequalities, 422
Lines
  parallel, 409
  perpendicular, 410

M

Magic number, 234
Mathematics and careers, 68, 316
Meaningful replacements, 311, 485
Missing terms, 198
Mixture problems, 460, 462, 463
Monomials, 193, 198
  as divisors, 235
  factoring, 249
  and multiplying, 211
Motion problems, 351, 469, 554
Multiples, 5
Multiplication. *See also* Multiplying.
  associative law of, 88
  commutative law of, 87
  with decimal notation, 21
  using the distributive law, 91
  using exponents, 177
  of fractional expressions, 12, 312, 315
  of polynomials, 211–214, 217–222, 230
  of radical expressions, 489, 492, 506
  of rational expressions, 312, 315
  of real numbers, 73–75
  using scientific notation, 188
  by zero, 74

Multiplication principle
  for equations, 117
  for inequalities, 156
Multiplicative identity, 85
Multiplicative inverse, 14, 80
Multiplicative property of zero, 74
Multiplying. *See also* Multiplication.
  exponents, 185
  by 1, 85, 312
  by -1, 97

N

Natural numbers, 3, 49
Negative exponents, 179
Negative integers, 49
Negative numbers, square roots of, 485
Negative square root, 483
Notation
  decimal, 19, 52
  exponential, 31, 175
  fractional, 9
  percent, 27
  scientific, 187
  set, 49, 155
Number line
  addition on, 59
  and graphing rational numbers, 51
  order on, 54
Numbers
  of arithmetic, 9
  complex, 485
  composite, 4
  factoring, 3
  integers, 49
  irrational, 52
  multiples of, 5
  natural, 3, 49
  negative, 55
  opposite, 61, 97
  order of, 54
  positive, 55
  prime, 3, 4
  rational, 50, 51
    nonnegative, 9
  real, 53
  signs of, 62
  whole, 9, 49
Numerator, 9

O

Odd integers, consecutive, 135
One
  equivalent expressions for, 10
  as exponent, 176

One (*continued*)
  identity property of, 9, 85
  multiplying by, 85
Operations, order of, 32, 99, 100
Opposite. *See also* Additive inverses.
  of a number, 61, 97
  of a polynomial, 204, 229
Order
  ascending, 196
  descending, 196
  on number line, 54
  of operations, 32, 99, 100
Ordered pairs, 381
Origin, 381

P

Pairs, ordered, 381
Parabolas, 561
Parallel lines, 409
Parallelogram, area of, 47
Parentheses, 32
  in equations, 127
  within parentheses, 99
  removing, 97, 98
Parking-lot arrival spaces, 484
Pendulum, period of, 504
Percent, problems involving, 143–146
Percent notation, 27
Perfect-square radicand, 485
Perimeter, 137
Perpendicular lines, 410
Pi (π), 176
Place-value chart, 19
Plotting points, 381
Point–slope equation, 405
Points, coordinates of, 381
Polygon, number of diagonals, 532
Polynomials, 193
  addition of, 203, 229
  additive inverse of, 204
  in ascending order, 196
  binomials, 198
  coefficients in, 195, 228
  collecting like terms (or combining
    similar terms), 195
  degree of, 197, 228
  in descending order, 196
  division of, 235
  evaluating, 193, 227
  factoring, *see* Factoring,
    polynomials
  missing terms in, 198
  monomials, 193, 198
  multiplication of, 211–214,
    217–222, 230

  opposite of, 204
  in several variables, 227
  subtraction of, 205, 229
  terms of, 194, 195
  trinomials, 198
  value of, 193
Positive numbers, 55
  integers, 49
Positive square root, 483
Power, 31, 175. *See also* Exponents.
  raising to a power, 185
Power rule, 185
Prime factorization, 4
  and LCM, 5
Prime numbers, 3, 4
Principal square root, 483
Principle of square roots, 535
Principle of squaring, 511
Principle of zero products, 289
Problem solving. *See also* Applied
    problems; Problems.
  five-step process, 133
  other tips, 138, 276
Problems. *See also* Applied Problems.
  accidents, daily, 199
  age, 458
  average, 163
  cost, total, 199
  falling object, distance and time of
    fall, 199, 540
  games in a sports league, 194, 297
  geometric
    area, 444, 553
    circumference, 552
    diagonals of a polygon, 532
    perimeter, 137
    triangle, 137, 296, 298, 553
  gravitational force, 361
  handshakes, 301
  height of a projectile, 566
  horizon, distance to, 514
  interest
    compound, 231
    simple, 47, 144
  lung capacity, 234
  magic number, 234
  medical dosage, 194
  mixture, 460, 462, 463
  motion, 351, 469, 554
  parking-lot arrival spaces, 484
  pendulum, period of, 504
  percent, 143–146
  planet orbits, 8
  price decrease, 145
  rent-a-car, 136, 452

revenue, total, 199
skidding car, speed of, 494
temperature conversion, 164
Torricelli's theorem, 549
wildlife population, estimating, 356
wind chill temperature, 508
work, 353
Product rule
  for exponential notation, 177
  for radicals, 489
Products. *See also* Multiplication;
    Multiplying.
  raising to a power, 186
  of square roots, 489, 492, 506
  of sums and differences, 219
  of two binomials, 212, 217
Projectile, height, 566
Property of -1, 97
Proportion, 354, 355
Proportional, 355, 413, 415
Proportionality, constant of, 413
Pythagoras, 517
Pythagorean equation, 517
Pythagorean theorem, 298, 517

Q

Quadrants, 382
Quadratic equations, 289, 529
  approximating solutions, 546
  discriminant, 545
  graphs of, 561
  solving
    by completing the square, 537
    by factoring, 290, 530, 531
    by principle of square roots, 535
    by quadratic formula, 543
  in standard form, 529
  x-intercepts, 564
Quadratic formula, 543
Quotient
  of integers, 79
  raising to a power, 186
  involving square roots, 497
Quotient rule
  for exponential notation, 178
  for radicals, 497

R

Radical equations, 511
Radical expressions, 484. *See also*
    Square roots.
  adding, 505
  conjugates, 507
  in equations, 511
  factoring, 489

meaningful, 485
multiplying, 489, 492, 506
perfect-square radicands, 485
product rule, 489
quotient rule, 497
involving quotients, 497
radicand, 484
rationalizing denominators, 498, 507
simplifying, 486, 489
subtracting, 505
Radical symbol, 483
Radicals, like, 505
Radicand, 484
  meaningful, 485
  negative, 485
  perfect-square, 485
Raising a power to a power, 185
Raising a product to a power, 186
Raising a quotient to a power, 186
Rate, 354
Ratio, 354. *See also* Proportion.
Rational equations, 345
Rational expressions, 311, 345. *See
    also* Fractional expressions;
    Rational numbers.
  addition, 329, 340
  complex, 365
  division, 321
  meaningful replacements, 311
  multiplying, 312, 315
  reciprocals, 321
  simplifying, 312, 315
  subtraction, 337
Rational numbers, 50, 51. *See also*
    Fractional expressions; Rational
    expressions.
  decimal notation, 52
Rationalizing denominators, 498, 507
Real numbers, 53
  addition, 59, 60
  division, 79–82
  graphing, 54
  multiplication, 73–75
  order, 54
  subsets of, 49
  subtraction, 65
Reciprocals, 14, 80
  and division, 15, 81, 321
  of rational expressions, 321
Rectangle
  area, 44
  perimeter, 137
Related equation, 422
Repeating decimals, 23
Revenue, total, 199

Reversing equations, 120
Right circular cylinder, area of, 227
Right triangles, 298, 517
Rise, 401
Roots
   of quotients, 497
   square, *see* Square roots; Radical
     expressions
Roster notation, 49
Rounding, 24
Run, 401

S

Scientific notation, 187
   converting from/to decimal
     notation, 187
   dividing using, 188
   multiplying using, 188
Second coordinate, 381
Set-builder notation, 155
Sets, 49
   empty, 437
   notation, 49, 155
   solution, 153
   subset, 49
Sidelights, 16, 68, 76, 316, 464, 508,
   556
Signs of numbers, 62
Similar terms, 195, 228. *See also* Like
   terms.
Simple interest, 47, 144
Simplest fractional notation, 10
Simplifying
   complex rational (or fractional)
     expressions, 365, 367
   fractional expressions, 312, 315
   fractional notation, 10
   radical expressions, 486, 489
   rational expressions, 312, 315
   removing parentheses, 97
Slant, *see* Slope
Slope, 401
   applications, 406
   from equation, 403
   of horizontal line, 402, 403
   of parallel lines, 409
   of perpendicular lines, 410
   point–slope equation, 405
   slope–intercept equation, 404
   of vertical line, 403
Slope–intercept equation, 404
Solutions
   of equations, 111, 387
   of inequalities, 153, 421
   of systems of equations, 435

Solving equations, 111. *See also*
   Solving formulas.
   using the addition principle, 112,
     123
   clearing of decimals, 126
   clearing of fractions, 126
   collecting like terms, 124
   containing parentheses, 127
   by factoring, 290, 530, 531
   fractional, 345
   using the multiplication principle,
     117, 123
   with parentheses, 127
   using principle of zero products,
     290, 530, 531
   procedure, 127
   quadratic
     by completing the square, 537
     by factoring, 290, 530, 531
     by principle of square roots, 535
     by quadratic formula, 543
   with radicals, 511
   rational, 345
   systems of
     by elimination method, 447
     by graphing, 436
     by substitution method, 441
Solving formulas, 149, 150, 361, 549
Solving inequalities, 153
   using addition principle, 154, 157
   using multiplication principle, 156,
     157
   solution set, 153
Special products of polynomials
   squaring binomials, 219
   sum and difference of two
     expressions, 219
   two binomials (FOIL), 217
Speed, 354
   of a skidding car, 494
Square, area of, 182
Square of a binomial, 219
Square roots, 483. *See also* Radical
   expressions.
   approximating, 483, 490, 500
   on a calculator, 483
   negative, 483
   of negative numbers, 485
   positive, 483
   of powers, 491
   principal, 483
   principle of, 535
   products, 312, 315
   quotients involving, 497
   of quotients, 497

of a sum, 492
using a table, 483
Squares, differences of, 273
Squaring, principle of, 511
Standard form of a quadratic
equation, 529
State answer, problem-solving
process, 133
Study tips, 16, 76, 464
Subscripts, 361
Subset, 49
Substituting, 44. *See also* Evaluating
expressions.
Substitution method, solving systems
of equations, 441
Subtraction
by adding inverses, 65
with decimal notation, 21
of exponents, 178
of fractional expressions, 13
of polynomials, 205, 229
of radical expressions, 505
of rational expressions, 337
of real numbers, 65
Sum
additive inverse of, 97
square root of, 492
Sum and difference of terms, product
of, 219
Supplementary angles, 455
Surface area, right circular cylinder,
227
Symmetry, line of, 561
Systems of equations, 435
solution of, 435
solving
by elimination method, 447
by graphing, 436
by substitution method, 441

T

Table of primes, 4
Temperature, converting Celsius to
Fahrenheit, 164
Terminating decimal, 52
Terms, 90
coefficients of, 195
collecting (or combining) like, 92,
195, 228
degrees of, 197, 228
like, 92, 195
missing, 198
of polynomials, 194
similar, 195
Torricelli's theorem, 549

Translating
to algebraic expressions, 45
to equations, 133
to inequalities, 163
Trapezoid, area, 361
Tree, factor, 4
Trial-and-error factoring, 255, 261
Triangle
area, 47
equilateral, 520
right, 298, 517
and Pythagorean theorem, 298
sum of angle measures, 138
Trinomial squares, 271
factoring, 272
Trinomials, 198
factoring, 255, 261, 267
True equation, 111
True inequality, 153

V

Value
of an expression, 44
of a polynomial, 193
Variables, 43
Variation
constant, 413
direct, 413
equation of, 413, 415
inverse, 415
and problem solving, 414, 416
Vertex, 561, 562
Vertical lines, 397
slope, 403

W

Whole numbers, 9, 49
Wildlife populations, estimating, 356
Wind chill temperature, 508
Work principle, 354
Work problems, 353

X

x-intercept, 395, 564

Y

y-intercept, 395

Z

Zero
degree of, 197
division by, 79
as exponent, 176
identity property, 9, 60, 85
multiplicative property, 74
Zero products, principle of, 289

PROFESSOR WEISSMAN'S SOFTWARE

Specially customized for **INTRODUCTORY ALGEBRA**, Sixth Edition, by **Keedy/Bittinger**

Build your mathematics skills with your own personal tutor! Professor Weissman's Software is an easy-to-use skill building program that you can use on your own home computer or in your school's computer labs.

You can select a specific topic and skill level, and Professor Weissman's Software gives you a series of randomly generated practice problems--You never get the same problem twice! If you get the answer wrong, the program leads you through a step-by-step solution to the problem--and then you can try again. The program will give you more challenging problems as you get more and more correct answers. It will also keep a record of your scores, so that you can track your own progress!

For more information and a free sample disk, call The Professor at 1-718-698-5219, Sunday through Thursday, 8-10 pm EST, or send in your order with the attached coupon for **33%-50%** off list price!

HARDWARE REQUIREMENTS: IBM PC or compatible with 384 K memory, CGA or HERCULES (color monitor not necessary), DOS 2.1 or higher.

IBM and **IBM PC** are registered trademarks of International Business Machines Corporation.

****Addison-Wesley accepts no responsibility for order fulfillment, customer service, or technical support. All inquiries should be directed to Professor Martin Weissman.

Order card: (tear here and place in envelope)

To order your customized version of **PROFESSOR WEISSMAN'S SOFTWARE** for use with INTRODUCTORY ALGEBRA, Sixth Edition by Keedy/Bittinger detach this card and send it with a check or money order payable to Professor Martin Weissman at P.O. Box 140612, Staten Island, NY 10314. Please be sure to indicate your disk preference on the reverse side.

Name_____

Street Address_____

City/State/Zip Code_____

Telephone (____)_____

School _____

TOTAL AMOUNT ENCLOSED $_____

TOPIC COVERAGE:

Disk #1

Inequality Symbols < >
Combine/Multiply Numbers
Remove Parentheses
Divide Signed Numbers
Order of OPerations #1
Order of Operations - Exponents
Simplify with Exponents
 Algebraic Substitution
Factoring
Distributing
Translating to Algebra
Solve Linear Equations
Number Word Problems

Disk #2

Simplify Algebraic Expressions
Word Problems: Coins
Word Problems: Rectangles
Solving Inequalities
Equations with Fractions
Equations with Decimals
Solving Proportions
Multiplying Binomials
Dividing POlynomials
Rectangular Coordinates
Points on Lines
Graphing a Line
Horizontal/Vertical Lines

Disk #3

Reducing Algebraic Fractions
Multiplying Algebraic Fractions
Dividing Algebraic Fractions
Add/Subtract Algebraic Fractions #I
Least Common Multiples
Add/Subtract Algebraic Fractions #2
Add/Subtract Algebraic Fractions #3
Integers as Exponents
Scientific Notation

Disk #4

Introduction to Polynomials
Multiplying Polynomials
Special Products
Factoring Polynomials
Differences of Squares
Trinomial Squares
Factoring Trinomials #1
Factoring Trinomials #2
Factoring Trinomials #3

Disk #5

Fractional Equations
Translating to Equations
Systems of Equations (substitution)
Systems of Equations (elimination)
Solving Word Problems #1
Solving Word Problems #2
Solving Quadratic Eqs by Factoring
Completing the Square

Disk #6

Square Roots
Radical Expressions
Multiplying Radicals
Simplifying Radicals
Roots of Quotients
Quotients of Roots
Add/Subtract Radicals
The Quadratic Formula
Radicals Quadratic Equations

WRITE FOR INFORMATION ABOUT OTHER AVAILABLE DISKS.

CHECK THE DISKS YOU WANT

_____disk #1 _____disk #2 _____disk #3 _____disk #4 _____disk #5 _____disk #6

CHECK THE TOTAL AMOUNT OF DISKS ORDERED

_____ I want 1 disk for $20 (reg. $30, I save 33%)*
_____ I want 2 disks for $38 (reg. $60, I save 37%)*
_____ I want 3 disks for $54 (reg. $90, I save 40%)*
_____ I want 4 disks for $68 (reg. $120, I save 43%)*
_____ I want 5 disks for $80 (reg. $150, I save 47%)*
_____ Send all 6 disks for $90 (reg. $180, I save 50%)*

CHECK THE DISK SIZE YOU WANT _____3.5" _____5.25"
*Include $5 per order for shipping and handling
*NY Residents add 8% sales tax
 Prices subject to change without notice